AF520613

AGROFORESTRY

Natural Resource Sustainability, Livelihood and Climate Moderation

AGROFORESTRY

Natural Resource Sustainability, Livelihood and Climate Moderation

Editors
OP Chaturvedi, A Venkatesh
RS Yadav, Badre Alam, RP Dwivedi
Ramesh Singh and SK Dhyani

National Research Center for Agroforestry, Jhansi - 284 003 (U.P.)

SATISH SERIAL PUBLISHING HOUSE
403, Express Tower, Commercial Complex, Azadpur, Delhi-110033 (India)
Phone : 011-27672852, Fax : 91-11-27672046
E-mail : info@satishserial.com, hkjain1975@yahoo.com
Website : www.satishserial.com

Published by :

SATISH SERIAL PUBLISHING HOUSE
403, Express Tower, Commercial Complex, Azadpur, Delhi-110033 (INDIA)
Phone : 011-27672852 Fax : 91-11-27672046
E-mail : info@satishserial.com, hkjain1975@yahoo.com

Reprint, 2019

ISBN 81-89304-64-X

Composed, Designed & Printed in India

Dr. A.K. Gogoi
Assistant Director General (Agronomy)
NRM Division
Indian Council of Agricultural Research
Krishi Anusandhan Bhavan - II
Pusa, New Delhi - 110 012
Ph: 91 -11 - 25848370 (O), Fax: 91 - 11-25848366
E-mail: akumargogoi@yahoo.com

Foreword

Climate change is one of the major problem that mankind is facing today in agriculture production and human health. Agroforestry practiced since ages has been recognized by the IPCC as having high potential for sequestering C as a part of climate change mitigation strategies. There is emerging evidence that agroforestry systems have strong potential to increase aboveground and soil C stocks to mitigate greenhouse gas emission. Considering a projection of agroforestry cover of 25 million hectare in next 25 years, the total C sequestration potential of these areas would be 118.4 million tones C ha^{-1} yr^{-1}. These potential could be further remarkably enhanced if appropriate agroforestry systems could be developed and adopted depending upon the specific agroclimatic zone which necessitates integrated research involving all the stakeholders and exploring all the opportunities as offered by the agroforestry. The present forest cover is about 20.7 per cent of total geographical area and the target fixed to increase tree cover to the extent of 33 per cent by 2012 can only be achieved through adoption of agroforestry. Restoration of degraded lands has a tremendous potential for C sequestration, improving soil quality and increasing productivity through afforestation involving silvipastoral system. Agroforestry has a huge potential to provide employment to rural and urban population through industrial application and value addition.

The present endeavour of the distinguished scientists Drs. O.P. Chaturvedi, A. Venkatesh, R.S. Yadav, Badre Alam, R.P. Dwivedi, Ramesh Singh and S.K. Dhyani in collecting, synthesing and editing the chapters contributed by learned experts on various aspects of agroforestry encompassing scope of the subject, C sequestration, climate mitigation, ecosystem development through agroforestry, economic aspects etc. in the form of the book

"Agroforestry : Natural Resource Sustainability, Livelihood and Climate Moderation" is highly praiseworthy.

I am confident that the book will generate great enthusiasm in planners, researchers, academicians, students and policy makers to make use of current state of the knowledge on various aspects of agroforestry enshrined in different chapters of the book for better planning, research, teaching and policy legislation.

January, 2009

(A. K. Gogoi)

Preface

Magnificent potentials of agroforestry for socio-economic development through enhanced livelihood security, better ecosystem services for natural resource conservation and improving system productivity, environmental security through carbon sequestration and mitigation of climate change have been recognized at the apex bodies of policy making at National and International levels namely the Planning commission of India, CGIAR, IPCC, UNFCC, LULUCF and UNEP. For the past few decades agroforestry research has made considerable progress which gave it a very convincing recognition for being treated it as one of the major areas of national importance. As no science is saturated with knowledge, continuing research efforts are essential for further tangible benefit out of it to improve the cause of agroforestry and thereby the contribution to the society.

One of the major new initiatives for better natural resource management practices is through agroforestry intervention for land use planning. Accomplishment of this involves diversification of management of land and crops. This needs better understanding of the mechanisms and reasons behind adoption of diversification measures. Moreover, keeping the implications of climate change and global warming in focus, various options are open for agroforestry to offer. These require to further strengthen the stake of agroforestry in cutting edge areas of relevant disciplines. Agroforestry has been increasing the expectations for improving the socio-economic conditions of the under-privileged countrymen, for increasing the green cover of our country, for bringing about 43 million ha non-arable degraded lands under agroforestry, for improving dry land agriculture, for mitigating and adaptation strategies for climate change and for many more such noble activities. This clearly indicates that how diversified knowledge is required to successfully achieve the mandate of agroforestry. Hence, prioritization, monitoring and implementation of all the essential activities relating to advancement of agroforestry assume much significance. Keeping these issue in view, the National Research Centre for Agroforestry, Jhansi, (U.P.), India developed the course module for lectures in winter school on *"Agroforestry for National Resource Sustainability, Livelihood Security and Climate Moderation"* which held from 4-24 February, 2009. The synthesis of all the lecture notes put together in the publication of the present compendium, which would be distributed to all the concerned organizations involving in implementation and monitoring of agroforestry research and development programmes in India besides the participants.

This book presents well focused topics of contemporary science for the greater interest of agroforestry. Agroforestry is an example of the integrated research of interdisciplinary subjects. Attempts have been made to include most of the relevant disciplines which are expected to be essential for their direct or indirect role in agroforestry development. Initial few chapters have been devoted for analyzing the past arena of agroforestry and emphasizing the scope and future challenges for it. We have covered articles from basic and applied aspects namely subject areas dealing with natural resource management for improving soil quality, fertility and carbon sequestration, climate change, soil water conservation and watershed development, water use efficiency, propagation techniques for quality planting materials, breeding strategies for MPTS through classical and biotechnological approaches, biofuel and food security, GIS and remote sensing, tree crop and environment interaction and statistical modelling, livestock and agroforestry, socio-economic aspects and value addition.

We feel that the authors have done justice in putting basic and applied aspects of research and development of agroforestry into various chapters. We appreciate the commitment of the contributing authors and thank all the contributors for complying with the time schedule. Hope this book would be very helpful to the students, scientists and policy makers who envisage greater role of agroforestry in natural resource management.

The editors acknowledge the Human Resource Division of Indian Council of Agricultural Research, New Delhi for providing the necessary grant for conducting the winter school and publishing this compilation in the form of edited book.

Editors

List of Contributors

A. Balasubramanian
Associate Professor
Department of Tree Breeding
Forest College and Research Institute
Tamil Nadu Agricultural University
Mettupalayam-641301 (T.N.)

A. Chandra
Pr. Scientist
Division of Crop Improvement
Indian Grassland & Fodder Research Institute, Gwalior Road
Jhansi-284 003 (U.P.)
Email: amaresh62@yahoo.com

A. K. Rai
Sr. Scientist
Division of Crop Production
Indian Grassland & Fodder Research Institute, Gwalior Road
Jhansi-284 003 (U.P.)

A. K. Shukla
Pr. Scientist
Division of Grassland Management
Indian Grassland & Fodder Research Institute, Gwalior Road
Jhansi-284 003 (U.P.)

A. K. Vasisht
IASRI, Library Avenue
Pusa campus, New Delhi-110 012
Email: amitvasisht@rediffmail.com

A. Venkatesh
Sr. Scientist (Forestry)
National Research Centre for Agroforestry
Near Pahuj Dam, Gwalior Road
Jhansi-284 003 (U.P.)
Email: drvenkatesh.a@gmail.com

A. Vijayakumar
Professor, Tamil Nadu Agricultural University, Coimbatore-641 003 (T.N.)

A.K. Handa
Sr. Scientist (Forestry)
National Research Centre for Agroforestry
Near Pahuj Dam, Gwalior Road
Jhansi-284 003 (U.P.)
Email: handaarun@rediffmail.com

Ajit
Sr. Scientist (Statistics)
National Research Centre for Agroforestry
Near Pahuj Dam, Gwalior Road
Jhansi-284 003 (U.P.)
Email: umaajitgupta@yahoo.co.in

Anil Kumar
Pr. Scientist (Plant Pathology)
National Research Centre for Agroforestry
Near Pahuj Dam, Gwalior Road
Jhansi-284 003 (U.P.)
Email: anilgargnrcaf@gmail.com

Ashok Shukla
Ph. D. Scholar
National Research Centre for Agroforestry
Near Pahuj Dam, Gwalior Road
Jhansi-284 003 (U.P.)

Badre Alam
Sr. Scientist (Plant Physiology)
National Research Centre for Agroforestry
Near Pahuj Dam, Gwalior Road
Jhansi-284 003 (U.P.)
Email: badrealam@gmail.com

C. Buvaneswaran
Scientist, Institute of Forest Genetics and
Tree Breeding, Coimbatore-641002 (T.N.)

C. K. Vajpayee
Technical Officer
National Research Centre for Agroforestry
Near Pahuj Dam, Gwalior Road
Jhansi-284 003 (U.P.)

D. R. Palsaniya
Scientist (Agronomy)
National Research Centre
for Agroforestry
Near Pahuj Dam, Gwalior Road
Jhansi-284 003 (U.P.)
Email: drpalsaniya@gmail.com

D. V. K. Nageswara Rao
Sr. Scientist, Division of Crop Production
Indian Grassland & Fodder Research
Institute, Gwalior Road
Jhansi-284 003 (U.P.)

G. P. Saravanan
Forest College & Research Institute
TNAU, Mettupalayam-641 301 (T.N.)

Garima Gupta
Ph. D. Scholar (Agroforestry)
National Research Centre for
Agroforestry
Near Pahuj Dam, Gwalior Road
Jhansi-284 003 (U.P.)
Email: garima811@gmail.com

Imtiyaz Ali
Sr. Research Fellow (RS-GIS Project)
National Research Centre for
Agroforestry
Near Pahuj Dam, Gwalior Road
Jhansi-284 003 (U.P.)

J. B. Singh
Sr. Scientist
Division of Crop Production
Indian Grassland & Fodder Research
Institute, Gwalior Road
Jhansi-284 003 (U.P.)
Email: jbs_igfri@rediffmail.com

J. P. Singh
Sr. Scientist
Division of Grassland Management
Indian Grassland & Fodder Research
Institute, Gwalior Road
Jhansi-284 003 (U.P.)
Email: jpsingh.igfri@gmail.com

K. Kareemulla
Pr. Scientist (Agricultural Economics)
Central Research Institute for Dryland
Agriculture, Santoshnagar, Hyderabad
Email: kalakareem@crida.ernet.in

K. T. Parthiban
Forest College & Research Institute
TNAU, Mettupalayam-641 301 (T.N.)

Kamlesh Kumar
Senior Research Fellow
Division of Grassland Management
Indian Grassland & Fodder Research
Institute, Gwalior Road
Jhansi-284 003 (U.P.)
Email: kam_kumaar@yahoo.co.in

K. D. Kaushik
Research Associate (RS-GIS Project)
National Research Centre for Agroforestry
Near Pahuj Dam, Gwalior Road
Jhansi-284 003 (U.P.)
Email: kdkaushik@gmail.com

M. Govinda Rao
Dean, Forest College & Research Institute
TNAU, Mettupalayam-641 301 (T.N.)

Munna Ram
Pr. Scientist (Soils)
National Research Centre for Agroforestry
Near Pahuj Dam, Gwalior Road
Jhansi-284 003 (U.P.)
Email: munnaram@gmail.com

N. Das
Pr. Scientist & Head
Division of Plant Animal Relationship
Indian Grassland & Fodder Research
Institute, Gwalior Road
Jhansi-284 003 (U.P.)

N. Kushwaha
Senior Research Fellow
Division of Grassland Management
Indian Grassland & Fodder Research
Institute, Gwalior Road
Jhansi-284 003 (U.P.)
Email: neeraj29aug84@gmail.com

Nishi Roy
Programme Coordinator
KVK, Bharari, Jhansi-284003 (U.P.)

O. P. Chaturvedi
Pr. Scientist (Agroforestry)
National Research Centre for Agroforestry
Near Pahuj Dam, Gwalior Road
Jhansi-284 003 (U.P.)
Email: opchaturvedi@mailcan.com

P. Masilamani
Associate Professor
Agricultural Engineering College and
Research Institute
Tiruchirappalli-621 712 (T.N.)

P. Rathakrishnan
Scientist (Forestry)
National Research Centre for Agroforestry
Near Pahuj Dam, Gwalior Road
Jhansi-284 003 (U.P.)

P. Sharma
Sr. Scientist, Division of Social Sciences
Indian Grassland & Fodder Research
Institute, Gwalior Road
Jhansi-284 003 (U.P.)

Parul Shukla
Senior Research fellow (NOVOD Project)
National Research Centre for Agroforestry
Near Pahuj Dam, Gwalior Road
Jhansi-284 003 (U.P.)

Pradeep Behari
Sr. Scientist
Division of Crop Production
Indian Grassland & Fodder Research
Institute, Gwalior Road
Jhansi-284 003 (U.P.)

R. Indumathy
NRCAF Colony
National Research Centre for Agroforestry
Jhansi-284 003 (U.P.)

R. K. Singh
Technical Officer
National Research Centre for Agroforestry
Near Pahuj Dam, Gwalior Road
Jhansi-284 003 (U.P.)

R. K. Tewari
Pr. Scientist (Horticulture)
National Research Centre for Agroforestry
Near Pahuj Dam, Gwalior Road
Jhansi-284 003 (U.P.)

R. P. Dwivedi
Sr. Scientist (Extension Education)
National Research Centre for Agroforestry
Near Pahuj Dam, Gwalior Road
Jhansi-284 003 (U.P.)
Email: rpdwivedi41@yahoo.com

R. S. Chaurasia
Senior Research Fellow
Division of Grassland Management
Indian Grassland & Fodder Research
Institute, Gwalior Road
Jhansi-284 003 (U.P.)
Email: rsc25.jhs@gmail.com

R. S. Tripathi
Pr. Scientist (Agricultural Economics)
Division of Technology Evaluation and
Transfer, Central Soil Salinity Research
Institute, Karnal- 01 (Haryana)

R. S. Yadav
Sr. Scientist (Soils)
National Research Centre
for Agroforestry
Near Pahuj Dam, Gwalior Road
Jhansi-284 003 (U.P.)
Email: yadav_rams@rediffmail.com

R. Sreenivasan
Tamil Nadu Newsprints and Papers
Limited, Kagithapuram (T.N.)

R. V. Kumar
Sr. Scientist
(Genetics & Plant Breeding)
National Research Centre for Agroforestry
Near Pahuj Dam, Gwalior Road
Jhansi-284 003 (U.P.)
Email: rvkumar4@rediffmail.com

Rajeev Tiwari
Technical Officer
National Research Centre for Agroforestry
Near Pahuj Dam, Gwalior Road
Jhansi-284 003 (U.P.)
Email: rajeevtiwari62@rediffmail.com

Rajendra Prasad
Pr. Scientist (Soils)
National Research Centre for Agroforestry
Near Pahuj Dam, Gwalior Road
Jhansi-284 003 (U.P.)
Email: drrajendraprasad@yahoo.co.in

Rajendra Singh
Technical Officer
National Research Centre for Agroforestry
Near Pahuj Dam, Gwalior Road
Jhansi-284 003 (U.P.)

Ram Newaj
Pr. Scientist (Agronomy)
National Research Centre for Agroforestry
Near Pahuj Dam, Gwalior Road
Jhansi-284 003 (U.P.)
Email: ramnewaj@rediffmail.com

Ramesh Singh
Scientist, SS (SWCE)
National Research Centre for Agroforestry
Near Pahuj Dam, Gwalior Road
Jhansi-284 003 (U.P.)
Email: rsingh_nrcaf@mailcan.com

S.A. Faruqui
Project Coordinator (AICRP FC)
Indian Grassland & Fodder Research
Institute, Gwalior Road
Jhansi-284 003 (U.P.)

S. K. Das
Sr. Scientist
Division of Crop production
Indian Grassland & Fodder Research
Institute, Gwalior Road
Jhansi-284 003 (U.P.)
Email: skdas1964@gmail.com

S. K. Dhyani
Director
National Research Centre for Agroforestry
Near Pahuj Dam, Gwalior Road
Jhansi-284 003 (U.P.)
Email: shivkdhyani@gmail.com

S. K. Mahanta
Sr. Scientist
Division of Plant Animal Relationship
Indian Grassland & Fodder Research
Institute, Gwalior Road
Jhansi-284 003 (U.P.)

S. K. Nag
Sr. Scientist
Division of Plant Animal Relationship
Indian Grassland & Fodder Research Institute, Gwalior Road
Jhansi-284 003 (U.P.)
Email: nagsk_67@rediffmail.com

S. Manivasakan
Department of Tree Breeding
Forest College and Research Institute
Tamil Nadu Agricultural University
Mettupalayam-641301

S. N. Pandey
Development Alternatives, Jhansi
Email: sn.1212@gmail.com

S. P. Ahalawat
Sr. Scientist
(Genetics & Plant Breeding)
National Research Centre for Agroforestry
Near Pahuj Dam, Gwalior Road
Jhansi-284 003 (U.P.)
Email: ahlawat@mailcan.com

S. Saravanan
Institute of Forest Genetics and Tree Breeding, Coimbatore-641002 (T.N.)

S. Vennila
Forest College & Research Institute,
TNAU, Mettupalayam-641 301 (T.N.)

Shabir Ahmed Dar
Research Associate
National Research Centre for Agroforestry
Near Pahuj Dam, Gwalior Road
Jhansi-284 003 (U.P.)
Email: shabbu_01@yahoo.com

Sharmila Roy
Sr. Scientist
Indian Grassland & Fodder Research Institute, Gwalior Road
Jhansi-284 003 (U.P.)
Email: roysharmilaigfri@gmail.com

Sultan Singh
Sr. Scientist
Division of Plant Animal Relationship
Indian Grassland & Fodder Research Institute, Gwalior Road
Jhansi-284 003 (U.P.)

Sunil Kumar
Sr. Scientist
Division of Grassland Management
Indian Grassland & Fodder Research Institute, Gwalior Road
Jhansi-284 003 (U.P.)

Sunil Kumar
Technical Officer
National Research Centre for Agroforestry
Near Pahuj Dam, Gwalior Road
Jhansi-284 003 (U.P.)

Sunil Kumar
Sr. Scientist (PC Unit)
Indian Grassland & Fodder Research Institute, Gwalior Road
Jhansi-284 003 (U.P.)

Suresh Kumar
Sr. Scientist
Division of Crop Improvement
Indian Grassland & Fodder Research Institute, Gwalior Road
Jhansi-284 003 (U.P.)
Email: sureshkumar3_in@yahoo.co.uk

Uma
Technical Officer
National Research Centre for Agroforestry
Near Pahuj Dam, Gwalior Road
Jhansi-284 003 (U.P.)
Email: uma41163@gmail.com

V. K. Gupta
Pr. Scientist (Genetics & Plant Breeding)
National Research Centre for Agroforestry
Near Pahuj Dam, Gwalior Road
Jhansi-284 003 (U.P.)

About the Editors

Dr. O.P. Chaturvedi was born on 1st January, 1956 at Barwan, Uttar Pradesh and completed M.Sc. from BHU, Varanasi and Ph.D. from Kumaun University, Nainital with specialization in Forestry/Forest Ecology. He has published more than 100 research papers in international and national journal of repute, one book, 14 book chapters and 25 popular articles. He is awardees of NAAS and National Institute of Ecology Fellowship. He is also the recipient of 1st K. G. Tejwani award for his outstanding contribution in the field of agroforestry research and NAAS recognition award-2008 for significant contribution in soil, water and environmental sciences. He has guided two M.Sc. students. He specializes in Forestry and Agroforestry research.

Dr. A. Venkatesh was born on 26th May, 1973 at Sattur, Tamil Nadu and completed his M.Sc and Ph.D. from TNAU, Coimbatore in Forestry. He got published 27 research papers, five books, 11 book chapters and seven popular articles. He has been awarded best thesis in Ph.D., ICAR-JRF and Prize Money by Government of Tamil Nadu. He has specialization in value addition in agroforestry and tree-crop interaction.

Dr. R.S. Yadav was born on 30th September, 1967 at Badhal, district-Jaipur, Rajasthan and did his M.Sc. (Agriculture) from MPUA & T, Udaipur and Ph.D. (Soil Science & Agricultural chemistry) from RAU, Bikaner. He has a good number of publications with 18 research papers, one book, eight book chapters and two popular articles. He is recipient of NRCAF best research paper award for 2007. He specializes in soil agroforestry research especially soil enrichment through trees, biomass decomposition and nutrient release in soil.

Dr. Badre Alam was born on 31st January, 1966 at Sakodda, West Bengal and completed M.Sc. from AMU, Aligarh and Ph.D. from IIT, Kharagpur with specialization in Plant Physiology. He got published 15 research papers, six book chapters and one popular article. He is recipient of University gold medal, BRP gold medal and Young Scientist award from ISPP. He has specialization in the field of environmental physiology, photosynthesis and climate change.

Dr. R.P. Dwivedi was born on 20th July, 1966 at Amritpur Kherwa, district-Banda, Uttar Pradesh and got his M.Sc and Ph.D. from IVRI, Izatnagar in Extension Education. He has a good number of publications in the form of 23 research papers, 15 book chapters and 23 popular articles. He is the recipient of Young Scientist award and best paper award of ISEE, IARI, New Delhi. He has guided five M.Sc. students. He specializes in the area of PRA and agroforestry extension.

Dr. Ramesh Singh was born on 1st July, 1970 at Varanasi, Uttar Pradesh and completed his M.E. from RAU, Bikaner and Ph.D. from GBPUA&T, Pantnagar in Irrigation & Drainage Engineering. He got published 17 research papers, eight book chapters and eight popular articles. He is the recipient of University gold medal for M.E. and best scientist and team award, NRCAF. His specialized area of research is watershed development and management, GIS, remote sensing and hydrological modeling.

Dr. S.K. Dhyani was born on 17th September, 1956 at Lansdowne, Uttarakhand and completed M.Sc. from H.N.B. Garhwal University and Ph.D. course work from M.S.U., USA and degree from North Eastern Hill University, Shillong with specialization in Botany. He has got published more than 100 research papers in international and national referred journals, 11 books, 41 book chapters and 27 popular articles. He has been bestowed with prestigious Fakhruddin Ali Ahmad award (1998), Dr. Rajendra Prasad Puruskar (2000) and Fellow of Indian association of Soil and Water Conservationist. His specialized area of research is agroforestry, soil and water conservation and watershed management.

Contents

□□□

Agroforestry: Natural Resource Sustainability, Livelihood & Climate Moderation © 2009
Editors: OP Chaturvedi, A Venkatesh et. al., pp. 1-11
Satish Serial Publishing House, Delhi

Chapter 1

AGROFORESTRY PAST, PRESENT AND FUTURE

S.K. Dhyani
National Research Center for Agroforestry, Jhansi-284 003 (U.P.)

Introduction

Agroforestry practices are very ancient but the organized agroforestry research in India began in 1979 at Imphal, Manipur when a Seminar on Agroforestry was organized by the Indian Council of Agricultural Research, New Delhi to discuss the perspectives of agroforestry research and development of agroforestry research in India. To implement the major recommendation of the aforesaid seminar ICAR launched the All India Coordinated Research Project (AICRP) on Agroforestry with 20 centers all over country in 1983. Further, National Research Centre (NRC) for Agroforestry was established on 8th May in 1988 at Jhansi to accelerate basic, strategic and applied research in agroforestry. At present there are 36 centers representing all the agro-climates of the country under AICRP on Agroforestry with project coordinating unit at NRC for Agroforestry, Jhansi. Agroforestry has received much attention from the researchers, policy makers, and others for its perceived ability to contribute significantly to economic growth, socio-economic upliftment and environmental sustainability. Significant progress has been made during the past in developing and understanding of interactions amongst the different components of agroforestry to achieve effective resource utilization and their ameliorative effects on soil and microclimate. The paper concentrates on the research, developmental efforts of agroforestry inteventions and their impact on energy security, socioeconomic upliftment, and environmental perspectives in India.

Agroforestry is an age old concept that has now been institutionalized. Farmers, especially those in the tropics, have a long tradition of raising food crops, trees and animals together as well as exploiting a multiple range of production from natural woodlots. It is now a sustainable land use system that maintains or increases total yield by combining food crops (annuals) with tree crops (perennials) and/or livestock on the same unit of land, either alternately or at the same time, using management practices that suit the social and cultural characteristics of the local people and the economic and ecological conditions of the area. Thus the term 'agroforestry' encompasses a diverse set of integrated land use practices employed for variety of purposes. The aim is to achieve synergy between the components, which leads to net improvements in one or more of a range of characteristics, such as productivity and sustainability.

Brief History

Broadly the agroforestry history in India can be studied in following categories:

Agroforestry During Vedic, Epic and Medieval Period

Man was basically a food collector during *Vedic* era and gradually acquired the knowledge of plant species yielding food, fruits, vegetables etc. The origin of agroforestry practice i.e. growing trees with food crops and grasses believed to happen during the Vedic, Pre Vedic period in India. The God's prime Rishi and Muni and Guru were residing in forest and derived food in the form of vegetables, dry fruits, sweetening medium medicine and gum, besides fuel from the forest. Further, they develop Ashrams and started growing fruit trees like mango, jackfruit, guava, mahua etc. along with flowers, vegetables and *Kutoo* (food grain) to meet their needs. Man gradually developed settled farming with the passage of time and continued to integrate tree components in his farming system.

The period of two great Indian epics (Ramayana and Mahabharat) is known as epic era. In both the epics the illustrations that Rishi and Muni were to grow on the land of Ashram (i.e agri-horti and horti-pasture, fruits (mango, banana, guava), (vegetables sweet potato), flowers and food grains. This clearly indicated that the practice could be considered as an agroforestry practice. Rearing of silkworm and lakh-worm was a popular profession during that period. It is evident from these epics that agroforestry practices such as agrisilviculture and agrihorticulture practices were in operation in their primitive state during epic era.

The agriculture was in its primitive state during medieval period. However, many agroforestry practices existed. From king Ashoka to Sher Shah Suri, the roadside avenue plantations are examples of keen interest of the rulers for environmental protection. The tribal dances performed around these trees are symbolic of the intimate ethnic bias to protect trees. Trees had considerable socio, ethnic and religious backing during this period.

Agroforestry During Recent Past

During the last 3-4 centuries spectacular improvements were made in tree plantation. The most important agroforestry practice is known from the Kangeyam tract of Tamil Nadu where *Acacia leucophloea* + *Cenchrus setigerus* in silvipasture system was perfected. Similarly in ravines of Yamuna and Chambal, trees, shrubs and bamboos with grasses were planted for rearing the milk producing Jamunapuri breed of goats and sheep. Scattered trees of *khejri* (*Prosopis cineraria*) and/or mehndi in association with bajara, jowar and chilly were grown in the semi-arid area of Tamil Nadu. Tree plantation continued as demarcation and control against wind erosion throughout the country. Plantation of *khejri* trees for various uses on farm was a common practice in Rajasthan.The concept of sacred groves and sacred forest even today is an extension of people's faith in conservation of biodiversity- a prerequisite for agroforestry. In the northeast, large cardomom (*Amomum subalatum*) was introduced in Alder (*Alnus nephalensis*) forests in Sikkim Himalayas by the local farmers more than two hundred years ago, which has spread to more than 48000 ha area in northeastern states at present (Dhyani, 1998). In the coastal areas, *Casuarina equisetifolia* and other trees were grown in association with crops on farmlands for cash and small timber. Live hedges were common as an agroforestry practice in which Mehndi, *Agave siselana, Euphorbia* species were common species. In paddy sowing areas, *Pongamia glabra* and *Sesbania grandiflora* were grown, lopped annually and their leaves applied to fields as green manure. Application of green manure to paddy field also was common in Madhya Pradesh and Uttar Pradesh. In Western Ghats *Terminalia* leaves were harvested, spread on land, burnt and then paddy, ragi and millets were sown. Multistorey homesteads / home gardens were in existence in Kerala, Karnataka, Tamil Nadu, Assam and later on in Andaman and Nicobar islands as an important agroforestry practice. This agroforestry practice is still followed in aforesaid states to fulfill the needs of fodder, fuel wood, fruit, vegetable, medicines etc. On the basis of nature of components following common Agroforestry systems / practices are prevailing in different agro-ecological regions of India:

1. Agri-silviculture (trees + crops)
2. Boundary plantation (trees on boundary + crops)
3. Block plantation (block of trees + block of crops)
4. Energy plantation (trees + crops during initial years)
5. Alley cropping (hedges + crops)
6. Agri-horticulture (fruit trees + crops)
7. Agri-silvi-horticulture (trees + fruit trees+crops)
8. Agri-silvipasture (trees+crops + pasture or animals)
9. Silvi-olericulture (trees + vegetables)
10. Horti-pasture (fruit trees + pasture and/or animals)
11. Horti-olericulture (fruit trees + vegetables)
12. Silvi-pasture (trees+pasture and/or animals)
13. Forage forestry (forage trees + pasture)
14. Shelter-belts (trees + crops)
15. Wind-breaks (trees + crops)
16. Live fence (shrubs and under- trees on boundary)
17. Silvi or Horti-sericulture (trees or fruit trees + silkworm)
18. Horti-apiculture (fruit trees + honeybee)
19. Aqua-forestry (trees + fishes)
20. Homestead (multiple combination of trees, fruit trees, vegetable etc).

Besides these systems, there are many more component combinations followed in different agroecological regions of India.

Thus the review of agroforestry practices indicate that the safest choice of agroforestry species have come from the native vegetation, which has a history of adaptation to local precipitation regimes. There are a number of tree crop combinations, which in turn reflect the differences in the climates and soil fertility of various regions in the country (Sharda et al., 2001). The examples of major trees in agroforestry practices are (1) *Grewia optiva* and *Alnus nepalensis* in northern region, (2) *Populus, Eucalyptus, Mangifera* and *Dalbergia sisoo* in indogangetic region, (3) *Azadirachta indica, Acacia nilotica* and bamboos in central India, (4) *Acacia nilotica, Prosopis cineraria* and *Zizyphus* in western India, (5) *Tectona grandis, Tamarindus indica* para rubber (*Hevea brasilensis*) and cashew nuts in southern region and (6) *Artocarpus, Azadiracta indica, Casuarina equisetifolia* and bamboos in coastal and island region. This would correspond to the high rainfall levels and long rainy seasons in the south, with drought periods comperatively longer for the regions to the west, indo-gangetic and central (Dhyani *et al.*, 2003).

Agroforestry Research in India

Agroforestry research is more than hundred years old in India beginning with introduction of shade trees in tea gardens, but organized research was initiated during fifties in few ICAR Institutes namely Central Arid Zone Research Institute (CAZRI), Jodhpur, Central Soil and Water Conservation Research and Training Institute (CSWCRTI), Dehradun, Indian Grassland and Fodder Research Institute (IGFRI), Jhansi, as fuel- fodder plantations on degraded lands or the silvipastoral system. However, India has been at the forefront since organized research in agroforestry started worldwide. In 1979 at Imphal, I[st] Seminar on Agroforestry was organised by the ICAR, New Delhi to accumulate and compile the data on the R&D of agroforestry in India. The All India Coordinated Research Project (AICRP) on Agroforestry with 20 centers all over country was launched in 1983 to implement the major recommendations of the aforesaid seminar. Further, National Research Centre for Agroforestry was established on 8th May in 1988 at Jhansi to accelerate the basic, strategic and applied research in agroforestry. At the initial phase agroforestry research also got financial support from different international agencies primarily USAID, UNDP and others, which helped in strengthening infrastructure and human resource component. At present there are 36 centers under All India Coordinated Research Project on Agroforestry with project coordinating unit at National Research Center for Agroforestry, Jhansi. These centers represent almost all the agroclimates of the country. In addition to ICAR, Indian council of forestry research and education (ICFRE) and its regional centres, private institutions and NGOs such as WIMCO, ITC, BAIF, IFFDC, West Coast Paper Mills Ltd, Hindustan Paper Mills Ltd, National Tree Growers Cooperatives are also engage in research and promotion of agroforestry in the country. Agroforestry research and education has been started in more than 15 State Agricultural Universities (SAUs). The major thrust of agroforestry research in the beginning was on the following areas,

1. Diagnostic survey and appraisal of existing agroforestry practices.
2. Collection and evaluation of promising tree species/cultivars of fuel, fodder and small timber.

3. Studies on management practices of agroforestry systems.
4. Environment, wasteland and community land development through agroforestry interventions, and
5. Integration of livestock, fishery, apiculture, lac etc. as component of agroforestry.

The research and developmental efforts undertaken during the last more than two decades have clearly demonstrated the potential of agroforestry for resource conservation, improvement of environmental quality, rehabilitation of degraded lands and providing multiple outputs to meet the day to day demand of the rural population. The diagnostic survey and appraisal of existing agroforestry practices in the country revealed that there are enumerable agroforestry practices prevalent in different agroecological zones (Dhyani and Chauhan 1994; Pathak *et al*, 2000; Sharda *et al*, 2001; Solanki, 2006). The review indicates that agroforestry systems or practices are widely based on nature and arrangement of the components and ecological or socio-economic criteria. In the documented agroforestry practices, trees serve as wind breaks and shelterbelts, delineate boundaries and provide shade, ornamentation and seclusion around homesteads. They supply not only poles, stakes, timber and fuel, but also cash crops, fodder, fruits and nuts, dyes, gums, resins, fiber and medicines. Fodder and food trees provide balanced diets during dry seasons, when other foods are scarce. Trees, with their deep rooting systems, consume moisture and nutrients from higher depth in the soil than arable and pastoral crops, and thus there is least competition among the different components. On the basis of the design (D&D) exercises, the prominent agroforestry system from different region of the country were enumerated (Table 1).

The second major aspect of the agroforestry research was collection and evaluation of promising tree species/cultivars of fuel, fodder and small timber. A number of germplasm has been collected and evaluated in arboretum established by different Centres of the project. About 184 promising tree species have been determined based on growth performance trials at these centers. The promising tree species identified include *Ulmus wallichiana, Ailanthus excelsa, Morus alba, Robina pseudoacacia* and *Grewia optiva* for Western Himalayan region, *Acacia auriculiformis, Alnus nepalensis, Bamboosa, Parasanthes falcataria* and *Gmelina arborea* for eastern Himalayan region; Poplar, Eucalyptus and *Dalbergia sissoo* for Indogangetic region, *Dalbergia sissoo, Acacia tortilis, A. nilotica, Ailanthus excelsa, Prosopis cineraria* and *Leucaena leucocephala* and *Azadirachta indica* for arid and semi arid regions; *Albizzia spp., Erythrina, Gliricidia, Acacia auriculiformis* for humid and sub humid regions, and *Casuarina equisetifolia, Toona ciliata, Grevillea robusta* for the coastal and island region. The efforts made so far has created voluminous database, which is strength. The information collected may be utilized for creating local and regional volume tables. Planning Commission has also identified six species for major thrust in agroforestry and Joint Forest Management (JFM) programmes in the country. The species are *Acacia nilotica, Bamboo species, Casuarina equisetifolia, Eucalyptus spp., Populus deltoides* and *Prosopis cineraria.*

Table 1: Prominent agroforestry systems in different regions of country.

Agroclimatic zone/region	Agroforestry system					
	Agri - silvi culture	Agri-horti-silvic ulture	Agri-horticu lture	Horti-pastor ral	Silvi-pastoral	Specialized
1.Western Himalayan	4	4	5	4	4	-
2.Eastern Himalayan	5	3	5	4	3	Jhum (shifting cultivation)
3.Lower gangetic plains	5	-	3	2	2	Energy
4.Middle gangetic plains	4	-	2	-	-	Homestead
5.Upper gangetic plains	5	-	5	-	-	Agri-horticulture
6.Trans-gangetic plains	5	-	5	-	-	Agri-hortisilviculture
7.Eastern plateau and hills	5	-	4	3	4	Block plantation
8.Central plateau and hills	4	-	5	4	4	Block plantation
9.Western plateau and hills	4	-	5	2	4	-
10.Southern plateau and hills	5	-	3	2	4	-
11.East coast plateau and hills	4	5	4	4	2	Home garden/ shelterbelt
12.West coast plains and ghats	4	5	4	2	2	Home garden
13.Gujarat plains and hills	4	2	4	3	5	Block plantation
14.Western dry area	5	3	4	-	5	Shelterbelt
15.The islands	4	5	4	1	2	Homegarden

(Solanki, 2006)

Due to tree selection and improvement efforts, *Populus, Dalbergia, Azadirachta, Acacia, Leucaena, Eucalyptus, Pongamia, Casuarina, A. mangium* x *A. auriculiformis* hybrids etc. are now available. There are Poplar 8 clones; Eucalyptus 1 clone; Jatropha 1 selection; Leucaena 1 introduction; Zizyphus (ber) 2 grafted, 2 selection; Aegle (Bael) 6 buded and Tamarind 2 selection have been developed by the research institutions under agroforestry programmes. Clonal Seed Orchards have been established for Shisham, *Acacia,* Neem etc.

On the basis of identification and evaluation of promising trees and D&D survey of the existing systems, agroforestry interventions were initiated in different agroclimatic regions. Primarily these systems are catagorised into agri-silviculture, agri-horticulture, agri-horti-silviculture, horti-pastoral, silvi-pastoral, and specialized systems. Among these systems, agri-silviculture followed by agri-horticlture, are the most prominent being practiced and advocates in the maximum agroclimatic zones. Home gardens, block plantation, energy plantation, shelterbelts and shifting cultivation are some of the specialized agroforestry systems developed by the research institutions. However the agroforestry systems developed and recommended by the Centres are location specific, besides in many situations these systems are not looked favorably by the farmers due to wide variations in the choice of species, crop preferences etc. and also some of the systems developed have not completed a full rotation for the tree component, therefore biomass production and returns are only on the basis of extrapolation. The stakeholders also require full package of practices, choice of crops with trees in temporal sequence and information on insect pest disease and effect of aberrant weather conditions, which are now of common occurrence.

Agroforestry initiatives in the country has greatest role in creating greater concern among the common man for the protection and preservation of trees, bushes and grasses. As such large amount of woods are now being produced from outside the conventional forestlands. Small landholdings and marginal farmers, through short rotation forestry and agroforestry practices are now providing the bulk of country's domestically produced timber products. This has a great significance as the extraction of timber from the state owned forests has been banned to reduce deforestation and to conserve the environment. The outcome of the efforts has resulted in considerable adoption of agroforestry practices nation wide. Some of the examples are enumerated below:

- Adoption of Eucalyptus, Poplar, Leucaena as important multipurpose agroforestry species by corporate sector (ITC, WIMCO etc.) in large scale plantations, which at present occupy more than 5 million ha.
- In Poplar and Eucalyptus based agroforestry systems the rotation in some states has come down to less than 5 years with improved clones as against short rotations of 6-8 years in the Indo-Gangetic regions.
- In north-western India the agroforestry systems are very prominent land use systems and about 70 million trees of Populus producing 10.40 million m^3 industrial wood annually are standing in U. P., Haryana, Uttrakhand, Punjab, H.P. and J&K with agricultural crops (equivalent to 1,40,000 ha pure plantation @ 500 trees ha^{-1}).

- About 25,000 ha equivalent plantation of poplar in north-west India and 5000 ha of Eucalypts in Andhra Pradesh are now being established every year in association with a number of agricultural, horticultural and medicinal crops with a rotation of 6-8 years.
- Rehabilitation of >3 million ha salt affected wastelands through agroforestry interventions.
- >13 million ETPs (Entire Transplantable Plants)/ year grown in nursery, though now demand is >15 million ETPs.
- Aonla based agroforestry system has been adopted in about 50,000 ha area in different parts of the country.
- Silvipastoral system on an average cycle of 10 years could generate 120 man $days^{-1}$ ha^{-1} year employment.
- Agroforestry practices have been intertwined in the various developmental programmes/ schemes like watershed development.

Focus on Future

Agroforestry is the only pragmatic alternative to meet the target of increasing forest cover to 33% from the present level of 20.68% by 2012. However, it will require appropriate research interventions, adequate investment, suitable extension strategies, harvest process technology, development of new products and market infrastructure. With these inputs the area under agroforestry can be increased from the current level of 7.45 million ha to a sustainable level of 25.36 million ha which will help in augmentation and stabilization of production and productivity, meeting basic needs of tree based produce, minimize ecological degradation and ensure sustainable management of land, water and biodiversity. A major role for agroforestry in near future will emerge in the domain of environmental services. This will need the development of mechanism to reward the rural poor for the environmental services such as biodiversity conservation, watershed protection and carbon sequestration that they provide to society (NRCAF, 2007).

In agroforestry, analogous programmes are being pursued by a number of other agencies such as the Indian Council of Forestry Research and Education (ICFRE), Wasteland Development Board, the Ministry of Environment and Forest etc, including the initiatives on biofuel and bioenergy in several Departments. However, there is little convergence and coordination among the various programmes. There is need to evolve an effective mechanism for synergistically linking and managing the linkages among the various programmes.

In view of changing scenario, the following thrust areas have been identified:

- Increase wood resources to meet the growing needs of households and industries.
- Genetic improvement of identified potential multi-purpose trees.
- Organic farming, mycorrhizae in relation to Agroforestry and integrated pestmanagement in Agroforestry.
- Developing decision support systems for replication of successful Agroforestry systems/ practices.
- Exploring unexploited and under-exploited species of high economic value, such as medicinal, aromatic, oil-yielding plants etc.
- Saline agriculture, including conjunctive irrigation with saline water, developing halophytic crops involving fish, shrimp culture and poultry with Agroforestry systems, genetic improvement of salt-tolerant plants, raising nursery with saline water and multiplication and conservation of useful genetic material.
- Environmental amelioration through the carbon sequestration, bioremediation and resource conservation.
- Enhance livelihood and employment avenues.
- Technology transfer and human resource development.
- Linkages with national and inter-national collaborative research programmes.

Conclusion

The analysis presented here is not based on a comprehensive treatment of all aspects of agroforestry research that has happened in all parts of India, but it gives a brief review of the agroforestry in the past, present and future. Owing to increased supply of wood in the market, there has been a significant increase in the number of factories /industries dealing with wood and wood based ventures. Such industries have promoted agroforestry (through Poplar and Eucalyptus) and contributed significantly in increasing area under agroforestry. On the whole, in addition to promoting indigenous agroforestry models, it appears that a great deal of research needs to be done to identify short rotation, high value species which suit the farmer's requirements of planting on marginal lands. It would probably be more realistic to select trees that could provide more cash benefit to farmers through their products, and to accept that in the longer term they will provide indirect environmental benefits arising from a more complex agro ecosystem.

References

Dhyani, S. K. 1998. Tribal alders- an agroforestry system in the northeastern hills of India. Agroforestry Today (ICRAF), 10(4): 14-15.

Dhyani, S. K., Sharda, V. N. and Sharma, A. R. 2003. Agroforestry for water resources conservation; issues, challenges and strategies. In: Agroforestry Potentials and Opportunities (Pathak, PS and Ram Newaj, Eds), Agribios (India) and ISA.

Dhyani, S. K., Sharda, V. N. and Samra, J. S. 2005. Agroforestry for sustainable management of soil, water and environmental quality; looking back to think ahead. *Range Management and Agroforestry*, 26 (1): 71-83.

NRCAF, 2007. Perspective Plan Vision– 2025. NRC for Agroforestry, Jhansi.

Pathak P.S. 2002. Common pool degraded lands: technological and institutional options,. In: Institutionalizing Common Pool Resources (Ed Marothia D.K.) Concept Publishing Company, New Delhi, India, Pp. 402-433.

Pathak.P.S., Pateria, N.M. and Solanki, K. R. 2000. Agroforestry Systems in India: A Diagnosis and Design approach. National Research Center for Agroforestry. ICAR, Pp. 166.

Sharda, V, N., Dhyani S. K. and Sharma, A. R. 2001. Role of agroforestry in disaster management. HPC (DM Plan) DOA&C, Govt of India, New Delhi, Pp. 37.

Solanki, K. R. 2006. Agroforestry- opportunities and challenges. In: Production Technology and Management of Agroforestry Models (Eds: S.K.Chauhan, S.S.Gill, H.N.Khajuria and R.Chauhan). Agrotech Publishing Academy, Udaipur, Pp. 14-15.

❑❑❑

Agroforestry: Natural Resource Sustainability, Livelihood & Climate Moderation © 2009
Editors: OP Chaturvedi, A Venkatesh et. al., pp. 13-25
Satish Serial Publishing House, Delhi

Chapter 2

CHALLENGES AND SCOPE OF AGROFORESTRY RESEARCH IN INDIA

S.K. Dhyani
National Research Center for Agroforestry, Jhansi-284 003 (U.P.)

Introduction

Agroforestry plays a vital role in the Indian economy by way of tangible and intangible contributions. Although agroforestry is an age-old practice in India, its popularization as a systematic farming system and as a development tool was initiated after the emphatic recommendation of National Commission on Agriculture, 1976. Large-scale initiatives under national and multinational aided programs during 1980s gave the much-needed impetus for agroforestry development. The foundation for organized agroforestry research was also laid around the same time. In the initial phase of agroforestry research and development, the thrust was on basic and strategic research, demonstrating the efficacy and motivating the people and other stakeholders for systematic integration of trees into conventional farming systems. Later on as agroforestry systems demonstrated their immense potential for industrialization, although to a limited scale in certain pockets of the country, the aspirations and expectations of all the stakeholders started growing. The onslaught of liberalization of Indian economy more or less coinciding with the globalization and WTO regime have changed the face and fate of all sectors and agroforestry is not immune to these developments. The impact was seen to be concerned in some form or the other, like highly unstable prices of poplar, teak boom, continued imports of wood for industrial uses like pulp and ply, abnormal crude oil prices, clean development mechanism etc. All these factors have a significant bearing on the direction and pace of agroforestry growth and development.

Agroforestry Research in India

India has been at the forefront since organized research in agroforestry started worldwide. In 1979, a seminar on agroforestry was organized by the Indian Council of Agricultural Research (ICAR), New Delhi at Imphal to accumulate and compile the data on the research and development of agroforestry in India. The All India Coordinated Research Project (AICRP) on Agroforestry was launched in 1983 to implement the major recommendations of the aforesaid seminar. Further, National Research Centre (NRC) for Agroforestry was established on 8th May in 1988 at Jhansi to accelerate the basic, strategic and applied research in agroforestry. At the initial phase, agroforestry research also got financial support from different international agencies primarily USAID, UNDP and others, which helped in strengthening infrastructure and human resource component. At present there are 36 centers under AICRP on Agroforestry with project coordinating unit at NRC for Agroforestry, Jhansi. These centers represent almost all the agroclimates of the country. In addition to ICAR, Indian Council of Forestry Research and Education (ICFRE) and its regional centres, private institutions and NGOs such as WIMCO, ITC, BAIF, IFFDC, West Coast Paper Mills Ltd, Hindustan Paper Mills Ltd, National Tree Growers Cooperatives, etc. are also engaged in research and promotion of agroforestry in the country. Agroforestry research and education has been started in 25 State Agricultural Universities (SAUs). More than 2000 scientists and technicians are engaged in agroforestry R&D at present. The R&D efforts undertaken during the last three decades have clearly demonstrated the potential of agroforestry for resource conservation, improvement of environmental quality, rehabilitation of degraded lands and providing multiple outputs to meet the day to day demand of the rural population. The outcome of the research interventions have been presented in this volume elsewhere.

Area under Agroforestry

The greatest role of agroforestry initiatives in the country has been in creating awareness among the common man for the protection and preservation of trees, bushes and grasses. However, there is always confusion on the statistics of the area under agroforestry in the country. As per one of the estimate about 25.72 million ha is under various types of tree plantations, which include agroforestry, social forestry, farm forestry and others (Table 1). Recently, Dhyani *et al.* (2006) reported 7.45 million ha under agroforestry. The task force on Greening India (Anonymous, 2001) has identified a potential of 10 million ha in irrigated lands and another 18 million ha in rainfed areas that could be developed through agroforestry on a watershed basis. There is scope for using modern tools and procedures like RS and GIS to ascertain the information. NRCAF has taken a lead in this regard recently.

Table 1. Area under various types of tree plantations in India.

S.No.	Plantation type	Area (million ha)
1	Agroforestry and social forestry	23.00
2	Externally aided social forestry	
	(a) Farm forestry / agroforestry	1.24
	(b) Village woodlots	0.50
	(c) Strip plantations	0.07
	(d) Rehabilitation of degraded forests	0.24
	Sub total	2.06
3	By NGOs / VAs	0.04
4	Tree growers co-operatives	0.04
5	Conservative forestry	0.19
6	Afforestation (windbreaks/shelterbelt), Under irrigation region development program of the Indira Gandhi Canal Project	0.40
	Total	25.72

Dhyani *et al.* (2007) has illustrated the potential of agroforestry contributions to sectors like fuel wood, timber, fodder, food, fruits, livestock and bioenergy. However, to realize the potential of agroforestry, appropriate research interventions, adequate investments, suitable extension strategies, harvest process technology, development of new products, value addition and market infrastructure is a must. With these inputs, agroforestry from its current 7.45 million ha area may be increased to 25.36 million ha by next two decades which will help in augmentation and stabilization of production and productivity, meeting basic needs of tree based produce, minimize ecological degradation and ensure sustainable management of land, water and biodiversity.

Fuel Wood : Fuel wood production from various sources was around 169 million t in 1990, which increased to 201 million t by 2001. More than half of this comes from the farmlands and community lands and agroforests. About 10 million rural people are estimated to be engaged in collection of fuel wood (Mahapatra, 2003). Besides this many more are engaged in the marketing of fuel wood in rural and urban areas. The fuel wood scarcity has been obviously felt in areas where forests are absent. Therefore, in such areas agroforestry has played its role to fulfill the firewood demand. Further, as the scope for extraction of firewood from the forests is decreasing gradually, the only alternative left for the people is to patronize agroforestry on farm and community lands. By promoting either silvipasture or pure woodlots, supply from wastelands could be enhanced.

Timber : The scenario of timber for domestic and industrial purposes is given in Table 2. Currently the gap between supply and demand is met through imports. The deficit of 16 million cum (7.23 million tonnes) as on date (2008) could be met by promoting fast growing species like *Eucalyptus,* Subabul, Poplar, *Acacia mangium* etc. and restricting imports. The value addition in processing and the employment generation are other additional contributions. The major contribution of agroforestry has been in the supply of small timber, where more than half of the demand is being met from agroforestry.

Table 2. Demand and supply of timber for furniture, agriculture & industry (Million cum.)

Particulars	2001	2006	2010*	2020*	2025*
Demand	73	82	89	107	116
Supply from forests	12	12	12	12	12
Supply from plantation and agroforestry	47	53	61	90	100
Deficit	14	17	16	5	4

(Note: 2.2 cum = 1 tonne) Source: Ganpathy (1997) and Saxena (1990).* estimates

Agroforestry for Soil Conservation and Environment Protection

Land degradation is a global phenomenon. The problem is acute in India with 120 million ha under wind/ water erosion, water logging, mine spoils, industrial polluted areas, etc. (NBSSLUP, 2008). With the establishment of National Rainfed Area Authority (NRAA) the problems of these areas are expected to be addressed with greater concern. Agroforestry plays a key role in keeping the soil resource productive which is one of the major sustainability issues. Agroforestry played a major role in the recent past in rehabilitation of wasteland such as desert and lands that have been degraded by salinization and ravines, gullies and other forms of water and wind erosion hazards. The example of rehabilitation of more than 3 million ha (out of total of 7.1 million ha) of salt affected soils, particularly the village level community lands, areas along road side, canals and railway tracts through agroforestry interventions have been mentioned elsewhere. The Tree Growers Co-operatives in Gujarat has come in a big way to implement agroforestry on farmlands and wastelands. Silvipastoral system is the most appropriate land use system for these degraded lands. A silvipastoral system comprising *Leucaena leucocephala* (K8) and *Cenchrus ciliaris* and *Stylosanthes hamata* has shown the way to rehabilitate some of the degraded and marginal land with red chalka soil in Andhra Pradesh. On an average the system has productivity (dry matter) of 6 t^{-1} $ha^{-1}yr$ from *Cenchrus ciliaris* and 6 t^{-1} ha^{-1} yr from *Leucaena leucocephala*, besides building up of soil structure. After 6 years, any cereal/ pulse / oilseed crop can be grown on the strip thus rejuvenated by *Cenchrus*- Stylo partnership by carrying out pollarding of *Leucaena leucocephala* trees (Rai and Handa, 2006). Similarly *Albizia lebbeck* –*Cenchrus*- Stylo and *Albizia procera* –*Cenchrus* - Stylo based agroforestry systems have recorded significant improvement in soil organic matter, available N, P and K (Deb Roy 1986). Singh *et al.* (1998) reported higher annual turn over of organic matter, total N and organic S under silvipastoral management in semi arid region. Research efforts made at NRCAF, Jhansi for over a decade has resulted in development of appropriate technologies for the rehabilitation of degraded lands of Bundelkhand region. Singh (2006), reported that *Prosopis* produced 160 t ha^{-1} air-dried firewood in 6 years when planted at 2m x2m spacing. This system improved the soil to such an extent that less tolerant but more palatable fodder species such as shaftal (*Trifolium resupinatum*) and senji (*Melilotus parviflora*) could be grown under mesquite trees after four and half year. The various agroforestry systems enumerated here are productive, remunerative and at the same time they are environmental friendly, e.g. the bacterial population was highest in tamarind based agrisilvihorti system (330×10^{6} CFU g^{-1}) followed by silvipastoral system with *Acacia nilotica* + *Cenchrus* grass. The soil organic carbon status was found to be highest under agrihorti system (0.90 %) followed by silvipasture system (0.76 %) at surface layer, which was minimum under fallow land. It shows that agroforestry systems have a positive impact on soil health and changes in fertility status.

Agroforestry for Carbon Sequestration

Agroforestry has importance as a C sequestration strategy because of C storage potential in its multiple plant species and soil as well as its applicability in agricultural lands and in reforestation. The potential seems to be substantial; average C storage by agroforestry practices has been estimated as 9, 21, 50 and 63 Mg C ha^{-1} in semiarid, subhumid, humid and temperate regions. For smallholder agroforestry systems in the tropics, potential C sequestration rate ranges from 1.5 to 3.5 Mg C ha^{-1} yr^{-1} (Montagnini and Nair, 2004). Agroforestry can also have an indirect benefit on C sequestration when it helps to decrease pressure on natural forests, which are the largest sinks of terrestrial C. Another indirect avenue of C sequestration is through the use of agroforestry technologies for soil conservation, which could enhance C storage in trees and soils. The analysis of C stocks from various parts of the world showed that significant quantities could be removed from the atmosphere over the next 50 years if agroforestry systems are implemented on a global scale (Singh *et al.*, 2007).

India's CO_2 equivalent emission form all energy, industrial processes, agricultural activities, land use, land use changes and forestry (LUC&F) and waste management practices for the base year 1994 were 1228.54 million t per year. If the area under agroforestry, afforestation and reforestation programme is optimized, then this land use will have maximum potential to C sequestration for next 25years. The area under plantation including agroforestry is expected to be 94.7 million ha. Out of that the area under forest is 69.70 million ha and in agroforestry 25 million ha. Thus the total C sequestration potential of these areas would be 118.37 million tones C ha^{-1} yr^{-1} in next 25 years. Looking to the present and future rate of CO_2 equivalent emission in next 25 years would be 1227.68 million tones C yr^{-1}. The forestry and agroforestry would sequester 9.3% atmospheric CO_2 into tree biomass. Steps proposed for terrestrial C management strategies and potential land use includes C sequestration and C conservation. For C sequestration the steps may include afforestation, reforestation and restoration of degraded and wastelands; improved tree management practices to increase growth rates; implementation of agroforestry practices/ systems on agriculture and pasture lands. For C conservation it may be conservation of biomass and soil C in existing sinks; improved lopping and harvesting practices, improved efficiency of wood processing; fire protection and more effective use of burning in both forest and agricultural systems; and for C substitution (substitution of fossil fuels): Increased use of biofuels, increased conversion of wood biomass into durable wood products for use in place of energy-intensive materials. Agroforestry practices such as: agri-silviculture or agri-horticulture systems for food and wood/fruit production; boundary and contour planting for wind and soil protection; Silvi-pasture system for fodder production as well as soil and water conservation; complex agroforestry systems, viz. multistrata tree gardens, home gardens, agri-silvi-horticulture and horti-silvi-pasture systems for food, fruits and fodder especially in hill and mountain regions and coastal areas and biofuel plantations are suitable for sequestering atmospheric C. C storage depends on several factors including climatic, edaphic, and socio-economic conditions. Perennial systems like home gardens and agroforestry can store and conserve considerable amounts of C in living biomass and also in wood products. Agrihortisilviculture and silvipastoral systems of agroforestry have been observed as the potential sinks for sequestering surplus C from the atmosphere. Since agroforestry increases the income of farmers, it reduces the incentive for future extraction from natural forest for income generation.

Agroforestry for Food Security

The country's food production has increased four folds in the past five decades, but recent improvements in food supply have been insufficient to fulfill the nutritional needs of the average person in the country. Increasing the production and consumption of protein-dense foods, pulses in particular, will be necessary if the country is to meet its protein needs. By virtue of diversity of the components in the agroforestry systems like food grains, vegetables, fruits, nutritional security to the communities could be ensured. A break through in agroforestry was achieved by introduction of *Poplar* in Punjab, Haryana and Uttar Pradesh during 1960. Later this tree became very popular due to fast growth with a short rotation of 6-8 years. Deciduous growth nature, amicability to pruning and clean bole renders the species ideal for agroforestry. It is a soft wood plant, hence ideal suited for match splint, paper pulp and plywood manufacturing. Similar compatible tree-crop combinations are also seen to a lesser extent in *Ailanthus, Eucalyptus, Dalbergia, Parkia* and other MPTS in different parts of the country. Initially for first 3-4 years any crop combination can be selected with trees, but once the canopy is fully developed apart from management of canopy, change of crops such as shade loving (e.g. rhizomatous crops, legumes, grasses or medicinal plants) will ensure adequate yields. If proper management techniques such as pruning, pollarding, lopping, thinning are followed and judicious crop sequencing is observed then the agroforestry practices will ensure adequate crop yields and overall economic benefits to the stake holders. Under agroforestry system, fruits owe special significance on account of short juvenile phase than timber trees and assure regular production for long period (up to 40 years). Experiences have indicated that agrihorticulture obviously due to assured returns and notional security is preferred by the farmers than agrisilviculture. In recent years, Aonla (*Emblica officinalis*) has been promoted under agrihorticulture system and already more than 50,000 ha area is under improved cultivars of the species. A number of industries utilizing its products are providing value addition as well as employment generation. Further, on wastelands there is plenty of scope for introducing fruit trees as hortipasture components for immediate monetary gains and higher biomass yield without hampering forage production. The estimated demand for food by 2025 will be 320 million t and food crops grown under agroforestry alone are likely to contribute about 25 million t yr^{-1} (NRCAF, 2007).

Agroforestry for Fodder Production

Trees and shrubs often contribute substantial amount of leaf fodder in arid, semi arid and hill regions during lean period through lopping/pruning of trees, popularly known as top feed. The fodder scarcity could be reduced through fodder cultivation on farmlands and through silvipasture on farm/community wastelands. Promotion of silvipasture contributes in two forms namely additional fodder supply that ensures proper rearing of livestock and environmental protection. Small ruminant farming based on goats and sheep is most ideal under agroforestry. This forage is usually rich in proteins, vitamins and minerals like calcium though in general low in phosphorus and crude fibre (Singh, 1990). Such top feeds species play an important role in human food security through their function as animal feed resource. The importance of top feeds increases with the severity of drought and progression of drought season. Other uses such as for live fencing are complimentary, as they encourage cultivation of the species and increase the availability of feed. The top feeds are considered very important in vegetation stabilization and sustained productivity of rangelands. They also play an important role as windbreaks and by providing shade for the grazing animal. The important ones are *Prosopis cineraria, Albizzia lebbek. Acacia* spp., *Leucaena leucocephala, Dalbergia sissoo, Ailanthus excelsa, Azadirachta indica, Acacia leucophloea etc.* The species such as *Aegle marmelos, Albizzia amara, Bauhinia* spp., *Butea monosperma, Cordial dichotoma (syn.C.myxa), Ficus spp., Hardwickia binata, Holoptelia integrifolia, Pithecellobium dulce (syn. Inga duluce), Morus alba and Moringa oleifera* have also been found to be useful as top feeds for livestock. *L. leucocephala* could be fed to goats to increase the rate of productivity per unit of land in terms of products including milk and meat. Thus leaves provide enough nutrients and can serve very well as a green fodder supplement. Besides providing green fodder, such leaves are also conserved in the form of hay and silage to supplement feed during scarcity periods. The mean forage production under silvipastoral systems is shown in Table 3.

Table 3: Mean forage production (dry) under different Silvipastoral system

Tree species	Pasture species	Pasture yield (t ha^{-1})
Albizzia lebbeck	*Cenchrus ciliaris*	7.4
	Chrysopogon fulvus	7.8
	Sehima nervosum	7.8
Dalbergia sisso	Natural pasture	2.9
Dichrostachys cinerea	*Cenchrus ciliaris*	4.3
	Chrysopogan fulvus	5.2
Leuceana leucophala	*Cenchrus ciliaris*	4.1
	Cenchrus setigerus	3.2

Source: Deb Roy, 1988; Singh and Pathak, 1992

Agroforestry for Energy Plantation

The main biomass energy sources in rural areas, which are being used in the households, include wood from forest, croplands and homesteads, cow dung and crop biomass. Bioenergy from tree biomass is one significant contribution from agroforestry. Species like *Prosopis juliflora* are being used for power generation in states like Karnataka and Andhra Pradesh. *P. juliflora* due to high calorific value of over 5000 Kcal is preferred in these power plants. An estimated 0.51 million ha area is under *P. juliflora*. Even if 25 percent of the area is meant for power generation leaving the rest 75 percent for fuel and charcoal, the bioenergy potential works out to 1000 MW. Similarly other tree species can be used for running the biomass based power plants. Thus a total of 5000 MW power could be produced from the biomass sources from trees under moderate conditions, which meets almost one third of the ultimate potential of 16000 MW from biomass. Promoting bioenergy through *P. juliflora* also encourages tremendous employment generation to the tune of 6.34 million mandays and 7.03 million woman days for fuel making in Tamil Nadu alone. The fuel wood potential of indigenous (*Acacia nilotica, Azadirachta indica, Casuarina equisetifolia, Dalbergia sissoo, Prosopis cineraria and Ziziphus mauritiana*) and exotic (*Acacia auriculiformis, A. tortilis, Eucalyptus camaldulensis* and *E. tereticornis*) trees revealed that calorific values ranges from 18.7 to 20.8 MJ kg^{-1} for indigenous tree species and 17.3 to19.3 MJ kg^{-1} for exotics. Species such as *C. equisetifolia, Prosopis juliflora, Leuceana leucocephala* and *Calliandra calothyrsus* have become prominent due to their potential for providing wood energy at the highest efficiency, shorter rotation and also their high adaptability to diverse habitats and climates.

Agroforestry for Biofuel Production

In India the energy demand is expected to grow at 4.8%. Further in view of the increasing gap between demand and domestically produced petroleum our dependence on import of oil will increase in the foreseeable future. It has been reported that the import of crude oil has gone up from 85 mmtpa to 147 mmtpa by the end of the last year correspondingly increasing the import bill from US$ 13.3 billion to US$ 15.7 billion. In this connection, fuels of biological origin have drawn a great deal of attention during the last two decades. Biofuels are renewable liquid fuels coming from biological raw materials and has proven to be good substitute for oil in the transportation sector as such biofuels are gaining worldwide acceptance as a solution for problems of environmental degradation, energy security, restricting imports, rural employment and agricultural economy. The potential tree borne oilseeds (TBOs) holding promise for biofuel are *Jatropha curcas, Pongamia pinnata,* Simarouba, *Azadirachta indica, Madhuca spp.,* etc. In agroforestry system, *Jatropha curcas* has been intercropped with annual crops such as cowpea, sesame, sunflower, French bean, black gram, green gram and groundnut etc. at various places. The promotion of the use of oils could also provide a poverty alleviation option in the rural areas. Farmers can use vacant, waste and marginally used land for growing such trees and benefit from the annual produce, which will add as their income. With the increased green cover the environment will also benefit greatly. The use of oils is also CO_2 neutral, which would mitigate green house effect. But the economics and viability of the Jatropha plantation and biofuel production are still at initial stage and will be governed by international market prices of crude oil as well as government policies.

Agroforestry for Socioeconomic Upliftment and Employment Opportunities

Dhyani *et al.* (2003) have highlighted the role of agroforestry products and environmental services to meet the subsistence needs of low income households and providing a platform for greater and sustained livelihood of the society. Increased supply of wood in the market has triggered a substantial increase in the number of small-scale industries dealing with wood and wood based products in the near past. Such industries have promoted agroforestry and contributed significantly to increasing area under farm forestry. Recognizing agroforestry as a viable venture, many business corporations, limited companies such as ITC, WIMCO, West Coast Paper Mills Ltd, Hindustan Paper Mills Ltd., financial institutes such as IFFCO have entered into the business and initiated agroforestry activities in collaboration with farmers on a large scale. One of the major contributions of agroforestry for the economy is the livelihood impact, both in terms of income and employment generation. Besides the existing agroforestry practices, there is a tremendous potential for employment generation with improved agroforestry systems. Dhyani and Sharda (2005) have indicated the potential of agroforestry for rural development and employment generation to the tune of 5.763 million human days yr^{-1} from Indian Himalayas alone. Agroforestry models adopted by the farmers in upper gangetic region and specially in Haryana are highly lucrative, therefore, attracting farmers in a big way. The agroforestry models include: (1) poplar (*Populus deltoides*)-sugarcane-turmeric block plantation model; (2) poplar-sugarcane-wheat-chari (*Sorghum bicolor*) block plantation model; (3) poplar-sugarcane-wheat-chari-potato-maize-bajra (*Pennisetum glaucum*) block plantation model; (4) poplar-sugarcane-potato-berseem (*Trifolium alexandrinum*)-chari block plantation model; (5) poplar-paddy-wheat boundary plantation model; and (6) poplar-sugarcane-wheat-paddy boundary plantation model. The net present value for the different agroforestry models on six years rotation varied from Rs. 26,626 to Rs. 72 705 ha^{-1} yr^{-1} whereas the benefit: cost ratio and the internal rate of return varied from 2.35 to 3.73 and 94 to 389%, respectively. Agroforestry has not only uplifted the socioeconomic status of the farmers but also contributed towards the overall development of the region (Kumar *et al.*, 2004). Similar examples are now available from other parts of the country. The economic analysis of the dominant agroforestry systems in different parts of the country reveal a B-C ratio ranging from 1.17 to 16.5 with a completion of one rotation (Anonymous, 2001). Increased supply of wood in the market has triggered a substantial increase in the number of small-scale industries dealing with wood and wood based industries in the near past. Such industries have promoted agroforestry and contributed significantly to increasing area under farm forestry. Recognizing agroforestry as a viable venture, many business corporations, limited companies have entered into the business and initiated agroforestry activities in collaboration with farmers on a large scale. Agroforestry got a major boost in northern India through the farmer friendly policies of an industrial plywood company, WIMCO ltd. This company has not only motivated the farmers to plant trees, such as poplar, on their farmlands but also created a big impact on the economy on these states by creating assured market for poplar wood. Similarly ITC Bhadrachalam Paper Board Ltd. promoted *Eucalyptus, Casuarina and Leucaena* in Andhra Pradesh and other southern states; West Coast Paper Mills Ltd promoted *Acacia mangium* x *A. auriculiformis* hybrid and Hindustan Paper Mills Ltd promoted Bamboo plantation.

Policy and Development Interventions Needed for Maximizing the Potential of Agroforestry

The suggested potential could only be tapped in the next two decades provided the following policy and development initiatives are taken up:

- Promotion of production and supply of quality planting material of agroforestry species through incentives and by promoting adequate infrastructure.
- Formulation and implementation of special line of scheme for credit to agroforestry.
- Rationalizing the tree felling and transit norms/rules
- Replicating the success of linkage models between farm and industry, by emphasizing public private partnerships.
- Relaxation in land leasing rules by the state governments for agroforestry.

References

Anonymous 2001. Report of the Task Force on Greening India for Livelihood Security and Sustainable Development. Planning Commssion, GOI, Pp. 231.

Deb Roy, R. 1986. Studies on some aspects of renewable energy (fuel) and pasture production on ecosystem of *Albizzia procera* and *Albizzia lebbeck* with grass and legume, Ph.d thesis Jiwaji University, Gwalior, M.P.

Dhyani, S. K., Sharda, V. N. and Sharma, A. R. 2003. Agroforestry for water resources conservation: issues, challenges and strategies. In: Agroforestry Potentials and Opportunities, (Pathak, PS and Ram Newaj, Eds), Agribios (India) and ISA.

Dhyani, S.K. and V.N, Sharda, 2005. Agroforestry systems as rural development options for the Indian Himalayas. *Journal of Tree Science*, 24 (1): 1- 19.

Dhyani, S.K., Kareemulla, K. and Handa, A.K. 2006. Agroforestry potential and returns across agroclimatic zones. NRCAF, Jhansi.

Dhyani, S.K., Samra, J.S, Ajit, Handa, A.K. and Uma. 2007. Forestry to Support Increased Agricultural Production: Focus on Employment Generation and Rural Development. *Agricultural Economics Research Review*, 20: 179-202.

Ganapathy. P.M, 1997. Sources of Non-wood fibre for paper, board and panels production: status, trends and prospects for India. Asia Pacific Forestry Sector Outlook study. Working paper series. Working paper no. APFSOS/WP/10 in Sustainable Forestry Management. Ford Foundation. New Delhi.

Kumar, R., Gupta, P, K. and Gulati, A. 2004. Viable agroforestry models and their economics in Yamunanagar District of Haryana and Haridwar District of Uttaranchal. *Indian Forester*, 130(2): 131-148.

Misra V.K. 2002. Greening the wasteland: experiences from the Tree Growers Co-operatives Project. In: Institutionalizing Common Pool Resources, (Ed. Marothia D.K.), Concept publishing company, New Delhi, India, Pp. 334-354.

Montagnini, F. and Nair, P.K.R. 2004. Carbon Sequestration: an under exploited environmental benefit of agroforestry systems. *Agroforestry Systems*, 61: 281-295.

Pathak. P.S., Pateria, N.M. and Solanki, K. R. (2000). Agroforestry Systems in India: A Diagnosis and Design approach. NRC for Agroforestry. ICAR, Pp. 166.

Rai, P. and Handa, A.K. 2006. Reclamation of degraded lands through agroforestry intervention. In: Production Technology and Management of Agroforestry Models. (Eds: S.K.Chauhan, S.S.Gill, H.N.Khajuria and R.Chauhan), Agrotech Publishing Academy, Udaipur. Pp. 64-65.

Sharda, V. N., Dhyani S. K. and Sharma, A. R. 2001. Role of agroforestry in disaster management. HPC (DM Plan) DOA&C, Govt of India, New Delhi. Pp 37.

Singh, G. 2006. Agroforestry models for rehabilitation of sodic soils. In: production technology and management of agroforestry models (Eds S.K.Chauhan, S.S.Gill, H.N.Khajuria and R.Chauhan), Agrotech Publishing Academy, Udaipur. Pp. 27-33.

Singh, G., Singh, H and Bhojvaid, P. P. 1998. Amelioration of sodic soils by trees for wheat and oats production. *Land degradation and management*, 9 (5): 453-462.

Singh, P. 1990. Agrisilviculture systems in India. In: shrubs and Tree Fodder for Farm Animals (Eds.C.Devendra), IDRC, Ottawa (Canada): 186-195.

Singh, S. 2000. Economics of Poplar with intercropping. M.Sc. Thesis, Department of Economics and Sociology, PAU, Ludhiana (unpublished).

Singh, V.P., Dhyani, S.K., Verchot, Lou and Lasco, R. 2007. Climate change and agroforestry. National Conference on Climate Change and Indian Agriculture. Organized by ICAR at NASC, New Delhi, Oct. 12-13, 2007.

Solanki, K. R. 2006. Agroforestry- opportunities and challenges. In: Production Technology and Management of Agroforestry Models (Eds: S.K.Chauhan, S.S.Gill, H.N.Khajuria and R.Chauhan). Agrotech Publishing Academy, Udaipur, Pp. 14-15.

Mahapatra, R, 2003. The multimillion dollar fuel wood trade is the last resort for India's poor. Infochange features. www.infochange.org.

NRCAF 2007. *Perspective Plan: Vision 2025*. NRCAF, Jhansi.

Saigal, S., H.Arora, and Rizvi, S.S. 2002. The new foresters. The role of private enterprises in the Indian forestry sector. IIRD, London. Pp. 151.

Saxena, N.C. 1990. Farm Forestry in North West India: Lessons from the 1980s. Studies of agroforestry in western U.P. Final report of APCF scheme.

❑❑❑

Agroforestry: Natural Resource Sustainability, Livelihood & Climate Moderation © 2009
Editors: OP Chaturvedi, A Venkatesh et. al., pp. 27-42
Satish Serial Publishing House, Delhi

Chapter 3

AGROFORESTRY RESEARCH PERSPECTIVES AND POTENTIAL

O.P. Chaturvedi and I.A. Khan
National Research Centre for Agroforestry, Jhansi-284 003 (U.P.)

Introduction

Agroforestry systems have existed since the very beginnings of plant domestication. There are innumerable examples of this traditional land use practices in many parts of the world (Nair, 1993). Agroforestry has a long tradition in India. The socio-religious fabric of the people of the country is interwoven to a very great extent with raising, caring for and respecting trees. Trees are integrated extensively in the crop-and livestock-production systems of the country according to the agro-climatic and other local conditions. Agroforestry is a farming system that integrates crop and/or livestock with trees and shrubs. The resulting biological interactions provide multiple benefits, including diversified income sources, increased biological production, better water quality, and improved habitat for both humans and wildlife. A traditional tree farm managed as single-purpose monocrop is not an agroforestry system. Neither is a woodlot when it's managed for wood products only. Agroforestry involves combining a tree planting with another enterprise-such as grazing animals or producing mushrooms-or managing a woodlot for a diversity of special forest products. For example, an agroforestry system might produce firewood, biomass feedstock, mulch, fodder for grazing animals, and other traditional forestry products. At the same time, the trees are sheltering livestock from wind or sun, providing wildlife habitat, controlling soil erosion, and the case of leguminous species fixing nitrogen to improve soil fertility. Agroforestry has vast potential not only to meet the demands of fuel, fodder, timber, medicine and other Non-wood Forest Products but also to enhance green cover, soil and water conservation and reduce pressure on forests. The objective of agroforestry is to take advantage of the complimentary relationships between trees, crops and livestock in such a way that the productivity, stability and sustainability of the total system are maintained. It has great potential in improving degraded and affected sites and has been recognized as a restoration agent, rehabilitation process, and bioremediation mechanism, in addition to increasing farm income.

Agroforestry systems have the potential to evolve through succession towards mature, productive system to form a mosaic of patches on landscape while producing marketable tree products for improved livelihoods. Agroforestry is increasingly being identified as an integrated land use that can directly enhance plant diversity while reducing habitat loss and fragmentation (Noble and Dirzo, 1997).

In India, the origin of agroforestry is believed to be since the Vedic period. In Ramayana 'Ashok Vatikia' and in Mahabharata 'Ashrams' depict the Agroforestry practices. The road side plantations by King Ashoka to Shersah Suri and ' Sacred groves' symbolizes the Agroforestry. In India, organized agroforestry research was started in 1983 with conceptualization and initiation of AICRP on Agroforestry after Imphal Seminar in 1979. The Agroforestry systems, collections, screening and selection of trees and shrubs for agroforestry including tree improvement and developing sustainable Agroforestry systems and their management. Considerable progress in Agroforestry research and extension has been made during last two and half decades.

Perspectives of Agroforestry Research

Webster's Ninth New Collegiate Dictionary defines research as *"investigation or experimentation aimed at the discovery and interpretation of facts, revision of accepted theories of laws in the light of new facts, or practical application of such new or revised theories of laws."* With such a broad view of research, agroforestry encompasses different perspectives. Perhaps there is no clear hierarchy of research levels and categories. But there are various dimensions and perspectives of research in agroforestry encircling a broad range of issues as summarized by Nair, 1993 (Table 1).

Table 1: Different perspectives of agroforestry research

Basis	Category/Operational unit	Example/Type of research
Organizational level	Ecosystem Farm/Plot Component Cellular/Molecular	Agroforestry-system design Field experiments MPT evaluation Biotechnology
Stage of technology generation	Exploratory Component/System mgmt. Prototype	Survey Plant-plant interactions Alley cropping
Subject	Biophysical Social, economic, political	Soil productivity Economic evaluation
Nature	Methodological Experimental	D & D, Stat. methods Plant- and soil management
Application of results	Basic Applied Strategic Adaptive	DNA, N_2-fixation process MPT improvement Genotype Soil-erosion control
Nature of questions addressed	What Why How	The result of growing crops near tree rows Why it happens? How it happens?
Place of research	On-station On-farm	On research stations On farmers fields

At the organizational level, agroforestry research could be dealing with the ecosystem (e.g., agroforestry systems design), the field/plot (e.g., management/adoptability investigations), the individual component (e.g., genotypes), or the cellular/molecular (e.g., biotechnology applications) level. Agroforestry investigations can also be considered in terms of the specific stages of their development in the process of technology generation. For example, the first step in the D & D procedure (prediagnosis and diagnosis) involve what can be termed as exploratory (or survey type of) research. Then there are specific management trials of components and systems (e.g., methods of planting, plant arrangements, or pest management). These trials lead to the development of prototype technologies, i.e., the products of research and synthesis, the practical performance of which have not been tested. Another way of looking at agroforestry research is based on the subject issues. Thus research could be on biophysical aspects or socioeconomic-policy issues. Research can also be methodological or experimental in nature. Methodological research includes development of methodologies for undertaking research in specific areas and subjects, i.e., developing procedural frameworks; experimental research involves testing of hypothesis according to one or more specific methodologies.

A distinction can also made between basic and applied research. Basic or fundamental research investigates processes and mechanisms; examples are research dealing with such issues as DNA, and fundamental processes of nitrogen fixation or photosynthesis. Applied research involves application of research results to solve specific problems. There are two categories of applied research in agricultural sciences; strategic and adaptive. Strategic research refers to innovative application of basic-research results to solve practical problems in the medium-to-long term; an example is the development of dwarf cultivars of cereals; another one is the development of alley-cropping technology. Adaptive research, on the other hand, refers to development of location-specific technologies for solving practical problems of an immediate nature; the use of alley cropping technology for soil conservation in a specific location is an example. Another perspective that is related to the concepts of basic and applied research is the nature of questions addressed in research. Thus there can be "what", "why", and "how" types of research. "What" research is mostly observational in nature; for example, what happens if a hedgerow is pruned according to a particular schedule of what happens if the crop is grown in association with the tree. The results obtained from such trials are highly site-specific. "Why" type of research tries to discover why the observed behavior happens, including how could it be different from what has been observed. The "what" type of research is mostly of applied nature, and undertaken by technicians, whereas the "why" and "how" type are of a more basic or strategic nature, and are in the domain of the research scientist.

Yet another way of looking at agroforestry research is based on the place of site of research, i.e., "on-station" or "on-farm". On-station refers to research conducted on research stations, whereas on-farm refers to the research in farmer's fields or other places outside research stations, with or without the involvement of the farmer or the land-user.

Agroforestry systems prevalent in India

Systems/Practices	Agro-ecological adaptations
A. Agrisilvicultural Systems	
1. Shifting cultivation	In tropical forest areas
2. Taungya	In all regions
3. Homegardens	Mainly humid tropics
4. Plantation based cropping systems	Mainly humid tropics
5. Scattered trees on farm lands regions	In all regions specially semi-arid and arid
6. Shelterbelts and wind breaks arid, alpine areas	In wind prone areas, all regions viz., coastal,
7. Boundary plantations and live hedges	In all regions
8. Woodlots for soil conservation lands	In hilly areas, along sea coast and ravine
9. Trees on rangelands	In all regions
10. Plantation crops with pastures and animals with less grazing pressures	Mostly humid and sub-humid regions
11. Industrial plantations with crops	Intensively cropped areas in north India
B. Agri-silvopastoral Systems	
1. Seasonal forest grazing	Semi-arid and mountainous ecosystem
2. Aquaforestry	Low lying lands
3. Agriculture with trees	All regions
C. Silvopastoral	
1. Silvipastures	Sub tropics and tropics with bioedaphic climaxes
2. Hortipastoral	In hilly orchards for soil conservation and forage.

Agroforestry Research in India

In India the organized agroforestry research began in 1979 at Imphal. In 1983, All India Co-ordinated Research Project on Agroforestry (AICRPAF) a large network project with 20 centers was initiated and later on more centers were added. At present there are 36 AICRPAF centers covering all the major agro-ecological zones of India with headquarters at National Research Centre for Agroforestry, Jhansi. During the last 25 years significant research work has been done on different aspects of agroforestry. The major research thrusts under this initiative have been on diagnostic survey and appraisal of existing agroforestry practices, collection and evaluation of MPTs for agroforestry, studies on management of agroforestry systems, environment, watershed, community land development and integration of different components in agroforestry. Based on D&D exercises following thrust areas were identified for research.

A. Collection and Evaluation of MPTS

Multipurpose trees are defined as all woody perennials that are purposefully grown to provide more than one significant contribution to the production and/or service function of a landuse system. Considerable attention is being paid in identification and collection of suitable germplasm and location of genetic resources through exhaustive network of seed sources, seed production areas and provinence trials, not only indigenous species but also promising exotics such as *Eucalyptus, Populus, Albizia, Paulownia, Acacia* etc. The priority tree species identified by AICRP on Agroforestry Centres and ICFRE for different regions are:

1. Himalayan Region

Grewia optiva, Morus alba, Celtis australes, Eucalyptus spp., *Alnus nepalensis, Acacia* spp., *Pinus spp., Parasenanthes, Falcataria, Albizia spp., Terminalia spp.*

2. Indo-Gangetic Plains

Populus spp., Acacia spp., Dalbergia sissoo, Anthocephalus cadamba, Casuarina equisetifolia, Prosopis chilensis, Eucalyptus spp.

3. Arid and Semi arid Regions

Prosopis chilensis, Azadirachta indica, Dalbergia sissoo, Leucaena leucocephala, Acacia spp., Bamboo, Hardwickia binata.

4. Humid and Sub-humid

Gmelina arborea, Acacia spp., Dalbergia sissoo, Bamboo.

5. Tropical Coastal and Island Region

Tamarindus indica, Azadirachta indica, Ceiba pentandra, Tectona grandis, Gliricidia spp.

Promising provenances, varieties/cultivars/clones of *Populus, Dalbergia sissoo, Azadirachta indica, Acacia nilotica, Eucalyptus spp., Leucaena leucocephala, Prosopis cineraria, Bombax ceiba, etc.* have been developed through selection and hybridization. Seed orchards of promising species like *Prosopis* at Karnal and Jodhpur, Babul at Hisar and Ludhiana, Neem at Jodhpur, Jhansi and Kushinagar, Poplar at Pantnagar and Ludhiana have been established for large scale multiplications. All the centres of AICRP on Agroforestry have started working on improvement of two allotted MPTs through collection and evaluation process.

B. Agroforestry Systems Research

The impact of agroforestry research/technologies generated in the last 25 years is visible in many areas. Recently, area under various tree plantations is 25.32 million ha and small holders have emerged as timber suppliers of 21st century. However, the assessment of impact of agroforestry is hampered by lack of rigorous procedure and methodologies.

The farmers of Punjab, Haryana, Western Uttar Pradesh and Terai belt of Uttaranchal Terai have adopted Poplar-Eucalyptus based agroforestry in a big way covering an area of 11.48 million ha. The poplar planted on agricultural fields and field boundaries have rotation of six-eight years. Punjab having only 2-3% area under forests, now has become a major wood surplus state and is supplying wood to Himachal Pradesh for making packaging boxes for fruits and also to some other Southern state for making paper. Poplar planted in agroforestry system is estimated to provide gross return of Rs.3.27 lacs per acre for seven years or a net discounted return of Rs.88,749 per acre in seven years at 12% rate of interest on investment. Without intercropping, the net discounted return from poplar growing is only Rs.53,476 per acre. The cost benefit ratio were 1:1.92 with intercropping and 1:2.13 without intercropping. The returns are substantially higher than raising sole agriculture crops. In northern Haryana the financial analysis of agroforestry systems showed net returns of Rs. 54,585; 11,2142 and 27,9080 per acre over five, six, and eight years, respectively. It works out to Rs. 10,917; 18,960 and 34,885 per acre per year. Whereas, net returns on the basis of cost of cultivation comes to Rs. 2,869 $acre^{-1}$ yr^{-1} from paddy-wheat rotation, generally followed in this region. Even with low prices of wood, poplar based agroforestry systems is more economical than monocropping of agricultural crops.

In poplar growing areas 30 million trees are producing 1.125 million m^3 industrial wood annually and around 25,000 hectare equivalent plantation are being established every year in association with number of agricultural and horticultural crops. In order to cover such acreage of plantation, 16 million ETPs are annually grown in nurseries and 13 million saplings are planted annually. This has created much needed employment to many people not only in plantation activities but in industries also.

In regions of high fertility and rainfall like Kerala, small farmers usually maximize returns from land through multistoreyed cropping where perennial crops such as coconut, arecanut, rubber and pepper are intercropped with seasonal and annual crops like tapioco, bananas, pulses and vegetables. Trees with multiple uses such as yielding fruit, fodder and mulch are preferred.

Improvement of degraded lands, poverty elevation, productive employment generation, fuelwood, timber and fodder production by the active participation of women, landless and disadvantaged section of the society was launched by Indian Farm and Fertilizers Co-operative (IFFCO) through its subsidiary called Indian Farm Forestry Development Co-operative (IFFDC) during 1986-1987. Leased wastelands of revenue departments of Govt. of Madhya Pradesh, democratically elected village bodies in Panchayat of Rajasthan and private lands in Uttar Pradesh were managed by the Cooperatives. Even a landless, or socially disadvantaged wage earner engaged for planting agroforestry spares a part of his/her wage to buy shares. During 1986-2004 about 26060 ha of wastelands was planted with multipurpose trees of 5.10 million. A total of 145 primary Farm Forestry Cooperative Societies with a membership of 28287 and 820 Self Help Groups were registered. A revolving fund of Rs.18.44 million was circulated among the SHG. Availability of fodder and credit promoted livestock rearing, conservation and amelioration of soil, water and biodiversity were important activities of the societies. Similarly, the Tree Grower Cooperative floated by National Diary Development Board has came in a big way to implement agroforestry on farmland and wastelands.

Prosopis juliflora in addition to provide fuelwood is also creating a large market of high value products. *P. juliflora* in Ramnathpuram, Distt. Of Tamil Nadu producing 15,000 tonnes of Charcoal which is transported to Chennai. Now, the charcoal making has become a cottage industry in parts of southern Tamil Nadu. As fuelwood and charcoal, *P.juliflora* offers employment to the tune of 6.34 million man days and 7.03 million women days per annum in southern Tamil Nadu. Apart from charcoal making, it has potential to be utilize in gassifiers for generating power units and thus, more prospects for improving livelihood. Three power plants in Andhra Pradesh have developed their captive plantation of *P. juliflora* for bioenergy. The commissioned biomass based power plant capacity in Karnataka and Andhra Pradesh was 266 MW with 46 power generation units in operation.

Ailanthus excelsa has an excellent market in Rajasthan and North Gujrat where its leaves are sold as fodder for goats, sheep and camel. Its timber has great demand in construction work and in plywood industries. The average, additional income received from planting *Ailanthus excelsa* is Rs. 7600 ha^{-1} yr^{-1}. The *Khejri* based agroforestry system existed in Rajasthan since time memorial. *Khejri* (*Prosopis cineraria*) green pods are being used as vegetables and green leaves (dried) called long used as cattle feed. This tree is commonly known as 'Kalpvarkhsa' and also declared as State tree of Rajasthan. The economic analysis of Khejri based agroforestry system revealed that from bajra alone gross income of Rs. 1600 ha^{-1} can be obtained but from the *Khejri* (30 trees ha^{-1}) inter cropped with bajra is capable of providing an income of Rs. 4600 ha^{-1}.

In Bihar Litchi is a major commercial crop and litchi based agrihorticulture system is widely spreaded in its growing areas. On a nine-year cycle of litchi based agri-horti system, establishment, management and harvest of fruits and intercultural operations of intercrops generate employment opportunities of 130-140 man days ha^{-1}. With development of suitable cropping pattern more and more farmers are adopting this diversified technology, which helps in generating employment and improving economy.

Silvipastoral system on an average cycle of 10 years could generate 120 mandays ha^{-1} yr^{-1} employment i.e. for 50 million ha - 20 million people yr^{-1} employment and 5 million people yr^{-1} for other ancillary activities employment.

In arid and semi arid tracts, there is about 40% land in degraded condition, which have possibility of being cultivated by tree/shrubs species like Ratanjot, Karanj, Neem, Mahua, Paradise tree and Jojoba for oil seed production. The total potential of tree borne oil seeds in India is 52.35 lakh MT and it would create employment to 0.44 million.

The un-measurable indirect impacts of agroforestry include improvement of soil quality for higher production, controlling desertification, checking soil erosion, providing year round employment to small and marginal farmers, landless labourers and particularly the rural women are also evidenced.

The changing climate of the earth has adverse effect on the sustainability poor due to loss of productivity of crops and can be improved through carbon sequestration by reducing CO_2level in the atmosphere. Agroforestry provide twin advantages by storing carbon through enhanced build-up of soil organic matter and increased wood biomass production that would absorb more CO_2 from air. Biomass production through multi species stands in agroforestry on degraded lands should be priority for providing sustainable environment. Poplar exhibits some of the highest CO_2 exchange rates and photosynthetic capacities among woody plants and therefore, forms a unique combination with agricultural crops for C sequestration in the atmosphere.

Potential of Agroforestry for Sustainable Development

Agroforestry systems have evolved with time in response to population pressure and depletion/degradation of non-renewable resources. The levels of productivity that can be achieved reflect the potential and the degree of the management of the resource base. High productivity comes only from systems where management intensities that are necessary for sustainability are attained without extensive depletion of resources. Forestry is faced with the challenge of meeting an increasing demand for goods, as well as for an expanding array of services, like clean water, soil conservation, and wildlife habitat, from a fixed or shrinking land base. Agroforestry is an important category of planted forests that has the potential to provide farmers, communities, and society-at-large with wide array of forest-related goods and services. Agroforestry can complement forestry sector efforts in sustainable forest management by providing a set of tree-based conservation and production practices for agricultural lands. Some important sustainability issues are:

(i) Biological diversity

There is not enough forested habitat remaining in some landscapes to support some species of plants and animals. Even when there are forest reserves, they may be too small to contain the habitat requirement of all species. In addition, most species have population that extends beyond reserve boundaries (Kramer *et al.* 1997). Agroforests are typically less diverse than native forests (but more diverse than agriculture), they do contain much greater number of plants and animals species than forest plantation (Murdiyarso *et al.*, 2002). Agroforestry provides ways of augmenting the supply of forest habitat and providing greater landscape connectivity. Where croplands occupy most of landscape, shelterbelts can be essential for maintaining plant and animal biodiversity, especially under a changing climate scenario. Agroforestry adds plant and animal biodiversity to landscapes that might otherwise contain only monocultures of agricultural crops (Noble and Dirzo, 1997). The complex agroforestry systems have been documented to contain over 100 plant species per field and support up to 180 bird species (Michon and de Foresta, 1990). In mature complex multi-strata agroforestry systems of Indonesia, plant diversity was in the order of 300 species ha^{-1}, while bird diversity was found to be 50 percent that in the original rainforest. In addition, almost all mammal species were still present at same level in these agroforestry systems (Thrupp, 1997).

(ii) Carbon Sequestration

Agroforestry systems hold considerable potential for improving C sequestration and storage in both the soil and the biomass. The *Gmelina arborea* based agroforestry system sequestered 31.37 Mg C ha^{-1} at the age of six years (Swamy *et al.,* 2003). Due to the integration of *Prosopis juliflora* and *Dalbergia sissoo* C flux in net primary productivity increased in a silvopastoral system (Kaur *et al.,*2002). Long rotation systems that use tree for windbreaks, border planting and overstory shade can sequester C for many decades. Over the 11 years since plantings, agroforests accumulated approximately 740 kg ha^{-1} yr^{-1} more C than forests and 520 kg ha^{-1} yr^{-1} more C than pastures (Sharow and Ismail, 2004). Out of the 15 million of world forest that is destroyed each year, the practice of shifting cultivation is reported to account for over 60 % (FAO, 1997). One of the environmental consequences of slash-and-burn farming is the loss of forest system C which in turn contributes to atmospheric change and impact upon global climates. The original forest contained 308 t C ha^{-1} and lost 220 t C ha^{-1} upon conversion to agriculture (Jean *et al.,* 1997). In degrade soils of the sub humid tropics, improved fallows agroforestry practices have been found to increase top soil C stocks up to 1.6 Mg C ha^{-1} yr^{-1} above continuous maize cropping (Mutuo *et al.,* 2005).

(iii) Soil

Agroforestry systems have the potential to contribute significantly to maintaining or improving soil and water quality. The degree to which these and other ecological functions cane be provided will depend on plant species composition and their physical structure both above and belowground. Agroforestry systems are generally perceived to be sustainable and to enhance soil properties. Growing trees in conjunction with annual crops or pastures is believed to provide a more thorough plant cover to protect the soil from erosion and a deeper or more prolific root system to enhance nutrient cycling. Appropriate agroforestry systems improve soil physical properties, maintain soil organic matter, and promote nutrient cycling. Research results during past two decades show that three main tree-mediated processes determine the extent and rate of soil improvement in agroforestry systems (Nair *et al.,* 1999). These are;

- increase N input through biological nitrogen fixation by nitrogen-fixing trees,
- enhanced availability of nutrients resulting from production and decomposition of substantial quantities of tree biomass, and
- greater uptake and utilization of nutrients from deeper layers of soils by deep-rooted trees.

(iv) Water

There is strong perception that water beneath trees is high. Rhoades (1995) reported increased soil water (4-53 % greater than in open) in the crop root zone beneath *Faidherbia albida* canopies. At high tree densities, the proportion of rainfall lost as interception by tree canopies and used for tree transpiration would exceed that saved by shading and stem flow, resulting in drier soil below the tree canopy. Agroforestry technologies have been shown to be effective in reducing water pollution from agricultural activities when they are well designed and properly located in a watershed (Dosskey, 2002). The buffers can stabilize stream channels and slow and reduce the transport of runoff to streams. This allows more time for infiltration of water and contaminants into the soil and increases the ability of the environment to degrade pesticides and animal waste products. Agroforestry practices are also being adapted to design best management practices to detain and treat storm water runoff from communities and restore ecological functions to watersheds.

(v) Productive Functions

Agroforestry and agroforests can be used to produce harvestable wood for fuelwood, pulp, saw timber, and veneer products. Similar to plantation forests can help relieve some of the pressure to harvest native forests. The potential for agricultural lands to augment the world wood supply is substantial (Watson *et al.*, 2000), and has the added benefits of bolstering on-farm income. Many agroforestry designs can also be used to produce non-timber commercial products. Agroforestry plantings mixed into and at the edges of forest plantations can be used to produce a wide array of products, like medicinal, ornamentals, and food products, which are compatible with wood production. This will also allow for greater structural diversity and the development of more diverse plant communities.

SWOT Analysis

a. Strength

- Variety of edaphoclimatic conditions to support diversity.
- Large area under degradation to support agroforestry.
- C fixation and sequestering role of trees.
- Organized research throughout the country.
- Availability of trained human resource.

b. Weaknesses

- Edaphoclimatic specific research inputs/infrastructure.
- Long gestation period of trees.
- Lack of wood based industries to affect farmers income in every region.
- Non-availability of quality seed and planting material.
- Availability and adaptation of very few transferable technologies.
- Policy about tree felling and transit awareness of farmers.
- Market imperfection and support price.

c. Opportunities

- Proper utilization of marginal agricultural and other degraded lands for agroforestry development and site amelioration.
- Designing of systems for improvement of land productivity to meet societal demands and land conservation.
- Development of small scale tree based industries for economic upliftment of farmers and employment generation.
- Bioremediation of polluted sites for improvement of soils and environmental quality.

d. Threats

- Increased pest disease complexes due to secondary hosts.
- Bird damage of crops.
- Unchecked growth of wild life.
- Shading effect of trees on yield of understorey vegetation.

Opportunities and Challenges

- Land use diversification through agroforestry and land management approaches for sustainable production.
- Basic research aimed at understanding process to support balanced WUE, NUE, IPM and below and above ground biodiversity for resource management.
- Integrating fisheries – tree crop production system and soil fauna.
- Studies on carbon sequestration, carbon balance and effort to minimize global warming.
- Improved planting material to the farmers.

- Ecological and economical modeling to value the social, cultural and employment (livelihood) systems for their sustained growth.
- Value addition through processing and marketing of tree product.
- Technology transfer.

Strategies to Meet Challenges

(i) Political Will

Strong political will and commitment at the level of Central and State Government for:

- Prevention of further degradation of well stocked forests.
- Restoration of forest growing stock and productivity to optimum levels.
- Adequate financial investments.

(ii) Improvement in Productivity

Substantial improvent in productivity of forest resources on sustainable basis through:

- Conserving the forest resources for meeting the national needs.
- Agroforestry practices to achieve productivity levels of 20-30 m^3/ha^{-1}.
- Reforestation of degraded forest lands and marginal agricultural lands.

(iii) Corporate Sector participation

Linking agroforestry to Corporate Sectors through PPP for:

- Production of quality planting material for plantations.
- Supply and marketing of raw materials to industry for value addition.

(iv) Local community co-operation

Co-operation of the local communities for sustainable development of forest resources absolutely essential. Therefore local communities should be involved in development of tree resources to:

- Conserve forests.
- Supply firewood and fodder on sustainable basis.
- Reduce illicit felling of trees.

(v) Liberalizations of Ceiling Limits / Restrictions

- Statutory ceiling limits on agricultural land holdings for wastelands should be liberalized.
- Tree planting and farm forestry plantations should be treated on par with commercial crops like rubber, tea and coffee etc.
- Restriction on felling of farm growing trees and movement of wood from farm plantations to markets/ factories should completely be abolished in all states.

References

Dixon, R.K. 1995. Agroforestry systems: sources or sinks of greenhouse gases? *Agroforestry Systems,* 31: 99-116.

FAO. State of the world's forest. FAO, Rome Italy. 1997. 200pp.

Jean, Kotto-Same, Paul, L., Woomer Moukan and Appolinaie Zapfack Louis, 1997. Carbon dynamics in slash-and-burn agriculture and land use alternatives of the humid forest zone in Cameroon. *Agriculture Ecosystem and Environment,* 65 (3): 245-256.

Jean, Kotto-Same, Paul, L., Woomer Moukan and Appolinaie Zapfack Louis, 1997. Carbon dynamics in slash-and-burn agriculture and land use alternatives of the humid forest zone in Cameroon. *Agriculture Ecosystem and Environment,* 65 (3): 245-256.

Kaur, B., Gupta, S.R. and Singh, G. 2002. Carbon storage and nitrogen cycling in silvopastoral system on a sodic soil northwestern India. *Agroforestry Systems,* 54: 21-29.

Kramer P., C.P. van Shaik, and J. Johnson. 1997. Lost stand - protected areas and the defense of tropical biodiversity. Oxford Univ. Press, Oxford, UK.

Michon, G., and H. DeForesta. 1990. Complex agroforestry systems and the conservation of biological diversity in harmony with nature. Proceedings of International Conference on Tropical Biodiversity. Kuala Lumpur (SEA MEO-BIOTROP).

Murdiyarso, D., Van Noordwijk, M., Wasrin, U.R., Tomaich, T.P. and Gillisom, A.N. (2002). Environmental benefits and sustainable land-use options in the Jambi transect. Sumatra, Indonesia. *Journal of Vegetation Science,* 13: 429-438.

Mutuo, P.K., Cadisch, G., Albrecht, A., Palm, C.A., and Verchot, L. 2005. Potential of agroforestry for carbon sequestration and mitigation of greenhouse gas emissions from soils in the tropics. *Nutrient-Cycling-in-Agroecosystems,* 71 (1): 43-54.

Nair, P. K. R., R.J. Buresh, D.N. Mugendi, and C.R. Latt. 1999. Nutrient cycling in tropical agroforestry systems: Myths and science. pp.1-31. In: L. E. Buck, J.P. Lassoie, and E.C. Fernandes. (eds), Agroforestry in Sustainable *Agricultural System,* CRC Press, Boca Raton, FL.

Nair, P.K.R. 1993. An introduction to Agroforestry. Kluwer Academy Publishers, Dordrecht, The Netherlands, Pp. 499.

Noble, I.R. and R. Dirzo. 1997. Forests as human-dominated ecosystems. *Science*, 7:522-525.

Rhoades, C. 1995. Seasonal pattern of nitrogen, mineralization and soil moisture beneath *Faidherbia albida* (syn. *Acacia albida*) in central Malawi. *Agroforestry Systems*, 29: 133-145.

Sharrow, S.H. and Syed Ismail, 2004. Carbon and nitrogen storage in agroforests, tree plantations, and pastures in Western Oregon, USA. *Agroforestry Systems*, 60 (2): 123-130.

Swamy, S.L., Puri, S and Singh, A.K. (2003). Growth, biomass, carbon storage and nutrient distribution in *Gmelina arborea* under an agrisilviculture system in substantial tropics of central India. New Forests.

Thrupp, L.A. 1997. Linking biodiversity and agriculture: challenges and opportunities for sustainable food security. World Resources Institute, Washington D.C.

Watson, R.T. I.R. Noble, B. Bolin, N.H. Ravindranath, J.D. Verarda, and Dokken, D.J. 2000. Land-use, Land-use Change and Forestry, IPCC Special Report. Cambridge University Press, Cambridge, Pp. 388.

□□□

Agroforestry: Natural Resource Sustainability, Livelihood & Climate Moderation © 2009
Editors: OP Chaturvedi, A Venkatesh et. al., pp. 43-54
Satish Serial Publishing House, Delhi

Chapter 4

TREE IMPROVEMENT FOR AGROFORESTRY SPECIES IN INDIA

V.K.Gupta
National Research Centre for Agroforestry, Jhansi-284 003 (U.P.)

Introduction

In most of the genetic improvement programme of forestry, increasing productivity of economic traits remained the major criteria of selection and improvement and on the other hand any variety of tree species developed for agroforestry systems other criteria are also taken into consideration besides productivity of economic traits. These are: competition between trees and understorey crops for natural resources like sunlight, water and nutrients and trees should not act as a host for pathogens responsible for spreading diseases of understorey crops. Hence, ideotype concepts for agroforestry species are quite different than forestry species. Poplar based agroforestry system in western Uttar Pradesh, part of Haryana and Punjab are well accepted by farmers due to shedding of leaves by poplar during winter season when there is more demand of sunlight to wheat crops. *Prosopis cineraria* and pearl millet system is quite common in Thar deserts because *P. cineraria* act as a wind breaker in Thar deserts. It results increase in grain yield of pearl millet.

Minimizing the risk to obtain food, fodder and fuelwood is also important objective of agroforestry in marginal and sub-marginal lands. Therefore, before starting any tree breeding programme, selection of tree species and deciding ideotype of a species are very important.

Ideotype

Ideotype has been defined as "a biological model which is expected to perform predictably leading to greater quantities and quality and quality of crop yield and tree products in toto in defined environment".

No Single Ideotype

- can be successful on all sites.
- can tolerate all type of management.
- can yield all type of products in same proportion.
- can provide all type of services.

Ideal Ideotypes

- to provide maximum economic returns.
- maximum land use in time and space.
- associative or non-competitive.

Ideotype for Agrisilviculture

- Medium to fast growing.
- Low branch habits.
- Good survival and adaptability.
- Geotropic angled roots rather than horizontally.
- Associative or less competitive with crops.
- Resistance to diseases and pests.
- Trees should not act as a common host for diseases and pests of crops.
- Crown should be round with small diameter.
- Single stemmed with clear bole height (> 7 m).

Ideotype for Silvipasture

- Medium to fast growing.
- Crown should be round with medium diameter.

- Multiple stems (3-4).
- Upright branch angle.
- High coppicing, lopping and pollarding habits.
- Good survival in harsh environments.
- Ability to produce green fodder during lean period.
- Leguminous species.
- Geotypic angled roots rather than horizontally.
- Resistance to disease and pests.
- Trees should not act as a common host for diseases and pests of crops.

Present Status

In agroforestry and forestry sectors, few and sporadic efforts have been made for tree improvement research in India. Most of the agroforestry trees species have not been even gone through simple selection for superior types (plus trees). Exploration, collection and conservation of germplasm of woody agroforestry species, selection for fast growth, clonal propagation, seed orchards and other advances in trees breeding research are still in fancy stage. At present, all the woody agroforestry species are propagated through seeds/ or propagules collected by unskilled labourers. Concept of plus tree selection, establishment of seed orchards and producing of quality seeds are still in air in India. Inadequate information is available for choice of species and seed sources, progeny trials, methodology, evaluation, breeding strategies etc. Little research has been conducted on species and provenances trials of agroforestry tree species in general. Perhaps *Leucaena leucocephala, Populus deltoides* and *Eucalyptus* are only agroforestry species which have been exploited fully for general improvement. In India, collection , evaluation and improvement have been carried out sporadically in *Leucaena leucocephala, Prosopis cineraria, Tecomella undulata, Acacia senegal, Acacia tortilis, Azadirachta indica, Populus deltiodes, Dalbergia sissoo, Acacia nilotica, Eucalyptus tereticornis, Eucalyptus camaldulensis, Albizia lebbeck* etc. Still quality seeds of agroforestry tree species are not available due to lack of coordinated efforts at national level and absence of seed certification polcy. There is a need to start All India Coordinated Project on Tree Improvement so that improved materials can be identify and mass multiplication is possible at national level.

Limitations of Tree Improvement Programmes

There are several limitations in the fast progress of tree improvement programmes which are as follows:

1. Long Duration Breeding Cycle: It results heavy investment on the programme and results will be obtained after long time.
2. Most of the Species are Cross Pollinated: It requires many generations to stabilise.
3. Inbreeding depression is high.
4. Reproductive biology is not known.
5. Lack of adequate skilled personnels and regular transfer of them.
6. Lack of coordination at national level.

Factors for Success of Tree Improvement Programme

Tree improvement programme rely on following points:

1. Selection of the species, or geographical sources within a species.
2. Determination of the amount, kind and cause of the variability within the species.
3. A complete package of desired qualities into improved individuals (breeding objectives).
4. Mass production of improved individuals for reforestation purposes.
5. Developing and maintaining a broad genetic base for further improvement of the populations.

Factors for Initiate Tree Improvement Programme

Since tree breeding programme is very costly affair and require very long time. It should be done when there is acute shortage of availability of suitable tree species in nature or some acceptable and economically viable tree species is suffering due to epidemic of disease or having characters that reduce market price of a product.

Breeding Aids

Following are some breeding aids that make the programme better and efficient and also can reduce time for successful achievement or evolving varieties:

1. Extent and type of variability present in the nature.
2. Heritability of the traits.
3. Correlation of juvenile and mature tree for economic traits.

4. Gene action for important and economic traits.
5. Genotype x environmental interaction.
6. Vegetative propagation methods.
7. Short breeding cycle.

Approaches in Tree Improvement

The principal and practices of plant breeding for tree are well established (Wright, 1976; Zobel and Talbert, 1984) and they apply equally to industrial plantations, farm plantation and community planting. Superior phenotypes are selected from the best populations and their breeding potential is evaluated in clonal selection or progeny tests on typical sites with typical managements. Superior genotypes of superior cross pollinated species are then propagated clonally and planted in special seed production areas or seed orchards where open or controlled pollination provides seed for pollination and if needed further selection, testing and breeding (Burley, 1980; Zobel, and Talber, 1984) should be done.

(A) Breeding for Immediate Need

Many times organisations with extensive planting programme need large quantities of improved seed immediately. The approach followed in these circumstances of immediate need will be some what different from that in which seed is required after few years. If seed is needed immediately for an operational programme, one viable approach is to choose outstanding phenotypes from natural stands or plantations and collect the seeds from them. Raise the nusery for the purpose.

(B) Breeding for Long Term

Following breeding methods are common for the improvement of crops.

1. Collection, evaluation and conservation of genetic resources of different geographical regions (provenances trials)
2. Selection in natural populations or planting stands (mass selection/plus tree selection)
3. Hybridization followed by selection

Tree improvement programme is very tedious due to difficulty in control pollination, long breeding cycle, traits are governed by complex genetic inheritance and high degree of genotype x environment interaction. Therefore, breeding methods require several generations to evolve a variety is not desirable in case of trees. Most of the trees are cross pollinated thus breeding methods requiring minimum generations/ time and no control on free pollination are generally recommended in case of tree species. Following breeding methods are popular in tree improvement programmes:

1. Mass selection (plus tree selection) and progeny testing.
2. Simple recurrent selection.
3. Hybrid variety.
4. Recurrent selection for general combining ability.

In order to provide promising species and provenances for planting, it is recognised that extensive systematic exploration and testing are required (Burley, 1980; Turnbull, 1983). This is emerged as a primary condition at a recent workshop on multi-purpose tree germplasm held in Washington, USA in June, 1983. Danger of susceptibility to pests and diseases is most common by cultivation of improved uniform materials as a mono crop and difficult to introduce resistance genes in trees. Therefore, broad genetic base in the form of genetic conservation is necessary to replace the provenance under cultivation immediately before creating alarming situations at national level which can delay or control onset of this stage. Collections should be properly evaluated under multi-location trials for economic and needful traits. Several varieties in tree have been developed simply by selection in natural stands.

Selection Methods

There are several methods which are available to the breeder depending types of information available. These selection methods are used in natural stands and unimproved plantation stands. Following are few important selection methods.

Mass Selection (Plus Tree Selection)

Mass selection involves choosing individuals solely on the basis of their phenotypes without any information about performance of ancestors, siblings, offsprings or other relatives. Mass selection works best for highly heritable traits where phenotype is a good reflection of the genotype or to provide improved seeds immediately. Often seed from this procedure will not yield large volume gains, although they sometimes greatly improved the quality and pest resistance and they can ensure good stability.

Family Selection

Family selection involves the choice of entire families on the basis of their average phenotypic values. There is no selection of individuals - trees values are used to compute family means. Family selection works best with traits of low heritability, where individual phenotypes are not good reflection of genotypes.

Progeny Testing

Progeny testing involves selection of parent trees based upon the performance of their progeny. It can be very precise selection method because it allows direct estimation of breeding values of lines in the selection process. Initial selection by progeny testing considerably lengthen the generation interval which means a critical loss in time.

Within Family Selection

Here individuals are chosen on the basis of their deviation from the family means, and family values *per se* are given no weightage when selection are made. This one gives rise slowest rate of inbreeding which is a major problems in most of the programmes. In practice within family selection is rarely used in tree improvement.

Recurrent Selection

Most tree improvement programmes are designed to provide continued gain through each of many cycles of improvement. The selection procedure that involves many cycles of selection and breeding is known as recurrent selection. The system most often use in forest tree improvement is known as simple recurrent selection. With this system, improvement programmes begin by selecting tree in natural stands or in unimproved plantation based upon their phenotypic values. Selected trees are then mated and their progenies are established in such a way that they can be used as a source of selection for the second generation improvement.

Traditional Strategy for Species and Provenances Testing

Ideally, the selection of species and provenances follow a systematic pattern of experiments. Although the pattern is usually not adapted in developing countries due to shortage of time and resources. The general sequence is as follows:

(i) Elimination Trial (ET)

The objectives of the ET stage are (1) to test the adaptability of species to the trial area quickly and low cost and (2) to identify most promising candidates and eliminate the least suited. They are generally unreplicated at each site but repeated at several locations. In general, small plots of 16 to 25 trees are used in square shape. Assessment parameters are generally survival, plant height and diameter at breast height. Survival is most important measurement in this trial. ET trials drastically reduces number of entries of the species or provenances or progenies.

(ii) Progeny Trial (PT)

The objective of PT to obtained comparative performance, growth and canopy of the promising progenies. Performance trial are normally scattered to cover a range of areas. Plots are usually larger (10x10 to 20x20 trees) and are replicated in randomised block design. Spacing is same as for normal plantations and survival, form, height and diameter are periodically assessed.

(iii) Provenance Trial

The main objective of this trial is to identify geographical seed sources whose performance for adaptability, growth, form and other desirable characters is comparatively superior.

Procedures to Reduce the Period of Breeding Cycles

Standard tree breeding method is both time consuming and expensive, and as a result, it is rarely used as such in developing countries. At least, 15-20 years are required to complete one cycle. Hence, it is too slow to meet the growing demand for quality seed for agroforestry programme (Turnbull, 1984). Hence there is a need to reduce the period of breeding cycle which is determined individually for each species due to differences in reproductive biology among species and the resources available for breeding. However, the following techniques may be considered:

(i) Shorting Evaluation Period

This may include use of juvenile-mature relationship that allow early prediction of performance at maturity.

(ii) Nursery Screening for Desirable Traits

Many traits of high heritability can be easily evaluated at nursery level by creating appropriate environment like disease resistance, acid tolerance, fodder quality etc.

(iii) Flower Induction Techniques for Late-flowering Species

Flowering may be accelerated by using gibberlines and through vegetative means like air layering, grafting, tissue culture. For example, in neem, through air layering, flowering and fruiting can be inducted in 9-months old saplings otherwise it takes 5-6 years.

Seed Orchard

There are two aspects of any successful tree improvement programme. The first related to obtaining immediate gain of desired products as rapidly and as efficiently as possible. This will be achieved by studying variability available in the nature. Maximum gains are achieved by the use of few very good genetic parents to supply planting stock for operational programmes. Lack of suitable seeds is one of the greatest deterrents to agroforestry. Second aspect of improvement programme is concerned with the long term need to provide the genetic base that is essential for continued progress over many generations.

All the tree improvement programmes must have an operational (production) and a development (research) phases. The two are closely linked yet they are requiring different approaches and philosophies. The operational phase produces quick economic gains from tree improvement and is the most easily understood and useful to the general public and to forest managers. Modest gains are often accepted in order to obtain improvement quickly. The main objective of developmental phase is to obtain and retain a broad genetic base and to combine desire characteristics into suitable trees that will be useful for future generations. No programme can be better without the the base of genetic materials upon which it is operated. Areas which are used for mass multiplication of improved genotypes for desirable traits are called seed orchards. These are used to get maximum genetic gain as quickly and least expensive. Requirement of seeds are of two types:

1. Meeting immediate seed needs.
2. Long term seed needs - seed orchards.

(A) Meeting Immediate Seed Needs

There are several methods that can be used to obtain genetically improved seeds for immediate planting. These are interim in nature in that they are used only until the more permanent seed orchard becomes in operation. Often seed from the interim procedure will not yield large volume gains although they some times greatly improve tree quality and pest resistance, and they can ensure good adaptability. Methods for production of immediate planting are discussed below:

(i) *Seed from Individually Good Phenotype*: One viable approach is to choose outstanding phenotypes from natural stands or plantation, mark the trees, and collect seed from them. Seed from good phenotypes in which the male is not known and is unselected will usually yield only limited volume improvement because of the low heritability for growth in most species.

(ii) *Seed from Proven Sources*: One of the most common methods obtaining large quantities of seed quickly is go to back to the original source or provenance that has been tested earlier and that has proven to be suitable. This can be a great success or an equally great failure. If the source information is correct and is indeed the same as reported for the initial successful planting, the method works well.

(B) Long - Term Needs - Seed Orchards

The standard method of producing genetically improved seed in operational quantities is to use the seed orchard approach. The method to prepare seed orchard consists of initially establishing orchards based on the phenotype of the parent trees and then rouging out the poor genotypes from the orchard based on progeny test results rather than waiting to establish an orchard only with parents that have already been tested for their genetic worth. One of important point to be considered to select the parents that flowering and pollen exchange among genotypes in the orchard will be uniform and equal.

Types of Production Seed Orchards

Seed orchards are generally put in two broad categories.

1. Vegetative orchards
2. Seedling Seed Orchards

(i) Vegetative Orchards

These are those which are establish through vegetative propagules as grafts, cuttings, tissue culture plantlets, or other methods. Vegetative orchards are the most common types use operationally.

(ii) Seedling Seed Orchards

These are established by planting seedlings followed by later rouging that will remove the poorest trees generally leaving the best trees of the best families for seed production.

Seed Orchard "Generation"

Seed orchards are commonly categorized by generation; that is first, second, or more advanced generation orchards depending upon how many cycles of improvement they represent.

The first - generation orchard usually results from selection from the natural stands or unimproved plantation, most often by using individual tree selection methods. The pedigrees of the parent trees are improved by rouging as determined by progeny testing. This type of orchards generally called 1.5 generation orchards.

Factors for Quality and Purity

1. Isolation distance.
2. Location.
3. Area under production.
4. Establishment.
5. Management.
6. Harvesting procedures.

Advantages and Limitations of Tree Improvement

(A) Advantages

One major advantage of genetic improvements in forest trees is that once a change is obtained, it can be kept over a number of generations. When improvements are made through silviculture, they will need to be done one more times at each rotation. Therefore, it is more economical. Improved provenances or genotypes can be kept for a very long time without any change.

(B) Disadvantages

1. The size of the trees creates problems in measurement, crossing and especially in seed collection.
2. Suitable areas are necessary for storage of desired genetic material and for testing.
3. Require long time and also expensive compared to crop breeding.

References

Burley, J. 1980. Selection of species for fuelwood plantation. *Forestry Review,* 59(2): 133-147.

Kedarnath, S. 1982. Plus trec selection - a tool in forest. In *Proc. of Symp. on Improvement of Orest Biomass,* Indian Society of Tree Scientists, Solan (HP), Pp. 13-19.

Turnbull, J.W. 1983. Tree seed supply - a critical factor for the success of agroforestry projects. In *Proceeding ICRAF/IBPGR/CFI/NAS Workshop on Multiple Purpose Tree Germplasm,* June 1983. Washington D.C.

Turnbull, J.W. 1984. Tree seed supply: a crtical factor for the success of Agroforestry project. In *Multipurpose Tree Germplasm*. ICRAF, Nairobi, Pp. 298.

Wright, J.W. 1976. *Introduction to Forest Genetics.*, Jhon Wiley and Sons New York, Pp. 463.

Zobel, B.J. and Talbert, J. T. 1984. *Applied Forest Tree Improvement,* Jhon Wiley and Sons, New York, Pp. 505.

❏❏❏

Agroforestry: Natural Resource Sustainability, Livelihood & Climate Moderation © 2009
Editors: OP Chaturvedi, A Venkatesh et. al., pp. 55-70
Satish Serial Publishing House, Delhi

Chapter 5

ECOSYSTEM DEVELOPMENT THROUGH AGROFORESTRY MEASURES

O.P. Chaturvedi and P. Ratha Krishnan
National Research Centre for Agroforestry, Jhansi-284 003 (U.P.)

Introduction

Soil degradation is a major threat to our food and environmental security. Many thriving civilizations vanished in the past because of inadequate attention to land care. Out of 328 m ha of total geographical area of India, more than 50% is considered as problematic where crop / forage production is very poor or negligible. These degraded lands are lying waste due to different constraints or unproductive / low productive due to mismanaged / improper and over utilization. Under these categorics, quite sizeable area of India is affected by water erosion, salinity, alkalinity, acidity and water logging. For addressing these concerns, proper land usc planning is a prerequisite. To facilitate such shifts, landuse should increase the production along with sustainability. One such technology evolved for this concern is "Agroforestry".

Agroforestry is a land use that involves deliberate retention, introduction, or mixture of trees or other woody perennials in crop/animal production field to benefit from the resultant ecological and economical interactions. In

India, agroforestry has been a way of life. Its evolution process was linked with the process of human evolution by which man has learnt the art of domesticating plant and animals after leaving hunting and gathering habit.

The concept of agroforestry during Vedic period was growing crops in between / boundary of trees whereas the present emphasis is on incorporating tree in the farm lands / bunds to increase the productivity. Even though, retaining trees on the farm fields had been an age old practice for fulfilling the fuel, fodder, timber and environmental requirements, its role in the degraded land is enormous and potential. Practicing agroforestry instead of sole cropping will improve the soil quality by addition of nutrients and organic matter to the soil, control of erosion, and better water use efficiency. In this juncture, agroforestry is identified as an alternate land use technology towards restoration of ecosystem by reclamation of degraded soils / lands, conservation of soil, water, and environment, livelihood security and mitigation of climate change.

Agroforestry for the Ecosystem Development

Because of the multifunctional possibilities and opportunities with agroforestry, it is explored for the past two decades for soil and water conservation and sustainable agriculture. The additional benefits like livelihood security and the climate mitigation gave new vista to agroforestry in the 21st Century. Integrated watershed approach should be given maximum attention to combat land degradation and environmental problems particularly in fragile and hilly areas (Dadhwal and Solanki, 2000). More over agroforestry has been an age old way of life of Indian farmers. Scattered trees on crop lands like *Prosopis cineraria* in Rajasthan, *Acacia nilotica* and *A. leucophloea* in South India and variety of trees due to religious importance are popular since ages. Importantly, agroforestry models are evolving as successful practices by farmer's interventions and used for the benefit of the society. Table 1 highlights the region specific challenge and changes observed due to practicing agroforestry.

Agroforestry Interventions to Develop Chemically Degraded Lands

A silvipastoral model comprising the growing of *Prosopis juliflora* + Karnal grass *(Leptochloa fusca)*, is highly successful for the development of community lands (Singh, 1995). In this model, *Prosopis* and *Leptochloa* are grown together for nearly 5 years. Later, when the surface soil is reclaimed, the grass is replaced by better forage such as *shaftal* (*Trifolium resupinatum*), *berseem* (*Trifolium alexandrium*) and *senji* (*Melitotus parviflora*). After 8-10 years of growing trees and forage, the soil is reclaimed to the extent that the normal arable crops may be raised either after removing the trees or in association with trees.

Table 1. Regional examples of challenges and changes due to agroforestry in India

Region	Challenge	Changes due to agroforestry
Himalaya (Kurukshetra)	Improvement of Sodic soils	Increase in microbial, tree biomass and soil carbon; enhanced N availability
Himalayas	Restoration of abandoned agricultural sites	Biomass accumulation (3.9 t ha^{1} in agroforestry compared to 1.1 t ha^{1} in degraded forests); improvement in soil physico-chemical characteristics; C sequestration
Western Himalayas	Reducing soil and water loss in agro-ecosystems in steep slopes	Contour tree-rows (hedgerows), reduced run-off and soil loss by 40 and 48% respectively (in comparison to 347 mm run-off, 39 Mg ha^{-1} soil loss per year under 1000 mm rainfall conditions)
Sikkim Himalayas	Enhancing litter production and soil nutrient dynamics	N fixing trees increase N and P cycling through increased production of litter and influence greater release of N and P; they help in maintenance of soil OM with higher N mineralization rates in agroforestry systems
Indo-Gangetic Plains	Biomass production and nutrient dynamics in nutrient-deficient and toxic soils	Biomass production (49 t ha^{-1} $decade^{-1}$)
Meghalaya	Enhancing tree survival and crop yield	Crop yield did not decrease in proximity to *Albizia* trees.
Western India (Rajasthan)	Compatibility of trees and crops	Density of 417 trees per ha was found ideal for cropping with pulses
Western India (Karnal)	Improvement of soil fertility of moderately alkaline soils	Microbial biomass C which was low in rice-berseem crop (96.14 $\mu g\ g^{-1}$ soil) increased in soils under tree plantation (109.22 $\mu g\ g^{-1}$ soil); soil carbon increased by 11-52% due to integration of trees and crops.
Central India (Raipur)	Biomass production in N and P stressed soils	*Azadirachta indica* trees were found to produce biomass in depleted soils
Central India	Soil improvement	Decline in proportion of soil sand particles; increase in soil organic C, N, P and mineral N
Southern India (Kerala)	Growing commercial crops and trees	Ginger in interspaces of *Ailanthus triphysa* (2500 tree ha^{-1}) helps in getting better rhizome development of the former compared to solo cropping

(Pandey, 2007)

Studies conducted at Pali (Rajasthan) on soil working in pits of the size 0.63 m^3 to 1.0 m^3 with manure @ 25 kg pit^{-1} and gypsum 3.6 kg pit^{-1} or auger holes of 150 cm depth and 15 cm diameter filled with original soil and 2 kg gypsum + 7-8 kg manure helped in improving soil environment and establishment of tree seedlings of *Acacia nilotica, Prosopis juliflora* and *Casuarina equisetifolia* (Jain, 1986) in salt affected wastelands. In addition to biomass production, trees on salt affected soil help in amelioration by improving physical, chemical and biological properties of the soils. Ameliorating effects of 20 years old tree plantations are reported in Table 2.

Table 2. Ameliorating effects of trees on alkali soil

Tree species	After 20 years	
	pH	**Organic carbon (%)**
Eucalyptus tereticornis	9.1	0.33
Acacia nilotica	9.0	0.55
Albizia lebbeck	8.6	0.47
Terminalia arjuna	8.1	0.58
Prosopis juliflora	8.0	0.58

(Original pH and OC % are 10.3 and 0.12, respectively) (Singh, 1995)

Agroforestry Interventions for Physically Degraded Lands

The physical deterioration of soil due to water logging or submergence / flooding has affected around 11.6 million ha land in India. Area occupying lower topographic positions, area long the rivers (diara lands) and those lying beyond levees (tal lands) are a few typical examples. Further, it has been estimated that nearly 40 million ha area of the country is flood prone. These lands are generally not remunerative for crop production but offer an alternative to increase the production through silvopastoral and hortipastoral systems. Growing of grasses under water logging may yield substantial green forage to meet the shortage of fodder to the animals. The species suitable for reclaiming such lands are *Casuarina equisetifolia, Eucalyptus camaldulensis, Syzygium cuminii, Acacia nilotica, Pongamia pinata, Leptochloa fusca, Brachiaria mutica, Paspalum spp., Veteveria zizanioides, Ziziphus mauritiana, Terminalia arjuna, Tamarix articulata, Atriplex spp.,* coconuts, red oil palms, etc.

It is estimated that nearly 3000 billion tones of mine overburden is dumped annually the world over and about 3,86,000 ha land per annum is disturbed by mining (Dadhwal and Katiyar, 1996). Mine spoils result in shifting mass of sand and rubble, denudation of forests, reduction in water holding capacity of reservoirs, rivers, streams, etc. Agroforestry is a viable option for rehabilitation of such degraded areas. Establishment of permanent vegetative cover of suitable trees, shrubs and grass can improve their deteriorated conditions and stop further degradation.

For the revegetation of mine spoils, plant species which are indigenous, hardy, have N_2 fixing ability, have aesthetics, provide quick coverage during establishment and are ecologically and economically acceptable should be preferred. Suitable species for revegetation of different kinds of mine spoils of the country are given in Table 3.

Table 3. Plant species suitable for revegetation of mine spoils

Mine spoil Category	Suitable Plant Species
Bauxite mined area of MP	*Grevillea pteridifolia, Eucalyptus camaldulensis, Pinus, Shorea robusta*
Coal mine spoils of MP	*Eucalyptus hybrid, Eucalyptus camaldulensis, Acacia auriculiformis, Acacia nilotica, Dalbergia Sissoo, Pongamia pinnata*
Lime stone mine spoils of Uttaranchal Himalaya	*Salix tetrasperma, Leucaena leucocephala, Bauhinia retusa, Acacia catechu, Ipoemea carnea, Eulaliopsis binata, Chrysopogon fulvus, Arundo donax, Agave americana, Pennisetum purpureum, Erythrina suberosa*
Rock-phosphate mine spoils of Mussoorie, Uttaranchal	*Pennisetum purpureum, Saccharum spontaneum, Vitex negundo, Rumex hastatus, Mimosa himalayana, Buddlea asiatica, Dalbergia sissoo, Acacia catechu, Leucaena leucocephala* and *Salix tetrasperma*
Lignite mine spoils of Tamil Nadu	*Eucalyptus species, Leucaena leucocephala, Acacia* and *Agave*
Mica, copper, tungeston, marble, dolomite, limestone, etc. Mine spoils of Rajasthan	*Acacia tortilis, A. Senegal, Prosopis juliflora,, Salvadora oleoides, Tamarix articulata, Zizyphus nummularia, Grewia tenax, Cenchrus setigerus, Cymbopogon, Cynodon dactylon, Sporobolus marginatus, D. annulatum.*
Iron ore wastes of Orissa	*Subabul,* local plant species, etc.
Haematite, Manganese spoils from Karnataka	*Albizia lebbeck,* local plant species, etc.

(Dadhwal and Solanki, 2000)

About 10 million ha in the Western Himalayan states of Jammu & Kashmir and Himachal Pradesh, lying between the Greater Himalayas and the mountains on the edge of the Tibetan plateau are a cold desert. The ever-increasing biotic pressures on the desert vegetation for fuel, fodder and grazing has led to serious ecological degradation. *Hippophae rhamnoides*, an important indigenous multipurpose shrub of the region, is used as fuel and fodder by the local inhabitants. One ha plantation of this shrub can meet the fuel wood needs of about 20 families (ICFRE, 1993). Recently some eucalyptus and poplar clones have been introduced as agroforestry species for revegetation of cold desert in Kashmir valley.

The agroforestry interventions aimed for restoration of ecosystem is occurred by the following measures.

a. By control of soil erosion

Check in soil erosion is the first step need to be addressed for the restoration of ecosystem. Table 4 compares the erosion rates under tropical forest, tree crops and some agroforestry systems. It clearly tells the potential of agroforestry practices in erosion reduction.

Table 4. Rates of erosion in tropical forest and tree crop systems

Land use system	Erosion (t ha^{-1} yr^{-1})		
	Minimum	**Medium**	**Maximum**
Multistorey tree garden	0.01	0.06	0.14
Natural rain forest	0.03	0.30	6.16
Shifting cultivation , fallow period	0.05	0.15	7.40
Forest plantations, undisturbed	0.02	0.58	6.20
Tree crops with cover crop or mulch	0.10	0.75	5.60
Shifting cultivation, cropping period	0.40	2.78	70.05
Taungya, cultivation period	0.63	5.23	17.37
Tree crops, clean weeded	1.20	47.60	182.90
Forest plantations, burned or litter removed	5.92	53.40	104.80

(Wiersum, 1984)

Vegetative barriers are cheap and effective as compared to mechanical measures on mild slopes in controlling runoff and soil loss (Narain *et al.*, 1998). Vegetative barrier of *Panicum maximum* have proved effective in controlling erosion in maize by reducing runoff by 28% and erosion by 48% at Dehradun (Bhardwaj and Khola, 1999). The barriers influence redistribution of sediment in the catchments resulting in deposition of 30-60 t ha^{-1} yr^{-1} sediment on contours and farming bunds / terraces naturally (Table 5).

Table 5 : Sediment deposition through vegetative barriers

Vegetative barrier	No. of years in existence	Sediment deposited (t ha^{-1})	Average deposition (t ha^{-1} yr^{-1})
Leucaena hedges in turmeric field	3	47.31	15.77
Leucaena hedges in maize field	3	183.98	61.33
Leucaena trees in maize field	9	256.54	28.50
Eucalyptus trees in maize field	9	185.58	20.62
Leucaena trees in turmeric field	9	90.06	10.01
Eucalyptus trees in turmeric field	9	103.70	11.52

(Narain *et al.*, 1988)

The performance of vegetative barriers is situation specific. For example *Vetiveria zizanioides* (Vetiver) grass barriers has neither been universally effective nor is accepted by the people (Bhardwaj, 1988). Therefore, barriers of native grass species are emphasized. Live bunds of *Panicum maximum*, *Eulaliopsis binata* and *Vetiveria zizanioides* grass reduced runoff by more than 18% and soil loss by more than 78% (Bhardwaj, 1990-91) as compared to cultivated fallow on 4% slope in Doon Valley. Guinea grass was found more effective than Bhabar and Vetiver. *Leucaena* barriers maintained as hedgerow also significantly reduces soil loss from 1.75 to 0.80 t ha^{-1} at Akola (Anon, 1993-94). In Bellary vertisols (Karnataka), vetiver reduced soil loss to an extent of 22.8% as compared to control (Jayaram and Raizada, 1994). In Shivaliks, the Eucalyptus hybrid - *Eulaliopsis binata* system showed negligible soil loss (0.07 t ha^{-1}) followed by *Acacia catechu - Pennisetum purpureum* (0.24 t ha^{-1}) and teak - Leucaena - *Eulaliopsis binata* (0.43 t ha^{-1}). The soil loss, run off and nutrient losses were less under agroforestry as compared to arable crops and fallow (Grewal, 1993).

Plantation of *Leucaena leucocephala* (Subabool) and *Eucalyptus* hybrid (Safeda) in paired contour and hedge rows of *Leucaena leucocephala* have been found to reduce 40-48% runoff and 12.09 t ha^{-1} soil loss in maize grown field on mild slopes upto 4%, erosion losses were brought to negligible by the block plantation of these species while the average rainfall during cropping season is 1002 mm. Eucalyptus tree-rows were more efficient than Leucaena when both were managed as sole tree. Eucalyptus tree - row barriers in grass plots further reduced the run - off and soil loss. However, in both the cases the tree products compensated the loss made by the system and gave economic viability (Narain *et al.*, 1998).

Table 6. represents the various kinds of agroforestry practices suitable for different climate and environmental conditions towards controlling the soil erosion.

Table 6. Agroforestry practices with potential for control of soil erosion

Agroforestry practices	Suitable climates	Important points
Plantation crop combinations	Humid to moist sub-humid climates	Densely planted combinations of agricultural plantation crops with MPTs appear to control erosion effectively on at least moderate slopes.
Multistorey including home garden	Mainly in Humid & moist sub-humid climates	Possess an inherent capacity to control erosion through combination of herbaceous cover with abundant litter.
Hedgerow intercropping, barrier hedge	Humid, sub-humid, possibly in semi-arid climates	A considerable apparent potential to combine erosion control with arable use on gentle to moderate slopes, more speculative potential on steep slopes.
Trees & erosion control structure	All climatic conditions	Supplementary use of trees stabilizes earth structures and gives production from land they occupy.
Windbreaks and shelterbelts	Semi arid zone	Proven potential to reduce wind erosion.
Silvipastoral practices	Semi arid, humid and sub-humid	Opportunities for inclusion of trees and shrubs as part of overall programms of pastoral improvement.
Reclamation forestry	All climatic conditions	Potential for planned design and development.
Combinations of the above in integrated watershed management	All climatic conditions	Substantial opportunities to include agroforestry with other major kinds of land use in integrated planning and management.

(Young, 1997)

b. By soil health improvement

It is well documented that agroforestry improves soil health. The trees improves soils by increasing inputs (Organic matter, nitrogen fixation, nutrient uptake), reducing losses (organic matter, nutrients) by promoting recycling and checking erosion, improving soil physical properties, including water holding capacity and beneficial effects on soil biological processes. Ameliorative effects of agroforestry landscape in marginal and degraded lands have been reported from Doon valley. A decrease in soil pH by 0.4 units, bulk density by 0.1 mg m^{-3} , improvement in prganic matter (OM) content by 0.16 % and water holding capacity by 15% was observed in valley of Uttaranchal (Dadhwal and Tomar, 1999). Low pH values and high OC beneath the tree cover than bare soil has also been observed by Aggarwal *et al.* (1975). For *Acacia albida,* cases of 50 – 100 % increase in organic matter and nitrogen under the canopy are known, together with increasing water holding capacity (Felker, 1978). In semi arid climates it is common to find higher soil organic matter and nutrient content under tree canopies than in adjacent open land (Table.7). Maize and sorghum in pot samples from soils under trees in Northern Nigeria grew 2 to 3 times faster than in soil with no trees; the order of fertility was *Azadirachta indica* > *Prosopis juliflora* = *Eucalyptus camaldulensis* > no trees (Verinumbe, 1987).

Table 7. Soil properties beneath trees - North – West India, Semi-arid climate

Available nutrients (kg ha^{-1})	Soil Depth (cm)	*Prosopis cineraria*	*Prosopis juliflora*	Open field
N	0 – 15	250	203	---
	15 – 30	193	212	196
P	0 – 15	22	10	8
	15 – 30	10	5	4
K	0 – 15	633	409	370
	15 – 30	325	258	235

(Aggarwal, 1980)

A study from Bundelkhand region revealed that cropping under trees almost double the OC of soils in three years. Leucaena based cropping increased the OC from 0.4 to 0.87% and sesbania based to 0.75 %. The tree cropping also improves N and P status of soil but marginally depress K (Hazra, 1994). Positive impact on soil fertility by perennial vegetation was also recorded in semiarid vertisols (Bellary region) where OC increased from 0.03 % to 2.16 % in ten years (Srivastava, 1994) with *Acacia nilotica*, the well preferred and adapted agroforestry tree species.

Leguminous shade trees benefit through shade and N which they provide to tea plants (Tejwani, 1994). A shade tree may add as much as 2500 to 5000 kg leaf, twig and pods and 63 – 126 Kg N, 18 – 36 kg P_2O_5, 22 - 44 kg K_2O, 32 – 64 kg CaO and 16 – 32 kg MgO ha^{-1} yr^{-1} to the soil (Gogoi, 1976). Therefore, leguminous trees play a vital role in fertility built up and enhancing production. In a study from Meghalaya, a sharp increase in exchangeable Ca, Mg, K and Na was noticed in all the agroforestry interventions over adjoining degraded lands. The exchangeable Al, the potential cause of infertility disappeared completely in natural grasslands within 10 – 15 years of agroforestry practice (Singh and Dhyani, 1996). Among the horticultural trees, *Emblica officinalis* Var: Kanchan was found highly remunerative (120 kg fruit $tree^{-1}$) causing only 30% yield reduction of black gram and also improved soil fertility (Solanki *et al.*, 1999).

Soil and water conservation measures coupled with planting of suitable pasture and tree species not only conserve natural resources but also improve productivity by 13 times and income generation by 3 times (Singh and Hazra, 1994). Productivity of some MPTs in silvipastoral systems on highly degraded lands in semiarid region of Central India was worked out. *Albizia amara* based silvipastoral system produced 7.95 t ha^{-1} yr^{-1} fodder followed by *Acacia tortilis* (7.6 t ha^{-1} yr^{-1}). The production of *Hardwickia binata* based silvipastoral system (800 trees ha^{-1}) didn't significantly affect the growth and yield of black gram in *Kharif* and mustard in *Rabi* even after six years (Solanki *et al.*, 1999).

In North –Eastern India (Meghalaya), the silvipastoral system comprising *Alnus nepalensis* and *Ananas comosus* and forage crop like *Panicum maximum* or *Setaria sphacealata* coupled with *Stylosanthes guyanesis* in 1 : 1 ratio sustained agroforestry system in areas having 30 – 60 % slope. Forage yield of 13.54 t ha^{-1} was obtained from the combination of Stylo and Setaria. In addition, 2.3 t ha^{-1} litter from Alnus and 4000 fruits ha^{-1} from pineapple were also obtained (Chauhan *et al.*, 1993). In another study from Barapani, sericulture based agroforestry system *Morus alba* + *Arachis hypogea* / and *Brassica juncea* generated net income of Rs. 2, 58,000 apart from ameliorating the fertility status of these degraded lands.

c. By ground water recharge

An integrated watershed management approach comprising mechanical, vegetation and agronomical measures applied for managing degraded lands in Aravali region of south-west Haryana reduced soil and run –off loss and increased ground water recharge. Tree species like *Acacia tortilis* and *Prosopis juliflora* planted in the watershed not only provide cover to denuded hills and wastelands but also provide fodder, fuel, timber which improve the socioeconomic status of the people (Singh *et al.*, 1994). In Tejpura watershed

(775.7 ha in Bundelkhand) about 231 ha m of rainwater was stored in 5 check dams which resulted in rise of ground water by 3 – 7 m (Hazra, 1998) and this increased under ground water table leading to the digging up of wells from five in number to present number of 315. This increased the irrigation to crop from initial 3.8 % to present 100 % and also increased cropping intensity (82 – 209 %). This also reduce run-off by 66 % and soil loss by 99%.

A review on the status of watershed research in India (Samra, 1997) reveals that integrated development reduced run-off, moderate flooding of down stream area and improve *in-situ* moisture conservation for increased biomass production. Besides improvement in crop yield, grasses and trees in the catchments help in ground water recharge and raise water table. Run-off and soil losses are substantially reduced when small watersheds with agriculture are replaced either by trees and grasses or with mechanical measures (Singh *et al.,* 1990).

d. By increasing water use efficiency

Planting of trees and shrubs on the landscape changes the surface energy balance, influence the surrounding microclimate, and has the potential to alter water use efficiency and productivity of adjacent crops (Cleugh, 1998; Brandle *et al.,* 2000). Agroforestry practices impact the water relations of the crop by affecting the loss of water through damaged leaves (Brandle *et al.,* 2008). On soils subject to wind erosion, wind breaks or other agroforestry buffers provide significant reductions in the amount of wind blown soil and subsequent abrasion of plant parts and cuticular damage (Kort, 1988). Loss of cuticular integrity or direct tearing of the leaves reduces the ability of the plant to control water loss (Miller *et al.,* 1995). Further out in the "quiet zone" the wind shelter reduce air turbulence and help conserve soil moisture by reducing evaporation from the soil. This may boost a crop's early growth, and increase its water use efficiency over the whole growing season (Helen Cleugh, 2003). However, trees and shrubs used in agroforestry practices themselves consume a portion of available water. So, in the area immediately adjacent to the barrier, competition for water between crops and trees has a negative impact on yield.

e. By offering livelihood security

There are some set of environmental and social conditions in which the potential for agroforestry is particularly high: densely populated, steeply sloping lands one such, frequently having problems of erosion, fertility decline, forest clearance and fuelwood shortage (Young, 1997). When formulating a participatory watershed management agenda the critical question "Why would a farmer want to think beyond the farm level?" have to be addressed (Turton and Farrington, 1998). Practicing agroforestry in the watershed area gives livelihood security to the dependent. The Garhkundar - Dabar watershed of Central India with agroforestry intervention can directly support 895 people, 2648 animal resources and secures livelihood of the dependent (Palsania *et al.*, 2008). It is possible through,

- Additional production along with diversified products.
- Sustained production with environmental amelioration.
- Microclimate enrichment, nutritious and healthy food.
- Social status along with motivation and confidence in life.
- Assured and increased employment.
- Earning and additional employment through value addition.

f. By the mitigation of climate change

Agroforestry has importance as a carbon sequestration strategy and increasing the efficiency of N use. Based on tree growth rates and wood production, and assuming ratios of tree-stem biomass to C content of 1 : 2, average carbon storage by agroforestry practices has been estimated to be 9, 21, 50 and 63 Mg C ha^{-1} in semiarid, sub-humid, humid and temperate regions (Schroeder, 1994). At a global scale it has been estimated that agroforestry systems could be implemented on 585 to 1275 X 10^6 ha^{-1} of technically suitable land, and these system could be store 12 to 228 Mg C ha^{-1} under the prevalent climatic and edaphic conditions (Dixon, 1995). For smallholder agroforestry systems in the tropics, potential Carbon sequestration rates ranges from 1.5 to 3.5 Mg C ha^{-1} yr^{-1} (Montagnini and Nair, 2004).

The leguminous components of agroforestry increase the N use efficiency and productivity. It is noteworthy that the non-nitrogen-fixing species, *Acioa* and *Alchornea* nevertheless contain substantial N, as does a species that is probably non-fixing, *Cassia siamea*. Up to 30 % of the N in pruning reaches the crop, the rest being lost by leaching and gaseous losses. Large amount of N can be fixed by hedgerows, e,g. 75 to 120 kg N ha^{-1} in six months by Leucaena (Mulongoy, 1986). That is why agroforestry was recognized by the IPCC as having a high potential for sequestering carbon as part of climate change mitigation strategies (Watson *et al.*, 2000). This kind of atmospheric deposition makes a significant contribution to nutrient cycling, greater in humid region than dry. Trees may increase rainfall but it reduces the wind speed and thereby provide preferential conditions for deposition of dust (Parker, 1983). A key question in this context is how small holders / watershed stake holders can benefit from carbon sequestration projects? (Montagnini and Nair, 2004).

Conclusions

Practicing agroforestry in the degraded lands will improve the soil quality, productivity, microclimate, water storage and use efficiency, status, health and livelihood of people. Even though the suitable agroforestry systems and models may vary with place to place, it assures the success of greening of land. This kind of integrated approach not only meet the needs for food, fodder, fuel, fibre, fertilizer, etc. but economic and environmental security is guaranteed for safe and protected habitat for humans and other living organisms by harmonization of proper land uses. It also capable to reduce the pressure on natural forests by supplying the demands of people and animals.

References

Aggarwal, R. K. 1980. Physico-chemical status of soils under " Khejri" (*Prospis cineraria* Linn.). In: Khejri (*Prospis cineraria*) in the Indian desert-its role in agroforestry, (H.S. Mann and S.K. Saxena, eds), Jodhpur, India: CAZRI, Pp. 32 - 37.

Aggarwal, R.K., Gupta, J.P., Saxena, S.K. and Muthana, K.D. 1975. Studies on soil physico-chemical and ecological changes under 14 years old desertic species of Rajasthan. *Indian Forester*, 102: 863 - 872.

Anon. 1993-94. Annual Report of the All India Coordinated Research Project on Dry land Agriculture, Main Centre, Akola (Maharashtra), India.

Bhardwaj, S.P. 1988. Effect of zero tillage and living mulch on rainfall disposition and erosion losses at 4% slope in Doon Valley. Annual Report, CSWCRTI, Dehradun, Pp.41-42.

Bhardwaj, S.P. 1990-91. Study on live bunds for soil and water conservation and maize yield on 4% slope in Doon valley. Annual Report, CSWCR&TI, Dehradun, Pp.41-42.

Bhardwaj, S.P. and Khola, O.P.S. 1999. Biological and mechanical measures for erosion control and crop production on 4 per cent slope. Soil & Water Conservation Annual Report, 1998-99, CSWCR&TI, Dehradun, Pp.22.

Brandle J.R., Xinhua Zhou. And Hodges, L. 2008. Agroforestry: enhancing water use efficiency. In: Encyclopedia of water science, (Stanley W. Trimble Eds.), CRC Press, New York, Pp. 17 - 19.

Brandle, J.R., Hodges, L. and Wight, B. 2000. Windbreak practices - in North American Agroforestry: In: Integrated Science and Practice, (Garrett, H.E. Rietveld, W.J. Fisher, R.F., Eds.), American Society of Agronomy, Inc.: Madison, WI, Pp. 79 - 118.

Chauhan, D.S., Dhyani, S.K. and Desai, A.R. 1993. Productivity potential of grasses in association with *Alnus nepalensis* and pine apple under Silvi-horti-pastoral system of agroforestry in Meghalaya. *Indian Journal of Dryland Agriculture Research & Development*, 8: 60-64.

Cleugh, H.A. 1998. Effects of windbreaks on airflow, microclimates and crop yields. *Agroforestry Systems*, 41, 55 – 84.

Dadhwal, K.S. and Katiyar, V.S. 1996. Land degradation and environmental pollution by mining and landslides. In: Soil management in relation to land degradation and environment, (Biswas, T.D., and Narayansamy, G.Eds.), Indian Society of Soil Science, New Delhi, Pp. 114 – 126.

Dadhwal, K.S. and Solanki, K.R. 2000. Agroforestry innovations for degraded lands, Pub: NRCAF, ICAR, Jhansi, Pp. 60.

Dadhwal, K.S. and Tomar, V.P.S. 1999. Enhancing production by managing tree roots in an agroforestry land use systems, SAIC Newsletter, Dhaka, Bangladesh, 9: 5&7.

Dixon, R.K. 1995. Agroforestry Systems: Source or sinks of greenhouse gases?. *Agroforestry Systems*, 31: 99-116.

Felker, P. 1978. State of art. Acacia albida as a complementary permanent intercrop with annual crops. Report to USAID, University of California, Riverside, USA: Pp. 133.

Gogoi, B.N. 1976. A review of research in shade tea in India. *Two and a* Bud, 23: 67-73.

Grewal, S.S. 1993. Agroforestry systems for soil and water conservation in Shivaliks. In : Agroforestry in 2000 AD for the semi-arid and arid tropics, National Research Centre for Agroforestry, Jhansi, India, Pp. 82-85.

Hazra, C.R. 1994. Forages from rainfed agroforestry in Bundelkhand region of India and its influence on soil fertility. In : Agroforestry systems for degraded lands, (P.Singh, P.S. Pathak and M.M. Roy Eds.), Vol.I, Oxford & I.B.H. Pub. Co. Pvt. Ltd., New Delhi, India, Pp.140-147.

Hazra, C.R. 1998. Management of rain water resources on watershed basis for sustained agricultural production – an experience of Tejpura Watershed (Jhansi). *Agricultural Situation in India*, 55 : 120-127.

Helen Cleugh. 2003. Trees for Shelter windbreaks for Australian farms, The JVAP research update series No.6, Pp. 15.

ICFRE, 1993. Annual research report for 1991 – 92, pp.81. Indian Council of Forestry Research and Education, Dehradun, India.

Jain. B.L. 1986. Afforestation and pasture development on salt affected lands. Proc. Summer Institute on Wasteland in arid and semiarid zones and technologies for their improved utilization, CAZRI, Jodhpur, India, Pp. 277 – 287.

Jayaram, N.S. and Raizada, A. 1994. Vegetative measures for erosion control in vertisols. *Indian Journal of Soil Conservation*, 22: 225-235.

Kort, J. 1988. Benefits of windbreaks to field and forage crops. *Agriculture Ecosystem and Environment*, 22: 23, 165 - 190.

Miller, J.M., Bohm, M. and Cleugh, H.A. 1995. Direst mechanical effects of wind on selected crops: A review, Technical report number 67; CSIRO Centre for Environmental Mechanics: Canberra, Australia, Pp. 68.

Montagnini, F. and Nair, P.K.R. 2004. Carbon sequestration: An underexploited environmental benefit of agroforestry systems. *Agroforestry Systems*, 61 & 62, 281 - 295.

Mulongoy, K. 1986. Nitrogen cycling in alley cropping systems. IITA Research Briefs 7. Ibadan, Nigeria, IITA, 3 - 5.

Narain, P., Singh, R. and Joshie, P. 1988. Studies on nutrient and water budget under different land use systems. Annual Report, CSWCRTI, Dehradun, Pp. 55-57.

Narain, P., Singh, R.K., Sindhwal, N.S. and Joshi, P. 1998. Agroforestry for soil and water conservation in western Himalayan valley region of India. *Agroforestry Systems*, 39: 191-203.

Palsania. D.R., Ramesh Singh, Tewari, R.K., Yadav., R.S., Dwivedi, R.P.,Kumar, R.V., Venkatesh, A., Kareemulla, K., Bajpai, C.K., Rajendra Singh, Yadav, S.P.S., Chaturvedi, O.P. and Dhyani, S.K. 2008. Socioeconomic and livelihood analysis of te people in Garhkundar - Dabar watershed of Central India. *Indian Journal of Agroforestry*, 10 (1): 65 - 72.

Pandey, D.N. 2007. Multifunctional agroforestry systems in India. *Current Science*, 92 (4): 455 - 463.

Parker, G.G. 1983. Throughfall and stemflow in the forest nutrient cycle. *Advances of Ecological Research*, 13: 57 - 133.

Samra, J.S. 1997. Status of research on watershed management. Presented at 173rd meeting of Governing body of ICAR, New Delhi.

Schroeder, P. 1994. Carbon storage benefits of agroforestry systems. *Agroforestry Systems*, 27: 89-97.

Singh, B.P. and Dhyani, S.K. 1996. An appraisal of indigenous landuse systems in Meghalaya for developing sustainable rainfed hill agriculture. In : IUFRO-DNAES International Meet on Resource Inventory Techniques to Support Agroforestry and Environmental Activities (Kohli, R.K. and Arya, R.K. Eds.), DNAES, Chandigarh, India, Pp. 61-72.

Singh, D.P. and Hazra, C.R. 1994. Rehabilitation of degraded wastelands through soil and water conservation and silvipasture at Gaharawa Watershed. In : Agroforestry systems for degraded lands (P.Singh, P.S. Pathak and M.M. Roy Eds.), Vol.I, Oxford & I.B.H. Pub. Co. Pvt. Ltd., New Delhi, India, Pp.54-62.

Singh, G. 1995. An agroforestry practice for the development of salt affected soils using *Prosopis juliflora* and *Leptochloa fusca*. *Agroforestry Systems*, 27: 61 - 75.

Singh, G., Sastry, G. and Bhardwaj, S.P. 1990. Watershed responses to conservation measures under different climatic regions of India. *Indian Journal of Soil Conservation*, 18 : 16-22.

Singh, P., Rao, M.S.R.M. and Yadav, K.S. 1994. Resource management for sustained production in Aravali region. *Indian Journal of Soil Conservation*, 22: 134-144.

Solanki, K.R., Bisaria, A.K. and Handa, A.K. 1999. A decade of research (1988-98) - National Research Centre for Agroforestry, Jhansi, Pp. 44.

Srivastava, A.K. 1994. Alternative land uses for sustained production in semi-arid vertisols. In : Agroforestry systems for degraded lands (P.Singh, P.S. Pathak and M.M. Roy Eds.), Vol.I, Oxford & I.B.H. Pub. Co. Pvt. Ltd., New Delhi, India, Pp.128-133.

Tejwani, K.G. 1994. Agroforestry in India. Oxford & IBH Publishing Co. Pvt. Ltd., New Delhi, Pp. 233.

Turton, C., Farrington, J., 1998. Enhancing rural livelihoods through participatory watershed development in India. ODI Natural Resource Perspectives 34, 1-4.

Watson, R.T., Noble, I.R., Bolin, B., Ravindranath, N.H., Verardo, D.J. and Dolken, D.J.2000. Land Use, Land use change, and Forestry. Intergovernmental Panel on climate change (IPCC). Special Report. Cambridge University Press New York.

Wiersum, K.F. 1984. Surface erosion under various tropical agroforestry systems. In C.L. O' Loughlin and A.J. Pearce, eds. Symposium on effects of forest land use on erosion and slope stability. Honalulu, Hawaii. USA: East - West Centre, 231 - 239.

Young, A. 1997. Agroforestry for soil management. 2nd Ed. C.A.B. International. Wallingford, UK, Pp. 320.

□□□

Agroforestry: Natural Resource Sustainability, Livelihood & Climate Moderation © 2009
Editors: OP Chaturvedi, A Venkatesh et. al., pp. 71-78
Satish Serial Publishing House, Delhi

Chapter 6

IMPORTANT AGROFORESTRY SYSTEMS DEVELOPED FOR DIFFERENT AGROCLIMATIC REGIONS OF INDIA UNDER AICRPAF

A.K. Handa, S.K. Dhyani and Uma
National Research Centre for Agroforestry, Jhansi-284 0003 (U.P.)

Introduction

The All India Coordinated Research Project (AICRP) on Agroforestry with 20 centers all over country was launched in 1983. At present there are 36 centers under All India Coordinated Research Project on Agroforestry with project coordinating unit at National Research Center for Agroforestry, Jhansi. These centers represent almost all the agroclimates of the country. The agroforestry systems developed and recommended by the Centres are location specific, besides in many situations these systems are not looked favorably by the farmers due to wide variations in the choice of species, crop preferences etc. and also some of the systems developed have not completed a full rotation for the tree component, therefore biomass production and returns are only on the basis of extrapolation. The stakeholders also require full package of practices, choice of crops with trees in temporal sequence and information on insect pest disease and effect of aberrant weather conditions, which are now of common occurrence. The important agroforestry systems for different agroclimatic regions of the India are presented in this chapter.

Out of 36 centres, 25 coordinating centres are located in different State Agricultural Universities and 11 in ICAR Institutes. The complete list of centres under the project is given below:

Table 1: Location, year of start and agroclimatic zone of the coordinating centres.

S.No.	Name of Centre and Place	Year of start	Agro-climatic zone
A)	**State Agricultural University based :**		
1.	Dr. YSPUH & F, Solan (H.P.)	1983-84	Himalayan
2.	SKUAS&T, Srinagar (J&K)	1983-84	Himalayan
3.	AAU, HRS Kahikuchi, Guwahati (Assam)	1983-84	Himalayan
4.	CSKHPKV, Palampur (HP)	2001-02	Himalayan
5.	PAU, Ludhiana (Punjab)	1983-84	Indo-Gangetic
6.	GBPUA&T, Pantnagar (Uttrakhand)	1983-84	Indo-Gangetic
7.	RAU, Pusa (Bihar)	1985	Indo-Gangetic
8.	NDUA&T, Faizabad (UP)	1985	Indo-Gangetic
9.	CSAUA&T, Kanpur (UP)*	1985	Indo-Gangetic
10.	BCKVV, Jhargram (W.Bengal)	1983-84	Humid & sub-humid
11.	OUA&T, Bhubaneshwar (Orissa)	1983-84	Humid & sub-humid
12.	BAU, Ranchi (Jharkhand)	1985	Humid & sub-humid
13.	IGKVV, Raipur (Chhattisgarh)	1996	Humid & sub-humid
14.	SDAU, Sardarkrushinagar (Gujarat)	1983-84	Arid & semi-arid
15.	RAU, ARS, Fatehpur-Shekhawati (Rajasthan)	1983-84	Arid & semi-arid
16.	CCSHAU, Hissar (Haryana)	1985	Arid & semi-arid
17.	MPKV, Rahuri (Maharashtra)	1985	Arid & semi-arid
18.	PDKV, COA, Nagpur (Maharashtra)	1985	Arid & semi-arid
19.	ANGRAU, Hyderabad (AP)	1985	Arid & semi-arid
20.	JNKVV, Jabalpur (MP)	1985	Arid & semi-arid
21.	UAS, Dharwar (Karnataka)	1983-84	Tropical
22.	TNAU, COF, Mettupalaym (TN)	1983-84	Tropical
23.	MAU, Parbani (Maharashtra)*	1985	Tropical
24.	KAU, Thrissur (Kerala)	1983-84	Tropical
25.	BSKKV, Dapoli (Maharashtra)	1985	Tropical
26.	TNVASU, Kattaupakkam (TN)	1996	Tropical
27.	UAS, Bangalore (Karnataka)	2001-02	Tropical

Contd...

Table 1. Contd...

B)	ICAR Institute based#		
1	CSWCR&TI, Dehradun (Uttrakhand)	1983-84	Himalayan
2	ICAR NEH Region Complex, Umiam, Barapani (Meghalaya)	1983-84	NEH region
3	ICAR NEH Region Complex, Imphal (Manipur)	1983-84	NEH region
4	ICAR NEH Region Complex, Gangtok, (Sikkim)	1983-84	NEH region
5	CSSRI, Karnal (Haryana)	1985	Indo-Gangetic
6	ICAR NEH Region Complex, Agartala (Tripura)	1983-84	Humid & sub-humid
7	ICAR RCER, HARP, Plandu, Ranchi (Jharkhand)	2003-04	Humid & sub-humid
8	CAZRI, Jodhpur (Rajasthan)	1983-84	Arid & semi-arid
9	IGFRI, Jhansi (UP)	1983-84	Arid & semi-arid
10	CRIDA, Hyderabad	1985	Arid & semi-arid
11	CARI, Port Blair (A&N Islands)	1983-84	Tropical

* Closed w.e.f. 01.04.2004 based on the recommendations of second QRT.
# Now working as only cooperative centres as scientists provided under this scheme have been merged as part of cadre strength of the institute.

Himalayan Zone

Agroforestry in hill regions requires special emphasis on the cultivation of indigenous tree species which have multiple economic and traditional importance. Various species of trees like *Acsandra, Bulyracea, Oleosa, Pyrus peshia, Rhododendron arboreum* etc. have plenty of nectar for the bee keeping. Species of *Grewia, Boeluneria* etc. have been utilized successfully by the inhabitants of Garhwal region in Uttar Pradesh for fibre and industry. Likewise trees viz. *Morus alba, M. serrata, Quercus lecuotrichophora* are utilised for sericulture purposes. In Western Himalayan zone, agroforestry practices vary due to elevation from sea, but silvipastoral practices are common to all. Of all the trees being used *Grewia optiva* upto mid hills and *Celtis australis* upto high temperate zones are important. On high temperate zones it is *a Robinia pseudoacacia* which is considered promising. Fruit trees get prominence in the farming system in some areas. In the central part, at lower altitudes agri-horti-system was found more prevalent. Mango, litchi and citrus were the preferred trees. Forests were used for grazing animals. Fuel wood deficits are common and about 25% farmers purchase it. Hortisilvimedicinal system namely peach + grewia + tulsi / mucuna followed by peach + morus + tulsi / mucuna are recommended for popularization in rainfed areas of sub-temperate sub humid and mid hills region of Himachal Pradesh in place of traditionally followed cereal based agri-horti-silviculture system. Bio-economic appraisal of horti-silvi medicinal system integrating peach, grewia and morus with tulsi and mucuna showed that peach + grewia + tulsi is highly economic combination followed by Peach + Morus + Tulsi. The net profit from each type was Rs. 1, 18,551 and Rs. 1, 18,820 per ha^{-1}, respectively. In Eastern Himalayas, large cardamom in association with *Alnus nepalensis* (Alder) and *Schima wallichi* (Chillowani) and citrus in association with maize/ginger, soybean/ginger is commonly practiced. Similarly potato and other vegetables often in combinations with pineapple, which is grown in the slopes in association with pine (*Pinus kysia*) is a very common practice in the Meghalaya. At Sikkim agricultural lands shaded by bamboos (*Bambusa nutans*) is effectively utilised by growing ginger, turmeric etc. In Assam, interspaces in coconut with rice seedlings and fodder crops provided net return of Rs 87,747/- ha^{-1} and also increased nut yield from 9,800 numbers ha^{-1} (sole coconut) to 13,890 numbers ha^{-1} (intercrops).

Indo Gangetic Plains

This is most productive rice-wheat regions of the country which geographically extends from Punjab in the West to West Bengal in the East. The region has maximum area under cropping and irrigation and supports massive human and livestock's population. The region has very poor natural forest cover. The deficits of timber, firewood, fodder etc. are also acute. The firewood demands are primarily met by burning crop residues and animals dung. The region shows almost all types of agroforestry practices with dominance of agri-silvicultural practice. The farm boundary plantation is one of the common and popular practices.

Poplars are today the most important among tree species constituting man-made forests in irrigated tracts of the upper Indo-gangetic plains demonstrating immense economic and industrial potential, while contributing sizeably to the well being of farmers. Poplar based agroforestry is dynamic and vibrant. The innovative farmers relentlessly experiment with cultivation of high value crops under the canopy of Poplars, refining agro forestry practices and rendering them ever more remunerative. Agriculture, which is in private sector in the State of Uttar Pradesh, Haryana and Punjab, today have more than parallels public sector forestry in creation of new timber resource. The extent of forest plantation raised in the public sector in above three States during the VIII Five Year Plan period were 68,035, 31,551 and

16, 987 ha respectively totaling to 1,16,573 ha or 23,314 ha yr^{-1}. The area planted with poplars in the agriculture sector during the same period was also of the order of 23,000 ha yr^{-1} but compared to forestry plantations, it was five to ten times higher in productivity. Some of the Poplar plantations in agriculture sector had mean annual increment of 50 m^3 ha^{-1} with average of 20 m^3 ha^{-1} against average of 4.5 m^3 ha^{-1} of forest plantations. The early sowing of wheat (10-11 Nov) in poplar plantation gave significantly higher grain yield (25 q ha^{-1}) as compared to the late sowings [end Nov (20.45 q ha^{-1}) and mid Dec (15.37 q ha^{-1})] in 4th, 5th and 6th years of poplar plantation. Among the six wheat varieties, the wheat variety PBW 502 (28.4 q ha^{-1}) and PDW 274 (26.8 q ha^{-1}) out yielded other varieties when sown on Nov.10-11. Under late sown conditions (End Nov) PBW 509 (22.8 q ha^{-1}) was better than others wheat varieties. The additional seed rate (25 per cent more than the recommended doze) increased wheat grain yield significantly (30.3 to 34.2 q ha^{-1} in 4th year) over the recommended dose.

Semi-Arid and Arid Zone

Though a large number of trees and grasses are grown on agricultural fields in arid areas of Rajasthan, the two most important woody components are *Prosopis cineraria* (Khejri) and *Zizyphus numularia* (Bordi). Depending on the amount of rainfall, 10 to 120 trees/ha is being grown along with common agricultural crops like *Pennisetum glacum* (Pearl millet), *Cyamopsis tetragonoloba* (Cluster bean), *Vigna mungo* (Moong bean) and *V. aconitifolia* (Moth bean). Grain production under *P. cineraria* was not significantly affected under wider tree spacing of 10 x10m but the yield of cluster bean was significantly affected when grown with *Acacia albida*. This is because of the horizontal spread of root system of *A. albida* utilizing moisture and nutrients more effective resulting in faster growth of the shoot compared to *P. cineraria* which has the capacity to penetrate its root deep during establishment years and subsequently show slow shoot growth at minimum interference with crop production. It was observed that *Alinathus excelsa* has no adverse effects on the crops sown as understorey if proper care is taken. For having better results, the spacing adopted should be at least 10m X 10m. Regular lopping of the leaves should be done to minimize shade effect on the crops and ploughing of the root zone of crops should be done to remove the lateral roots of the tree so that there will be no competition between tree and crop. To ensure the sustainable resource management the agroforestry is an important alternative system of farming in which trees and bushes are grown in association with arable crops. To maximize biomass production single, two-tier and multi-tier systems of fodder production were compared with natural pasture under semi-arid conditions. Multi-tier system (Ardu + Dichro + grass/ crop) provided maximum dry fodder and seed/ grains with the highest net returns (Rs. 6000.00) followed by two-tier (Rs. 5600.00) and single tier system (Rs. 3600.00).

Tropical Zone

Studies on red chalka soils showed that crops like arhar, castor, bajra and jowar associated with *Leucaena* yielded more grains as compared to sole crop with highest yield of arhar and least in jowar . In tree crop interaction studies, in 4 years old *Leucaena* planted at 7.5x2 m yield of sorghum was 53 per cent of the sole crop while aerial biomass of tree was 85 per cent of the sole crop at Hyderabad. They also recommended wider alley 7-8 m and pair rows (60 cm) of *Leucaena* in semi-arid tropics and leaf mulching better than using it as fodder. There was no depression in the yield of sorghum and castor in association with *Acacia albida, A. ferruginea* and *Prosopis cineraria* but marginal reduction in crop yield in *A. lebbek*. Higher transpiration rates were recorded in *A. albida* than *A. lebbek* as measured by heat pulse recorder.

Agrihorticultural systems revealed that in addition to getting about 40 kg of ber fruits it is also possible to get 100 kg ha^{-1} of horse gram or 450 kg ha^{-1} of cowpea grown in the interspaces. They also reported better performance of Stylo (4.2 t ha^{-1} yr^{-1}), Cenchrus (2.5 t ha^{-1} yr^{-1}) in association with *Leucaena* which provide fuel yield of 5 t ha^{-1} yr^{-1}. There was significant improvement in organic carbon in *Leucaen*a with *Leucaena-Stylo* compared to *Leucaena-Cenchrus*. In agrisilvihorticultural systems at Dharwar higher fruit yield of papaya in teak+papaya+sorghum combinations and least in association with grass or subabul instead of sorghum. Intercropping Subabul with *Agave* was found to be more remunerative than arable crops on marginal shallow soils at Bijapur (Karnataka). Fibre yield increased with increased plant population and had no significant effect on the Subabul plant growth. Better effect of Subabul and *Casuarina* on Rabi Jowar and Sunflower was least (upto 7-8 m) and was highest by the site of *A. nilotica* (20 m) followed by *Eucalyptus* (18 m). However, least run off was collected through the strips of *Leucaena* and *Casuarina* as compared to other species. Studies conducted at Rahuri revealed that in the 3rd year of the growth of the trees caused maximum reduction in grain yield on sunflower followed by red gram and pearl millet. To minimize the reduction loss in grain production the alley width was recommended at 10 m or more.

Cashew plantations in Goa and adjoining Western Ghats and coastal areas in general are grown on hill slopes and comprise self sown seedling progenies or in un-orderly planting ones with diverse genetic origin. Invariably such plantations have other trees such as teak, jackfruit, mango jamun, etc. density of trees varies from very high (400 trees ha^{-1}) to very low (40-50 plants ha^{-1}). To make such plantations more productive, orderly planting of more than two high yielding varieties in combination with other compatible tree species has to be meticulously planned to achieve sustainable agro-forestry. The major benefit that agro-forestry brings to sloping areas is its capacity to combine soil conservation with production. Further, it is feasible to grow several intercrops in cashew plantations like pine apple, papaya, drum stick, etc., for better utilisation of land during gestation period of the crop in the initial years. Coconut based systems are more prevalent in Kerala, Andhra Pradesh and Tamil Nadu. Black pepper, banana, pine apple, vanilla, ginger, turmeric, nutmeg, clove, tuber crops, betel vine, fodder crops, etc., are the common crop components in the systems found profitable in different agro-climatic situations depending on crop potential, agro-climatic conditions and the market.

Humid and Sub-Humid Regions

This is zone with characteristic high rainfall is divided into three categories viz. <1500 mm, 1500-2000 mm and > 2500 mm annually. The agroforestry practices vary from agrisilviculture to homestead with energy plantations. This is the region with high species diversity and many tree species are used by people. Species like *Casuarina,* Silver oak, teak, *Ceiba pentendra,* Jackfruit, *Acacia nilotica* and *Acacia planifrons* are common. Many fruit trees are used ranging from guava to coconut. Maize has given a grain yield of 10.2 q ha^{-1} when grown in association with *Acacia auriculiformis* where as Stylo recorded a fodder yield of 28.43 q ha^{-1} when grown in association with *Eucalyptus.*

Studies showed that *Gmelina arborea* could be a promising tree species in the agroforestry system particularly in the red and laterite zone of West Bengal. Crops like rice, groundnut, cowpea, blackgram and sesame have been found promising during rainy season and can be intercropped with Gmelina arborea having a plant density of 833 ha^{-1} i.e. 4 m × 3 m spacing. Among all the crops tested, groundnut produced maximum yield (1.55 t ha^{-1}). This system can be followed upto 8 years. Afterwards, number of gamhar tree should be reduced to 50 % (i.e. 4 m × 6 m spacing) by cutting to avoid shade effect. Then all the arable crops can be cultivated without hampering the yield. It was seen that *Dalbergia sissoo* and *Gmelina arborea* species based silvipasture system was more suitable for development and utilization of red lateritic wasteland in Chhatisgarh region. Association index between *Stylosanthes hamata* and *Chrysopogon fulvus* showed highest degree of association under *Dalbergia sissoo* plots (100%). Pasture crops grown under *Dalbergia sissoo* contain highest nutrient, crude protein and crude fibre with maximum share of grasses. Thus the combination of trees + *Chrysopogon fulvus* + *Stylosanthes hamata* was found suitable for restoring the organic matter in red lateritic wasteland.

Guava+sandal combination has given higher returns followed by guava+sissoo compared to guava+teak and guava+bija (*P. marsubium*). The black gram, rice and maize performed well under wider spacings of *Acacia auriculiformis*. Growth of *Eucalyptus* and *A. auriculiformis* was better with intercrops. Agrihorticulture system of litchi +cabbage and citrus+cabbage were found better in Orissa. Subabul and *Sesbania* were planted in interspace which yielded good return.

Conclusion

The retro spectral perusal of the agroforestry research exhibits its wide spectral potential in sustenance of agriculture as these systems provide food, fodder, fruit, vegetables, fuel wood, timber, medicines, fibre etc. from the same piece of land at a time which not only fulfils the demand of people but also elevate their socioeconomic status and standard of life. It is worth noting that agroforestry systems can be sources or sinks of green house gases depending on the components of the system and the methods through which these are established. Selection of agroforestry system and its management practices affects the spatial temporal flux of carbon and nitrogen reservoirs in soil and vegetation. Agroforestry systems may be managed to stabilize green house gases in three ways viz. sequestering carbon dioxide in plants and storing carbon and nitrogen in perennial vegetation and soils for long period, producing food, fodder, fuel, fibre etc. to suppress deforestation and sustained production of biofuels (fuel wood and crop stems, roots, leaves etc.) cuts down the combustion of fossil fuels which are the major source of green house gases. The apparent high potential of agroforestry systems lies in its applicability to control soil erosion, soil improvement, creating congenial and conducive microclimate for trees and understorey crops and reduction in the accumulation of green house gases in the atmosphere. In the present situation it is the first and foremost duty of ecologists, plant physiologists, agroforesters and policy makers to launch agroforestry programmes in states where suitable land for such system is available.

❑❑❑

Agroforestry: Natural Resource Sustainability, Livelihood & Climate Moderation © 2009
Editors: OP Chaturvedi, A Venkatesh et. al., pp. 79-89
Satish Serial Publishing House, Delhi

Chapter 7

BAMBOO BASED AGROFORESTRY SYSTEMS IN INDIA

S.P. Ahlawat, R.V. Kumar, V.K. Gupta and S.K. Dhyani
National Research Centre for Agroforestry, Jhansi-284 003 (U.P.)

Introduction

Bamboo is a promising renewable natural resource and its benefits are spread through employment provision, income generation and environment protection. It is well known for faster growth, multiple uses and aesthetic beauty. Bamboo grows naturally at heights ranging from sea level to 3500 m, in an extraordinary range of habitat - in almost 10 million hectares of forests land and on homestead and private plantations. Bamboos play an important role in the economy of Asia-Pacific region. They are multipurpose plants, with more than 1500 documented uses. The most important traditional uses include housing, food and material for handicrafts. Worldwide, more than 2.5 billion people trade in or use bamboo (INBAR,1999). Modern manufacturing techniques allow the use of bamboo in timber-based industries to provide bamboo flooring, board products, laminates and furniture. Bamboo is becoming a substitute for wood in pulp and paper manufacturing; about 25 per cent of the fibre used in the Indian paper industry each year comes from bamboo (FAO, 1994). Bamboo shoots are now an important food crop on the international market as well as locally and nationally. China is by far the leading exporter of bamboo shoot products, with an annual export value of nearly US$ 140 million (Feng Lu, 2001). Bamboo furniture is an expanding business in many countries. Worldwide, domestic trade and subsistence use of bamboo are estimated to be worth US$ 4.5 billion per year. Global export of bamboo generates another US$ 2.7 billion (INBAR, 1999). Yet international trade represents only a small proportion of total bamboo usage – domestic use is estimated to account for at least 80 per cent. Bamboo is thus a major world commodity.

Scope and Potential of Bamboo Based Agroforestry Systems

Bamboo as a woody grass is perfectly suited to agroforestry. Under agroforestry systems, it is suitable for intercropping, soil conservation, wind break, riparian filter, permaculture and yields value added products, timber, forages, edible shoots, fiber and craft wood. Bamboo based agroforestry models will provide higher economic returns to the farmers, improve the soil, fill the gap of national forest cover (33%) and provide raw material to the industry as well as for domestic use of the rural community. Economically its cultivation ensures IRR of more than 25%, almost equal to eucalyptus but returns are recurrent on annual basis after 5-6 years upto 30 years or more without recurring investments on plantations (Ashutosh, *et al.*,1996). Because of the use of various intercrops, products are obtained even in the early stages of plantation and the income would be much higher than any other system. Bamboo can be harvested every year from fourth year onwards and hence regular income will start much earlier than expected from any other woody component. Agroforestry practice will also benefit the bamboo plants by sharing the inputs of irrigation, manure and fertilizers, weeding etc. applied to agriculture crops, and hence bamboo quality and yield would be better than unmanaged plantations. So, the total returns can be much higher than other wood based agroforestry systems.

Many bamboo species grow quickly and are highly productive. For example the shoots of *Bambusa tulda* elongate at an average rate of 70 cm per day (Dransfield and Widjaja, 1995). *Dendrocalamus giganteus* also grow at the same rate and can reach 25 meters in three months. Annual productivity values mostly range between 10 and 20 t/ha/yr (Hunter and Wu Junqi, 2002). Bamboo stands may achieve a total standing biomass that is comparable to some tree crops (of the order of 20-150 t/ha) and therefore, they may sequester substantial amounts of carbon. However, their total carbon storage is likely to be lower than that of well-grown forest on favourable sites (Hunter and Wu

Junqi, 2002). Because the extensive rhizome system of bamboos lies primarily in the top layers of soil, they often play a major role in stabilizing soils on slopes and river banks, preventing erosion and land slips. This also makes them important in securing the hydrological function of catchments and rivers. All these properties show good scope for its cultivation under agroforestry and farm forestry.

Bamboo Based Agroforestry Models

In many Asian countries, bamboos are planted by farmers in courtyards, homesteads or small-scale farm lands and managed as groves or following some agroforestry model. Advanced agroforestry involves design principles, different models, and their components and structure. Most of the bamboos are cultivated for edible shoots and/or bamboo culms. Moso bamboo (*Phyllostachys pubescence*) is main species for agroforestry systems in China. Initially for 2-3 years of plantation, some agricultural crops like soybean or tuberose crops are planted as intercrops in bamboo plots, forming perfect agroforestry model. Certain bamboo species (e.g. *Phyllostachys pubescence* and *Phyllostachys edulis*) have formed the basis for major industrial development and have been domesticated into plantations in China. Fu Maoyi and Banik (1995) have reviewed different bamboo based agroforestry models in INBAR technical bulletin no. 8 and based on their functions and end-products following three groups i.e. bamboo-agriculture model, bamboo-fishery model and special purpose model were suggested.

Group 1: Bamboo-Agriculture (Agri-silviculture) Model

1. Bamboo + agricultural crops
2. Bamboo + conifer and/or broadleaf timber trees

Group 2: Bamboo-Fishery Model

1. Bamboo + fish pond

Group 3: Special Purpose Model

1. Bamboo + edible fungi
2. Bamboo + medicinal plants

From above three groups, following five bamboo based agroforestry systems can be established by combining one or more components with bamboo in a reasonable structure on a given site.

i. Bamboo + agricultural crops

This model exists until bamboo plants develop and occupy the whole site. Its final products will be culm timber or edible shoots. Bamboo is planted at varied spacing of 4 x 4 m to 8 x 8 m, depending upon the clump size. Intercropping of agricultural crops can be done for a maximum period of five years after bamboo planting or until bamboo covers the full space lying between lines and plants. If Intercropping of agricultural crops is to be done for more years than larger spacing of 10 – 15 meters is required. This is quite common in NE region, where ginger, turmeric and arrowroot is grown in jhum lands under shifting cultivation.

ii. Bamboo + conifer and/or broadleaf trees

In India, such model can be seen in Ziro valley of Arunachal Pradesh. The *Phyllostachys bambusoides*, a monopodial bamboo is grown with *Pinus kesiya* as well as pure stand. There are very few temperate bamboo species in the country that grows with conifer or broad leaved tree species. This can be established by either converting semi-naturally mixed stands or planting new ones. The ratio of bamboo to trees is important especially in semi-naturally mixed stands. Crops such as soybean, black gram, sweet potato, turmeric, zinger and vegetables can be intercropped within three years after planting. The planting time for bamboo and trees should be determined based on the growth rate of the tree species involved.

iii. Bamboo + crops + fish pond

This model is usually made on plain or wet lands where fish ponds are built. One to three rows of shoot-producing sympodial bamboos may be planted on the banks of the pond, and crops such as soybean and rye intercropped between bamboo clumps to form a complete food chain. Crops can be harvested only in initial stages until the bamboo occupies the full space. Bottom mud from the ponds may be dug out in winter and used as fertilizer for bamboo clumps.

iv. Bamboo + medicinal plants

For this model, the medicinal plants should be chosen to suit the climatic conditions and topography of the site. Some selected species those can tolerate shade, like *Withenia somnifera, Catheranthus roseus, Casia angustifolia, Aloe vera, Plantago ovata, Chlorophytum sp.* (safed musali) etc. can be grown as intercrop as they have good adaptability. Some others such as *Curnma domestica* need to be planted in valleys at altitudes of 800-900 meter. *Mentha arvensis* need slopes which get adequate sunshine.

v. Bamboo + edible fungi

There are a large number of edible fungi regarded as natural food rich in vegetable proteins. Among these, *Dictyophora tomentosa, Pleurotus ostreatus* and *Auricularia auriculajudoe* can be cultured in bamboo stands, which satisfy the fungi's need for humidity, shade and a fertile bed. A bed of decayed bamboo litter and cotton shells or rice husk placed evenly on the ground up to a height of 10 cm, is used as the substrate. Inoculation of *Dictyophora* sp. is done in September for varieties that grow in low temperature and in May-June for those that require a higher temperature. Harvest is after 4-8 months, depending on the fungus variety. *P. ostreatus* is inoculated in March and harvested two months later. *A. auriculajudoe* needs to be cultured in bags filled with the growth medium and hung on the bamboo. This system is reported from China.

Bamboo Based Agroforestry Systems in India

Scope for bamboo in agroforestry in India is very wide because of the uncertain weather conditions and increasing cost of labour involved in raising agricultural crops on marginal lands rendered the later option less attractive. A large amount of bamboo grows outside of forests particularly in homestead and farm fields. It has been effectively used as wind break and shelterbelts in costal areas to protect agricultural crops, plantations and habitations. They are planted with other tree species specific of the region, for example *Cocos nucifera* (Coconut), *Anacardium occidentale* (cashew nut) and *Eucalyptus* spp. as second storey crop (Jain and Rawat, 2008).

Bamboo in Homesteads and Shifting Cultivation Lands

It is basically a house hold plant species grown around houses and compounds, on farms and available spaces on farm holdings. In eastern, north eastern states and Kerala, bamboos are planted as an integral component of each homestead along with coconut, areca nut, betel vine and kadam etc. It forms a combination of timber trees, fruit trees, fuel wood trees, food crops and vegetables. In eastern Himalayan region, shifting cultivation is still a most common agricultural practice. During fallow period, density of bamboo increase and therefore, shrubs, trees and bamboo dominate in 25 years. Primarily, *Dendrocalamus hamiltonii, Bambusa tulda, B. khasiana*, and *Neohouzeuna dulloa* contributed to the biomass in the field upto 4.9 t/ha (Rao and Ramakrishna, 1990).

Some research work has been done on study of bamboo based agroforestry systems in India. A limited numbers of reports were found on bamboo based agroforestry system in the country and same are discussed here. Intercropping performance of bamboo with mango, cahewnut, kokum and rubber have been studied by Wagh and Rajput (1994). Study concluded that bamboo was the most profitable among the six horticultural crops grown in Konkan region in India. Shanmugavel and Francis (2000) studied intercropping performance of four crops viz. pigeon pea, soyabean, ginger and turmeric with *B. bambos* under Tamilnadu conditions. Intercropping of pigeon pea and soyabean were found more productive than ginger and turmeric. The Land equivalent ratio (LER) of intercropping of *B. bamboo* with pigeon pea and soyabean was equivalent to that of 1.2 ha or 1.1 ha under monoculture. Jha *et al.* (1999) reported that intercropping of soyabean with *M. baccifera* and *D. longispathus* is feasible on degraded *jhum* lands of Mizoram and gave better results than pure bamboo stands. Cultivation of soyabean (*Glycine max*) along with *Dendrocalamus strictus* has been found technically feasible and economically viable (Seshadri,1985). In Madhya Pradesh, the impact of bamboo based agroforestry models with short rotation agricultural crops namely soybean, niger, greengram, mustard, wheat, blackgram and pigeonpea were investigated. The model has played a significant role in soil management including checking soil erosion and improvement of soil fertility of degraded agricultural lands (Jamaluddin *et al.*, 1998). Lal (1995) has reported development of bamboo based nine agroforestry models for eco-rehabilitation and social upliftment of rural people in Allahabad district of U.P. The response of farmers and rural people of that area was encouraging and positive.

Most of the studies reported better economic returns from bamboo based agroforestry systems in comparison of sole cropping. Shanmugavel and Francis (2000) reported higher annual net returns (Rs. 13,300) when pigeon pea was intercropped in 1:1 rows at 3x3 m^2 spacing. Cost benefit analysis of bamboo plantation based on *D. strictus* at Gaul Pahari, Haryana revealed that this system yielded better economic returns (Rawat *et al.* 2002). Jayashankar *et al.* (1997) carried out the profitability analysis of three species of bamboo viz. *B. bambos, D. strictus* and *Thyrsostachys oliveri* grown in Kerala. Study revealed that all the three species of bamboo yielded B-C ratio higher than one, indicating high profitability. Among the three species *T. oliveri* gave highest returns.

Singh *et al.* (1992) studied the effect *B. nutans* on the yield of some agricultural crops at mid hills of eastern Himalaya. The bamboo was planted or allowed to grow along farm boundaries, drainage channels & uncultivated wastelands. Through experimentation it was found that agricultural land near bamboos can be effectively utilized for growing ginger, turmeric, cardamom, orchard grass and dinanath grass up to a distance of 11-15 m from bamboo rows. Rice, finger millets, soybean, nandi setaria and fine stylo were suitable crops beyond this distance. Further, the impact of 25-30 years old *B. nutans* clumps raised in agri-silviculture system on chemical properties of soil was studied (Singh *et al.*, 1992). The land near bamboo could be effectively utilized for growing ginger, turmeric, large cardamom, orchard grass and dinanath grass. The available phosphorus (P) increased whereas exchangeable K and Ca + Mg decreased with increasing distance from the bamboo row, soil pH and soil organic matter did not vary with distance. Singh and Kochar (2005) have studied the effect of clump density on nutrient uptake in 12.5 years old clumps of *B. pallida* and changes in soil properties. The plant density had significant influence on circumference of clump and characteristics of culms. The positive nutrient balance was reported under three plant densities (278, 204 and 156 clumps ha^{-1}). The floor litter added 79 kg ha^{-1} nutrients to the soil and drained nutrients 469 kg ha^{-1} yr^{-1}. EC and soil pH was improved and soil enriched by organic matter, Ca, Mg and Zn. Patil *et al.* (2004) analysed the effect of bamboo based agroforestry systems on soil profile and surface soil properties. Study revealed that organic carbon, carbon content of these soils ranged from 0.43 to 0.72%. Soil profile investigations revealed that all of these nutrients increased in bamboo based agroforestry sites. The organic C and organic matter content of these soils increased from 0.37 to 0.58% and 0.63 to 0.99%, respectively.

Bamboo Based Agroforestry System at NRCAF, Jhansi

In Bundelkhand region the bamboo has been found growing naturally in forests, field boundaries and homesteads. Mainly, *Bambusa bambos* and *Dendrocalamus strictus* are seen in the region. Both the species have better adaptability to the semi-arid conditions and degraded lands. Besides providing items of daily use to the rural poor it can be a good source of fodder for buffalo and cattle during lean season. Bamboo (*Dendrocalamus strictus* and *Bambusa bambos*) based agri-silviculture system was evaluated in rainfed red gravely and black soils of NRC for Agroforestry, Jhansi. The bamboo was planted at 6x6 m spacing. Performance of bamboo was better in red soils compared to black soil, and chickpea and blackgram were found ideal intercrops with bamboo. Intercropping did not exhibit any adverse effect either on survival or growth and yield of bamboo. In five years, clump height, canopy and number of culms in bamboo were higher under the agroforestry situation as compared to pure bamboo (NRCAF, 2004). Bamboo based agroforestry system gives output from harvest of bamboo from fourth year onwards, besides the yield of intercrops, so this system is more remunerative than any other systems. Recently a coordinated project on development of bamboo based agroforestry systems at six agroclimatic zones has been takenup with the financial assistance of National Bamboo Mission, New Delhi, in which five other research centres of All India Coordinated Research Project on

Agroforestry (AICRPAF) are included. In this project, besides developing the agroforestry systems at research fields, the on farm trials are also taken. The response of the farmers is very encouraging. In the first year, there is no reduction in yield of intercrops due to bamboo. The crops and bamboo species taken at different centers are given here.

Bamboo species and intercrops possible in agroforestry systems in different agroclimatic zones

Agroclimatic Zone	Research centre	Bamboo species	Possible Intercrops
Central Plateau and Hills Region	NRC AF Jhansi	*D. strictus,*	Gram, lentil and sesame
Eastern Himalayan Region	AICRPAF, Guwahati	*B. tulda* and *B. balcooa*	Ginger, turmeric, maize, mustard, chili, colocasia, sesamum,
Eastern Plateau & Hills Region	AICRPAF, Jhargram	*B. tulda, B. balcooa*	Chili, mustard, sesamum, niger , turmeric, black gram, groundnut,
Southern Plateau & Hills Region	AICRPAF, Dharwad	*B. bambos,*	Soyabean, mustard, sesamum, greengram, blackgram,cotton,
East Coast Plains & Hills Region	AICRPAF, Bhubneswar	*D. strictus,*	Arhar, Horse gram, sesamum, colocasia, cowpea, turmeric
West Coast Plains & Ghats Region	AICRPAF, Dapoli	*O. stocksii*	Finger millets , green gram, black gram, pigeon pea, niger,

Bamboo Species Suitable for Agroforestry and Farm Forestry

Different species are found in different agroclimatic conditions and their adaptability varies accordingly. They have different size and properties of culm. They have different uses and suitability for handicrafts, chipping, slivers, strips, laths, splitting, panel boards, composite boards, mats etc. Distribution pattern, ecological preference and uses of some the suitable species are given here.

Limitations

Besides so many advantages, similar to the trees there are some constraints of the agroforestry systems based on bamboos. The bamboos are tall plant which cast shade on under story crop and reduces the crop yield. The crops can be taken to a limited period until bamboo covers the entire soil. Otherwise, the spacing for bamboo planting is to be kept too wide, which is sufficient to grow crops at full maturity of bamboo clumps. Since new culms arise on the periphery of clump, the soil workings for intercropping may limit the bamboo

S. No.	Species	Ecological preference, altitude and distribution	Common uses
1	*Bambusa affinis*	Moist Tropical Region, Tripura	Making javelins, angling rod, furniture etc. (solid culm)
2	*Bambusa balcooa*	Moist Tropical upto 600m MSL, NE Region, West Bengal, Bihar	Building construction, Ladder, Agarbatti sticks, Edible shoots, Wood Chip industry,
3	*B. bambos* syn. (*B. arundinacea*)	Moist Tropical & semi-arid zone, Upto 1200m NE India, South India, Maharashtra	Building construction, Ladder, Tent pole, Baskets, Mats, Agriculture implements.
4	*B. nutans*	Moist Tropical, 600 to 1500m, Himalayan tracts, NE Region, W. Bengal	Construction, Pulp & paper, Basketry, Handicrafts
5	*B. pallida*	Moist Tropical, 700- 2000m MSL, NE Region, North Bengal	Construction, Mats, Basketry, Edible shoots, Agarbati sticks, Pulp & Paper
6	*B. polymorpha*	Tropical Zone, Madhya Pradesh, Kerala, Tamilnadu, NE- India,	Building construction, Pulp & paper, Agarbatti sticks, Fiber boards, Roofing, Edible shoots
7	*B. vulgaris* (suitable for saline & sandy areas)	Moist tropical zone, Upto 1200m, M.P., Chattisgarh, Bihar, U.P.,	Construction, Pulp, Basketry, Handicrafts, Hats, Food Grain containers
8	*Bambusa tulda*	Moist tropical zone, NE-Region	Construction, Pulp, Basketry, Mats, Toys, Handicrafts
9	*Dendrocalamus asper*	Tropical areas, (exotic sp.)	Edible shoots, Construction, Pulp, Basketry, Ladder,
10	*D. giganteus*	Moist tropical zone, 1200 m, Arunachal Pradesh Assam, Manipur, Nagaland, W. Bengal & NW Himalayas	Construction, Ladder, Pulp,
11	*D. strictus*	Semi-dry areas, upto 1000m, Widely through out India except NE Region	Construction, Pulp, Land reclamation etc.
12	*Melocanna baccifera*	Moist tropical, 600m, NE-region, Assam, Manipur, Meghalaya, Mizoram, Tripura, W. Bengal	Construction, Mats, Toys, Wall plates, screens, hangers, hats, baskets, food grain containers, pulp & paper, fruit edible
13	*Oxytenenthera stocksii*	Tropical zone, 800m, Goa, Karnataka, Northern Kerala	Construction, baskets, pulp

culms production. Its adventitious roots spread in the field and may interfere in intercultural operations. At the same time they are damaged by ploughing thereby limiting the growth of bamboo clumps. During extraction of culms the understory crops may also get damaged. Though leaf litter enrich the soil after decomposition but bamboo leaves create some hindrances in cultural operations of crops. Further, there may be sharing of common pests and diseases among the bamboo and understorey crops.

References

Ashutosh, S., Vishwakarma, P. and Nema, S. 1996 Financial analysis of bamboo based agroforestry models. *Indian Forester,* 122 (7): 543-546.

Dransfield, S. and Widjaja, E. 1995. Bamboos. Plant resources of South-East Asia. Backhuys Publishers, Leiden.

FAO 1994. Bamboo in Asia and the Pacific. Proceedings of the Fourth International Bamboo Workshop, 27-30 November 1991, Chiangmai, Thailand. FAO, Bangkok FORSPA Publication 6. Pp. 373.

Fu Maoyi and Banik, R.L 1995. Bamboo production systems and their management. In: Bamboo People and Environment: Propagation and Management (I.V. Ramanuja Rao, C.B. Sastry and E. Widjaja, Eds.), NBAR Tech. Report No. 8. Pub. INBAR, IPGRI and IDRC. 1: 18-33.

Fu, M.Y., Fu, J.H., and Fang, M.Y. (1993) Bamboo plays an important role in agroforestry management systems in China. *Journal of Bamboo Research,* 12 (2): 63-73.

Hunter, I.R. and Junqi, Wu. 2002. Bamboo biomass: An INBAR Working Paper. INBAR, China.

INBAR. 1999. Socio-economic issues and constraints in the bamboo and rattan sectors: INBAR's assessment. INBAR Working Paper 23. International Network for Bamboo and Rattan, New Delhi. Pp. 28.

Jain, S.S. and Rawat, G.S. 2008. Bamboo and agroforestry. In International Conf. on Improvement of bamboo productivity and marketing for sustainable livelihood, 15-17 April, New Delhi. National Bamboo Mission, GOI. India. Pp. 115.

Jamaluddin; Nema, S and Behari, B. 1998. Effect of VAM on growth and development of bamboos planted under agroforestry system. *Indian Forester, 124* (7): 516-523.

Jayasankar, B., Anitha, V. and Muraleedharan, P.K. 1997. Economics of bamboo cultivation in homesteads agroforestry system of Kerala. (Eds. Iyenkar, P.K) Proceedings of Ninth Kerala Science Congress, Trivandrum, 27-29 January 1997, KFRI, Peechi. Pp. 22-24.

Jha, L.K., Sahoo, U.K and Lalnundanga 1999. Prospects and promises of bamboo in Mizoram. The Proceedings of National Seminar on Processing and Utilisation of Plantation Timbers and Bamboo. (Eds. Damodaran, K; Aswathanarayana, B.S; Prasad, T.R.N; Shyamasundar, K; Padmanabhan, S) 23-24 July 1998 IPRITI, Bangalore. Pp. 49-59.

Lal, S. 1995. Rehabilitation of bamboos in Kaimur and Vindhyan Region, U.P. *Indian Forester*, 121(9): 786-791.

NRCAF 2004. Annual Report. National Research Centre for Agroforestry, Jhansi (U.P.) India. Pp 154.

Rao, K.S. and Ramakrishnan, P.S. 1990. Role of bamboo in secondary succission after slash and burn agriculture at lower elevations in north east India, In: Bamboo Current Research. Pub. KFRI, India and IDRC, Canada. Pp. 59-65.

Rawat, J.S., Singh, T.P and Rawat, R.B.S. 2002. Potentials of bamboos in agroforestry in India. Proceedings of the National Workshop on Policy and Legal Issues in Cultivation and Utilisation of Bamboo, Rattan and Forest Trees in Private and Community Lands, KFRI, Peechi. Pp. 38-44.

Seshadri, P. 1985. Intercropping of bamboo (*Dendrocalamus strictus* Nees) with soybean (*Glycine max* (L. Merrill) - An agroforestry study. Ph.D Thesis. Tamil Nadu Agricultural University, Coimbatore. Pp. 480.

Shanmughavel, P and Francis, K. 2000. Culm production of bamboos in an agroforestry system. *Journal of Non-Timber Forest Products*, 7(3/4): 194-198.

Singh, K.A., Patiram, Singh, L.N. and Roy, R.N. 1992. Effect of bamboo (*Bambusa nutans* Wall. ex Munro) shade on the yield of some agricultural crops at Mid Hills of Eastern Himalaya. *Indian Journal of Forestry*, 15(4): 339-341.

Singh, K.A and Kochar, S.K. 2005. Effect of clump density/ spacing on the productivity and nutrient uptake in *Bambusa pallida* and the changes in soil properties. *Journal of Bamboo and Rattan*, 4: 323-334.

Wagh, R. and Rajput, J.C. 1994. Comparative performance of bamboo with the horticultural crops in Konkan. Bamboo in Asia and the Pacific. Proceedings of the Fourth International Bamboo Workshop, 27-30 November 1991, Chiangmai, Thailand. International Development Research Centre, Canada and FORSPA, Bangkok, Thailand. FORSPA Publication 6, Pp85-86.

❑❑❑

Agroforestry: Natural Resource Sustainability, Livelihood & Climate Moderation © 2009
Editors: OP Chaturvedi, A Venkatesh et. al., pp. 91-111
Satish Serial Publishing House, Delhi

OPTIONS FOR NON-TIMBER FOREST PRODUCE FROM AGROFORESTRY SYSTEMS

A.Venkatesh
National Research Centre for Agroforestry, Jhansi-284 003 (U.P.)

Introduction

In general, minor forest produce reflects the products obtained from natural forest other than timber. This term is suitably replaced by non-timber forest produce. As per the Forest (Conservation) Act, 1980 and National Forest Policy, 1988, the dwellers of the forest only permitted to collect these forest produces. Reduction in forest cover owing to illicit felling, illegal encroachment of the forest, industrialization, infrastructure development etc., leads to vanishing of the valuable NTFP tree species and subsequently the livelihood source of the forest dwellers is in blink.

Considering the potential and great demand of the NTFP mainly natural resins and gums in the local and global market, the suitable host trees can be introduced in the out side forest area and promoted. This option will be highly benefit to the resource poor farmers by providing livelihood security besides creating additional employment.

Natural resins and gums are non-toxic biodegradable, low volume and high value produce. These can be further processed to add value in quality for higher returns. In some products, value addition through primary processing alone results in about three times higher returns. In India, lac and guar gum are processed but for most of other resins and gums processing and value addition is meager. In this chapter, the suitable tree species with respect to natural resins and gums in the agroforestry systems and its possible commercial / technical extraction are discussed.

Natural Resins

Natural resins are solid or semi-solid materials, usually a complex mixture of organic compounds called terpenes. These are insoluble in water but soluble in certain organic solvents. From two sources viz., insects (lac) and plants (rosin + dammar) the resins are obtained. Mainly used in paints, varnishes, food, pharmaceuticals, adhesives, lacquers, cosmetic, etc.

(i) Resin from Insect Source

Lac is the only natural resin from insect origin. It is the resinous secretion produced by the lac insect, *Kerria lacca* Kerr. (syn= *Laccifera lacca* Kerr; Homoptera: Tachardidae) which is formed as a protective test, or hard shell, around the insect's soft body throughout most of its life-cycle, and is the only natural resin of animal origin. The main use of lac until the 1960s was in the manufacture of gramophone records (Gupta and Guleria, 1982) but has since been replaced by vinyl. Today, due to its strength and excellent bonding properties as well its non-toxicity, it is used in wide range of industrial products including varnishes and polishes, pharmaceuticals, confectionery, adhesives, printing inks, electricals, leather and wood finishing, and for fruit and vegetable coating (Anon, 1995).

Lac Production in India

India is the main producer of lac and accounted for 85% of world lac production. Lac occurs most parts of India but presently the main areas of lac production in India are in the states of Chhattisgarh, Jharkhand, Bihar, M.P., West Bengal, Maharastra and Orissa. Chhattisgarh and Jharkhand are among the most productive states, accounting for 38 and 32% of the country's total out put of lac (Anon, 2007). Thailand has significantly increased its production of lac over last few decades and has emerged as India'a main competitor in the export market (Tewari, 1994). Smaller quantities of lac are produced by China, Vietnam, Burma, Malaysia (Krishnaswami and Singh, 1962; Tewari,

1994), Laos (Anon, 1990) and Pakistan (Sheikh, 1987). Total lac production of the country is 23,229 tonnes (Anon, 2007) and about 75% of lac is exported to over 100 countries. The export earning from lac and lac products during the year 2005-06 was around 160 crores. Only 15-20% of the total production is consumed internally in India (Rao and Singh, 1990; Pandey and Singh, 1992).

Production and export of lac in India in last 5 years

Year	Production (MT)	Export	
		Quantity (MT)	Value (Rs. Crore)
2001-02	NA	6363.59	86.34
2002-03	17,500	5021.18	78.46
2003-04	19,800	6632.00	129.20
2004-05	21,300	7301.53	165.88
2005-06	18,000	7280.58	159.62
2006-07	23,229	NA	NA

State wise production of sticklac in India during the year 2006-07

States	Production (MT)	Share of State (%)
Chhattisgarh	8,875	38.21
Jharkhand	7,490	32.24
Madhya Pradesh	3,164	13.62
Maharashtra	1,300	5.60
West Bengal	1,240	5.34
Orissa	440	1.89
Uttar Pradesh	450	1.94
Assam	130	0.56
Andhra Pradesh	60	0.26
Gujarat	50	0.22
Bihar	15	0.06
Meghalaya	15	0.06
Total	**23,229**	**100.00**

Trends in export of lac and its value added products

Name of product	Export in 2005-06		Change in last 5 years	
	Quantity (tones)	Value (Rs. lakh)	Quantity (tones)	Value (Rs. lakh)
Shellac	5120.97	8854.48	+ 11.14	+ 62.10
Seedlac	1325.29	2043.66	+ 143.26	+ 323.95
Bleached Lac	418.76	1126.01	+ 403.68	+ 725.16
Dewaxed Shellac	244.17	731.29	+ 164.05	+ 298.68
Aleuritic acid	162.35	2488.84	+ 166.15*	+ 268.21*
Gasket Shellac	4.54	6.50	- 66.17	- 51.78
Shellac Wax	4.50	11.30	+ 78.57*	+ 174.94*
100% E.O.Us	-	700.00	-	-
Grand Total	7280.58	15962.08	+ 30.24	+ 144.75

Diversity of Lac Insects and Host Plants

Lac insects require plant species for survival. More than 400 lac hosts have been observed to carry lac insects throughout the world. Of the nine genera and 87 species of lac insects reported from the world, two genera and 19 species are found in our country. Species belonging to genus *Paratachardina* produce a hard, horny substance, which is insoluble in alcohol. These are uni-voltine and are generally treated as parasites of economically important plants such as tea and sandal. However, recently, *Parachardina* spp. have been found to be potential bio-control agents for managing weeds and need to be nurtured as such. Lac insects under genus *Kerria* are generally bi-voltine except *K. lacca mysorensis* found on *jalari* (*Shorea talura*) and *K. sharda* on *kusum* (*Schleichera oleosa*) and raintree (*Albizia saman*), which are tri-voltine. Indian lac insect, *Kerria lacca* (Kerr), the most important and widely exploited insect for lac cultivation can further be distinguished into two strains or infra sub species forms, the *rangeeni* and *kusmi* on the basis of differences in life cycle, host preference and quality of lac produced. *Rangeeni* strain is characterized by unequal duration of bi-voltine life cycle and non preference of *kusum*, as a host, whereas *kusmi* strain by a more or less equidurational life cycle and preferring *kusum* as a host. Quality of resin produced by *kusmi* lac insect is superior in comparison to *rangeeni* lac insect. Significant quantitative and qualitative variation in various biological attributes of the lac insect, viz., yield of resin, frcundity, sex ratio and body colour have also been reported. The pigment present in the lac insect haemolymph (laccaic acid or lac dye) is non-toxic and finds numerous applications in textile, pharmaceutical and food industry. Qualitative and quantitative variations in lac insect dye have been observed showing crimson, yellow, albino and cream body colour.

Lac Insect Associated Fauna

The soft-bodied lac insects produce a resinous secretion, which protects them from adverse environment. In spite of this protective covering, the insects due to their sedentary habit are always a sitting prey to predators and parasites. Twenty two lac insect predators, 30 primary parasites, 45 secondary parasites and several fungal pathogens of lac insects as well as lac hosts, besides several other associated insects visiting lac insects for honey dew represent a rich bio-diversity of this ecosystem. Moreover, natural lac insect complex maintains a variety of other tree flora, macro fauna and soil micro-organisms.

Lac growing regions with various hosts

State	Important lac hosts reported
	Major lac growing region
Jharkhand	*Butea monosperma, Croton oblongifolius, Ficus* spp., *Protium serratum, Schleichera oleosa, Ziziphus mauritiana*
Chhattisgarh	*Acacica catechu, B. monosperma, S. oleosa, Z. mauritiana, Z. xylopyra*
West Bengal	*B. monosperma, Ficus* spp., *Samanea (=Albizia) saman, Z. mauritiana*
	Minor lac growing region
Orissa	*S. oleosa, Z. mauritiana*
Maharashtra	*Acacia catechu, A. nilotica, B. monosperma, S. oleosa, Z. mauritiana*
Gujarat	*A. catechu, B. monosperma, Z. mauritiana*
Uttar Pradesh	*A. nilotica, B. monosperma, Z. mauritiana, Z. xylopyra*
Assam	*Albizia lucida, Cajanus cajan, Grewia* spp., *Ficus* spp., *Flemingia macrophylla, Kydia calycina, Leea crispa*
Andhra Pradesh	*B. monosperma, Albizia saman, Peltophorum ferrugenium*
	Minor lac growing region presently
Rajasthan	*Acacia* spp., *B. monosperma, Ficus* spp., *Z. mauritiana*
Punjab	*A. nilotica, Ficus* spp., *Z. mauritiana*
Karnataka	*Acacia* spp., *Ficus* spp., *Shorea talura*
Tamil Nadu	*A. nilotica, B. monosperma, S. talura, Z. mauritiana*

Market value of lac and its products

S.No.	Particulars	Price (Rs./kg)
1.	Sticklac (Palas)	70-80
2.	Sticklac (Ber)	75-85
3.	Sticklac (Kusum)	100-110
4.	Seedlac (rangeeni)	160-180
5.	Seedlac (Kusmi)	180-210
6.	Shellac (Rangeeni hand made)	200-220
7.	Shellac (Rangeeni machine made)	190-220
8.	Lac dye	1,500,2000
9.	Lac wax	200-225
Value added products		
10.	Bleached lac (export quality	310-350
11.	Aleuritic acid	1,200-1,500
By Products		
12.	Mollamma	20-60
13.	Ghonghi	20-30
14.	Kiri	10-25
15.	Pashewa (Machine made)	25-35
16.	Pashewa (Hand made)	50-70

Lac processing centres in the country

Name of States	Districts	Centres
Jharkhand	Ranchi	Bundu, Khunti and Murhu
	West Singhbhum	Chakradharpur
	Garhwa	Garhwa
	Latehar	Latehar
	Simdega	Kolebira
	Palamau	Daltonganj
West Bengal	Purulia	Jhalda, Tulin and Balarampur
	Howrah	Howrah
Maharashtra	Gondia	Gondia
Chhattisgarh	Dhamtari	Dhamtari
	Kanker	Kanker, Sambalpur
	Rajnandgaon	Bandha Bazar
	Korba	Kathgora
	Janjgir	Sakti
	Bilaspur	Pendra

Conventional Method of Lac Cultivation

The trees are infested with lac insect crawlers by putting broodlac sticks along with leaves attached with it on the branches of new trees to be infested, when lac crawler emergence begins. These broodlac sticks are put here and there specially between forks of branches on new trees. The broodlac sticks remain on tree for a long time even after emergence is over. These sticks are collected when the farmers get time. The lac crop is harvested whenever encrustation is thick irrespective of crop maturity. Normally crop is not harvested completely but some amount of mature lac is left on the tree to act as summary of farmers' method, its demerits and its remedy by Scientific method of lac cultivation is given in Table 1. The scientific method of lac cultivation is as follows and involves several operations which are pruning of host tree, infestation of host with insect (Inoculation), removing used-up broodlac sticks, crop harvesting.

1. Pruning of Host Tree

The pruning of trees is an essential step for lac cultivation in order to have large number of new, succulent and healthy shoots at the time of infestation of trees with the insect. Regular pruning result into increase in size of tree and more area for infestation. The dead cracked and broken branches, which are not suitable are removed, otherwise the released lac insect may settle on such shoot and would die afterwards as these twigs can't supply required food (phloem sap) to lac insect. The pruning also help the trees to maintain its vigour and potential as it is being given rest when there will be no lac culture on trees.

Pruning time: Based on gestation period between pruning and infestation, we can divide the trees into two categories. First category are those trees which require pruning 6 months before infestation with lac insect (inoculation) and second category require 12-18 months before infestation. The former include hosts like *B. monosperma, Z. mauritiana, A. lucida* and *F. macrophylla, F. semialata*, etc. whereas latter include *A. auriculaeformis* and *S. oleosa*. The time of pruning for different species of host trees are given in Table 2.

Method of Pruning: The trees are pruned with axe, dauli, secateur or with tree pruner. The twigs more than 2.5 cm in diameter should not be cut but less than 1.5 cm diameter should be cut from the point of origin. The twigs 1.5 to 2.5 cm in diameter should be cut, leaving 1-1.5 ft. length from point of origin. In case of *A. auriculaeformis* twigs less than 7 cm diameter should be cut leaving 1.5 ft. length from point of origin. Similarly, all shoots of *Flemingia* sp. is cut 6 inch above the ground. While pruning the trees, precaution is taken to prune the trees lightly in order to maintain the shape of crown of host tree, umbrella like and allow plenty of space for growth of new shoots. Dry diseased and broken twigs are removed simultaneously. When trees are pruned for the first time, one should take care of himself, not to go on weak branches and in such cases thicker branches can be cut.

Table 2. Pruning time of hosts

Host	Period	Crop
Schliechera olelsa	Jan/Feb	Winter (Aghani)
	Jun/Jul	Summer (Jethwi)
Butea monosperma	Mid. Feb	Rainy (Katki)
	April	Summer (Baisakhi)
Zizyphus mauritiana	April	Summer (Baisakhi)
Albizia lucida	Jun/Jul	Summer (Jethwi)
	April	Summer (Baisakhi)
Flemingia semialata	Jan/Feb	Winter (Aghani)
	Jan/Feb	Summer (Jethwi)

2. Infestation of Host Tree (Inoculation)

Infestation of host trees with lac insect is by putting the bundles of broodlac (lac stick with mature mother cell) and allowing the young larvae (crawlers) to come out of this and settle on host plant. The period of infestation for various crops is given in Table 3. All the dead dried twigs from the trees must be removed before infesting the trees with lac insect.

Table 3. Inoculation and harvest schedule of various crops

Strain	Crop	Inoculation	Harvest	Duration
Kusmi	Jethwi	Jan/Feb	Jun/Jul	6 months
	Aghani	Jun/Jul	Jan/Feb	6 months
Rangeeni	Katki	Jun/Jul	Oct/Nov	4 months
	Baisakhi	Oct/Nov	Jun/Jul	8 months

Method of Inoculation : The unwanted portion from broodlac sticks like leaves and petioles and part of twigs, where there is no settlement of lac insect, removed with the help of Secateur. The broodlac stick of 15-20 cm length is cut. These sticks are bundled (about 100 gm) with plastic *sutli* is kept longer for tying on branches of tree. These bundles are tied on to the branches parallel of shoots on upper side. If broodlac is infested with insect predator and parasitoids which normally happen, then it would be better to remove or reduce its population. These can be done by physical and chemical treatment. In former method, the broodlac sticks are kept in 60 mesh (one inch linear length contains 60 numbers of holes) synthetic net bag (33x10 cm). Then these bags are tied on branches of tree. This allows lac insect crawlers to come out of synthetic net bag but lac insect predators and parasitoids are trapped inside, as these are larger in size. The new generation crop is expected to be devoid of insect predators and parasitoids. But practically it does not happen as if neighbourhood may not adopt this method and for new generation crop, predators and parasites may cross over from other field to this field. Normally 10-20 g of broodlac per metreis required for infestation of trees depending on host tree and crop. The broodlac rate for various crops on host trees is given in Table 4. For inoculating *kusum* trees, the broodlac bundles are tied on shoots of the interior of the crown in case of *jethwi* and exterior of the crown for *aghani* crops.

Table 4. Broodlac requirement for different crops.

Broodlac rate (g/meter shoot)	Crop	Host	Strain
20	*Jethwi + Aghani*	*Kusum*	*Kusmi*
	Aghani + Jetwi	*Kusum*	
	Aghani	*Ber, Khair*	
	Aghani	*Khair, akashmani Galwang*	
8-10	*Baisakhi + Katki*	*Palas*	*Rangeeni*
	Immature baisakhi (*ari*)	*Palas, ber*	
20	*Katki*	*Akashmani*	*Rangeeni*

3. Removing of Used-Up Broodlac (Phunki) Sticks

After complete emergence of lac larvae from broodlac, the use-up bundles (*phunki* bundles) are to be removed from trees. This is an essential operation in order to minimize attack of predators and parasitoids to new lac crop because this used-up broodlac sticks are the source of predators and parasitoids which normally emerge in large number after emergence of lac larvae from broodlac is over. This is also necessary to avoid wastage of lac after drying of lac encrustation on *phunki*-sticks which normally fall on the ground.

The *phunki* sticks are to be removed as soon as emergence of lac larvae is over. This period should not be exceeding more than three weeks after infestation. For its removal, there is no further need to climb on tree, instead, a small *dauli* mounted on top of a bamboo can be utilized for this purpose, which cut the plastic sutli and the bundles fall to the ground. These bundles are collected and lac encrustation is scraped either by simple knife or by lac scrapping machine. The scraped lac from this *phunki* sticks are to dispose to village market as early as possible or kep for 1-2 days under the sun in order to kill various stages of predators and parasitoids harbouring in lac encrustation.

4. Harvesting

The lac crop is harvested either as a mature lac or immature from tree along with host twigs. If the harvested lac is to be utilized for re-infesting the other trees i.e. to be used as a broodlac then only mature lac is harvested. The maturity is assessed by post-embryonic developmental stages of lac larvae inside the mother cell. Broadly speaking, it is to be cut from parent host tree when young - ones are ready to emerge out from mature crop. However, the immature lac can be cut at any time. Harvesting of immature lac should be avoided as this result into shortage of broodlac for raising next crop in coming days. The immature harvested lac is scraped, sold in market or kept for drying. The freshly scraped lac is spread on floor at a place where there is sufficient aeration and preferably dried in air.

The harvesting time of different crop is given in Table 3. Pruning is also carried out simultaneously while harvesting the crop.

Insect Pest Management

For a good harvest, lac pest management is essential as 30-40% lac crop is damaged by three major predators viz., *Eublemma amabilis* (white moth), *Pseudohypatopa pulverea* (Black moth) and *Chrysopa* spp. Use of synthetic net sleeves should judiciously be used for inoculation and three spraying of pesticides should be done as mentioned below. While spraying it should always be kept in mind that this period (spraying) does not coincide with the male emergence in any case.

Winter kusmi lac crop (Aghani)

First spray at 30 days of inoculation with endosulphan 35% EC @ 0.05% and carbendazim @ 001% solution. Second spray at 40 days of inoculation with dichlorvos @ 0.03% and carbendazim @ 0.01% solution. Third spray after 60 days of inoculation with dichlorvos @ 0.03% and carbendazim @ 0.01% solution. Application of carbendazim is done to protect the crop against fungal attack.

Table 1. Conventional method of lac cultivation its demerits and the Scientific method

Conventional method	Demerits	Scientific method
Trees are not selected or utilized for lac cultivation as per suitability of crop (Seasonal).	Heavy mortality of lac insect resulting into low productivity.	Trees are infested with lac insect as per suitability of lac crop season. Some trees are suitable for summer crop while others for winter crop.
Trees are not pruned. If pruned, then for the purpose of harvesting of lac crop and not for getting tender/succulent shoots	Due to hard and old shoots most of the lac insect die, as they are unable to pierce their proboscis inside the shoot as also scanty sap in old shoots. Due to lack of succulent shoots, most of the lac insect settles on petioles and with the leaf fall in the month of Feb./ March, the settled insects are wasted.	Pruning of trees at proper time and by efficient and accurate method.
Trees are infested with lac insect without removing dead and hard shoot prior to infestation	Living insect also settle on old and hard shoot and die afterwards.	Prior to inoculation, once again ensure removal of old, hard and dead shoot.
Broodlac is not quantified properly at the time of infestation.	Utilization of more than required broodlac result into either heavy mortality of lac insect or drying of shoots or both. Whereas less than required quantity will result into non-utilization of full potential of host trees.	Utilization of optimum quantity of broodlac for raising lac culture.
Improper placement of broodlac on the tree i.e. not according to ensuing season.	Resulting into head mortality of lac insect mainly settled on top strata of the host canopy, during summer season.	Broodlac bundles are to be placed on lower side of tree canopy during Oct./Nov.
Infestation of new trees without any consideration of good quality broodlac.	Emergence of nominal lac larvae and more attack of insect pest on new crop.	Give priority to quality broodlac for infestation. Use of 60-mesh synthetic net bag for enclosing broodlac sticks at the time of infestation.

Contd...

Table 1. Contd...

Repeated self-colonization of lac insect.	Reduction in yield potential of lac insect due to continuous exploitation of host tree. Increase in lac insect-pest population.	At most, self-colonization is carried out once only. Later on complete harvesting is to be carried out.
Trees are not given rest and some lac is always left on tree for further multiplication.	Reduction in yield potential of host tree.	Adoption of Coupe system, giving, rest to trees in a planned manner.
Broodlac sticks are left on trees for several weeks even after emergence of larvae is completed.	Carry over of insect pest from broodlac sticks to new crop. Lac encrustation from used-up broodlac sticks fall on the ground after automatic separation from host sticks, causing monetary loss.	Remove used-up broodlac sticks as soon as emergence is over and not later than 21 days after infestation.
Keeping used-up broodlac stick in open after removing from the host trees.	Transfer of enemy insect from use-up broodlac stick to new crop in large number.	Scrap the encrustation from sticks and dispose sticklac at the earliest possible or expose it under sun before storing.
Harvesting of immature lac crop without ensuring sufficient broodlac for next crop.	Shortage of broodlac for next crop.	Except in the month of April/ May, do not cut immature lac.
Production of broodlac and *aril ac* from the same tree.	Shortage of broodlac for next crop. Affects production of quality broodlac.	Take broodlac and immature lac from separate trees.
No measure to control insect pest.	Decrease in production. Reduction in quality of broodlac.	Adoption of IPM including use of insecticide at proper time of crop growth.
Storage of lac along with stick (without scrapping)	Quick breeding of insect predators and parasites resulting losses under storage condition.	Store lac only after scrapping or converting into seedlac (granulated form).
Harvesting of mature lac crop only after crawler emergence start in June/ July.	Wastage of a lot of lac insect crawlers due to high rate of emergence in June/ July.	Mature rainy season crop should be harvested 4-5 days prior to emergence of lac insect if needed.
Repeated use of broodlac from the same species of tree.	Yield gradually decreases.	Change host tree species frequently for sustained yield.

Schedule of operation for lac cultivation on Kusum

Year	Month	Coupes of kusum trees (70 trees/ha) A (14)	B (14)	C (14)	D (14)	E (14)
Ist Year	Jan-Feb	Pruning				
	June-July		Pruning			
2nd Year	Jan-Feb			Pruning		
	June-July	Inoculation			Pruning	
3rd Year	Jan-Feb	Partial harvesting →	Inoculation			
	June-July	Complete harvesting	Partial harvesting →	Inoculation		
4th Year	Jan-Feb		Complete harvesting	Partial harvesting →	Inoculation	
	June-July			Complete harvesting	Partial harvesting →	Inoculation
5th Year	Jan-Feb	Inoculation			Complete harvesting	Partial harvesting ↷ To- A
	June-July	Partial harvesting →	Inoculation			
		Complete harvesting	Partial harvesting →	Inoculation		
			Complete harvesting	Partial harvesting →	Inoculation	
				Complete harvesting	Partial harvesting →	Inoculation
					Complete harvesting	Partial harvesting ↷ To – A Complete harvesting

Rangeeni strain cultivation on palas or ber

Year	Month	Broodlac set-I	Immature (*ari*) set (*Palas/Ber*)	Broodlac set-II (*Palas*)
First	April	Pruning		
	Oct/Nov	Inoculation		
Second	April		Pruning	Pruning
	June/July	Self Inoculation		
	Oct/Nov	Brood harvesting →	Inoculation	Inoculation
Third	April	Pruning + *ari* collection	*Ari*-harvesting + Pruning	
	June/July			Self Inoculation
	Oct/Nov	Inoculation	Inoculation ←	Brood Harvesting
Fourth	April		*Ari*-harvesting + Pruning	Pruning + *ari* collection
	June/July	Self Inoculation		
	Oct/Nov	Brood harvesting →	Inoculation	Inoculation
Fifth	April	Pruning + *ari* collection	*ari*-harvesting + Pruning	

Summer kusmi lac crop (Jethwi)

First spray 40 days of inoculation with endosulphan 35% EC @ 0.05% and carbendazim @ 0.01% solution. Second spray at 80 days of inoculation with dichlorvos @ 0.03% and carbendazim @ 0.01% solution.

ii) Resin from plant source

Resins occur in special cavities or passages in a wide variety of plants. These represent exudation products of various essential oils. They normally ooze out through the bark, hardening on exposure. Commercial resins are collected from artificial wounds. Unlike gums, resins are insoluble in water but are more or less soluble in ordinary solvents such as ether, alcohol and turpentine. They are brittle, amorphous, and more or less transparent. They possess luster and burn with a smoky flame when ignited. The importance of natural resins has very much declined due to preparation of synthetic resins. The resins are widely used in the manufacture of paints and varnishes, water-proofing substances and soaps. They are used in medicine, pharmacy, for sizing paper, for incense, in preparation of sealing-waxes and many other products.

The resins, oleo-resins and gum resins obtained from trees are classified as pine resins and resins from broad-leaved species.

a) Pine Resins

The extraction of crude resins is confined to the gunus *Pinus*.The different pines tapped in the country are noted below:

Pinus roxburghii (chir pine) is the only species regularly tapped. Chir pine occurs in the Himalayan sub-tropical pine forests at elevations between 450-2,300 m, extending from Bhutan in the cast to Kashmir in the west. The tree occurs in abundance and yields the resin copiously.

Pinus wallichiana (kail) occurs in western Himalayas in moist and dry temperate forests between elevations of 2,000-3,000 m. The blue pine or kail produces resin less freely, but the quality of resins, is superior. With the development of communications and demand of resin, the specie is being taken up for resin tapping increasingly.

Pinus kesiya (khasi pine) is known to yield good quality resin, but due to its limited distribution only in khasi hills of Assam, make resin tapping uneconomic.

Pinus gerardina (chilgoza pine) occurs in remote; inaccessible localities at high altitudes and commercial tapping is not feasible.

b) Resin from broad leaved species

The gums and resins that exude from cracks or cuts and solidify upon exposure to air are described under the trade name of 'dammar'. The true dammar is obtained from *Agathis loranthifolius*, a conifer not found in India. It signifies the resins obtained from plants of Dipterocarpaceae and a few of Burseraceae. The principal dammars are as under:

Canarium strictum (white dhup) found in the forests of West Coast and Southern India. It yields the resin known as 'black dammar'. It is mainly used in varnishes, bottling wax, for plasters and for caulking boats.

Hopea odorata (thingan) yields, by natural exudation, a resin which is known as 'rock dammar'. The tree is found in Andaman Islands. It is used for the manufacture of varnishes, as a medicine for wounds, and sores and painting pictures.

Vateria indica (vellapine) a large evergreen tree occurs in the evergreen forests of Western Ghats. It yields the resin known as 'white dammar'. The resin when soft is known as 'piney resin', 'Indian copal' or 'dhupa' but when hard it is called dammar. It is used as a varnish, as an ointment for sores, an incense and for caulking boats.

Shorea robusta (sal) resin is known as 'ral'. When the tree is blazed the oleo-resin oozes out and on exposure hardens. It is opaque, brittle, pale yellow and aromatic. It finds its use as incense, a disinfectant, fumigant, for paints, and varnishes.

Oleo-Resin

Resins which contain a percentage of natural essential oils is called as Oleo-resin. These are volatile odoriferous oils, generally liquids but sometimes solids, which are found, secreted in the inter-cellular spaces, where they combine with the resins to produce exudations known as oleo-resins. Some of the important oleo-resins are described below:

Boswellia serrata (salai) is found in dry deciduous forests common on dry hills. The tree yields a gum oleo-resin which is known commercially as 'salai-gum' or 'guggal'. It is used as incense, for rheumatism and nervous diseases in medicine.

Dipterocarpus turbinatus (gurjan) a large tree of tropical forests of Assam and Andaman islands. It is a source of *gurjan* oil. It is used in lithographic inks, anti-corrosive coating for iron, medicinal purposes, preservation of timber and for caulking boats.

Kingiodendron pinnatum (piney) found in evergreen forests of Western Ghats from South Maharashtra to Kerala. The tree yields, a dark reddish-brown oleo-resin. It is locally used as wood varnish and for medicinal purposes.

Gum Resins

Gum resins are natural mixtures of gums and resins in variable proportions, therefore partly soluble in water. Have a characteristic flavour and taste. The trees yielding gum resins are described below:

Garcinia morella (gamboges) a small tree, found in the evergreen forests of Assam, West Bengal, and the Western Ghats. It is a source of gum-resin, 'gamboge'. It is used in varnishes, lacquer and metal work. It has also a medicinal use as a cathartic.

Commiphora mukul a small tree or shrub, occurring in the arid rocky tracts of Rajasthan, Gujarat and Karnataka. The resin has an aromatic taste and odour. It is used as incense, a fixative in perfumery, and in medicine.

Natural Gums

Gums and resins are exuded by plants, partly as a normal phenomenon and partly as the result of disease or injury to the bark of wood. These arise chiefly from the stem but some times also from the roots and even from the leaves and other parts of plants. A gum is a substance of more or less sticky nature or one which was at one time of a sticky or plastic consistency. Oozing out of tears or gum from tree bark is called as "gummosis. It is mainly due to the pathological response by the plants.

Acacia gums : Largely known as 'gum arabic' are derived from several species of the genus *Acacia*. The following are important species of *Acacia* yielding gum:

i) *Acacia catechu* (khair): The gum from the tree, though darker in colour, is of good quality and is regarded as a better substitute for true gum Arabic than babul gum.

ii) *Acacia nilotica* (babul): Indian gum Arabic or babul gum is from *Acacia nilotica*, and is of great commercial importance. The gum is used in calico-printing, dyeing and as a sizing material for silk and cotton and in the manufacture of paper. This is also used in preparing sweetmeats.

iii) *Acacia senegal*: The true gum Arabic of commerce comes from this species. It is small thorny tree found in Punjab, Uttar Pradesh and parts of Rajasthan. The production of true gum Arabic is negligible. The gum is usually mixed with other *Acacia* gums. It is used in medicine as a demulcent, in textile, mucilage, paste, polish and confectionery.

iv) *Acacia modesta*: This tree yields a fair quantity of gum which is highly valued in indigenous medicine and also in calico-printing.

Gum kino : Produced by *Pterocarpus marsupium* (bijasal), a large tree of Central and Southern India. The gum is of ruby red colour and contains about 75% tannic acid. It is used as an astringent in medicine.

Bengal kino : Product of *Butea monosperma* (palas). It resembles true gum kino and is used for similar purposes.

Jhingan gum : Obtained from *Lannea coromandelica*. It is used as mucilage, for making ink and confectionery. The tree is regularly tapped for a period of five years and then given a rest till the wounds are healed up.

Katira gum : The term 'gum-tragacanth' is used in commerce to denote a gum which is not soluble in water but which will absorb water and swell up into a mucilaginous mass. True gum-tragacanth is not produced by any Indian tree but *Sterculia urens, Cochlospermum religiosum* and *Bombax ceiba* produce gums slightly inferior to the true gum-tragacanth and pass under the name of gum katira. The gums are all very astringent and are used in medicine in case of diarrhea and dysentery, as an ingredient for face creams, cosmetics and for thickening ice creams.

Moringa gum : Obtained from *Moringa pterygosperma.* The gum is used locally in medicine.

Salai gum : Product of *Boswellia serrata.* It is a gum oleo-resin of a greenish yellow colour and has a pleasant aromatic smell when burnth.

Dhaura gum : Derived from *Anogeissus latifolia.* The gum is extensively used for sizing paper and in calico-printing.

Semla gum : From *Bauhinia retusa.* It is a clear gum resembling gum Arabic. It is used for eating, sizing cloth and paper and for water proofing terraced roofs.

Miscellaneous gums: Among the other gum yielding trees, the following may be mentioned: *Albizia chinensis, A. odoratissima, A. procera, Anacardium occidentale, Azadirachta indica, Bauhinia purpurea, B. racemosa, B. variegate, Chloroxylon swietenia, Feronia limonia, Mangifera indica, Terminalia alata,* and *T. bellirica.*

Tapping and Harvesting Gum Arabic

Gum Arabic-producing *Acacias* are highly drought-resistant. The gum is tapped from the bark of the trees during winter dormancy, coinciding with the tree's largest accumulation of carbohydrate reserves from the previous wet summer season. The trees begin shedding their leaves and entering dormancy after the end of the rains, generally in early November. Exudation of gum is reduced during the coldest winter months, so the best time for tapping is in late fall and early spring. Trees are only tapped once per year (Jamal and Huntsinger, 1993).

Trees are first tapped when they reach a height of about 4-5 ft. with a main stem of about 2 in or more in diameter – typically when they are 3-7 years old. Traditionally, a small bladed axe is used for tapping. A light blow is directed to the side of the branch and parallel to it so that the blade penetrates between the bark and the wood. The axe is then moved upwards and downwards to tear off the strip of bark, removing a strip of bark ¾ inch to 1-1/4 inches in width and 2-4 ft. in length. This is done for each main branch of the tree, tapping the young growing bark and not the main stem, because the bark soon becomes thick and fiberous and unfit for tapping. During the next year's tapping season, the opposite side of the branch is tapped (Awaudha, 1989). This gives the tree time to heal and re-grow its outer bark.

Once the branches are debarked or tapped the tree's reserve of carbohydrates rushes to seal the wounds in the form of liquid gum. The gum starts to solidify slowly as it comes in contact with oxygen, so the gum does not trickle as it does in latex or resins, but instead forms droplets at one or more points along the side of the stripped face. The gum exudes intermittently, forming a hard but slightly elastic skin on the outside. As more gum exudes the outer skin expands or cracks and the nodule grows in size to about 1-1.1/ 2 inches in diameter. When the outer skin becomes so hard that the liquid cannot force it to expand any further, the nodule ceases to grown in size and is ready for picking (Awouda, 1989). The first harvest of gum, the one that will eventually bring home most of the return each season, usually occurs 40 days after tapping. New nodules form after the first are picked, and within 10-15 days a second picking is possible. An average of four to five pickings over two months is common, although some farmers started that they may collect up to 15 times in six months.

Ethephon Induced Gum Production in *Acacia Senegal*

Bhatt and Ram (1990) standardized the procedure of ethephon (2-chloroethylphosphonic acid) induced gum production in *Acacia senegal.* They made holes measuring 5cm deepx 2.5 cm wide in the main trunk at about 1-1.5 m above the ground level using chisel and hammer. The opening slanted downwards and reached up to outer sapwood. Four ml; of ethephon solution (960 mg of the active substance) was introduced into the hole and it was covered with bee wax during April/May. An area measuring 2.5-4cm^2 of the bark, above the treated holes, was deliberately bruised to encourage increased gum production. After 4-8 days of treatment the trees started yielding gums and about 0.8-0.9 kg of good quality gum was obtained per tree.

Conclusion

Cultivation of lac not only provides livelihood to millions of lac growers but also helps in conserving vast stretches of forests, lac insects and associated biota as most of the lac hosts are in forest areas and farmers resist felling of these trees and protect them for lac cultivation. Thereby, lac culture plays a vital role in the protection of our bio-resources. Growing lac hosts for timber and fuel yields revenue in cycles of long years, whereas, cultivation of lac on these trees gives a return almost every year. Thus, lac growers give more importance to regular income from cultivation of lac over the years to one-time income from timber or fuel. Average net profit from one tree is Rs.109

for *palas*, Rs.202 to 1060 for *ber* and Rs.1320 for *kusum* per crop cycle. Lac also provides sustained high economic returns, generates employment opportunities and has potential to pave a strong foundation for lac-based rural cottage industries. About one million mandays are generated in the existing lac-processing factories. With increasing demand for natural products, e.g. in fruits and vegetable coating and as food colour, the time is ripe to introduce lac culture in the farming system and on idle lac host trees in the forests.

Extraction of gum from the available hosts in the farm lands / agroforestry system which provides additional out put and simultaneously enhances the value of the system. While the traditional method of gum extraction leads to uneconomical, since it consumes more labours and quantity of gum received is minimum. Hence, the improved techniques viz., ethephon induced gum production and timely operations of required activities will help to obtain more yield. Therefore, standardization of improved gum production procedures are needed for commercial exploitation of gums from trees.

References

Anon, 1990. Lao People's Democratic Republic Tropical Forestry Action Plan (First Phase). Ministry of Agriculture and Forestry, Government of Lao PDR.

Anon, 1995. Lac. Suter and Co., London.

Awouda, H.M. 1989. Out look from gum Arabic: production and supply. Report to the Gum Arabic Co., Khartoum. Sudan. Pp. 18.

Bhatt,J.R. and Ram,H.Y.M. 1990. Ethephon induced gum production in *Acacia senegal* and its potential value in the semi-arid regions of India. *Current Science* 59 (23): 1247-1250.

Campbell, M.H., Holtcamp, R.H., McCormick, L.H., Wykes, P.J., Donaldson, J.F., Gullan, P.J. and Gillespie, P.S. 1994. Biological control of native shrubs, *Cassinia* spp. using native scale insects, *Austrotachardia* and *Paratachardina* sp. (Hemiptera : Kerriidae) in New South Wales, *Plant Prot. Q.*, 9: 64-68.

Chauhan, N.S. and Mishra, Y.D. 1977. A new colour locus in *Kerria lacca* (Kerr). *Current Science* 46: 272-273.

Chauhan, N.S. and Teotia, T.P.S. 1973. A new variant in the lac insect, *Kerria lacca* (Kerr). *Entomological News letter*, 3: 33-34.

Das, B.B. 1990Present status of entomological research on lac in India and future strategies. Part 1. Present status of knowledge. *Indian Shellac* (Annual Number), 9-16.

Gupta, T. and Guleria, A., 1982. Non-wood Forest Products in India: Economic Potentials. IBH Publ., New Delhi, Oxford.

Jamal, A and Huntsinger, L 1993. Deterioration of a sustainable agro-silvo-pastoral system in the Sudan: the gum gardens of Kordofan. *Agroforestry Systems*, 23: 23-38.

Krishnaswami, S. and Singh, H.D., 1962. Lac cultivation in India : improved methods and extension service. In : Mukhopadhyay, B., Muthana, M.S. (Eds.), A Monograph on Lac. Indian Lac Research Institute, Ranch, Pp.145-155.

Mishra, Y.D., Sushil, S.N., Kumar, S. and Bhattacharya, A. 2000. Variability of lac productivity and related attributes of *Kerria* spp. (Homoptera : Tachardidae) on ber (*Ziziphus mauritiana*), *Journal of Entomological Research*, 24: 19-26.

Pandey, R.K. and Singh, R.P., 1992. Production and export of lac in India. *Agricultural Marketing* 34(4), 30-33.

Ramani, R. and Sharma, K.K. 1991. A review of some genetical aspects of lac insects. *Annals of Entomoogy.*, 9: 47-53.

Roonwal, M.L., Raizada, M.B., Chatterjee, R.N. and Singh, B. 1958. Descriptive account of the host plants of the lac insect, *Laccifer lacca* (Kerr) and the allied plants in the Indian region (Part 1 and 2), Indian Lac Cess Committee, Ranchi, Pp.140.

Shaoji, Gu. 1993. An investigation on the pathogens of both lac insects and their host plants. *Forestry Research*, 6: 711-713.

Sharma, K.K., and Ramani, R. 1999. An update on synoptic catalogue of lac insects (Tachardidae : Homoptera). *Journal of Bombay Natural History and Society*, 96: 438-443.

Sharma, K.K., Jaiswal, A.K. and Kumar, K.K. 2001. New record of fungi associated with Indian lac insect, *Kerria lacca* (Kerr). *Indian Journal of Entomology*, 63:369-371.

Sharma, K.K., Ramani, R. and Mishra, Y.D. 1997.An additional list of the host plants of lac insects, *Kerria* spp. (Tachardidae : Homoptera). *Journal of Non-Timber Forest Produce*, 4:151-155.

Tewari, D.N., 1994. Tropical Forest Produce. International Book Distributors, Dehradun.

Varshney, R.K. 1976.A check list of insect parasites associated with lac. *Oriental Insects*,10: 55-78.

Varshney, R.K. 1968. Further data on host-plants of lac insects (Tachardidae : Homoptera). *J. Bombay Nat. Hist. Soc.*, 65: 249-251.

Varshney, R.K. and Teotia, T.P.S. 196.7A supplementary list of the host plants of lac insects. *Journal of Bombay Natural History and Society*, 64: 488-511.

Varshney, R.K. 1977. Toxonomic studies on lac insects of India. *Oriental Insects, Suppl.*, 5: 97.

❑❑❑

Agroforestry: Natural Resource Sustainability, Livelihood & Climate Moderation © 2009
Editors: OP Chaturvedi, A Venkatesh et. al., pp. 113-123
Satish Serial Publishing House, Delhi

Chapter 9

DEVELOPMENT OF BIOFUEL AS RENEWABLE SOURCE OF ENERGY

R.V. Kumar, Parul Shukla, S.P. Ahlawat, D.R. Palsaniya and S.K. Dhyani
National Research Centre for Agroofrestry, Jhansi-284 003 (U.P.)

Introduction

Energy is the prime source which ultimately determines the carrying capacity on this earth. With the fast depleting natural resources and increasing demands on energy, the need to develop alternative fuel strategies has been felt globally. We are very much aware that petroleum sources are finite therefore the dependence on renewable energy sources has to increase. Of all the renewable energy sources, biomass accounts for over one third of the energy used in developed countries. The two important bio-fuels generated from biomass are bio-diesel and bio-ethanol. There are a number of plant species yielding oil and low and high molecular weight hydrocarbons, which can provide bio-diesel. For meeting the growing demand of biodiesel, developed countries are using edible oil seed crops but countries like India can not afford to use edible oils for the same. The reason being demand for edible oil is higher than its domestic production; there is no possibility of diverting this oil for production of bio-diesel. Fortunately there is a large junk of degraded forest land and un-utilised public land, field boundaries and fallow lands of farmers where non-edible oil-seeds can be grown.

India, a country with fast growing economy is facing the challenge of meeting the increasing demand for energy. In terms of energy requirement, it ranks sixth in the world. Its economy is projected to grow 7-8 percent over to next two decades and there will be a substantial increase in demand for oil to manage transportation and to meet various other energy needs. In the present situation, India imports approximately 70 percent of its petroleum requirement. The global annual requirement of petroleum products is approx. 17,000 MMT, out of which Indian requirement is estimated to be 120 MMT. Our domestic production of crude oil and natural gas is around 34 MMT, thus leaving a huge gap between demand and supply. The net import bill for the year 2006-07 was of the order of Rs. 1, 90,000 crores. If the alternative approaches are not worked out properly, it will be matter of serious concern for the country.

Alternative Approach

Many developed countries have active biofuel programme but basically they are dealing with edible oil crops like rape seed, sunflower etc. in Europe, soybean in USA, palm oil in Malysia etc. But country like India where edible oil is already in shortage, at present we can not think of biofuel programme based on edible oil. Secondly, in situations where expanded production of fossil fuel is unrealistic, there is immense need to think, how to utilize alternative or substitute energy sources. The environmental risks are also there in continuous use of fossil based fuels. So, substitutes are required which are technically and environmentally acceptable and economically competitive.

Biofuels

Biofuels are gaining importance in the backdrop of rapid depletion of the fossil fuel reserves, high prices and associated environmental degradation. Biofuels not only fulfill traditional domestic needs such as cooking, heating or lighting but also cater to new energy applications at a cost competitive with equivalent fossil based technologies. That is why researchers all over the world are looking for alternative fuels on sustainable basis. Biofuels are renewable source of energy derived from biological raw material. Biofuels -ethanol and biodiesel- are gaining worldwide acceptance as one of the solutions for problems of environmental degradation, energy security, restricting imports, rural employment and agricultural economy. World energy consumption and percent contribution by different source are presented in Table 1.

Table 1: World Energy Consumption (Million Tons)

Country	Oil	Gas	Coal	Nuclear	Hydro	Total
USA	943.1	595.7	573.7	192.1	56.8	2361.4
China	368.0	60.6	1311.4	14.2	109.3	1863.5
United Kingdom	78.2	82.3	39.2	14.1	2.1	215.9
Canada	102.3	84.6	30.4	21.1	83.3	321.7
India	128.5	36.2	208.0	4.0	27.7	404.4
India (%)	32	9	51	1	7	100
World	3952.8	2637.7	3177.5	622.0	709.2	11099.2

Source : BP Statistical Review of Energy, 2008

In India, there is a huge gap between annual requirement and domestic production of petroleum products. The huge gap between demand and supply is only met through import. The increasing trend of import burden is not a healthy sign and if the corrective measures are not taken in time, the India's valuable money will drain out in the import of petroleum products. Indian production and import of crude oils are presented in Table 2.

Table 2: Indian Production and Import of Crude Oils (Million Tons)

Year	Production		Import	
	Crude Oil	POL Products	Crude Oil	POL Products
1972-73	7	19	13	3
1980-81	11	24	16	6
1990-91	33	49	22	10
2000-01	32	96	74	9
2004-05	34	118	96	9
2005-06 P	32	120	99	12

P - Provisional POL - Petroleum, Oil, Lubricants

Source : Handbook of Statistics on Indian Economy, 2008

Bio-Ethanol

Ethanol is used as substitute fuel or as an additive. Raw material used for producing ethanol varies from sugar in Brazil, cereals in USA, sugar beet in Europe and molasses in India. Brazil uses ethanol as 100 percent fuel in about 20 percent of vehicles and 25 percent blend with gasoline in rest of the vehicles. USA uses 10 percent ethanol-gasoline blends whereas a 5 percent blend is used in Sweden. Australia uses 10 percent ethanol-gasoline blend and European Union uses 15 percent ethanol blended gasoline. In India, the bio-ethanol is produced from various crops viz., Sugarcane, Sugar beet, Sweet sorghum, Sweet potato, Cassava etc. The gasoline (Petrol) consumption during

2005-06 was 8.6 MMT, which is likely to increase 10-15 percent annually. The projected demand of petrol and estimated requirement of ethanol to replace different percent level of petrol is mentioned in Table 3.

Table 3: Demand of Petrol and estimated requirement of ethanol (MMT)

Year	Petrol Consumption	5%	10%	20%
2007-08	11.00	0.55	1.10	2.20
2008-09	12.50	0.62	1.25	2.50
2009-10	14.50	0.72	1.45	2.90
2010-11	16.50	0.82	1.65	3.30
2011-12	18.15	0.90	1.80	3.60

Source : Kureel, R.S., 2007

In India, there is possibility to use sugarcane as feed stock for the production of ethanol. 4 percent of the available sugarcane is to be diverted for ethanol production for 5 percent blending and 8 percent sugarcane for 10 percent blending (Kureel, 2007). However ethanol can be produced from other sources also. Cropwise ethanol production's presented in Table 4.

Table 4: Crop wise ethanol production

Crop	Ethanol yield ($l\ t^{-1}$)	Yield ($t\ ha^{-1}$)	Ethanol production ($l\ ha^{-1}$)
Sugarcane Juice	70	42	2940
Molasses	245	1.68	412
Sweet sorghum	60	45	2700
Sweet potato	125	25	3125
Cassava	180	25	4500
Sugar beet	110	35	3850
Maize Corn	360	4	1440
Potato	110	16	1760

Source : Kureel, R.S., 2007

Bio-Diesel

Biodiesel ia a fatty acid ethyl or methyl ester. Biodiesel is made from both edible and non edible oils as well as from animal fats. It is environment friendly fuel which has almost no sulphur, no aromatics and has about 10 percent in built oxygen. Biodiesel has the properties similar to petroleum diesel hence require very little or no engine modification for its use. Biodiesel is safe to handle and the flash points of biodiesel are higher than conventional diesel. It can be stored just like the petroleum diesel and hence does not require separate infrastructure. It can be easily blended with petroleum diesel and blending results in the reduction of unburnt hydrocarbons. Its higher Octane number improves the ignition quality even when blended in the petroleum diesel.

Jatropha

The country has enormous potential of oilseeds of tree origin like oil palm, mahua, neem, simarouba, karanja, ratanjyot, jojoba, cheura, kokum, wild apricot, bhikal, wild walnut, kusum, tung etc. which can be grown and established in the wasteland and varied agro-climatic conditions. These have domestic and industrial utility like agriculture, cosmetics, pharmaceutical, diesel and substitute, cocoa-butter substitute etc. Most of these TBOs are scattered in forest and non-forest areas and hardly 20% of the existing potential is being crushed and utilized that too of poor quality due to lack of awareness of their uses, collection and marketing etc. Among the various Tree Borne Non Edible Oilseeds, *Jatropha curcas* has been found more suitable for biodiesel production as it can be grown in varied agro-climatic conditions. As per the planning commission report, *Jatropha curcas* has been found most suitable for the purpose. It will use lands which are largely unproductive for the time being and are located in poverty stricken areas and in degraded forests. The cultivation of this species is proposed in degraded and marginal lands. So that there will be no competition for land available for food. Demand of diesel and estimated requirement of biodiesel for different blending levels are presented in Table 5.

Table 5: Demand of Diesel and estimated requirement of Biodiesel (MMT)

Year	Diesel Demand	Bio-diesel Requirement for blending		
		@5%	@ 10%	@ 20%
2006 -07	52.33	2.62	5.24	10.48
2011 -12	66.90	3.35	6.69	13.38
2012 -15	80.00	4.00	8.00	16.00

Souce : Kureel, R.S, 2007

Karanja

Karanja (*Pongammia pinnata*) can be grown very well in areas having an annual rainfall ranging from 500 to 2500 mm. In its natural habitat, the species tolerates a wide temperature range from slightly below 0°C to 50°C. Mature trees withstand light frost and water logging. This species grows to elevations of 1200 m, but in the Himalayan foothills is not found above 600 m. Pongamia can grow on most soil types ranging from sandy, gravelly and poor soils. It does not do well on dry sands, although it tolerates saline conditions, alkalinity and waterlogged soils. It is common along waterways or seashores, with its roots in fresh or salt water. Good growth can be obtained in soil with well drainage facility and assured moisture. Natural reproduction is profuse by seed and common by root suckers. It is now increasingly gaining popularity due to high oil yielding quality and as an alternative source of renewable energy. *Pongamia pinnata* has been found to be one of the alternative species due to its favorable attributes like its hardy nature, high oil recovery and quality of oil.

The yield performance of different non-edible oils as well as area required at different blending level is given in Table-6.

Table 6: Yield performance of different non-edible oils along with required area

Name of crop	Yield/ ha	Biodiesel /ha	Area requirement (lakh ha)			Yield of Biodiesel (lakh MT)		
			5%	10%	20%	5%	10%	20%
Jatropha	3.00	1.00	17.00	34.00	68.00	17.00	34.00	68.00
Karanja	2.00	0.50	5.00	10.00	20.00	2.50	5.00	10.00
Oil palm	FFB	4.00	1.00	2.00	4.00	4.00	8.00	16.00
Castor	1.00	0.45	4.00	8.00	16.00	1.80	3.60	7.20
Other TBOs	1.00	0.30	5.00	10.00	20.00	1.50	3.00	6.00
Total			**32.00**	**64.00**	**128.00**	**26.80**	**53.60**	**107.20**

Source: Kureel, R. S., 2007.

Government Initiatives

Agriculture being a state subject, the responsibility for the promotion of biofuel plantation rests with the state governments. While the authority for transfer or leasing of government lands rests with the Collector of each district, the nodal agency for processing of application differs in each state. In Rajasthan and Andhra Pradesh the nodal agency is the Department of Agriculture and Department of Rain and Shadow Area Development, respectively. In Tamilnadu, it is the Tamilnadu Watershed Development

Agency & Tamilnadu Wasteland Development Corporation and in Chhatishgarh, Chhatishgarh Biofuel Development Authority. Gujarat Agro Industries Corporation is the nodal agency in Gujarat. These state governments are trying their best to have a large plantation of biofuel species in their respective states.

Indian Council of Agricultural Research has identified first ever Jatropha variety, SDAUJ 1 (Chhatrapati) for commercial cultivation. The seeds contain 32 per cent oil with yield potential of 1000-1100 kg per hectare under rainfed condition. This variety is recommended for semi arid and arid regions with annual rain fall of 300-500 mm. This variety has been developed by Sardarkrushinagar Dantiwada Agricultural University, Gujarat.

Petroleum Conservation Research Association (PCRA) has established a Biodiesel Bank. Uttranchal government has established Uttranchal Biofuel Board and plantation of Jatropha is taking place in big way. Simultaneously, Chhatishgarh government has already established Chhatishgarh Biofuel Board and a large area has come under Jatropha plantation.

Earlier Efforts

Bio-ethanol Programme

With a view to reducing dependence on import bill by way of encouraging use of indigenous sources of energy to provide a supportive role to the farmers, the scheme of mandatory supplying 5 percent ethanol blended petrol (EBP) in nine major sugar producing states and four Union Territories w.e.f. 01.01.2003 was introduced. This was based on experiment in the three pilot projects, two in Maharashtra and in Uttar Pradesh during April-June, 2001. The Ethanol Blending Programme (EBP) could not succeed due to non availability of required raw material for production of ethanol. The annual requirement of ethanol with 5 percent blending is estimated to be 500 million tones.

Biodiesel Programme

A National Biodiesel Mission is being launched by the Planning Commission to cover 25 lakh hectare areas in the country, which will meet 5 percent replacement of the diesel requirement of the country. Government of India has fixed the target to replace 20 percent petrodiesel with biodiesel up to 2011-12 by producing 13.38 milliuon tones of biodiesel annually through plantation of Jatropha in 11.19 million hectare. The Ministry of Petroleum and Natural Gas has launched biodiesel procurement policy w.e.f. 01.01.2006

@ Rs. 25 per litre through state owned petroleum companies in 12 states. However, petroleum companies could not procure any quantity due to low price. However, recently the National Biodiesel Mission has been dropped.

Progress During 10th Plan

The scheme ' Integrated Development of Tree Borne Oilseeds ' has been initiated during the 10th Plan period to harness the existing potential and augment the future potential of TBOs like Jatropha, Karanja, Mahua, Neem, Jojoba, Wild Apricot, Cheura, Kokum, Simarouba etc. The sheme comprising of promotional programme such as development of elite nursery and model plantation, establishment of TBOs garden, transfer of technology programme through farmers and trainers training, Research and Development programme and back ended credit linked subsidy programme such as nursery raising and commercial plantation, establishment of procurement centres, installation of preprocessing and post processing equipments are being implemented.

Two National Network namely 'National Network on Jatropha and Karanja' and 'National Network on Wild Apricot and Cheura' were constituted by involving State Agricultural Universities, Institutions of CSIR, ICFRE, ICAR, CFTRI, TERI and IIT to address different researchable issues under the banner of National Oilseed and Vegetable Development Board, Gurgaon (Ministry of Agriculture).

Efforts by Private Organisations

Certain private sectors are taking very much interest in Jatropha plantation as well as production of biodiesel through establishment of biodiesel plant. General Motors (GM), Hindustan Motors (HM) and Tata Motors has shown keen interest in the biodiesel developed by Central Salt and Marine Chemical Research Institute (CSMRI), Bhavnagar using Jatropha seed. One of the private firm namely, Teena Group has announced to invest 130 Crore rupees in the cultivation of Jatropha in three to four years. This firm has planned to cultivate Jatropha in twenty thousand acre of land. In Maharashtra alone, this firm has started plantation in 1500 acres of land. Reliance Life Sciences, sister concern of Reliance group has already identified Kakinada area in Andhra Pradesh for plantation of Jatropha. This company is in touch with state authorities of Andhra Pradesh, Maharashtra, Tamil Nadu and Gujarat regarding aavailability of lands for its plantation in very large area. Cosmetic company Emami is also entering in Jatropha cultivation and this company has selected states of West Bengal and Orissa for its plantation.

National Biofuel Policy

The group of ministers headed by the Union Agriculture Minister has already approved the National Biofuel Policy drafted by Ministry of New and Renewable Energy (MNRE). The draft policy envisages utilization of a wide range of biocrops such as sugar cane, sweet sorghum, cassava, maize and tree borne oilseeds like Jatropha, Karanja etc. The Draft Policy also envisages setting up of "National Biofuel Development Board" to develop a road map for use of biofuels in petrol and diesel engines in a time bound manner besides taking appropriate policy measures. However, National Biofuel Policy assures that biofuel programme would not compete with food security and fertile farm lands would not be diverted for plantation of biofuel crops. The policy draft also deals with a number of issues like minimum support price for biofuel crops, subsidies for growers of biofuel crops, marketing of oil bearing seeds, subsidies and fiscal concessions for the biofuel industry and related aspects.

After award, in September, 2008 cabinet approved the National Biofuel Policy by fixing 2017 as the target for selling petrol mixed with one fifth of sugarcane extracted ethanol and diesel doped with non-edible oil with a view to cut the national dependence on expensive oil imports. Instead of setting up a National Biofuel Development Board, cabinet constituted a National Biofuel Coordination Committee headed by the prime minister.

Biofuel Programme at NRCAF, Jhansi

Research and development work on *Jatropha curcas* was started from November, 2003 at this Centre. So far the center has collected 284 accessions of *Jatropha curcas* from diverse agro-climatic zones of states like Uttaranchal, U.P., M.P., Chhattishgarh, Maharashtra, Rajasthan and Gujarat. These accessions are in the process of evaluation for growth, yield and oil content. Evaluation trials are being conducted at institute's farm as well as on farmers fields. 87 accessions have been recorded for more than 35 per cent oil content on seed basis. 128 accessions have been cryo preserved at NBPGR, New Delhi and accessions IC numbers have been issued. Four accessions have been sent for multilocation trial and few accessions are in the process for evaluation in multilocation trials so as to release as variety. 90 intra-specific crosses have been developed and being evaluated in fields conditions. Some of the germplasm has been assessed for biochemical activities (ADF, NDF, cellulose, hemicellulose, lignin, crude protein, peroxidase and polyphenol oxidase). Accessions are characterized on molecular level using different molecular markers. Study on reproductive biology and cytology is continued.

Experiments on cultivation of *J. curcas* under agroforestry systems to standardize the package of practices are in progress in experimental field. Experiments on seed viability, germination, storage conditions, clonal propagation and micropropagation and nursery techniques are being standardized for mass multiplication of elite genotypes

Similarly in Karanaj (*Pongammia pinnata*), so far the center has collected 143 accessions from diverse agro-climatic zones of states like Uttar Pradesh, Madhya Pradesh, Rajsathan and Haryana. These accessions are in the process of evaluation for growth, yield and oil content. Evaluation trials are being conducted at institute's farm as well as on farmers fields. 51 accessions have been recorded for more than 35 per cent oil content on kernel basis. Few accessions of early fruiting (2.5years) have been identified.

In Neem (*Azadirachta indica*), center has collected 220 accessions from diverse agro-climatic zones of states like Uttar Pradesh, Madhya Pradesh, Andhra Pradesh, Maharashtra and Orissa. Evaluation trials are being conducted at institute's farm as well as on farmers fields. 16 accessions have been recorded for more than 50 per cent oil content on kernel basis. Accessions are being evaluated for azadirachtin content also. Ten accessions have been sent for Multilocation trial in different states. Work of other Tree Borne Oil Seeds (TBOs) viz, Mahua and Simarouba are in progress at this centre. More than 25 ha area at this Centre is under plantation having different TBOs viz., Jatropha, Karanja, Neem, Mahua and Simarouba.

Conclusion

Fossil fuels have been the prime source of energy for the transportation and industrial sector for more than a century. However, their rapidly increasing consumption and consequent depletion of reserves clearly show that the end of the "Fossil Fuel Age" is not very far. For developing countries like India, rising world prices of crude oil and petroleum, gives maximum cause for concern. Among the many species, which can yield oil as a source of energy in the form of biodiesel, different non-edible oils have been found reasonably suitable to be covered under biofuel programme. Implementation of above programme will enable our country to replace 5-20 percent requirement of petrol and diesel with biofuel i.e. bioethanol and biodiesel without sacrificing area under agricultural feed stocks. Furthermore, these species provides many other products such as manure for the crops, methane for power generation and glycerol for industrial use besides, enormous advantage of carbon trading. The fruit pulp and the remaining de-oiled seed cake can be used for the production of bio-gas by anaerobic fermentation.

Slurry from bio gas plant can further be processed as manure. The slurry of the bio gas plant can be mechanically separated into concentrate of rich nutrients for using in plantation. The dried out slurry forms manure for the agricultural use. Both these products can provide a major substitute to chemical fertilizers and pesticides.. The investment in the sector would give boost to support rural based industry by providing raw material for extraction of oils, value addition, packing, transport, marketing etc. A lot of man power can be generated by successfully organizing this programme of plantation of biofuel species.

References

BP Statistivcal Review of Energy, June 2008

Handbook of Statistics on Indian Economy, RBI, 2008.

Kureel, R.S. 2007. Biofuels: Global and Indian Scenerio- Ppolicy Perspective. 4th International Biofuels Conference held at New Delhi India during Feb 1-2, 2007.

Kumar, R.V., Ahlawat, S.P., Gupta, V.K., Handa, A.K., Yadav, V.P. and Tripathi, Y.K. 2005. *Jatropha curcas*: The Biodiesel Plant. *SAIC Newsletter* 15 (3), Pp. 4 &6.

Kumar, R. V., Ahlawat, S.P., Handa, A.K. and Gupta, V.K. 2004. *Jatropha curcas*: Fuel of future. Employment News (20-26 Nov.), 29 (34): 1- 2.

❑❑❑

Agroforestry: Natural Resource Sustainability, Livelihood & Climate Moderation © 2009
Editors: OP Chaturvedi, A Venkatesh et. al., pp. 125-137
Satish Serial Publishing House, Delhi

Chapter 10

BIO-AMELIORATION OF SALT AFFECTED SOILS THROUGH AGROFORESTRY

R.S. Yadav, D.R. Palsaniya, Ramesh Singh, Munna Ram and Garima Gupta
National Research Centre for Agroforestry, Jhansi-284 003 (U.P.)

Introduction

Out of 329 Mha geographical area of the country about 187.8 million ha (57.1%) suffer from different limitations expressed to varying degrees and getting further degraded through natural and/or anthropogenic processes (Table 1). With no additional land source available for horizontal expansion of agriculture, India needs to find out viable technologies for utilization of existing land resources including the salt affected lands in order to meet future requirements for food, fodder and fuel.

Soil salinity and alkalinity have degraded about 7.41 mha of land in India (Sharma, 1998). A large acreage in the canal irrigated tracts of arid and semi-arid regions of the country suffers from water logging and salinity and the resultant loss to crop production is increasing at an alarming rate. Already a sizeable area has gone out of cultivation in almost all command areas, making the landscape devoid of any vegetation except a few hardy species and grass species. Large tracts of soils in India are confined to villages as communities lands, as public properties near cities, along the railway tracks and roads, and government lands reserved for specific purposes. Today these lands contribute least to the national output and present a picture of desolation. There is a dire need to revegetate these wastelands and prevent their further degradation. Agroforestry/tree plantation is an option of great promise for use of such lands, in view of the growing demand for fuel wood and fodder and also on environmental considerations. Establishment of a permanent vegetative cover will stop further degradation of these land resources. Establishment of tree plantations on salt affected soils sets in amelioration process by favourably affecting physical, chemical and biological properties of soil (Singh *et al.* 1993).

Table 1. Land degradation in India

Type of degradation	Area (M ha)	% of total area
Water erosion	148.9	45.3
Wind erosion	13.5	4.1
Chemical deterioration (Salinity etc.)	13.8	4.2
Water logging	11.6	3.5
Total	187.8	57.1

The Ameliorative Mechanism

Trees are known for their beneficial interactions with the surroundings in which they grow. Soil, which is an important natural resource, is influenced by trees in many ways. Establishment of tree plantations in salt affected soils sets in amelioration process by favourably altering physical, chemical and biological properties of the soil. The ameliorative effects are mediated through:

- Litter production and nutrient recycling, addition of organic matter and rise in the humus status of soil
- Initiation of biological activity consequent to litter addition in the otherwise almost lifeless mass of salt affected soil. The biological activity produces CO_2 in the soil which in combination with water mobilizes the insoluble calcium carbonate-
- $CaCO_3 + CO_2 + H_2O$ – – Ca $(HCO_3)2$
- (2Na) Clay + Ca $(HCO_3)2$ – – – Ca Clay + $2NaHCO_3$
- Reducing salt deposition in upper layers of the soil profile by cutting down evaporation losses of soil water.
- Lowering of water-table in waterlogged and salt affected areas as trees act as bio-pumps.

- Deep and sturdy root systems of trees open up the soil and improve water permeability facilitating leaching of salts.
- Exclusion of large amounts of salts during absorption by trees.
- Excretion of salts from the general plant surface.

Factors Affecting Amelioration Process

The degree to which the problematic soils get ameliorated by trees depends upon-

- The tree species.
- Their growth habit.
- Quantity and quality of litter.
- Planting density.
- Age of plantation.
- Ability to fix the N.
- Management practices.

Litter Production and Nutrient Cycling

Litter is formed of tree parts like dead leaves, twigs, flowers and fruits which fall on the ground. Upon decomposition, organic acids, essential nutrients and a few other chemical substances are released which help in soil reclamation. Accumulation of litter on the surface also checks weed growth, protects the soil from the erosive impact of rain and reduces surface runoff. Poor humus status of salt affected soils makes the litter production potential of forest species quite important.

In a four year study on *Acacia nilotica* and *Eucalyptus terreticornis* plantations raised on a highly alkali soil (pH_2 10.5, EC_2 3.98 dSm^{-1} and ESP 96 in the 0-15 cm surface layer), Gill *et al.* (1987) found that litter production in *Acacia* plantation was significantly higher than in *Eucalyptus* plantation of the same age and stocking rate. Variation in total annual yield during the observed period ranged from 1.03 to 1.13 t ha^{-1} in *Eucalyptus* and from 2.54 to 5.75 t ha^{-1} in *Acacia*. Concentrations of N, P, S and K were significantly higher in *Acacia* litter than in *Eucalyptus* (Table 2). The opposite was true for Ca, Mg, Na, Fe and Mn contents. Gill and Samra (1989) studied the nutrient recycling through litter production in *Leucaena leucocephala* var. K 8 and *Sesbania sesban* raised by direct seeding in soils of pH_2 8.4-8.5. Dry litter yields increased progressively with advancement of growth period in *Leucaena leucocephala*. But in *Sesbania sesban*, the increase was observed up to 42 months only. Amounts of important nutrients recycled through the litter are given in Table 3. From these data it is evident that trees add good amount of litter directly releasing essential nutrients being deficient in salf affected soils and improve crop productivity. Indirectly, availability of nutrients also increases by decreasing soil pH in these soil favourably.

Table 2. Mean concentration of nutrients (m mol kg^{-1}) in litter of *Acacia* and *Eucalyptus* from 1982 to 1986.

Nutrients	*Acacia*	*Eucalyptus*
N	243	160
P	59	20
K	213	100
Ca	134	171
Mg	147	159
S	72	49
Na	37	125
Fe	4.00	5.7
Mn	0.80	28.59
Zn	0.27	0.24
Cu	0.11	0.12

Table 3. Litter (dry) yield and amounts of important nutrients recycled in *Leucaena Leucocephela* and *Sesbania sesban* plantation

Item	*Leucaena leucocephala*	*Sesbania sesban*
Average litter (kg ha^{-1})	5406	5940
Nutrients recycled (kg ha^{-1})		
N	243	160
P	59	20
K	213	100
Ca	134	171
Mg	147	159
S	72	49
Na	37	125
Fe	4.00	5.79
Mn	0.80	28.59
Zn	0.27	0.24
Cu	0.11	0.12

Tree Species Suitable for Salt-affected Soils

Like agricultural crops, woody species also differ in their tolerance to alkali soil conditions at the genetic as well as the species level. The tolerance of plants is governed by factors like Na: K ratio of the shoots and the capacity to take up K under strong Na competition, selective absorption or retention of cations, mainly Na in the leaves, roots and shoots, tolerance to oxygen stress in the rhizosphere and nutritional stresses. Under saline conditions, only those plant species which can withstand high salt content and thrive under high water table conditions should be selected for planting. The promising crops, grasses and tree species for plantation under salt affected soils are listed according to their degree of tolerance (Table 4, 5 and 6). The suitable fruit species under such conditions are sapota, date palm, pomegranate, guava, Ber, custard apple and Aonla.

Table 4. Promising tree species for salt affected soils.

Alkali Soils	Saline Soils	Bio-drainage Species
Prosopis juliflora, Acacia nilotica, Tamarix articulata, Casuarina equisetifolia, Eucalyptus tereticornis, Pongamia pinnata, Terminalia arjuna, Prosopis alba, Dalbergia sissoo	*Prosopis juliflora, Tamarix troupi, Tamarix articulata, Acacia farnesiana, Acacia nilotica, Acacia tortilis, Casuarina glauca, Eucalyptus camaldulensis, Leucaena leucocephala*	*Eucalyptus tereticornis, Populus deltoides, Terminalia arjuna,Acacia auriculiformis, Syzugium cumini, Albizia lebbek, Dalbergia sissoo, Pongamia pinnata*

Table 5. Salinity tolerant crops.

High salt tolerant	Medium salt tolerant	Low salt tolerant
Sesbania, rice, sugarcane, Barley, sugar beet, berseem, oat, lucerne, fenugreek, senji	Castor, cotton, sorghum, wheat, Bajra, mustard, maize	Linseed, sesame, pulses, Sunnhemp, pea, gram

Table 6. Sodicity (Exchangeable Na^+) tolerance of some crops and grasses

Tolerant	Semi tolerant	Sensitive
Karnal grass, Rhodes grass, Para grass, Rice, Sugar beet	Barley, Wheat, Mustard Berseem, Oat, Sugarcane, Cotton, Bajra	Pulses, Maize, Groundnut

Soil Improvement

Salt-affected soils are very low in organic matter and biological activity. Tree growth provides a green cover and improves soil environment for enhanced biological activity, better water relations and increased organic matter and fertility status. Tree roots increases the carbon dioxide levels in the soil which help in mobilizing soil calcium carbonate and the releasing Ca exchange with Na on soil exchange complex. Further, the litter produced also hastens the process of reclamation.

High density plantation of *Sesbania sesban* resulted in marked decrease in pH_2 and EC_2 and improvement in organic carbon, total and available N in a growth period of about 30 months (Rao & Gill, 1988). The decrease in soil alkalinity and improvement in soil fertility was mainly confined to the surface 30 cm layer (Table7). *Sesbania sesban* produced 4 tones ha^{-1} yr^{-1} litter with a high mean N content of 2.3 per cent showing great promise for planting in semi-reclaimed alkali soils for firewood production and soil improvement.

Table 7. Changes in soil properties in 30 months in *Sesbania sesban* planted at selected spacing.

Spacing (cm)		Soil depth	pH_2		Organic carbon (%)		Total N (%)		Available N (kg ha^{-1})	
Rows	Plant	(cm)	1985	1988	1985	1988	1985	1988	1985	1988
50	100	0-15	8.80	8.75	0.71	0.82	0.08	0.09	74	109
		15-30	9.50	9.20	0.50	0.54	0.07	0.07	73	27
100	100	0-15	8.65	8.30	0.76	0.84	0.08	0.10	72	105
		15-30	9.35	9.15	0.49	0.58	0.06	0.07	66	67
200	100	0-15	8.61	8.50	0.82	0.90	0.08	0.10	76	105
		15-30	9.28	9.20	0.47	0.60	0.07	0.06	74	84
300	100	0-15	8.64	8.20	0.82	0.82	0.09	0.10	75	116
		15-30	9.25	9.00	0.50	0.64	0.07	0.07	71	77

Singh *et al.* (1997) evaluated agriculture, forestry and agroforestry practices in a moderately alkali soils in northwestern India over a period of six years at CSSRI, Karnal. The soil conditions were much improved in terms of the build up of soil organic matter, N, P and K when trees were associated with agricultural crops (Table 8). The improvement in soil condition with different trees was in the order: *Acacia* based system > poplar > *Eucalyptus* > agriculture (sole crops). The higher fertility status under *Acacia* and poplar trees results from higher annual litter yields and a better quality of the added material.

Mongia *et al.* (1998) observed that by growing *Prosopis juliflora, Acacia nilotica, Dalbergia sissoo* and *Casuarina equisetifolia* on sodic wasteland of Saraswati range, Kurukshetra, Haryana, for more than three years, pH, EC of the soil decreased, $CaCO_3$ lowered and organic matter increased (Table 9). A sharp decrease in soil pH and EC was observed within three years of growth under all the plantations, the decrease being more under *Prosopis juliflora* followed by *Acacia nilotica, Dalbergia sissoo* and *Casuarina equisetifolia.* However, pH and soluble salts increased in lower depths.

Table 8. Changes in some properties of 30cm soil in five years as affected by land use practices at Karnal, India.

Land use system	Soil properties					
	Initial					
	pH	EC(dSm^{-1})	OC (%)	Available nutrients (Kg ha^{-1}) N	P	K
Sole crops	9.26	0.56	0.22	65	14.3	238
Eucalyptus-based system	9.32	0.61	0.25	65	13.8	224
Acacia-based system	9.34	0.67	0.28	72	13.6	212
Poplar-based system	9.27	0.60	0.26	68	14.4	237
	After Five Years					
Sole crops	8.81	0.45	0.29	75	13.6	208
Eucalyptus-based system	8.65	0.32	0.37	86	18.5	259
Acacia-based system	8.71	0.41	0.48	103	31.9	286
Poplar-based system	8.47	0.35	0.43	93	21.7	321

The increase in soluble salts may be due to their translocation through leaching and lowering of pH can be related to the organic matter accumulation because of litter fall and their subsequent decomposition. Further, the table shows a general increase in organic C content throughout the soil profile under all the plantations, the increase being more in the surface layer and the rate of increase decreased with depth. The increase in organic C was maximum with *Prosopis juliflora* (3.2 g kg^{-1} soil) and the least in *Casuarina equisetifolia* (1.7 g kg^{-1} soil). As regards the available nutrients, available P declined while an increase in available K was observed under all plantations. The lowering of available P in the planted site may be due to downward movement of sizeable portions of P along with other soluble salts to the lower depths. The higher content of K may be due to release of K from the K-bearing minerals following reclamation and partly due to recycling of K on account of litter decomposition.

Table 9. Changes in soil properties after 40 months of trees plantations

Depth (m)	pH_2	EC_1 (dSm^{-1})	OC (g kg^{-1} soil)	Avail. P (mg kg^{-1} soil)	Avail. K (mg kg^{-1})	$CaCO_3$ (g kg^{-1})
			Initial soil characteristics			
0-0.15	10.7	2.7	0.5	43.0	118.6	11.4
0.15-0.30	10.7	3.3	0.3	28.3	115.6	14.8
0.30-0.45	10.2	1.1	0.2	11.0	28.9	23.2
0.45-0.60	9.8	0.5	0.2	7.5	16.6	22.9
0.60-0.90	9.6	0.4	0.3	7.3	28.9	18.2
0.90-1.20	6.4	0.6	0.3	14.0	13.7	2.7
			Prosopis juliflora			
0-0.15	9.3	1.1	3.2	24.7	249.0	10.0
0.15-0.30	9.8	1.9	2.6	30.3	160.3	12.5
0.30-0.45	9.9	2.4	2.0	34.6	133.2	20.5
0.45-0.60	10.4	2.6	1.8	28.0	127.9	20.3
0.60-0.90	10.6	2.9	1.5	12.6	128.5	18.0
0.90-1.20	10.8	3.1	1.0	10.8	115.6	21.0
			Acacia nilotica			
0-0.15	9.4	1.7	2.7	23.6	203.1	9.7
0.15-0.30	9.6	2.1	1.9	30.0	157.3	11.8
0.30-0.45	10.1	2.4	1.5	33.4	130.1	19.0
0.45-0.60	10.2	2.7	1.0	25.7	125.8	18.7
0.60-0.90	10.2	3.3	0.8	14.6	128.0	18.5
0.90-1.20	10.0	3.3	0.8	13.9	120.3	22.0
			Dalbergia sisso			
0-0.15	9.4	1.7	2.1	22.6	190.7	9.3
0.15-0.30	9.9	2.0	1.4	29.2	153.2	11.3
0.30-0.45	10.1	2.2	1.0	32.3	120.8	19.0
0.45-0.60	10.3	2.4	0.8	24.0	118.6	17.8
0.60-0.90	10.2	2.6	0.5	13.2	123.0	19.0
0.90-1.20	10.4	2.8	0.5	13.8	119.0	21.9
			Casuarina equisetifolia			
0-0.15	9.8	1.8	1.7	21.7	132.2	8.7
0.15-0.30	9.9	1.9	1.3	25.6	120.1	7.9
0.30-0.45	10.1	2.6	11.1	34.0	179.4	16.5
0.45-0.60	10.3	2.7	1.0	22.6	105.6	17.3
0.60-0.90	10.5	3.4	0.8	11.3	98.8	18.5
0.90-1.20	10.6	4.9	0.5	13.0	93.4	22.3

Based on investigation conducted at Karnal, Kaur *et al.* (2000) reported that on the basis of increased soil organic matter content and greater soil N availability, agrisilvicultural systems have been found to be ecologically sustainable land-use systems for utilizing moderately alkaline soils (Table 10).

Table 10. Soil carbon and nitrogen concentrations at different soil depths and distances from rows of tree (1, 2 and 3 m) in different systems.

Treatments	OC (%)			Soil N (%)		
	0-7.5	7.5-15	15-30	0-7.5	7.5-15	15-30
control	0.65	0.26	0.16	0.083	0.036	0.023
Rice-berseem	0.47	0.28	0.15	0.055	0.037	0.018
1 m Distance						
Acacia	0.68	0.30	0.21	0.083	0.037	0.021
Acacia+rice-berseem	0.77	0.36	0.27	0.090	0.043	0.033
Eucalyptus	0.48	0.19	0.14	0.058	0.029	0.017
Eucalyptus+rice-berseem	0.67	0.22	0.18	0.073	0.035	0.019
Populus	0.50	0.20	0.19	0.060	0.031	0.018
Populus+rice-berseem	0.71	0.36	0.28	0.088	0.042	0.025
$LSD_{(p<0.05)}$	0.034	0.038	0.033	0.010	0.011	0.008
2 m Distance						
Acacia	0.57	0.20	0.18	0.065	0.025	0.018
Acacia+rice-berseem	0.64	0.22	0.14	0.077	0.030	0.022
Eucalyptus	0.38	0.15	0.12	0.035	0.015	0.014
Eucalyptus+rice-berseem	0.55	0.19	0.15	0.050	0.021	0.017
Populus	0.42	0.16	0.14	0.049	0.019	0.015
Populus+rice-berseem	0.58	0.21	0.19	0.070	0.028	0.021
$LSD_{(p<0.05)}$	0.029	0.029	0.028	0.014	0.009	0.010
3 m Distance						
Acacia	0.48	0.13	0.11	0.057	0.019	0.016
Acacia+rice-berseem	0.50	0.15	0.13	0.064	0.023	0.019
Eucalyptus	0.29	0.14	0.11	0.026	0.012	0.011
Eucalyptus+rice-berseem	0.41	0.12	0.10	0.041	0.016	0.014
Populus	0.32	0.12	0.10	0.032	0.015	0.012
Populus+rice-berseem	0.49	0.14	0.12	0.061	0.020	0.017
$LSD_{(p<0.05)}$	0.044	0.029	0.028	0.012	0.007	0.006

Further it is evident from the data that the amelioration effect of trees was more conspicuous at 1 and 2 m distance from trees and in upper soil layers (0-7.5 cm). Maximum build up of organic matter content and soil N occurred under Acacia based system followed by *Populus* and *Eucalyptus* based systems. This might be due to organic matter inputs in the form of litter fall and fine roots from the tree species.

In a field experiment conducted at Agricultural Research Station, Gangavati, Karnataka on saline water-logged soils, Manjunatha *et al.* (2002) found that eight tree species showed better performance under saline and water logged conditions for their role in bio-amelioration (Table 11). *Dalbergia sissoo* contributed more to soil organic carbon (14.4 g kg^{-1} soil) and decomposition in soil followed by *Casuarina equisetifolia* (OC 12.6 g kg^{-1} soil) and Av. N 470 kg ha^{-1}). *Glyricidia maculata* with green manure value also contributed more organic carbon to soil (11.4 g kg^{-1} soil) and higher available NPK (560, 88 and 772 kg ha^{-1}, respectively) and remained comparable with *Casuarina equisetifolia*. In general, all the tree species contributed to available nutrient pool and enriched organic carbon status of the soil and improved soil environment.

Table 11. Bio-ameliorative effects of trees in saline soils.

Tree species	Available nutrients (kg ha^{-1})			OC (g kg^{-1} soil)
	N	P_2O_5	K	
Fallow	160.0	33.3	380	4.6
Acacia ferugenia	421.8	58.5	841	9.6
Albizzia lebbek	387.3	51.2	733	9.3
Casuarina equisetifolia	470.3	20.5	494	12.6
Dalbergia sissoo	435.7	65.8	841	14.4
Inga dulse	463.4	23.4	872	10.5
Eucalyptus hybrid	532.9	74.6	787	10.8
Pongamia pinnata	463.4	23.4	872	10.5
Glyricidia maculata	560.2	87.8	772	11.4

The two ever-green salt tolerant trees *Azadirachta indica* and *Pongamia pinnata* were tried at Banthra, Lucknow to reclaim soil over a period of nine years by Jain *et al.* (2002). The soil organic carbon status increased slightly under *A. indica* (2-4 times) in comparison to *P. pinnata* (1.7 times) stands. However, both the speciec enriched the soil organic carbon pool particularly in surface layer of 0-15 cm depth (Table 12). The most important factors, which contributed most significantly in this process, were litter fall, their decomposition, fine root mortality, root expansion, root exudates etc. Most of them are related to plant growth and productivity.

Although soil pH reduced considerably under both the species, yet its effect was more pronounced in *A. indica* stand particularly in surface soil of 0-15 cm depth, but the reduction in pH at lower depth of 45-60 cm was non-significant (Table 12). The restricted movement of the leachate through a hard

Kankar pan might have caused this effect. Electrical conductivity did not vary much between the control and planted soils.

Table 12. Properties of soil profile before and after nine years of plantation

Depth (cm)	Before plantation							
	pH	EC ($dS\ m^{-1}$)	OC ($g\ kg^{-1}$ soil)	ESP				
0-15	9.7	0.41	0.33	28.54				
15-30	10.4	0.41	0.31	31.26				
30-45	10.8	0.60	0.30	35.48				
45-60	10.9	0.61	0.27	40.16				
	After nine years of plantation							
	Azadirachta indica				*Pongamia pinnata*			
	pH	EC ($dS\ m^{-1}$)	OC ($g\ kg^{-1}$ soil)	ESP	pH	EC ($dS\ m^{-1}$)	OC ($g\ kg^{-1}$ soil)	ESP
0-15	8.62	0.512	3.21	19.30	9.05	0.485	2.38	23.48
15-30	9.10	0.436	2.25	23.80	9.35	0.464	1.72	27.30
30-45	9.49	0.495	1.72	27.48	9.72	0.422	1.18	31.05
45-60	9.85	0.532	1.21	32.76	10.12	0.318	1.00	37.50

Management of Trees in Salt Affected Soils

Management of trees in salt affected soils is an important practice to enhance the productivity and to accelerate the amelioration process. The following techniques can be adopted for this purpose.

- Treating pit soils with gypsum (3-4 kg/pit), FYM (25 kg/pit) and fertilizers.
- Augur hole planting (100-150cm deep and 20-30cm diameter augur hole).
- Sub surface planting (trenches/furrows of 30cm depth).
- Transplanting of older saplings.
- Light and frequent irrigation during initial years of establishment.
- Provide adequate drainage.
- Manage tree canopy depending upon growth of tree.

Summary

Experimental evidences accumulated so far have identified agroforestry/ tree plantation as an economically viable land use system for the utilization and improvement of salt-affected soils, especially the alkali soils of the Indo-Gangetic plains. In many salt-affected soils, where good quality of irrigation water is not available, afforestation with hardy species is the only practical solution for their proper utilization. Moreover, agroforestry/tree plantation is an alternative to check wide spread water logging and salinity in canal command areas. Agroforestry/tree plantation should not only be viewed as a source of wood/fuel, fodder and food production system, but also as an alternative form of biological reclamation of salt affected soils.

From results above, now it is well established that trees ameliorate salt affected soils by way of reducing soil pH and enriching soil with essential nutrients. Edaphic specific techniques and trees are available for adoption and reclaim the soils.

Future Research Thrust Areas

Based on data compiled for reclamation of salt affected soils through agroforestry/tree plantation, some areas of high priority for agroforestry research are:

- Agroforestry systems should be developed for salt-affected soils of the Indo-Gangetic plains, arid and semi-arid regions, black soils and coastal saline soils.
- Researches are needed to devise suitable silvipastoral models for canal induced saline areas for biomass production and for drainage and salinity control.
- Screening and determination of tolerance limit of promising species at the nursery stage and later on in the field is necessary. There is a need to improve plant types of tolerant tree species like *Prosopis* and *Acacia* using latest genetic and plant breeding tools.
- Practically no information is available on the tolerance limits of fruit trees and no technology has been developed for their cultivation in salt-affected soil. Efforts should be made to collect and screen germplasm of suitable fruit trees for their adaptation to salt, alkali and water logging stresses under different soil and climatic conditions.
- Identification of suitable inter-crops grown in association with salt-tolerant trees and quantification of tree-intercrop interaction in terms of competition for space, light, nutrients and water.
- Long term field experiments should be set up for comparative evaluation of agriculture, silviculture and agroforestry land use systems in terms of sustainability, biomass production, energy use and economic gains.
- Long term effects of tree growth on soil properties, organic matter build-up, nutrient recycling, hydrological status and ecological balance should continue to form an integral part of future researches in agroforestry.

- Simulation models of agroforestry systems in salt-affected soils should be developed.

References

Gill, H.S., Abrol, I.P. and Samra, J.S. 1987. Nutrient recycling through litter production in young plantations of *Acacia nilotica* and *Eucalyptus tereticornis* in a highly alkaline soil. *Forest Ecology and Management*, 22: 57-69.

Gill, H.S. and Samra, J.S. 1989. Evaluation of affirestation techniques and tree soil interactions in salt affected soils. Ann. Rep., CSSRI, Karnal, 150-157.

Jain, R.k., Singh, Bajarang, Tripathi, K.P. and Srrivastava, Neeta. 2002. Reclamation of a sodic soil through Afforestation with *Azadirachta indica* and *Pongamia pinnata. Journal of Indian Society of Soil Science,* 50(1): 147-148.

Kaur, B., Gupta, S.R. and Singh, G. 2000. Soil carbon, microbial activity and nitrogen availability in agroforestry systems on moderately alkaline soils in northern India. *Applied Soil Ecology*, 15: 283-294.

Manjunatha, M.V., Hebbara, M., Patil, S.G. and Minhas, P.S. 2002. Performance of multi-purpose tree species on saline-waterlogged soils. *Journal of Indian Society of Soil Science,* 50(1): 103-106.

Mongia, A.D., Dey, Pradip and Singh, G. 1998. Ameliorating effect of forest trees on a highly sodic soil in Haryana. *Journal of Indian Society of Soil Science,*46 (4) : 664-668.

Rao, D.L.N. and Gill, H.S. 1988. Nitrogen fixation and biomass production by *Sesbania* species. *Annual Report,* CSSRI, Karnal, 124-127.

Sharma, R.C. 1998. Nature, extent and classification. In: *Agricultural Salinity Management in India.* CSSRI, Karnal.

Singh, G., Singh, N.T., Dagar, J.C., Singh, H. and Sharma, V.P. 1997. An evaluation of agriculture, forestry and agroforestry practices in a moderately alkali soil in northwestern India. *Agroforestry System*, 37: 219-95.

Singh, G., Singh, N.T. and Tomar, O.S. 1993. Agroforestry in salt affected soils. *Bulletin No. 17*, CSSRI, Karnal, 65.

Singh, G., Yadav, J.S.P. and Singh, G.B. 2000. Multi-purpose tree species management. In: *Natural Resource Management for Agricultural Production in India* (J.S.P. Yadav & G.B.Singh eds.), Pp. 373-479.

□□□

Chapter 11

MINOR FRUIT PLANTS AND THEIR AGROFORESTRY PERSPECTIVES IN BUNDELKHAND REGION OF CENTRAL INDIA

D.R. Palsaniya, R.K. Tewari, R. Tiwari and C.K. Bajpai
National Research Centre for Agroforestry, Jhansi-284 003 (U.P.)

Introduction

Bundelkhand is one of the most backward regions of India as identified by planning commission. The region comprises of seven districts of U.P. and six districts of M.P. Physiographically, Bundelkhand is among the most disadvantaged regions of the country owing to undulated and rugged topography, highly eroded and dissected land, poor soil fertility and scarce underground water resources (Palsaniya *et al.*, 2008). Most of the population (>70%) derives their livelihood from agriculture which is mainly traditional and subsistence in nature. Their livelihood is further threatened by the unfavourable climatic conditions characterized by high temperatures and low and erratic rainfall. At the same time, the natural resources of the region are under immense pressure due to high animal and human population. During the last 17 years, the region has witnessed 7 years of severe drought (1990, 1994, 2002, 2004-07) leading to a heavy out migration to the extent of 48% of total population towards big cities in search of livelihood (Anonymous, 2008).

The situation is more alarming particularly in last 4-5 years in the countryside of Bundelkhand. The crop failures are frequent due to low and erratic rainfall and lack of rain water harvesting techniques. There is lack of alternate livelihood opportunities and sustainable land use systems like agroforestry. Under such circumstances, traditionally and naturally growing minor fruit plants can play an important role in the livelihood of the people of the region. Fruit plants which are not commercially cultivated by masses are commonly known as minor fruit plants.

Minor Fruit Plants of Bundelkhand Region and their Significance

Minor fruit plants of Bundelkhand agro-climatic conditions (semi-arid) are given below in Table 1.They naturally grow through natural regeneration in forest, wastelands, community lands, agricultural field bunds and near habitat. They are widely consumed either fresh or processed. Often, they have high nutritional values and are rich source of minerals, vitamins and important alkaloids (Table 2). Traditionally, they are important component of the dietary habits of the local people. They have high medicinal values. They are also the important source of fodder and fuel wood. Several minor fruit plants are well adapted to given agro-ecological and agro-climatic conditions as such commonly found on agriculture landscape as scattered plants. Many of them are quite remunerative and have good market in the region. These are hardy in nature and well adopted. Yadav (2003) while assessing the forest resources in Bundelkhand mentioned that they play important role in the socio economic life of the people in the region. Since most of them regenerate by seed, they invariably exhibit large variability in quality and quantity of produce. Palsaniya *et al* (2008) reported that the communities living in Bundelkhand have immense knowledge about plant wealth and are dependent on flora for their social and livelihood needs. During their study in Garhkundar-Dabar watershed of Bundelkhand they recorded 22 plant species used for edible purposes; seven plant species for extraction of gum, tannin, dyes and insecticidal properties; six plant species for making ropes, baskets, mats, brooms, fibers and thatching houses; three plant species for making country liquor; 12 plant species for fodder; five plant species for fencing and eight plant species for fuel wood, timber and making agricultural implements.

Some of the minor fruit plants like Aonla, Ber, Bael, and Imli have gained popularity in past two decades and now heading to become major fruit plants. There is vast potential to cultivate these minor fruits for proper nutrition and high remuneration. In light of growing population, good fertile lands may not be available for sole cultivation of minor fruits; hence, they are poised to seek shelter in croplands under agroforestry system.

Table 1. Minor fruit plants of Bundelkhand region

S. No.	Minor fruit plant	Scientific name	Family
1	Aonla	*Emblica officinalis*	*Euphorbiaceae*
2	Ber	*Zizyphus mauritiana*	*Rhamnaceae*
3	Jamun	*Syzygium cumini*	*Myrtaceae*
4	Mahua	*Madhuca latifolia*	*Sapotaceae*
5	Bael	*Aegle marmelos*	*Rutaceae*
6	Lasoda	*Cordia Myxa*	*Boraginaceae*
7	Imli	*Tamarindus indica*	*Leguminosae*
8	Jungle-Jalebi	*Inga dulsis*	*Leguminosae*
9	Sahejena	*Moringa oleifera*	*Moringaceae*
10	Chironjee	*Buchanania lanzan*	*Anacardiaceae*
11	Karonda	*Carissa carandus*	*Apocynaceae*
12	Ker	*Capparis decidua*	*Capparidaceae*
13	Jharberi	*Zizyphus nummularia*	*Rhamnaceae*
14	Sharifa	*Annona squamosa*	*Annonaceae*
15	Phalsa	*Grewia subinaequalis*	*Tiliaceae*

Table 2. Nutritional value of minor fruit plants

Minor fruit plants	Carbo-hydrates (%)	Protein (%)	Fat (%)	Ca (mg/ 100g)	P (mg/ 100g)	Fe (mg/ 100g)	Vit-A (mg/ 100g)	Vit-B (mg/ 100g)	Vit-C (mg/ 100g)
Aonla	14	0.5	0.1	50	20	1200	-	30	600
Ber	17.1	0.8	0.3	4	9	1.8	0.02	0.74	76
Jamun	19.7	0.7	0.1	20	10	1000	-	-	-
Mahua	22.9	1.4	1.6	45	22	1.1	512 IU	-	41
Bael	31.8	1.8	0.4	-	-	-	55	1.32	8
Lasoda	-	-	-	-	-	-	-	-	-
Imli	67.4	3.1	0.1	170	110	11	0.06	0.14	3
Jungle-Jalebi	-	-	-	-	-	-	-	-	-
Sahejena	3.7	2.5	0.1	30	110	5.3	-	-	-
Chironjee	12.1	19.0	59.0	279	528	8.5	-	2.7	5
Karonda	2.9	1.1	2.9	21	28	-	1619 IU	0.71	-
Ker	20.8	5.9	1.2	154	51	2	8.6	-	133
Jharberi	-	-	-	-	-	-	-	-	-
Sharifa	23.5	1.6	0.4	17	47	1.5	-	1.5	37
Phalsa	11.7	-	-	-	-	-	-	-	-

In India, agroforestry in its primitive form is being practiced since ages. The typical examples are scattered boundary plantations, Phoenix on paddy field bunds and Khejri plantations of arid Rajasthan. In recent past, with the introduction of fast growing short rotation industrial wood plants agroforestry has gained momentum. Among them, Poplar, *Eucalyptus, Casuarina, Ailanthus, Acacia mangium* are front runners. However, except *Eucalyptus* all others are localized in certain pockets. Expansion of industrial agroforestry is closely linked with the industrial development in a given region. Elsewhere, industrial agroforestry could not make much head way for want of ready market. As such, more than 80% (exact data is not available) agroforestry plantations are either scattered plants inside field (Khejri, Babul, Rohida in Rajasthan) or boundary plantations (Babul, Ber, Mahua, Shisham, Palas, Teak, Neem and others). Such plantations do not receive any care except protection. However, they flourish on resource sharing meant for main crop. On the other hand, crops grown in interspaces of planned plantations (agri-silviculture or agri-horticulture) receive due care in terms of manure/fertilizer, irrigation, weeding, plant protection, etc. as per their requirement. Preponderance of scattered or bund plantation agroforestry system in India is largely responsible for low productivity.

Most of the minor fruit plants find place in boundary plantation system besides those perpetuated in homestead. A survey of Datia district of Bundelkhand indicated that Ber constitute 40% of tree population where tree density was 11.2 tree ha^{-1}. Some of the special features of minor fruit plants of Bundelkhand region are given in Table 3.

Table 3. Special features of minor fruit plants of Bundelkhand region

Minor fruit	Special Features
Aonla	Hardy in nature, less browsed, richest source of vitamin C, widely consumed and marketed for pickle, medicine, chutney, etc.
Ber	Drought hardy, good coppiciability, browsable, wood suited for small agricultural implements, high heterogeneity in fruit, consumed fresh and dried.
Sharifa	Drought hardy, non-browsable, consumed fresh after ripening.
Kaitha	Drought hardy, thorny, non-perishable fruits used for chutney making
Jungle-Jalebi	Drought hardy, thorny, non-perishable fruit, consumed after ripening.
Mahua	Hardy in nature, huge tree, leaves very good top feed, flowers edible, made into country liquor, seed contains edible oil. It is life line of tribal economy in the region.
Chironjee	Slow growing, browsable, very deep rooted tree, fruits are highly priced as dry fruit.
Karonda	Very hardy, good coppicer, thorny, evergreen, good for fencing, bears inferior edible fruits. It can be top worked with improved varieties.
Jamun	Common along water courses, huge tree with edible fruits.
Bael	Hardy in nature, susceptible to water logging, non-perishable fruit ripening in April-May (lean period), highly medicinal, soothing effect.
Lasoda	Hardy, prolific bearer, medicinal, good for pickles.

Minor fruit plants have high potential for cultivation as they grow with little care and normally not stolen which is often big concern in fruit cultivation. Thorny minor fruit plants like Karonda, Ber serve the purpose of live fencing while others act as pole on boundary for brushwood or barbed wire fencing of field. Like other fruit plants, they are rich source of minerals and vitamins, thus, add to health of consumer. Many times these fruits are marketed hence; fetch cash to the farmers besides yielding fuelwood, timber, fodder, fibre, fence material, etc. They have aesthetic value and good means of in-situ germplasm conservation. Many minor fruits are processed as such they provide raw material for processing industry such as pickles made out of Lasoda, Ker, Karonda, Sahejena, Imli, etc. They also possess medicinal properties. Some of the minor fruits are rich in antioxidants; others contain alkaloids with high therapeutic value. These are known to people in nearby towns and find easy market through vegetable shopkeepers. Local people develop special taste and liking for minor fruits common in particular region.

Promotion of Minor Fruits

Promotion of minor fruits on farmer's field needs to be rigorously pursued in light of their importance. This warrants systematic efforts.

Selection of Superior Types and Varieties

Most of the minor fruit plants are seed propagated. Plants show very high variability with regards to quantity and quality of produce on account of open pollination and heterogeneity in population. Selection of high yielding plants with desired fruit quality can revolutionize their cultivation. Some of the desired fruit characters of different minor fruit plants have been given in Table 4.

Ensuring Availability of Quality Planting Material

Since beginning of fruit cultivation, availability of quality planting material remains major constraint in expansion of horticulture. Although, vegetative propagation techniques of several minor fruit plants have been well established, still quality planting material is not available to the farmers. Once a promising type is identified, it should be multiplied in bulk time to the type at large scale. Preferably, budded or grafted plants or cuttings should be prepared in polythene bags so that establishment of plants in field is good on account of undisturbed tap root system. Vegetative propagation techniques have been given in Table 5.

Table 4. Desirable characters of minor fruit plants for selection of superior types

Fruit Plant	Desirable traits
Aonla	Big fruit size, red colour blush on fruit, regular bearer, medium fibre and glossy fruits.
Ber	Big fruit size with small stone, tolerance to Ber fruit fly and powdery mildew, good keeping quality and high Total Soluble Sugars (TSS).
Karonda	Big fruit size, attractive red and white colour, heavy bearing.
Sharifa	Big and uniform fruit size, early ripening, high TSS, good keeping quality.
Kaitha	Big fruit size, precocious and high yielder.
Chironjee	Soft seed coat, precocious, prolific bearing.
Jamun	Big fruit size, synchromesh ripening, high TSS
Bael	Medium to big (1-2 kg) fruit size, few seeds, high TSS, prolific bearing, pleasant aroma, soft rind.
Jungle-Jalebi	Dwarf tree, prolific bearing, high pulp content, good flavour and attractive pulp colour.
Mahua	Precocious, bunch bearing, flower drop at down, high TSS in petal, prolific bearer

Table 5. Vegetative propagation techniques for various minor fruit plants

Fruit	Technique	Time
Aonla	Budding	June-July
	Grafting	July-August
	Bench grafting	February
Ber	Budding	July-October
	Bench grafting	June
Bael	Budding	July-October
Lasoda	Budding	July-August
Karonda	Seed	March
	(Top working of *C. spinarum* employing shield budding	May-June
Sharifa	Grafting	July-August
Kaitha	Seed	March-April
	Budding	July-August
Chironjee	In-situ grafting (Veneer)	July-August
Jamun	Budding	July-August
Jungle-Jalebi	Seed	March-April
Mahua	Air layering	September-October

Density Optimization

Under agroforestry system, tree density should be such that cropping goes unabated throughout the life period of fruit plants. This warrants wider spacing particularly row spacing. However, a density of 20-25 tree ha^{-1} can be safely recommended under agroforestry system on the line of Khejri plantations.

Agrotechniques

Agro-techniques of individual minor fruit plants need to be worked out particularly spacing, training and pruning needs, manure and fertilizer requirement, watering at critical growth stages. However, since most of them are hardy in nature, 30-40 kg FYM and 50-60 g N, 30-50 g P and K $plant^{-1}$ yr^{-1} in split dozes and watering once in a month during summer and once in two months during winter may be provided as starter doze. Protection to plant during initial growth stages against stray cattle particularly for browsable species holds key to success. Under agroforestry system, above requirements of minor fruit plants are met with the inputs used for crop. At later stages (bearing stage), separate provision for fertilizer and irrigation if any required can be made for tree alone.

References

Annonymus. 2008. Amar Ujala, Jhansi, January 12

The Wealth of India, A-Z, Council of Scientific and Industrial Research, New Delhi.

Lal, J. 2003. Minor forest produces resources in the economy of Uttar Pradesh and Uttranchal. National Seminar on Forest resource Management.15-17 Nov., CSAU&T Kanpur, Pp. 138 – 144.

Palsaniya, D.R., Singh, Ramesh, Tewari, R.K., Yadav, R.S., Dwivedi, R.P., Kumar,R.V., Venkatesh,A., Kareemulla, K., Bajpai, C.K., Singh, Rajendra, Yadav, S.P.S., Chaturvedi, O.P. and Dhyani, S.K.2008. Socio economic and livelihood analysis of people in Garhkundar-Dabar watershed of central India. *Indian Journal of Agroforestry*, 10 (1): 65-72.

Rai, R. and Nath, V. 2006. Socio-economic and livelihood pattern of ethnic group Baiga in Achanakumar Sal reserve forest in Bilaspur, Chhattishgarh. *Journal of Tropical Forestry*, 22 (3&4): 62-71

Yadav, D. 2003. Phytogeographical survey of assessment of forest resources in Bundelkhand region of Uttar Pradesh. National Seminar on Forest Resource Management, 15-17 Nov., CSAU&T Kanpur, Pp. 157.

❑❑❑

Chapter 12

BIOMASS, BIOENERGY, ENVIRONMENTAL AND LIVELIHOOD BENEFITS FROM *PROSOPIS JULIFLORA* (SWARTZ) D.C. IN INDIA

I.A. Khan, K. Karemulla and O.P. Chaturvedi
National Research Centre for Agroforestry, Jhansi-284 003 (U.P.)

Introduction

Fuelwood plays an important role in the rural economy of the developing countries. The importance of wood as an energy source is increasing dramatically, while its availability is a matter of serious concern. Over use of wood has resulted in rapid shrinkage of forest cover, ecological imbalance and potential economic disaster (Goel & Bhel, 1995).

Out of 329 million hectares geographical land area of the country, about 175 million ha suffer from different limitations expressed to varying degrees and are getting further degraded through natural or man made process (Singh & Dagar, 2005). Majority of these lands are treated as wasteland because of their low productivity due to soil-based constraints like water logging, salinity, sodicity, shallow depth, rocky substratum and sandy soils. Most of the alkali or sodic lands do not support a healthy vegetation cover, with the exception of some restricted natural species comprising only a few salt-tolerant trees and bushes. *Prosopis juliflora* is reported by many to have a high ability to grow under difficult conditions even in deteriorated soils and sites not suitable to many crops and trees (Deans *et al.*, 2003, Garg, 1992, Singh, 1998, Sood & Bhatia, 1998; Singh & Rathod, 2002, Tomar *et al*, 2003). It grows in all types of soils from sandy to saline-alkaline soils (Harsh & Tewari, 1998). It has proved to be most versatile species for afforesting shifting sand dunes, coastal dunes, river beds, saline-alkaline lands, eroded hill slopes, mine spoiled areas and other wasteland (Muthana & Arora, 1983). Initially, plantations of *P. juliflora* were established mainly for the purpose of conservation, but it has now become the main source of fuelwood in rural areas and also to large extent in urban areas, fulfilling more than 70% of the firewood requirements of rural people in the tropical arid and semi-arid region of the country (Harsh & Tewari, 1998).

P. juliflora is also playing a vital role in sustaining the livelihoods of the rural poor, including the landless, small farmers and artisans- the least vocal group of society. In rural areas, it is often the only source of small roundwood and dry season fodder, and provides the only income to many families (Pasiecznak *et al.*, 2001). It provides highly palatable and nutritious pods in large quantities, and the flowers produced good quality nectar for honey, which birds also consume. Ripe pods, which fall on the ground, are readily consumed by domestic live stock and wild ungulates (Mathur & Bhora, 1998). These are rich in protein and free sugars, which gives them a sweet taste (Muthana and Arora, 1983).

Distribution

Prosopis juliflora (Swartz) D.C. is commonly known as mesquite (U.S.A) or Vern, Vilayti babool, Pardesi babool (India). It is thorny, deciduous, large crowned and deep rooted bush or tree which grows up to 10 m height or more (Senthilkumar *et al.*, 2005, Gupta, 1993) depending upon the variety and climatic conditions, with a trunk more than 1 m (Rawat *et al*, 1992). It is a fast growing, hardy, drought resistant but frost tender species, with good coppicing ability (NAS, 1980), and has an extensive lateral root system with a stout tap root which can penetrate as deep as 20 m. (Sood & Bhatia, 1998). It is well adapted to warm and dry tropical climates and grows well in areas receiving 250-600 mm annual rainfall (Saxena, 1998). It is naturally distributed in west tropical and sub-tropical North and South America, very abundant from Buenos to Mexico and Texas and frequently planted also (Rawat *et al.*, 1992). It was first introduced in India in 1857 from Latin America and the first systematic plantations were carried out in 1876 in the Kaddapa area of Andhra Pradesh (Harsh & Tewari, 1998). Since then, the species was gradually introduced in many parts of the country including Gujrat, Rajasthan, Haryana, Punjab, Uttar Pradesh, Madhya Pradesh, Maharashtra and West Bengal (Gurumurthi *et al.*, 1984, Patel, 1986). Now the species has became acclimatized exotic in large parts of arid and semi-arid tropical India (Fig. 1)

Growth

The ecological amplitude of *P. juliflora* is very wide, excluding frost susceptible areas, and it can be grown in any soil type and can tolerate very severe drought and variable rainfall. It can be grown on the poorest soil, with low organic matter and nutrient status (Harsh & Tewari, 1998). Gupta and Balara (1972) noted respective shoot and root length of 6 cm and 21 cm after 1 week, increasing to 20 cm and 52 cm after 10 weeks, with equivalent increases in the number of secondary roots (37 after 10 weeks) in the desert waste land of Rajasthan. Goel (1987) recorded shoot and root length of 69 cm and 27 cm with 30 secondary roots after 8 weeks in alkali wastelands. While studying the growth and survival of different tree species on highly alkaline soils in Kurukshetra district of Haryana, Singh and Dagar (2005), reported a better growth and survival of *P. juliflora* than most of the other tree species studied. At the age of 2, 3.5, 5 and 7 years, the diameter at breast height (DBH) and height recorded was at par with *Pithsolobium dulce, Dalbergia sissoo, Leuceana leucocephla* and much better than *Cordia rohita, Anthocephalus cadamba, Pongamia pinnata, Bombax cieba* and *Tectona grandis* planted in deep and shallow augarholes. In waterlogged and highly saline areas where nothing else can be grown, this species thrives well and produces large quantities of biomass. In the Little Rann and Great Rann of Kutch, Gujrat it is the predominant species, with plant density of up to 2000 per hectare, depending on the rainfall pattern and available soil moisture (Harsh & Tewari, 1998). The tree has a wide range of adaptability and can be planted without any soil amendment (Patel & Pathak, 1986). It is an aggressive colonizer in areas where soil moisture availability is adequate. At the age of 1 year, the species had attained a height of 0.76 m and diameter of 0.93 cm (Singh *et al,* 1989) with a 163 kg ha^{-1} above ground biomass. Where as after 10 years of planting, 12.9 m height and 12.5 cm DBH was recorded under same soil conditions (Singh, 1998). Salt affected soils are extensively distributed both on coastal and inland areas of Gujrat. Maliwal (1999) reported the 5.48 m height, 19.4 cm DBH, with 14385 kg per hectare above ground biomass after seven years of planting in *P. juliflora* under these soil conditions. The growth potential of this tree species at different ages is presented in Table 1.

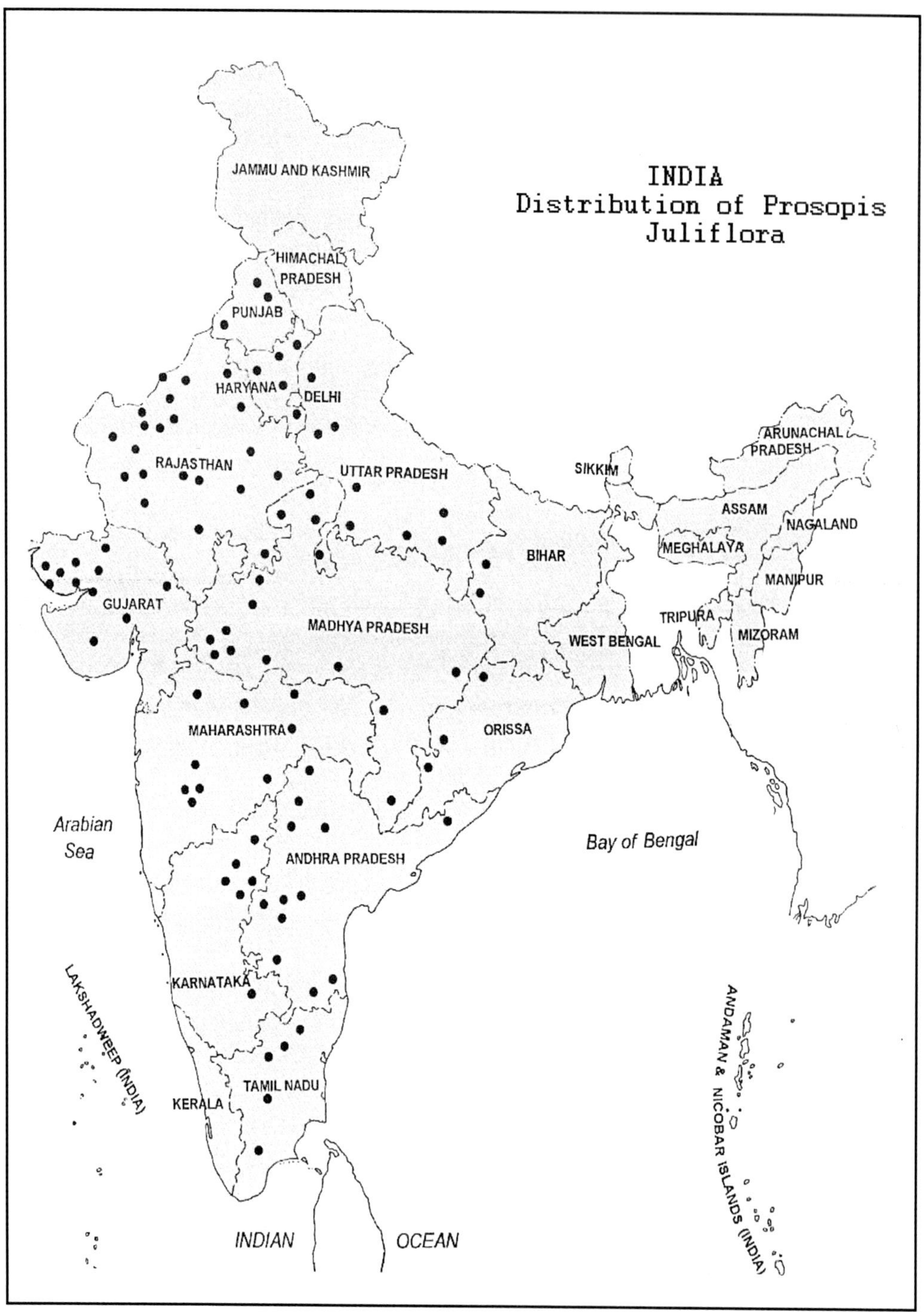

Fig. 1: Distribution of *Prosopis juliflora* in India

Table 1: Growth of *P. juliflora* in waste lands.

Age in years	Height (m)	Diameter (cm)	References
1	0.76	0.93	Singh *et al.*, 1989
2	0.86	1	Singh *et al.*, 1989
3	3.72[a] 2.00[b]	4.5[a] 3.6[b]	Singh and Dagar, 2005
4	3.2	3.6	Singh, 1996
7	5.48	19.4	Maliwal, 1999
8	8.2	7.8	Singh, 1996
9	6.7	12.0	Tomar *et al.*, 2003
10	12.9	12.5	Singh, 1998

[a] planted in deep auger holes
[b] planted in shallow auger holes

In natural habitat, two forms of *P. juliflora* are observed. The first is the large 'tree' form, with individuals often more than 8 m in height and having a clear single bole up to 3-4 m. Some trees of more than 15 m in height were observed near Jodhpur and in the Allen forest of Kanpur. However, 20-year-old trees in the artificially raised protective forest of *P. juliflora* in ravines near Kota on the hump top did not grow beyond 12.5 m height (Gupta, 1993). The second is the small 'bush' form, multi stemmed at ground level, although occasionally the multi-stemmed nature may be due to coppice regrowth. Though initially bush by habit, with forking at lower heights it can grow straight, if pruned from the base leaving one stem, (Harsh & Tewari, 1998).

Biomass Production

After one-year of growth the woody biomass of *P. juliflora* becomes suitable as a resource (Pasiecznik *et al.*, 2001). There are many reports on biomass yields, but the results have been found to vary considerably in different regions (Table 2). A study carried out by the Forest Research Institute (Dehra Dun) in Gandhinagar, Gujarat found that the utilizable dry matter biomass (stem & branches) of *P. juliflora* was 133 t ha^{-1} and the non-utilisable biomass (leaves and roots) was 35 t ha^{-1} after four years of out planting (Patel, 1986). Gurumurti *et al* (1984) reported a production of 113.25 t ha^{-1} when planted at 1.3×1.3 m spacing at 4 years, while Chaturvedi (1985) reported a harvest of only 6.7 t ha^{-1} from two year old plantation. Rao & Vasanthi (1986) recorded a biomass yield of 50 t ha^{-1} from Andhra Pradesh.

Table 2: Biomass production of *P. juliflora* at different ages.

Age (years)	Total Biomass (t ha^{-1})	References
0.5	3.53	Goel and Bhel (1995)
2	6.70	Chaturvedi (1985)
4	113.25	Gurumurti *et al.*, 1984
4	133.00	Patel (1986)
6	103.00	Singh (1998)
10	260.00	Singh (1998)

In Gujrat *P. juliflora* has been grown in semi-arid regions and its plantation getting momentum on waste and marginal lands owing to grave energy crises. Maliwal (1999) reported a total yield of 17.45 t ha^{-1} at the age of 7 years. The mean annual increments of 66 cm in height, 2.51 cm in collar diameter and 1.73 cm in breast height diameter was recorded, and a total yield of 50-60 t ha^{-1} after 10 years and 75-100 t ha^{-1} after 15 years (Vimal & Tyagi, 1986). Tentative biomass production in 10 years was 260 t ha^{-1} from *P. juliflora* (Singh, 1998). At the youngest age of just 6 months, only 3.53 t ha^{-1} biomass production was recorded. The mean annual production in different ages of plantations on varied edhapho-climatic conditions ranged from 2.50 to 33.25 t ha^{-1}.

Fuelwood Situation and Contribution

The total primary energy consumption in India in 1991 was 365-425 Mtoe (million tones of oil equivalent), with the share of biomass energy ranged from 36-46 %. Thus, wood fuel alone accounts for a share of about 20-30 % of total energy consumption in the country. For the year 2000, the consumption of fuelwood for South Asia was 397 million m^3 with India alone accounting for 297 million m^3 of the total consumption (FAO, 2002). The estimated fuelwood consumption in the country is about 380 million m^3 taking into account the illegal and unrecorded fuelwood collections. The incremental growth of Indian forests in terms of fuelwood is around 26.3 million m^3 (21 million tonnes) per year (FSI, 1995). However, the consumption of fuelwood in India is about five times higher than what can be sustainably removed from forest. Fuelwood meets about 40% energy needs of the country. Commercial and industrial units account for about 70% of the fuelwood. Pandey (2003) and Saxena (1997) reported that traditional fuel (fuelwood, crop residue and dung cake) dominates domestic energy use in rural India and accounts for about 95% of the total. The trends up to the year 2010 indicate that the consumption of fuelwood is likely to rise marginally in most countries. In India, however, the projected consumption of fuelwood will be 365 million

m^3 (FAO, 2002). The dependence on fuelwood for domestic usage varies greatly between the rural and urban household as per National Sample Survey Organization Report 2001. Use of dung and agricultural wastes as fuel is widespread in agriculturally prosperous regions with fertile soil and controlled irrigation, such as Punjab, Haryana, UP, and north Bihar, but continues to be the main domestic fuel in less endowed and power regions (NCAER, 1985).

Wood of *P. juliflora* is a very important source of domestic fuel for millions of people in many arid and semi arid zones of the world (Pasiecznik *et al.*, 2001), and in the arid and semi-arid tract of Rajasthan, Gujarat & Haryana more than 70% of the fuelwood demand of people is met by *P. juliflora* (Saxena, 1998). The fuel value of this species is also realized in Kutch (Gujarat), where the fuel requirement of the entire rural population of close to 1 million inhabitants is met by this species (Kareemulla *et al.*, 2005). The *P. juliflora* wood is hard with specific gravity of 0.7 (Tyagi & Singh, 1986), density 0.736 g cm^{-3}, ash 2.30%, calorific value 23.53 kilo jules per $gram^{-1}$, moisture 36.13% (Goel & Bhel. 1995), burns slow and evenly and holds heat well (Kanzaria & Varshaney, 1998). Bhat and Todaria (1990) reported that high calorific value; high density of wood, low water content is the most desirable characteristic for an ideal fuelwood. Among the different *Prosopis* species, *P. juliflora* is reported to be good firewood with high fuelwood value index of 2084 (Goel & Bhel. 1995). Because of its high heat value the wood is termed as 'wooden anthracite'. The wood of this species does not spit, spark or smoke excessively, and the smoke is never unpleasant. It must be noted, however, that its popularity is also linked to its ubiquity that is where trees are present, they are generally in large numbers and are often found on common land and are, thus, freely available to all sections of society (Pasizienik *et al.*, 2001).

The tree coppices readily and provides immediate fuel needs of the rural population. Pollarding in too close a felling cycle (6 months) showed less related out turn of wood, where ratio of twigs to small wood was more (Subbayan, 1980). Maximum fuel yield was from 5-10 cm mid dia pole. Weight per unit volume of stacked wood and air-dried matter contents increased with increase in diameter of fuelwood (Joshi & Prajapati, 1977).

Charcoal Production Potential

Charcoal is consumed mostly in urban areas and is often produced at considerable distance from the point of consumption. Firewood is bulky and expansive to transport whereas conversion to charcoal reduces the weight and increases the value of the product in energy and economic terms. *P. juliflora* is becoming a popular species in many parts of the India as a part of the farm forestry. The average firewood yield in Gujarat has been estimated to be approximately 30 t ha^{-1} per year dry matter, which if converted to charcoal would yield approximately 10 t. With about 1.75 million ha of *P. juliflora* plantations in Kutch alone, the potential for charcoal production from the region can be seen (Kanzaria an Varshney, 1998). The *Prosopis*-based Farm Forestry Programme has good potential in Gujarat not only for meeting energy needs, but also support the Wasteland Development Programme. *P. juliflora* is reported to produce a high quality charcoal, which has virtually replaced *Anogeisus pendula* as the main charcoal producing species in the arid-zone of India (Saxena, 1998). Approximately 3-6 kg of wood of *P. juliflora* is required to produce 1 kg of charcoal depending on the method used (Pascizenik *et al.*, 2001). Charcoal is manufactured in traditional or improved earth kilns, or less commonly in metal kilns. In India, forestry departments produce and market charcoal, predominately of *P. juliflora*, through special development corporation. From 1986-93, Gujarat state estimated that they had produce up to 88,00000 t of charcoal from *P. juliflora* wood with a maximum production of 30,00000 t during 1991-92, where as it was only 70,000 t in 1986-87 and 1992-93. Kareemulla *et al.*, (2005) reported a production of 152600 t of charcoal (Table 3) in 3 districts (Bhuj, Patan and Surendranagar) of Gujarat.

Table 3: Charcoal production potential of *P. juliflora* in Gujrat during the year 2005.

Name of the District	Estimated *P. juliflora* area (ha)	Estimated charcoal production (t)	Annual employment generation (million mandays)
Bhuj	88,225	1,03,800	0.44
Patan	23,037	13,8000	0.40
Surendranagar	59,710	35,000	1.02
Total	170972	276800	1.86

Use and Potential in Gasifiers

The biomass of *P. juliflora* has been found ideal for utilization in power generation through biomass gasification. Biomass gasification is basically conversion of solid biomass (wood, agriculture residues etc) in to a combustible gas mixture normally called 'Producer Gas' (or low Btu gas). The process involves partial combustion of biomass which produces combustible gases, Carbon Monoxide (CO) and hydrogen (H_2). After cleaning, the gas is finally fed in to a diesel generating set to run the engine or for direct heat application. The diesel generating set operates on dual fuel mode (typically 20% Diesel and 80% producer gas). Gasification of *P. juliflora* for powering water pumps for irrigation in India has been found profitable (Patel, 1986.). Boilers can also be fired directly, for example for the drying of agricultural products or grain processing. Many studies have suggested the use of *P. juliflora* as fuel for the generation of electricity by direct burning or via gasification (Parker 1982, Patel. 1986, Sutaria *et al.*, 1986). The biomass is low in Sulphure and is therefore not as polluting as other sources such as coal (Parker, 1982). Steam boilers have also been employed to convert wood energy directly in to mechanical energy for pumping of water (Rao & Vasanthi 1986). In India, many studies have shown the profitably of electrical generation from short rotation biomass plantation (Bose & Bandoyopadhay, 1986). There involves utilization of harvested wood in several ways, either by direct burning (Patel, 1986, Sutaria *et al.*, 1986), carbonization (Higorani, 1986, Raju, 1986), gasification (Rao & Vasanthy, 1986), pulverization (Felker, 1984). Gasification has been promoted as the most efficient method of converting *P. juliflora* wood energy to electrical energy (Patel, 1986). Approximately 1 kg of *P. juliflora* biomass generates 2.5 m^3 of gas, which can supply 3000 kilo calories of thermal energy (Rao & Vasanthi, 1986). Several small plants are in operation in northern India, producing 1-5 MW of electricity solely from *P. juliflora* biomass (Sutaria *et al.*, 1986). Karemulla *et al.*, (2005) reported that in Andhra Pradesh and Gujarat, *P. juliflora* is a major source of fuel for the power plants (Table 4), where as for the power plants in Karnataka presently it is a secondary fuel source. The three power plants studied in Andhra Pradesh have developed their captive plantations of *P. juliflora*. Thus, *P. juliflora* is the reliable source to meet such quantities of biomass and with further advancement of technology in gasification it will be used more efficiently to generate more electric power.

Table 4: Biomass Source and use by sample power plants in Andhra Pradesh, Karnataka and Gujarat.

State	Biomass based No. of Power Plants visited	Source of Biomass	Captive plantation of *P. juliflora*/MW (ha)	Status of Power plants
Andhra Pradesh	3	*P. juliflora*	75-500	Fully operational and supply to national grid
Karnataka	2	Rice husk, *P. juliflora*	Nil	Fully operational
Gujrat	2	*P. juliflora*	Nil	One unit closed and the other is a captive power plant

Employment Generation and Livelihood Support

P. juliflora is a species with significant potential for commercial exploitation and for enhancing livelihoods in the India's dry lands. When managed, it can be a very valuable source of commercial products, which can directly or indirectly improve livelihoods of rural poor. There is enough scope for exploring the possibility of utilizing the different parts such as the stem, root, bark, leaves, flowers and pods of *P. juliflora* for the benefits of the human population (Vimal & Tyagi, 1986). The species produces a good quality of wood, which is resistant to decay agents and displays distinctive straight or swirling grains that range from golden brown to dark reddish orange in color. The wood color is reported to compare favorably to the Indian Rose wood (Shisham), and equals or surpasses the physical and mechanical properties of other commonly recognized fine hardwoods in India (Tiwari *et al.*, 2000). The surface hardness of wood makes it ideal for furniture and flooring as it exhibits good tanning, finishing and polishing properties (Jain *et al.*, 2005). The timber is used for the construction of houses and the manufacture of farm implements throughout Gujarat state (Kanzaria & Varshaney, 1998). However, its use as timber is limited by its small size, but it can still provide a good substitute to rural people who can not afford to use other high quality and costly woods. Efforts have been made to examine the possibility of producing writing and printing papers from the bleached sulphate pulp of *P. juliflora* wood (Guha & Gupta, 1970, Vimal & Tyagi, 1986).

In Gujarat, the estimate made by Forest Department Corporation shows a considerable amount of employment generation from 1986 up to 1993 in charcoal processing. The man-days required in charcoal processing were 33,000 and 63,000 man-days in 1992-93 (Kanzaria & Varshney, 1998), and showing significant increase or decrease each year during this period. Kareemulla *et al.*, (2005) reported that in 3 districts (Bhuj, Patan & Surendranagar) of Gujarat, the Forest Department Corporation has estimated employment generation of 18.55 lakh man-days in charcoal processing (Table 3).

The tree flowers twice a year and the flowers of this species are very good source of nectar for honey bees (Kanzaria & Varshney, 1998). In parts of Gujarat where dense plantation exist, apiculture has developed, with a honey production of 300 tonnes during the 5 years period, and simultaneously during the process of purification, about 15 tonnes of beeswax was also obtained over the same period (Harsh & Tewari, 1998). The seed pods of *P. juliflora* are known to be good source of cattle feed, and the average yield per tree per year has been found to be 18.95 kg, with the main fruiting seasons being winter and summer (Kanzaria & Varshney, 1998). The pods contain

13% crude protein, 60% nitrogen free extracts, 0.15-0.4% phosphorus and 0.30-0.50% calcium on dry matter basis (Shukla *et al.*, 1984). Rao and Reddy (1983) did not observe any effect on the digestibility in the growing calves when offered 40% levels of pods in their concentrate ration. Shukla *et al.*, (1984) also found no adverse effect on the digestibility of various nutrients and also on nitrogen, calcium and phosphorus balance in Jersy × Kakrei F1 calves when offered *P. juliflora* pods at 0%, 10%, and 20% levels in their concentrate ration. *P. juliflora* pods provide a valuable feed for the livestock with an opportunity to the landless and small farmers for rearing animals.

Reclamation of Problematic Soils

The value of *P. juliflora* for reclaiming degraded saline soils has been widely acknowledged. The Planning Commission of Indias Action Plan for greening India identifies *P. juliflora* as one of the most promising agroforestry options for problematic and saline soils. When planted in saline land, it has been observed that the roots absorb the salt from the ambient soil solution and preserve it in the stem, and thus, the salt concentration in the soil is slowly reduced. The fallen leaves on decomposition also enhance the process of soil reclamation. By providing a vegetative cover, it also reduces soil erosion, and has proved beneficial in stabilizing coastal belts and dunes (Kanzaria & Varshney, 1998). It grows well in areas having high salinity or alkalinity levels, and can tolerate alkalinity as high as pH 9.5 (Singh *et al.*, 1993) or above 9.5 up to 10 and a soluble-salt content of up to 0.54-1.0 % (Yadav & Singh, 1970). Even in areas of high salinity, it can be planted by amending the soil with farmyard manure and gypsum. The litter falling on the ground adds to the humus content of salt affected soils. The organic acids from the decomposed litter react with the calcium carbonate in the soil and release calcium which substitutes for sodium in the exchangeable complex and thus helps in the reclamation of alkali soils (Singh, 1998). The plantations of *P. juluiflora* on sodic soils in Uttar Pradesh have reduced the pH levels by a fraction (0.3) to a significant level (2.0) in about 17 years besides improving organic carbon by 2.3-9.1 g/kg (Kareemulla *et a*l, 2005). There are many other reports showing a significant decrease in soil salinity (Table 5) and increase in organic carbon (Maliwal, 1999, Singh, 1998, Singh & Dagar, 2005). Bhojvaid and Timmer (1998) reported that *P. juliflora* growth altered microclimate, improved soil moisture status, organic C, total N, extractable P, exchangeable Na levels that contributed to the reclamation of sodic sites. Shah and Vora (1965) reported that *P. juliflora* plantation decreased the salinity and sodicity considerably in absence of shallow saline aquifer. Similar results were also observed by Agarwal *et al., (*1976) and Maliwal *et al.*, (1991). It is also observed that *P. juliflora* could be used to decontaminate heavy metal polluted soils in view of its ability to accumulate various heavy metals like Cupper, Nickel, Chromium, Aluminum, etc (Senthilkumar *et al.*, 2005).

Table 5: Effect of *P. juliflora* on alkali soil properties.

Soil Properties	pH	EC (dS m^{-1})	OC (%)	ESP	References
Initial	10.3	2.2	0.18	-	Singh, 1998
After 6 years	9.3	0.46	0.43	-	
Initial	10.3	-	0.12	-	Singh, 1998
After 20 years	8.03	-	0.58	-	
Fallow	-	0.99	0.72	17.7	Maliwal, 1999
After 7 years	-	0.38	0.82	6.7	
Initial	10.45	2.17	0.03	90	Singh & Dagar,2005
After 7 years	9.9	0.84	0.22	61.5	

Singh (1998) observed that *P, juliflora* can yield 5-8 t ha^{-1} of air dried leaf litter after 4-6 years, containing 2.2% nitrogen, 0.2-0.4% phosphorus, 1.5-1.9% potassium and sodium content generally less than 0.2%. The annual turnover of macronutrients to the soil through litter would be 88-132 kg h^{-1}, 8-16 kg Phosphorus per hectare and 60-76 kg Potassium per hectare. Such nutrient addition through litter fall will raise the fertility level of salt affected soils, which are otherwise deficient in essential plant nutrients. There are also some reports about the presence of growth regulators (Triacontanol) in the plant parts of this tree species, particularly in leaves (Khan *et al.*, 1987, Khan, 1998). The triacontanol was found to be effective in increasing the growth and water uptake of rice seedlings (Reis *et al.*, 1977), and also to promote germination, growth and stubble sprouting in corn and barley. The presence of such a plant growth promoting substance in the leaves of *P. juliflora* may increase the growth of plants in *P. juliflora* based agroforestry system, that are affected by leaf litter fall. *P. juliflora* is also reported to help in reclaiming salt affected soils more effectively than other trees, such as *Acacia nilotica, Eucalyptus tereticornis, Terminalia arjuna* and *Albizia lebbeck* (Singh, 1998, Singh & Dagar, 2005).

Prospect in Future Bioenergy

In terms of total energy consumption, presently India is the fifth-largest energy consumer in the world (B.P, 2005). On a per capita basis, however, this consumption is very small as compared to that of the world average. India faces enormous challenges as it aspires to reach rates of economic growth of 8 %-10 % per annum. Nearly 57 % of Indian population has no access to electricity as per the Census Survey of 2001. Forty three percent have no access to clean cooking fuel and continue to burn biomass. Continued huge electricity peak shortage of 12.3 % and energy shortage of 7.7 % and oil import dependence of 75 % (Bhojvaid, 2006) have forced to think on alternative renewable resources, such as wood based bioenergy program on sustainable basis to reduce the gap between demand and supply in the country.

P. juliflora wood is the potential single most important natural resource which could be utilized as fuelwood, charcoal and production of biomass based electricity from gasifiers. The area occupied by this tree is reported to be more than 5 million hectare in India. It produces 721.65 million tonnes of biomass annually. If we assume that 50% biomass (360.82 m t.) is consumed for the fuelwood and remaining 25% (180.41 million tonnes) each for the charcoal production and power generation, number of households which can obtain the fuelwood from *P. juliflora* can be calculated. The daily consumption of firewood by a medium sized household is about 7.5 kilogram (Chaturvedi & Singh, 1992). The 50% of existing biomass produced can support the 329 million families for their fuelwood need and nearly 132 families for fuelwood from 1 ha of area on sustainable basis. Each medium size family will consume 257 Kilo Watt hour energy on monthly basis. According to the Pasizenek *et al.*, (2001), approximately 3-6 kilogram of *P. juliflora* wood is required to produce 1 kilogram of charcoal. Therefore, the total charcoal of 7.5-15 million tonnes can be produced from the 180.41 m t biomass distributed in 1.25 million hectare area.

With a conversion factor of 1 Mega Watt hour = 3.6 Giga Joules and assuming a 35% heat conversion efficiency with an operating load of 50%, the total energy requirement for operating a 10 Mega Watt hour power plant is 205.71 Giga Joule. To run this power unit round the clock through out the year would require 1802019.6 GJ of energy. The energy stored in the 25 % biomass of 5 million hectare area under *P. juliflora* with four year rotation period can therefore operate 589 power plants of 10 Mega Watt hours on the sustainable basis of 4 years rotation, and about 212 hectare area is required for 1 Mega Watt hour power generation. It is also reported that the culturable wasteland in the country is about 175 million hectare. If only 25% of it (43.75 million hectare), is planted with *P. juliflora*, can produce 6315.75 million tonne more biomass, and if 11 million hectare (25 % area), will be planted for the power plants, can support additional 5150 power units of 10 Mega Watt hour through out the year on sustainable basis. Similarly 2884 million more families can get the firewood for daily use from 50% biomass (3157.87 million tonnes.), and about 65.5-131.25 million tonnes of charcoal can be produced by utilizing the remaining 25% biomass of total additional production from the wastelands.

The place of *P. juliflora* in rural development of India has formerly brought up by Reddy (1978) with his thoughts.

"The rebellious sands are subdued and the inhospitable soils are colonized.

The dreary scene of dry districts is changed to that of green belts.

The bleak tree-less landscape is painted with splash of brown, green and yellow.

The monotony is broken for the traveler and the sheep and goats munch and crunch on the proteinous pods.

The rural folk, whose lands were getting buried under drifting sands, are grateful to the Forester and Prosopis, and the poor folk who had no fuel to burn in their hearths now have Prosopis.

They collect the fuel in their leisure and sell in towns for a decent price".

References

Aggarwal, R.K., J.P, Gupta, S.K, Saxena and Mathana, K.D. 1976. Studies on soil physico-chemical and ecological changes under twelve year old 5 desert tree species of Western Rajasthan. *Indian Forester*, 102: 863-872.

B.P (Britis Petroleum). 2005. Statistical Review of World Energy. 54 th edn. London: BP. http://www.bp.com.

Bhat, B.P. and Tadoria, N.P. 1990. Fuelwood characteristics of some mountain trees and shrubs. *Biomas,* 21: 233-238.

Bhojvaid, P.P and Timmer, V.R. 1998. Soil dynamics in an age sequence of *Prosopis juliflora* planted for sodic soil restoration in India. *Forest Ecology & Management*, 106 (2-3): 181-193.

Bhojvaid , P.P. 2006. Biofuels: Towards a greener and secure energy future. TERI press. The Energy and Research Institute Darbari Seth Block, New Delhi, Pp. 273.

Bose, R.K. & S.K, Bandoyopadhyay. 1986. Economics of energy plantations in alkali soils of Indian semi-arid regions. *Biomass*,11: 51-60.

Census of India. 2001. New Delhi: Office of the Registrar General, Govt of India.

Chaturvedi, A.N. 1985. Biomass production on saline soils. A case study. Proceedings of the Bio-energy society. First Conventional Symposium, New Delhi, Pp 15-20.

Chaturvedi, O.P. and Singh, J.S. 1992. Accumulation, transfer and dissipation of energy in Chir Pine forests. *Biomass and Bioenergy*, 6 (6): 363-367.

Deans, J.D., J, Diagne.,J, Nizinski., D.K, Lindley., M, Seck., K, Ingleby. & R.C, Muro. 2003. Comperative growth, biomass production, nutrient use and soil amelioration by nitrogen-fixing tree species in semi-arid Senegal. *Forest Ecology & Management*, 176 (1-3): 253-264.

FAO. 2002. Forest products yearbook database. Role of wood energy in Asia. Food and Agricultural Organization, Rome, Italy.

Felker, P. 1984. Economic, environmental, and social advantages of intensively managed short rotation mesquite (*Prosopis spp.*) biomass energy farms. *Biomass*, 5: 65-77.

FSI. 1995. Fuelwood, Timber and fodder from Forests of India, Forest Survey of India, Ministry of Environment & Forest, Dehradun.

Garg, V.K. 1992. Litter production and its nutrient concentration in some fuelwood tress grown on sodic soil. *Biomass & Bioenergy*, 3 (5): 323-328.

Goel, V.L. 1987. Performance of some firewood species in nursery of alkali wastelands. *Indian Forester*, 113: 792-797.

Goel and Behl, H.M. 1995. Fuelwood production potential of six Prosopis species on an alkaline soil site. *Biomass & Bioenergy*, 8 (1): 17-20.

Guha, S.R.D. and Gupta, R.K. 1970. Production of writing and printing papers from *Prosopis juliflora. Indian Forester*, 96 (6): 429-432.

Gupta, R.K. 1993. Multipurpose Trees for Agroforestry and Wastelan Utilization. Oxford and IBH Publishing Co. Pvt. Ltd. Pp. 562.

Gupta and Balara, G.S. 1972. Comparative studies on the germination, growth and seedling biomass of two promising exotics in Rajasthan desert (*Prosopis juliflora* (Swartz) DC. and *Acacia tortilis* (Foresk.) Hayne ssp. tortilis). *Indian Foreste*, 98: 280-285.

Gurumurhti, K., Raturi, D.P. and Bhandari, H.C. 1984. Biomass production in energy plantations of *P. juliflora, Indian Forester*, 110: 879-894.

Harsh, L.N. and Tewari, J.C. 1998. Prosopis in the arid region of India: Some important aspects of research and development. Proceeding of a conference held at the Central Arid Zone Research Institute Jodhpur, Rajasthan India, Nov, 21-23 (1993). Publishe by the Prosopis society of India and the Henry Doubleday Research Association.

Hingorani, N. L. 1986. Power from biomass. In: Patel, V.J., The Role of Prosopis in Wasteland Development. Javraybhai Agroforestry Centre, Surendrabag, Gujrat, India, Pp. 1-8.

Jain, S.H., Hamant Kumar., Arya Rajana. and Srivastava, R.L. 2005. Protection and efficient utilization of plantation grown lesser-known timbers of arid region in India- *Acacia tortilis, Prosopis juliflora* and *Prosopis cinaria* . poster paper presented in the 36 th Annual meeting Banglore, India.

Joshi, P. and Prajapati, M.C. 1977. Yield characters of Prosopis juliflora on humps of chambal ravines, Kota. Annual Progress Report, CS and WCR and T Inst. Dehradun, Pp. 79-81.

Kanzaria, M.K. and A.K, Varshney. 1998. *Prosopis juliflora*- its uses. Proceeding of a conference held at the Central Arid Zone Research Institute Jodhpur, Rajasthan India, Nov, 21-23 (1993). Published by the Prosopis society *of* India and the Henry Doubleday Research Association (1998).

Kareemulla, K., O.P. Chaturvedi and Dwivedi, R.P. 2005. Bioenergy from *Prosopis juliflora* Economics and Environmental Evaluation. Annual Report.2005-2006 NRCAF. National Research Centre for Agroforestry, Jhansi, India, Pp 61-63.

Khan, H.A. 1998. A plant growth regulator from *Prosopis cinaria* and *Prosopis juliflora.* Proceeding of a conference held at the Central Arid Zone Research Institute Jodhpur, Rajasthan India, Nov, 21-23 (1993). Published by the Prosopis society of India and the Henry Doubleday Research Association (1998).

Khan, H.A ., V. Lodha., and Ghanim, A. 1987. Triacontanol from the leaves of *Prosopis cinaria* and *Prosopis juliflora.* Trans ISDT. 17: 29-32.

Maliwal, G.L. 1999. Reclamation of saline and sodic soils through *P. juliflora. Indian Journal of Forestry*, 22 (2): 132-135.

Maliwal, G.L., C.M.Ndiadara., S.H. Punjani. and C.L. Patel,. 1991. Changes in soil properties as influenced by *Prosopis juliflora* and crop covers in salt affected soils. *Annals of Arid Zone*, 30: 57-60.

Mathur and H.C. Bhora. 1998. Nutritive value and palatability of *P. juliflora* pods : A review. Proceeding of a conference held at the Central Arid Zone Research Institute Jodhpur, Rajasthan India, Nov, 21-23 (1993). Publishe by the Prosopis society of India and the Henry Doubleday Research Association (1998).

Muthana, K.D. and Arora, G.D. 1983. *P. juliflora* (Swartz). DC. A fast growing tree to bloom the desert. CAZRI Monograph No. 22. Central Arid Zone Research Institute. Jodhpur, India. Pp. 19.

National Sample Survey organization (NSSO). 2001. Energy used by Indian Households, NSS. 55 th Round, July 1999- June 2000. Report No. 464 (55/1.0/6). National Sample Survey organization (NSSO), Department of Statistics, Government of India, New Delhi.

NCAER. 1985. Domestic fuel survey with special reference to kerosene, Vols. 1 and 2. National Council of Applied Economic Research, New Delhi.

Pandey, D. 2003. Fuelwood studies in India: Myth and Reality. Published in G. Kohlin (ed). Fuelwood-Crises or Balance? Proceedings from Workshop at Goterborg university (2003).

Parker, H.W. 1982. Mesquite as biomass residue. Pp. C1-C11. In: Mesquite utilization. (Ed. H.W. Parker), Texas Tech university, Lubboc, Texas, USA.

Pasiecznik, N.M., Felker, P., Harris, P. J. C., Harsh, L.N., Cruz, G., Tewari, J.C., Cadoret, K. and Maldonado, L.J. 2001. The *P. juliflora-Prosopis palida* Complex: A Monograph. HDRA, Conventry, U. K., Pp. 162.

Patel, G.B. and Pathak, S.R. 1986. *Prosopis* experiment on coastal saline lands of Bhavnagar, In: Role of *Prosopis* in Wasteland Development (V.J Patel, ed.), P KCS1. Surendrabag, India: Jivrajbhai Patel Agroforesty Centre, Gujarat.

Patel, V.J. 1986. Economics of energy plantations on wasteland. In: Patel, V.J (ed). The Role of Prosopis in Waste land Development, Javrajbhai Patel Agroforestry Centre, Surenderabag, Gujrat, India.

Raindernath, N.H. and Hall, D.O. 1995. Biomass, Enegy and Environment, Oxford University Press-New Delhi.

Raju, G. 1986. Bio-mass conversion technology. In: The Role of Prosopis in Wasteland Development (Patel, V.J. Eds.), Javraybhai Agroforestry Centre, Surendrabag, Gujrat, India, Pp. 1-8.

Rao, K.S. and Vasanthi, K.S. 1986. Economics and promotion of biomass gasifier for water pumping. pp. KSRI-KSRI 13. In: Patel, V.J., The Role of *Prosopis* in Wasteland Development. Javraybhai Agroforestry Centre, Surendrabag, Gujrat, India.

Rao, N. S. R. and Reddy, M.P. 1983. Utilization of *Prosopis juliflora* pods in the concentrate feed of cattle and sheep. *Indian Journal of Animal Science*, 53: 365-372.

Rawat, M.S., Uniyal, D.P. and Vakshasya, R.K. 1992. *Prosopis juliflora* (Swartz) DC: Fuel, Fodder and Food in arid and semi-arid areas: Some observations and suggestions. *Indian Journal of Forestry*, 15: 164-168.

Reddy, C.V.K. 1978. *Prosopis juliflora*, the precious child of the plant world. *Indian Forester*. 104: 14-18.

Reis, S.K., Sweely, C.C. and Leavit, R.A. 1977. Triacontinol: a new naturlay occurign plant growth regulator. *Science*, 195: 1339-1340.

Saxena, S.K. 1998. Ecology of *P. juliflora* in the arid regions of India. Proceeding of a conference held at the Central Arid Zone Research Institute Jodhpur, Rajasthan India, Nov, 21-23 (1993). Publishe by the Prosopis society of India and the Henry Doubleday Research Association (1998).

Saxena, N.C. 1997. The Fuel Scenario and Policy Issues in India. Regional Wood Energy Department Programme in Asia, FAO Regional Office for Asia and the Pacific, Bangkok.

Senthilkumar, P.W., S.P.M. Prince., S. Sivakumar. and C.V. Subharaam. 2005. *Prosopis juliflora*- A green solution to decontaminate heavy metal (Cu and Cd) contaminated soils. *Chemosphere*, 60: 1493-1496.

Shah, R.K. and Vora, I.C. 1965. *Prosopis juliflora* DC. For improving saline soils, *Vidya*, 8: 210-214.

Shukla, P.C., Talpada, P.M. and Pande, M.B. 1984. Prosopis juliflora pods- a new cattle feed source. Technical Bulletin. 3: Pp. 84. Animal Nutrition Dept., G.A.U., Anand, India.

Singh, G. 1998. Practice for raising Prosopis plantation in saline soils. Proceeding of a conference held at the Central Arid Zone Research Institute Jodhpur, Rajasthan India, Nov, 21-23 (1993). Published by the Prosopis society of India and the Henry Doubleday Research Association (1998).

Singh, G.1996. Effect of site preparation techiniques on *P. juliflora* in an alkali soil. *Forest Ecology and Managemet,* 8: 267-278.

Singh, G. and Dagar, J.C. 2005. Greening sodic lands: Bichian Model. Tech. Bull No. 2/ 2005, Central Soil Salinity Research Institute Karnal-132001 (India), Pp 51.

Singh, G., I.P. Abrol and Cheema, S.S. 1989. Effect of gypsum application on Mesquite (*Prosopis juliflora*) and soil properties in an abandoned sodic soils. *Forest Ecology and Managemet,* 29: 1-4.

Singh, G and Rathod, T.R. 2002. Plant growth, biomass production and soil water dynamics in shifting dunes of Indian Desert. *Forest Ecology Management,* 171 (3): 309-320.

Singh, G., N.T. Singh.,and J.C. Dagar. 1993. Mesquite for development of salt affected soil. *Wasteland News,* VIII: 7174.

Sood, R. and S. Bhatia. 1998. Biomass production and portioning of *Prosopis juliflora* on a degraded ravenous site. Proceeding of a conference held at the Central Arid Zone Research Institute Jodhpur, Rajasthan India, Nov, 21-23 (1993). Published by the Prosopis society of India and the Henry Doubleday Research Association (1998).

Subbayan, R. 1980. Studies on management of *Prosopis uliflora* for fuel in Bellery. Annual Progress Report. pp. 86-88. CS and WCR and T Inst., Dehra Dun.

Sutaria, J.N., R. Andharia, V.J. Patel. 1986. Thermal power station gas on fire wood from *Prosopis chilensis* in four talukas of Bhavnagar District. pp. 1-8. In: Patel, V.J., The Role of Prosopis in Wasteland Development. Javraybhai Agroforestry Centre, Surendrabag, Gujrat, India.

Tewaria, J.C., Harris, P.J.C., Harsh, L.N., Cadoret, K. and Pasiecznik, N.M. 2000. Managing *Prosopis juliflora* (Vilayti baboul). A technical manual. Pp 96. CAZRI, Jodhpur, India and HDRA, Conventry, U.K.

Tomar, O.S., Minhas, P.S., Sharma, V.K.., Singh, Y.P.., and Gupta, K.R. 2003. Performance of 31 tree species and soil conditions in a plantation established with saline irrigation. *Forest Ecology and Management,* 177: 333-346.

Tyagi, P.D. and Singh, L. 1986. *Prosopis Juliflora* in bioenergy. In: The Role of Prosopis in Wasteland Development (Patel, V.J. Eds.), Javraybhai Agroforestry Centre, Surendrabag, Gujrat, India, Pp 15-20.

Vimal, O.P. and Tyagi, P.D. 1986. Fuelwood from wastelands. Yatan Publications, New Delhi.

Yadav, J.S.P. and Singh, K. 1970. Tolerance of certain forest species to varying degree of salinity and alkalinity. *Indian Forester,* 96: 587-599.

❏❏❏

Agroforestry: Natural Resource Sustainability, Livelihood & Climate Moderation © 2009
Editors: OP Chaturvedi, A Venkatesh et. al., pp. 165-183
Satish Serial Publishing House, Delhi

Chapter 13

ROLE OF VESICULAR ARBUSCULAR MYCORRHIZAL FUNGI IN ESTABLISHMENT OF AGROFORESTRY SYSTEMS ON MARGINAL LANDS

Anil Kumar and Ashok Shukla
National Research Centre for Agroforestry, Jhansi-284 003 (U.P.)

Introduction

Plant roots provide an ecological niche for many of the microorganisms that abound in soil. German Botanist A.B. Frank (1885) introduced the Greek word *mycorrhiza,* which literally means "fungus root". Vesicular arbuscular mycorrhizal fungi (VAMF) are an important group of soil-borne microorganisms that contribute substantially to the establishment, productivity, and longevity of natural or man-made ecosystems (Harley and Smith, 1983). These fungi form symbiotic association with most terrestrial plant families (Trappe, 1977). In natural ecosystems mycorrhizal fungi can colonize much of the root system. Colonization is restricted to the root cortex and does not enter the vascular cylinder. The symbiosis is so well balanced that, although many of the host cells are invaded by the fungal endophyte, there is no visible tissue damage and under certain conditions it enhances the growth and vigour of the host plant. Because most economically important plants form mycorrhizae, the subject is currently attracting much attention in agricultural, horticultural, and forestry research. There are different types of mycorrhizae. The main types of mycorrhizae and the host plants in which they are commonly associated are given in Table 1. As the present paper is concerned with VAM fungi, information available in literature on its benefits, occurrence, taxonomy, ecology, culturing techniques and their role in agroforestry are being reviewed and being presented here, in short.

Table 1: Different types of mycorrhizae and their hosts

Types of Mycorrhizae	Host plant/families
Ecto	Pinaceae, Fagaceae, Betulaceae, Leguminaceae, mostly Caesalpinoid legumes, Salicaceae, Tiliaceae, Rosaceae, Juglandaceae, etc.
Arbuscular	Majority of the plants including those important in agriculture, horticulture, pasture and tropical forests
Ericoid	Ericaceae and Epacridaceae
Orchidaceous	Orchidaceae
Arbutoid	Arbutus and Monotropha

Due to the extensive network of external hyphae, which function as plant rootlets and increase P uptake, VAM fungi are considered as beneficial microorganisms. Some of the benefits conferred by the VAM fungi are as follow:

Benefits of Mycorrhizae

Mycorrhiza plays a very important role on enhancing the plant growth and yield due to an increase supply of phosphorus to the host plant. Mycorrhizal plants can absorb and accumulate several times more phosphate from the soil or solution than non-mycorrhizal plants.

Mycorrhiza increase root surface area for water and nutrients uptake. The use of mycorrhizal bio-fertilizer helps to improve higher branching of plant roots, and the mycorrhizal hyphae grow from the root to soil enabling the plant roots to contact with wider area of soil surface, hence, increasing the absorbing area for water and nutrients absorption of the plant root system. Therefore, plants with mycorrhizal association will have higher efficiency for nutrients absorption, such as nitrogen, phosphorus, potassium, calcium, magnesium, zinc and copper and also increase plant resistance to drought. Benefits of mycorrhizal bio-fertilizer are as follows:

- Allow plants to take up nutrients in unavailable forms or nutrients that are fixed to the soil, especially phosphorus which are immobile in the soil. In addition, mycorrhiza help to absorb other organic substances that are not fully soluble for plants to use, and also help to absorb and dissolve other nutrients for plants.
- Enhance plant growth, improve crop yield and increase income for the farmers.
- Reduce the use of chemical fertilizer.
- Improve plant resistance to root rot and collar rot diseases. Mycorrhizal association in plant roots will help plant to resist root rot and collar rot diseases caused by other fungi.
- Enhance water transport in plants, decrease transplant injury and promote establishment of plants in wasteland. Mycorrhizal plants recover better following moisture stress, more efficient in water use and frequently have higher root-shoot ratio than non-mycorrhizal plants.
- Increase plant growth by 20-30% in terms of bio-mass.
- Enable the plants to grow and survive better under various stress conditions, like those in coal spoils, sand dunes, saline-alkaline soils, eroded- degraded sites and lands polluted by industrial wastes.

Occurrence

VAM fungi have the widest host range and distribution of all the mycorrhizal associations. It is estimated that about 90% of vascular plants normally establish mutualistic relationship with VAM fungi. VAM fungi have been observed in 1000 genera of plants representing some 200 families. There are at least 300,000 receptive hosts in the world flora (Kendrick and Berch, 1985) and there are about 120 species of VAM fungi (Schenck and Perez, 1987). If the hosts are divided up evenly among the fungi with no overlap in host range, each fungus would have more than 2500 potential partners. We know that host ranges overlap extensively, suggesting that some individual VAM fungi may well have access to thousands of hosts (Kendrick and Berch, 1985).

According to Gerdemann (1975), it is easier to list most plant families that do not form VAM than to list those that do. Families not forming vesicular arbuscular mycorrhizal fungi include Pinaceae, Betulaceae, Orchidaceae, Fumariaceae, Commelinaceae, Urticaceae, and Ericaceae (Table 2). Families that rarely form vesicular arbuscular mycorrhiza include the Brassicaceae, Chenopodiaceae, Polygonaceae, and Cyperaceae. Families that form both ectomycorrhizae and vesicular arbuscular mycorrhizal fungi include Juglandiaceae, Tilliaceae, Myrtaceae, Salicaceae, Fagaceae, and

Caesalpiniaceae (Gerdemann, 1975). Important crops with VAM fungi include wheat, maize, all millets, potatoes, beans, soybeans, tomatoes, apples, oranges, grapes, banana, castor, tobacco, tea, coffee, cocoa, sugarcane, mango, asparagus, rubber, cardamom, pepper, etc. Most of the tropical rain forest trees are arbuscular mycorrhizal (Janos, 1983). Harley (1969) has listed the gymnosperms in which VAM fungi have been observed. They are found in Pteridophytes (Cooper, 1976) and Bryophytes (Parke and Linderman, 1980). Recently, arbuscular mycorrhizal colonization has been reported in floating (Bagyaraj *et al*, 1979) and submerged aquatic plants (Clayton and Bagyaraj, 1984). Usually arbuscular mycorrhizae are confined to the roots, they have been reported in diverse structures such as the modified leaves of water fern *Salvinia cucullata* (Bagyaraj *et al*, 1979), fruiting peg of peanut (Graw and Rehm, 1977) and modified scale like leaves and rhizomes of ginger and canna (Selvaraj *et al*, 1986).

Table 2: Important plant families and tree genera known to form mycorrhizal associations

Family	Genus	ECM	VAM
Betulaceae	Alnus *(alder)*	√	√
	Betula *(birch)*	√	√
Caesalpinoideae	Cassia	√	
	Non N-fixing trees	√	√
Casuarinaceae	Casuarina	√	√
Dipterocarpaceae	Shorea	√	√
Ebenaceae	Diospyros		√
Fagaceae	Fagus *(beech)*	√	
	Quercus *(oak)*	√	
Juglandaceae	Carya *(hickory, pecan)*	√	
Meliaceae	Cedrela		√
	Khaya *(African mahogany)*		√
	Swietenia *(mahogany)*		√
Mimosoldeae	Acacia		√
	Albizia		√
Myrtaccae	Eucalyptus	√	√
Paplionoideae	Dalbergia *(rosewood)*		√
	Pterocarpus		
Rosaceae	Malus *(apple)*		√
Rutaceae	Citrus *(orange, lemon)*		√
Sallcaceae	Salix *(willow)*	√	√
Commercial crops	Caria *(papaya)*		√
	Coffee *(coffee)*		√
	Hevea *(rubber)*		√
	Most palms		√

VAM fungi, in addition to their widespread distribution throughout the plant kingdom, are also geographically ubiquitous and occur in plants growing in arctic, temperate, and tropical regions (Mosse *et al*, 1981). In general, VAM population is more in cultivated soil compared to virgin soil (Mosse and Bowen, 1968). They are mostly seen in the top 15-30 cm of soil, and their numbers decrease markedly below the top 15 cm (Redhead, 1977). They are normally not found in depths beyond the normal root range of plants (Mosse *et al*, 1981). The distribution of species of VAM fungi varies with climatic and edaphic environment as well as with land use. For example *Acaulospora laevis* is common in Western Australia (Abbott and Robson, 1977) and New Zealand (Mosse and Bowen, 1968) but occurs less frequently in soils of Eastern Australia (Mosse and Bowen, 1968). *Glomus* species appear to have widest distribution. *Gigaspora* and *Sclerocystis* species are more common in tropical soils. *Acaulospora* seems to be better adapted to soils with pH <5.0. In fact, certain VAM fungi have been linked to particular kinds of soil: *Glomus mosseae* with fine textured, fertile, high pH soils, *Acaulospora leavis* with coarse-textured, acid soils, and *Gigaspora* species with sand dune soils (Kendrick and Berch, 1985).

Classification of VA Mycorrhizal Fungi

VAM fungi, being obligate biotrophs, do not grow on synthetic media and hence are classified according to morphological characteristics of the spores formed in the soil. One hundred and twenty species have been described by Schenck and Perez (1987), and they are all grouped in a single family, the Endogonaceae. The family is considered to be in an order of its own, the Endogonales tentatively placed in the class Zygomycetes in the subdivision Zygomycotina (Benjamin, 1979). The seven genera, *Endogone, Glomus, Sclerocystis, Entrophospora, Acaulospora, Scutellospora and Gigaspora,* presently constitute the Endogonaceae. Of these, only *Endogone* forms zygospores characteristic of Zygomycetes, while others probably lack sexual reproduction, species of *Endogone* do not form VAM fungal association. The other six genera form vesicular arbuscular mycorrhizal associations.

The main features of VAM fungi are summarized by Trappe and Schenck (1982) and Hall (1984). Hall and Abbott (1983) compiled photographic transparencies illustrating diagnostic features of the Endogonaceae. Walker (1983) has suggested a standardized terminology for use in description of endogonaceous spore walls involving a stylized graphic representation of the wall structure ("murograph"). This concept is recommended for future descriptions of new species. Recently, Schenck and Perez (1987) published a manual for the identification of VAM fungi.

Glomus, the most common genus of VAM fungi, has over 50 species, which form globose, ellipsoid or rather irregularly shaped spores that range from 20-400 μm. These spores are thick-walled (up to 30 μm) and hyaline, yellow, red-brown, brown or black. Spores are attached to a single subtending hypha (or to two, three, or more in exceptional species) and produced in the soil near plant roots, at the soil surface, or occasionally in roots, and they may form singly, in groups of a few to many, or in large aggregates called sporocarps. Sporocarps are most commonly found in undisturbed forest communities with perennial plants and accumulated organic material. Sporocarpic species, when grown on plants in pot culture, usually produce spores singly or in small aggregates. Some *Glomus* species (e.g. *Glomus lacteum*) are not known to form spores in sporocarps, while others form spores outside the plant root only infrequently (e.g. *Glomus intraradices*). The spores of most species germinate by emergence of a germ tube directly through the subtending hypha.

Sclerocystis species produce chlamydospores that are very similar to those of *Glomus*, except that they tend to be clavate rather than globose. All *sclerocystis* species form spores only in sporocarps up to 700 μm in diameter. The spores, which can be up to 125 μm long, develop in a single layer, radially arranged around a central plexus of sterile and sporogenous hyphae. The sporocarps may group together in masses up to several cm in diameter on the soil surface, on leaves, twigs, and mosses. There is no published information on the germination of *Sclerocystis* spores.

Spores of the genus *Acaulospora* appear to form laterally on the neck of a small thin-walled saccule, called variously a mother spore, a vesicle, a hyphal terminus, a soporiferous saccule, or a sporogenous saccule (Walker, 1987). When mature, the spore does not display a subtending hypha because the vesicular structure that gives rise to it loses its cytoplasm and collapses during development of the spore. The spores are globose or ellipsoid, from <100 to >400 μm in diameter, and hyaline, yellow, or reddish-brown. The surface of the sporewall may be ornamented with pits, projections of various shapes, folds, spines, or reticulations, and the wall is up to 12 μm thick. One species of *Acaulospora* is now known to produce sporocarps, but in other species, spores are normally found in the soil of the rhizosphere, *Entrophospora* is more or less similar to *Acaulospora* except that the spore is formed inside the parent hypha just below the vesicle.

Scutellospora spores can be greater than 600 μm in diameter, and the smallest ones are just under 200 μm in diameter. Spores develop terminally on a bulbous hyphal suspensor, which remains attached at maturity and may bear short lateral projections. Spores are usually globose but can be ovoid, obovoid, pyriform or irregular. Spores are hyaline, white, yellow, pinkish, grey-green, or brown. Spore wall structure is composed of at least two wall groups, with one or more flexible membranous or coriaceous walls in the inner group or groups. The genus is characterized in particular by its mode of spore germination. Germination is by means of one or more germ tubes produced near the spore base from a germination shield formed within or upon a flexible inner wall. Thin walled, knobby, or broadly papillate auxiliary cells (vesicles) are borne in soil on straight or coiled hyphae, formed singly or in clusters. *Gigaspora* produces spores which are superficially similar to those of *Scutellospora* spores. The striking difference is the germination of spores, which is by direct growth of one or more germ tubes through the spore wall. It is possible that this genus evolved from *Scutellospora* (Walker, 1987).

Another fungus, or group of fungi, that forms endomycorrhizae has been referred to in the literature as "the fine endophyte," since its hyphae are <4 um wide, compared to 10-20 μm for typical VAM fungi. This anomalous endophyte has been described as *Glomus tenue*, though it does not resemble the other species of *Glomus*. Its spores are 10-15 μm in diameter, and when young are hyaline and indistinguishable from the vesicles that form in the roots, when mature, however, they are pigmented. Because of their small size as compared to those of other VAM fungi, the spores of *Glomus tenue* are not often recorded in nature, though mycorrhizae of this type are common, particularly in dry regions.

Morphology and Histology

VA-mycorrhizal roots are surrounded by an extensive loose hyphal network that may extend into the soil for as much as 1 cm. hyphae also grow along root surface tightly appressed to the epidermis. However, they are never abundant enough to form a fungal mantle comparable to that of ectotrophic mycorrhizae. Vesicles and large thick walled spores are borne on hyphae in the soil. Hyphae produce appressoria on the root surface prior to penetration. Mycorrhizal infection produces either very little or no modification in the external morphology of roots.

Histology of mycorrhizal roots can easily be observed by examining whole mounts under relatively low power of a compound microscope. The hyphae penetrate the epidermal cells of young roots behind the meristematic region. The fungus grows throughout the cortex, but it does not invade the endodermis, stele or root meristem. Cells containing chloroplast are not infected. Shortly after infection, the fungus forms arbuscules within cortical cells. Vesicles are inter or intracellular, depending on the host species. Some species of the endophyte produce vesicles that remain relatively thin walled. In others the vesicles become thick walled, and do not differ in any important respect from chalmydospores produced in the soil. Vesicles function as food-storage organs, or if they develop into thick-walled chlamydospores, they may function as reproductive structures.

Production of VAM Inoculum

Identification of efficient indigenous strains of VAM fungi along with suitable host and substrate is a pre-requisite for mass production of VAM inocula. Plant growth period of 2-3 months gives a large crop of mycorrhizal spores to produce sizable amount of inoculum. VAM inoculum can also be produced under sterile environment by using hydroponic cultures and transferred organ culture techniques. Although, these techniques have not reached commercialization due to certain limitations.

Inoculation Methods

For successful mycorrhizal infection, the inoculum is placed in the root zone of actively growing plants, which are not heavily fertilized. The quantity of infection by the introduced endophytes depends on placement, inoculum potential, inoculum density, type of inoculum and environmental factors. Various methods proposed for placement of the inoculum are as follows:

Layering or Pads

Inoculum is placed in layers or pads beneath seeds such that the roots penetrating the inoculum will get infected. Presently, it is one of the most effective methods of inoculating plants with VAM fungi. Layering on small and large scale can be done by using manual planters and tractor drawn seeders, respectively.

Banding or Side Dressing

In this method, VAM inoculum is placed to sides of seedling or seeds in the zone of root proliferation. Banding mycorrhizal inoculum 5 cm deep and 5 cm away from 2 month old citrus seedling was found to be as effective as placing the inoculum directly on the roots at transplanting and more effective than seed pelleting.

Mixing Soil with Inoculum

Mixing mycorrhizal inoculum with soil is the most natural method for inoculating plants. However, it requires large amount of inoculum to obtain rapid infection. Generally, this method is used for inoculating plants in nursery.

Seed Pelleting

In this method, inoculum is extracted from a pot culture by wet sieving and decantation, root retained on 1 mm sieve are fragmented and added to the inoculum. Five ml aqueous solutions (1% w/v) of methyl cellulose and calcium carbonate are mixed with 35 ml of mycorrhizal sievings. This material is poured over seeds (amount depending upon the crop species), mixed thoroughly and seeds are dried.

Pre-inoculation of Transplanted Seedlings

In this method, inoculum is placed in the seedbeds and seedlings are transplanted when these get adequately infected. The method is useful with crops like onion and tomato.

Ecology

Factors Affecting VAM Symbiosis in Field

Initial studies have shown that temperature, moisture, pH, amendments and some management practices like fallowing, cropping pattern, fertilizers and pesticides may all affect the conduciveness of soil to mycorrhizal symbiosis. In general, long fallowing, excessive secondary tillage, high inputs of fertilizers and pesticides especially broad-spectrum fumigants, substituted aromatic hydrocarbons and benzimidazoles leads to a decline in the inoculum and VAM colonization. Further, maintaining a plant cover on soil help in building up the inoculum of VAM in soil. Since, crop species differ in association with VAM fungi, selection of appropriate crop/ cultures in the cropping sequences can be expected to help raise the population of VAM in soil.

Biological Interactions with VAM Fungi

In 1904, Hiltner coined the term *rhizosphere* to denote the region of the soil subjected to the influence of plant roots and characterized by intense microbial activity. Many workers have subsequently shown that the microflora of the rhizosphere differs both quantitatively and qualitatively from that in the soil beyond the influence of the root (Parkinson, 1967, Bowen, 1976, Bagyaraj *et al*, 1972). The rhizosphere effect was greatest with bacteria followed by actinomycetes and fungi. The increased microbial activity in the rhizosphere has been attributed to the extra nutrients available in the region. The extra nutrient supply comes from the root exudates, sloughed-off parts of the root cap, mucigel, root hair residues and abraded epidermal cells (Rovira, 1979 and Tinker, 1980). The microflora of the rhizosphere could influence plant growth in many ways as all plant nutrients pass through this region (Rovira, 1967 and Nye, 1977).

Of the various microorganisms colonizing the rhizosphere, VAM fungi occupy a unique ecological position as they are partly inside and partly outside the host. The internal phase of a mycorrhizal fungus does not encounter competition and antagonism from other soil microorganisms and has an ensured source of nutrients from the host. This advantage enables them to achieve a large and more functional biomass in more intimate contact with the root and thus increases their chances of exerting a greater effect on plants than other microbial species restricted only to the rhizosphere.

Certain generalizations and conclusions about the interactions of VAM fungi with other soil organisms can be drawn. VAM markedly improve nodulation and nitrogen fixation by legume bacteria mainly by providing high phosphorus required for the fixation process. Mycorrhizal plants also allow introduced populations of beneficial soil organisms like *Azotobacter* and phosphate solubilising bacteria to maintain higher numbers than around non-mycorrhizal plants and to exert synergistic effects on plant growth. In general, VAM fungi decrease the severity of root diseases. Summary of some important interactions between VAM fungi and beneficial soil organisms are presented in Table 3.

Table 3: Summary of important interactions between VAM fungi and beneficial soil micro-organisms.

Groups	Species	Interactions
N-fixation bacteria	Rhizobium *sp.*	*Increased VAM infection level, nodulation, N-fixation and plant growth*
Free living N-fixation bacteria	Azotobacter *sp.* Beijerinkia *sp.* Clostridium *sp.* Azospirillum *sp.*	*Increased VAM infection level, AM spore production and bacterial population*
Phosphate solubilizing bacteria	Pseudomonas *sp.* Agrobacterium *sp.*	*Good bacterial population in rhizosphere of mycorrhizal plants*
Phosphate solubilizing fungi	Aspergillus niger Penicillum funiculosum	*Availability of P increased by P-solubilizing fungi & uptake increased by VAM*

VAM Fungi and Agroforestry Systems

The role of VAM fungi in enhancing plant growth and yield, resistance to drought and salinity and tolerance to pathogens is well documented, but little is known about their interactions with agroforestry systems. Studies are being carried out in this regard at National Research Centre for Agroforestry, Jhansi. The results obtained so far showed the presence of VAM fungi in roots of 19 forestry trees, 7 horticultural trees, 10 field crops and 8 fodder crops. The colonization index varied from 20.4 to 95.9% (Kumar, 1992). Local soils were found to be rich in VAM spore populations. *Glomus* was predominant genus, apart from this *Acaulospora*, *Gigaspora* and *Sclerocystis* were also present (Kumar *et al*, 2006, Jha *et al*, 2008 and Kamalvanshi *et al*, 2008). Cultures of 11 *Glomus* and 2 *Acaulospora* species were purified from tree rhizosphere. Inoculation of tree seedlings (*Acacia nilotica, Pongamia pinnata, Jatropha curcas, Eucalyptus tererticornis, Albizia procera, Dalbergia sissoo, Leucaena leucocephala, Emblica officinalis, Zizyphus mauritiana, Buchanania lanzan* and *Cordia dichotoma*) and intercrops (*Phaseolus mungo, P. vigna, Zea mays, Triticum aestivum, Cicer arietinum* and *Pisum sativum*) with suitable VAM species recorded substantial increase in plant growth and phosphorus (P) uptake (Kumar *et al*, 2007). This resulted in a benefit comparable to high input levels of P in *P. mungo, T. aestivum, E. tererticornis* and *A. procera* at much lower P concentrations (Shukla *et al*, 2008).

In agroforestry systems, trees acted as VAM inoculum reservoir for intercrops (Kumar *et al*, 2007). Cross inoculation test performed by using VAM fungi isolated from trees showed that these could colonize the roots of important rainy season (*P. mungo, P. vigna* and *Z. mays*) and winter (*C. arietinum, P. sativum* and *T. aestivum*) intercrops (Kumar *et al*, 2007). VAM fungi also cross infected the plant species under agroforestry situations. Intercropping increased VAM activity in rhizosphere of trees (Kumar *et al*, 2007). Tree shade reduced the rate of mycorrhization of intercrops, especially during winter months. VAM inoculations enhanced the growth and phosphorus uptake of intercrops in shade and the tree canopy management increased this advantage. Use of excessive shading (25% of full sun or more) in nurseries reduced mycorrhization of tree seedlings (Shukla *et al*, 2008). VAM activity in terms of spore population and mycorrhizal diversity was more in light soils as compared to heavy soils. Pruning/ canopy management of trees reduced their VAM colonization. Excessive stagnation of water during rainy season in plantations (*E. officinalis* and *Hardwikia binata*) was found to be harmful for colonization of plants. *Glomus* was more adapted to such conditions in *E. officinalis* plantations. Inoculation of intercrops (*P. mungo* and *T. aestivum*) and tree seedlings (*E. tererticornis* and *A. procera)* with a suitable VAM consortium was found more beneficial than individual VAM fungi in *P. mungo, T. aestivum, E. tererticornis* and *A. procera.*

Conclusions

Mycorrhizal symbiosis has lately created a keen interest among biological scientists of the country involving mycologists, soil scientists, agronomist, plant physiologists, plant geneticists, environmentalist, foresters and microbiologists alike. Major efforts are now being concentrated towards the production of bulk inoculum of mycorrhizal bio-fertilizers for use in forestry, horticulture, floriculture, agriculture, etc. While many ectomycorrhizal fungi have been brought in pure cultures, VAM fungi till now, defied all attempts to bring them in pure culture on synthetic media. Many efforts are being made all over the world to decode the genetic barrier to grow these fungi on synthetic media. In India, though fundamental studies on molecular and biochemical level are scarce, much headway has been achieved in selecting suitable VAM fungi for different crops and for different site conditions (Table 4).

Table 4: List of efficient vesicular arbuscular mycorrhizal fungi for different plant species*

Plant Species	Efficient mycorrhizal fungus/fungi	Mycorrhizal fungi tested	Parameters evaluated
Agriculture crops			
Allium cepa (Onion)	*Gigaspora margarita*	*G. margarita, G. calospora*, and a known VAM fungus	Leaf number, dry matter, P-uptake, bulb yield
	G. calospora	*G. margarita, G. calospora*, and a known VAM fungus	Root colonization
Amorphophallus paeonufolius	*Glomus mosseae* + *G. aggregatum*	*G. mosseae, G. aggregatum, Gigaspora albida, Pisolithus tinctorius*	Tuber yield/ plant
Ananas comosus	*Glomus fasciculatum*	*G. fasciculatum, Glomus* sp., *Acaulospora* sp., *Scutellospora* sp.	Plant growth, nutrition
Acacia nilotica	*Glomous mosseae* + *Gigaspora gilmorei*	*G. mosseae, G. constrictum, Gigaspora gilmorei*, and their combinations	Collar diameter, root -shoot dry weight and length

Contd...

Table 4. Contd...

Acacia auriculaeformis	*Glomus etunicatum, G. macrocarpum*	*G. etunicatum, G. macrocarpum, G. fasciculatum, G. mosseae*	Height, diameter, biomass, P-uptake
Acer saccharum (Sugar maple)	*Glomus macrocarpum*	*G. macrocarpum, G. clarum, G. epigaeum, G. mosseae*	Leaf number, mean lateral root number
Azadirachta indica (neem)	*Glomus mosseae*	*Glomus mosseae, G. fasciculatum,* and seven other VAM fungi	Plant height, stem girth, biomass, P-content, zinc concentration, root colonizations
Cajanus cajan	*Glomus clarum*	*G. fasciculatum* and seven other VAM fungi	Plant growth
	Glomus etunicatum	*G. etunicatum, Gigaspora margarita*	Shoot dry matter, nutrient uptake
Capsicum	*Glomus intraradices*	*G. intraradices,* indigenous mixed culture, commercial inoculum (Mycorise & C)	Fruit yield
Carica papaya	*Glomus fasciculatum*	*G. fasciculatum, Glomus* sp., *Acaulospora* sp., *Scutellospora* sp.	Plant growth, nutrition
Cacocasia esculenta	*Glomus mosseae+ G. aggregatum*	G. mosseae, G. aggregatum, Gigaspora albida, Pisolithus tinctorius	Tuber yield / plant
Cucumis sativus	*Glomus caledonium*	*G. caledonium, Glomus* sp.	P-uptake
Eleusine coracana	*Glomus caledonicum*	*G. caledonicum, G. mosseae, G. fasciculatum, G. epigaeum (G. versiforme), Gigaspora calospora, G. margarita*	Mycorrhizal efficacy, root colonization
Fragaria xananassa (micropropagated)	*Glomus versiforme* 36-366	*G. versiforme* (2 isolates), *G. mosseae, G. etunicatum*	Yield, production of runners
Glycine max	*Acaulsopora scrobiculata*	*A. scrobiculata, Glomus intraradices,* and four other AM	Plant growth, crop yield

Contd...

Table 4. Contd...

Lycopersicum esculentum	*Glomus etunicatum*	*Glomus etunicatum, G. mosseae*	Shoot dry weight, plant height
cv. Manapal	*G. mosseae*	*Glomus etunicatum, G. mosseae*	Root colonization
cv. Walter	*G. mosseae*	*Glomus etunicatum, G. mosseae*	Shoot weight, plant height
Manihot esculenta (cassava)	*G. fasciculatum*	*G. fasciculatum*, G. *mosseae*, G. *constrictum, Glomus etunicatum Acaulsopora morrowea*	Root colonization, plant weight, shoot and root dry weight
Manihot esculenta (cassava)	*G. mosseae + G.aggregatum*	*G. mosseae, G.aggregatum, Gigaspora albida, Pisolithus tinctorius*	Tuber yield/ plant
Medicago sativa	*Glomus intraradices*	*G. intraradices, G. mosseae, Gigaspora margarita*	Root colonization, fungal sporulation
Musa acuminata	*Glomus fasciculatum*	*G. fasciculatum, Glomus* sp., *Acaulospora* sp., *Scutellospora* sp.	Plant growth, nutrition
Oryza sativa cv. Prakash	*Glomus intraradices*	*G. intraradices, G. fasciculatum*	Grain yield
Orzya sativa (upland rice)	*Acaulospora spinosa*	*Acaulospora spinosa, A. scrobiculata*	Plant biomass, grain yield, root colonization
Persea Americana	*Glomus fasciculatum*	*G. fasciculatum, Glomus* sp., *Acaulospora* sp., *Scutellospora* sp.	Plant growth, nutrition
Polianthes tuberosa (rajnigandha)	*Glomus intraradices*	*G. intraradices,* indigenous mixed cultures, commercial inoculum (Mycorise)	Fruit yield, spike length
Trifolium (Clover)	*Acaulospora trappei*	*A. trappei, A. laevis, Acaulospora* sp.,	Root colonization
Triticum aestivum wheat (var. Swift)	*Glomus intraradices*	*G. intraradices, Gigaspora margarita*	Plant yield, number of grains/spike

Contd...

Table 4. Contd...

Vigna mungo (black gram)	*Glomus epigaeum* (*G. versiforme*)	*G. epigaeum, Acaulospora spinosa, A. morrowae*	Root colonization, shoot dry weight, N and P concentrations
Vigna radiata (green gram)	*Glomus epigaeum* (*G. versiforme*)	*G. epigaeum, Acaulospora spinosa, A. morrowae*	Root colonization, shoot dry weight, N and P concentrations
Vigna unguiculata	*G. etunicatum*	*G. etunicatum, Gigaspora margarita*	Shoot dry matter, nutrient uptake
Vigna unguiculata cv. Katumanu K80	*G. etunicatum*	*Glomus etunicatum, G. claroideum, Glomus* sp.	Top dry weight
Vigna unguiculata *cv.* California No. 5 black eye	*G. claroideum*	*Glomus etunicatum, G. claroideum, Glomus* sp.	Top dry weight
Zea mays (pioneer 3905)	*Glomus etunicatum*	*G. etunicatum, G. mosseae, G. aggregatum, G. versiforme*	Leaf mass, protein concentrations
Nephrolepis exaltata var. whitmanii (fern) micropropagated plants	*Glomus intraradices, G. clarum*	*G. intraradices, G. clarum, G. versiculiferum, G. versiforme*	Root infection
Tree Crops			
Malus domestica (apple)	*Glomus mosseae*	*G. mosseae, G. macrocarpum*	Root colonization, plant height, plant growth
Hevea brasiliensis	*Glomus clarum, Glomus* sp.	*G. clarum, Glomus* sp., *A. spinosa, Gigaspora margarita, Entrophospora colombiana, Glomus* sp., and three other species	Dry weight
Tectona grandis	*Glomus leptotichus*	Not mentioned	Seedling growth parameters
Zizyphus mauritiana	*Glomus fasciculatum*	*G. fasciculatum, G. mosseae, Scutellospora calospora*	Growth, root colonization, nutrient uptake
Citrus jambhiri (rough lemon)	*Glomus fasciculatum*	*G. fasciculatum, G. mosseae, G. etunicatum*	Plant height, top weight, root weight, chlamydospores in

* Cited from Singh, 2001.

Special attempts are being made to determine the role of VAM fungi in ameliorating soils polluted with industrial wastes. Subjects on which more information is needed are as follows:

- Distribution of mycorrhizal fungi under varying edapho-climatic conditions.
- Plant growth responses and mycorrhizal dependency.
- Role of mycorrhiza on difficult/ disturbed sites.
- Tripartite biological interactions involving rhizosphere flora, mycorrhizal fungi and plant roots.
- Physiology and biochemistry.
- Management of pests, mycorrhizae and biocides for increased plant productivity.
- Soil/ water pollution and mycorrhizae.

References

Abbott, L.K. and Robson, A.D. 1977. Growth stimulation of subterranean clover with vesicular arbuscular mycorrhizas. *Australian Journal of Agricultural Research,* 28: 639-649.

Bagyaraj, D.J. and Rangaswami, G. 1972. Studies on the rhizosphere microflora of *Eleusine coracana* under the influence of foliar chemicals sprays. I. Quantitative incidence of microorganisms. *Madras Agriculture Journal,* 59: 517.

Bagyaraj, D.J., Manjunath, A. and Patil, R.B. 1979. Occurrence of vesicular arbuscular mycorrhizas in some tropical aquatic plants. *Transactions British Mycological Society,* 72:164-167.

Benjamin, R.K. 1979. Zygomycetes and their spores. in B Kendricks (ed) The Whole Fungus. National Museums of Canada, Ottawa. pp. 573-616Bowen, G.D. and Rovira, A.D. 1976. Microbial colonization of plant roots. *Annual Review Phytopathology,* 14: 121-144.

Clayton, J.S. and Bagyaraj, D.J. 1984. Vesicular arbuscular mycorrhizas in submerged aquatic plants of New Zealand. *Aquatic Botany,* 19: 251-262.

Cooper, K.M. 1976. A field survey of mycorrhizas in New Zealand ferns. *New Zealand Journal Botany,* 14:169-181.

Frank, A.B. 1985. Uber die auf Werzelsymbiose beruhende Ernahrung gewisser Baume durch unterirdische. Pilze. *Ber. Dtsch. Bot. Ges,.* 3: 128-145

Gerdemann, JW. 1975. Vesciular arbuscular mycorrhizae. in JG Torrey and Glarkson DT (eds) *The Development and Function of Roots.* Academic Press, London, Pp. 576-591.

Graw, D. and Rehm, S. 1977. Vesicular-arbuscular Mycorrhiza in den Fruhttragern von *Arachis hypogea. L.Z. Acker Und Pflanzenbau,* 144: 75-78.

Hall, I.R. and Abbott, L.K. 1983. Photographic slide collection illustrating features of Endogonaceae. *Transactions British Mycological Society,* 17: 54-55.

Hall, I.R. 1984. Taxonomy of VA mycorrhizal fungi. in CL Powell and Bagyaraj DJ (eds) VA mycorrhiza. CRC Press, Boca Raton, FL, Pp. 57-94.

Harley, J.L. 1969. The Biology of Mycorrhiza, 2nd ed. Leonard Hill, London

Harley, J.L. and Smith, S.E. 1983. Mycorrhizal Symbiosis, 1st edn. Academic, London

Hiltner, L. 1904. Uber neure Erfahrungen und Probleme auf dem Gebiet der Bodenbakteriologie und unter besonderer Beruksichtigung der Grundingung und Brache. *Arb Dtsch Landwirtsch Gesellschaftswiss,* 98: 59-78.

Janos, DP. 1983. Tropical Mycorrhizas. In SL Sutton, TC Whitemore and AC Chadwick (eds) *Tropical Rain Forest: Ecology and Management.* Blackwell Scientific Publications, Oxford, Pp. 327-345.

Jha, A, Kumar, A., Saxena, R. K., Kamalvanshi, M. and Shukla, A. 2008. Occurrence of Arbuscular Mycorrhizal Fungi in Selected Agroforestry Tree Species of Bundelkhand Region. National Symposium on Agroforestry Knowledge for Sustainability, Climate Moderation and Challenges Ahead. Dec.15-17. NRCAF, Jhansi. 60.

Kamalvanshi, M., Kumar, A., Jha, A. and Shukla, A. 2008. Occurrence of arbuscular mycorrhizal (AM) fungal species in *Pongamia pinnata, Jatropha curcas* and *J. gossypifolia.* National Symposium on Agroforestry Knowledge for Sustainability, Climate Moderation and Challenges Ahead. Dec.15-17. NRCAF, Jhansi, Pp. 167.

Kendrick, B. and Berch, S. 1985. Mycorrhizae: Applications in Agriculture and Forestry. In : CW Robinson (ed), Comprehensive Biotechnology, Vol. 4. Pergamon Press, Oxford, Pp.109-150.

Kumar, A. 1992. Occurrence of VAM among some plant important in agroforestry under semiarid conditions. *Agroforestry Newsletter,* 4(1): 7.

Kumar, A., Hashmi, S., Shukla, A. and Jha, A. 2006. Colonization of some important intercrops by Vesicular Arbuscular Mycorrhizae isolated from minor fruit crops. National Symposium on Agroforestry for Livelihood Security, Environment Protection and Bio fuel Production. Dec.16 -18. NRCAF, Jhansi, Pp. 20.

Kumar, A., Hashmi, S. and Shukla, A. 2006. Occurrence of vesicular arbuscular mycorrhizae in *Buchanania lanzan* and *Cordia dichotoma. Range Management & Agroforestry,* 27 (1): 84-86.

Kumar, A., Shukla, A. Hashmi, S., Jha, A. and Kamalvanshi, M. 2007. Studies on arbuscular mycorrhizal fungi in agroforestry. *Agroforestry Newsletter* January- March, 2008, 20 (1) Pp. 5.

Kumar, A., Hashmi, S., Shukla, A. and Chaturvedi, O.P. 2007. Effect of intercropping on inoculum potential of arbuscular mycorrhizal fungi in agroforestry systems. *Annals of Arid Zone,* 46 (1): 31-36.

Kumar, A., Shukla, A., Hashmi, S. and Tewari, R.K. 2007. Effect of trees on colonization of intercrops by vesicular arbuscular mycorrhizae in agroforestry systems. *Indian Journal of Agricultural Sciences,* 77 (5): 291-298.

Mosse, B. and Bowen, G.D. 1968. The distribution of Endogone spores in some Australian and New Zeland soils and in the experimental field soil at Rothamsted. *Transactions British Mycological Society,* 51: 485-492.

Mosse, B., Stribley, D.P. and Le Tacon, F. 1981. Ecology of mycorrhizae and mycorrhizal fungi In M Alexander M (ed) Advances in Microbial Ecology. Plenum Press, New York, Pp. 5: 137-209.

Nye, P.H. 1977. The rate limiting step in plant nutrient absorption from soil. *Soil Science,* 123: 292.

Parke, J.L. and Linderman, R.G. 1980. Association of vesicular-arbuscular mycorrhizal fungi with the moss *Funaria hygrometrica. Canadian Journal of Botany,* 58: 1898-1904.

Parkinson, D. 1967. Soil microorganisms and plant roots in A Urges, F Raw (eds) *Soil Biology*. Academic Press, London, Pp. 449.

Redhead, J.F. 1977. Endotrophic mycorrhizas in Nigeria. Species of the Endogonaceae and Ttheir distribution. *Transactions British Mycological Society,* 69: 275-280.

Rovira, A.D. and McDougall, B.M. 1967. Microbiological and biochemical aspects of the rhizosphere in: AD McLaren, GH Peterson (eds) *Soil Biochemistry*. Marcel Dekker, New York, Pp. 417-493.

Rovira, A.D. 1979. Biology of soil-root interface in JL Warley, SR Russel (eds) *The Soil-Root Interface*. Academic Press, London, 145-160.

Schenck, NC and Perez, Y. 1987. *Manual of the identification of VA Mycorrhizal Fungi.* University of Florida, Gainesville, Florida, USA, Pp. 1-245.

Selvaraj, T., Kannan, K. and Lakshminarasimhan, C. 1986. Vesicular-arbuscular mycorrhizal fungi in roots and scale-like leaves of *Canna indica* L. (Cannaceae). *Current Science,* 55: 728-730

Singh, S. 2001. Mycorrhizal dependency, Part 1: Selection of efficient mycorrhizal fungi. *Mycorrhiza News,* 13 (1): 2-16.

Shukla, A., Kumar, A., Jha, A., Chaturvedi, O.P., Prasad, R. and Ajit 2008. Effect of shade on arbuscular mycorrhizal colonization and growth of crops and tree seedlings in Central India. *Agroforestry Systems* (in press).

Shukla, A., Kumar, A., Jha, A., Kamalvanshi, M. and Gupta, G. 2008. The effects of phosphorus application on growth and arbuscular mycorrhizal colonization of crops and tree seedlings. National Symposium on Agroforestry Knowledge for

Sustainability, Climate Moderation and Challenges Ahead. Dec.15-17. NRCAF, Jhansi, Pp. 41.

Tinker, P.B. 1980. Role of rhizosphere microorganism in phosphorus uptake by plants in FE Khasawneh, EC Sample and EJ Kamprath (eds) *The Role of Phosphorus in Agriculture*. American Society of Agronomy, Madison, Wisc. Pp. 617-647.

Trappe, J.M. 1977. "Biogeography of hypogeous fungi: Trees, mammals and continental drift" In HE Bigelow and EG Simmons (eds) *Abstr. 2nd Int. Mycol. Congr. Inc.* University of South Florida, Tampa, Pp. 675.

Trappe, J.M. and Schenck, N.C. 1982. Taxonomy of the fungi forming endomycorrhizae. A. Vesicular-arbuscular mycorrhizal fungi (Endogonales) in NC Schenck (ed) Methods and Principles of Mycorrhizal Research, *American Phytopathological Society Publication*, St. Paul, MN, Pp.1-10.

Walker, C. 1983. Taxonomic concepts in the Endogonaceae: Sproe wall characteristics in species description. *Mycotaxon*, 18: 443-455.

❑❑❑

Agroforestry: Natural Resource Sustainability, Livelihood & Climate Moderation © 2009
Editors: OP Chaturvedi, A Venkatesh et. al., pp. 185-200
Satish Serial Publishing House, Delhi

Chapter 14

NUTRIENT RECYCLING AND SOIL FERTILITY MANAGEMENT IN AGROFORESTRY SYSTEM

Munna Ram, R.S. Yadav and Sunil Kumar
National Research Centre for Agroforestry, Jhansi-284 003 (U.P.)

Introduction

The inevitable consequence of farming is to diminish the natural fertility of soils whether natural or acquired. The provision of an adequate supply of nutrients to crops has always been a leading objective of soil management. The former practice of supplying nutrient deficiencies simply by adding fertilizers has given place to the new approach of integrated fertility management. This is directed at using low to moderate quantities of fertilizer inputs to maximum efficiency, in combination with biological methods of soil management. It is in making more efficient use of nutrients, both from natural sources and from fertilizer, that one of the greatest potential contributions of agroforestry to sustainable land use lies. This potential rests on the basis that natural forest ecosystems achieve a high degree of recycling, through the repeated taking up of nutrients from the soil by plants and their return by decomposition of litter and that it should be possible to make use of this potential in managed systems of agroforestry.The meaning of efficient use can be scanned narrowly or broadly. Fertilizer use efficiency is the proportion of nutrients in added fertilizer which find their way into a crop, and fertilizer response is the crop-yield increment per unit of fertilizer. More generally, efficient use requires two conditions. i) the degree of recycling depends on individual nutrient ions should pass through plants many times before they are lost to the plant-soil ecosystem and ii) There should be relatively high plant growth responses per unit of nutrients taken up by plants leading to nutrient balance, soil physical conditions and adequacy of the water supply, together with the biological efficiency of the plant cultivar. The problems now being encountered in soil nutrient management include negative nutrient balances in low input systems (nutrient mining), reduced responses to fertilizers in medium to high input systems, the cost of fertilizer and hence their limited availability to the poor, and the environmental problems of soil acidification and water pollution. The widespread nature of these problems means that the potential benefits if improvements can be brought about through agroforestry are extremely high.

Plant requires sixteen essential and three beneficial elements for their normal growth and they are conventionally split into three group's namely primary nutrient (Nitrogen, Phosphorus and Potassium), secondary nutrient (Calcium, Magnesium and Sulphur), micronutrients (zinc, iron, manganese, copper, boron, molybdenum and chloride) and beneficial elements (sodium, cobalt and vanadium). Nitrogen and phosphorus is most frequently the limiting nutrient in tropical and temperate soils. Except on lands newly clear from forest, there is nearly always a substantial initial response to nitrogen fertilizer. Phosphorus deficiency is found on highly weather soils, where it remains only in small quantities, and on strong acid soils, where mineral phosphorus gets fixed on the clay minerals and exchangeable Iron and Aluminium. Phosphorus deficiency also commonly appears after number of years of cultivation, as soil reserves become depleted. Potassium deficiency is less often limiting except under root crops, while sulphur appear locally, where it is lacking in the soil parent material. Deficiency in micro nutrient are most likely to appears when major shortages have been remedied by the fertilizer. Nutrient is necessarily removed in harvest, of both crops and trees, and, if they are not present in the soil parent material, recycling can reduce but not replace, the need of the input.

The Nutrient-Recycling

One of the main principles of soil management in agroforestry is to make the best use of its resource-conserving and resource-sharing potentials. In natural and derived ecosystems, living things interact with the abiotic environment in clear tropic structures enabling the flow of the nutrients. The efficiency with which nutrients are captured by and flow through living things is directly related to biomass production in natural system and harvested products in managed systems. Nutrient cycling refers to transfer of nutrients from one component to another in the soil-plant-animal-environment system. Under natural woodland and forest ecosystems, a large quantity of nutrients circulate annually between plants and the soil; in a mature (climax) community the plant-soil system as a whole is in equilibrium and the inputs and outputs of nutrients are small. A high-input agricultural system, in contrast, has inputs through fertilizer balanced by large losses through accelerated natural processes (leaching, gaseous losses and erosion) and removal in harvest. The element of recycling, through decomposition of root and, perhaps, crop residues is relatively small. The recycling hypothesis states that, by the inclusion of trees, agroforestry systems can achieve a condition intermediate between these extremes, by reducing losses and increasing plant uptake.

This can be expresses quantitatively as the nutrient recycling ratio, treating the plant-soil complex as a system. Gains and losses are external to the system; uptake and return are internal transfers, respectively soil to plant and plant to soil. Treating the organic and mineral parts of the soil together and indicating where process affect certain nutrients only; flows in the plant-soil nutrient cycling are:

Gains to Systems

This includes addition through i) Atmospheric rain and dust, ii)Fertilizers, iii)Organic additions (compost, manure, etc), iv) Symbiotic and Non-symbiotic Nitrogen fixation, v) Deep capture (uptake by roots of trees from below the rooting depth of crops) and vi) Rock weathering (excluding nitrogen).

Losses from the Systems

i) Erosion, ii) Leaching, iii) Harvest (including for fodder), iv) Removals not treated as harvest i.e. grazing, v) Gaseous losses: De-nitrification and volatilization (nitrogen), vi) Burning (nitrogen, sulphur) and vii) Net fixation on to clay minerals (phosphorus)

Internal Flows

Uptake : Uptake by plants from soil.

Return : Mineralization of litter, Humification of litter, through fall and stem flow, Ash from burning.

For any Specified Nutrient

Nutrient balance (kg ha^{-1} $year^{-1}$) = Gains-Losses.

Nutrient recycling ratio (NRR) of a system could be worked out by using %NRR = Uptake + Return ÷Gains + Losses + Uptake + Return X 100.

The nutrient balance of the system is the gains minus losses, as kg ha^{-1} yr^{-1} of the nutrient, and may be positive or negative. The recycling ratio is the average of the internal flows (uptake and return) as a proportion of the external plus internal flows, expressed as a percentage. The recycling ratio would be 100% in a closed system, with neither gains nor losses, and 0 % if there were no internal recycling. In rotational systems, both these indices can be averaged over a cycle of land use.

Mature natural forest and woodland have nutrient balances close to zero and can attain recycling rates as high as 90%. For example, for the phosphorus cycle of a rain forest in Panama, the flows (kg phosphorus ha^{-1} yr^{-1}) were measured as gains = 1.0, losses= 0.9, uptake =11.0, return= 11.9, a phosphorus recycling ratio of 92% (Golley *et al.*, 1975). In savannahs, the trees act as a stabilizing influence, having a slower nutrient cycle than the grasses (Swift *et al.*, 1989). In natural ecosystems, the capacity of trees roots to scavenge nutrients and concentrate these beneath their canopies results, over time, in a mosaic of islands of higher fertility, which influence community structure and ecosystems junction (Rhoades, 1997).

In contrast, the nitrogen budget for a fertilizer cereal crop, ignoring smaller flows, might be of the order of (kg N ha^{-1} yr^{-1}):

Gains:	Fertilizer	100
Losses:	Leaching, etc.	50
Harvest		100
Uptake:		150
Return:	Root residues	50

The nitrogen is 100-150 = -50 kg N ha^{-1} yr^{-1}, and the recycling ratio (200/450) x 100 = 44%.

To reduce harvest conflicts with the objective of managed land use systems and increases in nutrient use efficiency through agroforestry can therefore be achieved primarily by:

* Increasing the cycling of nutrients from tree litter and pruning, via the soil, into crops.
* Reducing losses by leaching
* Reducing losses by erosion.

On sloping land without erosion control, the nutrient budget is dominated by losses in eroded soil. Where erosion is moderate to severe, these may amount to an order of 10-100% kg ha^{-1} of nutrients. The large savings in nutrients can be made by controlling erosion. Since leaching is a loss in the course of the tree-to-crop pathway, recycling and leaching can be considered together. A recent review of evidence for the various processes of nutrient recycling has been given by Buresh and Tian (1997).

Recycling Paths

Short-term and Longer-Term

There are two paths for nutrient recycling from plant to soil and back into plants: the short-term path, via mineralization of plant residues, and the longer-term path, via humification and storage as soil organic matter. The short term or mineralization path takes place within the time span of the growth of annual crop, in the tropics largely within the first three months. In the longer- term or humification path, nutrients stored as organic molecules are released slowly, with a half life of the order of 10 years in the tropics and up to 50 years in the temperate zone.

In agroforestry, the short term path concerns nutrient transfer from tree residues to crops grown in the same year. This involves questions of plant residue decomposition, leaching loss and synchrony between nutrient release and uptake. This aspect has tended to dominate recent agroforestry research. The longer-term path, in which nutrients recycled by trees may reach crops in subsequent years, should not be ignored. It is one of the two major roles played by soil organic matter, along with that of maintaining soil physical properties. Root residues are an important component in both paths, so there are further links with the functions of roots.

Increasing Inputs to the Plant-Soil System

Nitrogen Fixation

Nitrogen-fixing trees can substantially increase nitrogen inputs to agroforestry system. Biological nitrogen fixation takes place by non-symbiotic and symbiotic means. Non-symbiotic is carried out by soil organisms not dependent on plants. It is important in natural ecosystems but amounts of nitrogen fixated are small in relation to the greater requirements of agro-ecosystems. Its rate varies with the organic matter status and therefore microbiological activities of soil. Symbiotic fixation occurs through associations of plant roots with nitrogen-fixing bacteria. The symbioses are:

- Between many leguminous species and *Rhizobium* or *Bradyrhizobium*, forming nodules on roots.
- Between a small number of non-luminous genera and *Frankia*.

A key feature of biological nitrogen fixation is that additional nitrogen is brought into soil-plant system at no cost, in two respects: there is no fertilizer cost, and the loss from the atmosphere is negligible. It is well established that *L. leucocephala* and *Sesbania rostrata* can fix 100-500 kg N ha^{-1} yr^{-1}.Other fast-growing species, including *Sesbania sesban, Gliricidia sepium, Albizia lebbek. Acacia mangium* and the non-legume *Casuarina equisetifolia*, fix amounts of the order of 50-100 kgN ha^{-1} yr^{-1}.On the basis of the nitrogen balance study, irrigated *Prosopis glandulosa* in California appeared to have fixed 700kg ha^{-1} in seven years (Abrams *et al.* 1990).

Increasing Nutrient Uptake

Trees can increase nutrients inputs to agroforestry systems by retrieval from lower soil horizons and weathering rock.

In addition of N fixation, two processes have the potential to increase nutrient inputs to the soil-plant cycle: nutrient removal from weathering rock, and increased inputs from rain and dust. Uptake from the deep horizons and weathering rock is highly plausible, although difficult to prove since minerals in weathering rocks cannot be leveled. Indirect evidence is that tree roots penetrate weathering rock and some tree will grow where there is almost no soil. Retrieval of N from deep soil horizons has been demonstrated on an Oxisol (ferralsol) and an alfisols (acrisol) under a bimodal-rainfall, moist sub humid climate of Kenya where nitrate N at 50-200 cm depth decreased during the growing season under both natural bush fallow and planted *S. sesban* fallow, but not under maize (Hartemink *et al.*, 1996; Buresh and Tian, 1997).

Atmospheric deposition makes significance contribution to nutrient cycling, greater in humid regions than dry. It comprises nutrients dissolved in rainfall (wet deposition) and those carried in dust (dry deposition). Trees do not increase rainfall, but they increase infiltration and thus, presumably, the input of dissolved nutrients. Trees also reduce wind speed and thus favour deposition of dust.

Plant Residue Quality and Decomposition

The decomposition of tree litter and pruning can substantially contribute to maintenance of soil fertility.

The addition of plant residues from the tree component to soil play critical role in agroforestry system. These residues may be in the form of litter - natural leaf fall - or pruning - fresh materials cut from the tree. Litter has a lower nutrient content than pruning, since plants at senescence translocate nutrients into the fruits or ground storage organs before dropping leaves.

When litter decomposes, through the action of soil fauna, the nutrients contained within it may follow two paths: humification or mineralization. Nutrient held in the humus are, like those of residues, protected against leaching; they are subsequently released by the slow but steady process of humus mineralization. The mineralized nutrients pass into the soil solution. They then become available for uptake by plant roots, but the same become subject to loss from the plan-soil system by leaching, gaseous losses and fixation on to clay minerals.

Most natural ecosystem achieves high nutrient–use efficiency, through a complex balance between litter decay rates and nutrient uptake. This balance is not so well achieved for agro-ecosystems. In order to increase the nutrient uptake and reduce the losses, it is desirable to synchronize the release of nutrient by litter decomposition with the requirements of crops; the greater the degree of synchrony between release and requirements, the higher should be the proportion of nutrients taken up by crops. It is also important to avoid N immobilization, which occurs when litter with a high carbon/nitrogen ratio is added directly to soil. This is the reason for composting to reduce the C/N ratio prior to its addition to the soil. Immobilization is much less of a problem in the tropics than in temperate zone, owing to faster decay rates, but it can still occur, e.g. with maize Stover. So composting, it dose not appear to be necessary in the tropics for most tree residues

Agroforestry systems offer many opportunities to manipulate the timing of litter decay and nutrient release. This can be achieved through:

- Selecting tree species with differing rates of litter decomposition;
- Controlling the timing of addition of pruning to the soil, through adjusting the timing of pruning, tree cutting or bringing of plant residues from external sources;
- Pretreatment: addition of fresh materials or composting;
- Controlling manner of addition i.e. placed on the ground surface or ploughed/hoed into the soil.

Tree-species selection is influenced by a variety of considerations, while the timing of pruning is often determined by the need to reduce shading of young crops. The concept of 'quality' of plant residues refers to their relative content of nutrients (especially N), lignin and polyphenols; the C/N ratio and the content of sugars, cellulose and hemicellulose may also be taken in to account. Residues of high quality, (high in N, low in lignin and polyphenols); decay and release nutrients rapidly, giving short-term release of nutrients to meet peaks in plant requirements. Low-quality residues: low in N, high in lignin and/or polyphenols; decay slowly, giving extended release of nutrients, protected against leaching until mineralized; also of value for prolonged soil cover as mulch (Swift *et al.*,1979; Cadisch and Giller,1996). The provisional guidelines for measure the residue quality (Palm, 1995) suggest that N per cent or polyphenol / N (PP/N) may serve as a guide to short-term nutrient release verses immobilization and lignin plus polyphenol / N (LI+PP)/N as an index for longer-term release pattern. If the nitrogen in plant residues is less than 1.74%; N immobilization will occur. If lignin concentration is higher than 15% or polyphenols higher than 3%, there is likely to be reduced or delay nitrogen release because polyphenols binds nitrogen in resistant compound.

The rate of decomposition can be expressed by the litter decomposition constant, K_{lit} (commonly called 'k' but here designated K_{lit} to distinguish it from the decomposition constant for soil humus). This is not, in fact, a constant since the early stages of decomposition proceed more rapidly than later; the half-life of litter, i.e. time taken for 50% loss in mass, is a simpler alternative measure (Anderson and Ingram, 1993). The rate of change of surface litter accumulation in a natural ecosystem, dL/dt, is given by:

$$dL / dt = A - (L \times K_{lit})$$

Where A = annual litter additions and L = accumulation surface litter. If K_{lit} less than 1.0, then the mean residence time of litter is less than one year. This is the case with most ecosystems in tropics.

Absolute rates of the decomposition vary widely with climate, faster for warmer and wetter condition, and hence litter decomposition constants are useful only for different species in the same environment or vice versa. Decomposition is more rapid for buried residues than on soil surface (Lehmann, *et al.* 1995 and Mafongoya, *et al.* 1997). The most useful practical suggestion, however, is too add mixture of pruning from two or more trees of differing quality. The high-quality residues meet the peak demand from young crops, while those of low quality extend the supply through the growing season.

Role of Agroforestry in Soil Improvement

Agroforestry system being ecologically balanced and sustainable ecosystems, trees play many roles as wood components. Besides providing useful products such as fuel wood, fruit, edible seeds and fodder, they also minimize nutrient drain due to leaching and soil erosion, restore nutrient losses from the ecosystem and perform other key environmental services. Thus, they play a major role in maintaining soil fertility through nutrient cycling. They serve as windbreaks or shade trees on pasturelands and in crop fields, help to improve the microclimate, so that animals and plants have better chances of survival. Trees posses long taproots for anchorage and wide spreading surface from the lower strata. Some trees are N fixers and they produce nitrate that can be cycled from decomposed leaves in to the cropping system along with the pumped nutrients. The nutrient demands for trees are much lower as compared with agriculture crops. The presence of mycorrhizal roots in most of the trees makes them successful on moisture starved and nutrient poor soil. Tree plays both productive and protective role in the system and soil conservation undoubtedly is the most important service function performed by the trees. Through originally equates with control of erosion, soil conservation now connotes with maintenance of soil fertility (Young, 1987a & b).

Tree and Shrub for Soil Improvement

Trees for soil improvement should posses certain qualities to sustain the fertility status and there are: 1) a high rate of production of leafy biomass, 2) a high rate of nitrogen fixation, 3) a dense network of fine roots either with abundant feeder's roots or a capacity for mycorrhizal association, 4) the existence of some deep roots, 5) a moderate to high balanced nutrient content in foliage, 6) an appreciable nutrient content in the root system, 7) absence of toxic substances in the foliage and root exudates, 8) for soil reclamation, a capacity to grow on poor soils and 9) low invasiveness.

Fifty-five tree and shrub species, belonging to 32 genera, are identified which have a potential to maintain or improve soil fertility. Species with particularly high potential included: *Acacia albida, Acacia tortilis, Calliandra calothyrus, Casuarina equisetifolia, Erythrina poeppigiana, Gliricidia sepium, Inga jinicuil, Leucaena leucocephala, Prosopis cineraria* and *Sesbania sesban.*

Mechanisms Involve in Soil Fertility Improvement

1. through increasing inputs (organic matter, N fixation and nutrient uptake)
2. by reducing losses (organic matter, nutrients) by promoting recycling and checking erosion
3. Improving soil physical properties, including water-holding capacity
4. Beneficial effects on soil biological processes.

Nutrient Content

The characteristic feature of deciduous tree species are to translocate nutrients from leaves to perennial organs well before the leaf fall, and nutrients in living leaves are usually higher than in litter. Thus the nutrient transfer to the soil will differ between pruning of green leaves and litter fall. Nitrogen fixing trees in this respect are far superior to non- nitrogen fixing trees. N fixing trees and shrubs, growing with in practical agroforestry systems are capable of fixing about 50-100 kg N ha^{-1} yr^{-1}. Major role of tree is to improve nutrient cycling. Mechanisms are uptake from lower soil horizons, reduction of leaching loss by tree- root systems balanced nutrients supply, and improvement in the ratio between available and fixed minerals. Agroforestry system promotes closer nutrient cycling than agriculture system, 1) through uptake of soil nutrients by tree root systems and recycling them as litter, including root residue and 2) by helping synchronizing nutrient released with crop requirements by controlling the quality, timing and manner of addition of plant nutrients.

Tree Roots-Role in Soil Fertility

There has recently been increasing recognition of the importance of roots as a component of primary production. Root biomass of trees is typically 20-30% of total plant biomass (or 25-43% of above ground biomass). However, net primary production of roots is substantially more than standing biomass, owing to the turnover of fine roots. Roots form an appreciable store of nutrients, and since they are almost invariably returned to the soil, constitute a substantial element in nutrient recycling. Tree root systems, together with their associated mycorrhizae, improve the efficiency of nutrient cycling, defined as the ratio between plant uptake and losses by leaching and erosion. They also contribute to soil physical properties. The key to making use of root and mycorrhizal systems in agroforestry lies in maximizing these positive effects whilst reducing tree crop competition for moisture and nutrients. There is a clear need for more knowledge of root growth and functioning in agroforestry systems.

Changes in Soil Properties in Agroforestry

When trees are planted as components in agroforestry systems they bring about improvement in soil fertility by improving physical properties. Sreemannarayana *et al*, 1994 studied influence of *Albizia lebbek* on soil properties reported that soil organic matter improved significantly over the initial value. Kater *et al.* (1992) reported that soils under trees were slightly richer in organic matter and several cations compared to adjacent tree less sites. Comparative study of different land use systems was made with regard to soil fertility improvement was least in fallow land and was highest in *Albizia lebbeck* based system and *Acacia nilotica* based silvipastoal system. Nutrient inventory of soil in agroforestry system for predicting the land degradation and related aspects form some basis for the appraisal of nutrient flow within the ecosystem, although the type of vegetation and agro-climatic conditions play equal role in moderating the nature and magnitude of nutrient cycling. In silvipastoral system, the improvement in fertility is as encouraging when leguminous species are included as pasture crop components. This help in reducing the infiltration rates appreciably through their root proliferation (Giri Rao *et al.*, 1998) besides improving organic matter status of the soils. The concentration of nutrients in the litter decreased with increasing age of stands (Faize, Mohsin and Babu Ram, 2002). Many acid loving tree species reduce the soil acidity by the accumulation of aluminium in above and below ground plant parts. Afforestation has been successfully employed as a means of reclaiming saline and alkaline soils. Plantation of *Acacia nilotica* and *Eucalyptus tereticornis* in Karnal, India, lowered pH of topsoil from 10.5 to 9.5 in five years and of electrical conductivity from 4 to 2 dSm^{-1}, has been reported, but with tree establishment assisted by additions of gypsum and manure (Gill and Abrol, 1986; Grewal and Abrol, 1986). Soil studies conducted at NRCAF, Jhansi clearly revealed the beneficial role of trees in croplands in improvement and/or maintenance of soil fertility. Integration of *Tectona grandis, Albizia procera* and *Azaderchta indica* with black gram-wheat sequence increased the organic carbon to 5.7, 6.0 and 6.4 g kg^{-1} soil, respectively over to 5.4 g kg^{-1} soil, under cropping only. Similarly, different MPTs based silvo-pastoral system also brought noticeable build up in organic carbon from 3.9 to 6.5 g kg^{-1} and total N from 0.43 to 0.67 g kg^{-1} soil, compared to the state before the experiment began (Rai *et al.*, 2001). In a comprehensive study in semi arid conditions of Rajasthan, Yadav (2005) reported significant and marked increase in soil organic carbon, total and available pool of N, P and K; soil microbial biomass C, N and P and soil enzymatic activity under different trees based traditional agroforestry systems. He further reported that extent of enrichment of soil was linked to the quantity and quality of litter fall added by the trees.

Impact of Intercrops on Soil Fertility Changes in Agroforestry

A combination of woody perennials and agricultural crops may cause considerable improvement in soil physical conditions and fertility. They may also influence the hydrological characteristics and help in control of erosion from sloping sites. The productivity, fertility and erosion under *Cassava* based agroforestry systems were studies by Ghosh *et al.* (1989). It was reported that due to inclusion of *Cassava* with *Eucalyptus* and *Leucaena,* there was increase in phosphorus and potassium status of soil and reduction in the runoff and soil loss.

Agroforestry Practices in Relation to Soil Fertility

Practices with Substantial Positive Effects on Soil Fertility

1. Improved tree fallow, 2. Trees on cropland, 3. Plantation crop combinations, 4. Home gardens, 5. Hedgerow inter-cropping, 6. Trees on erosion control structures, 7. Wind breaks and shelterbelts, 8. Biomass transfer, 9. Trees on rangeland or pastures, 10. Wood lots with multipurpose management, 11. Reclamation forestry leading to multiple uses.

Practices with Smaller Positive or Neutral Effects on Soil Fertility

1. Boundary planting 2. Plantation crops with pastures

Practices with positive or negative effects on soil fertility : Shifting cultivation

Practices with neutral or negative effects on soil fertility : Taungya

Adverse Effect

Trees can have directly adverse effects on soil properties, whilst other consequences arise when they are grown in association with herbaceous plants. Leaving aside shading, a major problem at the tree/crop interface but unconnected with soils, the main soil related problems that can arise are as under:

1. Loss of organic matter and nutrients in tree harvest: In agroforestry, the soil improving potential of trees is greatly reduced if both foliage and wood are harvested for fodder and fuel wood.
2. Nutrient competition between trees and crops: It is desirable that trees in agroforestry should have rooting systems, which penetrate deeply but have limited lateral spread.
3. Moisture competition between trees and crops: In the semi-arid and dry zones, moisture competition is possibly the most serious problem in agroforestry.
4. Production of substances, which inhibit germination or growth: It is well known phenomena that tree leaves and roots of many species contain certain chemicals that would not promote the crop growth, though their positive influence on soil is not to be forgotten. The entire chemical such as benzoic and para coumaric acids when present in the large quantity can create problem on the germination of various crops.

5. Acidification by trees which produce moor-type humus: This is known problem in conifer plantations of the temperate zone.

Fertility Management in Agroforestry

Application of fertilizers to forest trees in agroforestry systems is an expensive process, and it suits to trees of some economic value as seen in horticultural tree based systems. In such cases selection of fertilizer should be done very carefully. An ideal fertilizer should be cheaper to produce a unit of plant food, give high growth response, maintain fertility and improve physico-chemical properties of soil (Maselkar, 1990). In general hardwoods require more nutrients than conifers and fast growing trees require more nutrients than slow growing.

Conclusion

Agroforestry systems promotes more closed nutrient cycling than agriculture and more efficient use of nutrient by means of 1) increasing nutrient inputs, through nitrogen fixing trees and nutrient uptake from deep soil horizons, 2) reducing leaching losses through tree root and mycorrhizal systems and 3) recycling nutrients through the decomposition of litter, pruning and root residues.

Both the short-term pathway, via mineralization of decomposing plant residues and longer term pathway via humification, contribute to nutrient recycling. In the mineralization of litter, agroforestry offers excellent opportunities for synchronization of nutrient release from decomposing litter with the requirements of growing crops through: 1) selecting tree species with differing rates of litter decomposition, 2) controlling the timing of litter addition to the soil and 3) controlling pretreatment of litter and manner of addition.

Dense, mixed agroforestry technologies, including multistrata systems and perennial crop combinations, can achieve high rates of nutrient recycling, while technologies with lower proportions of trees can substantially improve recycling and nutrient use efficiency. Agroforestry systems that include animals offer a third pathway for nutrient recycling via fodder- animal-manure-soil. Taking into account both economics and environmental sustainability, the brown manure system of feeding pruning as fodder to animals may sometimes be preferable to the green manure system of adding them to the soil. While agroforestry system can provide adequate N for moderate yields of cereal crops (2 to 4 t ha^{-1}), additional fertilizer N is needed to achieve the potential yields of the season and improved varieties. Agroforestry system can not recycle adequate P, hence inorganic P fertilizer have to be applied even for moderate yield, where soil P availability is low.

References

Anderson, J.M.and Ingram, J.S.I. 1993. Tropical soil biology and fertility. A Handbook of methods, 2nd edn. CAB International, Wallingford.

Abrams, M.K., Jarrell, W.M., Smith, H.A. and Clark, P.R. 1990. Nitrogen accretion in soil and biomass production by trees *Prosopis* species. *Agroforestry Systems,* 10: 93-97.

Borthakur, D.N. and 9 others 1979. Agroforestry based farming system as an alternative to humming. Shillong, India: *ICAR Research Complex*, Pp. 32.

Buresh, R.J.and Tian, G. (1997) Soil improvement by trees in sub-Saharan Africa. *Agroforestry Systems,* 38: 51-76.

Cadsch, G. and Giller, K.E (eds) 1996. Driven by Nature: Plant litter quality and decomposition. CAB International, Willingford.

Das, D.C. 1980. Some aspects of shifting cultivation related to soil and water conservation. *Indian Journal of soil conservation,* 8: 53-59.

Faiz Mohsin and Babu Ram 2002.Six month study on litter production, changes in soil chemical properties and productivity under Poplar intercropped with wheat. *Indian Journal of Agroforestry*, 4(2): 81-87.

Ghosh, S.P., Mohan Kumar, B., Kabeerapumma, S. and Nair, G.N. 1989. Productivity, Soil fertility and soil erosion under *Cassava* based agroforestry system. *Agroforestry System*, 8: 67-82.

Gill, H.S. and Abrol, I.P. 1986. Salt affected soils and their amelioration through afforestation. In: Amelioration of soil by trees: a review of current concepts and practices (R.T. Prinsley and M.J. Swift, eds), London: *Commonwealth Science Council,* Pp. 43-56.

Giri Rao, L.G., Joseph, B. and Sreemannrayana, B. 1998. Effect of in application on biomass yield of *Stylosanthes hamata* and *Cenchrus* in *Acacia nilotica* based silvipastoral systems. *Range management & Agroforestry,* 19 (2): 173-178.

Golley, F.B., Mcginnis, J.T., Clements, R.G. et al. 1975. Mineral Cycling in a Tropical Moist Forest Ecosystem. *University of Georgia, Athens, Georgia, USA.*

Gourou, P. 1948. *Les pays tropicaux*. Paris: *Presses Universities*, Pp. 157.

Grewal, S.S. and Abrol, I.P. 1986.Agroforestry for alkali soils: effect of some management practices on initial growth, biomass accumulation and chemical composition of selected tree species. *Agroforestry Systems*, 6: 221-232.

Gupta, J.P., Rao, G.G.S.N., Rao, B.V.R. and Gupta, G.N. 1983. Soil drying and wind erosion as affected by different types of shelterbelts planted in the desert region of western Rajasthan. *Journal of Arid Environment*, Pp. 58-63.

Hertemink, A.E., Buresh, R.J., Basir Jama and Janssen, H.B. 1996. Soil nitrate and water dynamics in sesbania fallows, weed fallows, and maize. *Soil Science Society of America Journal, 60*: 568-574.

Kang, B.T. 1993. Alley cropping: past achievement and future directions. *Agroforestry Systems*, 23: 141-155.

Kang, B.T. and Duguma, B. 1985. Nitrogen management in alley cropping systems. In: Nitrogen Management in Farming System in Humid and Sub-Humid Tropics (Kang, B.T. Van der Heida, J. Eds.), *Institute of Soil Fertility, Haren, Netherland*, Pp.269-284.

Kang, B.T., Grimme, H. and Lawson, T.L. 1985. Alley cropping sequential crop maize and cowpea with *leucaena* on a sandy soil in Southern Nigeria. *Plant and Soil*, 85:267-276.

Kang, B.T., Wilson, G.H. and Lawson, T.L. 1984. Alley cropping: A stable alternative to shifting cultivation IITA, *Ibadan, Nigeria.*

Kater, L.J., Kante, H.S. and Budelman, A. 1992. Kariete and Parkia biglobosa associated with crops in south Mali. *Agroforestry Systems*, 18: 89-105.

Lal, R., 1989a. Agroforestry systems and soil surface management of a tropical alfisol: I soil moisture and crop yield. *Agroforestry Systems*, 8: 7-29.

Lal, R., 1989b. Agroforestry system and soil surface management of a tropical alfisol: II water runoff, soil erosion and nutrient loss. *Agroforestry Systems*, 8: 97-111

Mselkar, A.R. 1990. Forests companion, Published by Jugal *Kishore And Company* (Pub. Div.) Dehradun, Pp. 604.

Lehmann, J., Schroth, G. and Zech, W. 1995. Decomposition and nutrient release from leaves, twings and roots of three alley-cropped tree legumes in central Togo. *Agroforestry Systems*, 29: 21-36.

Mafongoya, P.L., and Nair, P.K.R. 1997. Multipurpose tree pruning as a source of nitrogen to maize under semiarid conditions of Zimbabwe.1.Nitrogen recovery rates in relation to pruning quality and method of application. *Agroforestry Systems*, 35: 31-46

Mishra, P.K., Srinivas Sharma, Rao, K.V. and Singh, H.P. 1989. Potentiality of water harvesting and use as micro levels in semi arid alfisol- a case study. Proceeding of National Seminar on Water Resource Development and Management of Sustainable Crop Production. WALMTARI, Hyderabad.

Mulongoy, K. 1986. Nitrogen cycling in alley cropping system. IITA Research Briefs 7, Ibadan, Nigeria, IITA: 3-5.

Palm, C.A., Salazar, A.A., Szott, L.T. and Fernandes, E.C.M. 1995. Long-term performance of alley cropping on acid soils of the Amazon basin. In: Kang *et al.* (1995), q.v: Pp.379-392

Rao, M.R., Ong, C.K., Pathak, P. and Sharma, M.M. 1991. Productivity of annual cropping and agroforestry system on a shallow alfisol in semi arid India. *Agroforestry Systems*, 15: 51-63.

Singh, R.P. and Osman, M. 1995. Alternate land use systems for dry land. In: Sustainable development of dry land agriculture in India, Singh, R.P. (Ed.) Scientific Publishers, Jodhpur, Pp. 375-398.

Singh, R.P., Ong, C.K. and Saharan, N. 1989a. Microclimate and growth of sorghum and cowpea in alley cropping in semi arid India. *Agroforestry Systems*, 9: 259-274.

Singh, R. P., Ong, C.K. and Saharan, N. 1989b. Above and below ground interactions in alley cropping in semi-arid India. *Agroforestry Systems*, 9: 259-274.

Sreemannarayana, B., Giri Rao, L.G. and Joseph, B. 1994. Evaluation of MPT and their influence on soil fertility improvement. *Range management and Agroforestry*, 15(2): 199-202.

Sreemannarayana, B., Giri Rao, L.G. and Joseph, B. 2003. Nutrients returns and withdrawals from trees in agriculture system. Paper presented at *6th Agriculture Science Congress* held at IISS, Bhopal, 6-8 February.

Swift, M.J., Heal, J.W., and Anderson, J.M. 1979. Decomposition in Terrestrial Ecosystems. Blackwell Scientific Publishers, Oxford, UK.

Swift, M.J., Frost, P.G., Campbell, B.M. et al. 1989. Nitrogen cycling in farming systems derived from savanna: perspective and challenges. In: Ecology of Arable Lands – Perspectives and Challenges (Clarholm, M. and Berstrom, L. Eds), Kluwer, Dordrecht, Pp.63-76.

Von Carlowitz, P.G. 1986c. Recommendations for the design and establishment of demonstration trails at the Ethiopian Center for Community Forestry and Soil Conservation. ICRAF Working paper 41.Nairobi: ICRAF, Pp. 89.

Yadav, R.S. 2005. Organic matter dynamics and nutrient availability under traditional agroforestry systems in semi arid conditions of Rajasthan. *Ph. D thesis, Rajasthan Agricultural University*, Bikaner.

Young, A. 1987a. The potential of Agroforestry for soil conservation. Part II: maintenance of soil fertility. ICRAF Working paper 43.Nairobi: ICRAF, Pp. 135.

Young, A. 1987b. The potential of agroforestry for soil conservation and sustainable land use. In J. Kozub ed. Land and water resource management. Washington, DC: Economics Development Institute of the World Bank, 301-17.

Young, A. 1989. Agroforestry for soil conservation, CAB International Wallingford, OXON (U.K.), Pp. 276.

Young, A. 2005. Agroforestry for Soil Management, 2nd edition, CAB International Wallingford, OXON OX10 8DE (U.K.).

□□□

Agroforestry: Natural Resource Sustainability, Livelihood & Climate Moderation © 2009
Editors: OP Chaturvedi, A Venkatesh et. al., pp. 201-216
Satish Serial Publishing House, Delhi

Chapter 15

CARBON SEQUESTRATION POTENTIAL IN DIFFERENT LAND USES AND OPPORTUNITIES OFFERED BY AGROFORESTRY

Ram Newaj and Shabir Ahmad Dar
National Research Centre for Agroforestry, Jhansi-284 003 (U.P.)

Introduction

The world's climate has always been varied naturally. The people believe that a new kind of climate change is now under way. Its impact on people and ecosystems are to be drastic. Level of carbon dioxide (CO_2) and other green house gases in the atmosphere have risen steeply since the industrial revolution. Concentrations have increased mainly because of the use of fossil fuels, deforestation and other anthropogenic activities. Like a blanket around the planet, green house gases stop energy escaping from the earth's surface and atmosphere. If levels rise too high, excessive warming can distort natural patterns of climate. About 30 gases produced by human activity have been identified as contributing to the greenhouse effect. The main ones are CO_2, Methane (CH_4), Nitrous oxide (N_2O), Hydroflurocarbon (HFCs), Perflurocarbons (PFCs) and Sulphur hexafluoride (SF_6). The contribution of CO_2, methane and nitrous oxide is 55, 18 and 6%, respectively, of the overall global warming effect arising from human activities. India currently ranks fifth in C emissions in the world, behind only the United States, China, Russia and Japan and currently account for about 4.2% of the world's total fossil fuel-related CO_2 emissions. In the past decade alone, India's C emissions have increased by about 60% and are about nine times higher than they were forty years ago. Much of this increase is due to India's increasing utilization of its coal resources for power generation. Emission from coal made up 69% of the total emission in 2002 followed by petroleum at 27%. C emissions are forecast to grow by about 3.3% annually through 2020, which would cause India to displace Japan as forth-greatest C emitter by 2010 (TERI 2001). Originally CO_2 has been believed to be the main pollutant but the list of pollutant now include gases such as halocarbons, methane, nitrous oxide and ozone as well aerosols. Aerosols are of two types, suspended such as black C and reflective such as sulphate and nitrates. Black C (BC) emissions are a product of incomplete combustion from coal, diesel oil bio fuels and biomass; these are particularly very large in China and India. R. Parthasarathy in a statement related to climate change (Hindu, New Delhi edition Dec. 2002) stated that reducing the amount of BC aerosols, in addition to having human health benefits, actually help diminish the intensity of flood in south China and also drought in north China. Similar considerations may apply to India, which has experienced recent droughts.

The idea about mitigating climate change by sequestration through forest conservation and management was discussed as early as in the 1970s. But it was only in 1992, that several countries agreed to the United Nations Framework Convention on Climate Change (UNFCCC), with the major objectives of developing national inventories of greenhouse gas emission and sinks, and reducing the emission of greenhouse gases (FAO 2001). At its third meeting in 1997 in Kyoto, Japan, the participating countries, including the US, agreed for the Kyoto Protocol, to reduce greenhouse gas emission to 5% or more below 1990 levels by 2012. The protocol provides a mechanism by which a country that emits C in excess of agreed-upon limits can purchase C offsets from a country or region that manages C sinks. Initially there was no agreement as whether forests could be considered as C sinks, but the potential role of forest conservation and management to decrease greenhouse gases in the atmosphere was soon recognized. Globally, forests contain more than half of all terrestrial C, and account for about 80% of the C exchange between terrestrial ecosystems and the atmosphere. Forest ecosystems are estimated to absorb up to 3 Pg (3 billion tones) of C annually. In recent years, however, a significant portion of that has been returned through deforestation and forest fires. For example, tropical deforestation in the 1980s is estimated to have accounted for up to a quarter of all C emission from human activities (FAO 2003). Decrease in deforestation helps to preserve current C reservoir and afforestation helps holding C for longer time. During the decade 1981-90, land-use changes in the tropics accounted for CO_2 emission of about 1.6 G t yr^{-1} (G t yr^{-1}, 1 Giga t $=10^9$ t) the other hand, terrestrial vegetation assimilated approximately 1.8 G t of C per year during the same period. In sum, the C balance shows that in the 1980s the terrestrial vegetation in the tropics acted as a net sink of C (Bhadwal and Singh, 2002).

The consequences of climate change will be felt across the world and only limited attempts were made to model crop productivity in response to climate change in India (Rao *et al.* 1995, Gadgil 1995). But it is expected that agricultural crop production might be significantly affected by the predicted changes in climate and atmospheric CO_2. The probable effect of higher CO_2 concentration in atmosphere may increase plant photosynthesis and result in higher crop yield (Kimball 1983) and also increase in net primary production in tropical forest ecosystems (Mingkul and Woodard 1998). However, a rise in temperature may reduce crop yields by hastening plant development by modifying water and nutrient budgets and by increasing plant stress (Long 1991). The net effect of increased CO_2 and climate change on crop yield, thus depend on local conditions. While warmer summer air temperature might be beneficial to crop production in the temperate latitudes, where the length of the growing season and frost-free period would increase, warmer temperatures exert negative effects on crop maturity in those regions where temperature and water stress limit crop production. Therefore, vast areas of the arid and semi-arid regions of India, where agriculture is mostly rainfed will have probably, strongly adverse effect of global warming. Lal *et al.* (2001) projected that 5 to 25% decline in winter rainfall and 10 to 15% increase in monsoon rainfall over India during the 2080s, which is significant and may lead to droughts during the dry months and more intense rainfall spell during the wet season. Large increase in grain sterility of cereal and legume crop caused by climate change raises an alarming food security issue, increasing the world's challenge to feed itself in the coming decades. The extent of the sterility threat to root and tuber crops, pasture and tree species is unknown. If rates of rice yield decrease due to thermal stress are broadly validated and assuming the range in temperature increase in the latest IPCC data (0.14-0.58^{0}C per decade), yield of tropical grain crop could decrease by 5-11% by the year 2020 and by 11-46% by 2050 (CGIAR and Inter Centre Working group on Climate Change 2004).

Carbon Sequestration Potential in Different Land Uses

C sequestration potential differs with the kind of land use. It is a proven fact that forest ecosystems are the best way to sequester C. However, considering the huge human population and degraded areas in developing countries like India, the immediate need is to provide food and under such circumstances, much of the land cannot be spared for increase in forest cover. Also, landholding of most of the farmers in developing countries is less and they cannot spare land to permanent tree cover, as it will not provide immediate benefit. Finding low-cost methods to sequester C is emerging as a major international policy goal in the context of global climate change. Therefore, the need is to assess the potential of different land-management options which can fulfill both environmental and economic goals. C sequestration depends upon biomass production capacity, which in turn depends upon interaction between edaphic, climatic and topographic factors of an area.

Globally 700 million ha of land might be available for C conservation and sequestration programs (345 million ha for plantation and Agroforestry, 138 million ha for slowed tropical deforestation and 217 million ha for natural and assisted regeneration of tropical forests (Brown, 1999). The global potential of biological mitigation has been estimated at 100 Gt C (cumulative) by 2050, equivalent to about 10 % to 20% of projected fossil fuel emissions during that period (IPCC, 2001). The largest potential is in the subtropical and tropical regions, but realization of this potential will depend on land and water availability and rates of adoption (Watson *et al.*, 2000, IPCC, 2001). The largest opportunities for biological mitigation in tropical countries cannot be considered in isolation of broader policies in forestry, agriculture and other sectors. Barriers to reaching the potential level of mitigation include: (i) lack of funding and human and institutional capacity to monitor and verify mitigation efforts and out comes, (ii) food supply requirements, (iii) people living off the natural forests, (iv) existing incentives for land clearing. (v) population pressure and (vi) switch from forests to pressures because of demand of meet (IPCC, 2001).

Bloomfield and Pearson (2000) present a review of estimates of the potential of LUCF activities to offset greenhouse gas emissions. Some of their figures are presented in Table 1. Brown *et al.* (1996) estimated that by 2050, plantation in tropical countries have the potential to capture as much as 16.4 Gt C whereas agroforestry has the potential to capture 6.3 Gt C. The estimates of Trexler and Haugen (1994) are much lower than this, especially for agroforestry (Table1), indicating that there is a large degree of uncertainty regarding estimates of potential C sequestration at a global scale.

Afforestation and reforestation of degraded forests and wastelands offer attractive opportunities. But the mitigative capacity may be weak, and enough land and water may not be available (IPCC, 2001). Also semi-subsistence farmers manage much of the land in the tropics and shifting cultivators, so their willingness to participate in biological mitigation projects may be an important factor (de Jong *et al.*, 2000).

Table 1. Potential for C sequestration in tropical countries in Gt C by 2050

Production system	Texler and Haugen (1994)	Brown *et al.* (1996)
Plantations	2.0 – 5.0	16.4
Agroforestry	0.7 – 1.6	6.3
Forest re-growth	9.0 – 23.0	11.5 - 28.7

In the terai zone of West Bengal, Koul and Panwar (2008) compared the C sequestration potential of seven land uses under vogue, namely fallow land, agriculture field, tea garden, agrihorticulture agroforestry system, plantation of *Dalbergia sissoo*, plantation of *Terminalia arjuna* and natural forest of *Shorea robusta*. They reported that fallow land and agricultural field sequester 5.86% and 4.73% C respectively, compared to the natural forest of *Shorea robusta*. However, agroforestry systems, viz. tea garden and agrihorticulture contributed 24.24% and 9.09% C respectively. The agrihorticulture system while sequestering C also provides agricultural crops and other economic gains, including C credits, and hence seems to be the best option. The potential of C storage of tree + crop-based system can be further increased using improved planting materials of perennial components.

Agroforestry and Its Potential for Carbon Sequestration

Agroforestry is a viable alternative to prevent and mitigate climate change. Agroforestry was recognized by IPCC as having high potential for sequestering C as part of climate change mitigation strategies (Watson *et al.*, 2000). It can increase and stabilize agricultural yields and reduce soil erosion (Prinsley, 1990). The biomass can provide fuelwood, foods, basic construction materials, shade, medicines, etc., and thereby decreases pressure on natural forests. Further more, it may allow land to be taken out of fallow rotation in shifting cultivation systems; for example, one hectare of land sustainably managed with agroforestry could replace 5- 10 ha of land under shifting rotation slash and burn (Trexler, 1993). Agroforestry in urban areas can provide local biomass and help the public to recognize the usefulness of tree planting. In the U.S. and Europe alone, 50 M t C yr^{-1} could be absorbed by urban tree planting (Kulp, 1990). Indirect effects, such as substitution for fossil fuels, could prevent the release of 17 M t C yr^{-1} worldwide (Evans, 1992). Agroforestry is ideal option to increase productivity of wastelands, increase tree cover out side the forest, and reduce human pressure on forests under different agro-ecological regions of India. An IPCC special report (IPCC, 2000) indicates that conversion of unproductive croplands and grasslands to agroforestry have the best potential to soak up atmospheric C. In agroforestry, soil restoration process involves recovery of organic based nutrients cycle through replenishment of soil organic matters, about half of which is C.

In India, 107.43 million ha lands, once biologically productive have been rendered unproductive due to several degradation processes (MOA, 1994). These soils are characterized by low soil organic C (SOC), low soil quality, and low biomass productivity. Restoration of these soils is a high priority for economic and environmental reasons. Environmentally, restoration of biological productivity of these soils will improve water quality by reducing transport of sediments and sediment - borne pollutants, and mitigate green house effects by C immobilization in the biomass and sequestration in the soil. The technological options for biotic and soil C sequestration include afforestation of marginal soils, restoration of degraded ecosystems, establishment of bio-energy plantations with a large potential for biomass production, establishing perennials with a deep and prolific root system, growing species with high cellulose and other resistant materials, and developing land use systems characterized by a high NPP. Similarly, strategies for soil C sequestration include adoption of conservation tillage and mulch-farming techniques, maintenance of soil fertility, soil and water conservation, adoption of complex rotations. Increasing SOC content at 0.01% per year to 1 m depth (for a bulk density of 1.5 Mg m^{-3}), restoration of 1 x 10^9 ha of soil in the tropics has a potential to sequester C at the rate of 1.5 Pg yr^{-1} (Lal and Kimble, 2000) and greenhouse effect mitigation. Bhattacharya *et al.* (2006) has recently worked out about 10 Pg SOC stock in 0-30 cm soil depth in arid tropics of India. They observed that the soils in these areas revisited after 25-30 years indicate an overall increasing trend in SOC and the soils respond to controlled management level and are not depleted in SOC.

The main goal of management of C emission avoidance is to conserve existing C pool in forest vegetation through options such as controlling deforestation or logging, protecting forest in reserves, changing harvesting regimes and controlling other anthropogenic disturbance such as fire and pest outbreaks. Unfortunately, most forest soils are not suitable for profitable agriculture and quickly become unproductive. The felling of trees for commercial and domestic wood products is mostly unregulated and beyond the forests' ability to replenish itself. Similarly, the grazing of livestock in forested areas is often beyond their carrying capacity. Going by the potential for economic exploitation, it would appear that 90% of the forests are performing the critical functions of protecting fragile watersheds and are not fit for commercial exploitation (Dhyani *et al.*, 2006). Due to shift in the National Forest Policy of India harvesting from forests has now practically been banned with social benefits mainly flowing from the protective and environmental functions of the forest apart from meeting the subsistence needs of the communities living close to the forests.

The most if not all agroforestry systems have the potential to sequester C. With adequate management of trees under agroforestry systems, a significant fraction of the atmospheric C could be captured and stored in plant biomass and in soils. The average C storage by agroforestry practices has been estimated as 9, 21, 50 and 63 Mg C ha^{-1} in semi-arid, sub-humid, humid and temperate regions. For small holder agroforestry systems in the tropics, potential carbon sequestration ranges from 1.5 to 3.5 Mg C ha^{-1} (Montagnini and Nair 2004). In agroforestry systems C sequestration is a dynamic process and can be divided into two phases. At establishment, many systems are likely to be source of GHGs (loss of C and N from vegetation and soil). Then follow a quick accumulation phase and a maturation period when tonnes of C are stored in the boles, stems, roots of trees and in the soil. At the end of the rotation period, when the trees are harvested and land returned to cropping (sequential system), part of the C would be released back to the atmosphere (Dixon *et al.*, 1994). Therefore, effective sequestration can only be considered if there is a positive net C balance from an initial stock after a few decades (Felter *et al.*, 2001).

C storage in plant biomass is only feasible in the perennial agroforestry systems (perennial–crop combinations, agroforests, wind-breaks, boundary plantations etc.), which allow full tree growth and where the woody component represents an important part of the biomass. One comparative advantage of these systems is that sequestration does not have to end at wood harvest. C storage can continue way beyond if boles stem or branches are processed in any form of long lasting products (Roy, 1999). For the estimation of these effects, a decomposition rate of wood products of between 1 and 2% has been suggested for the tropics (Findlay, 1985; Roy, 1999). Alternatively the wood can serve as fuel, in which case an important part of the plant-stored C returns to the atmosphere.

Agroforestry trees improve land cover in agricultural fields in addition to providing C inputs (root biomass, litter and prunings) to the soil. This has often reduced soil erosion, which is a crucial process in the soil C dynamics.

But, one must be aware of the fact that soils have a finite sink capacity of 0.4 – 0.6 Pg C yr^{-1} over 50-1000 years (Paustian *et al.*, 2000; Ingram and Fernandes 2001). In above ground and soil C together, 1.1 – 2.2 Pg C could be sequestered annually over 50 years, which, as estimate suggest, would offset about 10-15% of the current annual C emissions (Dixon, 1995). Looking at this contribution, it becomes clear that agroforestry alone cannot solve the current climatic problems, but can only be one among a range of strategies. However, the implementation of agroforestry could be justified for many other reasons. First, increasing soil C greatly benefits agricultural productivity and sustainability. Second, given the improbability of obtaining any single mitigating method, adding modest contributions together appear to be a more realistic way of achieving CO_2 reduction targets (Paustain *et al.*, 1997). Third, the financial cost of C sequestration through agroforestry appears to be much lower (approx. \$1.69 Mg^{-1} C, median \$13 Mg^{-1} C) than through other CO_2 mitigating options. Economic analysis showed that these costs could be easily offset by the monetary benefits from agroforestry products and trading in C credits.

Carbon Stocking in Different Agroforestry Systems

According to the recent projections, in India the area under agroforestry will increase substantially in the near future (NRCAF, 2006). Undoubtedly, this will have a great impact on the flux and long term storage of C as the inclusion of trees in the agricultural landscape often improves the productivity of system while providing opportunities to create C sinks (Winjum *et al.*, 1992; Dixon *et al.*, 1993; Krankina and Dixon, 1994; Dixon, 1995).

The amount of C sequestered largely depends on the agroforestry system being practiced. Other factors influencing C storage in agroforestry systems include tree species, system management, environment and socio-economic aspects. The C storage potential of agroforestry systems in different regions of the world varies from 12 to 228 Mg C ha^{-1} (Table 2).

In an agrisilvicultural system in the arid region of India, eight-year study indicated soil organic carbon content (SOC) decreased with time in control plot with a loss of SOC by 56%. However, integration of trees allowed a loss of SOC only by 3.2% in *Emblica officinalis*, 22% in *Hardwickia binata*, and 35.5% in *Colospermum mopane* plots, indicating greater sequestration of C in *E. officinalis* plot and least in *C. mopane* plots. The study suggests that agroforestry is an important strategy to sequester C not only in the form of biomass but also in soil and may therefore help to maintain soil productivity (Singh, 2005). The range of sequestration was 5 (*C. mopane*) to 13 Mg C ha^{-1} *(E. officinalis)*.

Table 2. Potential C storage for Agroforestry systems in different eco-regions of the world

Country	Eco-region	System	Mg C ha^{-1}	Reference
Africa	Humid tropic high	Agrosilvicultural	29-53	Dixon *et al.*,
South America	Humid tropic low	Agrosilvicultural	39-102[a]	1993
	Dry low land		39-195	
Southeast Asia	Humid tropical	Agrosilvicultural	12-228	Krankina
	Dry lowlands		68-81	and Dixon,
Australia	Humid tropics low	Silvopastoral	28-51	1994
North America	Humid tropics high	Silvopastoral	133-154	Schroeder,
	Humid tropics low		104-198	1994
	Dry lowlands		90-175	
Northern Asia	Humid tropical low	Silvopastoral	15-18	Winjum *et al.*, 1992

[a]Carbon storage value were standardize to 50 year rotation

In India, a number of studies (Table 3) have indicated that the tree component in agroforestry has a capacity for biomass production at least as great as that of natural vegetation. It is possible to design agri-silvicultural systems in which the organic matter loss under the crop component is matched by a gain under the tree component. Das and Itnal (1994) reported that organic C content was about double in agri-horticultural and agroforestry systems as compared to sole cropping. Similarly, alley cropping is a promising agroforestry technology for humid and sub-humid tropics. The hedges (mostly legumes) are pruned periodically during the crop-growth and provide biomass which when added to soil acts as mulch and provides nutrients to soil. The lopping is also used as forage. Lal (1989) observed changes in SOC and nutrient status under different management system. He found that over a period of six years (12 cropping seasons), the relative rates of decline in the status of nitrogen and organic C was much less under alley cropping of *Leucaena* and *Gliricidia* as compared to normal arable crops. In high rainfall coastal areas the beneficial interactions of mixed cropping and mixed farming components on soil fertility have been reported in Sri Lanka in terms of soil physical, chemical and biological properties.

The impact of agroforestry systems on soil fertility has also been shown by workers from other coconut growing areas in terms of higher organic matter content, total nitrogen, available phosphorus and potash in the top soil, and improved microbial activities in the system (Varghese *et al.*, 1978; Bavappa *et al.*, 1986; Liyanage, 1994; Dagar, 1995 a,b). Quantity of litter fall, its chemical composition, nutrient addition and change in chemical composition of soil were studied by Singh *et al.* (1989) under agroforestry systems involving *Populus deltoides* and *Eucalyptus* hybrid tree with intercrops of aromatic grasses *Cymbopogon martinii* and *C. flexuosus* in the tarai tract of Kumaon hills. On an average, dry litter production of *P. deltoides* was 5 kg $tree^{-1} yr^{-1}$ where as of *E.* hybrid 1.5 kg $tree^{-1}yr^{-1}$. Under the canopies of these two trees SOC was enhanced by 33 to 83% and available nitrogen by 38.1 to 68.9% over control in 0-15 cm soil layer. There was significantly higher fertility build up under *P. deltoides* than *E.* hybrid.

Table 3. SOC content in some agroforestry systems

Region	Agroforestry practices	SOC ($g\ kg^{-1}$ soil)	Reference
Semi-arid	Alley cropping-		Kang *et al.,*
Tropics	*Leucaena leucocephala* (pruning retained)	10.7	(1990)
	L. leucocephala (pruning removed)	6.5	
	L. leucocephala (pruning + 80 kg N ha^{-1})	11.9	
Dryland	Sole cropping	3.9	Das and Itnal
region	Agroforestry	7.2	(1994)
	Agrihorticulture	7.3	
	Agrisilviculture	4.7	
Bay Islands	Monoculture systems (Coconut)	9.6	Liyanage and
and Kerala	Mixed farming with coconut	16.3	Dassanayake
	Agroforestry with coconut and cocao/ coffee/pepper/clove/cinnamom	12-14.6	(1993)

In agrisilviculture system at semi-arid regions of U.P., *Dalbergia sissoo* at age of 11years was able to accumulate biomass 48 to 52 t ha^{-1} which corresponds to 24 to 26 t C ha^{-1}. In another study at the same place C dynamics was studied by Ram Newaj *et al.* (2006) under agrisilviculture involving three pruning regimes and two crop sequences (Table 4). C accumulation in tree biomass was 23.61 to 34.49 t C ha^{-1}, respectively in different pruning regimes in blackgram-mustard crop sequence. C accumulation in herbaceous layer ranged from 0.28 to 0.56 t C ha^{-1}, respectively in different pruning regimes under blackgram-mustard crop sequence. C content in tree and crop was slightly higher in greengram-wheat crop sequence. C storage capacity of different agroforestry system under different agro- ecological regions have been given in Table 4.

Table 4. Total C storage under agroforestry systems in different regions of the country

Region	Agroforestry system and components	Total C storage (t C ha^{-1})	References
Semi-arid region	Silvipastoral system (age 5 yrs)*		Rai *et al.*, 2001
	Acacia nilotica + natural pasture	9.5-17.0	
	A. nilotica + established pasture	19.7	
	Dalbergia sissoo + natural pasture	12.4	
	D. sissoo + established pasture	17.2	
	Hardwickia binata + natural pasture	16.2	
	H. binata + established pasture	17.0	
North-western India	Silvipastoral system (age 6 yrs)		Kaur *et al.*, 2002
	Acacia/Dalbergia/Prosopis + *Desmostacya*	6.8-18.55	
	Acacia/Dalbergia/Prosopis + *Sporobolus*	1.5-12.32	
Central India	Block plantations (age 6 yrs) *Gmelina arborea*	24.12-31.12	Swamy *et al.*, 2003
Arid region (Rajasthan)	Agrisilvicultural system (age 8 yrs)		Singh, 2005
	Emblica officinalis + *Vigna radiata*	12.7-13.0	
	Hardwickia binata + *Vigna radiata*	8.6-8.8	
	Colophospermum mopane + *Vigna radiata*	4.7-5.3	
Semi-arid region	Agrisilvicultural system (age 11 yrs) *Dalbergia sissoo* + crop	26.02	NRCAF, 2005
Semi-arid region	Agrisilvicultural system (age 5 yrs)		Ram Newaj *et al.*, 2006
	Albizia procera + blackgram-mustard	34.77	
	Albizia procera + greengram-wheat	35.13	
North-western Himalayas	Silvipastoral system	2.17	AICRPAF, 2006
	Agrihortipastoral	1.15	
	Hortipastoral	1.08	

Dry matter production in agroforestry system of mandarin grown in association with *Albizia* and mixed species was studied in Mamlay watershed in the south district of Sikkim in the eastern Himalaya by Sharma *et al.* (1995). Data show that *Albizia*, mandarin and crop accumulated 13879 kg biomass ha^{-1} in watershed area and in this way the system can store 6939 t C ha^{-1} in tree and crop biomass.

Kaur *et al.* (2002) studied the C storage in 6 years old silvo-pastoral systems on a sodic soil in northwestern India. The total C storage in trees + *Desmostachya* systems ranged from 6.80 to 18.55 t C ha^{-1} and 1.5 to 12.3 t C ha^{-1} in the case of *Dalbergia sissoo* + *Sporobolous marginatus* and *Prosopis juliflora* + *S. marginatus*. *Acacia nilotica* could not survive along with *Sporobolous* under sodic conditions of the soil due the adverse effect of water logging and frost.

Adoption of silvipastoral system in degraded land in semiarid regions of UP, India was able to accumulate biomass 18.91 to 22.25 t ha^{-1} in natural pasture and 32.20 to 35.01 t biomass ha^{-1} in established pasture. The C storage in the system ranged from 1.89 to 3.45 t C ha^{-1} in silvipasture and 3.94 t C ha^{-1} in pure pasture (Rai *et al.*, 2001). In another study, biomass production from natural pasture was 2.1 to 3.6 t ha^{-1} and from established/improved pasture was 2.0 to10.4 t ha^{-1} under silvipasture in different years. It indicates that 1.8 to 3.5 times increase in biomass due to adoption of silvipasture system in degraded lands (Rai, 1999).

The realization of this potential, however, will depend on land and water availability and rates of adoption (Watson *et al.*, 2000; IPCC, 2001). These opportunities for biological mitigation cannot be considered in isolation of broader policies in forestry, agriculture and other sectors. Barriers to reaching the potential level of mitigation include: (i) lack of funding and human and institutional capacity to monitor and verify mitigation efforts and out comes, (ii) food supply requirements, (iii) people living off the natural forests, (iv) existing incentives for land clearing, (v) demographic pressure and (vi) shift from forests to other use to meet the demands of other sectors.

References

AICRPAF, 2006. All India Coordinated Research Project on Agroforestry, Report, NRCAF, Jhansi.

Bahttacharyya, T., Chandran, P., Ray, S.K., Mandal, C., Pal, D.K., Venugoplan, M.V., Durge, S.L., Srivastava, P., Dubey, P.N., Kamble, G.K., Sharma, R.P., Wani, S.P., Rego, T.J., Ramesh, V. and Manna, M.C. 2006. Estimation of Carbon Stocks in Red and Black Soils of selected benchmark spots in semi-arid tropics of India. *Global Theme on Agro-ecosystems Report* No. 28. Patancheru, India. ICRISAT, Pp. 86.

Bavappa, K.V, Kailasam, C., Khader, K.B.A., Biddappa, C.C., Khan, H.H., Kasthuri, Bai, K.V., Ramadasan, A., Sundararaju, P., Bopaiah, B.M., Thomas, G.V., Misra, L.P., Balasimha, D., Bhat, N.T. and Bhat, K.S. 1986. Coconut and arecanut based high density multispecies cropping systems. *Journal of Plantation Crops*, 14 (2): 74-87.

Bhadwal, S. and Singh, R. 2002. Carbon sequestration estimates for forestry options under different land –use scenario in India. *Current Science*, 83 (11): 1380-1386.

Bloomfield, J and Pearson, H.L. 2000. Land use, land use changes, forestry and agriculture activities in the Clean Development mechanism: estimates of greenhouse gas offset potential. *Mitigating and Adaptation Strategies for Global Changes*, 5: 9-24.

Brown, S. 1999. Opportunities for mitigating carbon emission through forestry activities. Winrock International, 1611 N Kent St, Suit 600, Arlington, VA 22209, Pp. 7.

Brown, S. Sathaye,J., Cannell, M., and Kauppi, P. 1996. Management of forests for mitigating of greenhouse gas emissions. In: Climate Change 1995 (Eds. Watson, R.T., Zinyowera, M.C. and Moss, R.H.), Impact Adaptation and Mitigation of climate Change: Scientific-technical Analysis. Cambridge University Press, NY.

Dagar, J. C. 1995a. Organic amendments in the reclamation and management of problem soils. In: *Organic Agriculture* (Ed.Thampan, P.K.) , Peekay Tree Crops Development Foundation, Cochin, Pp. 145-171.

Dagar, J. C. 1995b. Agroforestry systems for the Andaman & Nicobar Islands. *International Tree Crops Journal,* 8: 107-128.

Das, S.K. and Itnal, C.J. 1994. Capability based land use systems: role in diversifying dryland agriculture. *In: Soil Management for Sustainable Agriculture in Dryland Area – Bull. Indian Society of Soil Science,* 16: 92-100.

De Jong, B.H.J., Tipper, R. and Montoya-Goez, G. 2000. An economic analysis of the potential for carbon sequestration by forests: evidence from Southern Mexico. *Ecological Economics*, 33: 313-327.

Dhyani, S. K., Kareemulla, K. and Handa ,A. K. 2006. Agroforestry potential and returns across agroclimatic zones. NRCAF, Jhansi.

Dixon R.K., Brown S., Houghton R.A., Solomon A.M., Trexler M.C. and Wisniewski, J. 1994. Carbon pools and flux of global forest ecosystems. *Science,* 263: 185-190.

Dixon, R. K. 1995.Agroforestry systems: sources or sinks of greenhouse gases? *Agroforestry Systems,* 31: 99-116.

Dixon, R.K., Andrasko, K.J., Sussman, F.A., Lavinso, M.A., Trexler, M.C. and Vinson, T.S. 1993. Tropical forests: their past, present and potential future role in the terrestrial carbon budget. *Water Air Soil Pollution,* 70: 71-94.

Dixon, R.K., Brown, S., Houghton, R.A., Solomon, A.M., Trexler, M.C. and Wisniewski, J., 1994. Carbon pools and flux of global forest ecosystems. *Science*, 263, 185-190.

Evans, J. 1992. Plantation Forestry in the Tropics. 2nd Edition. Carendon Press, Oxford.

FAO 2001. State of the world's forests 2001. Food and Agricultute Organization of the United Nations. Rome Italy, Pp. 181.

FAO 2003. State of the world's forests 2003. Food and Agricultute Organization of the United Nations. Rome Italy, 126pp.

Fellter, C., Albrecht, A., Blanchart, E., Cabidoche, Y.M., Chevallier, T., Hartmann, C, Eschenbrenner, V., Larre-Larrouy, M.C. and Ndandou, J.F. 2001. Soil Carbon sequestration in tropical areas: general considerations and analysis of some edaphic determinants for Lasser Antilles soils. *Nutrient Cycling and Agroecosystem,* 61: 19-31.

Findlay, W.P.K., 1985. Preservation of Timber in the Tropics.Marinus/Junk Publishers, Dordrecht, Pp. 273.

Gadgil, S. 1995. Climate change and agriculture – an Indian perspective. *Current Science,* 69: 649-659.

Ingram, J.S.I and Fernandes, E.C.M. 2001. Managing carbon sequestration in soils: concepts and terminology. *Agriculture Ecosystem and Environment,* 87: 111-117.

IPCC, 2000. Special report on land Use, Land Use Change and Forestry. Summary for Policy Makers. Geneva, Switzerland. Pp. 20.

IPCC, 2001. Climate Change 2001. Mitigation, A Special Report of the Intergovernmental Panel on Climate Change. Cambridge Univeristy Press, NY. (http:www.ipcc.ch/pub/tar/wg2/index.htm).

Kang, B.T., Reynolds, L. and Atta-Krah, A.N. 1990. Alley farming. *Advances in Agronomy*, 43: 315-359.

Kaur, B. Gupta, S.R. and G.Singh 2002. Carbon storage and nitrogen cycling in silvopastoral systems on a sodic soil in northwestern India. *Agroforestry Systems*, 54: 21-29.

Kimball, B. A. 1983. Carbon dioxide and agricultural yield: an assemblage and analysis of 430 prior observation. *Agronomy Journal,* 75: 779-786.

Koul, D. N. and Panwar, P. 2008. Prioritizing land-management options for carbon sequestration potential. *Current Science*, 95(5): 258-263.

Krankina, O.N., Dixon, R.K. 1994. Forest management options to conserve and sequester terrestrial carbon in the Russian Federation. *World Resource Review,* 6: 88-101.

Kulp, J. J. 1990. The photosystem as a sink for carbon dioxide. EPRI Report EN-6786. Electrical Power Research Institute, Palo Alto, CA, USA.

Kursten, E., 2000. Fuelwood production in agroforestry systems for sustainable land use and CO_2 mitigation. *Ecologial Engineering*, 16: 69-72.

Kursten, E., Burschel, P., 1993. CO_2- mitigation by agroforestry. *Water Air Soil Pollution,* 70: 533-544.

Lal, M., Nozawa, T., Emori, S., Harasawa, H., Takahashi, K., Kimoto, M., Abe-Ouchi, A., Nakajima, T., Takemura, T. and Numaguti, A. 2001. Future Climate change: Implications for Indian summer monsoon and its variability. *Current Science,* 81 (9): 1196-1207.

Lal, R. 1989. Agroforestry systems and soil surface management of a tropical alfisol. Arts I-VI. *Agroforestry Systems,* 8(1): 1-6; 7-29; 8(2): 97-111; 8(3): 197-215; 217-238; 239-242.

Lal, R. and Kimble, J. M. 2000. Tropical ecosystems and the global C cycle. *In: Advances in Soil Science: Global Climate Change & Tropical Ecosystems,* CRC Press Roca Raton, Pp. 3-32.

Liyanage, M de S. and Dassanayake, K.B. 1993. Experience in coconut based farming system in Sri Lanka. In: Nai, M.K. et al. (eds.). *Advances in Coconut Research and Development*. Oxford & IBH Publ. Co. Pvt. Ltd., New Delhi, Pp. 358-367.

Liyanage, M. de S. 1994. Coconut based agroforestry in Sri Lanka. *In*: *Trees and Tree Farming* (Ed. Thampan, P.K.), Peekay Tree Crops Development Foundation, Cochin, Pp. 05-122.

Long, S.P. 1991. Modification of the response of photosynthetic productivity to rising temperature by atmospheric CO_2 concentration: Has its importance been underestimated? *Plant Cell and Environment,* 14 (8): 729-739.

Mingkui, C. and Woodard, F. I. 1998. Dynamic responses of terrestrial ecosystem carbon cycling to global climate change. *Nature,* 367:133-138.

Ministry of Agriculture, 1994. *Draft Report of Land Degradation in India,* Soil and Water Conservation Division, department of Agriculture and Cooperation, MOA, New Delhi.

Montagnini, F. and Nair, P.K.R. 2004. Carbon sequestration: an underexploited environmental benefit of agroforestry systems. *Agroforestry Systems,* 61: 281-295.

NRCAF, 2005. Annual Report, NRCAF, Jhansi

NRCAF, 2006. Perspective Plan – Vision 2025, NRCAF, Jhansi, U.P.

Paustian, K., Andren , O., Janzen, H.H., Lal, Smith, P., Tian, G., Tiessen, H., van Noordwijk, M. and Woomer, P.L., 1997. Agricultural soils as a sink to mitigate CO_2 emissions. *Soil Use Management,* 13: 1-15.

Paustian, K., Six, J., Elliott E.T. and Hunt H.W. 2000. Management options for reducing CO_2 emissions from agricultural soils. *Biogeochemistry,* 48: 147-163.

Prinsley, R.T. 1990. Agroforestry for Sustainable Production: Economic Implications. Commonwealth Science Council, London.

Rai, P. 1999. Growing greener and increasing the production of wastelands. *Wastelands News* XIV: 49-50.

Rai, P., Yadav, R.S. Solanki, K.R., Rao, G.R. and Rajendra Singh. 2001. growth and pruned biomass production of multipurpose tree species in silvipastoral system on degared lands in semiarid region of Uttar Pradesh, India. *Forest Tree and Livelihood,* 11: 347-364.

Ram Newaj, Shanker, A.K. and Yadav, R.S. 2006. Carbon and nitrogen dynamics in *Albizia procera* based agroforestry system. Annual Report of ICAR ad-hoc Scheme, NRCAF, Jhansi (unpublished data).

Rao, D.G., Katyal, J.C., Sinha, S.K. and Srinivas, K. 1995. Impacts of climate change on sorghum productivity in India: simulation study. In: Climate Change and Agriculture: Analysis of Potential International Aspects, pp.325-337. American Society of Agronomy, Specual Publication 59, Madison, WI, USA.

Roy, C. 1999. Options techniques et socio-economiques des emissions de CO_2 et d'augmentation des stocks de carbone. *CR Acad Agric. France,* 85: 311-320.

Schoeder, P.1994. Carbon storage benefits of agroforestry. *Agroforestry Systems,* 27: 89-97.

Sharma, R., Sharma, E. and Purohit, A.N. 1995. Dry matter production and nutrient cycling in agroforestry system of mandarin grown in association with Albizia and mixed tree spcies. *Agroforestry Systems,* 29: 165-179.

Singh, G. 2005. Carbon sequestration under an agrisilvicultural system in the arid region. *Indian Forester,* 131(4): 543-552.

Singh, K., Chauhan, H.S. Rajput, D.K. and Singh, D.V. 1989. Report of a 60 month study on litter production, changes in soil chemical properties and productivity under poplar (*P. deltoides*) and *Eucalyptus (E. hybrid*) interplanted with aromatic grasses. *Agroforestry Systems,* 9: 37-45.

Swamy, S.L., Puri, S. and Singh, A.K. 2003. Growth, biomass, carbon storage and nutrient distribution in Gmelina arborea Roxb. Stands on red lateritic soils in central India. *Bioresource Technology,* 90: 109-126.

TERI, 2001. Teri Eenergy Data and Directory Year Book 2000/2001. Tata Energy Research Institute (TERI),New Delhi.

Trexler, M.C. 1993. Mitigating Global Warming Through Forestry: A Partial Literature Review. Report to GTZ for Enquette Commission, German

Trexler, M.C. and Haugen, C. 1994. Keeping it Green: Tropical Forestry Opportuinities for Mitigating Climate Change. World Resources Institute, Washington DC.

Varghese, P.T., Nelliat, E.V. and Balakrishnan, T. 1978. Beneficial interactions of coconut cacao combination. Proceedings *PLACROSYM* - 1, Indian Soc. Plant Crops, Kasargod, Pp. 383-392.

Watson, R.T., Noble, I.R., Bolin, B., Ravindranath, N.H., Verardo, D.J. and Dokken, D.J. (eds.) 2000. Land Use, Land Use Changes and Forestry : A Special Report of the Intergovernmental Panel on Climate Change. Cambridge University Press, NY.

Winjum, J.K., Dixon, R.K. and Schroeder, P.E. 1992. Estimating the global potential of forest and agroforest management practices to sequester carbon. *Water, Air and Soil Pollution,* 64: 213-228.

❑❑❑

Agroforestry: Natural Resource Sustainability, Livelihood & Climate Moderation © 2009
Editors: OP Chaturvedi, A Venkatesh et. al., pp. 217-229
Satish Serial Publishing House, Delhi

Chapter 16

BIODRAINAGE TECHNOLOGY FOR SALINE AND WATER LOGGED LANDS

A. Balasubramanian and S. Manivasakan
Forest College and Research Institute, Mettupalayam-641 301 (T.N.)

Introduction

The irrigated agriculture across the world has increased the water availability to many cultivable areas. This has resulted with construction of water storage facilities like reservoirs etc... Ultimately increased the ground water and shallow ground water is the problem in many command areas. Shallow groundwater tables and associated salinity problems have become dominant features in agricultural areas around the world. These problems have been caused by increasing pressures on land resources caused by rising populations, especially in irrigation areas. Increasingly, large areas of land having historically deep groundwater tables are now in need of some form of water table control. In many places management strategies have been developed to address this problem. These often focus on engineering approaches such as deep open ditches, vertical drainage (groundwater pumping) or horizontal subsurface drainage.

The total irrigated area in the world is 255 million ha of which more than two-third lies in Asia. About 20 per cent of the irrigated land has been rendered saline due to waterlogging. Each year, an additional area of about 1.5 million ha. of irrigated land gets affected by secondary salinisation due to waterlogging, and thereby looses its productivity.

In India, the waterlogging problem is expected to increase 5 fold over the next 30 years, threatening the livelihood of one million farm families and having a significant negative impact on food grain production of India as a whole. The agricultural production losses due to waterlogging and salinity are over Rs. 860 million in the year 2000. The research evidences proved that conventional drainage system can reduce water table only by realizing the heavy cost (Cazs, 2002). It is also realized that the existing waterlogging area is still large and is expected to increase with the present status of irrigation systems, besides the population growth increases pressure to bring more land under irrigation. As a whole the net increase of waterlogging problem will be realized in India in the near future. In order to effectively address this problem with less cost, the new system called "Biodrainage" is advocated which will be given due emphasis in the next five year plan (Mona Metha, 2002). Much of the waterlogged areas can be reclaimed by growing tolerant tree like *Eucalyptus* (Calder, 1994).

Biodrainage relies on vegetation, rather than engineering mechanisms to remove excess soil water through evapotranspiration. It is often considered attractive because it requires only an initial investment in site development (planting of a "biodrainage crop") and (potentially) returns a benefit when the biocrop is harvested for fodder, wood or fibre. In addition, under some management scenarios, viz. certain cropping systems and slightly saline conditions, it might offer limited scope to achieve nutrient and/or salt-balance through removal of biomass, thus alleviating the problem of the disposal of polluted drainage effluent from the biodrainage crop area by reducing volumes and improving the quality of the effluent. Advocates of biological drainage plantations often emphasize the low-cost aspects for the removal of excess water (SMEDEMA 1997). Moreover, biodrainage tree plantations do not demand the installation of physical structures, but can be implemented by farmers themselves who in addition may gain marketable products such as fodder and wood when adequate tree species are chosen.

Biodrainage Definition

"Biodrainage is the process of removing excess surface and sub surface water in water logged areas using vegetation".

Aim of Biodrainage

The aim of biodrainage is to remove excess groundwater through the process of transpiration by vegetation. This is achieved by enhancing the transpiration capacity of the landscape by introducing high-water use vegetation types in large enough areas to balance recharge/discharge processes to maintain groundwater balances below the root zone of the agriculture crops.

Requirements to be Met by Bio-drainage

For biodrainage to be effectively adaptable, the following requirements must be met:

(a) Water balance : The quantity of water removed from the ground water annually should equal the quantity of recharge.

(b) Salt balance : The quantity of minerals removed annually should be nearly equal to the quantity of mineral import.

(c) Area under plantation : Irrigation is practised primarily to promote agriculture, horticulture, dairy etc. Therefore in term of economic returns afforestation or agro-forestry should be comparable with that from other alternative uses of land. If it is not so, afforestation may still be justified, on considerations of the environmental and drainage benefits.

(d) Water for plantations : Under ideal situation, trees in afforestation area on full development should be able to draw most of their requirement of water from the ground water table, so that surface irrigation water can be put to other productive uses. If, and so long it is not possible, plantation trees would need some irrigation water. They may also need some water periodically to leach down salts from the root zone, if and when the salinity levels approach threshold limits.

(e) Ground water quality : The quality of ground water, when the water table approaches the root-zone of trees, should be such as can be tolerated by the plant species, otherwise the trees would need to be supplied irrigation water.

(f) Effect on lowering ground water table : Trees can lower the ground water table directly underneath the plantation area, to a depth upto which the tree roots can extend. This can be upto 15 m from ground surface or even more. To be effective as a drainage measure, the ground water table must be lowered in the irrigated area to a minimum critical depth (say 2 m below ground level), at the farthest point from the edge of the plantation area.

Bio-Drainage

Dukhovny and Umarov (2000) classify reclamative regimes as follows:

Automorphous regime	-	Irrigation water not connected with ground water. Infiltration is free
Semi automorphous regime	-	Ground water backs up irrigation water but almost doesn't feed plants
Semi hydromorphous	-	Ground water feeds plants prevailing over irrigation water
Hydromorphous	-	Plants are mainly fed by ground water

Under the bio-drainage scheme, the irrigated crops (cotton etc.) would grow under automorphous regime while the tree plantations to provide drainage would be under the hydromorphous regime. The tree plantations are to be raised in blocks and strips adjacent but outside the blocks of irrigated areas.

Plants can remove water from the soil either (1) directly from the saturated zone below the water table, (2) from the unsaturated capillary fringe above the water table or (3) from unsaturated top soil layers after rainfall or irrigation. Scenarios (1) and (2) result in water table control; scenario (3) recharge control. In scenario (3) leaching is unimpeded; when water application exceeds plant water demand, leaching will take place. In scenarios (1) and (2) leaching becomes restricted and salt accumulation processes begin to occur. This happens especially where water tables are shallow, as is often the case in irrigation areas. A final equilibrium salinity level will establish, depending on applied water salinity, soil hydraulic conductivity, hydraulic gradients (vertical and lateral) and vegetation type (salt tolerance).

Scientific Basis of Biodrainage

In natural environments the components of the hydrological system, i.e. rainfall, evapotranspiration, change in soil-water storage and drainage, are (on average) in balance. Periods of high rainfall might temporarily result in increased drainage flows, a rise in the groundwater table and/or soil moisture storage, then over a period of about 5-10 years equilibrium is established. Vegetation plays a vital role in the evapotranspiration and soil-water storage components of this balance.

Seepage

The increased seepage scenario prevails and the development of land for either rainfed or irrigated agriculture generally increases groundwater recharge rates; the new landscape systems "leak", showed that in a semi-arid region in southern Australia (average rainfall 250-300 mm yr^{-1}), the recharge rate beneath native Eucalyptus spp. was <0.1 mm yr^{-1}.

Consumptive Water Use of Plants

Studies in Australia (Greenwood *et al.*, 1985) suggested that the rates of transpiration and groundwater uptake by trees underlain by relatively shallow (5-8 m below surface) water tables, were very high, exceeding the annual evaporation from pasture (~400 mm) by a factor 3-6 (1,200-2,300 mm yr^{-1}). Potential water use differences between species are also a topic of discussion.

Salt Balance

One of the major factors determining the sustainability of plant productivity (and thus plant water use) processes is salt balance. If the salts moving into the rootzone are not either (1) taken up by the vegetation and harvested or (2) removed from the rootzone by leaching, the vegetation is doomed to succumb to salinity. Therefore, salt balance considerations are crucial to the viability of biodrainage. In the case of *Eucalyptus* species, it reduces to about one-half of potential when the water salinity increases to about 8 dSm^{-1} (Oster *et al.*, 1999).

Water Table

An important consideration when discussing biodrainage issues is the definition of "water table". The water table can be defined as "the upper surface of a zone of saturation, where the body of groundwater is not confined by an overlying impermeable formation". Perched groundwaters are typically shallow, small in extent and often fresh (George *et al.*, 1997). The distinction between water tables and piezometric pressures should be kept in mind when analysing soil water movement under vegetation.

Consideration in the Development of Biodrainage Systems

Water Balance

Biodrainage plantations should be able to extract groundwater volumes equal to the net recharge. The water balance is to be maintained such that the water table is kept below the rootzone.

Plantation Area

The biodrainage plantation area should be kept as small as possible. Agriculture (particularly irrigated agriculture) is practiced primarily to produce high-value crops.. Particularly in arid and semi-arid regions, dryland areas surrounded by irrigated land could be earmarked for tree plantations without loss of productive resources.

Drawdown of Water Table

Crops, including trees, act as biopumps; they depress the water table directly underneath plantation areas and consequently lower the water table in the surrounding area. The drawdown effect under trees/crops depends on the tree/crop's water use, the rate of recharge in the surrounding area, the hydraulic conductivity of substrata and the depth to deeper barrier layers. Biodrainage plantings should be established in blocks or strips and spaced to keep water table levels in the irrigated farmland in between the plantings below the rootzone. The harvesting of the biodrainage plantations would need to be planned in such a manner that the "drainage" function is not lost (thinning regimes).

Economic Aspects

The growing of biodrainage trees and crops requires a different operational management approach than the growing of agricultural crops. Up-front costs associated with planting and maintenance precedes the income from harvesting by many years. Some form of contract growing, based on annual payments might have to be considered to make the system acceptable to landholders.

Social Acceptance

The introduction of new crops such as tree plantations affects rural social societies. New markets might have to be developed, security arrangements differ from those for normal crops (illegal pruning or cutting for fuelwood) and fires could destroy the results of many years of labour in a single day. Active participation of local communities in the development of tree plantation based biodrainage systems is extremely important to overcome problems and ascertain that the benefits of the biodrainage systems are reaped to the maximum extent.

Biodrainage Research Work at FC & RI, Mettupalayam

Forest College and Research Institute, Mettupalayam is pioneer in Biodrainage research. First project on "Screening superior genotypes of Eucalyptus for Biodrainage through ecophysiological approaches" for biodrainage efficiency funded by Ministry of Water Resources is operating at this institute since 2004. This is the model project given for the entire South India. This project is basically working to identify 47 superior genotypes of *E. tereticornis.* However preliminary screening has narrowed down 27 best performing phenotypes. Now these 27 phenotypes are in On Farm Trails (OFTs) in two different locations *viz.,*

1. Tamil Nadu Rice Research Institute, Aduthurai and,
2. Agricultural Engineering College and Research Institute, Kumulur.

These clones were subjected both for growth characteristics as well as biodrainage efficiency related ecophysiological characters *viz.,*

1. Transpiration
2. Photosynthesis
3. Inter cellular CO_2 concentration
4. Stomatal conductance

The biometrical estimation of the clones revealed that among the 26 clones and one seed source, seven clones viz., MTP-6 (147.89 cm), MTP-4 (144.04 cm), MTP-1 (139.16 cm), MTP-5 (131.20 cm), MTP-2 (116.33 cm), *E. saligna* (108.82 cm) and MTP-7 (107.61 cm) recorded significantly higher height. On the contrary, very low performance of the clones in terms of height was observed in 11 clones viz., TCR-2 (63.35 cm), SGc-2 (62.40 cm), ITC-286 (59.99 cm), PD-6 (58.23 cm), EU-1(54.32 cm), EU-2 (51.74 cm), NGL-2 (51.61 cm), ALR-1 (48.22 cm), ALR-2 (48.00 cm), PD-9 (46.58 cm) and PD-10 (43.13 cm). Similar observation was observed among the clones for basal diameter and volume index. A moderate performance of all the biodrainage traits was observed in the clones MTP-3, MTP-8, SG-1, SGc-1, SGc-3, KK-1, KK-2, EC-I and ALR-5.

The clonal efficiency to fit for biodrainage efficiency was identified by measuring ecophysiological traits viz., transpiration rate, stomatal conductance, photosynthetic rate and Inter Cellular CO_2 Concentration. The estimation of ecophysiological traits revealed that transpiration rate was ranged from 0.21 m mol m^{-2} s^{-1} (ALR-5) to 0.61 m mol m^{-2} s^{-1} (MTP-1) at 3 MAP. During 6 MAP, seven clones viz., MTP-1, MTP-5, MTP-3, MTP-6, MTP-2, MTP-4 and *E. saligna* recorded significantly high transpiration rate of 0.75, 0.71, 0.70, 0.69, 0.66, 0.64 and 0.63 m mol m^{-2} s^{-1} respectively. The lowest transpiration rate was observed in three clones viz., PD-9, ALR-3 and EU-1 with the value of 0.28, 0.30 and 0.30 m mol m^{-2} s^{-1} respectively.

The stomatal conductance estimation revealed that significantly higher stomatal conductance was observed in five clones viz., MTP-5 (51.88 m mol m^{-2} s^{-1}), MTP-2 (50.18 m mol m^{-2} s^{-1}), *E. saligna* (48.26 m mol m^{-2} s^{-1}), MTP-6 (48.25 m mol m^{-2} s^{-1}) and MTP-1 (48.22 m mol m^{-2} s^{-1}). On the contrary, significantly lowest stomatal conductance was exhibited by ALR-5 (31.30 m mol m^{-2} s^{-1}).

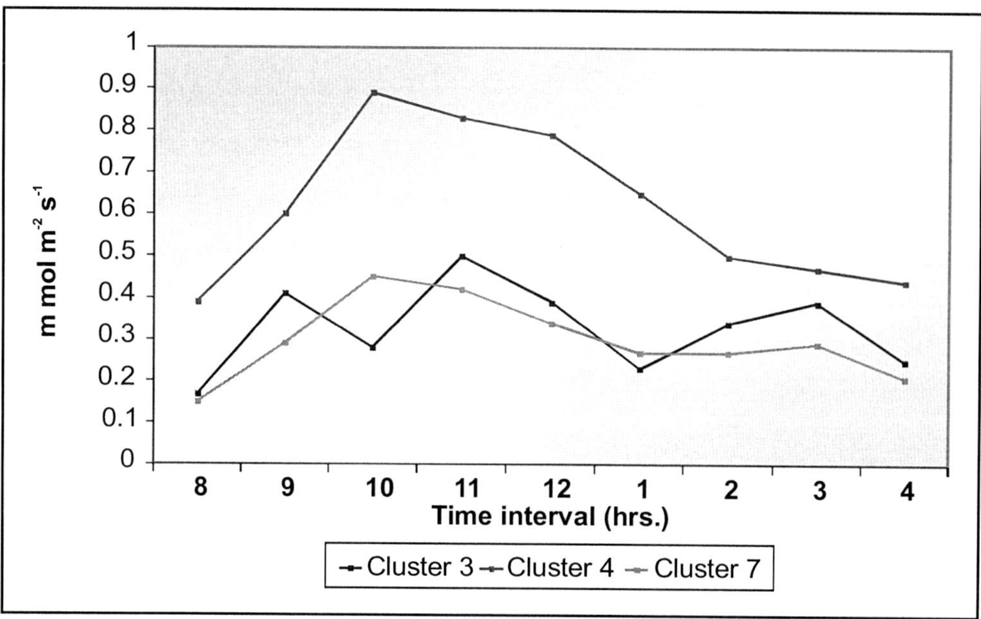

Fig. Cluster means for diurnal variation in Transpiration rate (Balasubramanian *et al.*, 2006)

The result revealed that the Inter Cellular CO_2 Concentration was ranged from 298.25 to 189.99 ppm. The highest level of Inter Cellular CO_2 Concentration was registered by MTP-4 (298.25 ppm) followed by MTP-5

(296.21 ppm) and the lowest performance was showed by the clone PD-6 which was accounting the Inter Cellular CO_2 Concentration of 189.99 ppm. Similar to transpiration rate and stomatal conductance, photosynthetic rate also exhibited same trend of results. Photosynthetic rate was significantly higher for the clones viz., MTP-1, MTP-2, MTP-5, MTP-3, MTP-4 and MTP-6 with the value of 0.55, 0.51, 0.48, 0.47, 0.47 and 0.47 m mol m^{-2} s^{-1}. The lowest photosynthetic rate was accounted by ALR-2 (0.24 m mol m^{-2} s^{-1}).

The genetic divergence studies using multivariate analysis helped to genetically group the 26 clones and one seed source into seven distinct clusters. The cluster mean revealed that high productive clones viz., MTP-1, MTP-5, MTP-6 and SG-1 were present in the cluster IV. The low productive clones viz., SGC-3, PD-9, ALR-2, ALR-5 and TCR-2 were present in the cluster VII.

The ecophysiological characterization of clones revealed that highest transpiration rate was accounted by the clonal members in the cluster IV (0.653 m mol m^{-2} s^{-1}) followed by clones in cluster VI (0.553 m mol m^{-2} s^{-1}). The lowest transpiration rate with cluster mean value of 0.320 m mol m^{-2} s^{-1} was exhibited by cluster VII. On the contrary, the lowest clusters mean value of 32.573 m mol m^{-2} s^{-1} and 0.286 m mol m^{-2} s^{-1} respectively for stomatal conductance and photosynthetic rate were observed in the cluster VII. The lowest Inter Cellular CO_2 Concentration was observed in the cluster III (201.90 ppm).

The ecophysiological mechanism that present in the clones were studied in terms of diurnal variation for the traits viz., transpiration rate, stomatal conductance, photosynthetic rate and Inter Cellular CO_2 Concentration. The diurnal variation study revealed that highest transpiration throughout the day was observed among the clonal members present in the cluster IV by registering the peak transpiration rate of 0.89 m mol m^{-2} s^{-1} at 10.00 AM. The same cluster showed very high transpiration rate of 0.79 m mol m^{-2} s^{-1}even at 12.00 Noon. The lowest transpiration rate through out the day was observed among the clonal members in the cluster VII. A moderate transpiration rate was observed among the clonal member in the cluster III and V. It is also interesting to note that all the clusters have maintained steady transpiration rate throughout the day without prominent midday depression. However, transpiration remarkably varies among the clusters. It is also important to note that all the cluster members varied in their ability to maintain different levels of photosynthetic rate, stomatal conductance and Inter Cellular CO_2 Concentration but none of the cluster showed any prominent midday depression for all the traits under consideration.

Biodrainage Work in India

India is making great effort in biodrainage research work next to China and Australia. MoWR, Govt. of India is organizing biodrainage research across the country through Indian National Committee on Irrigation and Drainage (INCID). The following research organizations are undertaking biodrainage research in the country.

1. Forest College and Research Institute, TNAU, Mettupalayam
2. Haryana Agricultural University, Hissar
3. Tropical Forest Research Institute, Jabalpur
4. Arid Forest Research Institute, Jodhpur
5. WTC for Eastern Region, Bhuvaneswar

Besides, the ministry has formed biodrainage working group which is actively co coordinating and implementing biodrainage research in consultation with INCID.

Application in Pollution Abatement

Biodrainage technology is the best technology that can be conveniently used for abatement of surface and ground water pollution. Many river pollution can be abated through floating bed techniques. Similarly sewage farms, effluents from dye, chemical and paper industries can be handled through biodrainage technique. This adds advantage of handling pollution at the source itself, which will greatly reduce the pollution of ground and surface water. In addition the yield of trees planted for biodrainage purpose is also adding revenue at the time of harvestable maturity.

Tree Species for Biodrainage

Characteristics of Species

Any tree/plant species that can with stand water logging can be advocated for biodrainage purpose. However the species that are having the following characters are desired for biodrainage utility.

1. Withstand prolonged water logging.
2. High transpiration rate.
3. High water use efficiency.
4. Perennial in nature.
5. Should not contain allelo chemicals.
6. Salt tolerant / resistant.
7. Evergreen.
8. Broadleaved

Suitable Species

Many tree species are having biodrainage suitability. However the following tree species are recommended for biodrainage,

1. *Eucalyptus tereticornis*
2. *Eucalyptus camaldulensis*
3. *Eucalyptus saligna*
4. *Terminalia arjuna*
5. *Anthocephalus cadamba*
6. *Acacia nilotica*
7. *Pongamia pinnata*
8. *Bambusa bambos* (shorter period of inundation)
9. *Prosopis juliflora*

Tree water use of various species reported in arid and semi arid regions

Region	Species	Method	Water use	Unit	Source
Turkmenistan	*Elaeagnus* spp.	Destructive	138	1	Eliseev, 1939
	Morus alba	Gravimetric	275	day^{-1}	
	Populus spp.		346	$tree^{-1}$	
	Salix spp.		379		
Pakistan	*Eucalyptus* spp.	potential ET	137-143	1	Roitzsh & Marsur
	M. alba		153	day^{-1}	Cited by Bhutta &
	Populus spp.		45	$tree^{-1}$	Chaudhry, 2000
Western USA	*Populus* spp.	n/a	650-2350	mm	
	Salix spp.		770-1350	$year^{-1}$	Cited by Bhutta &
	Tamarix spp.		2200-2800		Chaudhry, 2000
Australia	*Eucalyptus* spp.	Heat pulse	300	mm	Morris *et al.*, 1998
	E.globulus	Hydrologic	262-595	yr^{-1}	Hingston *et al.*, 1995
	Corymbia	analysis	675		Theiveyanathan &
	maculata	n/a			Benyon, 2000

(Khamzina *et al.*, 2005)

Conclusion

Biodrainage is the best tool to be introduced to the waterlogging areas by integrating biodrainage with agriculture system. Though it is an effective system the research evidences in this new vista is limited in many areas like choice of tree crops, root zone salinisation etc. However this system is advocated for its inbuilt advantage including less cost, additional income and environmentally friendly nature.

References

Asia Khamzina, John P.A. Lamers, Bart Wickel,Yulduz Djumaniyazova and Christopher Martius. 2005. Evaluation of Young and Adult Tree Plantations for Biodrainage Management in the Lower Amudarya River Region, Uzbekistan. ICID 21st European Regional Conference 2005 - 15-19 May 2005 - Frankfurt (Oder) and Slubice - Germany and Poland

Balasubramanian, A., Umadevi, M., Dhixiya Deve, K.R., Dhayalan, K., Srimathi, P. and Paramathma, M. 2006. Assessment of clonal variation for biodrainage efficiency. In: National seminar on Agroforestry for livelihood security, environment protection and biofuel production, held at National Research Centre for Agroforestry, Jhansi, during 16-18 December, 2006.

Bhutta, M.N. and Chaudhry, M.R. 2000. Biological control of waterlogging. Proceedings of the 8th ICID International Drainage Workshop, New Delhi, India. Vol. IV, Group V: Pp. 33-45.

Calder, I. 1994. Eucalyptus water and sustainability report. ODA Forest Series No.6.

Cazs. 2002. www.cazs,ban.gor.ac.UK/salinity/index.html.

Dukhovny, V.A and Uma Rov, P.D. 2000. "The land salinization and drainage problem in Central Asia" paper presented at the 8th ICID International Drainage Workshop, New Delhi - Vol. I – Pp. 71, 79-84

Eliseev, L.V. 1939. Transpiration of woody species in irrigated conditions of Turkmen SSR. Forestry J 5. (in Russian) Forestry and Forest Products. No. 31: Pp. 5.

George, R.J., McFarlane, D.J. and Nulsen, R.A. 1997. Salinity threatens the viability of agriculture and ecosystems in Western Australia. *Hydrogeology Journal*, **5**: 6-21.

Greenwood, E.A.N., Klein, L., Beresford, J.D. and Watson, G.D. 1985. Differences in annual evaporation between grazed pasture and Eucalyptus species in plantations on a saline farm catchment. Journal. *Hydrologist*, 78: 261-278.

Heuperman, A.F., Kapoor, A.S and Denecke, H.W. 2002. Biodrainage - Principles, experiences and applications. Food and Agriculture Organization of the United Nations, Rome.

Hingston, F. J., Galbraith, J. H. and Dimmock, G. M. 1995. Evaluating the effects of soils and climate on the productivity of *Eucalyptus globulus* plantations on contrasting sites in south-west of Western Australia. Final Rep. Project, CSF-41A, RIRDC, Pp. 67.

Kapoor, A.S. 2003. A new approach to overcome the waterlogging and salinity problems in the aral sea basin. Paper No. 053. Paper presented at the 9th International Drainage Workshop, held at Utrecht, The Netherlands during September 10 - 13, 2003.

Mona Mehta. 2002. Next five year plan to focus or biodrainage plantations. The Times of India, 18.

Morris, J., Mann, L. and Collopy, J. 1998. Transpiration and canopy conductance in a eucalypt plantation using shallow saline groundwater. *Tree Physiology*, 18: 547-555

Oster, J.D., Macedo, T.F., Davis, D. and Fulton, A. 1999. Developing sustainable reuse and disposal of saline drain water on Eucalyptus. Department of Environmental Sciences, UC Cooperative Extension, University of California, Riverside, USA.

Theiveyanathan, V. and Benyon, R.G. 2000. Planting trees for sustainable rice production. In Onwood; research updates from CSIRO.

❏❏❏

Agroforestry: Natural Resource Sustainability, Livelihood & Climate Moderation © 2009
Editors: OP Chaturvedi, A Venkatesh et. al., pp. 231-240
Satish Serial Publishing House, Delhi

Chapter 17

APPROACHES AND RESOURCE CONSERVATION STRATEGIES FOR SUSTAINABLE MANAGEMENT OF AGROFORESTRY INTERVENTIONS

Ramesh Singh, D.R. Palsaniya, R.K. Tewari, R.S. Yadav and R.K. Singh
National Research Centre for Agroforestry, Jhansi-284 003 (U.P.)

Introduction

Survival of the mankind on earth is essentially dependent on three basic resources-soil, water and vegetation. A no-care attitude and gross negligence coupled with over exploitation of these resources to meet the requirements of increasing population, 1.4 per cent per annum, have taken the problem to the threatening dimension. Average soil erosion loss in India is 16.35 tones ha^{-1} of which 20 per cent is lost to the sea (Dhruva Narayana *et al.*, 1983). On the water front, India is blessed with one of the highest precipitation (about 400 million ha- m) in any country of this size, but because of its wide space and time variation agricultural production suffers severely. Out of 400 million ha-m precipitation in the country, 70 m ha-m is consumed as evapotranspiration, 215 m ha-m infiltrates into soil, whereas 115 m ha-m is lost as runoff laden with sit and nutrient resulting into drought and land degradation in the catchment and floods and siltation of reservoirs in the downstream. Harvesting of 25 per cent of this 115 m ha-m runoff through water harvesting structures can provide three irrigations to the entire rainfed area, which constitute about 66 per cent of cultivated land in our country (Samra *et al.*, 1998). Harmonized area statistics on wasteland/degraded lands is depicted in Table 1. These areas can be rehabilitated through agroforestry interventions with watershed approach.

Watershed Approach

Watershed is a naturally occurring geohydrological unit draining to a common point, hence it is an ideal unit of development process. Watershed development needs integrated development of non-arable land, arable land, rainwater, vegetation, livestock, local materials, common property resource and human resource in a participatory process (Samra and Dhyani, 2001). Agroforestry interventions are the only way out for rehabilitation of degraded lands. These interventions are very much effective in conserving natural resources. Grewal (1993) compared some agroforestry interventions for eroded lands of Shivalik and reported minimum runoff, soil and nutrient loss from *eucalyptus* based silvipastoral system (Table 2). Watershed approach provides an opportunity for rational utilisation and management of land, water and vegetation. Following are the essence of watershed approach:

- Land use planning based on its capacity.
- Construction of relevant soil and water conservation structures.
- Rain water harvesting and efficient water management for drought proofing.
- Draining out excess runoff safely.
- Maintaining good vegetative cover over ground to minimise erosion of soil and water and maximize water retention in soil.
- Maximizing productivity per unit area, per unit time and per unit of water.
- Optimizing income of the people through integrated crop-livestock-tree-labour complex over years and to minimize the risk due to aberrant weather conditions.

Measures for Erosion Control in Arable Lands

Erosion control measures are also called as soil conservation practices/ measures/structures and can be grouped under biological measures and engineering or mechanical measures. Biological measures are adequate when the land slope is less than 2 per cent. Mechanical measures for arable lands and their adaptability are depicted in Table 3.

Table 1. Harmonized area statistics of wastelands/degraded lands of India (million-ha)

S. No.	Type of Degradation	Arable Land	Open Forest land (<40% canopy)	Data Source
1.	Water erosion (> 10 t ha^{-1} yr^{-1})	73.27	9.30	Soil loss map, CSWCR&TI
2.	Wind erosion (Aeolian)	12.40	-	Wind erosion map, CAZRI
3.	Chemical degradation	17.45	7.23	National salt affected soil map, CSSRI, NBSS&LUP, NRSA
4.	Physical degradation	1.07	-	Wasteland map of NRSA
	Total	104.19	16.53	
	Grand Total (Arable land and Open forest)	120.72		

(Source: Anonymous, 2008)

Measures for Erosion Control in Non-Arable Lands

Non-arable lands are major source of runoff and sediment resulting into devastating affect in the lower reach. Some of erosion control measures for non-arable lands are as follows:

- Trenching
- Stone wall
- Silt detention tank
- Diversion drain
- Toe wall/toe drain
- Inceptor drain
- Stream training
- Retaining wall
- Revetment
- Waterways
- Wattling
- Drop structure
- Drop inlet structure
- Chute structure

Table 2: Runoff, Soil and nutrient loss from different land use systems in Shivalik region

Land use systems	No. of years of observations	Runoff	Soil loss (t ha^{-1})	Nutrient loss (kg ha^{-1})	
				N	K
Eucalyptus -Bhabar grass	6	0.05	0.07	0.46	0.90
Acacia catechu - forage grass	3	2.0	0.24	6.97	0.52
Leucaena – Napier grass	3	4.4	0.28	6.60	1.20
Teak - Leucaena - Bhabar	3	3.3	0.43	2.08	0.55
Eucalytus - Leucaena -Turmeric	5	2.6	0.59	2.47	0.73
Poplar-Leucaena - Bhabar	5	4.8	1.54	5.90	1.10
Sesamum -rape seed	3	20.5	2.69	42.50	3.00
Cultivation fallow	3	23.0	5.65	51.30	5.00

Runoff

Design of any erosion measures is mainly dependent on runoff from a catchment. Therefore, estimation of probable runoff from any catchment is the first step towards designing and construction of soil and water conservation measures. Runoff is drainage of precipitation from a catchment, which flows out through its natural drainage system. After the occurrence of infiltration and other losses from the precipitation, the excess rainfall flows out through the small natural channels on the land surface to the main drainage channel. Such types of flows are called surface flows. A part of the infiltrated rainwater moves parallel to land surface as subsurface flow, and reappears on the surface at certain other points. Such flows are called interflows. The other part of the infiltrated water percolates downwards to ground water, and moves laterally to emerge in depressions and rivers, and joins the surface flow. This type of flow is called the subsurface flow or ground water flow.

Runoff is a rainfall driven process and depends upon biophysical characteristics of the catchments. Runoff estimation includes its volume and peak rate of flow volume. In designing spillways and outlets or waterways, peak rate of runoff is required while for assessing the storage in earthen dam, tanks and ponds etc. the estimates of runoff volumes are required. Another important variable of interest in drainage line treatment is the flow velocity that is required for determining scour pattern in the river bed and along the banks. In this way, various characteristics parameters of runoff are required for the design of soil and water conservation structures.

Table 3. Suitability of various measures for arable lands

S. No.	Measures	Land slope (%)	Soil depth	Rainfall (mm)
I.	Bunding			
	1. Narrow based terracing			
	a. Contour bunding	<8	Shallow & light	<600
	b. Graded bunding	<8	Shallow & light	>600
			Heavy (clay)	<600
	2. Broad based ter racing			
	a. Level or absorptive type	<8	Shallow & light	<600
	b. Graded or disposal or channel type	<8	Shallow & light	>600
			Heavy (clay)	<600
II	Trenching			
	1. Contour trenching			
	a. Continuous	<8	Medium to deep	<1500
	b. Staggered	<8	Shallow	<800
		<33	Medium to deep but well drained	<2000
	2. Graded trenching	<33	Medium to deep but well drained	2000-3000
III	Terracing			
	a. Level	<33	Medium to deep	<2500-3000
	b. Inward sloping	<33	-do-	-do-
	c. Outward sloping	<33	Shallow	<1200
	d. Puertorican type	<33	Shallow to deep	<1500
IV	Stonewall			
	a. Contour	<33	Shallow to medium	<1500
	b. Graded	<33	-do-	1500-2500
V	Zing Terracing	<33	Shallow to medium	<1200
VI	Conservation bench terracing	<33	-do-	<1200

Methods of Runoff Computation

There are many methods available for runoff estimation. The most commonly adopted methods are the Curves Number method of the Soil Conservation Service of the USA for estimation of both excess runoff volume and peak flow rate and the rational method for the peak flow rate. Many empirical relationships are also widely used for estimation of flow rates.

Curve Number Method

The curve number of method was developed for the determination of the rainfall excess of agricultural watersheds, on per storm basis.

$$Q = \frac{(p - 0.2S)^2}{P + 0.8S}$$

Where, Q = direct runoff, mm

P = storm rainfall, mm

S = a parameter for surface retention

The parameter S is defined as

$$S = \frac{25400}{CN} - 254$$

Where, CN = an arbitrary curve number varying from 0 to 100

The amount of rainfall (P) is also affected by duration. For design considerations, maximum runoff volume is required. It is established that minimum storm duration for flood estimation can be taken as six hours. But in certain conditions design rainfall for greater durations can also be taken.

Antecedent rainfall conditions are explained in Table 4. The curve number CN for Antecedent rainfall condition II are given in Table 5.

Procedure for Using SCS Curve Number Method

1. Divide the watershed area into uniform hydrologic soil cover groups.
2. Find out the antecedent moisture condition (AMC) based on 5 days total rainfall in previous days.
3. Find out the CN from Table 5.
4. Determine weighted curve number

$$CN = \frac{CN_1 x A_1 + CN_2 x A_2 + CN_3 x A_3 . + ... CN_n x A_n}{A_1 + A_2 + A_3 + ... + A_n}$$

5. If applicable, convert CN for the AMC II to other AMC from conversion factors given in Table 6.
6. Compute S
7. Compute direct runoff using the equation.

Example

Determine the runoff from a watershed of 60 ha from following giving observations:

Depth of rainfall, 100mm

Antecedent rainfall condition. AMC I

Row crop, good condition in 40 ha

Woodland, good condition in 20 ha

Solution : The curve number data from Table 5 are available for AMC-II. Since the present condition is AMC I, therefore, the curve number need to be appropriately converted to this condition from AMC II, by using Table 6.

Type of crop	Curve number at	
	AMC II	AMC I
Row crop, good	82	82 x 0.8 = 65.6
Woodland, good	55	55 x 0.65 = 35.75

$$\text{Weighted curve number} = \frac{(656 \times 40) + (35.75 \times 20)}{60}$$

CN = 55.65

S = (25400/CN)- 254 = 202.42 mm

$$\text{Therefore, } Q = \frac{(100 - 0.2 \times 202.42)^2}{100 + 0.8 \times 202.42}$$

$$= 13.5 \text{ mm}$$

The runoff in response to 100 mm of rainfall will be 13.5 mm.

Table 4 : Antecedent Rainfall Conditions

Condition	Description	5-day antecedent Rainfall, mm
I	Low moisture condition, wilting point to dry Average conditions	<35
II	Average conditions	35-52.5
III	Wet moisture conditions, heavy rainfall prior to the day under consideration	>52.5

Rational Method

This method is based on the assumption that constant intensity of rainfall is uniformly spread over an area, and the effective rain falling on the most remote part of the basin takes a certain period or time, known as the time of concentration (Tc), to arrive at the basin outlet. This method is used for a estimation of peak rate of runoff. It is expressed as

$$Qp = \frac{CIA}{360}$$

Where, Qp = peak discharge in m^3/sec.

C = Runoff coefficient

I = Design intensity for a particular return period and duration equal to time of concentration, mm/h

A = Area of the watershed, ha

Return Period

The return period, sometimes called recurrence interval, is defined as the period of time within which the depth of rainfall for a given duration is expected to be equalled or exceeded atleast once. Recommended return periods for various types of structures are given in Table 7.

Intensity, Duration and Frequency Relationship

In order to find the design intensity for a particular location or design, intensity duration-frequency relationships have been developed. A general expression for rainfall intensity is given as:

$$I = \frac{KT_r^a}{(Tc+b)^n}$$

Where, I is rainfall intensity (cm hr^{-1}), T is return period (year), Tc is the time of concentration (hour), and K, a, b, n are the constants for a specific location.

Table 5: Curve Numbers for Antecedent Rainfall Conditions II

Landuse or cover	Treatment or practice	Hydrologic condtion	Hydrologic soil group A	B	C	D
Fallow	Straight row	-	77	86	91	94
Row crop	Straight row	poor	72	81	88	91
	Straight row	Good	67	78	85	89
	Contoured		70	79	84	88
	Contoured		65	75	82	86
	Terraced		66	74	80	82
	Terraced		62	71	78	81
Small grain	Straight row		68	76	84	88
	Straight row		63	75	83	87
	Contoured		63	74	82	85
	Contoured		61	73	81	84
	Terraced		61	72	79	82
Close -seeded	Terraced	Good	59	70	78	81
Legumes or	Straight row		66	77	85	89
rotation	Straight row		58	72	81	85
Pasture or range	Contoured		64	75	83	85
	Contoured		55	69	78	83
	Terraced		63	73	80	83
	Terraced	Good	51	67	76	80
		Poor	68	79	86	89
Woodlots		Fair	49	69	79	84
		Good	39	61	74	80
	Contoured	Poor	47	67	81	88
Roads	Contoured	Fair	25	59	75	83
	Contoured	Good	6	35	70	79
		Poor	45	66	77	83
		Fair	36	60	73	79
		Good	25	55	70	77
			74	84	90	92

Table 6 : Factors to convert N for condition II to other conditions

Curve number for condition II	Factors to convert Curve Number for condition II to Condition I	Condition II
10	0.40	2.22
20	0.45	1.85
30	0.50	1.67
40	0.55	1.50
50	0.62	1.40
60	0.67	1.30
70	0.73	1.21
80	0.79	1.14
90	0.87	1.07
100	1.00	1.00

Table 8: Values of parameters for intensity duration return period relationship for different zones of India.

Zone	K	a	b	n
Northern Zone	5.92	0.162	0.50	1.013
Central Zone	7.47	0.170	0.75	0.960
Western Zone	3.98	0.165	0.15	0.733
Southern Zone	6.31	0.153	0.50	0.950

Time of Concentration

For determination of the time of concentration, the most widely used formula is the equation given by Kirpich (1940), which is given as :

$$Tc = 0.01947L^{0.77}S^{-0.385}$$

where,

Tc = Time of concentration (min)

L = Maximum length of travel of water (m)

S = Slope of the drainage basin

References

Anonymous. 2008. Harmonization of wasteland/degraded lands datasets of India. Wasteland News, XXIII, No. 4, Pp. 50-53.

Dhruva Narayana V.V. and Babu Ram. 1983.Estimation of soil erosion in India. *Journal of Irrigation and Drainage Engineering,* 109 (4): 419-434.

G. Das, 2004. Hydrology and Soil Conservation Engineering. Prentice-Hall of India, New Delhi, Pp. 489.

Grewal, S.S. 1993. *Agroforestry systems for soil and water conservation in Shivaliks.*In: Agroforestry in 2000 AD for the semi arid and arid tropics. NRC for Agroforestry, Jhansi, Pp. 82-85

Samra, J.S. and Narayan, Pratap. 1998. *Soil and Water Conservation.* In: 50 Years of Natural Resource Management Research (Eds. G. B. Singh and B. R. Sharma), ICAR, New Delhi, Pp. 145-176.

Schwab, G.O., R.K. Frevert, T.W. Edminster and Barnes, K.K. 1981. Soil and Water Conservation Engineering, John Wiley, N. York.

Ram Babu, Tejwani, K.G., Agarwal, M.C. and Bhushan, L.S. 1979. Rainfall intensity-duration-return period equations and nomographs of India, Bulletin No.3, CSWCRTI, Dehradun.

Samra, J.S. and Dhyani, S. K. 2001. *Role of trees in watershed development.* In: Agroforestry (Eds. Sunil Puri *et al.*), ICAR, New Delhi.

❑❑❑

Agroforestry: Natural Resource Sustainability, Livelihood & Climate Moderation © 2009
Editors: OP Chaturvedi, A Venkatesh et. al., pp. 241-261
Satish Serial Publishing House, Delhi

Chapter 18

TREE-CROP INTERACTIONS AND THEIR MANAGEMENT IN AGROFORESTRY

D.R. Palsaniya, Ram Newaj and R.S. Yadav
National Research Centre for Agroforestry, Jhansi-284 003 (U.P.)

Introduction

In agroforestry, tree and agricultural crops are combined together and they compete with each other for growth resources such as light, water and nutrients. Tree-crop interaction is simply how the components interact and affect each other in a given environment. More specifically, the biological potentials and restraints of agroforestry systems and environmental responses of plant components within them, is tree-crop interface (Huxley, 1985). Tree-crop interactions occur when more than one individual are present in a biological system and when resources are limited. The resources sharing by the components may result in complementary or competitive effects depending upon the nature of the species involved in the systems, the manner in which they are grown and the climatic factors. Tree-Crop Interaction is an important aspect in agroforestry. The positive (complementary) and negative (competitive) effects of tree in agroforestry systems have been realized by field observations. Many of the quoted examples of soil fertility improvement are based on traditional systems that have been established for many years. *Faidherbia albida* in western Africa and *Prosopis cineraria* in Rajasthan, India are the best examples for improving soil fertility and less competitive effect with crop. Similarly, Poplar based agroforestry system is a promise in some part of Haryana, Punjab, and Uttar Pradesh to increase the productivity. Shelterbelts and windbreaks are well established examples of microclimatic improvement that helps to improve the crop yield where high wind and sand movement are major constraints in cropping. Several other benefits of trees have been realized for soil conservation, environmental protection, nutrient cycling, weed control, sustainability and these benefits may be direct or indirect. But, there is still an urgent need for research to acquire more scientific evidences before attempting the formidable task of translating the promise of agroforestry in to sustainable land use.

Competition between trees and crops for the same limiting growth factors is more obvious when they are grown in close proximity. However, the extent of below ground competition is often not apparent. In agroforestry system, competition between tree and crop is usually a significant factor, which has a negative impact on extension of agroforestry. Quantification of competitive effect for soil moisture, light and nutrients in tree-crop system can play an important role in designing of an agroforestry and can suggest the management practices to maximize the complementarity and minimize the competition.

Factors Affecting Tree-Crop Interactions

The nature and quantum of tree-crop interactions and their effects are not uniform and depends upon many factors like:

- Nature of the species grown.
- Age and size of trees.
- Density of components.
- Management practices- spacing, training, pruning, irrigation, fertilization, etc.
- Environmental factors.

In case of erect growing and sciophyte (shade loving) plants, the extent of tree-crop interaction and the resultant effects will be less as compared to spreading type and heliophyte (sun loving) plants. Similarly, the adverse effect will be more in case of aged and large sized of trees. Further, higher the density of components (tree and crop), higher will be the negative effect on the under storey crop and vice-versa. The management practices like training, pruning, irrigation, fertilization, etc., environmental factors and availability of water, nutrients and light also affect the tree-crop complementarities or antagonism.

Types of Tree-Crop Interactions

Based on the resources sharing by the components, tree-crop interaction may be complementary or beneficial or positive or synergistic in one hand and inhibitory or competitive or negative or antagonistic on the other. Tree-crop interactions may also be as above ground and below ground interaction. Interaction may be among crop components, among tree components and between tree and crop components, the latter is more important in agroforestry. In a legume and non-legume based intercropping system, part of the nitrogen fixed in the root nodule of the legume may become available to the non-legume component is an example of complementarity. Further, the complementarities in resource capture and sharing under mixed cropping/intercropping system may be spatial or temporal. In temporal complementarity, the leaf canopies of component crops may occupy different vertical layers with taller component tolerant of strong light and high evaporative demand and shorter component favoring shade and high relative humidity. Multistoried cropping in coconut and planting shade trees in cocoa and tea plantation uses this principle. In similar manner under agroforestry system, exploitation of different layers of soil by the root systems of trees and crops may lead better utilization of resources with much less competition. In temporal complementarity, the yield advantages provided by the mixture cannot always are explained by more effective use of growth resources at specific times. Indeed, there are substantial opportunities for temporal complementarity if species make their major demands on available resources at different times, thereby reducing the possibility of competition. For example, the usually advantageous intercropping of the slow-growing pigeonpea with fast-growing cereals such as maize or sorghum produces only limited increase in total light or water use (Natarajan and Willey, 1980) but nevertheless provides intercrop advantages of 40-70%.

Manifestation of Tree-Crop Interactions

Tree-crop interaction may be manifested in terms of effects on following parameters of the components of the agroforestry.

- Effect on growth.
- Effect on yield and productivity.
- Effect on quality.
- Effect on intercultural operations and other agronomic practices.

Based on the above manifestation which noticed on the basis of experimental evidences and whose manifestation may be beneficial or harmful, tree-crop interactions in agroforestry may be divided into two categories. These are:

Complementary Effects

1. Increased productivity.
2. Improved soil fertility (organic matter addition and N-fixation).
3. Nutrient recycling.
4. Soil and water conservation.
5. Microclimate improvement.
6. Anchorage to climber crops.
7. Crop protection from cold and heat waves and frost (wind break and shelterbelt).
8. Bioamelioration/Phytoremediation.
9. Sustainability.

Competitive Effects

1. Competition for light, water, nutrients space and CO2 (above and below ground)
2. Allelopathy (+ve and –ve)
3. Insects-pests and diseases
4. Birds (+ve and –ve)

Table 1: Main effects of tree-crop interaction and their evidences

Effects	Direct	Indirect	Source
Increased productivity	+	+	Ong, 1991
Improved soil fertility	+	+	Kang *et al.*, 1990
Nutrient recycling	+	+	Szott *et al.*, 1991
Soil and water conservation	+	+	Lal, 1989
Microclimate improvement	+	+	Wiersum, 1991
Competition	-	-	Monteith *et al.*, 1991
Allelopathy	0	-	Ong *et al.*, 1991
Weed control	0	+	Tian and Kang, 1994
Sustainability and stability	0	+	Rizvi, 1991
Pest and diseases	0	+, -	Young, 1991

Positive effects are indicated by (+) and negative by (-): where evidence is not available it is indicated by (0).

Complementary Effects

Increased Production and Productivity

There are ample evidences to show that the overall (biomass) productivity of an agroforestry system is generally greater than that of an annual system although not necessarily greater than that of a forestry or grassland system. The basis for the potentially higher productivity could be due to the capture of more growth resources e.g. light or water or due to improved soil fertility. Competition, which is a negative effect in this context, is usually a significant factor in tree-crop system, even when there is evidence of increased combined productivity by both components. Some evidences on increase in productivity under various agroforestry systems are given in following paragraphs.

Poplar (*Populus deltoides*) has become a popular species in agroforestry system. Deciduous nature of poplar allows agricultural crops to grow under it without adversely affecting the yield. Up to first 2 years, the returns obtained from sugarcane+poplar combination are higher than any other crop, as the preparatory and cultural operation for both of these are complementary. However, third year onward, shade-loving crop like turmeric and ginger can be grown under it. During the winter (Rabi) season wheat can be grown till its harvest (Jagdish Chander, 1998; Sharma *et al.*, 2001; Sharma and Singh, 1992). In another study, under agrisilvicultural system having poplar (clone G-3) as tree component and wheat crop as an intercrop, a substantial reduction in wheat yield (23.3%) under three-year old poplar planting was observed. Pruning after the third year permitted some recovery in yield. It should be noted that the losses in wheat crop were amply compensated by the increment put on by the tree component where poplar planted in January/February1985 were harvested and disposed off by the farmers in August/September1999, i.e. at a rotation of six and half year (Ralhan *et al.*, 1992).

Poplar tree matures for harvest after 7-8 years having a girth of 90-100cm. One tree of 90 cm of girth provides 5-6 quintals of fresh bole wood and each quintal of wood fetches about Rs. 400 in the open market. One tree after 8 years can provide a handsome price of about Rs.2000. In one acre, 160 trees fetch about Rs.3, 20,000. On an average, a farmer can earn about Rs. 40,000 $acre^{-1}$ yr^{-1}. The income from crops grown in association with poplar is additional. In fact, in intercropping the growth of poplar significantly increased compared to their monoculture (Toky, 1997).

Singhal and Panwar (1992) studied on poplar based agroforestry system under eight years cropping scheme (poplar spacing 4 m x 5 m, 500 plants ha^{-1}). The crop rotation followed is sugarcane-main crop (1st year), sugarcane coppice (2nd year), wheat-jowar fodder (3rd to 6th years) and wheat -fallow (7 th to 8th years). The other crops included in the system were maize, potato, mustard, soybean, lentil, and turmeric, fodder crops and aromatic herbs. The commercial value of tree was calculated at the rates prescribed by M/s WIMCO which was Rs. 530/tree for economic analysis. The yearly expenditure and return were discounted at 12% interest rate (Table 2). The net present value (NPA) per hectare per year was Rs. 29,955.00 with benefit/cost of 4.45. The high benefit/cost ratio of 4.45 obtained in this case may be due to increased number of trees per hectare (500 plants ha^{-1}) and use of fast growing clones.

In an agroforestry system integrating grey gum (*Eucalyptus teriticornis*) for pole wood on border rows of 40 x 40 m plot, K-8 *Leucaena leucocephala* for organic mulch and manure in paired rows along the plots of shade loving turmeric. Grey gum poles (380) harvested after 6.5 years gave income of Rs.18, 380. The agroforestry system generated cash return of Rs.5642 ha^{-1} yr^{-1} compared with Rs.2997 from rainfed crops and thus proved economically more viable than normal system of growing field crops (Grewal *et al.*, 1995). Another study (1983 to 1992) conducted on nine (15 m x 90 m) runoff plot at 4% slope to compare the production efficiency of *Leucaena leucocephala* and *Eucalyptus* hybrid based agroforestry as well as monocropping landuse system in warm sub humid climate of the western Himalayan region of India. Treatment for the first sequence were monocropping system of *Leucaena, Eucalyptus, Chrysopogon fulvus* grass and maize-wheat rotation and alley cropping system of grass and crop at 4.5 and 10.5 m alley widths with paired contour tree rows of *Leucaena and Eucalyptus*. In second sequence, alley width increased to 22.5 m in 1989, grass was replaced by turmeric and paired contour rows of *Leucaena* hedges were introduced in monocropping system of grain crop and turmeric. Production of biomass was much lower near the tree rows than in mid alleys. Managing *Leucaena* as contour hedgerows eliminated crop yield reduction in alleys. Performance of grass and turmeric in alleys was not found to be satisfactory.

Table 2: Cost benefit analysis of one ha poplar (*P. deltoides*) plantation at spacing 4 m x 5 m with intercropping in North-Western Uttar Pradesh

Year	Cost of cultivation (Rs.)		Total working cost (Rs.)	Discounted @ 12% interest (Rs.)	Gross benefits (Rs.)	Discounted @ 12% interest (Rs.)
	Poplar	Crop				
I	8550.00	10596.00	19446.00	17362.00	20960.00	18714.00
II	900.00	6896.00	7796.00	6960.00	20960.00	18714.00
III	1640.00	7640.00	9290.00	8294.00	16315.00	14566.00
IV	1890.00	7640.00	9530.00	8508.00	16315.00	14566.00
V	1890.00	7640.00	9530.00	8508.00	16475.00	14709.00
VI	1890	7640.00	9530.00	8508.00	12180.00	10875.00
VII	--	6240.00	6240.00	5571.00	8775.00	7834.00
VIII	--	6240.00	6240.00	5571.00	234052.00	208950.00
Total	17070.00	60532.00	77602.00	69282.00	3460005.0	308928.00

i. Net present value @ 12% interest Rs. 29955/- ha^{-1} yr^{-1}.
ii. Benefit /cost ratio @ 12% interest Rs.4.45.

Biomass production from trees adequately compensated the crop yield reductions. Land equivalent ratio of agroforestry land use was comparable or even better than monocropping systems indicating suitability of this system for the western Himalayan valley region (Pratap Narain *et al.,* 1998). Deb Roy (1991) reported that, in marginal soils, a silvipastoral system involving subabul *(Leucaena leucocephala)* and stylo (*Stylosanthes hamata*) as a pasture legume provides 8 to 9 tones of nutritious dry matter per hectare in a year and enrich the soil. At an interval of 4 or 5 years, stylo may be replaced with *Cenchrus ciliaris* grass. This system is more remunerative than growing crops on marginal soils.

A large number of tree-crop combinations, particularly of nitrogen fixing trees (NFTs) with sorghum, groundnut, caster and pulses were evaluated in Alfisols and Vertisols. Though the NFTs improve the soil fertility, the excessive competition for moisture between the tree component and arable crops did not make this system popular with the farmers. However, *Acacia albida* and *Hardwickia binata* have been found to be compatible with arable crops (Singh *et al.,* 2000).

Agroforestry gives more income to the farmer per unit area of land than pure agriculture or forestry. Several studies in different parts of the country suggested that agroforestry is more profitable to farmers than agriculture or forestry for a particular area of land (Singhal and Panwar, 1992; Mathur and Sharma, 1983; Mathur *et al.*, 1984; Jagadish Chander, 1998; Ralhan *et al.*, 1992; Tokey, 1997). Plantation of kinnow at 4 m x 4 m spacing and Bhabar at 50 cm x 50 cm after minor leveling at Relmajra provided early return to the farmers. The system was highly profitable and quickly paid off the leveling cost (Samra *et al.*, 1995). In another study, a silvipastoral system involving *Eulaliopsis binata* in association with *Bauhinia purpurea* or *Grewia optiva* appear to be the most promising to sustain biomass production from the degraded bouldery riverbed lands of Doon valley (Samra *et al.*, 1999).

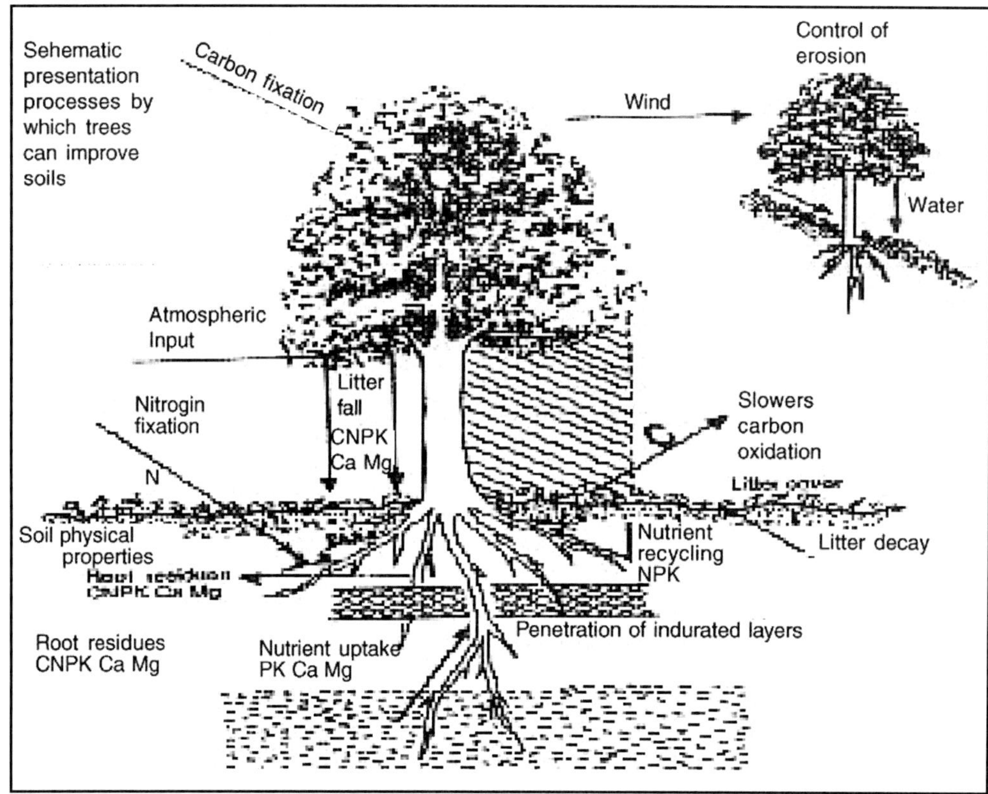

Fig. 1 Processes by which trees improve soil (Young, 1889).

Soil Fertility Improvement

The primary objective of soil conservation is to improve / maintaine soil fertility. To achieve this, control of erosion, maintenance of organic matter, maintenance of physical properties, organic matter addition and maintenance of nutrient is necessary. In this way, agroforestry system constitutes sustainable land use and help to improve soils in a number of ways as shown in figure 1

(Young, 1889). Some of these beneficial affects are apparent in experiments carried out in different parts of the country. Tree crowns and roots protect the soil from the impact of raindrops and create a specific microclimate in the particular area (Pradhan, 1973). Agroforestry practices play an important role in improving the fertility of the soils. Increase in amount of N and P and other macro and micro-nutrients are reported under the growth of 14 year old plantation of *Prosopis cineraria* as compared to a bare site of *Prosopis juliflora* of the same age (Shankarnarayan, 1984). It is reported that forage species produce higher biomass under *Prosopis cineraria* tree canopy due to a high fertility status (Singh and Lal, 1969). There are evidences that soil under the trees was richer in organic matter and nutrients than those in the surroundings and without tree cover. For *Acacia albida*, 50 to 100% increase in organic matter and nitrogen under canopy are recorded together with increased water holding capacity (Felker, 1978).

Table 3: *In-situ* litter accumulation and their percentage contribution and nutrient concentration in leaf litter.

Component	*In-situ* litter accumulation (t ha^{-1})	Nutrient conc. of leaves (%)	
Leaves	3.25	N	0.58
Large twigs	0.75	P	0.04
Small twigs	1.25	K	0.41
Bark	0.5	Ca	1.55
Total	5.75	Mg	0.59

In semi-arid India, it is also widely believed that shading by the tree is responsible for poor yield of associated crops, although the available evidence clearly indicates that the real problem is competition for moisture. It should, however, be recognized that alley cropping is very sound technology for improving the fertility of degraded low-activity clay soil in the tropics and the key to its success lies in the use of pruning as green manure or mulch (Ong *et al.*, 1991). An experimental evidence under agrisilvihorticultural system (Aonla+ Leucaena + blackgram) in rainfed conditions showed that after nine years 65 to 109.4% organic carbon was increased under below canopy and 28.1 to 62.5% under open canopy as compared to initial value (0.32%). The organic carbon per cent under the canopy of Aonla is higher under the canopy than open canopy owing to complete falling of Aonla leaves mostly limited to below tree canopy (Ram Newaj *et al.*, 1999).

The influence of *Eucalyptus citriodora* plantation on the physical properties of the soil was observed due to the differential organic matter addition, root density, root length and soil faunal activity. Contribution of 12 years old *Eucalyptus citriodora* plantation with tree density of 1600 tree ha^{-1} towards *in-situ* litter accumulation and nutrient concentration of leaves (Table 3) was carried out by Balamurugan *et al.* (2000).

In a silvipastoral study, the performance of different tree (*Albizia lebbek, Grewia optiva, Bauhinia purpurea* and *Leucaena leucocephala*) and grass species *Chrysopogon fulvus* and *Eulaliopsis binata* on changes in soil fertility over the 14 year period of experimentation from 1980 to 1993 revealed that soil pH decreased markedly from 7.3 in 1979 to 6.3 in 1993. There was a marginal decrease in organic carbon and available K content over the 14 year period. On the other hand, total N and available P increased appreciably by 50.0 and 74.5%, respectively over the initial status. The large improvements in soil organic carbon, total N and available K were noticed in the surface layer only. The mean increase in 0-5 cm depth in 1993 over the 0-15 cm depth in 1979 was nearly two-fold in organic carbon and available K, and more than three-fold in total N. However, all these parameters were much lower in 5-15 cm depth that observed in the 0-15 cm depth at the beginning of the study. This was because the residues of leaf fall and other left over material remained lying on the soil surface and there was no incorporation into the deeper layers (Vishwanatham *et al.*, 1998).

Samra and Singh (2000) observed an increase in soil organic carbon status of surface soil by 0.39 % to 0.52% under *Acacia nilotica* + *Sacchrum munja* and 0.44% to 0.55% under *Acacia nilotica* + *Eulaliopsis binata* after five years and suggested that *Acacia nilotica* + *Eulaliopsis binata* are conservative but more productive and less competitive with trees and suitable for eco-friendly conservation and rehabilitation of degraded lands of Shivalik foot hills of subtropical northern India. Rehabilitation of degraded forests is possible through afforestation by adopting integrated land use planning with soil and water conservation measures on watershed basis (Samra *et al.*, 1996). At farmers' field in semi arid condition of Rajasthan, significantly larger amounts of organic carbon, available N, P and K were found (Table 4) below the canopy of MPTs having a crown diameter of 8m in the traditional agrisilviculture system compared to the tree less fields (Yadav *et al.*, 2008). Soil pH was significantly lower in MPTs based agrisilviculture system.

Table 4: Soil pH, OC (g kg^{-1} soil) and available NPK (μg g^{-1} soil) under four MPTs in an agrisilviculture system (n = 20)

MPTs	pH	OC	N	P	K
P. cineraria	7.67	4.91	138.3	15.93	180.1
D. sissoo	7.71	4.43	114.5	14.20	162.4
A. leucophloea	7.84	4.21	106.4	11.56	150.8
A. nilotica	7.87	3.82	99.67	13.09	151.2
No tree	8.29	2.51	69.98	7.72	106.5

Soil Conservation

In agroforestry, multistrata practices simulate the protective nature of natural forest. Purposefully planted cover can take several forms that can differ in their ability to counter soil and water loss. Vegetative cover is the key to arrest runoff and soil loss. It has been estimated that soil erosion is accelerated to more than 900 times when tree plantation were clean weeded or the litter was burnt. In a semi-arid Alfisols degraded land, established pasture reduced runoff to 26% and soil loss to 8% (Pathak, 2000). In another study, agroforestry system recorded runoff 2.7%, soil loss 0.54 tones ha^{-1} yr^{-1}, N loss 2.1 kg ha^{-1} yr^{-1} and K loss 0.7 kg ha^{-1} yr^{-1} with mean monsoon season rainfall of 899 mm compared with runoff 25.4%, soil loss 3.21 tones ha^{-1} yr^{-1}, N loss 12.5 kg ha^{-1} yr^{-1} and K loss 4.3 kg ha^{-1} yr^{-1} from the field crops raised under same condition (Grewal *et al.*, 1995).

By alternate land use systems like silvipasture, agrisilviculture and agrihorticulture, the productive and protective benefits from watershed management are considerably higher than the investment, benefit: cost ratios ranging from 1.92: 1 to 7.1:1 with 48 to 99 % reduction in runoff and 81 to 98% soil loss besides reducing the migration of population from 26.6% before the implementation of WSM programme to 9.3% during the project period (Samra *et al.*, 1999). On sloppy lands, contour hedge rows have consistently been demonstrated to be highly effective in controlling soil erosion even in as little as 18 months (Lal, 1989; Maclean *et al.*, 1992). The woody hedgerows provide a semi-permeable barrier to surface movement of water, while mulch from the trees reduces the impact of raindrops on the soil and minimizes splash and sheet erosion (Young, 1989).

Nutrient Cycling

This is one of the most important hypotheses. It is based on the capacity of tree root systems to trap nutrients in the soil solution that would otherwise be lost by leaching and to recycle them through litter to the soil surface. There are implications for systems with high or low inputs. For low-input systems, the hypothesis states that nutrient cycling can become highly efficient, so that fallowing will only be necessary after long period. Alternatively, under conditions of weak leaching, continuous cultivation might be possible, with nutrient removal in crop harvest compensated by natural inputs. Under high-input systems, the hypothesis states that the ratio of nutrient recycling to nutrient losses will be greater for agroforestry than for agricultural systems, with consequent economic benefit through more efficient fertilizer use.

The studies on nutrient cycling showed that wherever, Shisham and Subabul were introduced with crops, the organic carbon content increased significantly over treatments of pure crop and trees alone. The organic carbon status in Subabul + maize treatment was 0.72%. In shisham + maize, it was 0.61%, while in crop alone it was 0.48% only (Table 5). In another study, Sharma *et al.* (1997a&b) reported that the N_2-fixing tree species increased the N and P cycling through production of more aboveground litter and influenced greater release of these nutrients. N_2-fixing species help in maintenance of soil organic matter levels with higher N-mineralization rate as land use change natural forest system to agroforestry systems with sparse tree population. Various MPTs based traditional agroforestry system at farmers' field in semi arid condition of Rajasthan adds good amount of available nutrients. Yadav *et al.* (2008) reported that *P. cineraria* (having a canopy diameter of 8 m) based system recycled 34.9 kg N, 2.86 kg P and 11.7 kg K ha^{-1} through 1325 kg ha^{-1} of litterfall with a mean tree density of 26 trees ha^{-1} (Table 5).

Table 5: Estimated total nutrient release (kg ha^{-1} yr^{-1}) under different tree based agrisilviculture systems

MPTs	N	P	K
P. cineraria	34.97	2.86	11.78
D. sissoo	11.08	0.94	3.51
A. leucophloea	12.98	0.98	4.17
A. nilotica	11.28	0.88	3.66

Poplar based agroforestry for an alluvial soil under irrigated condition in western Uttar Pradesh revealed that total amount of 913.8 kg ha^{-1} (520.9, 283.3 and 109.7 kg ha^{-1} between 0-3, 3-6 and 6-9 m distance from tree line, respectively) and 1291.8 kg ha^{-1} (593.9, 451.9 and 245.3 kg ha^{-1} between 0-3, 3-6 and 6-9 m distance from tree line, respectively) leaf litter was added in the soil through 3 and 4 year old plantation. Considerable amount of nutrients was recycled to the soil through leaf fall. The analytical results of leaf litter showed that the quantity of nutrients added into the soil decreased in the order of Ca>N>Mg>K>P. The nutrients return to the soil through litter fall followed the same trend that amount of litter fall at different distances. Moreover, the maximum addition/return of nutrients was estimated near the tree line, where maximum quantity of litter fall was observed .It was estimated that on an average, 11.9 kg ha^{-1} N (6.1, 3.9 and 1.9 kg ha^{-1} between 0-3, 3-6 and 6-9 m distance from tree line, respectively), 2.5 kg ha^{-1} P (1.3, 0.8 and 0.4 kg ha^{-1} between 0-3, 3-6 and 6-9 m distance from tree line, respectively), 7.9 kg ha^{-1} K (4.0, 2.6 and 1.3 kg ha^{-1} between 0-3, 3-6 and 6-9 m distance from tree line, respectively), 21.3 kg ha^{-1} Ca (10.8, 7.1 and 3.4 kg ha^{-1} between 0-3, 3-6 and 6-9 m distance from tree line, respectively) and 8.9 kg ha^{-1} Mg (4.5, 3.0 and 1.4 kg ha^{-1} between 0-3, 3-6 and 6-9 m distance from tree line, respectively) was added into the soil through litter fall (Sharma *et al.*,2001).

Table 6: Effect of trees on fertility and moisture status of soil

Treatment	OC (%)	Available nutrients (kg ha^{-1})			Moisture status (%)	
		N	P	K	0-15 cm	15-30cm
Sole crop	0.48	252.5	20.2	280.5	11.1	13.7
Subabul alone	0.46	180.2	12.5	250.0	11.6	12.2
Shisham alone	0.40	192.5	12.5	260.7	7.1	7.2
Shisham + crop	0.61	265.3	23.5	260.7	10.6	10.8
Subabul + crop	0.72	283.5	25.5	320.5	8.3	10.0

Source: AICRPAF Annual Report, 1999-2000

Microclimate Improvement

The use of trees as shelterbelts and windbreaks in areas that experience high wind or sand movement is well-established example of microclimate improvement that resulted in improved yields. Increased agricultural production due to windbreaks and shelterbelts in India has been widely reported (Rao and Sita Ram, 1980). Establishment of micro-shelterbelts in arable lands, by planting tall and fast-growing plant species such as cluster bean on the windward side, and shorter crop such as vegetables in the leeward side of tall plants helped to increase the yield of lady's finger by 41% and of cowpea by 21% over the control (Venkateswarlu, 1993). In spite of good results the community shelterbelts in arable lands are not very popular with farmers, especially as the technique normally cut across their field boundaries.

Experimental shelterbelts of *Acacia nilotica* and *Dalbergia sissoo* were established over 120 km in square blocks at the Central Mechanized Farm of Central Arid Zone Research Institute (CAZRI), Jodhpur and at Suratgarh in Bikaner District, Rajasthan. Reduction in wind speed was greater in the monsoon period (July-September) than the summer period before the monsoon period (April-June). There was a 5-14% reduction in evaporation from April to July. In general, the use of shelterbelts brought about a 50% reduction in the magnitude of wind erosion. In the *Cassia siamea* shelterbelts, the loss was 184.3 kg ha^{-1} while in bare soil (without shelterbelt), it was 346.8 kg ha^{-1}. The evidence for the beneficial effects of shade trees depends on the nature of the under storey crops. The clear effect is reported for crops that require shading for normal growth, e.g. black pepper, turmeric, cacao (Nair, 1984). In studies carried out in an agroforestry system, *Acacia tortilis* (7 year old) and Guar crop at Jodhpur indicated that relative humidity recorded beneath the tree canopy during the active cropping season of Guar was found to be 7% more than in the open (Ramkrishna and Shastri, 1977). This will, in turn, help for better growth of the crops.

Sustainability

The sustainable land use is that which maintain an acceptable level of production and at the same time conserves the basic resources on which production depends, thus enabling production to be maintained. The utility of agroforestry practices and their symbiotic effects on better growth and yield of the crops is widely recognized. Agroforestry as subject of scientific investigation has assumed wider recognition in view of the need to maximize production on the basis of sustainable land management (Singh *et al.*, 1998).

Inhibitory Effects

Competition (reduction in crop yield)

It is true that trees do compete with agricultural crops for nutrients, soil moisture and light, but the planting is so arranged in respect of location, spacing, and direction that perhaps and hopefully, tree cause the minimum adverse effect on crop. Harper (1977) most thoroughly examined the importance of understanding the mechanism of plant interaction. While describing the nature of interactions, he focuses that plant may influence its neighboring species, not only adding or removing of some factor, but by affecting conditions such as temperature, light or wind movement and by altering the balance between beneficial and harmful insect. Rao *et al.* (1991) investigated the tree-crop interface in the hedgerow intercropping. The results of four years from 1984 to 1987 did not show any effect of row orientation, and similarly, no effect was seen on crop rows due to their location on the windward or leeward side of hedge. The tree-crop interaction was positive on the first crop row in the first year because *Leucaena* grew slowly, but depressed the yield of the first 4-6 rows (1.8 to 2.7 m from hedgerow) in subsequent years. The negative effect of *Leucaena* was noted more on sunflower in a relatively dry year than on sorghum in other years.

Singh *et al.* (1989) studied the above and below ground interaction in alley cropping in semiarid region of India and results revealed that growth and yield of crop decline from 150 to 30 % of sole crop as the distance from the hedgerows decreased from 5 to 0.3 m. The presence of root barrier had a marked effect on crop growth and completely eliminated any reduction in crop yield, although shading by hedgerow reached 30 to 85% of full sunlight. There was some modification of the microclimate in the alleys but the changes were not great enough to significantly influence the crop yield. The studies on tree-crop interaction have shown that the yield of unirrigated wheat (grown in Rabi) and paddy (grown in Kharif) crops decreased consistently and progressively when *Grewia optiva*, *Morus alba* and *Eucalyptus* hybrid were grown along with them at 4 x 5 m spacing. Over a period of 13 years, there was an average loss of 29, 29 and 28 % in yield of wheat when grown in association with *Grewia optiva*, *Morus alba* and *Eucalyptus* hybrid respectively (Khybri *et al.*, 1992).

In semiarid and arid conditions, since there is insufficient soil moisture even for crop growth, *Acacia nilotica* interferes severely with crop growth. In addition, its dense canopy and shade also affect crop growth adversely. The yield of irrigated wheat crop is also reported to decrease significantly up to 4 m distance from *Acacia nilotica*. The yield of wheat increased gradually 4-15 m away from the trees, it was significantly higher when the distance from trees was more than 15 m (Sharma, 1992). Chauhan *et al.* (1995) observed the effect of mulberry (*Morus alba*), poplar (*Populus deltoides*), safed siris (*Albizia procera*) and subabul (*Leucaena leucocephala*) on the yield of rainfed wheat at Dhaulakuan District Sirmour in Himanchal Pradesh. Tree were planted in August, 1987 in 3 blocks in randomized block design with 2.5 m between trees and 25 m between rows (in N-S orientation), with trees of each species planted in group of 8. Wheat (var. Sonalika) was planted in Oct.1992. In the case of Safed Siris, Subabul and the no tree (control) blocks at 10 and 8-m distances from the trees gave a higher wheat yield and those at 6, 4 and 2-m distances a lower yield.

Allelopathy

Certain organisms or plants produce some chemical substances which inhibit the growth and yield of neighbouring plants or organisms. Such chemicals are known as allelochemicals and the process is known as allelopathy. Allelochemicals are either directly secreted by the roots or foliage of the plants or produced from the decaying portion of the plants. In agroforestry too, certain plants like *Eucalyptus* are known to have allelopathic effect on the under storey crops.

Management of Tree-Crop Interaction

Since the performance of the agroforestry, particularly the intercrop depends largely on the type and magnitude of tree-crop interaction, hence, management of tree-crop interaction is very important aspect in agroforestry for its success. Management of tree-crop interaction means enhancing the complementary effect and minimizing the competitive effect by following strategies.

- Selection of species- sciophyte, erect growing.
- Proper density, spacing and geometry.
- Canopy management- Training and Pruning (initiation and intensity).
- Root management.
- Input management- water, nutrients, etc.

At NRCAF, Jhansi, Ram Newaj studied tree-crop interaction in *Albizia procera* based agroforestry system with two crop sequences namely blackgram-mustard and soybean-wheat during 2000-2005 with eight different treatments namely, planting of tree allow to grow normally + crop, pruning of tree up to 70 % plant height + crop, treatment 1 + soil barrier + crop, pruning of tree up to 70 % plant height + soil barrier + crop, treatment 1 + irrigation as per requirement of crop, treatment 2 + irrigation as per requirement of crop, sole tree and sole crop. He observed that the tree growth was better in soybean-wheat crop sequence than blackgram-mustard crop sequence. The height of trees was 7.7, 6.80 and 3.59% less under pruning up to 70% tree height, soil barrier and one or two irrigation given to the crop over naturally growing, without soil barrier and irrigation given to the crop as per their requirement. Similar effect of these management practices were also observed in case of dbh and crown diameter. The tree growth was observed in order of: irrigation>pruning>soil barrier. As far as effect of different treatments on growth, yield and yield attributes of crops is concerned, he noted that after a year of tree planting (2001), tree did not affect the germination, plant height, number of branches/plant and number of pod/plant of intercrops (blackgram and soybean) but after two years, tree had an adverse effect on growth, yield and yield attributes of crops. The effect of tree on crop reduced with increasing distances from tree base. Among different treatments, pruning of tree up to 70% plant height had better growth, yield and yield attributes over tree allowed grow normally. Similarly, soil barrier had higher values of growth, yield and yield attributing characters in blackgram and soybean over without barrier. Similar results were also obtained in case of Rabi crop (mustard and wheat). In these crops the role of irrigation was higher during initial year (fist two year) but after that pruning of tree up to 70% plant height took a lead over other treatments. Pruning of tree along with irrigation as per their requirement of crop had a significant effect on growth, yield and yield attributes of intercrops.

Yield benefit due to pruning was less during first two years but the benefit of irrigation was more during this period. Yield benefits due to pruning increased gradually with increasing age of trees. Pruning of tree is essential to get regular crop yield from tree-crop system and there is a need to study the amount of canopy pruning with intensity so that tree growth as well as under storey crop yield should no be affected. Soil barrier is another option to reduce competitive effect between tree and crop but tree roots will again be active on upper surface of soil after three to four years. However, soil barrier is not a permanent solution to minimize belowground competition between tree and crop. The beneficial effect of soil barrier on yield increase in crops was comparatively less than pruning and irrigation.

Future Research Thrust and Areas of Interest in Planning Studies in Tree-Crop Interaction

- Selection of suitable MPTs.
- Tree density, spacing and geometry studies.
- Canopy management- pruning intensity and initiation studies.
- Root management- rooting pattern, root barriers, tillage practices (deep, shallow) studies.
- Input management studies- irrigation, nutrients, moisture conservation techniques.
- Insect pests and diseases based studies.
- Microclimate and climate based studies.

Conclusion

Individual plants in crop mixture interact with their neighbours for growth resources. In a mixture plant may influence its neighbours by changing their environment by adding or removing some factor from the environment. In agroforestry, yield reduction in crop is the removal reaction through competition between tree and crop/grass for limited growth resources. Density is the primary component of competition and for optimal tree density needs considerably more research. Management aspect such as root and shoot pruning, irrigation, soil barrier needs more research to maximize complementarity and minimize competition.

References

Annual Report. 1999-2000. All India Coordinated Research Project on Agroforestry, 45.

Balamurugan, J., Kumaraswamy, K. and Rajarayan, A. 2000. Effects of *Eucalyptus citriodora* on the physical and chemical properties of soil. *Journal of Indian Society of Soil Science*, 48(3): 491-495.

Chauhan, V.K., Sood, S.K., Bhargava, J.N. and Mishra, V.K. 1995. Effect of different trees on the yield of rainfed wheat. *Annals of Forestry*, 3(2): 147-151

Deb Roy, R. 1991. Agroforestry to meet our food, fodder and fuelwood needs. *Indian Farming*, 40 (10): 18-21.

Felker, P. 1978. State of the art: *Acacia albida* as a complementary permanent intercrop with annual crops. Report to USAID. Riverside, California, USA: University of California, 133.

Grewal, S.S., Singh, Kehar and Juneja, M.L. 1995. Conservation and production potential of an agroforestry systems integrating grey gum (*Eucalyptus tereticornis*), white popinac (*Leucaena latisiliqua*) and turmeric (*Curcuma longa*). *Indian Journal of Agricultural Sciences*, 65(3): 191-195.

Harper, J. L 1977. *Population Biology of Plants*, Academic Press, Orlando, Florida, 892.

Huxley, P.A. 1985. The tree/crop interface or simplifying the biological/environmental study of mixed cropping agroforestry systems. *Agroforestry Systems*, 3: 251-266.

Jagdish Chander. 1998. Poplar - an ideal species for agroforestry. *Indian Farming*, January, 17-22.

Khybri, M.L., Gupta, R. K., Sewa Ram and Tomer, H.S.1992. Crop yield of rice and wheat grown in rotation as intercrops with three tree species in outer hills of Himalaya. *Agroforestry Systems*, 17: 193-204.

Lal, R. 1989. Agroforestry systems and surface management of a tropical Alfisols: (1) Soil moisture and crop yields. *Agroforestry Systems*, 8: 7-29.

Maclean, R. H., Listinger, J.A., Moody, K. and Watson, A.K. 1992. The impact of alley cropping *Gliricidia sepium* and *Cassia spectabilis* on upland rice and maize production. *Agroforestry Systems,* 20: 213-228.

Mathur R.S. and Sharma, K.K. 1983. Poplars in India. *Indian Forester,* 109: 519-628.

Mathur, R.S., Sharma, K.K. and Ansari, M.V. 1984. Economics of *Eucalyptus* plantations under Agroforestry. *Indian Forester,* 110: 171-201.

Nair, P.K.R. 1984. *Soil Productivity* Aspect of Agroforestry, ICRAF, Nairobi, Kenya.

Natarajan, M and Willey, R.W. 1980. Sorghum-pigeonpea intercropping and the effects of plant population density. 1. Growth and Yield, 2. Resource Use. *Journal of Agricultural Science and Conservation,* 95: 51-58 and 59-65.

Ong, C.K., Corlett, J.E., Singh, R.P. and Black, C.R. 1991. Above and below ground interactions in agroforestry systems. *Forest Ecology Management,* 45: 45-57.

Pathak, P.S. 2000. Agroforestry: a tool for arresting land degradation. *Indian Farming,* 49 (11):15-19.

Pradhan, I.P. 1973. Preliminary study of rainfall in interception through leaf litter. *Indian Forester,* 99: 440-443.

Pratap Narain, Singh, R.K., Sindhwal, N.S. and Joshi, P. 1998. Agroforestry for soil and water conservation in the western Himalayan Valley Region of India. *Agroforestry Systems,* 39: 192-197.

Ramakrishna, Y.S. and Sashtri A.S.R.A.S. 1977. Microclimate under *Acacia tortilis* plantation. *Annual Progress Report,* Central Arid Zone Research Institute, Jodhpur, India, Pp. 9-70.

Ram Newaj, Shukla, S.K. and Yadav, R.S. 1998-99.Varietal evaluation of aonla (*Emblica officinalis*) in agrisilvihorticultural system under rainfed conditions on marginal lands. *Annual Report,* NRCAF, Jhansi, Pp. 44-46.

Ralhan, P.K., Singh, A. and Dhanda, R.S. 1992. Performance of wheat as intercrop under poplar (*Populus deltoides* Bartr.) plantations in Punjab (India). *Agroforestry Systems,* 19: 217-222.

Rao, M .R , Sharma, M.M., and Ong, C.K. 1991. A tree/crop interface design and its use for evaluation the potential of hedgerow intercropping. *Agroforestry Systems,* 13: 143-158.

Rao, M. and Sita Ram. 1980. Influence of shelterbelts on annual crops. *Proceedings of second Forestry Conference held at Forest Research Institue (FRI), Dehradun, India,* 525-529.

Samra, J.S. and Singh, Shiv Charan, 2000. Silvipasture systems for soil, water and nutrient conservation on degraded lands of Shivalik foot hills (subtropical northern India). *Indian Journal of Soil Conservation,* 28 (1): 35-42.

Samra, J. S., Vishwanatham, M.K. and Sharma, A.R. 1999. Biomass production of trees and grasses in silvipasture system on marginal lands of Doon valley of north-west India 2. Performance of grass species. *Agroforestry Systems*, 46: 197-212.

Samra, J.S., Bansal A.K., Sikka, S.P., Mittal S.P. and Agnihotri, Y. 1995. Resource conservation through watershed management in Shivalik foot hills. *Bulletin* T-28/ C-7, CSWCRTI, Research Centre, Chandigarh.

Samra, J.S., Pratap Narain and Dadhwal, K.S. 1996. Rehabilitation of degraded forest through integrated land use planing with soil and water conservation measures. *Indian Journal of Soil Conservation*, 24 (1): 10-18.

Samra, J.S. Dhyani, B.L. and Sharma, A.R. 1999. Problem and prospects of natural resource management in Indian Himalayas- A base paper. Hill and mountain Agro-ecosystem Directorate, NATP, CSWCRTI, Dehradun, Pp. 146.

Shankarnarayan, K.A. 1984. Influence of *Prosopis cineraria* and *Acacia nilotica* on soil fertility and crop yield. In Shankarnarayan K.A. (Ed.). Agroforestry in Arid and Semi-arid Zones, Central Arid Zone Research Institute (CAZRI) Publication No.24, Jodhpur (Raj.), India, Pp. 59-66.

Sharma, R., Sharma, E. and Purohit A.N. 1997a. Cardamom, mandarin and nitrogen-fixing trees in agroforestry systems in India. Himalayan Region. I Litterfall and decomposition. *Agroforestry Systems*, 35: 239-253.

Sharma, R., Sharma, E. and Purohit A.N. 1997b. Cardamom, mandarin and nitrogen-fixing trees in agroforestry systems in India. Himalayan Region. II Soil Nutrient Dynamics. *Agroforestry Systems*, 35: 255-268.

Sharma, K. K. 1992. Wheat cultivation in association with *Acacia nilotica* (L) Wild ex Del. Field bund plantation- a case study. *Agroforestry Systems*, 17: 43-51.

Sharma, K.K. and Singh, R.K. 1992. Studies on tree-crop interaction in *Populus deltoides* "G-3" bund plantation under irrigated conditions. *Indian Forester*, 118 (2): 102-108.

Sharma, N.K., Samra, J.S. and Singh, H.P. 2001. Poplar (*Populus deltoides*) based agroforestry systems for an alluvial soil under irrigated condition in western Uttar Pradesh. *Indian Forester*, 127 (1): 61-69.

Singh, H.P., Venkateswarlu, B. and Dixit Sreenath. 1999-2000. Sustainable land use practices for combating desertification, *National Action Plan - India, Working Group 2 Draft Report*, CRIDA, Hyderabad, 45.

Singh, R.P., Ong, C.K. and Sharma, N.1989. Above and below ground interactions in alley cropping in semiarid India. *Agroforestry Systems*, 9: 259-274.

Singh, K.S. and Lal, P. 1969. Effect of *Prosopis cineraria* and *Acacia arabica* trees on soil fertility and profile characteristics. *Annals of Arid Zone*, 8: 33-36.

Singh, Mahendra, Arrawatia, M.L. and Tewari, V.P. 1998. Agroforestry for sustainable development in arid zones of Rajasthan. *International Tree Crops Journal*, 9:203-212.

Singhal, R.M. and Panwar, B.P.S. 1992. Commercial approach to poplar (*Populus deltoides* M.) based agroforestry system in North- Western Uttar Pradesh. *Van Vigyan,* 30 (1): 29-38.

Toky, O.P. 1997.Poplar an economy booster and eco-friendly agroforestry tree. *Agroforestry News Letter,* 9: 2-3.

Venkateswarlu, J. 1993. Problems and prospects in desertification control - Role of Central Arid Zone Research Institute. In: *Desertification and its Control in the Thar, Sahara and Sahel Regions* (Sen, A.K. and Kar, A., eds.). Jodhpur, India: Scientific Publishers, Pp. 249-267.

Vishwanatham, M.K., Samra, J.S. and Sharma, A.R. 1998. Biomass production of silvipasture systems on gravelly lands of Doon valley. *Bulletin* No.T-39/D-26, CSWCRTI, Dehradun, Pp. 68-69.

Yadav, R.S., Yadav, B.L. and Chhipa, B.R. 2008. Litter dynamics and soil properties under different tree species in a semi-arid region of Rajasthan, India. *Agroforestry Systems,* 73: 1-12.

Young, A. 1989. Agroforestry for Soil Conservation. CABI-ICRAF, Nairobi, Kenya.

❑❑❑

Agroforestry: Natural Resource Sustainability, Livelihood & Climate Moderation © 2009
Editors: OP Chaturvedi, A Venkatesh et. al., pp. 263-276
Satish Serial Publishing House, Delhi

Chapter 19

POTENTIALS OF CONTRACT FARMING IN AGROFORESTRY

D.R. Palsaniya, R.S. Yadav, Ramesh Singh
O.P. Chaturvedi, R.V. Kumar and R.K. Tewari
National Research Centre for Agroforestry, Jhansi-284 003 (U.P.)

Introduction

At present, Indian agriculture is said to be at cross roads. Gone are the days when farming was considered as a novel, profitable and prestigious occupation. While presenting the data of National Sample Survey to Rajya Sabha on May 19, 2006, Shri Kanti Lal Bhuria, Minister of State in Ministry of Agriculture, Government of India stated that 27 per cent farmers maintain that farming is a non-profitable occupation, 8 per cent maintain that it is a risky profession and 5 per cent feel that it does not suit them as it carries no social prestige. The survey also revealed that small and marginal farmers are moving away from agriculture and taking up other means of livelihood. Thus, 40 per cent of the farmers of the country want to leave agriculture if provided with suitable alternate livelihood option. Large numbers of farmers are committing suicides every year. The core agricultural issues that threaten future development arise from following challenges:

- First, the challenge of ramping up agricultural growth and ensuring food security.
- Second, the challenge of inclusive growth and sustainable livelihood.
- Third, the challenge of energy security and
- Finally, the challenge of environmental security and climate change.

The convergence of demand for food, fodder, fuel, furniture and fibre will lead to an unprecedented pressure on natural resources like land, water and forests. Meeting this demand in an economically viable, socially acceptable and inclusive and environmentally sustainable manner is the greatest challenge that confronts us today. Unless we find solutions to these multiple challenges, we stand the risk of not only perpetuating social discontent and consequent unrest, but also of rendering future growth unsustainable.

Further, over many decades, the world has pursued a growth model primarily based on exploitation of precious natural resources without any material to conserve and replenish. The launch of "Green Revolution" in India through the use of improved varieties & hybrids, assured availability of inputs like irrigation and fertilizers and minimum support prices cover has resulted in a resounding success in terms of food security, increased farm income, enhanced employment opportunities for rural landless with a safeguard to our public distribution system. This success, however, came at some costs like increased inter regional, inter crop and inter personal disparities. There was only a minimal or marginal role for the Private sector in green revolution. The Essential Commodities Act became a mean of ensuring Govt. procurement while the MSP became a benchmark reference pricing, rather than a farmer safety net. The productivity level has reached a plateau and started declining. It also promoted mono-culture, over fertilization and excessive water usage which resulted in soil degradation. The subsidized exports to liquidate stock created a further distance between the international market and domestic reality. Thus, in-toto, this approach has put the needs of future generations at a significant risk. One solution to these multiple problems lies in the adoption of contract farming model for agriculture in general and agroforestry in particular and to succeed in this context, contract farming needs to go well beyond its simple definition.

What is Contract Farming?

Contract farming is being seen as one option for creating a role for the private sector in agricultural development. It is obvious that innovative public-private-people partnership will have to be forged to make a meaningful difference in respect to common issues of inclusive growth, sustainability, livelihood security, energy security, food security and environmental security. Corporates, on their part, are uniquely placed to play a constructive role in this process. Positioned, as they are, in the frontline of engagement with civil society, corporates have the wherewithal, including the more crucial managerial resources, to deliver social projects efficiently at a lower incremental cost.

Corporates engaged in processing and marketing of agricultural products are entering into contract with the farmers wherein the farmers are provided inputs and required expertise to plant the contractor's crop on his land on predefined terms and conditions. They buy back the product at a pre decided price. The terms and conditions of the contract are well defined and spelt out in advance and a consent letter is obtained from both the parties. Contract farming has manifold advantages for every stakeholder- farmer, contractor, consumer and the society as a whole in terms of income, security, productivity, quality, employment, etc.

Why Contract Farming?

Contract farming is the need of the hour in Indian agriculture. There is an urgent need to radically change the way we look at our agriculture particularly in the post-WTO scenario which opened new marketing opportunities and challenges. In the post-WTO scenario, besides a gamble in the hands of monsoon, agriculture is also becoming a gamble in the hands of market. Agriculture in general and agroforestry in particular are also important from the angle of ecological and livelihood security and also ensuring the country's national sovereignty. So, we have to look at agriculture and agroforestry from modern perspectives. Contract farming is being viewed as a necessity due to following:

- To increase private sector participation in agriculture through enhanced investment and greater public private partnership.
- To bring about a market focus in terms of crop selection by Indian farmers.
- To promote processing and value addition in agricultural products.
- To meet the opportunities and challenges aroused in post-WTO era.
- To reduce the load on the central and state level procurement and public distribution system.
- To generate a steady and sustainable source of income at the individual farmer level.
- To meet the requirement for quality products and greater consumer satisfaction.

- To generate gainful employment in rural areas, particularly for landless agricultural labour and to flatten as far as possible, any seasonality associated with such employment.
- To reduce migration from rural to urban areas.
- To promote rural self-reliance in general by pooling locally available resources and expertise to meet new challenges.
- To promote greater and newer technology in agricultural production system.

Thus, contract farming is necessary in the modern agricultural perspective where new markets, newer marketing strategies and new thinking to boost Indian agriculture are necessary. Broad based contract farming programs can be one possible solution to the current challenges.

Advantages of Contract Farming

Contract farming has manifold advantages for every stakeholder- farmer, contractor, consumer and the society as a whole. Contract farming helps eliminate the chain of middlemen and brings the producers and processors face-to-face. For the processing firm, it is an important step in supply chain management. Consumers are able to get the products of required quality at affordable price. A country can get large foreign exchange by exporting these value added products.

Advantages for Farmer

- Assured market and support price for his produce.
- Assured supply of quality inputs like seeds, fertilizers, pesticides, etc.
- Exposure to world class mechanized agro technology.
- Better remunerative price for his produce as there are no middlemen involved.
- Efficient, timely and almost free of cost technical guidance at doorstep.
- Financial support in cash or kind.
- Assured supplies of healthy disease free saplings, agricultural implements and technical bulletins.

Advantage for Company/Contractor

- Uninterrupted and regular supply of raw material.
- Required quality in the raw material can be ensured.
- Protection from fluctuation in market pricing.

- Long term planning made possible.
- Concept can be extended to other crops and areas.
- Backward market integration is possible with assured supply.
- Building of long term commitments is possible.
- Assured and dedicated supplier base.
- Generates goodwill for the organization.
- Residual toxicity level, which is very much required for export, is ensured.

Problems in Contract Farming

- The main problem in contract farming is difficulty in honoring the contracts. There is no credible enforcement mechanism in place for this purpose. In case of short duration crops, especially vegetables, farmers tend to divert the produce to the open market rather than supply to the contractor when the prices are high. Similarly, in long duration crops such as plantation crops, the firm often fails to honour the contracts as they know that farmers have no alternative but to sell the products to them at lower prices. The latter often succumb to the firms diktats.
- Small size of land holdings are also a major hindrance in the adoption of contract farming in India.
- Need to contract with a larger number of farmers, thereby increasing risk.
- Lack of a comprehensive crop insurance scheme.

Contract Farming Potentials in Agroforestry

Apart from food processing and value chain based agricultural growth, one of the most powerful means of bringing about transformational change in rural India lies in the relatively untapped agroforestry sector, and its allied areas across the entire wood based value chain. Agroforestry is a dynamic, ecologically based, natural resources management system that, through the integration of trees on farms and in the agricultural landscape, diversifies and sustains production for increased social, economic and environmental benefits for land users at all levels (ICRAF, 2006). It is an age old practice where farmers have been developing and managing agroforestry systems by nurturing trees on their farms, pasture lands and homesteads. In case of annual crops the private benefits are high as compared to social benefits while in agro-forestry, social benefits are relatively high as compared to private benefits. Traditionally, agroforestry produces a wide variety of products such as timber, fuel wood, fruits, fodder, fibers, vegetables, spices, resins, and medicines, primarily to meet household needs and also to generate some income through sale in local markets. The supply of timber from natural forests has declined considerably. At present, imports of forest raw materials, promoted by industrial and trade liberalization, are filling the gap. The wood based industry is growing rapidly with increasing demand for furniture, housing, construction material, packaging, agricultural implements, sports goods, plywood, veneer, matches, and so on. The total wood based imports for various uses today is already over US$ 1 billion. Depending on the regeneration of wood and the changing demand profile, India is likely to face a wood deficit of 20 to 70 million cubic metres by 2020. Certainly, the case of agroforestry is obvious. Non timber forest products, also lends itself to medicinal and herbal applications, and is a very important source of sustenance for forest dwellers. It is estimated that over 75 % of forest export revenue comes from these sources. Approximately 40% of the forest products are supplied from outside forest areas. Similarly, the fuel wood supply from agroforestry is over 95 per cent. Most of the timber requirement is being met from trees outside the forests. Furthermore, it provides employment to millions of people. Taken together, the wood based value chain, including plantations, has the potential to create sustainable livelihood for nearly 100 million people. Agroforestry also supports a number of handicrafts, including wooden jewellery and gift items, potentially another revenue stream in appropriate areas.

The societal benefits of agroforestry outweigh that of many other sectors. One of the most important is in its role as an instrument of climate change mitigation. Potentially, by greening even 10 million hectares of degraded land, nearly 500 million MT of CO_2 can be sequestered. This implies a compensatory sequestration for as much as 500 million MWh of electricity generated from fossil-fuels. Agroforestry also contribute to other ecological benefits in terms of arresting top soil erosion and enabling in greater ground water recharge.

It is obvious that innovative public-private-people partnership will have to be forged to make a meaningful difference in respect to common issues of inclusive growth, sustainability, livelihood security, energy security, food security and environmental security. Corporates, on their part, are uniquely placed to play a constructive role in this process. Positioned, as they are, in the frontline of engagement with civil society, corporates have the wherewithal, including the more crucial managerial resources, to deliver social projects efficiently at a lower incremental cost. The fullest potentials of the agroforestry can only be realized through corporate involvement in it.

Agroforestry Based Industries and their Contract Farming Potentials

A number of industries like paper and pulp, plywood, particle board, fibre board, furniture, housing and construction material, agricultural implements, sports goods, packaging and matches box and value addition industries are based on wood which is contributed from forestry or agroforestry, where contract farming can be promoted. In India, paper is manufactured from a variety of raw materials like wood, recovered paper, bagasse, wheat straw, rice husk, etc. Wood is the main raw material for paper in India and the paper industry's per annum demand for wood is expected to grow from 5.2 million tones in 2000 to 13.2 million tones by 2020. The government source accounts for 20 percent, mostly from bamboo while non-government source covers 80 percent of the total supply. At present, India is deficit in paper production and nearly 50 percent of the industry's requirements are met through imports, which are increasing. Current imports amount to 2 million tones, worth US$500 million. Owing to rapid depletion of our forests a need was felt to conserve the natural forest resources by encouraging reconstituted wood products such as Plywood, Hardboard, Particleboard and Medium Density Fibre board (MDF) to meet the rising demand of wood from consumers including individuals, Railways, Defence, Furniture and Laminate manufacturers, builders etc. This led to a greater scope for agroforestry. Given the major environmental advantage of wood as a C sink, greater usage of wood in furniture, housing and construction material is a better alternative than the use of cement and steel which have much larger ecological footprints. By 2010-12, more than 500 million sq.ft. of space is expected to be built in India, mostly in urban areas. Even a limited use of wood for door and window frames, flooring, wooden cabinets etc. would mean a demand for wood products valued at US$ 3 billion. With greater usage of wood as a predominant material for housing and construction material, in rural and urban areas, this potential can increase exponentially. Today, the Indian furniture market is estimated at US$ 8 billion and employs around 3, 00,000 people. Wooden home furniture is the largest segment. Imports of furniture are growing at a fast clip with a rate as high as 60 percent.

Trees are also a valuable source of bio chemicals. The cellulose part of the structure, used primarily in pulp and paper industry, represents only 20 to 30% of the total, and the balance can be a valuable resource for extraction of bio chemicals with wide applications from food to medicine. These non timber forest products are also very important source of sustenance for forest dwellers and contribute over 75 % of total forest export revenue.

Most of the sports goods, packaging materials and matches boxes are made from wood and their demand in future is likely to increase indicating greater potential of agroforestry to produce required raw material.

A major deterrent to competitiveness and growth is the lack of raw material in the form of wood fibre. Unavailability of sufficient raw material prevents mills from growing in size to achieve economies of scale and attain global benchmarks. As an alternative, many major mills have begun sourcing their raw material from agro-forestry sources on contract farming basis. This may boost agroforestry adoption in India.

Marketing Issues in Agroforestry and their Solution through Contract Farming

Markets for tree products are important to integrate farmers into a market economy. In these markets, however, farmers encounter competition, and policy constraints make it difficult for them to compete. The cumbersome and costly bureaucratic procedures lead the farmers to depend on intermediaries to market their produce. Relatively low returns to agro-forestry largely result from inappropriate policies, complicated transit rules, unfavorable institutional environment and bureaucratic hindrances. In most Indian states, farmers can only harvest and transport wood products after a long and cumbersome process to obtain permits. Further, there is a ban on timber export and restrictions on interstate movement of high quality wood. Marketing is an important issue which affects the decision making and ultimately determines the adoption of agroforestry. Wood marketing in India has several unique and distinctive features as compared to food grain and cash crop marketing which can be summarized as following:

- Absence of any price control mechanism.
- Inadequate marketing information.
- Poor market access and the farmers contact to local buyers only.
- Large number of middlemen and their high margins.
- Credit is not available on the basis of wood resources.
- Rigid forest and environmental laws wherein the farmer's rights to trees on their own lands are vague and ill defined.
- Poor value addition infrastructure.
- Lack of wood storage facilities and
- Poor grading or quality standards.

By adopting agroforestry on contract farming basis much of these problems can be solved easily. Corporates are better placed and suited for these roles as they are rich in resources like value addition infrastructure, transportation and storage facilities, money and man power, brand name and recognition, etc. They are more equipped for market analysis and understanding of market forces for designing and making effective marketing strategies. By understanding the unique and distinctive features of wood marketing in India and competitive forces within agroforestry, market strengths, weaknesses, opportunities and threats can be identified and successful strategies for adoption of contract farming can be developed.

With the careful analysis, we can say that enough commercial opportunities exist for farm communities to transform their traditional agroforestry systems towards market orientation. Most farmers produce small quantities of tree based products primarily for household consumption with limited market sales, which they often sale at the local markets. Forest product producers and gatherers often cannot sell what they have produced to the processors directly. Thus, large processors are unable to get sufficient quantities of raw materials. This is an indication of a frequent mis-match between demand and supply and is believed to be due to lack of sufficient production of quality products and unavailability of timely good market information with both suppliers and buyers for correct decision-making. By adopting contract farming, they can able to achieve this objective.

The experience of WIMCO India Ltd and ITC Ltd

WIMCO India Ltd operates in Uttar Pradesh in northern India where the company raises fast-growing poplars as raw material for its mills, which produce matches, veneer, pulp and paper, fuel, lumber, pencils, packaging, joinery and light furniture. WIMCO has been motivating landowners to plant poplars with wheat, lentils, sugar cane, potatoes and a wide variety of other crops. WIMCO buys the tree crop from the farmer at the market rate or at a predetermined price (whichever is greater). Through an extension programme the company supplies clonal planting stock and advises farmers on annual per ha returns under different crop/tree combinations. It also carries out an ongoing tree improvement programme, continuously introducing new varieties or provenances for testing. Trees are bred for narrow crowns and early leaf-fall to facilitate capture of sunlight by crops. This type of cooperation with local landowners has generated an immense amount of good will for the company and has significantly increased the income level of local farmers.

Similarly, ITC Ltd. also planted trees over 80000 ha of land in Andhra Pradesh and Orissa under its social and farm forestry programmes which has generated over 35 million man days of employment. The wood produced is being used for the production of eco friendly, state-of-the-art elemental chlorine free paper and paperboard- the only one of its kind in India. Because of the green cover created by it, the company has become C positive and earning foreign exchange through C credit trading. The company is also doing research for production of clonal saplings that are disease resistant and grow faster even under harsh conditions.

Sound Contract Farming Strategies for Agroforestry

Agriculture, as it stands today, does not adequately reward long term objectives of building a sustainable economy. A more enduring solution will emerge only when policy makers and market participants assign a larger priority to promoting contract farming in a way that does not compromise the interests of the future for short term gains of the present. By understanding the above scenario of contract farming prospects in agroforestry, successful contract farming strategies can be developed. Promotion of wood based rural enterprises based on contract farming will not only encourage agroforestry adoption but also to total income and employment in the rural areas. Possibilities for a variety of rural enterprises based on agroforestry lies not only growing and harvesting, but also value-added processing, packaging, and transportation. Economic viability is an important aspect of agroforestry development. Potential agroforestry based entrepreneurs should complete four evaluations before investing in a new venture. These include a personal evaluation, a resource evaluation, a market evaluation, and a project feasibility evaluation. Following suggestions can be adopted to develop a sound contract farming strategy in agroforestry.

Government Policy Support Through Institutional Reforms

Wood marketing in India is governed by rigid forest and environmental laws wherein the farmer's rights to trees on their own lands are vague and ill defined. Many provisions of the forest-related legislations in India have acted as serious disincentives to tree farming on private lands. The spirit of existing legislations in this sector is geared primarily to curtail the felling of trees, as opposed to facilitating the growing of renewable forests. The import of timber under Open General License (OGL) and the inconsistencies in inter-state timber trade/transit rules have upset the wood production by farmers. This calls for evolving appropriate policy packages to popularize agroforestry, covering aspects such as harvesting, processing, marketing, and utilization of farm-grown wood. Adequate policy protection is much needed in agroforestry related investments because of long gestation periods involved in yielding adequate returns on investment. For example, maturing of teak wood could take as long as a quarter century, while many other forms would need a growing-harvesting cycle of 5 to 10 years. Therefore, policy support and financing mechanisms will necessarily have to be longer term oriented. What is equally important is to protect such policies from any adverse mid-course changes that could impact the very viability of these long term investments. The following policy support can be provided to promote contract farming in agroforestry.

- Leveraging the ICAR and university systems to provide region specific crop solutions and making them part of public information domain.
- Facilitating import of varieties/hybrids/clones for contract farmers as growth will be led by productivity enhancement and market orientation.
- Incentivizing agriculture research scholars in pursuing their studies in contract farming in synergy with both farmers and private sector.
- Making purchase interference by a third party in a contract farming program, a cognizable offence and establishing a quasi judicial system of contract enforcement.
- Single tier regulation for contract farming at the state level.
- Contract farming organizations may be allowed to take out realistic and deregulated crop insurance policies.

Government Fiscal Support

Custom duty on most wood imports is presently at 5 %, despite a bound rate of 40 % committed under WTO. This does not encourage investment in agroforestry in India. Tariff policy to promote agroforestry will have to enable a larger corporate involvement in assisting farmers with R&D, extension services and linkages to markets. The direction of the current tariff policy, with lower applicable rates, is to make imported wood more attractive than the felling of trees in Indian forests. However, such a policy prevents livelihood creation in the country and exports that potential to the countries selling these products. Similarly, the environmental benefits that result from growing of such plantations accrue to those countries. The following fiscal support may be extended to agroforestry to harvest its potential.

- Collecting no taxes from food and wood processors involved in contract farming and compelling them to invest in rural infrastructure and farmer upliftment to the extent of tax saved.
- Offering 150% deduction on investments made in the creation of extension services for participating farmers linked to procurement of output.
- Legislation needs to be clarified in order to determine whether or not it is permissible to procure agricultural and agroforestry produce directly from the farmers.

- No taxes or duties on import of agricultural equipment and machinery to be used in registered contract farming programmes.
- Abolish all fees, taxes, cess, duties, levies on procurement effected by a registered contract farming programme.

Providing Infrastructural Facilities

Providing infrastructural facilities in rural areas can promote the development of industries which in turn encourage taking up of agroforestry based ventures. The basic requirements of water, power, transportation, communication, roads, storage, value addition, etc. can be provided to promote agroforestry based ventures.

Extension and Agroforestry Management

Extension helps to empower and motivate farmers to enhance and diversify the productivity/ profitability of their agroforestry systems and also in strengthening their capacity to seize market opportunities, both existing and developing. Extension agents can help in greater adoption of agroforestry on contract farming basis by educating the farmers about its provisions. Further, sound agroforestry management is required to produce a raw material of desired quality and quantity. Knowledge of project goals, commercial requirements, planting site conditions, and the range of potentially suitable species is essential for such sound strategies. The productivity of such products in agroforestry systems can be improved by adopting suitable agrotechniques and enhancing management skills of the farmers.

Simple Carbon Credit Trading Policy for Agroforestry

Agroforestry is being seen as one of the most important instrument of climate change mitigation. The Prime Ministers council on Climate Change recently unveiled its National Action Plan. As part of 8 missions announced, the National Mission for a Green India envisages the afforestation of 6 million ha to enhance forest and tree cover to 33%. Agroforestry will be an important component of this. The plan has further suggested in its Mission for Enhanced Energy Efficiency, a market based mechanism that would certify energy savings that could be eventually traded. In a similar fashion, to promote biomass energy from agroforestry, the differential in cost of generation between the biomass energy and the least cost generation from fossil fuel based energy sources could also be reimbursed in the form of energy credits, which can be traded. Needless to state that an appropriate regulatory frame work would be required to mandate purchase of energy credits by polluting industries. By adopting contract farming in agroforestry, the benefits of the global C trading can be extended to farmers as corporates are well worsed with such mechanisms.

A Lot Can Be Done Despite of Absence of a Legal Framework

- Maintain a proper database and record of farming activities and the contract.
- Incentives, rewards and public recognition to the best performers.
- Publicizing the names of defaulters in the locality.
- Farmer encouraged to set own targets.
- Clearly allocate quantities for fresh market.
- Repeat defaulters are not considered again.
- Maintain a high motivation level.
- The social stigma usually suffices as a disincentive to default.
- Promote "ownership" of the business and build long term loyalty.

Conclusion

There is an urgent need for evolving appropriate policy packages to popularize contract farming in agroforestry. Concerted efforts to promote wood plantations in farmer's field through contract farming in India could potentially change the fortunes of small and marginal farmers and consequently the opportunities in rural India. It would be desirable to extend to agroforestry all the benefits applicable to agriculture, so that farmers can be encouraged to cultivate their private lands including wastelands with equivalent benefits like price support, unrestricted movement of such produce, long term credit support on priority basis, and so on. To promote public-private partnership and enable larger corporate involvement in renewable agroforestry, and to enhance linkages for commercial application, industry must be given appropriate incentives. Simple C credit trading policy should be developed for the agroforestry sector separately. India, given the diverse agro climatic zones, can be a competitive producer of a large number of crops and multi purpose tree species (MPTs) and contract farming offers one possible solution.

References

Anonymous, 2008. The Hindu, August, 2, 2008.

Anonymous, 2004-2009. The Foreign trade policy 2004-2009, GOI, New Delhi.

Anonymous, 2007. Current affairs, competition wizard, Career Classics, New Delhi.

Anonymous, 2008. India 2008, Publication Division, GOI, New Delhi.

Gupta, S.K. 2002. Contract farming for rural development. *Kurukshetra*, February 2002, Pp. 13-19.

Palsaniya, D.R., Yadav, R.S. and Tewari, R.K. 2008. Marketing issues and constraints in agroforestry. Abstract: National symposium on "Agroforestry knowledge for sustainability, climate moderation and challenges ahead, December 15-17, 2008, NRCAF, Jhansi, Pp. 197.

□□□

Agroforestry: Natural Resource Sustainability, Livelihood & Climate Moderation © 2009
Editors: OP Chaturvedi, A Venkatesh et. al., pp. 277-287
Satish Serial Publishing House, Delhi

Chapter 20

INDUSTRIAL AGROFORESTRY THROUGH CONTRACT FARMING

K.T.Parthiban[1], S.Vennila[1], R.Seenivasan[2]
G.P.Saravanan[1] and M.Govinda Rao[1]
[1]Forest College & Research Institute, Mettupalayam - 641 301 (T.N.)
[2]Tamil Nadu Newsprints and Papers Limited, Kagithapuram (T.N.)

Introduction

Contract farming is an effective and efficient system of success for production and consumption systems of agricultural and allied sectors. It is an essentially agreement between unequal parties, which include the growers, processors and consumers (Eaton, 1998, Gahukar, 2007). Contract farming is viewed as essentially benefiting the promoters or user agencies by enabling them to obtain the necessary raw material on sustainable basis (Gupta, 2002). In the era of market liberalization, globalization and expanding agribusiness, there is a danger that small scale farmers will find difficulty in participating in the market economy (Jackson and Cheater, 1994). In many cases, small farmers could become marginalized as larger farms become increasingly necessary for profitable operations. This system of contractual farming existed from time immemorial but the success in many sectors of agricultural production is of recent origin.

In contrast to agriculture, the production in forestry is of long rotation in nature. Besides, the forests of the country are under acute socio- economic pressure and till recently, the deforestation rate in the country was estimated to be 1.5 million ha yr $^{-1}$. Currently the forest area in the country is around 23.57 % and Tamil Nadu state is around 17.41% (FSI, 2005), which is low against the mandated requirement of 33%. The less forest area coupled with low productivity of Indian forest has ushered in a large gap between demand and supply of both domestic and industrial wood requirements (Parthiban and Govinda Rao, 2007).

The biggest challenge faced by wood based industries is the raw material shortage. A conservative forest policy coupled with promotion of farmers/ industries linked plantation activities on under-utilized cultivable and marginal agricultural lands will help to mitigate the crisis. This necessitates a business farm forestry model, in order to expand the area under farm and agro forestry plantations through multi-stake holder's participation. A multi functional agroforestry systems in India has already been elaborately indicated (Pandey, 2007).

The success of industrial wood plantation schemes and the related plantation establishment is widely questioned. The reasons for the failure are numerous; but the key reasons are non-involvement of local people, lack of quality planting stock, lack of assured buy back / minimal support price etc. The middlemen and local contractors in the existing marketing system play a significant role and in most of the cases, they decide the harvesting time and also price fixation. Absence of efficient production to consumption system coupled with price incentives and assured buy back agreement is a major constraint faced by the tree growing farmers. This constraint on industrial agroforestry can be overcome through augmenting the existing supply chain system by linking the farmers directly with industries through contract farming system. The success of WIMCO seedling Ltd. for popular in North-Western states and ITC for clonal eucalyptus are the resultant initiatives of mult-istakeholder's partnership on quality planting stock, marketing, finances, etc. (Pande and Pandey, 2004; Dhiman, 2008; Karmarkar and Haque, 2008). Hence the quad - partite model contract farming system through public private partnership has been introduced in the state of Tamil Nadu involving various levels of stake holders and its success is discussed in this manuscript.

Pulp Wood Industries

There are 39 paper industries in the state of which, two industries are wood based and the remaining industries are based on agricultural waste or other sources of fibre. The two industries viz., Tamil Nadu News Prints and

Papers Limited (TNPL), Karur and Seshasayee Paper Board (SPB), Erode use predominantly hardwoods like Eucalyptus and Casuarina as raw material. These two paper industries required nearly 4.0 lakh tonnes of wood pulp and most of the raw material requirement is met from the Tamil Nadu Forest Plantation Corporation and partly from farm lands. Currently, both the industries have expanded their paper production capacity, which ultimately resulted in the wood pulp demand of nearly 8.0 lakh tonnes per annum against the actual availability of around 3.5 to 4.0 lakh tonnes of wood pulp. Industries have developed contract farming system through technological support from Forest College and Research Institute of TNAU at Mettupalayam, Coimbatore.

Existing Supply Chain

Currently there exist two supply chain systems. Viz., Forest department supply and farmer's supply. The major raw material supply comes from the Tamil Nadu Forest Plantation Corporation (TAFCORN), which accounts for nearly 1.5 lakh tonnes of wood pulp supply (TNPL, 2008). The remaining raw material comes from the farm lands through the traders. The supply chain of farm forestry is depicted below

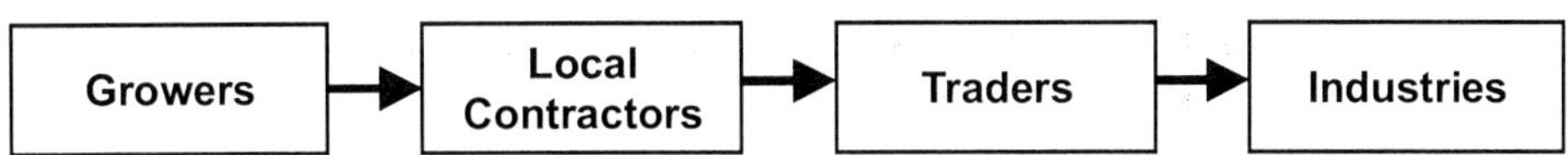

In the existing supply chain, growers sell the whole plantation on the acreage basis not taking into consideration the actual volume of the growing stock, which resulted in minimal returns to the growers. The local contractor's buy the plantations, fell them and classify them into poles, pulpwood and firewood. Only the wood size of 1 to 2 inches alone is supplied to the industries and other sizes are sold for pole purpose by the traders, which resulted in uncertain and erratic pulp recovery and quality (SPB, 2008). This necessitated development of strong supply chain with a successful contract farming model.

Contract Farming Model

To meet the above needs and increase the area under pulp wood plantation through farm forestry, the two paper industries in association with Tamil Nadu Agricultural University has developed a tripartite and quad-partite model for promotion of pulp wood based contract farming system. Through the system, it is intended to produce quality and sustained raw material through a strong supply chain.

Tri-partite Model

This model incorporates industry, growers and financial institutions (Fig 1). Under this system, the industry supplies quality planting material at subsidized rate and assures minimum support price of Rs. 2000 per tonnes or the prevailing market price which ever is higher. The financial institutions viz., Indian Bank, State Bank of India and Syndicate Bank provide credit facilities to the growers at the rate of Rs. 15000 to 20000 per acre in three installments. For credit facilities, a simple interest rate at 8. 5 per cent is followed and the repayment starts after felling. The contract farming system extends no collateral security for loan amount up to Rs.1, 00,000/- per farmer.

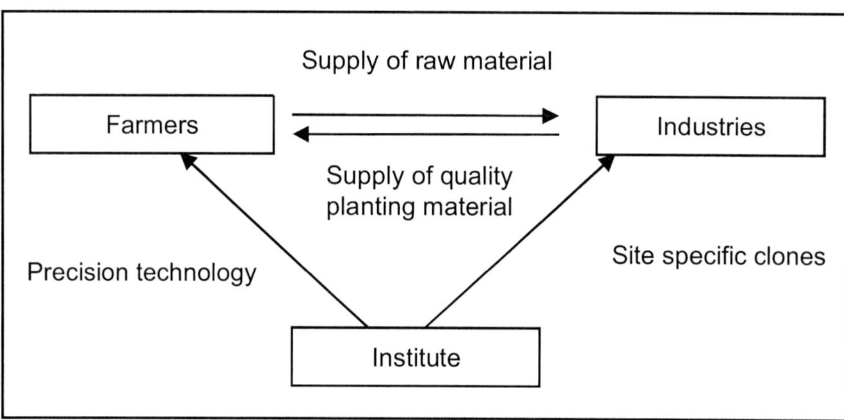

Fig 1. Tri-partite Model contract tree farming

Quad-Partite Model

This system is similar to tri-partite model barring the involvement of research institute (Fig 2). In this system, research institute particularly Forest College and Research Institute (TNAU) play a significant role for technological advancements through varietal development and also to advice site specific precision silvicultural technology to the growers. A pre and post-plantation scientific advice helps to develop human resources through on and off institute mode to farmers and plantation staff of the industries.

Similarly, the industries mass multiply the potential genetic materials identified by the research institute in a decentralized manner and supply them at subsidized costs. The industry also facilitates felling and transport at their own costs, which resulted in strong linkage between industry and the farmers. The industry also help to repay the loan amount after felling of farm grown raw materials there by help the financial institutions for timely repayment, which resulted in strong institutional mechanism for sustainability of the contract tree farming system in the state. Hooda and Hooda (2005) emphasized to provide suitable packages to provide credit facility and marketing to recap benefits of agroforestry/ farm forestry to reduce poverty in rural areas.

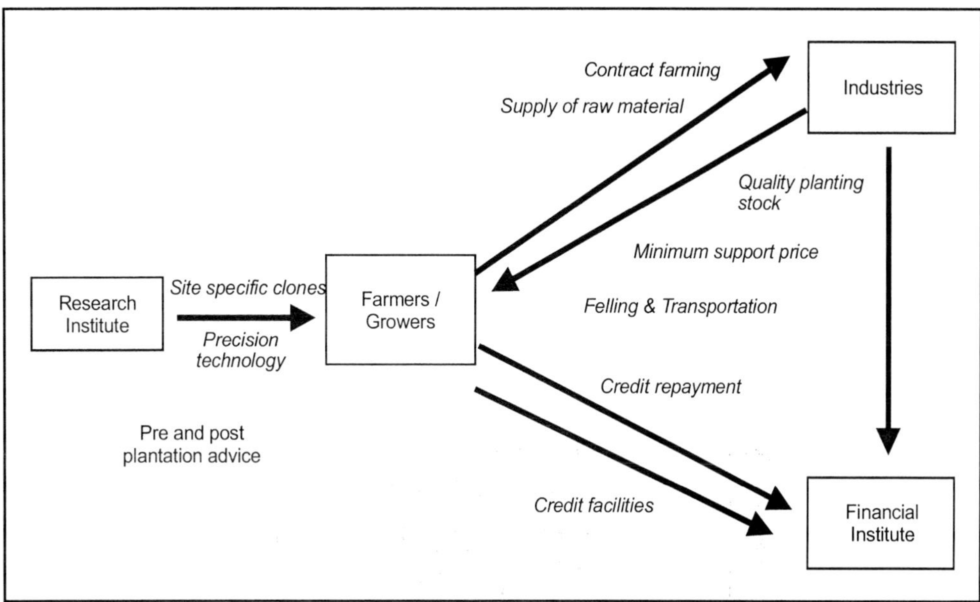

Fig. 2. Quad-Partite Model contract tree farming

Technological Support

Forest College and Research Institute, TNAU, Mettupalayam has established research linkages with both the pulp wood industries in order to develop high yielding short rotation varieties along with precision silvicultural technology for augmenting productivity and to address the site specific problems. Accordingly, short rotation clones in *Casuarina* species have been developed, which could be harvested in less than 30 months with an average yield of around 100 tonnes/hectare. Similarly, high yielding clones in *Eucalyptus tereticornis, E. camaldulensis, E. europhylla* and their inter-specific hybrids have been developed. These elite planting materials coupled with precision silviculture have been successfully transferred to farm land condition with industrial participation through a contract farming system developed and promoted by the institution.

The promotion of industrial wood plantation on the farm lands in Tamil Nadu was followed through following three approaches to develop industrial wood plantations under contract farming system.

Farm forestry	:	The farmers raise pulpwood plantations in their farm lands as block plantations
Agroforestry	:	The farmers raise pulpwood plantations as the major crop coupled with suitable intercrops
Captive plantations	:	The industries develop pulpwood plantations in the farm lands through lease or through benefit sharing mechanism.

Planting Material

Currently pulpable species viz., *Casuarina, Eucalyptus, Leucaena, Melia dubia* and various *Bamboo* species are promoted in the state. However, the farmers in the state mostly prefer *Eucalyptus* for dry lands and *Casuarina* under irrigated conditions. In both the species, superior genetic resources both through progenies and clones have been developed for commercial adoption.

Both the pulp wood industries have developed centralized facilities for mass multiplication of *Casuarina* and *Eucalyptus* through clonal approaches using rooting of coppice shoot cuttings (*Eucalyptus*) and through sprigs (*Casuarina*). The clonal technology developed by Tamil Nadu Agricultural University for mass multiplication of *Casuarina* and *Eucalyptus* has been transferred to industry for large scale adoption and multiplication. The clonal planting materials are supplied to all stake holders at subsidized rates through contract farming system.

Industrial Wood Plantations Covered Under Contract Farming

In order to have better monitoring and improve working efficiency with minimum logistic expenses, contract tree farming activities were confined to a compact area i.e., around 200 km from the mill.

As on date, two industries have concentrated in 16 Districts of the state and covered nearly 40,000 acres under the contract tree farming system. Among two industries, TNPL, a state government owned industry has established more area under contract farming than SPB.

Contract Tree Farming Established by TNPL

During the last three years, the industry has established an area of 19349 acres under industrial wood plantations viz., *Casuarina* and *Eucalyptus*. Out of the total 19349.64 acres, 91% of the area (17641 acres) is raised under farm forestry projects in association with small and medium farmers through contract farming. Under farm forestry programme, 52 % of area (14601 acres) was planted with *Eucalyptus* and the remaining 3040 acres with *Casuarina* (Table 1; Fig 3).

Table 1. Species wise plantation programme in TNPL

Particulars	2004-05	2005-06	2006-07	Total
	Area planted (acre)			
Eucalyptus seedlings	1705.55	4447.39	3806.82	9959.76
Eucalyptus clones	88.85	1310.57	3242.39	4641.81
Casuarina	96.3	296.57	2647.10	3039.97
Total (acre)	1890.70	6054.53	9696.31	17641.54

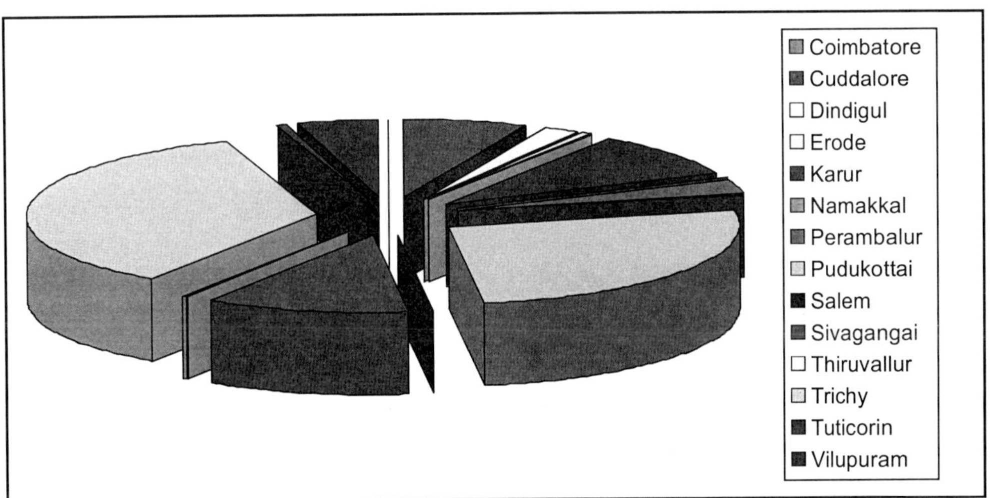

Fig 3. District wise plantation programme through TNPL

Contract Tree Farming Established by SPB

The industry has initiated contract farming system during the year 2005 and in the last two years, it covered an area of more than 10,000 acres in 16 districts of the state. This industry promoted two pulpable species viz., *Casuarina* and *Eucalyptus* depending on the locality. Among two species, the industry covered around 6,017 acres under *Eucalyptus* and 4,339 acres under *Casuarina* plantations (Table 2).

Table 2. Species wise plantation programme in SPB

Particulars	2005	2006	Total
	Area planted (acre)		
Eucalyptus	4933	1084	6017
Casuarina	2026	2313	4339
Total (acre)	6959	3397	10356

The pulp wood based contract farming system established by the two pulp wood industries in the state indicated that farmers in Trichy, Pudukkottai, Sivagangai and Karur districts are more interested for pulpwood plantations preferably with *Eucalyptus* plantations. In these districts, *Eucalyptus* was preferred under rainfed conditions. However, the farmers in Cuddalore, Tanjore, Villupuram, Vellore, Nagapattinam and Thiruvarur districts preferred *Casuarina* based pulpwood plantations due to higher returns from the tree coupled with traditional acceptability of *Casuarina* in the coastal regions. In total, areas of 27,997 acres have been covered under contract tree farming system (Table 3; Fig 4).

Table 3. Species wise plantation programme in TNPL& SPB

Particulars	2004-05	2005-06	2006-07	Total
	Area planted (acre)			
Eucalyptus spp.	1794.4	10690.96	8133.21	20618.57
Casuarina	96.3	2322.57	4960.1	7378.97
Total (acre)	1890.7	13013.53	13093.31	27997.54

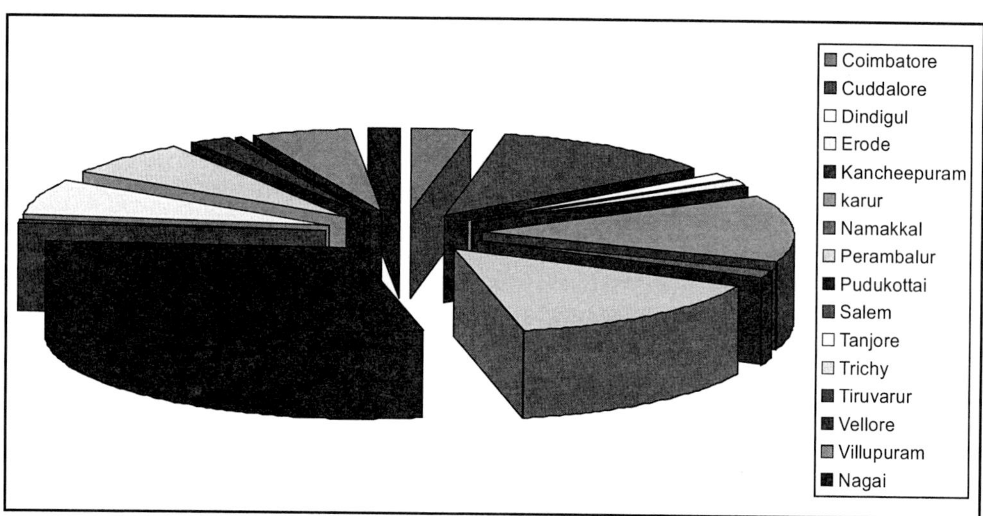

Fig 4. District wise plantation programme through SPB

Conclusion

Contract farming through public private partnership has long-term benefits for both the grower and the purchaser, provided that their long-term association is mutually complementary. Contract farming, as it existed then was different from the design and functioning of contract farming method today. Tree farming is gaining momentum in the state through the institute-industry linked public participation. The pulp wood industries in the state of Tamil Nadu promoted pulpwood based contract farming system and contributed significantly for the successful establishment of *Eucalyptus* and *Casuarina* based pulp wood plantations. The industries have adopted tri and quad-partite contract farming models and these systems proved successful in promotion of farm forestry across the state. The tree based contract farming system, industry provided quality planting material in terms of seedlings and clones coupled with strong buy back system. Besides, the contract farming system also facilitated availability of credit facilities, which promoted more number of small farmers entering into pulp wood plantations. This contract farming system has proved lucrative in the state, which resulted in successful industrial agro and farm forestry model.

Acknowledgement

The authors profusely thank the ICAR for funding the World Bank aided National Agricultural Innovation project on "A Value chain on Industrial Agroforestry in Tamil Nadu" to TNAU, where in the current study formed a part of objective of the scheme.

References

Dhiman.R.C. 2008. Marketing farm grown exotic tree species. In: Exotics in Indian Forestry. (Eds. Chauhan, S.K; Gill S.S; Sharma S.C and Chauhan. R). Agrotech. Publi. Acad. Udaipur, Pp. 787-805.

Eaton, C.S. 1998. Adaptation performance and production constraints of contract farming in China. (unpubl. Ph.D. thesis), Department of Geography, University of Western Australia, Perth.

FSI, 2005. State of Forest Report- 2003, Forest Survey of India, Dehradun. Pp. 134.

Gahukar, R.T. 2007. Contract farming for organic crop production in India. *Current Science,* 93 (12): 1661-1663

Gupta, S. K. 2002. Contract farming for rural development. *Kurukshetra,* 50 (4): 13-20

Hooda,A.K. and Hooda, N. 2005. Poverty reduction for small farmers through agroforestry: overcoming risk aversion. The Commonwealth Forestry Conference, Colombo, Srilanka (Feb.28 to March5, 2005), Pp. 13.

Jackson, J.C. and Cheater, A.P. 1994. Contract farming in Zimbabwe: case studies of sugar, tea, and cotton. *In* P.D. Little & M.J. Watts, eds. Living under: contract farming and agrarian transformation in sub-Saharan Africa, Madison, University of Wisconsin Press, Pp. 140-166.

Karmakar, K.G. and Haque, M.S. 2008. NABARD's initiatives in funding clonal eucalyptus under short rotation forestry. In: *Exotics in Indian Forestry.* (Eds. Chauhan, S.K; Gill S.S; Sharma S.C and Chauhan. R). Agrotech. Publi. Acad. Udaipur, Pp. 246-254.

Luna, R.K. 1989. Plantation forestry in India, International Book Distributor, Dehra Dun, Pp. 506.

Pande.S.K. and Pandey, D. 2004. Impact of incentives on the development of forest plantation resources in India. In: What does it take? The role of incentives in forest plantation development in Asia and the Pacific. (Eds. Enders, T and Durst, P.B.). FAO regional office for Asia and the Pacific, Bangkok, Pp. 81-102.

Pandey, D.N. 2007. Multifunctional Agroforestry systems in India. *Current Science,* 92 (4): 455-463.

Parthiban, K.T. and Govinda Rao, M. 2007. *Eucalyptus* based pulpwood programme in Tamil Nadu. Paper Presented at the Pulpwood Workshop organized by TNPL at Trichy, 18th May, 2007.

SPB, 2008. Plantation Status Report, SPB, Erode.

TNPL, 2008. Plantation Report. TNPL, Karur.

Contd...

Two year old Eucalyptus clonal plantation - Trichy District

One year old Casuarina (progeny) plantation - Trichy District

❏❏❏

Agroforestry: Natural Resource Sustainability, Livelihood & Climate Moderation © 2009
Editors: OP Chaturvedi, A Venkatesh et. al., pp. 289-316
Satish Serial Publishing House, Delhi

Chapter 21

SEED TECHNOLOGICAL ASPECTS FOR DRY LAND AGROFORESTRY

P. Masilamani[1], A. Vijayakumar[2], S. Saravanan[3]
C. Buvaneswaran[3] and A.Venkatesh[4]

[1]Agricultural Engineering College & Research Institute
Tiruchirappalli - 621 712 (T.N.)
[2]Tamil Nadu Agricultural University, Coimbatore - 641 003 (T.N.)
[3] Institute of Forest Genetics and Tree Breeding
Coimbatore-641002 (T.N.)
[4]National Research Centre for Agroforestry, Jhansi-284 003 (U.P.)

Introduction

In dry lands of arid and semi arid regions, a judicious and sustainable mixture of land use and natural resource management (NRM) strategies have to be evolved to maintain ecological, economic and social security for the farming community. The arid and semi-arid lands are characterized by low and erratic rainfall, intense solar radiation and high wind velocity. The evapo-transpiration exceeds precipitation during most parts of the year. The arid regions of India cover an area of 317,090 km^2 and lie at 24^0-29^0 N lat. and 70^0-76^{0}E long. (Shankaranarayan et al., 1987). The basic remedial measures in such dry lands should aim at erosion control coupled with conservation of soil moisture and nutrients. Agroforestry practices can be one such bio-remedial measure. Depending on the type of land, resources availability and specific user need, any one of the agroforestry practices such as energy plantations, silvi-pasture, agri-silviculture, silvi-horticulture, homestead gardens woodlots, etc. can be envisaged.

Other traditional agroforestry practices includes retaining or growing of trees like *Acacia nilotica, Acacia leucophloea, Azadirachta indica, Pongamia pinnata, Mangifera indica, Sygyzium cuminii, Prosopis cineraria* as scattered in their fields to have green manure, fuel wood, timber for agricultural purposes. *Ailanthus excelsa, Melia dubia, Thespicia populnea, Morus alba, Bombax sp.,* and *Lannea coromandelica* are the fast growing indigenous tree species getting importance in the recent past.

Acacias and *Albizias* being the most suitable tree components in dry land agroforestry, important seed technological interventions are related to breaking of physical dormancy exerted by the presence of hard seed coat. Details on this seed technological intervention and other important seed technological aspects related to dry land agroforestry practices are reviewed and discussed in this chapter.

Selection of Seed Trees

Seedlings inherit the qualities of their parents, therefore seeds should be collected from trees with characteristics that are preferred by the eventual tree planters. The following guidelines for selection of suitable trees for seed collection may be adopted.

1. All seed trees should be healthy, vigourous, and should have good form.
2. Trees intended for fodder should have abundant foliage and a bushy crown.
3. Trees intended for fuel wood can have plenty of branches.
4. Trees intended for timber should be straight and without too many thick branches.
5. If a species has multipurpose (or) more than one use, it will be necessary to compromise on these characters.
6. Collect seeds from middle-aged trees to matured trees, immature trees as well as over matured trees should be avoided, since seeds from them may be of low viability.
7. Seeds should be collected from dominant and co-dominant trees when trees are in block.
8. Avoid collecting seeds from isolated trees particularly in species where self-polinated seeds are poor performing in nature like *Casuarina.*
9. Tree stands containing numerous poorly formed individuals should not be considered for seed collection. Such trees are generally abnormal or diseased (Mapier and Robbins, 1989).

Age of Mother Tree

Seed production depends mainly on climatic conditions, but is also affected by the age of the tree and manner of growth. Few trees set seed after two to three years; others may set seeds after 20-30 years. Seeds are borne by trees that have sufficient maturity, and, it is believed that, in most species, there is an optimum age of seed bearing. Seeds from trees just beginning to bear are often of poor quality. There is some evidence that the age of the mother tree influence seed and seedling attributes by Olsen (1932) and Swaminathan *et al.* (1994). Therefore, it is important to know the optimum age of the tree for seed collection and effective seedling production.

Swaminathan *et al.* (1994) reported that in *Ceiba pentandra,* seeds were collected from 4, 5, 6 and 7-year-old stand, and investigated the physical and physiological parameters. It was observed that seeds collected from seven year- old stand gave more 100 seed weight (5.85 g), higher germination (76 per cent), germination energy (5.86) and germination value (7.69) compared to seeds collected from other age classes. Similarly, Masilamani *et al* (2000) reported an increase in germination percentage, seedling dry matter and seedling vigour of seeds from 20 year old trees compared to neem seeds from the trees of various other ages (Table 1).

Table 1. Effect of age of neem mother trees on germination and seedling growth

Age (yr)	Germination (%)	Days taken for seedling emergence	Root length (cm)	Shoot length (cm)	Dry matter production (mg)	Vigour index
5	68.4^{d} (55.79)*	8.0^{ab}	11.8^{ab}	16.7^{ab}	81^{ab}	1955^{c}
10	70.6^{c} (60.68)	8.5^{ab}	12.4a	16.2^{b}	59^{b}	2169^{b}
15	80.6^{b} (63.87)	9.5^{b}	11.2^{ab}	18.2a	70^{ab}	2375^{ab}
20	88.2^{a} (69.91)	7.5a	10.5^{ab}	17.3^{ab}	93^{a}	2453^{a}
25	60.8^{f} (51.24)	9.5^{ab}	9.6^{ab}	16.3^{b}	73^{ab}	1629^{d}
30	64.00^{c} (53.13)	9.5^{b}	9.1^{b}	17.2^{ab}	73^{ab}	1684^{d}

*Figures in parenthesis represent Arc sine transformation. Means followed by same letter in a column are not significantly different at 0.05 level by DMRT. (Masilamani *et al.*, 2000)

Suresh (1994) reported that 100 seed weight was highest (17.1 g) from 11 year- old trees of *Acacia nilotica*. This gradually declined with decrease of tree age 15.23 (10 yr), 10.66 (6 yr) and 9.26 (3 yr). Maximum germination (90%) was recorded by 9 and 11 year- old trees, which was followed by 81 per cent (8 yr). There was a drastic reduction to 56% and 34% in 7 and 3 years -old trees respectively. Seedling growth was also highest from seeds of 11-year-old trees. Prasad and Kandya (1992) recommended that the seeds should be collected from mature or nearly mature trees. Over mature or immature trees should be avoided, since seeds from them may be of low viability. This is not always true in all the tree species, where as in *Santalum album*, age of a tree has no effect on viability and germination (Manonmani, 1997). While our results suggest that tree age may significantly influence seed quality characteristics, further study is needed in other important native tree species of dry land agroforestry.

Time and Season of Seed Collection

The time of seed collection depends on the local edapho-climatic conditions through the natural ranges of each species. Progress of ripening and dispersal time of seeds need to be closely monitored. In a number of species, for example *Tectona grandis* and *Lagerstromia lanceolata* most legumes, seeds ripen in a fairly definite time. In such cases, dry seed/ fruits must be collected as soon as possible. Bhardwaj and Chakraborty (1994) reported that the *Terminalia bellirica* and *T. chebula*, when the seeds were collected during first fortnight of January and sowing was done at last week and third week of March gave better germination of 52.2 percent and 44.4 percent, respectively. *Dalbergia sisoo*, profusely seeding almost every year. Collection time of seed is different from place to place (Table 2).

Table 2: Seed collection months of *Dalbergia sisoo* in different states.

State	Seed collection time
Assam	December - January
Bihar	November - March
Orissa	November - March
Punjab	December - February
Uttarpradesh	January - February
West Bengal	December - Mid - February

(Tewari, 1994)

The fruiting season of teak in India is between November and March / April. Collection of fruits during the later part of the season gives better results. Studies conducted on the effect of time of collection on germination of teak fruits from Top slip source in Tamil Nadu revealed that fruits collected in March-April gave better germination percent than those collected in other months. The natural fall of teak fruits take place in March - April (Gurudev Singh *et al.* 2001). Seeds with the highest germination capacity and with the least damage to the seed are found at the end of the season (Hedegart, 1975). In Cauvery delta zone of Tamil Nadu, neem seeds were collected during June-July (normal bearing season) gave maximum germination of 80.8% and seedling vigour of 2002 as compared to seeds collected during off season (December-January) with germination of 53.2% with seedling vigour of 1467 only. It is concluded that collection of neem seeds should be collected during the months of June-July for raising of seedlings (Masilamani *et al.*1998).

Methods of Collection

There are two general methods of collecting the fruits and seeds of forest trees in India.

Firstly, the collection of naturally - fallen fruits or seeds from the ground surface, as followed in *Tectona grandis, Gmelina arborea* and several *Dipterocarpus* and most other heavy seeds / fruit.

Secondly, direct collection of fruits from felled or standing trees, as followed in *Cassia fistula, Casuarina equisetifolia, Simarouba glauca (*Plate 1, 2 & 3*) Emblica officinalis, Shorea robusta, , Eucalyptus, Cassia siamea.*

Though ground collection seems to be simple and easy method, there are certain disadvantages in ground collection.

- Seed if not collected immediately after shedding from the trees, may be attacked by insects, fungi and animals (e.g.) *Acacia spp.*
- Seeds having short viability lose germinability quickly. Once the seeds have been shed, any delay in collection may lead to collection of non-viable seed (e.g.) *Bassia latifolia, Pongamia pinnata* etc. (Khullar *et al.* 1991).

Collection of seeds from standing trees also poses some practical problems. Specially designed ladders, tree cycles etc., have been designed for use in collecting seed from difficult trees. However, where as the ladders cannot be safely used in inhospitable terrain, the use of tree cycle needs exceptional energy, skill and training.

Acacia catachu pods must be picked by hand directly from the trees before the pods open and scatter their seeds, pods laying on the ground might already damaged by *Bruchus bilincatopygus* or other insects (Tewari, 1995). *Masilamani et al* (1995) reported that in *Prosopis juliflora* pods collected from natural forest floor showed heavy microbial population (fungi, bacteria and actinomycetes). The seeds extracted from infected pods were lesser in germination (21.0%) and vigour than to seeds from uninfected pods (90% germination with higher seedling vigour).

Seed Extraction

Extraction denotes the procedure of physically releasing and separating the seeds from the enclosing fruit structure. Extraction of seeds in tropical species is generally done close to the site of collection. After the collection of fruit, seeds should be extracted in the shortest possible time to avoid damage to seeds, as well as to facilitate storage and sowing. The main rationale of extraction is to reduce bulk, ease handling, enhance germination and improve storability of seeds and lower the risk of transfer of diseases or pests between countries, this is achieved by removing material that may harbour disease and pests. Different methods of extraction are available for both dry fruits and fleshy fruits (Table 3).

Table 3 : Summary of extraction methods for various fruit types

Fruit type	Extraction procedure
Dry dehiscent fruits/ pods/ follicles / capsules/ cones, e.g.. *Eucalyptus* and most Leguminosae	Drying → shaking / tumbling
Dry indehiscent fruits, e.g. indehiscent pods of *Acacia nilotica*	Drying → threshing
Fleshy fruits with very thin pulp, e.g. *Vitex* spp. and *Ziziphus* spp.	Drying Soaking → maceration → washing
Fleshy fruit with soft pulp. e.g. Ficus	Soaking → fermentation →washing Soaking → maceration → washing
Fleshy fruits with soft, fibrous pulp, e.g. *Gmelina*	Soaking → maceration → washing→ Abrasion / polishing
Fleshy fruits with felty pulp, e.g. *Tectona grandis*	Soaking → abrasion / polishing

Seed Extraction from Dry Fruits

This is very diverse group. Examples are follicles that split open down one side, pods from legumes (e.g. *Acacia sp, Albizia sp*) that spilt down two sides and capsules of Eucalyptus that split down three (or) more sides. From extraction point of view, dry fruits can be divided into two groups; the truly dehiscent fruit that will split open and expel the seed when dried (e.g. *Leuceana leucocephala* and *Cassia siamea* etc.) and the indehiscent fruit that will remain closed after drying (e.g. *Albizia lebbeck, Cassia fistula, Cassia grandis, Delonix regia, Peltophorum pterocarpum and Prosopis juliflora* etc.) In case of dehiscent fruits, pick the fruit before split *open* and spread on the cement floor or on trays and exposed to the sun (or) artificial drier, these fruit dried too fast, they will have to be extracted or threshed after drying.

Lot of scope for threshing of leguminous fruits / pod through thresher (Fig.1) or dehuller (Fig. 2). The major Advantages are

1. Complete removal of the seeds from the fruit / pod is achieved.
2. Minimum labour requirement
3. Increased the efficiency of operation
4. Maximum seed extraction with minimum cost
5. A beneficial scarification of seeds may be made due to rubbing action.
6. Since the thresher winnows the threshed seeds, the chaff and other lighter impurities are separated. Hence the quality of the seeds is improved.

Disadvantages

1. It may damage the seeds causing loss of viability and vigour, if the machine is not adjusted properly.
2. Effective and economical only if large quantity of seeds is to be threshed.

In *Albizia lebbek,* seeds were extracted by different treatments, the results revealed that seed extraction with pliable stick gave 100 percent seed recovery but involved more than 80 man hours for splitting by hand and 30 man hours for beating with pliable stick, for extracting 50 kg of pod (plate 4). Where as threshing with commercial paddy thresher gave 100 per cent seed recovery and the same quantity of pods could be extracted within 18 minutes (Table 4). The extracted seeds were evaluated for germination test. Prior to germination, the seeds were scarified with sulphuric acid (200 ml kg^{-1} of seed) for 20 minutes and untreated seeds served as control. Germination recorded at 25 days after sowing revealed that seeds extracted through commercial paddy thresher gave maximum germination of 13% followed by beating with pliable stick and splitting with hand (9% and 5% respectively).

The higher germination might be due to fact that the rubbing of seeds against the internal surface of a thresher cylinder caused beneficial scarification of seeds (Masilamani *et al.* 1999). Sulphuric acid treated seeds recorded more than 95% germination irrespective of the extraction methods.

Table 4 : Effect of seed extraction methods on man-hours requirement, cost, seed recovery (%) and germination (%) of *Albizia lebbeck*

Treatments	Seed recovery (%)	Cost of seed extraction kg^{-1} (Rs)	Comparative time taken for operation	Germination (%) Control	Sulphuri acid
Splitting with hand	100	29.90	100 % (ref)	5	95
Beating with plyable stick	100	10.83	37.5 %	9	96
Paddy Thresher (18 minutes)	100	5.00	<1%	13	96

(Masilamani *et al.* 1999)

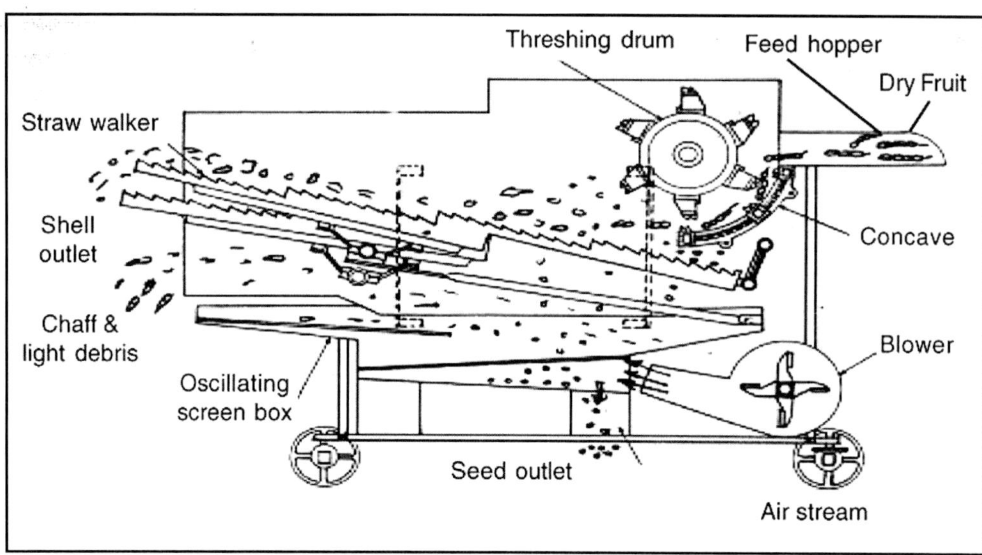

Fig. 1. Principle of a Commercial Thresher

Some seeds may be extracted by light rubbing or tumbling, where as others may need a series of hard mechanical, chemical (or) acid treatments. e.g. *Prosopis juliflora.* Seed extraction from *Prosopis juliflora* is a challenging task. The pods have tough peri carp, waxy exocarp, spongy mesocarp and a cartilaginous endocarp, enclosing the seeds in tomentaceous segments. As the cartilaginous inner lining of the pod does not allow the seed to escape easily, due to that the natural regeneration through seeds in the normal way is time consuming.

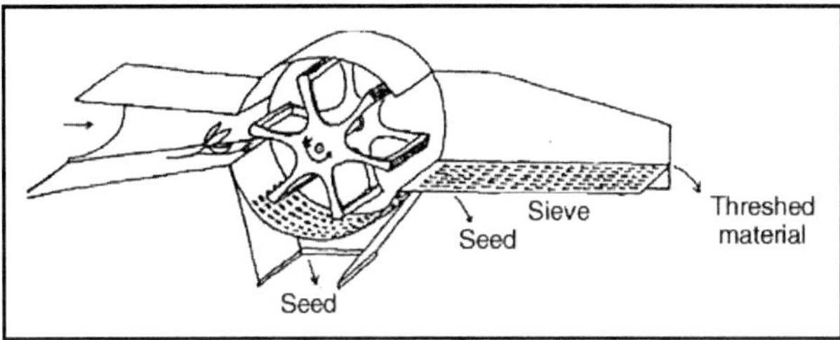

Fig. 2: Principle of a Dehuller

The pods were subjected to eleven extraction methods and assessed for germination and vigour (Table5). The results indicated that, the seeds collected from goat droppings recorded the highest germination (82.5%) followed by

Table 5: Influence of seed extraction methods on seed germination in *Prosopis juliflora.*

Treatments	Germination %	Root length (cm)	Shoot length (cm)	Vigour index
1. Steam (121°C)				
10 minutes	50 (45.0)	6.2	6.5	635
20 minutes	40 (42.1)	5.8	6.0	472
30 minutes	35 (39.2)	6.0	5.8	413
2. Hot water (90°C)				
10 minutes	25 (29.9)	6.0	5.6	290
20 minutes	10 (18.0)	5.8	5.3	111
30 minutes	0	0	0	0
3. Cold water soaking				
24 hours	15 (22.6)	5.8	6.2	180
4. Commercial $H_2 So_4$				
20 minutes	0	0	0	0
30 minutes	60 (50.8)	6.2	5.7	714
5. Goat droppings				
	82.5 (76.8)	6.0	7.0	1073
6. Termite feeding				
4 days	0	0	0	0
8 days	65 (55.8)	8.0	6.3	930
SEd	2.63	0.87	0.867	74.78
CD (p=0.05)	5.43	1.80	1.79	154.35

(Figures in parentheses represent Arc sine transformation)
(Masilamani and Vadivelu, 1997)

the seeds of termite fed pods (65.0%) (Masilamani and Vadivelu, 1997). The highest percentage of germination of seeds from the pods fed to goats might be due to fact that the seeds while passing the digestive system might have got scarified through the organic acids like acetic acid, propionic acid, succinic acid, butric acid, lactic acid etc. produced by bacteria *viz., Bacterioides succinogens, Ruminococeus flavifaciens, R. albus, Butyrovibrio fibrisolveros* etc., those are present in the ruminant fluid.

Saxena and Khan (1974) reported that soaking of pods of *Prosopis juliflora* in diluted NaOH for 30 minutes followed by washing in tap water and rubbing with cloth in the best treatment for removing seed form the septa. It is however not necessary to have absolutely clean seed, without any covering of the pericarp since it does not affect the germination much, also the cost of extraction becomes high. Brown and Belcher (1979), dried the pods of *Prosopis* in a forced air oven for 18 hr at 32°C, later subjected to electric scarification for 10-15 seconds gave good results.In some cases, the only option may be extracted each seed by hand e.g. *Hardwickia binata* and *Pongamia pinnata.*

Casuarina equisetifolia seeds are normally extracted by spreading the cones on canvas or plastic sheets in the sun until the bracteoles are open. Shaking and sieving can then separate the seeds. However, sun drying is less reliable and takes more time in climate or periods of the year with high relative humidity. To overcome these problems, study was carried out to compare the different method of drying for seed recovery and germination. The cones were dried under sun, shade and in oven maintained at 40°C for 5 days. The result revealed that cone dried under hot air oven at 40°C for 1 day gave maximum seed recovery (87.6%), 100 seed weight (176 mg) and germination (58%) compared to cone dried with shade and open sunlight (Umarani *et al.* 1997).

Seed Extraction from Fleshy Fruits

Species with juicy and puply fruits are potentially troublesome. Soon after collection the pulp begins to soften and various fermentation process may start. After soaking or decomposition the fruit pulp and fruit skin are separated from the seed. Several methods, often used in combination, are applicable. Selection of the most appropriate method depends on fruit type, quantity of fruits to be processed and equipment available (Table 6).

Table 6: Seed extraction methods.

Extraction method	Type of fruit / seed and extraction procedure
1. Individual manual extraction	Very small seed lots, seeds with extremely fragile seed coats or large fruits. Individuall y extracted seeds may be washed and cleaned e.g. *Syzygium cumini*.
2. Washing in deep bowls or drums	Small seed lots and small seeds extracted effectively. Separation of pulp and seeds is done by flotation in excess water; seeds or stones remain at the bottom while pulp tend to ascend to the surface where they may be skimmed off e.g. *Ficus* species.
3. Washing on wiremesh screens	Very small seeds or seeds with fragile seed coats. Fruits are rubbed manually on the screen, the mechanical impact can easily be adjusted when extracting fragile seeds e.g. *Anthocephalus cadamba*
4. Cement mixture with abrading material	Fruits with sticky pulp. Abrading material (gravel) rubs against the fruits. The pulp is gradually torn off. The mixture must be separated in to three fractions after extraction viz. seed, abrading material and fruit pulp. e.g. *Vitex keniensis* and *Cordia* spp.
5. Mechanical depulping	For large quantities of fruits or where the flesh tend to remain firm; fruits with relatively hard seed coats or stones. Depulper mechanically abrades the fruit pulp by rupturing and squeezing against or between its mechanical parts and the pure seeds were separated e.g. *Azadirachta indica* and *Gmelina arborea*

The fruit of many species such as *Azadirachta indica and Simarouba glauca* fleshy and pulpy. It is necessary to remove the pulp immediately at maturity to prevent loss of germination capacity. In neem, several methods are available for extracting the seeds from the fruits viz., the fruits were depulped in a coffee depulper, fruits soaked in water for six hours, then depulped by rubbing them over a soft wire mesh and the pure seeds were separated. All the afore said methods were laborious and time consuming. Masilamani (2003) conducted a study to find out the different mesocarp removal treatments on germination and initial seedling vigour of Neem. The seeds soaked in concentrated sulphuric acid (50 ml kg^{-1}) for 3 minutes was found to have higher percentage of mesocarp removal influencing germination and initial seedling vigour attributes of neem, with lesser time and cost (Table 7; plate 5).

Table 7 : Effect of mesocarp removal on germination and initial seedling vigour of neem.

Treatments	Percentage of mesocarp removal	Germination (%)	Days taken for seedling emergence	Root length (cm)	Shoot length (cm)	Drymatter production (mg seedling^{-1})	Vigour Index
T_1 Control	0 (0.98)	70.0 (56.85)	9.0	9.3	15.5	87	1734
T_2 Hot water (60°C soaking for 10 min.)	68.1 (55.61)	91.3 (76.42)	7.7	11.0	18.0	99	2622
T_3 Hot water (60°C soaking for 20 min.)	73.2 (66.03)	91.0 (73.45)	7.3	11.5	17.5	84	2634
T_4 Hot water (60°C soaking for 30 min.)	83.1 (65.73)	89.3 (74.31)	7.7	11.5	18.0	89	2623
T_5 Ferment for 12 hr at 30°C	38.4 (38.27)	89.3 (71.19)	6.3	13.7	21.0	121	3064
T_6 Ferment for 24 hr at 30°C	65.5 (53.67)	79.3 (72.15)	7.7	11.4	18.2	82	2350
T_7 Ferment for 36 hr at 30°C	58.6 (49.95)	90.0 (72.32)	6.3	12.2	20.0	101	2889
T_8 Soaking in 4% HCl (25ml/kg) for 10 min.	71.7 (57.91)	68.0 (55.59)	7.7	11.5	18.0	104	1991
T_9 Soaking in 4% HCl (25ml/kg) for 20 min.	76.6 (61.07)	80.0 (63.49)	7.7	12.5	19.2	104	2534
T_{10} Soaking in 4% HCl (25ml/kg)for 30 min.	79.5 (63.13)	93.3 (78.92)	7.3	12.0	20.0	108	2975
T_{11} Soaking in 10% sodium carbonate for 12 hr	42.4 (40.62)	77.3 (61.60)	7.7	11.0	19.0	85	2283
T_{12} Soaking in 10% sodium carbonate for 24 hr	43.6 (41.29)	82.0 (65.04)	6.3	11.0	20.0	102	2497
T_{13} Soaking in 10% sodium carbonate for 36 hr	46.3 (42.90)	85.3 (67.63)	7.3	12.0	21.0	108	2827
T_{14} Soaking in conc. H_2SO_4(50ml/kg) for 3 min.	94.5 (76.52)	98.6 (85.91)	5.3	11.4	21.0	102	3169
T_{15} Soaking in conc.H_2SO_4 (50ml/kg) for 5 min.	93.5 (75.23)	97.3 (82.36)	5.3	10.5	18.4	85	2821
T_{16} Soaking in conc. H_2SO_4(50ml/kg) for 10 min.	92.3 (74.29)	97.3 (80.73)	5.3	11.8	19.0	84	2973
T_{17} Soaking in conc. H_2SO_4 (50ml/kg) for 15 min.	96.2 (78.71)	95.3 (77.84)	5.3	11.3	18.4	87	2850
T_{18} Soaking in conc. H_2SO_4 (100ml/kg) for 3 min.	95.2 (77.57)	76.0 (60.97)	5.3	10.6	16.0	85	2015
T_{19} Soaking in conc. H_2SO_4 (100ml/kg) for 5 min.	97.3 (80.61)	95.3 (79.79)	5.3	10.8	18.0	91	2742
T_{20} Soaking in conc.H_2SO_4(100ml/kg) for 10 min.	98.5 (83.06)	86.3 (68.55)	5.3	10.3	19.0	87	2523
T_{21} Soaking in conc. H_2SO_4(100ml/kg) for 15 min.	100 (89.34)	76.0 (60.70)	5.3	10.2	20.0	88	2319
SEd	0.241	4.865	0.431	0.617	0.794	8.409	175.590
CD (P=0.05)	0.49	9.79	0.88	1.24	1.60	16.92	353.28

(Figures in parentheses indicate arc sine values) (Masilamani, 2003)

Palaniswamy and Masilamani (2000) developed hand operated roller type neem fruit pulper (Fig. 3 & 4) and its efficiency and the cost economics of the unit is compared with manual pulping (Table 8). The results revealed that seed extraction with manual and fruit pulper gave 99.0 and 95.48% pulping efficiency. While calculating the cost of extraction of 1 kg of fruit, it was 20 paise for fruit pulper followed by manual pulping (40 paise). Germination recorded at 21 days after sowing revealed the seed extraction through manual pulping gave maximum germination and seedling vigour. The rate of extraction of seed by manual pulping is 6 kg hr^{-1} when compared to 30 kg hr^{-1} in mechanical pulping. The seeds extracted through fruit pulper exhibited reduced germination, which might be due to, at the time of fruit pulping, the fruit mesocarp not completely removed. This mesocarp flesh having water-soluble germination inhibitors, might be responsible for lowering of germination. In order to maximize the germination of pulped seeds, it is recommended that technology must be developed for complete removal of mesocarp simultaneously along with the removal of fruit skin at the time of depulping.

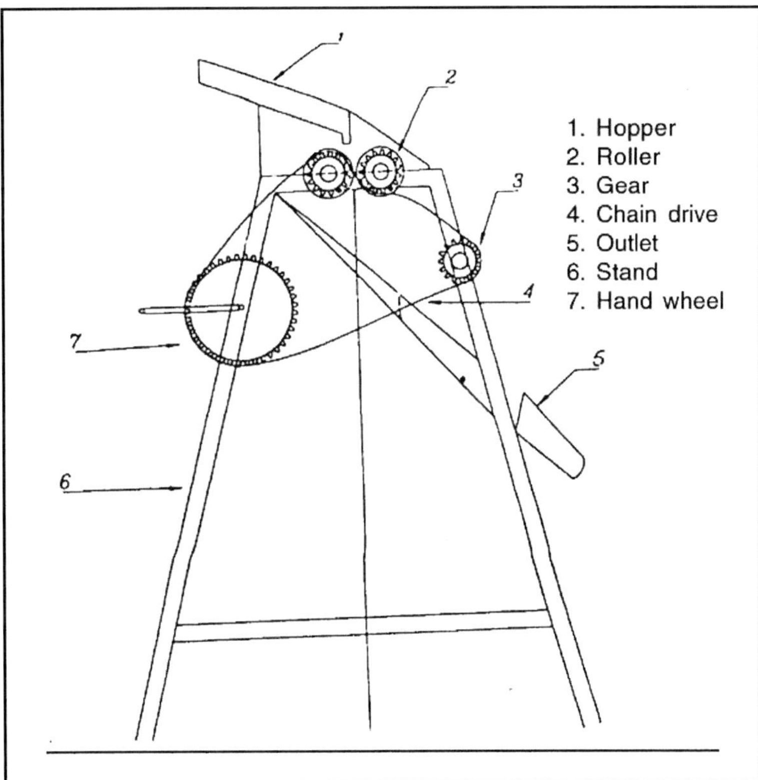

Fig. 3 : Side view of Neem Pulper

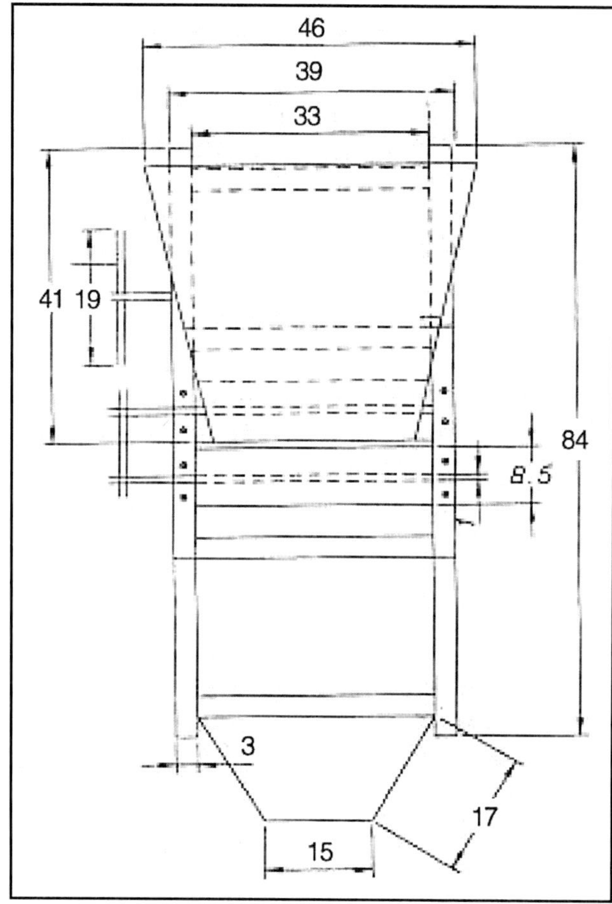

Fig. 4 : Plan of Neem Pulper (All dimensions in cm)

Table 8 : Effect of pulping methods on cost requirement, pulping efficiency, germination and seedling vigour of neem

Treatments	Pulping efficiency (%)	Cost of seed extraction kg^{-1} (Rs.)	Quantity of fruit pulping hr^{-1}	Cost requirement of pulping/ hr^{-1} (Rs.)	Germi nation (%)	Seedli ng vigour
Mechanical pulping	95.45	0.20	30 kg	8.34	71.4	2035
Manual pulping	99.0	0.40	6 kg	16.90	88.2	2452

(Palaniswamy and Masilamani, 2000)

In amla (*Emblica officinalis*) alternate soaking and drying of fruits in 20% sodium chloride solution upto three cycles could extract seeds. The fruit pulp after treatment was easily extractable and the nuts on sundrying expelled seeds. In jamun (*Syzygium cumini*) heaping fruits for a day made the separation of seed easier. Longer duration of heaping damaged the seed. In ber (*Ziziphus mauritiana*) removal of flesh with a sharp knife and thorough washing with water followed by soaking in lime solution for 30 min, made the extraction easier than drying under sun or water washing of fruits. These extraction techniques not only made the extraction easier but also made the edible part economically useful (amla and ber). The extracted seeds recorded higher seed quality in terms of germination, seedling vigour and storability (Sasthri *et al.* 1999). In *Tamarindus indica,* seed extraction from fruit was quickest using the thresher (1.5 min to produce 1kg seed) and slowest using the knife (15 min to produce 1kg seed). However, the thresher resulted in the highest percentage of damaged seed (31%). Manual maceration gave the lowest percentage of damaged seed (0.7%), was quick (3.5 min to produce 1 kg seed) and required the lowest number of fruits to produce 1kg seeds. Seeds extracted using the knife or concrete mixture showed highest seed germination percentage (92%), and seedling vigour. Finally they concluded that seed extraction using a concrete mixture or knife or by manual maceration are recommended (Chamshama *et al.*1998).

Fleshy fruits are also adopted to being ingested by animals. Seeds and stones are often left cleaned and intact after ingestion although in some species a relatively large amount of seeds may be digested. Both ants and termites may readily and efficiently clean fruits and seeds of sweet material, both moist and dry. In the Philippines, the pods of *Samanea saman* are piled up in a dark place and exposed to termite attack. The termites consume the fruit parts only, and after a short time the seeds will be left clean (Seeber and Agpaoa, 1976).

Seed Drying

After extracting the seeds from the fruit / pod, drying is essential before the seed can be safely stored until use in the nursery or packing for storage/ transport. The moisture content of most of the seeds for storage ranges between 10-12 percent. Development of all type of storage fungi (*Aspergillus sp* and *Penicillium sp*) and storage insect (Internal feeders – *Sitophilus oryzae, Rhizopertha dominica, Callosobruchus aculates , C. chinensis, C. pisorum, Sitotroga cerealella, Caryedon serratus, Trogoderma granarium* and external feeders – *Tribolium castaneum, Oryzaephilus surinamensis, Cryptolestes ferrugineus, Plodia interpunctella, Ephestia cautella, Corcyra cephalonica, Liposcelis divintorius, L.*

transvallensiss) in seed is retorted in the moisture content lower than 13%. (Mohan and Sundarababu ,1999; Jakar and Duhan,2001; Kashyap *et al*,2001). Increase the moisture content from 15 to 24% increase in storage fungi, they cause reduction in germination and seedling vigour; discoloration, shrinkage and shriveling of seeds; production of heat and mycotoxin; reduction in nutritional value and bio-chemical constituents of seed and development of mustiness and caking and total decay of seeds.

In case of storage insect pest, the seed quality is affected on the basis of insect damage to endosperm or embryo. If embryo portion is damaged, seed fails to germinate, seed may germinate and develop into a seedling but the vigour of the seedling will depend up on the extent and intensity of the damage. The degree and type of insect infestation and insect species involved depends upon the tree seed and prevailing ecological conditions.

To avoid the above said storage fungi and insect pest infestations, proper precautions should be taken both at harvest time and during post harvest operations like removal of the inert matter such as dried stick, leaves, soil clod, broken seeds, other tree seeds and weed seeds. Further, the dried, sound, cleaned and undamaged seeds should be treated with insecticide (Malathian 5 D @ 250 g / 100 kg of seeds (or) carbaryl 200 g / 100 kg of seeds) along with fungicide (like captan (or) thiram 250 to 300 g / 100 kg of seed (or) Bavistin 200 g / 100 kg of seeds). The seeds treated with both insecticide and fungicide should be packed in moisture vapour - pervious (or) moisture vapour resistant (or) moisture vapour proof containers and labeled properly. Extreme care is required to ensure that treated seeds are never consumed for human or animal food. Treated seeds should be clearly labeled as being dangerous, if consumed.

Pre Sowing Seed Treatment

The seeds of many native plants have built-in dormancy mechanisms, which protect them from germinating in times of droughts. In the wild, seeds will lie dormant until the proper conditions for growth occur. But in cultivation, the grower must become familiar with several simple pre-sowing seed treatment methods, which will unlock the dormancy mechanisms and stimulate quicker, more consistent germination.

In the species *Albizia lebbeck*, poor germination in the nursery is a problem due to seed coat dormancy; to over come this, seeds are treated with concentrated H_2SO_4 for different durations ranging from 5 to 30 min. After treatment, they are washed in running water (Bimlendra and Toky, 1993). Improvement in germination through acid scarification was reported by several authors (Srimathi *et al.*, 1991:, Zodape, 1991). Padma *et al.* (1996)

reported that scarification of seeds with H_2SO_4 for 30 minutes found to increase the germination and vigour status of the seedlings.

According to Natarajan (1999), the germination differed significantly among different duration of acid scarification. There was a significant increase in germination as the duration of scarification increases up to 30 minutes and thereafter it decreased. The higher germination of 88 per cent was obtained at 30 minutes acid scarification, which was followed by 25 minutes (82 per cent). A germination of 27 per cent was recorded in non-scarified control. The longest root of 13.0 cm and shoot of 11.3 cm was recorded at 35 min. and 15 min. scarification respectively. Higher vigour index of 1886 was recorded in seeds scarified for 25 minutes, while higher dry matter production of 0.241 g seedling^{-5} was recorded in seeds scarified for 30 minutes (Table 9).

Table 9 : Standardization of acid scarification treatments in Albizia lebbeck

Duration of acid scarification	Germination (%)	Root length (cm)	Shoot length (cm)	Vigour index	Dry matter production (g seedlings^{-1})
0 min	26.5	105	9.4	528	0.120
5 min	46.5	12.6	10.0	1049	0.118
10 min	55.7	10.7	8.9	1091	0.108
15 min	64.8	9.4	11.3	1336	0.117
20 min	77.5	11.9	9.8	1678	0.165
25 min	81.8	12.3	10.8	1886	0.221
30 min	88.3	12.2	8.4	1818	0.241
35 min	69.5	13.0	9.3	1543	0.212
Mean	63.8	11.6	9.7	1366	0.163

Soaking of seeds in aqueous solutions of thiourea (0.2 %) or KNO_3 (0.3 %) for 12 hr enhanced the germination percentage (Roy, 1992). The enhancement in germination due to KNO_3 may be attributed to an increase in cytochrome oxidase activity, which could break the post harvest dormancy. Thiourea enhances cytokinin activity thereby promoting seed germination. Illango *et al.,* (1999) also reported that soaking of scarified seeds of *Albizia lebbeck* in IAA 100 ppm for 12 hours or 3 % KCl increased the germination and seedling vigour (Table 10).

Soaking the seeds in boiled water for 24 hr and subsequent cooling also improved germination. The seeds, which showed rapid imbibition, germinated rapidly, a gentle but sure indication on the quality of seed being used. For treatment of large quantities of seed to overcome seed coat dormancy, they are tumbled or churned in a concrete mixer with sharp gravel mixer or sand or in a special drum with abrasive material (Kemp, 1975). This was tried for 15 minutes and germination was improved.

Table 10 : Effect of growth stimulants on germination and seedling growth of *Albizia lebbeck*

Treatments	Germination (%)	Vigour index	Root length (cm)	Shoot length (cm)
Control	54.5	1558	23.9	11.3
IBA 100 ppm	75.5	1840	36.4	15.1
IAA 100 ppm	80.5	2118	27.6	11.3
Succinic acid 1%	55.8	1586	25.2	10.3
KCl 1%	83.5	2771	31.5	10.3
CD (P=0.05)	2.91	235	3.18	1.55

In Ber (*Zizyphus mauritiana*), the nut/stone is used as the seed due to the reason that the true seed can not be easily extracted from the hard endocarp. Sharma and Kore (1990) and Grice (1996) also opined that the germination of ber seed is slightly difficult due to the hard endocarp and requires about 3-4 weeks to germinate. The germination time could also be reduced to one week by cracking the endocarp carefully. However, for easier and improved germination, Krishnan and Kulasekaran (1984) recommended soaking the seeds in GA_3 500 ppm or 5 per cent sucrose solution or thiourea 500 ppm for *Zizyphus mauritiana*. Ashok and Prasad (1997) also expressed that seed treatment with GA_3 at 500 ppm resulted in the highest germination of 96 per cent followed by 94 per cent and 92 per cent with 200 ppm + 5 per cent sucrose or thiourea at 500 ppm, respectively. Ghosh and Sen (1988) compared the various physical and physiological seed treatments for maximizing the germination of *Zizyphus mauritiana* and reported that soaking in 10 per cent thiourea for 24 hr could maximize germination. But, Murthy and Reddy (1989) recommended soaking of seeds in a low concentrated solution of 1 per cent thiourea for maximizing the seed germination of *Zizyphus mauritiana* cv. Umran. Sharma and Kore (1990) reported that the germination of ber could be improved by simple water soaking of seeds and germination could be improved to 60-98 per cent with low dose of gamma irradiation. Singh (1990) expressed that the germination of freshly harvested seeds was found to improve by soaking for 6 min. in sulphuric acid and in one-year old seed, soaking for 24 h in water followed by stratification in moist sand for 4 days improved the germination of seed. Srimathi (1997) reported that soaking of the seeds of ber (cv Umran) in 2% thiourea for 24 hr enhanced the germination to 85 per cent, while the germination was only 10% in control.

With regard to *Hardiwickia binata*, the seeds are readily germinating but could also be improved by several seed management practices as the germination capacity and plant percent is low and ranges between 17 and 32% respectively. Soaking of seed in cold water for 24 hr prior to sowing reported to give satisfactory results (Raynov, 1940). Pathak and Gupta, (1984) said that ascorbic acid (500 ppm) improved germination and reduced mortality of seedlings. Suresh *et al.*, (1994) revealed that pre-sprouting of seed improved the germination and survival capacity of seeds (Table 11). Masilamani, (1992) also revealed that when stored seeds were sown, the seeds could be soaked for 24 hr in cold water and pre-sprouted by placing in wet gunny bags for 48 hr for removal of dead seeds. Tiwari *et al.* (1999) evaluated the effect of 3 plant growth regulators (IAA, IBA, GA_3) at 6 concentrations (1,2,5,10,20 and 100 ppm) on germination percentage and plant percentage of, *Hardwickia binata* and found that IAA at lower concentrations (1-20 ppm) increased both germination and plant percentages up to 10% in *Hardwickia binata.* Higher concentrations of IAA had an inhibitory effect. IBA only promoted marginal increases in germination and plant percentages at lower concentrations and was inhibitory at higher concentrations and the effect of GA_3 was similar to IBA. Mamatha *et al.* (2000) revealed that seed treatment with *Trichoderma spp.* reduced the incidence of mycoflora and enhanced the germination and vigour.

Table 11 : Effect of pre sowing treatments on germination and seedling growth in *Hardwickia binata*

Seed treatment	After 15 days	After 90 days	
	Mean Germination (%)	Mean Survival (%)	Mean Height (cm)
Untreated	38	61	7.1
Soaked	70	76	11.5
Pre sprouted	89	82	14.3
CD (P=0.05)	17.6	NS	0.41

Similarly, in Emblica officinalis, Bajpai and Shukla (1990) obtained the highest percentage of germination (92.5%) and seedlings with best root system from seeds soaked in 500ppm GA_3 for 24 hour. Srimathi (1997) recommended stratification treatment with sand moistened to 60 percent with 0.5 percent KNO_3 solution at 5 °C for 10 days, which improved germination of amla seeds to 85 percent, since the seed possess morpho-physiological dormancy.

In Simarouba glauca, the seeds / nuts are non dormant which initiates its germination on 12-15 days after sowing. Soaking the seeds in cold water for 24 hours can break hard endocarp and facilitate quicker and enhanced germination.

In an experiment on *Cassia siamea* in Thailad, 1-2 minutes soaking in 85°C warm water, or submersion at 85°C with subsequent cooling in the water for 12-36 hr gave a germination percentage of 82-89. Longer duration of soaking at 85°C slightly decreased germination percentage. Soaking from 1 to 3 minutes in water at 95°C caused rapid reduction of viability; it was 71% after 1 minute, 47% after 2 minutes and 40% after 3 minutes. Hence, for this species a brief exposure to high temperature or prolonged exposure to 85°C apparently caused heat damage (Kobmoo and Hellum, 1984).

Rai *et al.*, (1986) reported that repeated scarification with conc. H_2SO_4 gave upto 95% germination. Todaria and Negi (1992) reported that *Cassia siamea* is of medium dormancy with percentage seed imbibition between 13.8% to 21.7%. Mechanical scarification was superior to either chemical (H_2SO_4) or biological treatment with rumen liquor.

The effect of various seed pretreatments and seed inoculation with isolates of *Azotobacter chroococcum* and *Rhizobium* were tested on the germination and growth performance of *Cassia siamea*. Treatment with *Azotobacter choococuum* isolate SM4 increased the germination percentage significantly. This inoculam also increased the leaf area, weight of plant and dry matter accumulation (Meshram *et al.*, 1994).

Padma *et al* (1996) suggested that Seeds of *Cassia siamea* responded well to acid treatment, for 12 minutes by exhibiting 60 and 99% germination of freshly harvested and one year stored seed respectively, whereas 4 to 5% germination was recorded in untreated control. Hot water soaking for 4-12 minutes of *Cassia siamea* seed increased germination in freshly harvested (21-46%) and year stored seed (12-95%). However Balkrishnan and Singh (1995) reported that freshly collected small seeds with light brown showed maximum germination and vigour index, but couldn't withstand hot water soaking beyond 6 minutes. Similarly, one year stored seed soaked in hot water for 5 minutes attained 95% germination and subsequently as the soaking period increased the germination gradually decreased.

A comparison was made of hot water and manual scarification treatment of *Cassia siamea*. Small part of the seed was cut off opposite the hilum and the seed was than soaked in water for 24 hr. It is concluded that while hot water treatment is more practical for large-scale plant production, manual scarification should be used in small nurseries (Napier and Napier, 1987).

The seeds pretreated with concentrated H_2SO_4 recorded more germination compared to untreated ones. Softening of the seed coat through scarification with concentrated H_2SO_4 would have facilitated the cell membranes to imbibe water, which would have provided the proper condition for the germination of the seed. This might be the probable reason for the higher germination (Masilamani, 1992).

Conclusions

To recapitulate, Traditional agroforestry systems based on species of *Acacia, Albizia* and other drought tolerant leguminous trees have proved to be reasonably effective in dry land agroforestry systems. Other traditional agroforestry practices includes retaining or growing of trees like *Azadirachta indica, Pongamia pinnata, Acacia nilotica, Mangifera indica, Sygyzium cuminii* as scattered in their fields to have green manure, fuel wood, timber for agricultural purposes. *Ailanthus excelsa, Melia dubia, Thespicia populnea, Morus alba, Bombax sp.,* and *Lannea coromandelica* are the fast growing indigenous tree species getting importance in the recent past.

Important seed technological interventions to improve seed and seedling qualities of these hardy species are related to

- Selection of superior mother trees for seed collection following the prescribed guidelines. For instance, excepting for Neem, Pungam, Mahua, not much work has been carried out in selection of superior mother trees in other native trees for dry land agroforestry systems.
- Understanding on the right age of the mother trees for seed collection. As mentioned earlier, over mature or immature trees should be avoided, since seeds from them may be of low viability and the seedlings from such seeds may have less vigour.
- Knowledge on the right time of seed collection. The studies conducted in the species in dry localities glaringly revealed that there is difference in germination behaviour among seeds collected in different season.
- Method of seed collection and extraction. It is observed that the care taken to address maximum precautionary points during seed collection and extraction will pay through the vigorous offspring's and remunerative plantations.
- Breaking of dormancy in seeds. Most of the dry land tree species possess physical dormancy exerted by the presence of hard seed coat. Suitable pre-sowing seed treatment methods may be adopted to unlock such physical dormancy and to stimulate quicker, more consistent germination.

Good seed handling is an essential complement to genetic improvement. Therefore various steps in seed handling operations viz., collection, extraction, processing, seed treatment, packing, storage and seed sowing operations should be meticulously planned and synchronized to improve (or) maintain seed quality, particularly for tree species to be grown under harsh site conditions as under dry land agroforestry systems.

References

Ashok, P. and Prasad, U. 1997. Effect of Gibberellic acid (GA_3) on germination and seedling growth of 3 host plants of tasar silkworm. *Indian Journal of Serculture,* 36: 153-154.

Bajpai, P. N. and Shukla, H. S. 1990. Aonla. In: Fruits-Tropical and subtropical. (Bose, T. K. and S. K. Mitra Eds.), Naya Prakash publication, Calcutta, Pp. 759-767.

Balakrishnan and Singh, U. 1995. Effect of seed size and colour on germination and seedling growth in five tree species. *Advances in Horticulture and Forestry,* 4: 199-204.

Bhardwaj, S.P and Chakraborty, A.K. 1994. Studies on time of seed collection, sowing and presowing seed treatments of *Terminalia bellirica* Roxb and *Terminalia chebula* Retz. *Indian Forester,* 120: 430-439.

Bimlendra, K, and Toky, O.P. 1993. Breaking of seed coat dormancy with acids in Bose, T. K. and S. K. Mitra. Naya Prakash publication. Calcutta. Pp. 592-615.

Brown, F.M. and Belcher, E. 1979. Improved techniques for processing *Prosopis* seed. *Tree Planters Notes,* 30: 19.

Chamshama, S.A.O, Mugasha, A.G., Msanga, H.P, Gowela, J.P. and Nshubemuki,I. 1998. The effect of different seed extraction methods and presowing treatments on the physiological quality of *Tamarindus indica. Annals of Forestery,* 6: 135-147.

Ghosh, S. N. and Sen, S. K. 1988. Effect of seed treatment on germination, grandiflora and Zizyphus mauritiana, two invasive shrubs in tropical woodlands of Northern Australia. *Australian Journal of Ecology,* 21: 324-331.

Grice, A.C. 1996. Seed production, dispersal and germination in C*ryptostegia grandiflora and Zizyphus mauritiana,* two invasive shrubs in tropical wood lands of Northern Australia. *Australian Journal of Ecology*, 21:324-331.

Gurudev singh, B., Arjunan, M.C., Manimuthu, L., Mahadevan, N.P. and Masilamani, P. 2001. Teak seed research problems and prospects. In: *"Genetics and Silviculture of Teak"* (Mandal, A.K and Ansari, S.A. Eds.), Published by International Book Distributors, Dehradun, Pp. 41-53.

Hedegart, T. 1975. Seed collection of teak. In: FAO / Danida training course on forest seed collection and handling II : FAO, Rome, Pp. 274 – 279.

Illango, K., Mallika Vanangamudi, K., Venkatesh, A., Rai, R. S. V. and Balaji, S. 1999. Effect of growth stimulants on seed germination and seedling vigour in *Albizia lebbeck* (L) Benth. *Seed Research,* 27 (2): 188 – 190.

Jakar, S.S. and Duhan, J.C. 2001. Deterioration of seeds by storage fungi and their management. Training manual on "Seed Health". Department of Seed Science & Technology. CCS Haryana Agricultural University, Hisar. Pp. 108-124.

Kashyap, R.K., Dahiya, B.S. and Khan, M.S. 2001. Insects and seeds store. SST Bulletin. No.15. Department of Seed Science and Technology, CCS Haryana Agricultural University, Hisar, Pp. 23.

Kemp. 1975. Pretreatment and principle of nursery handlings. FAO/DAMDA training course on forest.

Khullar, P., Thapliyal, R.C.,Beniwal, B.S., Vakshasya, R.K. and Sharma, A. 1991. *Forest Seed*. ICFRE publication. Dehradun, Pp. 82-94.

Krishnan, B.M. and Kulasekaran, M. 1984. Studies on seed germination in wild ber. *Z. rotundifolia. South Indian Horticulture,* 32:153-154.

Kobmoo, B. and Hellum, A.K. 1984. Hot water and acid improve the germination of *Cassia siamea* Britt Seeds. *The Embryon.* 1:27-33.

Mamatha, T., S. Lokesh and Rai, V.R. 2000. Impact of seed mycoflora of forest tree seeds on seed quality and their management. *Seed Research,* 28 : 59-67.

Manonmani, V. 1997. Investigation on seed collection, storage and germination of teak (*Tectona grandis*) and (*Santalum album*) Ph.D., Thesis, Tamil Nadu Agricultural University, Coimbatore, Tamil Nadu.

Mapier, I and Robbins, A.M.J. 1989. *Forest Seed and Nursery practice in Nepal.* Nepal U.K. forestry research project, Forestry research division, Department of Forestry and plant research. Babar Mahal. P.O.Box. 3339, Kathmandu, Nepal.

Masilamani, P. 1992. Production, Processing and Storage technology for seeds of *Cassia siamea* Lamk, *Hardwickia binata* Roxb and *Prosopis juliflora* swartz DC. M.Sc.(Ag.) Thesis. Tamil Nadu Agricultural University, Coimbatore, Tamil Nadu.

Masilamani, P and Vadivelu, K.K. 1997. Effect of seed extraction methods on germination and vigour of Honey mesquite. *Madras Agricultural Journal,* 83 : 512-514

Masilamani, P., Parthiban, K.T., Rajeswari , E., and Tilak M. 1995. Effect of seed microflora on germination and vigour of Honey mesquite (*Prosopis juliflora* swartz DC). Paper presented in "National Seminar on Seed Pathology" held at University of Mysore during March 23-25 Abstract, Pp.25.

Masilamani, P., Annadurai, K., and Chinnusamy, C. 1999. Seed extraction using commercial paddy thresher. *Agro India,* 3: 30.

Masilamani, P., Annadurai, K., and Marimuthu, R. 1998. Influence of season of fruit collection on germination and seedling vigour in Neem (*Azadirachta indica).* Paper presented in IUFRO Seed Symposium on "Recalcitrant seeds" held at Kuala Lumpur, Malaysia, during October 12-15.

Masilamani, P., Bhaskaran, M., Chinnusamy., C. and Annadurai, K. 2000. Influence of age of mother tree on germination and initial vigour of neem (*Azadirachta indica). Journal of Tropical Forest Science.*12: 188-190.

Masilamani, P. 2003. Collection and extraction of forest tree seeds. In " Forest Tree Seeds and Nursery Management (Current Trends)" edited by Singh, V and Lavania, S.K. Published by Bishen Singh Mahendra Pal singh publishers, Dehra Dun. Pp. 27- 63.

Meshram, S.O., Joshi, S.N., Pande, J. S. S., Gaikaward, S. J. and Jawarkur, A. S. 1994. Microbial technology for raising seedlings of fast growing trees. *Indian Journal of Forestry*, 17: 243-248.

Mohan, S. and Sundarababu, P.C. 1999. Stored product pests and their management. TNAU, Publications. Pp. 125.

Murthy, B.N.S, and Reddy Y.N. 1989. Temperature dependence of seed germination and seedling growth in ber (*Zizyphus mauritiana* lamk) and their modification by presowing treatments. *Seed Science and Technology*. 18:621-627.

Napier, I. and Napier, I. A. 1987. Pre germination treatment of *Cassia siamea* and *Leucaena leucocephala* seeds. *Banko-Jankari*, 1:5-6.

Natarajan, S. 1999. Investigations on seed source variations, standardization of seed testing. Procedures and nursery techniques in *Albizia lebbeck* (L.) Benth. Ph.D. Thesis, TNAU, Coimbatore.

Olsen, D.S. 1932. Germination capacity of seed produced from young trees. *Journal of Forestry*, 30: 871.

Padma, V., G. Satyanarayana and Reddy, B.M. 1996. Studies on pre-sowing seed treatments in three species of Cassia. *Seed Res*earch, 24:51-54.

Palaniswamy, P.T and Masilamani, P. 2000. Development and testing of Neem fruit pulper. *Journal of Non Timber Forest Products*, 7: 211-213.

Pathak, P.S. and Gupta S.K. 1984. Seed germination and seedling growth of Hardwickia, binata Roxb. *Nitrogen fixing Tree Research Reports*, 3:25-26.

Prasad, R. and Kandya, A.K. 1992. *Handling of forestry seeds in India*. Associated publishing company, New Delhi. Pp. 45-144.

Rai, S. N., Nagaveni , H. C and Padmanabha, H.S. A. 1986. Germination and viability of some tree species. *Van Vigyan*, 24: 8-12.

Raynov, E.W. 1940. A preliminary guide to the germinative capacity in the united provinces of seeds of certain common species and a few exotics together with estimates of quantities of seed required for 1000 plants. U.P. For. Dep. Bull. 14, Pp. 20.

Roy, M.M. 1992. Effect of seed pretreatment with potassium nitrate and thiourea on germination of *Albizia lebbeck* (L.) Benth. *Indian Journal of Forestry*, 15:356-357.

Sasthri, G., Srimathi, P., Malarkodi, K and Karivaratharaju, T.V. 1999. Seed extraction techniques on seed and seedling quality characteristics in few arid zone fruit trees. Abstract proceedings of the National Symposium on "Forestry Towards 21st Century", Coimbatore, Tamil Nadu. September 27-28, 1999. Pp. 137.

Saxena, S.K. and Khan, W.A. 1974. A quick method for obtaining clean seeds of *Prosopis juliflora*. *Annals of Arid Zone*, 13: 269-272.

Schmidt, L. 2000. *Guide to Handling of Tropical and Subtropical Forest Seed*. Danida Forest Seed Center Publications, Denmark. Pp.143-182.

Seeber, G. and Agpaoa, A. 1976. Seed extraction In: Agpaoa, A., Endarigan, D., Festin, S., Gumayagay, J., Hoenninger, Th., Seeber, G., Unkel, K., and Weidelt, H.J (eds); *Manual of reforestation and erosion control for the Philippines*. Eschborn. Pp. 509-511.

Shankaranarayan, K.A., Harsh, L.N and Kathju, S. 1987. Agroforestry in the arid zones of India. *Agroforestry Systems,* 5: 69-88.

Sharma, V.P. and Kore, N.V. 1990. Ber. In: Fruits-Tropical and subtropical. (Bose,T.K. and S.K. Mitra Eds), Naya Prakash Publication. Calcutta. Pp. 592-615.

Singh, V. B. 1990. Fruits of North East Region. Wiley Eastern Region, New Delhi.

Srimathi, P. 1997. Research focus on seed collection, processing and storage of amla (*Emblica officinalis* Gaertn.), jamun (*Syzygium cuminii* Skeels) and ber (*Zizyphus mauritiana* Lamk). Ph.D. Thesis, Tamil Nadu Agricultural University, Coimbatore

Srimathi, P., Rai, R.S.V and Surendran C. 1991. Studies on method of seed collection and optimum duration for acid scarification in a few tree legumes. *My Forest,* 27: 57-61

Suresh, K.K. 1994. Effect of seed size through grading and source on seedling quality in *Acacia nilotica*. Personal Communication.

Suresh, K.K., Swaminathan, C and Rai, R.S.V. 1994. Nursery efficiency by use of presprouted seeds in *Hardwickia binata* Roxb. *Indian Journal of Forestry,* 17: 356-358.

Swaminathan, C., Srimathi, P and Sivagnanam, K. 1994. Effect of tree age on germination and seed and pod quality in *Ciba pentandra* (L.) GAETRN. *Indian Journal of Forestry,* 17: 46-48.

Tewari, D.N. 1994. *A monograph on Dalbergia sisoo* Roxb. International Book Distributors, Dhera Dun. India. Pp. 316.

Tewari, D.N. 1995. *A monograph on Khair (Acacia catechu willd.)*. International Book Distributors, Dehra Dun, India. Pp. 127.

Tewari, P.K., Ravi, A., Savita, K., Agrawal, R. and Kiran, S. 1999. Studies on the Effect of growth regulators on germination percent and plant percent of some leguminous forest tree species. *Journal of Tropical Forestry,* 15:196-201.

Todaria, N.P. and Negi, A. 1992. Pre treatment of some Indian cassia seeds to improve their germination. *Seed Science and Technology,* 20:583-588.

Umarani, R. Bharathi, A and Karivaratharaju, T.V. 1997. Effect of methods of cone drying on seed recovery and seed quality of *Casuarina equisetifolia*. *Madras Agricultural Journal,* 84: 31-32.

Zodape, S.T. 1991. The improvement of germination of some forest species by acid scarification. *Indian Forester,* 117:61-66.

Contd...

Plate 1. Pod Collection in *Cassia fistula Cassia fistula* (Mature fruits)

Plate 1. Cone Collection in *Casuarina equisetifolia Casuarina equisetifolia* (Mature fruits)

Plate 3. Fruit Collection in *Simarouba glauca Simarouba glauca* (Mature fruits)

Plate 4. Manual and Mechanical Threshing of *Albizia lebbek* pods

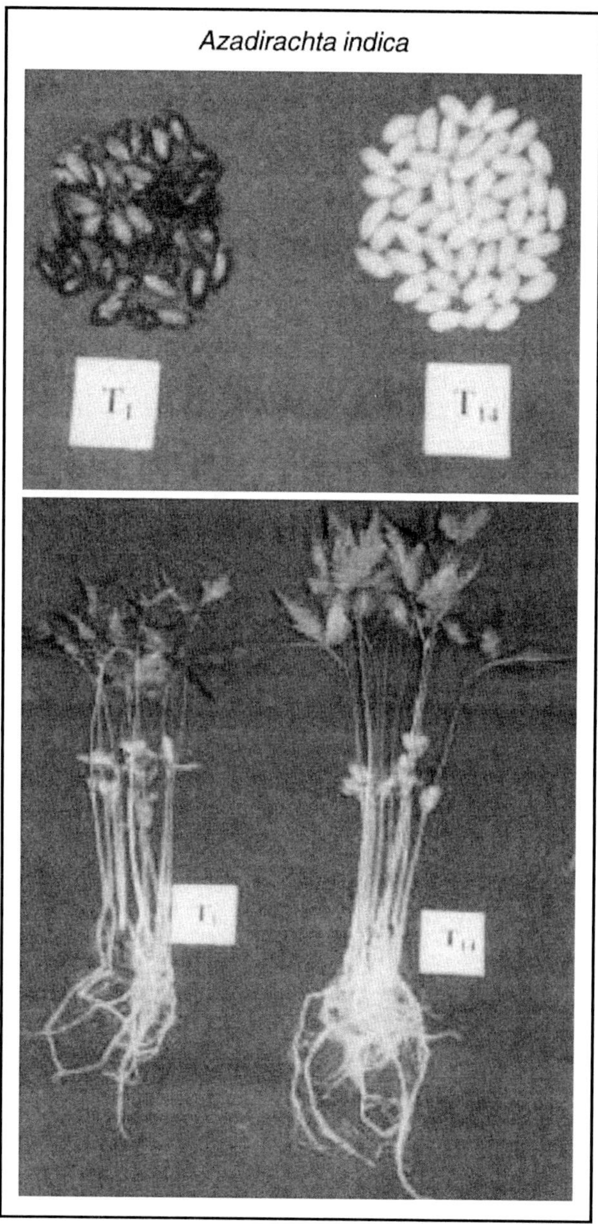

Plate 5. Mesocarp removal of Neem Seed Germination on Mesocarp removal T1 - control T14 - Treated with Sulphuric acid

Agroforestry: Natural Resource Sustainability, Livelihood & Climate Moderation © 2009
Editors: OP Chaturvedi, A Venkatesh et. al., pp. 317-334
Satish Serial Publishing House, Delhi

Chapter 22

LITTERFALL, DECOMPOSITION AND NUTRIENT RELEASE AND ITS IMPLICATIONS IN AGROFORESTRY INTERVENTIONS

R.S. Yadav, D.R. Palsaniya, R.K. Tiwari and Garima Gupta
National Research Centre for Agroforestry, Jhansi-284 003 (U.P.)

Introduction

Agroforestry is in practice since time immemorial over the globe and consists of integration of woody plants with crop or pastures to take advantage of the soil fertility improvement by trees. Researchers and planners are increasingly recommending agroforestry systems as a sustainable form of land use for augmentation of biomass production in the agricultural systems (Nair, 1984; Kang and Wilson, 1987 and Anonymous, 2000b) and recently for mitigation of climate change by way of sequestering C in both standing biomass and soil. Litterfall and decomposition in this alternate use system are the two major processes responsible for nutrient recycling and in the maintenance of organic matter in soil. Trees can maintain or enhance soil fertility by adding N through N_2-fixation, recycling nutrients through litterfall or pruning, or importing nutrients through biomass transfer systems. These nutrients become available to crops through decomposition of the tree prunings and litter. The bound organic forms of N and other nutrients are not available to plants unless they are decomposed and released as inorganic form.

Only about 20% of the N released from tree prunings or litter is taken by the current crop (Palm, 1995; Giller & Cadisch, 1995). Much of the remaining 40% to 80% of the applied organic N is incorporated into soil organic matter (SOM) (Hagger *et al.*, 1993). Because of the slow release of N and P from SOM organic inputs have a greater residual effect on soil fertility than do inorganic fertilizers. The rate of decomposition is mainly regulated by edaphic, climatic variables and resource quality (Singh & Gupta, 1977; Thomas & Askawa, 1993, Yadav *et al.*, 2008). High quality organic inputs (high N but low lignin and polyphenol contents) will release nutrients rapidly. In contrast, low quality organic inputs (poor in N and high in polyphenol & lignin) will release nutrients slowly or even immobilize (Palm, 1995). This has practical implications for attaining synchrony between nutrient release and crop nutrient demand and, hence, for efficiency of nutrient use. This review with few examples attempts the current state of knowledge on the litterfall, quality and decompostion pattern of organic inputs in tree based ecosystems. Authors have also put their views and most promising area of future research in the chapter.

Litter Defined

Litterfall is a fundamental process in nutrient cycling and it is the main means of transfer of organic matter and mineral elements from the vegetation to the soil surface (Vitousek & Sanford, 1986, Regina *et al.*, 1999). Litter is a general term for senescent plant parts. It differs from green foliage in that much of the soluble C compounds and nutrients are translocated from leaves during senescence, prior to abscision and leaf fall. As a result, the quality and decomposition rates of litter and foliage from the same plant can be quite different (Table 1).

Table 1. Generalized comparison of the carbon and nutrient quality of green foliage and litter of the same material.

Parameter	Foliage	Litter
Cellulose	More	Less
Lignin	Less	More
Soluble C	More	Less
N	More	Less
P	More	Less

Collection of Litterfall

Litter collection may be made for a period of one year using quadrants or litter traps of convenient size. Traps should be fixed 30-50cm above the ground. Litter can also be collected by fixing quadrants. The contents of the traps may be collected at monthly/bimonthly intervals depending on the purpose. Litter can be further seperated into different components viz.- leaf, twigs, reproductive parts etc. The samples then can be oven dried at 80^{0}C until weight is constant and the litterfall is computed on unit area basis for each month and species. Monthly litterfall could be summed to obtain the total annual litter yield.

Litter traps should be preferred for litter collection in scattered trees (low density), while quadrants can be used in collecting litter in forest plantation and uniformly high tree density plantation in different agroforestry systems. In scattered tree plantation especially found in traditional agroforestry systems, wind velocity and wind direction plays an important role in influencing litter production. In such cases, number of litter traps should be sufficient enough to harvest maximum and accurate litter production.

Importance of Litterfall

Litter from plants are requisite to ecosystem stability and ecological functions. The roles that litter plays in ecosystems were recognized as early as the 1850s and have been documented by a large number of studies worldwide (Sayer, 2006). Litter contributes to forest and agro-ecosystem mainly by nutrient and carbon turnover during litter decomposition and thus maintaining biogeochemical cycling in the ecosystems. Litter cover acts as a protective layer for maintaining soil physical properties like retention of soil moisture (Ginter *et al.*, 1979), buffering against soil temperature and compaction change (MacKinney, 1929), and soil conservation from erosion or leaching (Mo *et al.*, 2003). It also provide habitats and substrates for soil fauna (Attignon *et al.*, 2004) and flora (Ruf *et al.*, 2006).

Litterfall Production

Litter production is a function of many factors. Quantum of litter production varies with tree species, species composition, stand age, stand density, edaphic and climatic factors. It is natural that trees at higher age and density yield more litter. Also trees grown on land capability class I and II may yield more litter as compared to those grown on non arable lands due to good site characteristics.

In a review, Kumar (2007) reported a range from 1.02 (*Eucalyptus tereticornis*) to 14.5 Mg ha^{-1}yr^{-1} (*Pinus caribaea*) for mean annual litterfall of 49 tropical species. George (1982) estimated annual litter production in age sequence of *Eucalyptus* hybrid plantation per hectare basis (Table 2). Considering three plantations individually a total of 3377, 3801 and 6207 kg ha^{-1} of litter is produced at ages of 5, 7 and 10 years respectively indicating the effect of age. Ten year old plantation is producing almost double the amount of litter than that of 5 year old plantation with almost the same density while there is only a little difference between litter production at age 5 and 7 years. Of the total litter, the contribution of various components in these three plantations varies. Leaf litter contributes between 66% to 76% of the total litter; twigs between 16% to 24% and bark between 5% to 10%.

Table 2. Litter production in Eucalyptus hybrid plantation

Age	Leaf		Twig		Bark		Total
(Years)	Kg ha^{-1}	%	Kg ha^{-1}	%	Kg ha^{-1}	%	Kg ha^{-1}
5 (1167)	2567	76.0	647	19.2	163	4.8	3377
7 (1167)	2511	66.1	897	23.6	393	10.3	3801
10(1133)	4642	74.8	969	15.6	596	9.6	6207

Populus deltoides produced significantly higher litter than *Eucalyptus hybrid* (Singh *et al.*, 1989). Litter production increased with advancement in the age of both the trees. Total litter production tree^{-1} of *P. deltoides* was about 3.5 time than the *E. hybrid* (Table 3). *E. hybrid* is an evergreen species, low litter production may be due to the longevity of the leaf. During entire 60 month cycle, *P. deltoides* produced, on an average, 12.5 t litter ha^{-1} while *E. hybrid* only 11.5 t ha^{-1}.

Sandhu *et al.* (1990) made experiments on a 2-yr old *Leucaena leucocephala* stand to measure the litterfall. Annual litterfall was 10 t ha^{-1} and was maximal in the dry summer months. The seasonal peaks were in June and December (15 and 11% of total litterfall, respectively). Leaves comprised 88% of the total litterfall.

Litter dynamics and associated nutrient turnover were studied in 4-5 year old silvo-pastoral systems involving four fast growing MPTs (*Leucaena leucocephala, Casuarina equisetifolia, Acacia auriculiformis* & *Ailanthus triphysa*). Annual addition of litter ranged from 1.92-6.25 Mg ha^{-1} and showed variation between different systems (George & Kumar, 1998). Timing of the large pulse of deposition generally coincided with the dry period (Feb. to May) and the period of reduced litterfall with that of the wet season (June to Oct., Table 4). In scattered trees at farmers field, Yadav *et al.* (2008) estimated an annual litter production of 50.9, 54.2, 41.5 and 35.8 kg ha $^{-1}$ in *P. cineraria*, *D. sissoo*, *A. leucophloea* and *A. nilotica* having a canopy diameter of eight meter, respectively indicating inter species variation in litter production. Irrespective of tree species, maximum litter production was obtained in winter season.

Table 3 Periodic litter fall and addition of nutrients to soil via litter of *E. hybrid* and *P. deltoides* at CIMAP, Pantnagar, India.

Duration (months)	Litter fall (dry wt. t ha^{-1})		Nutrient addition (kg ha^{1} yr^{-1})					
			N		P		K	
	E. hybrid	*P. deltoides*	*E. hybrid*	*P. deltoides*	*E. hybrid*	*P. deltoides*	*E. hybrid*	*P. deltoides*
24	0.66	1.50	6.1	15.3	1.6	5.4	3.0	11.1
36	2.43	2.68	17.0	22.5	5.8	10.7	10.7	18.0
48	3.80	3.80	19.8	25.8	7.2	11.8	15.2	22.8
60	4.60	4.50	22.1	25.2	9.2	15.7	17.0	26.1

Table 4. Annual litterfall in four fast growing multipurpose tree spp. grown in silvo-pastoral systems in a humid region of Central Kerala, India.

Species	Annual litterfall (Mg ha^{-1}Year^{-1})		
	Mean	Min.	Max.
A. auriculiformis	6.27	3.72(Aug)	10.26(Feb)
C. equisetifolia	2.31	1.16(Oct.)	4.83(Feb)
L. leucocephela	2.30	0.98(Jun)	3.91(Feb)
A. triphysa	1.92	0.70(Aug)	3.42(Mar)

Resource Quality and Decomposition

The resource (substrate) quality is defined by the organic constituents and nutrient content of the material. The parameters of resource quality should be such so as to use to assess the potential decomposition and nutrient release patterns of organic materials. The recommended parameters of organic materials for quality characterization are-

Soluble carbon
Cellulose
Lignin
Polyphenols
Protein-binding capacity
Total N
Total P
NDF-N

Lignin is considered by many workers to be the most important component for determining the rate of decomposition. It is of low quality and provides little or no energy to decomposers until the late stage of decomposition. Basically, high quality leaves will decompose faster than low quality leaves.

Methods of Litter Decomposition Study

Diverse efforts have been made over the years to quantify the decomposition process. Among the methods or approaches, following tecniques have been widely adopted in contemporary studies.

- The perfusion method
- Litterbag tecnique
- Tracer tecnique
- Measurement of soil respiration

Litterbag technique is most widely used method to study the litter decomposition in the fields due to its simplicity in use. Litterbags of convenient size and mesh are get prepared of nylon. Plant material is placed in these bags, labellled and installed in the fields as per requirement and purpose of study. Litterbags are retrieved periodically for mass loss and further calculations and interpretation.

Litter Decomposition

Decomposition is a basic process in the control of nutrient cycling and formation of soil organic matter (Berg & McClaugherty, 2002). Unless there is release of nutrients from litter via decomposition, litter cover and tree pruning could deposit to a level causing dis-functioning of the ecosystem. Two strategies can be used to manipulate litter decomposition in order to improve nutrient use efficiency- 1) regulate the release of nutrients to improve the synchrony of nutrient supply with crop demand, and 2) provide a more favourable environment for plant growth. The two strategies relate to the short term improvement of nutrient availability and the longer term improvement of SOM (Table 5). The long term effect of building up SOM pools may be more important in some cases than short term nutrient release patterns. The effect of organic inputs on both of these strategies is highly dependent on the quality of the inputs.

Table 5: Beneficial roles of leaf prunings and litter from agroforestry trees.

Benefit	Timescale	Resulting from	Mediated by
Improved nutrient uptake	Short-term	Increased N availability	N release during decomposition
		Increased P availability	Reduced P fixation due to organic acids
		Enhanced root proliferation	Reduced Al toxicity due to complexation
Reduced nutrient losses	Long-term	Smaller leaching losses	Increased CEC Increased nutrient uptake
Improved moisture availability	Long-term	Increased moisture retention	Increased SOM
Reduced soil erosion	Long-term	Increased infiltration Increased soil cover Increased soil stability Reduced runoff	Increased SOM Surface litter Increased SOM Increased SOM

Sandhu *et al.* (1990) examined the litter decomposition in the field using leaf, twig and fruit fractions of above-ground litter of *Leucaena*. Roots were also studied. Loss of mass was in the order fruit> Root> twig> leaf (Table 6). The initial mass loss was high and there was rapid loss in the rainy season (July-September).The initial N content played a significant role in decomposition. Each fraction of litter decompose at a specific rate and is variably dependent on the resource quality (C-to-N) or one of the enviromental factors.

Raizada & Srivastava (1991) studied the rate of litter decomposition of Poplar leaves by litter bag method. The weight loss in the first six months (July-Dec.) was nearly 53% of the original weight, with a relatively high decay rate during July, Aug. and September (Table 7). However, the rate slowed down in the next six monthss, with a loss of only 10% more in the dry weight remaining. Thus, over a one year period, about 63% of the initial dry matter within the litter bag was lost. Further, dry weight loss of confined litter, in a year was more closely related with monthly rainfall (r = 0.63) than with temperature (r = 0.40).

Table 6 : Percentage of initial ash-free mass (g) remaining in decomposing substrate at each retrieval date

Months	Leaf	Twig	Root	Fruit
1986				
Oct	83.3	75.10	78.0	75.2
Nov	75.3	63.0	66.1	60.4
Dec	64.5	58.3	60.5	46.2
1987				
Jan	57.6	53.4	56.4	36.7
Feb	52.0	49.8	52.8	29.2
Mar	48.4	46.4	49.3	22.4
Apr	45.6	43.8	46.1	16.5
May	42.4	41.1	42.9	12.0
June	39.3	38.5	40.8	
July	36.0	35.5	36.6	
Aug	31.8	29.9	27.4	
Sep	25.1	18.5	14.5	

Table 7 : Dry weight loss and climatic variable during the study period

Months	Elapsed days	Wt. remaining (g)	% Wt. lost	Max temp. (°C)	Min. temp. (°C)	Rainfall (mm)
June 86	1	15.00	0.00	33.7	23.2	282.5
July 86	30	12.41	17.22	29.9	22.3	620.3
Aug. 86	60	8.63	42.46	29.9	22.4	603.4
Sep. 86	90	7.75	48.30	29.6	19.7	147.8
Oct. 86	120	7.69	48.74	27.3	13.8	101.6
Nov. 86	150	7.15	52.34	25.3	8.9	7.4
Dec. 86	180	7.06	52.94	20.3	4.8	39.0
Jan.87	210	6.64	55.74	20.7	4.4	24.8
Feb. 87	240	6.60	56.00	23.0	7.2	48.7
Mar. 87	270	5.71	61.90	26.9	10.6	34.4
Apr. 87	300	5.70	62.00	32.2	13.0	49.5
May 87	330	5.53	63.10	31.7	15.2	141.4
June 87	360	5.41	63.94	36.1	20.1	116.2

Pande (1999) examined the decomposition rate of four tree species viz. sal, teak, eucalyptus and pine. Leaf litter decomposition followed the order: sal (1.67) > teak (1.65) > pine (1.35) > eucalyptus (1.34, Table 8). In general, values for higher decomposition rate were observed during rainy season and lowest during winter except pine and sal (Table 9).

Table 8 : Decomposition constants and time required for the one half and 95% of the original dry weight for tropical plantations

Species	Component	k	Time required (Years) Half life (0.693/k)	95% (3/k)
Shorea robusta	Leaf	1.67	0.41	1.80
	Twigs	0.95	0.73	3.17
Tectona grandis	Leaf	1.65	0.42	1.81
	Twigs	0.97	0.71	3.08
Eucalyptus spp.	Leaf	1.16	0.59	2.57
	Twigs	0.82	0.84	3.64
Pinus roxburghii	Leaf	1.35	0.81	2.23
	Twigs	0.86	0.80	3.46

Table 9 : Decomposition constants (k) of leaf and twig litter in different seasons

Species	Component	Winter (Nov.-Feb.)	Summer (March-June)	Rains (July-Oct.)
Shorea robusta	Leaf	0.206	0.612	0.85
	Twigs	0.210	0.167	0.57
Tectona grandis	Leaf	0.239	0.502	0.91
	Twigs	0.183	0.204	0.57
Eucalyptus spp.	Leaf	0.284	0.278	0.61
	Twigs	0.216	0.256	0.35
Pinus roxburghii	Leaf	0.208	0.439	0.61
	Twigs	0.181	0.128	0.56

George & Kumar (1998) reported that mass loss of litter, decay rate coefficient (k) and half-lives varied among multipurpose trees (Table 10). On the whole residual litter mass decreased exponentially with time, except for *Ailanthus* which exhibited a linear pattern. Of the four tree species studied, only *Casuarina* and *Leucaena* litter lost mass completely during the experimental period. Rapid mass loss occurred in *Casuarina* needles, which had a half life of one month. In scattered tree plantation at farmers field under irrigation, litter decomposition rates also differed with tree species. Litter decay follow the trend: *P. cineraria* (k_D=0.0174) > *D. sissoo* (k_D =0.0150)) > *A. leucophloea* (k_D =0.0121) > *A. nilotica* (k_D =0.0119).

Table 10 : Monthly decay rate coefficients (k) and half lives of decomposing litter of four multipurpose tree spp. grown in silvo-pastoral systems in a humid region of Central Kerala, India.

Species	k	R^2	n	SE	Half life ($t_{0.5}$) (month)
A. auriculiformis	0.416	0.874	124	0.727	1.6
C. equisetifolia	0.666	0.908	61	0.476	1.0
L. leucocephela	0.510	0.502	75	1.156	1.3
A. triphysa	0.143	0.806	124	0.325	4.8

Indices for Predicting Nutrient Release

Many researchers have developed predictors or indices of mass loss and N release. The indices or predictors include-

Nitrogen (N) concentration

C : N ratio

Lignin

Polyphenol to Nitrogen (PP : N)

Lignin to Nitrogen (L : N)

Polyphenol + Lignin to Nitrogen (PP + L) : N)

All of the indices are apparently valid but depend on the time course under consideration and the type of materials. The (PP + L) : N) ratio of less than 10 will release N.

Nutrient Release

Nutrient release is a consequent process of decomposition of litter or any other plant material. Recycling of C and nutrients during decomposition is a key ecosystem process (Swift *et al.*, 1979, Cadish & Giller, 1997) that has major control over the C cycle, nutrient availability and consequently plant growth and community structure (Wardle, 2002, Bardgett, 2005). In the natural ecosystems, organic matter decomposition synchronizes with plant growth and C and other nutrients are utilized in the system with maximum efficiency (Gregorich & Janzen, 1998). In agro-ecosystems, however, no single organic material can be expected to release N in perfect synchrony to match the plant demand (Berg, 2000). Nevertheless, materials high in N and low in lignin and polyphenol release a large proportion on N faster, in advance of the main period of crop demand, while materials high in lignin or polyphenol release only a smaller proportion of their N initially and the remainder is released at a continuous slower rate.

Annual above-ground litter N content was high (256 kg ha^{-1}) with an annual release of 208 kg ha^{-1} i.e. 81% of the total litterfall in *Leucaena* (Sandhu *et al.* 1990). The rate of N release (Table 11) was greatest from fruit litter and the relative loss rate was high throughout the study. N concentrations had decrease in all components by the end of study. Except of roots, the release of N per unit area was highest during the first month for all the components.

Table 11 : Percentage of initial ash-free mass (g) remaining in decomposing substrate at each retrieval date

Months	Leaf	Twig	Root	Fruit
1986				
Oct	65.9	75.1	103.7	71.8
Nov	68.0	57.7	84.9	57.3
Dec	61.3	52.5	81.0	42.7
1987				
Jan	54.7	51.1	47.0	34.5
Feb	52.4	47.5	53.0	30.9
Mar	47.6	44.2	65.7	21.3
Apr	43.7	4.5	76.7	7.5
May	40.4	40.9	71.7	4.4
June	36.7	36.9	68.0	
July	27.3	37.1	54.7	
Aug	23.6	30.5	41.3	
Sep	19.5	20.2	21.7	

Palm & Sanchez (1991) determined N realese pattern from the leaflets of 10 tropical legumes and rice straw in a lab experiment. Three patterns of net N mineralization during the 8- weeks (Table 12). One pattern exhibited by the control soil, rice straw and leaves of *Inga edulis* and *Inga* spp. Pattern II showed initial net mineralization followed by low but positive net mineralization rates represented by *C. reticulata, C. Cajan, L. leucocephala* and the two *Desmondium* spp. Pattern III is that shown by *G. sepium, Erythrina* spp. and *A. saman* in which net mineralization was higher than the control during the first 2 weeks and then again at 4-8 weeks. *G. sepium* had the higher net mineralization, 54 m g N g^{-1} soil and *D. ovalifolium* the lowest, 15.9 m g N g^{-1} soil.

Maharudrappa (2000) reported that incubation of litter of different tree spp. enhanced nutrient availability. The available N decreased slightly at 30 days of incubation compared to control (Table 13). However, available N increased at the end of 60 days of incubation. This indicated temporary immobilization of nutrients by microorganisms. Unlike N, the release of P to the available pool from litter was recorded at 30 days of incubation. Mineralization of organic matter played a major role in P release for plant growth. The decomposition of litter material with narrower C : P ratio are likely to increase the available P as compared to those with higher C : P ratio. Incubation of litter also increased available K in all the treatments. The release of K to soil was more dependent on the quantity and quality of the litter. Among spp., teak recorded significantly higher values than other tree spp. The increase in available K may be attributed to the fact that K is not strongly bound in organic structures, unlike that of N and S.

Table 12 : Characteristics of organic additions, changes in $(NH_4^+ + NO_3^-)$– N and 8-week cumulative mineral N

Treatment	N	Lignin (%)	Phenolic	$(NH_4^+ + NO_3^-)$– N Week 1	2	3 (μ g N g^{-1} soil)	4	8	Total
Control				0.47	2.93	8.70	4.17	3.74	24.01
G. sepium	3.74	7.8	1.02	12.12*	15.64*	9.30	3.72	13.23	54.00*
Erythrina spp.	3.52	9.7	1.04	8.67*	10.86*	11.58	2.06	15.10*	48.25*
A. saman	3.20	6.3	1.48	1.84	9.05*	10.86	0.70	18.23*	40.68*
I. edulis	3.18	16.3	3.43	1.47	0.92	6.07	4.27	8.89	21.62
O. sativa	1.13	6.2	0.69	0.33	3.04	6.75	0.85	12.33	23.31
Inga spp.	2.97	17.9	1.61	-2.17	3.50	6.32	2.55	12.01	22.21
L. leucocephala	3.94	5.2	2.94	-4.59	4.79	7.54	9.22	8.23	25.19
C. reticulata	2.50	10.8	2.20	-9.44*	6.85	8.64	0.58	18.06*	23.52
C. cajan	3.48	10.2	3.34	-5.38*	4.71	4.57	4.18	10.67	18.74*
D. gyroides	2.99	11.7	1.91	-4.40*	4.75	3.22*	6.59	7.38	17.53*
D. ovalifolium	2.37	9.1	3.61	-7.00*	4.13	3.01*	3.78	11.99	15.91*

Table 13 : Available N, P and K status (kg ha^{-1}) of soils incubated with different litters

Treatments	N		P		K	
	30 days	60 days	30 days	60 days	30 days	60 days
Control	252.39	261.53	13.84	15.15	215.67	228.33
Eucalyptus 2.5 t ha^{-1}	248.92	283.68	13.95	15.57	222.56	236.08
Eucalyptus 5.0 t ha^{-1}	245.83	301.13	14.20	16.09	230.16	245.84
Eucalyptus 7.5 t ha^{-1}	243.53	325.46	14.55	16.85	238.07	253.66
Eucalyptus 10.0 t ha^{-1}	242.65	343.85	14.79	17.35	247.18	261.04
Teak 2.5 t ha^{-1}	252.14	306.92	14.50	16.95	235.66	243.06
Teak 5.0 t ha^{-1}	249.14	345.36	14.90	17.83	258.38	265.28
Teak 7.5 t ha^{-1}	248.26	378.65	15.30	18.52	281.54	292.33
Teak 10.0 t ha^{-1}	246.92	403.82	15.75	19.32	301.36	312.38
Casuarina 2.5 t ha^{-1}	247.09	279.96	14.10	15.72	223.18	238.08
Casuarina 5.0 t ha^{-1}	245.33	299.21	14.50	16.15	232.18	247.28
Casuarina 5 t ha^{-1}	242.00	318.33	14.85	16.97	240.58	255.79
Casuarina 10.0 t ha^{-1}	241.48	339.08	15.35	17.48	249.36	264.94
Acacia 2.5 t ha^{-1}	250.98	302.76	14.30	16.76	225.38	241.46
Acacia 5.0 t ha^{-1}	248.46	338.16	14.80	17.46	236.68	252.48
Acacia 7.5 t ha^{-1}	247.89	373.48	15.15	18.09	247.94	259.56
Acacia 10.0 t ha^{-1}	245.62	399.43	15.50	18.89	259.56	269.09

Absolute amount of N, K and Ca released were greater in legumes than in the grasses due to high initial concentration in legumes (Thomas & Askawa, 1993). Kenjale *et al.* (1994) reported decomposition of leaf litter from seven forest tree species and release of plant nutrients in soil. It was observed that leaf litter increased the available nutrients in the soil after decomposition as compared to control (Table 14). More release of available N was observed through teak, karanj and shivan may be due to faster rate of decomposition of leaf litter of these tree species.

Isaac & Nair (2002) compared decomposition pattern of jack leaf litter at the two sites using weight loss as the index of decay. The weight loss (%) from the decomposed leaves is presented in Table 15. The time taken for litter decay varied between the two sites. The decay revealed to be much faster beneath the tree compared to that in the open. It took 14 fortnights in the open for disappearance of 95% of the initial litter weight, whereas, the same took only 10 fortnights in the shaded condition. This clearly brings to light the strong influence of the site characteristics on the rate of decomposition. Further, the decay rate constants (k) were computed from the linear equation for the two sites. These were 1.381 and 1.941 on the open and beneath tree canopy, respectively in accordance with the time taken for 95% decay. Higher the value, faster is the rate of decay.

Table 14 : Available major nutrients and exchangeable cations in soil after decomposition of leaf litter

Plant spp.	Available N Kg ha^{-1}	Available P_2O_5 Kg ha^{-1}	Available K_2O Kg ha^{-1}	Exchangeable Ca me $100g^{-1}$	Exchangeable Mg me $100g^{-1}$
Gmelina arborea	326.04	25.47	315.08	5.39	3.78
Terminalia tomentosa	333.74	25.93	248.44	4.95	4.80
Tectona grandis	361.71	30.78	236.62	5.29	3.54
Casuarina equisetifolia	302.93	29.31	243.93	5.03	4.47
Acacia auriculiformis	310.64	31.42	221.17	4.33	4.50
Eucalyptus hybrid	290.10	31.19	311.69	4.68	3.15
Pongamia pinnata	329.89	30.23	264.26	5.12	2.63
Control	279.83	20.60	204.40	3.89	2.05

Table 15 : Decomposition (%) and decay constants computed from linear equation

Period (fortnight)	Open		Beneath tree canopy	
	Residual wt. (g)	Decomposition (%)	Residual wt. (g)	Decomposition (%)
0	20.00	0.00	20.00	0.00
1	19.20	3.99	18.33	8.83
2	17.22	13.86	16.15	19.50
3	15.93	20.32	14.71	26.41
4	15.16	24.20	12.36	38.17
5	13.14	34.26	7.54	62.28
6	11.37	43.13	6.28	68.59
7	10.19	48.02	5.30	73.46
8	6.46	67.67	3.16	84.16
9	5.23	73.85	2.89	85.51
10	4.04	79.79	0.59	97.04
11	2.51	87.42	-	-
12	1.97	90.14	-	-
13	1.06	94.68	-	-
14	0.66	96.67	-	-
Decay constant	1.38		1.94	

Conclusion

Agroforestry research and development will go a long way in present scenario of global climate change and wide gap in demand and supply of tree produce. This necessitates more studies on litter production, its decomposition and consequent nutrient release for better understanding of ecosystem functioning and application of this understanding in soil nutrient management. From the above brief results and discussion, it may be concluded that-

- Litterfall production varies with the age of plantation, tree density, species composition, nature & type of spp. and season of litterfall. More the litterfall, higher will be nutrient addition potential of the tree spp. on decomposition
- Rate of decomposition of litter also varies with the decomposition of substrate quality, climatic variables and edaphic factors
- Nutrient release by litter depends upon their concentration, nutrient in question, components of litter, season, tree spp., nutrient ratios, rate of decomposition etc.

Above conclusions were mainly drawn from the studies in forest ecosystems. Tree based agro-ecosystem is more complex than forest ecosystem requiring different approaches under a given situation. Agro-ecosystem especially agroforestry is more dynamic than the forest ecosystem. Farmers manipulate the agro-ecosystem to get better harvest of the crop. Sequential studies on litter production, decomposition and subsequent release can go a long way to synchronize the nutrient release and crop demand as an important aspect for organic farming. Here it is worth mentioning that trees in addition to nutrient recycling by way of litterfall and/or pruning addition, plays an imporatnt role to maintain the already fragile ecological balance.

Future Research Thrust

From the above discussion, it is observed that considerable thrust in future research areas of tree pruning, litter production and its decomposition should be given to-

- Relative importance of below- and above- ground organic matter inputs on nutrient cycling.
- Residual effects on nutrient release and long- term changes in soil fertility resulting from application of pruning biomass of different quality and periodical addition of litterfall.

- Develop ways of managing the decomposition rate of applied biomass to optimize rates of short- and long- term nutrient release.
- Synchronization of nutrient release from tree prunings and litter to meet the crop demand.
- Litter production in agroforestry should be recorded till tree attains maturity.
- Role of management practices for both crop and tree in agroforestry in litter production and decomposition should be quantified for better nutrient use efficiency.

References

Anonymous, 2000b. National Agricultural Policy. Department of Agriculture & Cooperation, Ministry of Agriculture, Government of India, Pp. 4.

Attignon, S.E.; Weibel, D., Lachat, T., Sinsin, B., Nagel, P. And Peveling, R. 2004. Leaf litter breakdown in natural and plantation forests of the Lama forest reserve in Benin. *Applied Soil Ecology*, 27: 109-124.

Berg, B. 2000. Litter decomposition and organic matter turnover in northern forest soils. *Forest Ecology and Management*, 133: 401-417.

Berge, B. and McClaugherty, C. 2002. Plant litter decomposition humus, Springer-verlag, New York.

George, M. 1982. Litter production and nutrient return in Eucalyptus hybrid plantations. *Indian Forester*, 109: 253-260.

George, S. J. and Kumar, B. M. 1998. Litter dynamics and cumulative soil fertility changes in silvopastoral systems of a humid tropical region in central Kerala, India. *International Tree Crops Journal*, 9: 267-282.

Giller, K.F. and Cadisch, G. 1995. Future benefits for biological nitrogen fixation. An ecological approach to agriculture. *Plant and Soil*. 174: 255-277.

Ginter, D.L., Mcleod, K.W. and Sherrod, C. 1979. Water stress in longleaf pine induced by litter removal. *Forest Ecology Management*, 2: 13-20.

Haggar, J.P., Tanner, E.U., Beer, J.W. and Kass, D.C.L. 1993. N dynamics of tropical agroforestry and annual cropping systems. *Soil Biology Biochemistry*, 25: 1363-1378.

Isaac, S.R. and Nair, M.A. 2002. Litter decay: Weight loss and nitrogen dunamics of Jack leaf litter on open and shaded sites. *Indian Journal of Agroforestry*, 4: 35-39.

Kang, B.T. and Wilson, G.F. 1987. The development of alley cropping as a promising agroforestry technology. In: Agroforestry: A decade of development (H.A. Steppler and P.K.R. Nair, eds.). ICRAF, Nairobi, Pp. 227-243.

Kenjale, R.Y., Chavan, K.N. and Chavan, A.S. 1994. Recycling of nutrient elements by some forest tree speceis of Konkan in Maharashtra. II- Leaf-litter decomposition and release of nutrients in laterite soil of Konkan. *Van Vigyan*, 32: 7-14.

Kumar, B.M. 2007. Litter dynamics in plantation and agroforestry systems of the tropics-A review of observations and methods. In: Ecological basis of agroforestry (D.R. Batish, R.K. Kohli, S. Jose and H.P. Singh, eds.), CRC Press, USA.

MacKinney, A.L. 1929. Effects of forest litter on soil temperature and soil freezing in autumn and winter. *Ecology*, 10: 312-321.

Mafongoya, P.L., Giller, K.E. and Palm, C.A. 1998. Decomposition and nutrient release patterns of tree prunings and litter. *Agroforestry Systems*, 38: 77-97.

McClaugherty, C. and Berg, B. 1987. Cellulose, lignin and nitrogen concentration as rate regulating factors in late stages of forest litter decomposition. *Pedobiologia*, 30: 101-112.

Maharudrappa, A., Srinivasmurthy, C.A., Nagaraja, M.S., Siddaramappa, R. and Anand, H.S. 1999. Decomposition rates of litter and nutrient release pattern in a tropical soil. *Journal of Indian Society Soil Science*, 48: 92-97.

Mo, J., Brown, S., Peng, S. and Kong, G. 2003. Nitrogen availability in disturbed, rehabilitated and mature forests of tropical China. *Forest Ecology Management*, 175: 573-583.

Nair, P.K.R. 1984. Soil productivity aspects of agroforestry. ICRAF, Nairobi, Pp.164.

Palm, C.A. 1995. Contribution of agroforestry trees to nutrient requirements in intercropped plants. *Agroforestry Systems*, 30: 105-124.

Palm, C.A. and Sanchez, P.A. 1991. Nitrogen release from the leaves of some tropical legumes aas affected by their lignin and polyphenolic contents. *Soil Biology Biochemistry*, 23: 83-88.

Pande, P.K. 1999. Litter decomposition in tropical plantations: impact of climate and substrate quality. *Indian Forester*, 126: 599-608.

Raizada, A. and Srivastava. M.M. 1991. Leaf litter decomposition and pattern of nutrient release in a *Populus deltoides* Marsh. Plantation. *Van Vigyan*, 29:175-179.

Ruf, A., Kuzyakov, Y. and Lopatovskaya, O. 2006. carbon fluxes in soil food webs of increasing complexity revealed by 14C labeling and 13C natural abundance. *Soil Biology Biochemistry*, 38: 2390-2400.

Sandhu, J., Singh, M. and Ambasht, R.S. 1990. N release from decomposing litter of *Leucaena leucocephala* in the dry tropics. *Soil Biology Biochemistry*, 22: 859-863.

Sayer, E.J. 2006. Using experimental manipulation to assess the roles of leaf litter in the functioning of forest ecosystems. *Biological Review*, 81: 1-31.

Singh, K., Chauhan, H.S., Rajput, D.K. and Singh, D.V. 1989. Report of a 60 month study on litter production, changes in soil chemical properties and productivity under poplar (*P. deltoides*) and Eucalyptus (*E. hybrid*) interplanted with aromatic grasses. *Agroforestry Systems*, 9: 37-45.

Thomas and Askawa, 1993. Decomposition of leaf litter from tropical forage grasses and legumes. *Soil Biology Biochemistry*, 25: 1351-1361.

Vituosek, P. and Stanfiord, R.I. 1986. Nutrient cycling in moist tropical forest. *Annual Review of Ecology and Systematics*, 17: 1137-1167.

Yadav, R.S., Yadav, B.L. and Chhipa, B.R. 2008. Litter dynamics and soil properties under different tree species in a semi-arid region of Rajasthan, India. *Agroforestry Systems*, 73: 1-12.

Young, A. 1989. Agroforestry for Soil Conservation. CAB International, Wellingford, UK.

❑❑❑

Agroforestry: Natural Resource Sustainability, Livelihood & Climate Moderation © 2009
Editors: OP Chaturvedi, A Venkatesh et. al., pp. 335-347
Satish Serial Publishing House, Delhi

Chapter 23

ADVANCES IN EVALUATION OF WATER USE EFFICIENCY IN CROPS AND TREES IN THE CONTEXT OF CLIMATE CHANGE

Badre Alam
National Research Centre for Agroforestry, Jhansi-284 003 (U.P.)

Introduction

It would be most appropriate to start with the slogan that great agricultural scientist and the father of the Indian Green Revolution, Dr.M.S.Swaminathan reiterates:

"More Crop Per Drop".

Water is precious for sustaining life and it is essential for plant growth as it is required for cellular functions and acquisition of CO_2 by plants. Soil-Plant-Atmosphere-Continuum (SPAC) mediates the water uptake of the plants through an integrated process where transpiration plays a crucial role. Thus, dual role of transpiration for water uptake and water loss by the plants is very conspicuous. The water loss through stomata is critically linked with the prime event of gas exchange (CO_2 and O_2) that ultimately leads to the growth of the plants. Thus, water use efficiency (WUE) is defined as the ratio of dry matter production to water consumed or as the ratio of assimilated

CO_2 to transpired water (Farquhar and Richards, 1984). Sustainable crop productivity under water-limited conditions and economically saving irrigating water are the two most essential agenda that need urgent attention for agriculture to meet various challenges including the implications of Climate Change and increasing population.

The significance of quantifying water requirement of crops and the WUE was felt long back in early twentieth century (Briggs and Shantz, 1913). It was demonstrated that plant growth and biomass production depend on the quantity of water used (WU) for growth as well as WUE. Subsequently, simplest quantifying expression was given as TDM (Total dry matter) = WU x WUE (Passioura, 1976, 1986). WUE is highly variable among genotypes and crop species which emphasizes for further efforts for its efficient use for crop improvement (Rao *et al.*, 1993; Ebdon *et al.*, 1998). This is also influenced by the prevailing environment and availability of water. Factors which strongly regulate the function of stomata, the central points of water loss are major determinants of water requirement and water use. Most probably due to this, many earlier studies have claimed that the C_4 species typically have a WUE of 4 to 5.5 g kg^{-1} compared to the C_3 species for WUE of 1.5 to 3.0 g kg^{-1} (Sheshshayee *et al.*, 2003). This, perhaps, explains for relatively better productivity of C_4 species than the C_3 species in dry tropics where water is limited. Thus broadly evaluation of WU and WUE emerged from the perspective of agronomic and physiological interest. From the agronomic point of view, this was straight forward, simple and direct measure of dry matter, yield and the water used. But, this is not feasible always as it depends on gravimetric quantification. However, this technique holds importance form experimental perspective and it is still in use where other advanced techniques are not accessible. To overcome the problems associated with gravimetric analysis (e.g. lysimetric; weighing platform with rain-out shelter etc.), many advanced techniques have been emerging to precisely determine the WUE in crops. Improvement in determining the WUE in perennials namely trees is felt essential as the conventional techniques were not ideally suitable to these. For this, scientists have put forwarded efforts to exploit the conservative principles of the gas exchange process through stomata in plants which ultimately led to develop a number of advanced techniques which are serving as surrogate tools for determining WUE (Farquhar *et al.*, 1982; Farquhar and Richards, 1984; Hall *et al.*, 1990; Condon *et al.*, 2002; Sheshshayee *et al.*, 2003).

Understanding WUE across the genotypes in the context of climate change assumes much importance as variation in WUE largely depends on three factors: 1) the impact of variation in WUE on crop growth rate 2) the impact of variation in WUE on the rate of crop water use and 3) how growth and water use interact over the crop's duration to produce grain yield (Condon *et al.*, 2002). Comparative significance of these factors will vary depending on the crops being grown and the type of the cropping environment. This holds true on trees as well which are also bound to face the consequences of climate change with elevated atmospheric CO_2, increasing temperature and water limiting environments. Attempts have been made in this paper to discuss most important and advanced approaches for determination of WUE from the physiological perspectives and their significance in the context of Climate Change.

Simple way to understand WUE

The term WUE is commonly used to express the ratio of total dry matter production to evapotranspiration (Richards, 1991). It can be expressed in the following ways:

$$\text{WUE (dry matter)} = \frac{\text{TE}}{1+\text{Es}/\text{T}} \qquad \text{.............. [1]}$$

where,

$$\text{TE (transpiration efficiency)} = \frac{\text{Above ground dry matter,}}{\text{Transpiration}}$$

Es = water lost by evaporation from the soil surface, and T= water lost through transpiration by the crop.

From equation [1], it is clear that crop WUE can be increased either by increasing TE or decreasing the magnitude of Es. The components of the equation [1], depends on rainfall distribution. So, if a crop is dependent on water stored in the soil, and precipitation (rainfall) during the growing period is low, then increasing TE provides the highest opportunity to enhance WUE. However, if a crop depends on in-season rainfall then decreasing Es could result in the largest gains. Decreasing Es solely depends on crop management practices and it is not difficult but gain is up to a limit. But, scope of increasing WUE through TE is unlimited and trials are still on by Plant Physiologists, Plant Breeders and Biotechnologist in this direction. This can be easily realized from the following model hypothetical example suitable for pot experiments (Table 1).

Water use is a function of evaporative demand and leaf area. There is little can be done to alter evaporative demand, although there may be possibility for managing crop phenology and therefore the timing of crop growth, but there are a large avenues to genetically reduce leaf area development. This can be explained with the findings of substantial genetic variation for tillering, xylem vessel diameter, leaf dimensions and stomatal or cuticular water loss. In the case of tillering in wheat there is a major gene on chromosome 1AS that inhibits tillering (Richards, 1988). The expression of this gene varies with climate and genetic background and so it provides substantial scope for the regulation of tillering and therefore leaf area. This gene can result in plants with primarily only productive tillers without significantly compromising vegetative growth (Richards, 1988).

Understanding the principle of Gas exchange and Intrinsic-WUE in plants

There is genetic variability in the important trait TE and to harness the benefit out of it for crop improvement for WUE, it is important to understand the underlying physiological mechanism(s) contributing to it. It is well known that the stomatal conductance (Gs) or mesophyll efficiency (Gm) substantially influence the intercellular CO_2 partial pressure (Ci) and hence contribute to the variation in TE (Johnson *et al.*, 1995; Virgona and Farquhar 1996). Instantaneous intrinsic WUE at the leaf level is defined as the ratio of the rates of net CO_2 uptake (A) and transpiration (T). Both CO_2 uptake in photosynthesis and water loss through transpiration occur almost simultaneously and exclusively through the stomata of the leaf as diffusion process. So, the rates of these are determined by the conductances to diffusion of CO_2 and water vapour through the stomata and the concentration differences in CO_2 and water vapour between the ambient atmosphere outside the leaf boundary layer and inside the substomatal cavities (Farquhar and Sharkey, 1982). This can be understood from the following expressions.

$$A = \frac{Gc(Ca\text{-}Ci)}{Pa} \qquad \text{...... [2]}$$

Where Gc is the stomatal conductance to CO_2, Ca is the atmospheric pressure of CO_2 and Ci is the intercellular pressure of CO_2, Pa is the atmospheric pressure. Similarly,

$$T = \frac{Gw\ (ei\text{-}ea)}{Pa} \qquad \text{......... [3]}$$

Where Gw is the stomatal conductance to water vapour, ei and ea are the water vapour pressures inside the leaf and in the atmosphere, respectively.

Table 1. Model example (hypothetical) for understanding the gravimetric estimation of WUE.

S No.	Soil moisture Maintained at 100% FC	D-1 (g) A	D-2 (g) B	D-3 (g) C	D-4 (g) D	ET (g)	E (g)	CWT (g) (ET-T)	DMi (g)	DMf (g)	DMdif (DMf-DMi) (g)	WUE (g/Kg) (DMdif/CWT)
1	30	200	300	250	400	1150	40	1110	1.5	5.0	3.5	3.153
2	31	400	350	400	300	1450	40	1410	1.6	4.5	2.9	2.057
3	30	300	200	350	200	1050	40	1010	2.0	6.0	4.0	3.960

Description of the parameters in Table 1. : The hypothetical example is suitable for pot container grown experiments under rain-out shelter facility. The initial weighing of the empty container, soil, and the amount of water the soil can hold in its field capacity (%) and initial dry biomass (DMi) will be recorded in designated day. Weighing of the pots with plant will be taken from days D-1 to D-4 along with the weighing of pots without plants everyday. The difference in weight of the pots (with plants) on the second day will be replaced with water. ET = Evapotranspiration (A+B+C+D); E = Evaporation (Differential Weight of the pots with soil but without plants); CWT= Cumulative water transpiration; DMi =Initial dry biomass; DMf=Final dry biomass; DMdif = Differential dry biomass (DMf-DMi); WUE = Water Use Efficiency (g kg^{-1}).

From the equations [2] and [3], instantaneous intrinsic WUE (I_{WUE}) at the leaf level (A/T) can be determined as follows:

$$\frac{A}{T} = \frac{Ca\ (1 - Ci/Ca)}{1.6v} \qquad \ldots\ldots [4]$$

Where v is the vapour pressure difference i.e. (ei-ea) and 1.6 is the empirically determined ratio of the diffusivities of water vapour and CO_2 in air.

From the equation [4] it is clear that instantaneous intrinsic WUE (I_{WUE}) at the leaf level (A/T) will be determined by the genotypic variation in Ci/Ca and or stomatal conductance as it influences Gw and A. Hence, this obviously depicts that estimation of A vs. Ci curve through Photosynthesis measurement (net CO_2 uptake) assumes utmost importance for determining WUE across the genotypes including crops and trees.

The determination of I_{WUE} across the genotypes captures central focus of research for dry land agriculture and also for mitigating climate change. Similar to A/T, another parameter A/Gs i.e. ratio of net CO_2 uptake (A) to stomatal conductance (Gs) could indicate I_{WUE} in plants. In addition to this, through A/Gs ratio, genotypes may be grouped for their dependence on stomatal or non-stomatal limitations and or for mesophyll efficiency (Jacob and Lawlor, 1991; Jacob and Lawlor, 1992). Net CO_2 assimilation rate and its association with leaf water relation has been established in a field experiment with natural rubber trees (Alam *et al.*, 2006a).

Only limitations of I_{WUE} are that it can not indicate the effects on integrated time scale and it is limited to single leaf level. Considering these, several advanced approaches are being emerged. However, importance of gas exchange parameters for estimating WUE is always there and it is essential for getting precise measure regarding genotypic differences in WUE (Udaya Kumar *et al.*, 1998). Net CO_2 assimilation rate and its association with leaf water relation have been well established.

Carbon Isotope Discrimination in Plants and WUE

Use of stable isotopes has emerged as a powerful tool for understanding many physiological processes including photosynthesis and WUE. Most elements of biological interest have two or more stable isotopes with one form will be in greater abundance than other. Accordingly, it is widely reported in literature that in nature average terrestrial abundance of the stable isotopes of major elements namely Carbon- ^{12}C (98.89%), ^{13}C (1.11%); Oxygen- ^{16}O (99.759%), ^{17}O (0.037%), ^{18}O(0.204%) and Nitrogen- ^{14}N (99.63%), ^{15}N(0.37%).

The overall abundance of ^{13}C relative to ^{12}C in plants tissue is commonly less than in the carbon of atmospheric carbon dioxide, indicating that carbon isotope discrimination occurs in the incorporation of carbon dioxide into plant biomass. As the isotopes are stable, the information inherent in the ratio of abundance of carbon isotopes, presented following convention as $^{13}C/^{12}C$, is invariant as long as carbon is not lost (Farquhar *et al.*, 1989). Various contributions have been made to our understanding of carbon isotope discrimination in plants since this area was extensively reviewed by O' Leary (1981). How this is important to determine WUE are discussed.

Variation in the ratio $^{13}C/^{12}C$ is the consequences of "isotope effects," and the detailed has been reviewed by Farquhar *et al.* (1989). As per Farquhar *et al.* (1989), isotope effects, denoted by α, are also called fractionation factors because they result in fractionations of isotopes. For understanding the carbon isotope discrimination in plants, mass spectrometric (Isotope Ratio Mass Spectrometer) estimation of the isotopic composition of the sample in comparison to a recognised standard is used. According to Farquhar *et al.* (1989), it is expressed in the following equations:

$$\delta p = (Rp - Rs)/\ Rs \text{ or } \delta p = (Rp/Rs) - 1 \qquad \text{.......... [5]}$$

Where dp is the isotopic composition of the sample and Rp and Rs are the molar abundance ratio, $^{13}C/^{12}C$ in sample and standard (PDB, the Pee Dee Belemnite standard) respectively. For simplifying this bio-physical relationship, Farquhar *et al.* (1989) derived the following equation for estimating Carbon Isotope Discrimination (Δ) in plants:

$$\Delta = (\delta a - \delta p)/\ (1 + \delta p) \qquad \text{................ [6]}$$

Where, da is the isotopic composition in air (generally the $\delta^{13}C$ value of the air is taken as -8 %o per mill) and δp is the isotopic composition in plant sample with reference to the standard (PDB, the Pee Dee Belemnite standard) and also where $\delta^{13}C = \{R\ (sample)/\ R\ (standard) - 1\} \times 1000$ %o (it is expressed as per mill). The standard is carbon dioxide obtained from "PDB" a limestone from the Pee Dee formation in South Carolina. The non dimensional units of $d^{13}C$ are expressed as "per mill" or %o. Hence more simply:

$$\Delta\ ^{13}C = [(Ra/Rp) - 1] \times 1000 \qquad \text{.......... [7]}$$

where, Ra is the value of $^{13}C/^{12}C$ ratio in the atmosphere and Rp is the value of $^{13}C/^{12}C$ ratio in plant material. Following convention and convenience, units of $\Delta^{13}C$ are expressed as per thousand (%o), i.e. the fractional difference from unity is multiplied by 1000. There are numerous studies that indicate $\Delta^{13}C$ has negative association with WUE in plants (Condon *et al.*, 1990; Udayakumar *et al.*, 1998; Condon et al., 2002; Zhao *et al.*, 2007). This means that lesser the $\Delta^{13}C$ higher the WUE and this is robust for many C_3 species.

Enrichment of ^{18}O in Leaf Biomass as a Measure of Time-Integrated Leaf Transpiration Rate

Evaporation will enrich water at the sites of evaporation in the leaf since the $H_2^{18}O$ vapour has a lower vapour pressure being heavier than $H_2^{16}O$ and because $H_2^{18}O$ diffuses more slowly than $H_2^{16}O$. This principle has been used to analyze $\delta^{18}O$ in leaf tissue and cellulose as in case of $\delta^{13}C$.

$$\Delta\,^{18}O = \frac{\delta p\,^{18}O - \delta s\,^{18}O}{1+\delta p\,^{18}O} \qquad \ldots\ldots\ldots\ldots [8]$$

where, $\delta p^{18}O$ is the ^{18}O isotopic composition of the plant sample corrected against the international standards i.e. Standard Mean Ocean Water (SMOW) and ds ^{18}O is the ^{18}O isotopic composition of the source water (irrigation water).

It has been shown that $\delta^{18}O$ can also be used for a measure of time-averaged transpiration rate (Sheshshayee *et al.*, 2005). In many studies it has been shown robust association with $\Delta^{13}C$ (Barbour and Farquhar, 2000; Condon *et al.*, 2002; Sheshshayee *et al.*, 2005).

Advanced technique for Online estimation of ^{13}C and ^{18}O discrimination

Sophisticated techniques are also available to simultaneously determine leaf gas exchange and isotope discrimination by linking an advanced well equipped system for estimating gas exchange to a mass spectrometer (Griffiths *et al.*, 2007; Cousins *et al.*, 2008).

Alternate Approaches to Quantify Water Use Efficiency in Plants

Specific Leaf Area (SLA)

Specific Leaf area is the ratio of leaf area to leaf dry weight. Since SLA is often considered as an indirect measure of leaf expansion, higher SLA will be taken as higher leaf area per unit biomass. This will provide larger surface area for transpiration. But as obviously it will reduce the leaf thickness, the capacity of photosynthesis would be smaller. As earlier defined the Transpiration Efficiency (TE) in equation [4], then with higher SLA, TE will decrease. Thus, it is used as a good indirect measure for WUE (Wright *et al.*, 1994).

Plant Minerals and Ash Content

During the process of transpiration, there is a continuous uptake by the roots from the soil. Nutrients and several ions namely K, silicon etc. also get an entry along with the mass flow. There are numerous reports that showed the association of these components with WUE (Merah *et al.*, 1999).

Recent use of Chlorophyll Fluorescence techniques

Simultaneously measuring gas exchange and pulse amplitude modulated chlorophyll fluorescence techniques are being emerged to monitor the photosynthetic performance of plants non-invasively. Major developments in the instrumentation for measuring chlorophyll fluorescence from intact plants and improvements in the understanding of how changes in plant fluorescence characteristics relate to physiological performance have led to a widespread use of chlorophyll fluorescence measurements in plant physiological studies (Baker and Rosenqvist, 2004; Alam *et al.*, 2005; Alam *et al.*, 2006a).

Evaluation of WUE in Trees from Real Time Data on Canopy Transpiration

Over and above the aforesaid techniques, for more precision in case of perennials, mainly trees, estimation of canopy transpiration for a long term basis assumes much significance to determine WUE. The scientific quest of plant physiologists and ecophysiologists over the decades has been fulfilled to a great extent after having received the techniques of determining sap flow rate of trees and the system for data logging (Zweifel *et al.,*, 2001; Zweifel *et al.*, 2006). To give an idea of how really sap flow data apparently represent the rate of water transpired by trees is presented (Fig.1) taking an example of data collected in rubber trees in a humid tropical climate of Kerala, India (Alam *et al.*, 2006b unpublished data). Thus once, an estimate of water transpired by a tree is known over a period of time, then from the biomass data based on empirical or allometric assessment, it will be simple to evaluate WUE or water demand of a tree.

WUE is a complex phenomenon and it now has gathered a central point of investigation in the context of climate change where severe water limitations and warmer atmosphere are anticipated. Rapid development of technical knowledge and instrumentation and the sincere quest of the scientists to solve the problem of man kind would definitely help keep the Earth safe, green and productive despite the climatic constraints.

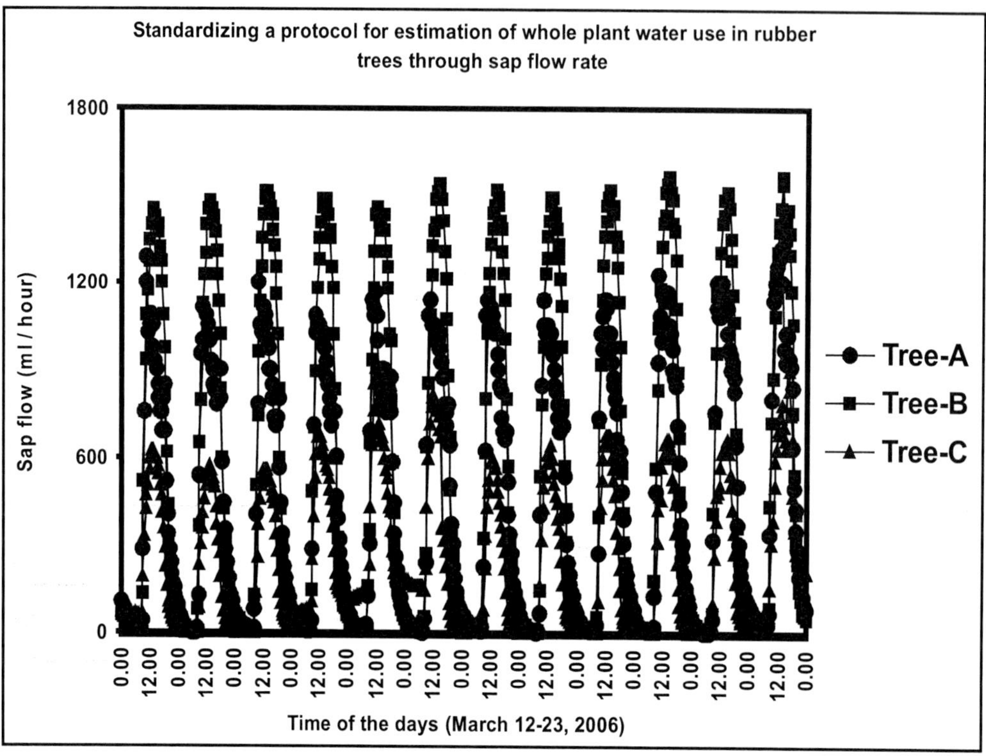

Fig.1 : Example of estimation of sap flow rate using Sap Flow Measurement system with data logger in rubber trees in a humid tropical climate of Kerala, India.

References

Alam, B., Nair, D.B. and Jacob, J. 2005. Low temperature stress modifies the photochemical efficiency of a tropical tree species *Hevea brasiliensis*: Effects of varying levels of CO_2 and photon flux density. *Photosynthetica,* 43(2): 247-252

Alam, B., Lakshmanan, R., Nair, D.B. and Jacob, J. 2006a. Changes in leaf water relations and photosynthetic activity due to drought in mature natural rubber plants grown in North Kerala. *Journal of Plantation Crops,* 34(3): 625-630.

Alam, B., Nathan, K.A. and Jacob, J. 2006b. Annual Review Meeting -2006 of RRII, Kottayam (Limited distribution).

Baker, N.R. and Rosenqvist, E. 2004. Application of chlorophyll fluorescence can improve crop production strategies: an examination of future possibilities. *Journal of Experimental Botany,* 55(403): 1607-1621.

Barbour, M.M. and Farquhar, G.D. 2000. Relative humidity and ABA-induced variation in carbon and oxygen isotope ratios of cotton leaves. *Plant, Cell and Environment,* 23: 473-485.

Briggs, L.J. and Shantz, H.L. 1913. The relative water requirement of plants: in *Investigations in great plains in 1910 and 1911* (USDA Bureau of Plant Industries Bulletin), Pp. 284.

Cousins A.B., Badger, M.R. and von Caemmerer, S. 2008. C4 photosynthetic isotope exchange in NAD-ME- and NADP-ME-type grasses. *Journal of Experimental Botany,* 59 (7): 1695-1703.

Condon, A.G., Farquhar, G.D. and Richards, R.A.. 1990. Genotypic variation in carbon isotope discrimination and transpiration efficiency in wheat. Leaf gas exchange and whole plant studies. *Australian Journal of Plant Physiology,* 17: 9-22.

Condon, A.G., Richards, R.A., Rebetzke, G.J. and Farquhar, G.D.2002. Improving intrinsic water-use efficiency and crop yield. *Crop Science,* 42: 122-131.

Ebdon, J.S., Petrovic, A.M. and Dawson, T.E. 1998. Relationship between carbon isotope discrimination, water use efficiency and evapotranspiration in Kentucky blue grass. *Crop Science,* 38: 157-162.

Farquhar, G.D. and Sharkey, T.D.1982. Stomatal conductance and photosynthesis. *Annual Review of Plant Physiology,* 33: 317-345.

Farquhar, G.D., O'Leary , M.H. and Berry, J.A. 1982. On the relationship between carbon isotope discrimination and the intercellular carbon dioxide concentration in leaves. *Australian Journal of Plant Physiology,* 9: 121-137.

Farquhar, G.D. and Richards, R.A. 1984. Isotopic composition of plant carbon correlates with water-use efficiency of wheat genotypes. *Australian Journal of Plant Physiology,* 11: 539-552.

Farquhar, G.D., Ehleringer, J.R. and Hubick, K.T. 1989. Carbon isotope discrimination and photosynthesis. *Annual Review of Plant Physiology Plant Molecular Biology,* 40: 503-537.

Griffiths, H., Cousins A.B, Badger, M.R. and von Caemmerer, S. 2007. Discrimination in the dark. Resolving the interplay between metabolic and physical constraints to phosphoenolpyruvate carboxylase activity during the crassulacean acid metabolism cycle. *Plant Physiology,* 143: 1055-1067.

Hall, A.E., Mutters R.G., Hubick, K.T. and Farquhar, G.D. 1990. Genotypic differences in carbon isotopic discrimination by cowpea under wet and dry field conditions. *Crop Science,* 30: 300-305.

Jacob, J. and Lawlor, D. 1991. Stomatal and mesophyll limitations of photosynthesis in phosphate deficient sunflower, maize and wheat plants. *Journal of Experimental Botany,* 42(241): 1003-1011.

Jacob, J. and Lawlor, D. 1992. Dependence of photosynthesis of sunflower and maize leaves on phosphate supply, ribulose-1,5-bisphosphate carboxylase/oxygenase activity, and ribulose-1,5-bisphosphate pool size. *Plant Physiology,* 98: 801-807.

Johnson, R.C., Huehlbauer, E.F., and Simon, C.J. 1995. Genetic variation in water use efficiency and its relation to photosynthesis and productivity in lentil germplasm. *Crop Science,* 35: 457-463.

Merah, O., Deleens E. and Monneveux P. 1999. Grain yield, Carbon isotope discrimination, mineral and silicon content in durum wheat under different precipitation regimes. *Physiologia Plantarum,* 107: 387-394.

O'Leary, M.H. 1981. Carbon isotope discrimination in plants. *Phytochemistry,* 20: 553-567.

Passioura, J.B. 1976. Physiology of grain yield in wheat growing on stored water. *Australian Journal of Plant Physiology,* 3: 559-565.

Passioura, J.B. 1986. Resistance to drought and salinity: Avenues for improvement. *Australian Journal of Plant Physiology,* 13: 191-201.

Rao, R.C.N., Williams, J.H., Wadia, K.D.R., Hubick, K.T. and Farquhar, G.D. 1993. Crop growth, WUE, and CID in groundnut (Arachis hypogea L.) genotypes under end season drought conditions. *Annals of Applied Botany,* 122: 357-367.

Richards, R.A.1988.A tiller inhibitor gene in wheat and its effect on plant growth. *Australian Journal of Agricultural Research,* 39: 749-757.

Richards, R.A.1991. Crop improvement for temperate Australia: future opportunities. *Field Crops Research,* 26: 141-169.

Sheshshayee, M.S., Bindumadhava, H., Shankar, A.G. and Udayakumar, M. 2003. Breeding strategies to exploit water use efficiency for crop improvement. *Journal of Plant Biology,* 30(2): 253-268.

Sheshshayee, M.S., Bindumadhava, H., Ramesh, R., Prasad, T.G., Lakshminarayana, R. and Udayakumar, M. 2005. Oxygen isotope enrichment ({Delta}-18O) as a measure of time-averaged transpiration rate. *Journal of Experimental Botany,* 56 (422): 3033-3039.

Udayakumar, M., Sheshshayee, M.S., Nataraj, K.N., Bindu Madhava, H., Devendra, R., Hussain, I.S.A and Prasad, T.G. 1998. Why breeding for water use efficiency has not been successful. An analysis and alternate approach to exploit this trait for crop improvement. *Current Science,* 74(11): 994-1000.

Virgona, J.M., and Farquhar, G.D. 1996. Genetic variation in relative growth rate and carbon isotope discrimination in sunflower in relation to photosynthetic capacity. *Australian Journal of Plant Physiology,* 23: 227-236.

Wright, G.C., Rao, R.C.N and Farquhar, G.D. 1994. Water use efficiency and carbon isotope discrimination in peanut under water deficit conditions. *Crop Science,* 34: 92-97.

Zhao, L., Xiao,H., Liu, X., Li, J. and Xiao, S.2007. Foliar Carbon isotope discrimination and nutrients in two desert plant species. *Polish Journal of Ecology*, 55(1): 57-66.

Zweifel, R. Item, H., Ha¨sler, R. 2001. Link between diurnal stem radius changes and tree water relations. *Tree Physiology*, 21: 869–877.

Zweifel, R., Zeugin, F., Zimmermann, L., Newbery, D.M. 2006. Intra-annual radial growth and water relations of trees: implications towards a growth mechanism. *Journal of Experimental Botany*, 57: 1445–1459.

❑❑❑

Agroforestry: Natural Resource Sustainability, Livelihood & Climate Moderation © 2009
Editors: OP Chaturvedi, A Venkatesh et. al., pp. 349-357
Satish Serial Publishing House, Delhi

Chapter 24

SCOPE FOR VALUE ADDITION IN AGRI-HORTICULTURE

A.K. Shukla and Sunil Kumar
Indian Grassland and Fodder Research Institute, Jhansi-284 003 (U.P.)

Introduction

India is bestowed with varied agro-climatic conditions suitable for growing a large variety of horticultural crops. Therefore, almost all types of fruits can be grown in one or the other parts of the country. India is the second largest producer of fruit and vegetable in the world. In India horticultural crops occupy about 6.7 % of gross area and contribute about 18% of gross value of agricultural output and 52 % of export earnings in agriculture. Unfortunately, Indians in general do not get the basic daily requirement of fruits and vegetables as per Indian Council of Medical Research (ICMR) recommendation of 80g fruits and 300g vegetables or their products per person per day. The major reason for non-availability of per capita recommended quantity of fruit and vegetables are the increasing population pressure of the country and high degree of post harvest loss in these perishable commodities, which has been estimated to the tune of 20-30 per cent in various commodities under different conditions. In India, quanta vise loss in these commodities is to the tune of 10 million tones for fruits and 22 million tones for vegetables, amounting to a total revenue loss of > 5000 crores annually. If one considers the loss in terms of both quality and quantity of produce, the loss exceeds 50,000 crores. In arid and semi arid areas, situation is more alarming owing to harsh climatic conditions (favours rapid loss) and poor post harvest infrastructures and facilities such as on farm pre-cooling, packaging, storage, transportation, processing, etc. The introduction of proper post harvest techniques and value addition procedures would result in getting 50,000 crores back into the countries economy. Like other developing countries, India needs to increase the quantity of food available for its rapidly growing population. There are various ways to combat this problem. One is to produce more food; another is to conserve whatever is produced. In the past, much emphasis has been put on growing more food, while the post harvest aspect has been generally ignored. Undoubtedly we have to grow more food for the fast growing population of our country not only by increasing the area under cultivation, but also by applying promising technologies. Horticultural produce supplies essential food items of daily use. There has been significant progress in increasing the area under these crops since independence. Also exotic, high yielding and good quality varieties have been introduced in India. A major and often neglected step towards offering a greater volume of nutritious food is to prevent its loss between the time of harvesting and consumption. .

Processing plays a significant role in reducing post harvest losses of fruits and vegetables and thereby ensuring their increased availability besides better returns to farmers. However, at present only about 2 per cent of the annual production of fruits and vegetables is processed as against 30 per cent in Thailand, 70 per cent in Brazil and 80per cent in Malaysia. Value addition to food in India is 7 per cent as against 23per cent in China and 188 per cent in UK. According to CII- Mckinsey Report, the value added processed foods in India are of the value of Rs 80000 crores out of the estimated value of food industry of Rs. 250,000 crores.. The share of agricultural food industry in the world could be $1 trillion. In India it could easily be of the order of Rs. 30,000 crores per year.

Optimum Maturity Standard Horticultural Crops

There are several visual, physical and biochemical criteria to determine the optimum stage of maturity in fruits and vegetables are visual indices like change in fruit colour (ber, pomegranate, phalsa, Mango, pineapple,), fruits/ pods tenderness (khejri, *gonda*, karonda, ker, drum stick), flesh colour (bael, date palm, tamarind, water melon pomegranate), Vitamin C content (aonla and ber), sugar:acid ratio (ber, phalsa and tamarind) and TSS (ber, date palm, water melon)

Maturity standards of fruits and vegetables for value addition

Name of crop	Maturity standard/physical symptoms	Potential use in value addition
Mango	Yellowing of fruits, falling of some fruits	Canned products, Jam, jelly, squash, juice etc.
Ber	Full ripe stage colour change from green to yellow green	Preservation through salt, sugar and dehydration
Aonla	Full developed, glossy shining with hard and fibrous stone	Preservation through salt, sugar, dehydration and Ayurvedic medicines
Pomegranate	Change in skin colour, sweet arils, acidic bold seeds with full ripe stage	Dehydration (Anardana), preservation in sugar
Bael	Change in colour and flovour	Preservation in sugar and medicinal uses
Phalsa	Red colour of fruit, better sugar- acid blend	Juice, Squash etc
Khejri	Tender pod (15-20 days old)	Dehydration, salt preservation
Ker	Mature green with soft seeds	Preservation with salt and dehydration
Moringa	Tender pods with soft seeds	Preservation with salt, dehydration and value addition of recepies.
Karonda	Full mature with hard texture	Preservation with salt and sugar
Water melon	Full ripe stage, metallic sound with deep red flesh colour, drying of the tendril just above the fruit	Preservation with sugar and dehydration of seeds

Harvesting Operation

Optimum time of harvesting is very important aspect for value addition of agricultural and horticultural crops. Stage and method of harvesting depends on the nature of commodity, however, morning harvesting is found better in almost all the crops. Time of harvesting significantly affects the fruit quality and its shelf life. Morning time is usually preferred because of more cool and turgid fruits, which can be disposed or processed during the day. Successive fruit ripening in many fruits particularly in non-climacteric types (ber, phalsa, watermelon), makes the harvesting operation most tedious. In such crops, fully ripened fruits are harvested. If they are harvested by shaking the plants (ber and phalsa), a lot of produce is damaged or plucked immature, therefore hand picking of individual fruits is advisable. In date palm, whole bunches are harvested and the individual berries are separated or entire bunch is packed. In many fruits and vegetables harvesting of individual fruit is done to avoid the damage.

Method of harvesting in important fruits and vegetables

Name of crop	Method of harvesting
Ber	Individual ripe fruit or by shaking of branches
Mango	Individual fruit picking
Guava	Individual fruit picking
Pomegranate	Individual fruit picking
Aonla	Shaking of twigs or individual fruit picking
Date palm	Cutting of whole bunch
Ker	Selective picking of individual tender fruit
Gonda	Picking of whole bunch
Phalsa	Individual fruit picking
Bael	Individual fruit picking
Grape	Cutting of bunch
Tomato	Individual fruit picking
Pineapple	Individual fruit harvesting
Custard apple	Individual fruit picking
Jamun	Individual fruit picking

The main constraints for the development of industry for value addition are non availability of raw material, limited duration of availability and uncertain demand in both domestic as well as export market besides techno-economic deficiencies, lack of consistent quality, preference for fresh fruits and vegetables and high requirement of working capital. In addition, the following constraints also need attention:

a. Lack of suitable varieties for processing.

b. Improper market infrastructure.

c. Lack of adequate domestic demand for processed product.

d. Lack of product wise quality standards, insurance and reliability.

e. Difference in products for local, domestic and export market

f. Product standards, malpractices and safety assurances.

g. Lack of mission-oriented network for value addition of a particular crop commodity in a defined region of its production keeping in view the related producers, technocrats, administrators, and industry and market managers.

The value addition deals with operations right from fruit maturity (both physiological and horticultural) to processing and waste utilization. These all factors play direct or some time indirect role in value addition in horticultural commodities. The value addition in broader term signifies the steps and series of operations like delineation of criteria for maturity, pre harvest treatments to reduce post harvest losses, techniques of harvesting to minimize "on farm" losses, standards for grading and packing for distance transportation, post harvest treatments and conditions of storage to improve shelf life, processing techniques to develop more useful product and utilization of waste to develop byproducts.

Present Status of Value Addition

Common processed fruit products available in the Indian markets are pulps and juices particularly from citrus fruits, mango and pineapple. Now products such as tamarind pulp and *Garcinia indica* juice are becoming popular. Ready-to-serve beverages like nectar are being prepared from phalsa, kinnow, lime, bael, aonla, etc. Fruit squashes, jams and preserves from aonla and bael are commonly sold in the market. Ketchup, sauces and pickles are also common products. Canned fruits and vegetables are available in the market in semi-processed form. Of course dehydrated products are being produced since long. The main value addition methods for disposal of the produce are:

1. Dehydration : Fruits of *boradi* and *jharberi* are sundried. Fruits of ker, pods of khejri (*sangari*) and cluster bean and leaves of methi are also sundried. Grated snapmelon (*khelara*), slices of peeled kachri are also sundried. Red chilli is sundried. In all these cases the quality of raw material is variable and drying methods are often not properly standardized. Rose petals are dehydrated and marketed in some products. The methods need to be standardized and the scope of such dehydration could be a enlarged to a number of flower crops.

   Possibilities for value addition through dehydration- There are much possibilities for dehydration of fruits and vegetables as it is an easy and economic technique.

   Fruits such as ber, aonla, bael, ker, lasoda, anardana, dates etc. can be dehydrated on commercial scale sangria and khokha from khejri and dehydrated quality products from vagatables such as cluster bean, chilli, mushroom, garlic, potato, kachri, snapmelon, tinda, methi, mint, senja, etc. can be made. Products of medicinal value from senna, aloe, isabgol, shankhpuspi, neem giloy, etc. could have great demand. Production of dehydrated seeds from the crops like Kumat, mateera, sunflower, methi, cumin, ajwain etc and of seed kernels by using decorticators in watermelon and mateera, tumba, snapmelon, cucumber, muskmelon pumpkin etc. have great potential

2. Seeds and Magaz (seed kernels) : Seeds of kumat watermelon (mateera), tumba, cumin, ajwain, methi, etc. are collected and marketed. Magaj of mateera, cucumber and muskmelon is produced manually and also by using special decorticators. The methods as well as the quality are variable and no standards are in practice.
3. Pickles and Murabbas : There is a considerable traditional know-how on these aspects. Pickles, mainly in oil, of different types are prepared and marketed. This know-how needs to be used to develop standard methods using good raw material of aonla, ker, lasoda, karonda, bael, chilli, kachri, aloe, drumstick, muskmelon, etc.
4. Fruit Juices : Fruit juices from fruit crops such as pomegranate, phalsa, kinnow, lime, mateera, toddy palm, etc. are served in selected market places to the consumers. There is need to develop standard products in disposable packs and in recycled bottles to enlarge the scope of such value addition. Ready-to-serve beverages from fruits such as pomegranate, bael, phalsa, lime, watermelon and snapmelon , neera etc. should become fairly popular.
5. Jam and Jelly : The majority of pulpy fruits are being utilized for either directo consumption or preserved product like jam and jelly, like Aonla, bael, guava, pineapple mango etc.
6. Marmalade : fruits of citrus group are being used for preparation of marmalade.

Modern Processing Techniques

i. Freeze-dried fruits and vegetables- Freeze-dried fruits and vegetables retain their structural integrity and colour and reconstitute rapidly even in cold water. Frozen to a temperature of –10^{0}C-40^{0}C and then moisture is removed from the solid matrix under very high vacuum of 100-200m or negligible atmosphere at low temperature of 40-60^{0}C to moisture content of about 2%. These products being hygroscopic need to be properly packed.

ii. Minimally processed cut fruits and vegetables- Trimmed and cut fruits and vegetables are preserved without causing significant change in fresh like properties.The trimmed, peeled, cot pieces are non-thermally treated by dipping in a soak solution, generally consisting of potassium metabisulphite (0.01-0.9%) sodium benzoate (0/03-0.08%), citric arid (0.5-1.2%), ascorbic acid (200-300 ppm) and salt (3-6%) at 20-25^{0}C for few hours or at 5^{0}C for several hours. Soaked products are drained and surface dried in basket centrifuge (1000 rpm). The product is the exposed to UV light for 10 minutes in clean room and packed in flexi polymeric sachets and stored at 5^0C. Ionizing radiation at 105Gy have also been used to check spoilage.

iii. Intermediate moisture (IM) fruit pieces- It is an alternate to tradition dehydration and canning. Moist infusion of fresh fruits and vegetables in soak solution containing ingredients such as glycerol sucrose, salt, potassium sorbate, potassium/sodium metabisulphite, dextrose, pectin, etc. with or without partial dehydration through osmotic process giving final moisture contents of 30-40% and 12-13.5%.

iv. Hurdle technology (HT) stabilized fruits and vegetables- Heat leathality and process temperature, chill temperature, pH, acidity, radiation dose, redox potential, preservation, competitive microflora and storage temperature are the factors (hurdles) that control microbial activity in foods. Generally two or more of such hurdles are used in combination to provide hurdle stabilization none of which is individually capable to prevent spoilage unless used at their extreme levels. Extreme levels spoil the nutritive and eating quality of products. Extreme levels spoil the nutritive and eating quality of the products.

v. Ready-to-eat and cost effective substitutes for canned products are being developed.

vi. Curries are being packed in pouches for ready-to-eat market.

Requirements for Value Addition

1. Production and access to raw material
2. Low cost technologies.
3. There should be strong infrastructural support
4. Market intelligence
5. Training facilities in processes technologies should be made available.
6. The value addition options must meet the WTO standards relating to price, quality and safety to meet the international needs.

Researchable Issues

- Change in the population dynamics, economic condition, life styles and eating habits have been created a very good market demand of processed and value added products. Further, increased health consciousness among the large section of consumer community of the processed products have been putting their demand for processed item low in sugar, salt and oil content, because of the increased problems of diabetes, high blood pressure, cordiac problem and obesity. Therefore, the processing and value addition of fruit and vegetable need a systematic research for development of quality products with enhanced nutritional and medicinal value.

- Product diversification is another important researchable issue of the value addition. These aspects of value addition are directly related to maturity standards and can be achieved through utilization of a crop for value addition at different maturity stages.
- Drying is one of the easiest and old method of fruit and vegetable preservation and almost all fruits and vegetables are suitable for dehydration. However, there is still problem of oxidation and discoloring of the dried products with low shelf life. These problems are needed to be addressed through systematic research. Further, there is good scope of utilization of dehydrated products into processing and value addition either alone or in combination with other fruits and vegetables.
- Research and developmental efforts are needed to develop more than one maturity standard for a single crop for utilization into various processed items and value added products. Therefore, it could be possible to delineation of criteria for maturity standards specific to the product development.
- It is apparent that the chemical constituents, maturity and ripening of the fruits is under strict genetic control. Therefore, a systematic research are needed to categorized all available varieties, improved strains and germplasm according to their maturity period and suitability to processed item for quality products.
- Biotechnological approaches have been used for isolation of genes responsible for endogenous ethylene biosynthesis and incorporation into the commercial varieties for controlled maturity and ripening. Moreover, synthesis of amino acids, enzymes, aroma, coloring additives and vitamins can be biologically synthesized through plant tissue culture techniques. Therefore, these research efforts will ensure utilization of agri-horticultural crops in processing and value addition with enhanced quality for global market opportunities.

References

Bakshi, A.K. 2002. Strategies for post harvest handling, storage and processing of subtropical fruits. Manual of Summer school on Recent advances in production, protection and hostharvest management of subtropical fruits, CISH, Lucknow,

Mangal, J.L. and Siddiqui, S. 2000. Post harvest technology of vegetables present status and future strategies. In: Emerging Scenario in vegetable research and development (Eds. G. Kaloo and Kirti Singh). Research periodicals and book publishing house India. Pp. 227-241.

Pareek O.P. and Shukla, A.K. 2006. Future scope for value addition in arid region. In Prospects and Dimensions for Utilization of Arid Foofds (Eds. Goyal and Sharma). Pp. 309-314.

Roy, S.K. 2002. Potentiality of post harvest management and processing of subtropical fruits. Manual of Summer school on Recent advances in production, protection and host harvest management of subtropical fruits, CISH, Lucknow, June, 3-23, 2002.

Singh, D. and Shukla, A.K. 2004. Maturity standard for value addition in Horticultural crops. Lecture delivered in winter school at RAU, Bikaner.

❑❑❑

Agroforestry: Natural Resource Sustainability, Livelihood & Climate Moderation © 2009
Editors: OP Chaturvedi, A Venkatesh et. al., pp. 359-367
Satish Serial Publishing House, Delhi

Chapter 25

LAND USE DIVERSIFICATION WITH SPECIAL REFERENCE TO AGROFORESTRY

R.K. Tewari, D.R. Palsaniya and Rajeev Tiwari
National Research Centre for Agroforestry, Jhansi-284 003 (U.P.)

Introduction

Since independence, India made high strides in agriculture as a result, inspite of ballooning population pressure, country became self sufficient in food gain production in seventies. This was possible due to increase in area under cultivation, irrigation resources, improved seeds, use of chemical fertilizers and agrochemicals for control of insect, pest and diseases. Further, human and animal population continued to increase unabated but agriculture production reached a plateau and did not increase at the same pace. Retrospection of our agriculture development indicated that increase in production during green revolution period came mostly from irrigated areas which occupy only one third of cropped area in India. Remaining two third areas under dry land agriculture had yet to realize its potential. The general policy of government had so far been a) supply of inputs like HYV seeds fertilizers, & agrochemicals at reasonable rates, (b) supply of credit, electricity, canal water at subsidized rates and (c) fixing of minimum support price for important food grains and other crops alongwith procurement system for wheat and rice only. This lead to increase in area under wheat and rice crops which have least risk of yield as well as price. As such mono cropping of rice and wheat became major land use through out the country. Over years, mono cropping lead to over exploitation of ground water, water pollution due to constant use of chemical fertilizers, weedicides and pesticides which in turn started adversely influencing environment quality , productivity and sustenance of animal and plant biota. Meanwhile, remuneration from crop production in comparison to other land uses /occupations minimized on account of increase in input cost and static prices of food commodities. In nineties with opening up of Indian economy globally, input cost further increased and government was under pressure to reduce subsidy on agrochemicals. However, globalizations opened up avenues for trading agricultural commodities globally and thereby ensure high returns to the farmers. Fortunately, India is bestowed with varying climatic conditions which offer scope for cultivation of variety of crops and plant species. This prompted our visionary planners to advocate land use diversification in tandem with global demand for specific commodity. As such, diversification of land use envisaged 1) Combating soil health degradation, 2) sustaining land productivity 3) checking water table depletion in some areas and water logging in some other areas, 4) improving water quality through checking chemical pollution of water, 5) maintaining ecological balance and 6) increasing remuneration from farming through export of agro based commodity in demand (Sangwan, 2002).

In order to achieve the goals of diversification of land use, government its organizations and farmers have to walk hand in hand. There should be sound policies with governments to induce the farmers to adopt desired land use and dedicated organization and teams of workers to implement the policies and assured market / logistics to ensure remuneration to farmers. This requires application of knowledge in terms of technical know-how, market intelligence and international relation at government level, organization level and framers level. Acreage planning for individual crops and trees at country/state level may help in land use diversification and controlling the output of concerned crops so as to ensure proper returns to the farmers.

Current Land Use

Current land use of country (2002-03) is given in Table 1. Out of 328.7 M ha geographical are 43.5 M ha (14.2%) is not available for agriculture. Forest land occupies 69.1 M ha (23.6%) area which can not be diverted to other uses for the sake of primarily meeting timber demand and secondly environmental conservation concerns. Another 27.4 M ha (9%) is not cultivated. Remaining 166.1 M ha (54.1%) area is under pressure for producing food grains, oilseed, pulses, industrial cash crops, fruits and vegetables. Out of net sown area (132.9 M ha), only 53.1 M ha is irrigated. Cropping intensity is around 130%. Total area under agriculture is likely to reduce further, on account of urbanization, industrialization, expansion of roads, special economic zones and habitations. The situation warrants vertical growth in productivity to meet food demand as practically no scope is left for increasing area under crop. Vertical increase in productivity is possible by increasing cropping intensity and alternate land use planning. In the past 50 years , best efforts in country could hardly raise cropping intensity by another 20% so we should not expect a miracle. Area under fruits (4.8 M ha) and vegetables (6.3 M ha) is almost static over years. There is hardly any fertile land available for sole orchard.

Table 1 : Land use classification of India (2002-03)

Land use	Area (M ha)	%
Geographical Area	328.7	-
Reporting Area for land use	306.1	93.1
Forest	69.1	22.6
Not available for cultivation	43.5	14.2
Other non cultivable Area	27.4	9.0
Culturable Area	166.1	50.5
Net sown Area	132.8	40.4
Net irrigated Area	53.1	16.3
Cropping intensity (%)	**132.1**	

Compulsions Leading to Diversification

Small Holding Size

In India majority of farmers are small holders, per capita land availability (net sown area) which was less than 0.4 ha in 1950 has come down to less than 0.2 ha only. Sustenance of farmers from such small holdings is not possible unless land use is diversified.

Table 2 : Distribution of land holdings

Category	% holders	Area (M ha))
Small to Marginal (<2 ha)	78.0	32.2
Medium (2-10 ha)	20.4	50.4
Large (>10-ha)	1.6	17.4

Slow Agriculture Growth Rate

Growth rate in agriculture (2004-05) is variable and slow. Even some of the states registered negative growth rate (Table 3). The growth rate indicated average 2% increase in agriculture growth. This is grossly insufficient looking to increase in demand- supply gap of various food, fodder, feed, fibre, energy, fruits & vegetables etc. The scenario warrants immediate attention.

Irrigation

In the country net irrigated area is pegged at 53.07 M ha which accounts for 40.6% of net sown area. Gross irrigated area is 70.67 M ha. About 80% of available irrigation water directed to only two crops i.e. rice and wheat where productivity has reached at plateau. Most of the perennial rivers in past few years have gone dry during summers, thus, limit further creation of water resource. Under such situations land use diversification has become mandatory to enhance productivity. It is estimated that if one hectare of land under rice-wheat is diverted to maize-mustard, the saved water can irrigate at least another 3 hectare area under maize-mustard and can enhance productivity by around 30% in monetary terms.

Table 3: State wise agriculture growth rate (2004-05)

State	Growth Rate (%)	State	Growth Rate (%)
Bihar	4.4	Rajasthan	2.0
Himachal Pradesh	3.8	Karnataka	1.4
Maharashtra	3.3	Kerala	1.4
West Bengal	2.7	M..P.	1.2
J & K	2.6	Punjab	1.1
Assam	2.4	A.P.	1.0
Haryana	2.3	Orissa	(-) 0.02
Gujarat	2.3	T.N.	(-) 0.06
U.P.	2.3		

Source: National Commission on Farmers, Report submitted by Dr. M.S. Swaminathan, to Govt. of India on 23.11.07

Employment

In India 56% work force is employed in agriculture which contributes 18% to the GDP (2006-07) wages for agriculture labor in real terms have remained static & low since 1997 (Annon, 2007), unemployment give rise to several socio-economic conflicts, hence, due emphasis and corrective measures are immediately required.

Productivity

Productivity of all food grains in India (2004-05) was 1.7 t ha^{-1} yr^{-1}. During this period productivity of fruits and vegetables was 10 and 13 t ha^{-1} yr^{-1} respectively where as that of rice and wheat was 2.0 and 2.7 t ha^{-1} yr^{-1}. The vast deference in productivity suggests strong need for crop diversification in the country (Table 4). World bank report suggests that India produces 1/10th of world vegetables and 1/16th of fruits at 1/3rd to 1/2 of global price but India's share in world market is hardly 0.5% (fruits) and 1.7% vegetables. Indian vegetable prices are 53% and fruits 63% cheaper in international market than that of other countries. This explains vast potential in diversification of Indian agriculture and thereby bringing in prosperity to poor farmers (Annon, 2007).

Table 4: Area, production and productivity of some of crops in India.

Crop	Area (M ha)	Production (MT)	Productivity t ha^{-1} yr^{-1}
Total Food grains	12.16	204.62	1.7
Rice	43.7	85.6	2.0
Wheat	26.3	71.3	2.7
Fruits	4.9	49.3	10.0
Vegetables	6.3	8.3	15.0

Land Use Diversification As a Tool

Land use diversification can increase productivity and profitability with available land and water resources. It can be used as a tool to generate year round employment as agriculture is basically seasonal activity. It can earn foreign exchange and ensure nutritional security. Alternate land uses minimize risk in agriculture due to ever changing climate and weather vagaries. They can contribute to environmental quality through higher and efficient carbon sequestration.

Variety of Products

Land use diversification ensures variety of products and thereby balance market force. Glut in market of one product and scarcity of other can be avoided through land use diversification. This will check sharp fluctuations in prices of commodities.

Year Round Employment

In Indian condition variety of crops, fruits and vegetables can be cultivated throughout the year. Fruit plants being perennial in nature and variable in time of fruiting give year round employment for rural work force. Associated activities like harvesting, transport, processing and marketing etc. open up new avenues for employment generation.

Nutritional Security

Nutritional security of country men is at stake due to less availability of fruits, vegetables and pulses which are main source of nutrition. Land use diversification will increase availability of nutritious fruits and vegetables at affordable cost to rural poor.

Soil Health

Soil health in past few decades has deteriorated badly to the extent that many good fertile ands are not even responding to chemical fertilizer due to continuous imbalance use, lack of organic matter addition and soil erosion from frequently tilled lands. Deforestation has played havoc. In order to restore land productivity and soil health, due emphasis should be given on organic farming, retention of permanent vegetation and balanced use of macro as well as micro nutrients. Extent of degraded lands in India is given in Table 5.

Table 5: Extent and type of land degradation in India

Type of land degradation	Area (M ha)
Water eroded	73.6
Wind eroded	12.9
Saline and Alkali	7.2
Forest degraded land	35.9
Total	**129.6**

Source: Agricultural Research Data book, 2006

Pre-requisites for Diversification

Land use diversification needs thorough understanding of natural resources and economic status of individual farmer. Diversification strictly follows farming system approach at micro level and acreage planning at country or state level. Assured market is lone driving force to persuade farmers. Technical know-how is available with research organizations. However, there is a need to transfer the knowledge to field functionaries and intern to end user, i.e., farmer. Assured availability of quality seed and planting material at reasonable prices and approachable distance holds the key to success. Logistic support to transport and store the produce with minimal spoilage, developing cool chain facility, refrigerated, transportation etc. need to be provided . Small scale processing units need to be encouraged.

Planning Diversification

Planning diversification at macro level can be done on acreage basis in light of global market. This can be done at state level or for agro-ecological zones. At micro level, farming system approach needs to be fallowed. For common property resources, watershed may be regarded as unit for land use diversification. Following step may be helpful.

Land Capability Assessment

Land use Planning should follow land capability. Class-1 lands should go under cash crops like flowers, vegetables, potato, sugarcane, tobacco, spices and high value cereals, pulses and oil seeds. Class-II lands may be developed under intensive cropping under staple food. Class-III and IV lands may be suggested for alternate land use viz. Agrihorticulture, agrisilviculture etc. Non arable lands from class-VI to VII may be developed under silvipasture or hortipasture. If topography and water availability permit, pisciculture can be advocated. Lac culture, sericulture, apiculture can also be accommodated in non arable lands.

Socio-economic Status

Socio-economic status of farmer is deciding factor in adoption of land use. Resource poor farming may be suggested to adopt low input requiring crops and plants whereas resource rich farmers may opt for high capital requiring crops and techniques. Certain social groups have traditional knowledge of vegetables, flowers, and fruits in particular, such farmers may be encouraged. Credit facility if required can be made available at reasonable interests.

Market

Specific markets for farm produce have developed in different parts of the country. Considering market demand, land uses have changed. Increase in area under polar around Yamuna Nagar in past two decades is unique example of market driven land use change. There is constant increase in demand of herbal medicines, flowers and fruits, globally. Such opportunities may be cashed upon by diversifying land use. However, it should be in tandem with transport and storage facility.

Diversified Systems

Diversification envisage 'get away' with macro cropping to escape risk of crop failures on account of insect, pest and disease epidemics, weather vagaries, water pollution arising from constant use of chemical fertilizes, insecticides / pesticides /fungi cide/herbicide, land degradation etc. further, low remuneration from existing crops due to high input costs and opportunities in cultivation of export oriented crops are major forces in land use diversification.

Diversification is not only needed for crops but also warrants introduction of alternate land uses. Over year several agroforestry systems have been evolved for different areas. Nair (1985) classified agroforestry systems on the basis of structure, function, component and socio-economic factors. Out of several classifications of agroforestry system, one based on components is given below. Following common alternate land use systems are prevailing in India.

1. Agrisilviculture (Crops+ Timber trees)
2. Agrihorticulture (Crop + Fruit tree)
3. Energy plantation (Tree + Crop during initial stages)
4. Alley cropping (Hedges + Crop)
5. Agrisilvihorticulture (Tree + Fruit tree + Crop)
6. Agrisilvipasture (Crop + Pasture + Animal tree)
7. Silvipasture (Tree + Pasture)
8. Hortipasture (Fruit tree + Pasture)
9. Forage forestry (Forage tree + Pasture)
10. Shelter belts (Tree + Crop)
11. Wind breaks (Tree + Crop)
12. Live fence (Shrubs on boundary)
13. Homestead (Fruits tree + Tree + Spices + Vegetables in Homegardens
14. Entomo forestry (Apiculture, sericulture)
15. Acqua forestry (Tree + Fish pond)

In addition, several combinations involving components like medicinal plants, flowers, spices, fruit trees, plantation crops, tuber crops, trees etc. may be developed as per edapho-climatic suitability.

Further, agroforestry systems have been classified on the basis of arrangement of trees. Common systems are as follows:

1. Block Plantation
2. Boundary Plantation
3. Field Bund Plantation
4. Field Plantation

References

Anonymous 2006. Agricultural Research Data Book, 2006 IASRI, (ICAR) New Delhi.

Anonymous 2007. Financial Express, 21.4-2007

Nair, P.K.R. 1985. Classification of Agroforestry System. *Agroforestry Systems*, 3: 97-128

Sanwan, SS 2002. Diversification in agriculture: How to do it. *Kurukshetra,* February, Pp. 20-22.

Swaminnathan, M.S. 2007. National Commission on Farmers. Report submitted to Govt. of India.

Agroforestry: Natural Resource Sustainability, Livelihood & Climate Moderation © 2009
Editors: OP Chaturvedi, A Venkatesh et. al., pp. 369-383
Satish Serial Publishing House, Delhi

Chapter 26

LIVESTOCK PRODUCTION UNDER AGROFORESTRY LAND USES IN INDIA

N. Das and S.K. Mahanta
Indian Grassland and Fodder Research Institute, Jhansi-284 003 (U.P.)

Introduction

The use of trees and shrubs as a source of feed for livestock is not new. Plato, in his Critias and Timaeus, cites that there are many places in the world where only trees can be grown and the local livestock are completely dependent on trees for their survivability. Since Biblical times, the carob has been recognized as a source of food, mainly for humans, but it is well recognized in parts of the world as a source of fodder for animals. Carob tree grows on rocky or stony outcrops, or clefts of rocks where no grasses could ever get a hold (Duke 1981). In England and some parts of northern Europe, Elms were used as a source of fodder until the mid-1800s. They formed an integral part of the farming system. The elm's large volume of canopy contains much needed proteins, carbohydrates and minerals sought by livestock (Lamb 1979). Atacama Desert in Chile is a desert covered where no grass grows as the top soil is covered with a salt crust up to one metre thick. The meagre rainfall, just a few millimetres a year, does not promote the growth of grasses anyway.

Holes about a metre or so across, and a couple of metres deep, are excavated through the salt crust for the tree plantation. *Prosopis tamarugo* trees have been planted. Despite very scanty rain, the trees receive abundant water from the water table below the surface. These trees provide about six tonnes of leaves and pods per hectare each year which can support twenty sheep or goats per hectare (Felker 1981, Felker et al. 1984). In Peru, *Prosopis juliflora* tress are grown in thousands of hectares for livestock. On maturity, the trees provide six or seven tonnes of high protein pods per hectare in a year (Felker 1981). South Africa has been developing trees of the species *Prosopis* and *Gleditsia*, with reported yields of edible beans of up to twenty tonnes per hectare per year. In Africa, *Acacia albida* trees are grown for centuries as a source of feed for animals. Acacias, being legumes, are able to convert, by means of rhizobium bacteria, atmospheric nitrogen into usable nitrates. These trees can thus grow on nitrogen-deficient soils. There perhaps no limit to grow trees for fodder. Hill tops that cannot be ploughed or planted to grasses, can carry trees (Douglas & Hart 1978; Lamb 1979). Steep slopes that are unsuited to ploughing will grow trees. Low rainfall areas with annual, irregular rainfall of less than 200 millimetres per year can grow some of the chenopods, or carobs, which will yield several tonnes of dry matter per hectare every year. Unlike pastures, there are few limits for growing at least some species of trees or shrubs for fodder. In some parts of the semi-arid and arid regions of India, few tree species grow in soil that is too infertile, and too saline to permit almost any other type of vegetation. Rainfall can be as little as 150 millimeters fall per year, yet the trees in these area produces tones of dry matter per hectare and the survivability of local sheep and goats are mainly dependent on these tress.

The major challenge in rain-fed areas is not only to eke out the most from every millimeter of rain, but to utilize the reserves of water deep below the surface; this can only possible by plantation of trees. Many of the trees and shrubs that are suitable for fodder are far less affected by drought (Douglas and Hart 1978). The feed yield from fodder trees - even grown under adverse conditions and in poor soils - can match, and often exceed, that from pastures grown under good conditions (Lamb 1979). Agroforestry is a good alternative to pastures and grasses. It involves the use of fodder trees and shrubs - trees that are permanent, and which will replace annual grasses; trees which will provide their own nitrogenous fertilizers; trees which will extend their roots deep beneath the ground to the water table. Agroforestry is the use of trees that give high feed yields all year round, every year, even where grasses cannot grow (Douglas and Hart 1978). Although there may be wild claims about the potential and yields from some fodder trees (Oates and Clarke 1987), there is no doubt that, because trees are deep rooting compared with grasses, they will tap minerals and water deep in the soil.

Owing to increase in population of human and livestock, there is increasing demand of food as well as fodder, particularly in developing countries like India. Each year the farmers of the world must now attempt to feed 81 million more people irrespective of weather conditions. It is important to note that there will be 19% decline in crop land per head by the end of this century due to population explosion. In general, there is acute shortage of food and fodder in developing countries. However, the situation in India is more alarming than in other countries of the third world. Asia posses the smallest area of the "potentially arable land" (672 million ha compared with 681 million ha in South America and 734 million ha in Africa) and has the highest population. Out of this, 590 million ha (83%) is already under cultivation. Hence, there is very limited scope for diverting arable land from crop production for human consumption to fodder production for animal consumption. To meet the ever increasing demand of animal products, we have to utilize the land not conventionally considered arable for animal farming. A management system like agroforestry, therefore needs to be devised that is capable of producing both food and forages from marginal agricultural land and is also capable of maintaining and improving the quality of producing environment (Rai and Handa, 2008). The challenge before us is how to make vast barren soils of India productive within a few years and provide a plentiful supply of fresh, highly nutritious and palatable feed for the animals. Many of the trees that are planted as sources of fodder modify the microclimates beneath their canopies; they reduce water loss from the soil, and reduce evaporation from nearby fields, and from dams; they reduce soil erosion; they lower the temperatures the animals must endure. There will be cost for growing trees, but gain will be substantially more than the cost.

On the basis of nature of components the following are common agroforestry systems prevailing in different agro-ecological regions of India:

1. Agri-silviculture (trees+crops)
2. Boundary plantation (tree on boundary + crops)
3. Block plantation (block of tree+ block of crops)
4. Energy plantation (trees+crops during initial years)
5. Alley cropping (hedges+crops)
6. Agri-horticulture (fruit trees+crops)
7. Agri-silvi-horticulture (trees+fruit trees+crops)
8. Agri-silvipasture (trees+crops+pasture or animals)
9. Silvi-olericulture (tree + vegetables)

10. Horti-pasture (fruit trees+pasture or animals)
11. Horti-olericulture (fruit tree + vegetables)
12. Silvi-pasture (trees+pasture/animals)
13. Forage forestry (forage trees+pasture)
14. Shelter-belts (trees+crops)
15. Wind-breaks (trees+crops)
16. Live fence (shrubs and under- trees on boundary)
17. Silvi or Horti-sericulture (trees or fruit trees+sericulture)
18. Horti-apiculture (fruit trees + honeybee)
19. Aqua-forestry (trees + fishes)
20. Homestead (multiple combination of trees, fruit trees, vegetable etc).

Out of these 20 types of agroforestry, animal is included as component only in three types viz. Agri-silvipasture, Horti-pasture and Silvi-pasture. The use of livestock under agroforestry can be a controversial topic. It is often claimed that animals such as cattle, goats and sheep are associated with ecological degradation and deforestation. There is no doubt that the faulty use of animals and poor management leads to many environmental degradation, but it is also the fact that the livestock factor is merely one of the many factors responsible for soil degradation. Whenever animals are separated from their natural environment (as generally practiced), the beneficial cycle of animals and their environment is broken. It has been observed, especially in the semi-arid and arid regions, that animals such as cattle, goats and sheep are associated with ecological degradation and deforestation. If planned and managed properly, appropriate kinds of animals can be key components in sustainable farming systems. Agroforestry systems that incorporate animals with tree crops can enhance important cycles of nature such as nutrient cycling and balancing of insect populations while reducing energy intensive management techniques. The key is to integrate the natural needs, behaviors, and products of animals with the environment provided by the agroforestry system in a way that maximizes the benefits to the animals and to the system as a whole.

Andrews (1998) listed the following characteristics of ideal fodder trees:

- Have high growth rate: many fodder trees fit well within this criteria;
- Cause minimal soil disturbance: generally fodder trees need no soil disturbance after establishment;

- Prevent nutrient loss from the soils: because of their permanence, fodder trees help to prevent nutrient losses;
- Require little or no need for irrigation: fodder trees can tap deep water tables out of the reach of short-rooted pastures. They may need a little irrigation till they are established;
- Have forage yields comparable to pastures.
- Have the ability to fix nitrogen to reduce the dependence on artificial fertilizers: legumes fodder trees fix atmospheric nitrogen by use of bacteria and convert it into nitrates that the plants can use.
- Have high protein content in their edible portions improve livestock nutrition. the protein content of legume tree leave will be high - often around 20 to 25 percent, comparable to lucerne;
- Provide their feed in a short time. Some trees require possibly seven or eight years before the pods and seeds are freely available. But then, some pasture improvement programs can take up to fifteen years to achieve.
- Palatable: Livestock are selective about what fodder they will eat
- Free of toxic substances: Some fodder species contain toxic substances that can be consumed only in small amounts. Leucaena sp., for example, contain mimosine which, in large dosed, is harmful to animals. It must, therefore, be used in combination with other fodder sources
- Sprout well: Recover rapidly after pruning

Animals in Agroforestry

In agroforestry animal and plant are complementary to each other; animal gets food from plant and in return, it spread manure, and dig and/or scratch in the soil. Many of the products and behaviors of animals, for example manure, are considered problems or pollutants in modern conventional production of e.g. poultry and livestock. Elevitch and Wilkinson, (2008) outlined this complementary aspect.

Products and Services Received of Animals for Agroforestry

- Grazing/weed maintenance can enhance tree growth by controlling grass competition for moisture, nutrients and sunlight.
- Cleaning of fallen fruit/nuts, other organic wastes (livestock)
- Spreading nutrients in the form of their manure (livestock)
- Scratching and digging habit of pig help to prepare land for planting.

- Well managed grazing provides economical control of weeds and brush without herbicides, maintains fire breaks and reduces habitat for gnawing rodents. Insect control is done by poultry
- Fertilizer applied for forage is also used by trees.

Products and Services of Agroforestry Systems for Animal Husbandry

- Favorable habitat
- Access to a diverse diet
- Access to medicinal plants and insects
- Natural population densities
- Some forage species tend to be lower in fiber and more digestible when grown in a tree-protected environment.
- Trees that provide shade or wind protection can have a climate-stabilizing effect to reduce heat stress and wind-chill of livestock.
- Protection from trees can cut the direct cold effect also by 50 percent or more and reduce wind velocity by as much as 70 percent.
- Livestock require less feed energy, so their performance is improved and mortality is reduced.

Economic Advantages of Integrating Animals in Agroforestry

- Less reliance on outside inputs
- Reduced human maintenance
- Better plant health through natural fertility cycling
- Creates a land management system that produces marketable products while maintaining long-term productivity.
- Economic risk is reduced because the system produces multiple products, most of which have an established market.
- Production costs are reduced and marketing flexibility is enhanced by distributing management costs between timber and livestock components.

Environmental and Aesthetic Benefits

- Increase wildlife diversity when livestock are introduced in the forest and improve water quality.
- The forage protects the soil from water and wind erosion, while adding organic matter to improve soil properties.

- Provide an attractive landscape with an aesthetically pleasing "park-like" setting.
- In contrast to concentrated livestock operations, silvopastoral systems are less likely to raise environmental concerns related to water quality, odors, dust, noise, disease problems and animal treatment.

Disadvantage

But the same integration may be disastrous if an animal is placed in the wrong place at the wrong time. For example, the wild pigs destroy new plantings and damage young trees in forest. The same pigs can be used successfully for farm forestry in controlled paddocks to graze and consume farm residues, and to provide meat for sustenance. The disadvantages of animal integration can be checked if we take the following measures.

- Animals must be carefully selected for particular systems. The introduction and management of animals must be carefully thought out, to ensure that the right kinds of animals are introduced to the right place, and at the right time. Animals those are more appropriate to their local environment, culture, and market demands need to be introduced. For some very fragile ecosystems, animals may not be appropriate at all.
- Animals require tending and management of grazing
- Food and water may need to be supplemented at times
- Health care required
- Requires a special set of skills and knowledge

Impact of Animal Grazing on Plant Vegetation

Component interactions include the effect of one component of a system on the performance of the other components as well as the system as a whole. In agroforestry system, site, vegetation, animals and their management and pastoral community are the major components. These components interact with each other over a period of time and yield plant and animal products. Site provides necessary physical base and mechanical support for plants and animals. Plants obtain their vital inputs such as light, oxygen, carbon dioxide, water, plant nutrients, etc. from the site. Site determines variety, extent and productivity of plants. The plants in return not only improve the site but also provide feed and shelter to animals. The animals provide organic manure to plants, help in seed dispersal, soften the seed coat, bury the seed in the soil, loosen and open up the soil, decompose litter, eat away leading shoots and thus promote tillering etc. The animal consume live as well as dead plant parts and build their biomass in the form of meat, hides, skins, wool, bones, etc. which is the end product of range ecosystem (Quraishi *et al.*, 1993). The major positive and negative impacts at the tree pasture interface (TPI) and the tree animal interface (TAI) in a silvopasture system involving animal component are given in Table 1 (Nair, 1993).

Table 1: The major positive and negative impact at the tree pasture interface (TPI) and the tree animal interface (TAI)

Impact	At the TPI	At the TAI
Positive	Shading trees (stress reduction) Biomass contributions Water conservation Soil conservation	Shading Manure deposition
Negative	Light competition Nutrient competition Water competition Allelopathy	Phytotoxins Browsing damage Trampling Disease/pest hosts

Animals protected by windbreaks can be healthier and more productive. Productivity of farm animals is optimal for a comfort zone of a relatively narrow temperature range. Animals experiencing excessively hot or cold conditions require more energy to maintain basic metabolism and thus have less energy available to increase body weight or to produce meat, milk or wool. In other words, more feed is required to counter environmental stresses. Under adverse conditions, livestock performance is inefficient relative to the quantity of feed consumed. Experiments with sheep and cattle have shown that strong wind and rain double the energy requirement of animals for maintenance. It is well known that dairy production is improved by providing sheltered conditions. In southern Victoria it has been calculated that the provision of shelter can increase milk production by 30%. Ten per cent of this is due to greater efficiency of conversion of feed, and 20% due to the greater amount of feed available (Fitzpatrick 1994).

Productivity of Animals in Agroforestry Systems

Plantation Crop Systems

The principal plantation crops where integration of livestock is possible are rubber, oil palm and coconuts. Keeping in view he unique quality of coconuts, compared to most other plantation crops, is the relatively constant and bright light environment over the life of the crop (60-80 years) emphasis is being given on the integration of cattle with coconuts. Under storey pastures can be grazed on a semi-permanent basis. In Malaysia, sheep are grazed in rubber and oil palm plantations; the productivity of sheep was moderate under young rubber (2-5 years) provided palatable leguminous cover crops were grazed, but low under mature rubber where light levels had fallen to <20% transmission of light. There is potential to increase productivity by using shade-tolerant forage species and by altering the conventional rubber planting system to a hedgerow system (Gutteridge and Shelton, 2008).

Forestry Systems

In the temperate countries (southern Australia and New Zealand), successful sheep based forest grazing systems have been developed. Sheep are preferred because they are less inclined to browse the trees than cattle. Stocking rates vary between 12 and 25 animals ha^{-1} in the early stages of tree development, but it is reduced by at least 50% by year 10. But in the tropics, forest grazing systems have been less successful. In the Solomon Islands, pastures and cattle under Eucalyptus deglupta plantations at densities of 130 stems ha^{-1} could not be sustained in the long term. In a young plantation (trees 2 - 3 years old), a stocking rate of 2.5 beasts ha^{-1}, giving live weight gains of 0.45 kg hd^{-1} day^{-1} was achieved; however, by year 7 with almost complete tree canopy cover, stocking rates had to be reduced to less than 1 beast ha^{-1}, producing only 0.25 kg hd^{-1} day^{-1} live weight. Substantial damage to the trees due to bark stripping by the cattle also occurred.

Horticultural Systems

A number of fruit tree species such as mangoes, kapok, tamarind, cocoa and cashews have some potential for integration with livestock when the trees are young which facilitate penetration of sun rays. But necessary precautions are needed to prevent damage to the palatable leaves of the crop. There is also problem of the low light profile when the trees reach maturity. However, a specialized agrosilvopastoral horticultural system was reported in Bali, Indonesia where there is vanilla production under coconuts—the vanilla orchid is supported and shaded by the tree legumes *Gliricidia sepium* or Erythrina sp. The shade level is regulated by lopping the branches of shade trees which are fed to livestock.

Alley Farming Systems

In alley cropping, rows of trees are planted closely together between rows of agricultural crops. These hedgerows can reduce soil erosion, and, when pruned regularly, provide valuable fuel and fodder. It was observed that the surplus foliage from 1 ha of Leucaena leucocephala and Gliricidia sepium from alley farming could provide half the daily fodder requirements for 29 goats. In the dry season fallow period, livestock can be allowed direct access to the cropping area to graze crop residues and browse the trees. In Malawi, a 0.25 ha leucaena alley farm produced enough leaf material to feed three dairy cattle @ 3 kg dry matter day^{-1} during the dry season. In lowland humid regions of the tropics where farm size is often small, the amount of forage generated from alley farming is very low. However, in semi-arid regions, the additional forage from both prunings and crop residues may be the major incentive attracting farmers to use the alley farming technique.

In India

Rail *et al.*, 1994 conducted a grazing trial of 241 days duration was conducted on one year established silvopasture consisting of *Albizia amara* and *Leucaena leucocephala* as tree, *Dichrostachys cineria* as shrub, *Chrysopogon fulvus* as grass, *Stylosanthes hamata* and *S. scabra* as pasture legume and natural *Sehima-Heteropogon* grass to record the comparative performance of small ruminants (sheep and goats). Study revealed (Table 2) that goats and sheep that grazed on 1 ha of silvopasture gained in their body weight @ 28.6 and 2.1 g hd^{-1} day^{-1}, respectively, while animals grazed on 2 ha of natural grassland showed that goats gained in their body weight @ 10.8 g hd^{-1} day^{-1} and sheep lost in their body weight @ 27.4 g hd^{-1} day^{-1} even after supplementation of 1.5 kg hd^{-1} day^{-1} green subabul leaves.

Table 2: Performance of goats and sheep grazed on silvopasture and natural grassland

Attributes	SilvopastureNatural grassland			
	Goats	Sheep	Goats	Sheep
Initial weight (kg)	21.6	43.1	21.6	42.6
Final weight (kg)	28.5	43.6	24.2	36.0
Total weight gain/loss (kg)	6.9	0.5	2.6	-6.6
Daily weight gain/loss (g)	28.6	2.1	10.8	-27.4
Weight gain/loss (kg/ha)	41.4	3.0	7.8	-19.8

In another study (Table 3), the growing lambs and kids (age between 6 to 12 months) grazed on silvopasture, gained weight @ 157.2 and 105.6 kg hd^{-1}, respectively, while lambs and kids grazed on natural grassland, the gain was 59.1 and 37.8 kg hd^{-1}, respectively in a total grazing period of 478 days. This showed that sheep and goats gained weights (kg hd^{-1}) more than 2.5 times when grazed on silvopastures over natural grasslands. On an average, gain in body weight of sheep and goats grazed on silvopasture was higher at the tune of 33.0 and 39.9%, respectively over natural grasslands. Better performance of sheep and goats on silvopasture was due to availability of nutritious herbage from tree leaves and pasture legumes. Gain in body weight was also observed in sheep (18.3 g hd^{-1} day^{-1}) and goats (28.4 g hd^{-1} day^{-1}) grazed on a silvopasture system for a period of 180 days (Sharma and Ogra, 1990).

Table 3: Performance of kids and lambs grazed on silvopasture and natural grassland

Attributes	SilvopastureNatural grassland			
	Kids	Lambs	Kids	lambs
Initial weight (kg)	11.8	13.4	11.8	13.4
Final weight (kg)	29.4	39.6	24.4	33.1
Total weight gain/loss (kg)	17.6	26.2	12.6	19.7
Daily weight gain/loss (g)	36.8	54.8	26.3	41.2
Weight gain/loss (kg hd^{-1})	105.6	157.2	37.8	59.1

A silvipasture system involving five species of the tree viz., *Acacia tortilis, Albizia lebbeck, A. amara, Leucaena leucocephalh and Hardwckia binata* with under storey of *Sehima nervosum* and *Heteropogon contortus* dominant grassland community was established and evaluated under mixed herd grazing with cattle, sheep and goat for productive and reproductive efficiency of the animals for 55 months (Upadhyay and Pathak, 1994). Growing Tharparkar (*Bos indicus*) heifers gained on an average about 500 g hd^{-1} day^{-1} (including compensatory growth) during monsoon period without any supplemental feeding. The concentrate supplementation at the rate of about 1 percent of the body weight during dry months led the animals to gain on an average about 100 kg live weight per animal per year. The breeding efficiency of sheep was 0.9 lambs/ ewe/year while goat produced on an average 1.8 kids doe^{-1} yr^{-1}. It was again suggested that proper selection and density of the tree species, their lopping management, conservation of forage specially in the form of silage fortified with legumes or urea etc. can enhance further the productivity of animals on silvipastoral system.

Sharma *et al.*, 2001 reported the findings of a long term (1990-1994) study was carried out on comparative performance of small ruminants on silvopasture and natural grassland under in situ grazing condition and their impact on plant vegetation at research farm, National Research Centre for Agroforestry, Jhansi. The plant composition (number of plants) of *Sehima nervosum, Iseilema laxium, Themeda guadrivalvis* and forbs increased from 1990 to 1994 after about 3 years of grazing round the year by sheep and goats (Table 4). While percentage of *Heteropogon contortus*, other grasses and legumes decreased during 1994 as compared to 1990, the decreased in per cent composition of *H. contortus*, other grasses and legumes might be due to round the year grazing of animals and as such most of the annual species could not form the mature seeds.

Table 4: Impact of grazing on plant vegetation under natural grassland system

Species	Composition (%)		Dominance (%)	
	1990	1994	1990	1994
Sehima nervosum	20.38	23.67	47.92	52.45
Heteropogon contortus	23.62	14.16	21.78	27.02
Iseilema laxum	0.53	1.05	0.78	1.09
Themeda guadrivalvis	2.38	29.64	1.82	8.27
Other grasses	26.46	7.48	17.66	3.28
Legumes	14.10	7.50	4.29	1.19
Forbs	12.41	16.28	5.69	6.62

Per cent dominance of all the perennial grass species increased during 1994 as compared to 1990. It revealed that there was no harmful effect of grazing on the health of perennial grasses. The per cent basal cover of other grasses and legumes decreased considerably during 1994 than that of 1990. This showed that the vegetation was more utilized by animals as well as mostly being of an annual nature, more damage occurred during the grazing period. Increase in per cent composition as well as dominance during 1994 in case of forbs showed the aggressiveness of these species as most of species were less palatable. Decrease in per cent composition and dominance of other grasses and legumes may be attributed to their more palatability and high grazing frequency. The plant population of *C. fulvus* increased during 1994 four times as compared to 1990 (Table 5), while decrease in plant population of *S. hamata* and *S. scabra* was recorded to the tune of 67 and 93.4 per cent, respectively after three years of grazing of sheep and goats. Thus, the results showed that there was no adverse effect of grazing on *C. fulvus* stand and as such it can tolerate grazing for many years with optimum stocking rate of animals. In case of *Stylosanthes* species, it can be attributed that these species may not withstand grazing for many years. However, *S. hamata* seemed to be more tolerant of grazing than *S. scabra*. Thus, for getting nutritious forage reseeding of *Stylosanthes* is required every fourth year.

Table 5: Impact of grazing on grass and legume population under silvopastoral system

Species	Plant population/m^2	
	1990	1994
Chrysopogon fulvus	2.0±0.0	8.0±1.4
Stylosanthes hamata	10.6±5.1	3.5±10.6
Stylosanthes scabra	10.7±6.0	0.7±0.2

Among the tree components, survival of *A. amara* was excellent and *D. cinerea* was medium, while *L.leucocephala* was very poor during 1994 (Table 6). Low survival of *L. leucocephala* was due to attack by sheep and goats during summer months for peeling the bark.

Table 6: Impact of grazing on survival of tree component under silvopastoral system

Species	Plant population m^{-2}	
	1990	1994
Albizia amara	100.0±0.0	92.5±7.0
Dichrostachys cinerea	93.3±2.5	64.0±11.9
Leucaena leucocephala	96.6±0.8	11.5±5.7

In this long term study, it was also observed that adult sheep and goats maintained their body weight grazed on silvopasture while supplementation through tree leaves required to the animals grazed on natural grassland. The body weight of growing lambs and kids increased by 33 and 40 per cent, respectively, when grazed on silvopastoral system over natural grassland in a total grazing period of 478 days. Further. It was observed that double number of animals can be reared on silvopasture than that of natural grassland as the grazing area of silvopasture system was one hectare and it was two hectare in case of natural grassland. As regards to health of pasture components, it was observed that there was no any adverse effect of round the year grazing on the perennial grasses, while legumes drastically decreased in both established silvopasture and natural grassland. Out of three tree species, L. leucocephala affected severely as survival was only 11.5 per cent during 1994. Thus, the studies revealed that silvopastoral system was better than natural grassland for higher body weight gain of sheep and goats as well as rearing of more number of livestock

Conclusions

Notwithstanding a substantial potential for animal production in livestock-based agroforestry systems exists but very little efforts are taken in our country for designing of suitable system to improve animal productivity. High levels of production are only possible through the use of high quality fodder-tree legumes when livestock production is the major economic activity. Evaluation of the available legume varieties of trees will broaden the resource base available and promote the expansion of these systems into a wider range of environments. The animal production potential of forest and plantation crop grazing systems is relatively low and improvements in these systems will come largely from the identification and selection of herbaceous forage species tolerant of lower sunlight conditions. Improvement is also possible through changes in plantation management practices such as tree spacing and pruning regimes. In all systems the service role of the trees in providing shade and shelter must be taken into account as this aspect is often undervalued in assessing the productivity of agroforestry systems. Comparative studies conducted on silvopasture and natural grassland have proved that production performances of animals are better in silvopasture kind of agroforestry system under semi-arid conditions and the same can be exploited to augment livestock productivity. But animals must be carefully selected for particular systems.

References

Andrews, G.J. 1998. The use of trees and shrubs for livestock production. *Journal Australian Biologist*, 11 (2), June 1998.

Douglas, J.S. and Hart, R.A. 1975, Forest Farming; Towards a Solution to Problems of World Hunger and Conservation, Rodale Press, Emmaus Pa.

Duke, J.A. 1981. Handbook of Legumes of World Economic Importance, Plenum Press, New York.

Felker, P. 1981. Uses of tree legumes in semiarid regions. *Economic Botany*, 35 (1): 174–179.

Felker, P. and Bandurski, R.S. 1979. Uses and potential uses of leguminous trees for minimal energy input agriculture. *Economic Botany*, 33 (2): 172–182.

Felker, P., Clark, P.R., Osborn, J.F. and Cannell, G.H. 1984. Prosopis pod production - Comparison of North American, South American, Hawaiian, and African Germplasm in young plantations. *Economic Botany*, 38 (1): 36–51.

Lamb, R. 1979. World without Trees, Magnum Books, London.

Oates N. and Clarke, B. 1987. Trees for the Back Paddock, Goddard & Dobson. http://www.grahamandrews.com/fodder_trees.htm

Elevitch, C. and Wilkinson, K. 2008. Animals in Agroforestry. Adopted from http://www.agroforestry.net/overstory/overstory35.html.

Fitzpatrick, D. 1994. Money Trees on Your Property: profit gained through trees and how to grow them. Inkata Press, Sydney.

Gutteridge, R. and Shelton, M. 2008. Productivity of animals in Agroforestry systems. Adopted from http://www.agroforestry.net/overstory/overstory37.html.

Klopfenstein, N.B., Rietveld, W.J., Carman, R.C., Clason, T.R., Sharrow, S.H., Garrett, G. and Anderson, B.E. 2008. Silvopasture: An agroforestry practice. Adopted from http://www.agroforestry.net/overstory/overstory36.html.

Nair P.K.R. 1993. An introduction to agroforestry. Kluwer Academic Publishers, London.

Quraishi M.A.A., Khan,G.S. and Yaqoob, M.S. 1993. Range management in Pakistan. Kazi Publications, Lahore.

Rai, P., Debroy, R., Verma, N.C. and Rao, G.R. 1994. Performance of small ruminants on silvipasture and natural grassland under grazing condition. In: Agroforestry systems for degraded lands (Eds. Singh, Panjab, Pathak, P.S. and Roy, M.M.). Oxford & IBH Publishing Co. Pvt. Ltd., New Delhi.

Rai. P. and Handa, A.K. 2008. Agroforestry. http://www.agroforestry.definition.concept.Jhansi. pdf.

Sharma, K. and Ogra, J.L.1990. Influence of continuous grazing by goats and sheep under three tier pasture on their performance and constituent vegetation. *Range Managemnet and Agroforestry*, 11: 137-144.

Sharma, N.K., Singh, R.P., Singh, M. and Harsh, L.N. 2001. Silvopasture management in hot arid and semi-arid ecosystems. Agrotech Publishing Academy, Udaipur, India.

Upadhyay, V.S. and Pathak, P.S. 1994. Silvipastoral system evaluation for secondary production in Central India. In: Agroforestry systems for degraded lands (Eds. Singh, Panjab, Pathak, P.S. and Roy, M.M.). Oxford & IBH Publishing Co. Pvt. Ltd., New Delhi.

❑❑❑

Agroforestry: Natural Resource Sustainability, Livelihood & Climate Moderation © 2009
Editors: OP Chaturvedi, A Venkatesh et. al., pp. 385-400
Satish Serial Publishing House, Delhi

Chapter 27

AGROFORESTRY BASED LIVESTOCK PRODUCTION FOR LIVELIHOOD SECURITY

Purushottam Sharma
Indian Grassland and Fodder Research Institute, Jhansi-284 003 (U.P.)

Introduction

Agroforestry as a traditional land-use adaptation may potentially support livelihood improvement through simultaneous production of food, fodder and firewood as well as mitigation of the impact of climate change (Pandey, 2007). According to him, the contribution of agroforestry systems in India is to:

i. biodiversity conservation;
ii. yield of goods and services to society;
iii. augmentation of the carbon storage in agroecosystems;
iv. enhancing the fertility of the soils, and
v. providing social and economic well-being to people.

The role of agroforestry in animal husbandry is mainly to improve crop productivity vis-à-vis fodder availability, feed and fodder availability from Agroforestry systems, leaf meal and vegetative shelter. It estimated that fodder requirement of India is 26 million tonnes annually. To meet the requirement, it is of vital importance to enhance the fodder production by improving top feed tree species. The tree species besides producing fodder, yield sufficient fuel wood also. Fodder tree species such as *Prosopis cineraria, Acacia nilotica, Albizia lebbec Azadirachta indica,* and *Dalbergia sissoo* etc. are native to Indian sub-continent. These species have wide variation with respect to their growth rate, biomass production, nutritive value and size of fruits. Tree leaves are being increasingly used to provide fodder for livestock, as they have a number of unique characteristics which make them attractive for both smallholder and largescale livestock enterprises. On an average, tree leaf fodder production of 0.2 to 2.0 ton /ha/year can be obtained up to 50% pruning height of the trees every year from the various agroforestry systems of different agroclimatic zones. The tree leaves contain 8-33 per cent crude protein, 1-19 per cent ether extract, 11-50 per cent crude fibre, 36-66 per cent nitrogen free extract, 22-57 per cent neutral detergent fiber, 0.2-3.0 per cent calcium and 0.1-0.3 per cent phosphorus. The digestibility coefficient of dry matter, crude protein, crude fibre, ether extract and nitrogen free extract in tree leaves ranged from 40-75, 28-83, 24-82, 32-65 and 51-85 per cent, respectively" (Rai *et al* ., 2007).

The role of agroforestry is more important in arid and semi arid regions of country. Arid and semi arid regions of the country are affected by frequent occurrence of drought, misery, hardships, large scale migration of livestock and human beings. Under difficult environments, livestock plays an important role in the livelihood of human beings. In periods of drought people look towards their livestock to overcome the difficult time, but it is this sector, which gets least attention in the fight for survival. Governments often resort to short term measures providing immediate relief to human population through food aid and through programmes like food for work. The Food Corporation of India (FCI) arranges food grains but the mute animals are most affected. The drought preparedness manuals do list that efforts should be made to make available sufficient quantity of fodder for milch animals and looking after their health of during the drought period. The Railways also helps in, to transport fodder to the affected areas free of cost. Very often the periphery of the drought affected regions do not have sufficient fodder to spare which makes its procurement and transport very difficult, inspite of the willingness of the Railways and the Animal Husbandry departments. Therefore, it is essential that mechanism for mitigating the hardship to animals and restoring the livelihood of people be created in the zones of vulnerability by enhancing the productivity of land through employing improved technologies like Agroforestry, agri-silviculture, creation of shelter belt, management of crops, using drought resistant crop verities, improving the productivity of wastelands and degraded forests, strip cropping in arid regions, meeting the demands of peri urban livestock through integrating perennial grasses and legumes etc. besides creation of fodder banks.

In arid and semi-arid region tree fodder are very important for ruminants because of (Rai *et al.*, 2007)-

- higher protein and mineral contents specially during dry months.
- important sources of high quality feed for grazing ruminants.
- help to strengthen a plant animal-soil ecosystem.
- important source for browsing and grazing.
- helps the herbivores especially the goats to extend as well as meet their feed preferences.
- the leaves serve as a concentrate to goat and sheep population at a very low cost or virtually free to both sedentary and marginal farmers.
- tree leaves like babul *(Acacia nilotica)*, jackfruit *(Artocarpus heterophyllus)*, pipal *(Ficus religiosa)*, plantain *(Musa paradisica)*, siris *(Albizia lebbek)* and subabul *(Leucaena leucocephala)* contained higher levels of Ca, Cu, Zn and Fe.

The discussion is mainly focused to identify the extent of deficit and the strategies to augment the forage resources through agroforestry in the arid and semi arid regions of the country. The important aspects included are agroforestry to improve crop productivity vis-à-vis fodder availability, feed and fodder availability from Agroforestry systems, leaf meal and vegetative shelter.

The Dry Zones

The arid region covers the Cold mountainous arid region of Ladakh, Hot Western plain region in the states of Rajsthan, Gujrat, Punjab and Haryana; and the Hot Deccan Plateau in the states of Karnataka and Andhra Pradesh. Similarly, the semi arid region has been classified into 5 sub regions namely, North Central Highland, Central Malwa Highland, Deccan Plateau, Telengana Plateau and Eastern Ghats & Deccan Plateau. The states falling under arid and semi arid regions have varying level of agricultural intensification brought about by irrigation projects and hence do not have the same degree challenges and opportunities.

Livestock

The dry zone has about 172 million total livestock of a total of 480 million in the country, which is 36% of the total. The buffaloes are more in arid and semi regions with a total population of 41% of the total against 26% cattle in a geographical area of 37%. Similarly the dry zones have 62% of total sheep and 34% of total goats in the country. This indicates the greater dependence of human beings on livestock for their livelihood. The arid region has 782 livestock per 1000 human beings whereas the semi arid region 535.The cold arid region of Ladakh has over 3 livestock per person and the arid Western Rajasthan 1 animal per person. Hence it is evident that livestock plays an important role in mitigating the adverse effects of drought in dry regions. Therefore it is imperative that drought proofing measures should have an important component to augment the feed and fodder supply in dry zones of the country to overcome the misery of human beings during periods of drought.

The livestock pressure (ACU ha^{-1}) in the dry zone of the country is 0.66. In arid region it is 0.38 and in semi arid region 0.79 ACU ha^{-1}. Within the arid region, the parts of Punjab and Haryana have close to 1 ACU ha^{-1} because the intensity of cropping is very high and over 80 % of the area under holding which makes available sufficient crop residues and agricultural by products available for animal feeding. On the other hand supporting even 0.3 ACU ha^{-1} in parts of Rajasthan and Gujrat becomes a difficult task as these regions are resource poor with vast wastelands and degraded lands (Table 1).

Table 1: Livestock pressure (ACU ha^{-1})

Sn	Region	Cattle ACU	Buffalo ACU	Sheep ACU	Goat ACU	Total ACU (,000)	ACU ha^{-1}
1.	Arid Region	5800	6063	1443	1380	14686	0.38
2.	Cold Arid Region	21	0	13	26	61	0.01
3.	Hot Western Plain Arid Region	5038	5086	1242	1251	12617	0.41
4.	Hot Deccan Plateau Arid Region	741	977	188	102	2008	0.51
5.	Semi Arid Region	32353	24932	3921	4101	65307	0.79
6.	North central highland	7739	11930	393	1337	21400	1.00
7.	Central Malwa highland	5717	3474	115	619	9924	0.78
8.	Deccan Plateau	9564	4301	595	1073	15533	0.70
9.	Telangana Plateau & E. Ghat	4133	3672	1831	443	10079	0.66
10.	E. Ghats & Daccan Karnataka Plt.	5200	1555	986	629	8371	0.75
11.	Grand TOTAL	38153	30995	5364	5481	79993	0.66
12.	INDIA	143683	74392	8627	16012	242714	0.77

To feed the above mentioned livestock population in dry zones, the approximate amount of nutrients needed is as follows-

Region	DM (million t yr^{-1})	TDN (million yr^{-1})	CP (million t yr^{-1})	DCP (million t yr^{-1})
Dry zones	186.40	79.83	12.88	7.42
INDIA	564.97	238.74	39.35	22.45

Improving Fodder Availability Through Enhancing Soil Fertility and Water Use Efficiency

Ecologically sound agroforestry systems such as intercropping and mixed arable-livestock systems can increase the sustainability of agricultural production. Agroforestry can enhance soil fertility and restore soil organic matter by alternate land-use systems such as agroforestry, agro-horticultural, agro-pastoral and agro-silvipasture. A field experiment to study N_2 fixation efficiency suggests that planting of stem-cuttings and flooding resulted in greater biological N_2 fixation, 307 and 209 kg N ha^{-1} by *Sesbania rostrata* and *S. cannabina* respectively which can be used as a green manure. Through a combination of mulching and water conservation, trees in agroecosystems may directly enhance crop yields of coarse grains. For instance, in the arid region of Haryana, the effect of *Prosopis cineraria, Tecomella undulata, Acacia albida* and *Azadirachta indica* on the productivity of *Hordeum vulgare* (barley) was found to be positive. *P. cineraria* enhanced grain yield by 86.0%, *T. undulata* by 48.8%, *A. albida* by 57.9% and *A. indica* by 16.8% over the control. Biological yield was also higher under trees than that in the open area. Soils under different tree canopies were rich in organic carbon content, moisture availability and nutrient status. Recent studies have found that multiple-use species such as *Bambusa nutans* have the potential to help in soil nutrient binding during restoration of abandoned shifting agricultural lands (jhum fallows) in northeastern India under *B. nutans*. A comparison of jhum cultivation and agroforestry suggests that the latter is an option to address the challenges of slash-and-burn26. A study of nutrient cycling, nutrient use efficiency and nitrogen fixation in *Alnus*–cardamom plantations in the eastern Himalaya found that nutrient standing stock, uptake and return were highest in the 15-yr-old stand. Annual N fixation increased from the five-yr-old stand (52 kg ha^{-1}) to the 15-yr-old stand (155 kg ha^{-1}) and then declined with advancing age. The agroforestry systems have the potential for improving water use efficiency by reducing the unproductive components of the water balance (run-off, soil evaporation and drainage). Integration of persistent perennial species with traditional agriculture also provides satisfactory drainage control to ameliorate existing outbreaks of salinity. Agroforestry systems can also be useful for utilization of sewage-contaminated wastewater from urban systems. Yield reductions may also be compensated in the long run by microclimate modification (Pandey, 2007).

The perennial tree based systems like natural grassland, silvipasture systems, agroforestry promote more biodiversity in soil as compared to comparison seasonal crops. Thus agroforestry systems have beneficial effect on soil biology which in turn increases yield of crop.

Vegetative Shelter

Under Indian conditions, the basic principle that should guide the dairy farmers or manager embarking on building a farm is to reduce heat gain and promote heat loss from the structure of animal house by radiation and conduction during summer. In rainy season, housing should promote evaporative cooling, thus saving animals from excess humidity resulting from precipitation. During winter, the structure especially those parts that come in constant physical contact with animals, like floor and walls should not get too cold and should give protection from cold winds. The loose housing system for dairy cows and buffaloes has generally been advocated but here too sufficient modifications are required for different regions and seasons of country the study of whose has been neglected.

A new housing system that will promote vegetation will be less stressful to the animal and will be eco-friendly especially for dairy buffaloes. It will be a sort of natural environment in which animal may found itself in zone of comfort. The possibility is that in vegetative shelter system the climatic stress will be considerably less resulting in better productivity. Thus vegetative shelter

- Improve microclimate i.e Temperature, THI and vapour pressure
- Physiological reaction i.e. Rectal temperature (morning, evening), skin temperature, respiration rate pulse rate. Increase in these parameter indicate more day time heat gain. Feed Intake and Efficiency of utilization i.e. Total feed intake (kg animal^{-1} day^{-1}), Feed intake (g wkg^{-1}), Feed intake (g) per kg FCM produced etc.
- Milk Yield and Composition i.e. average weekly milk yield and FCM Percent milk composition should be improved. To reduce the impact of hot weather on dairy cattle productivity, it is recommended to keep cows in barns with large side walls and roof opening, shade, fans and water sprinkler system, mistress cooling, sprinkling and fan evaporative cooling system in hot summer to increased daily milk yield. Thee milk yield had negative correlation with temperature humidity index (THI) and rectal temperature of cows.

- *Behavior* : Behavior of dairy cows and buffaloes was studied by Sharma and Singh, (2002). The analysis of shelter seeking behaviour revealed that excessive protection of closed housing resulted in considerable panting. No panting was observed in loose housing cows. It was observed that if temperature was less i.e., in summer and rainy season in night, open area was more preferred by cows. In loose housing cows Tree-shade was the third preference. By observing behavioural observations, it can be concluded that cows were more comfortable in loose housing in summer and rainy season while in winter season cows housed loose with central shed were more comfortable.

Leaf Meal

Leaf meal of legume forages such as *stylosanthes, lucerne* and top feed rich in protein such as *leucaena, sesbania, gliricidia* etc. are important ingredients o replace feed concentrates for livestock. Leaf meal can be easily produced and is profitable. Leucaena leaf meal (LLM) can replace mustard cake to meet up to 45% to the protein requirement. *Sesbania grandiflora* leaf meal can completely replace GNC nitrogen. Stylo meal prepared from *S. sebrana* species is rich in amino acids particularly lysine, cysteine and methonine and can be incorporated in the diet of livestock including poultry. Stylo meal was incorporated in the diets of sheep replacing concentrate mixture at 50 and 100% level. The leaf meal provides sufficiently enough nutrients for replacement of concentrate mixture in the ration of ruminants with reduced feed cost.

Fodder from Agroforestry

The overall requirement of livestock in dry zone have been estimated to be 186 million tones of dry matter, 79 million tones of TDN, 288 million M Cal, 12.9 million tones CP and 7.4 million tones of DCP.

Rao *et al* (2004) estimated feed availability in India. It was assumed that farmers aim at maximizing their returns from livestock, subject to feed availability. For our purposes, such an exercise helps in classifying zones on the basis of feed availability per livestock unit. The average feed availability from all sources on dry matter basis was 1.6 t per livestock unit. They further observed that crop residues were the most important feed source, contributing 45-70% of the total feed on dry matter basis, except in Zones 13 and 5c, where grasses from grazing lands provided more than 80% of the feed. The contribution of grasses from grazing lands was relatively high in the humid (MR, LI) region compared with all other regions. Fodder from forage crops was important in the irrigated zones with a few exceptions. The share of forage crops in total feed on dry matter basis was highest in Zone 2, which incidentally had low irrigation levels. In the rainfed areas with minimal irrigation facilities, fodder crops should be introduced in novel ways as, for instance, on field bunds, orchards and off-season planting, without affecting the main crop production. The availability of agro-industrial byproducts was above average in many SAT zones and in two humid zones.

The availability of grazing lands from forests, pastures, grassland and uncultivated arable land has been declining mainly due to encroachment of area under forests, pastures and public land by a rapidly growing population, and redistribution of common lands under poverty elevation programs; in part, a reduction in their biomass output due to degradation of forests and other grazing lands. The intensity of these phenomena and their impact again varied between different zones. They seem likely to affect ovine (which depend primarily on open grazing and very little on crop residues) more than bovine, and areas where animals are reared under nomadic or transhumance systems.

During the last 20 years, feed availability per livestock unit increased in nine zones and remained constant in others. Besides the irrigated zones, the increase was also in SAT (LR, LI) region due to an impressive growth in area under wheat and rice, with expansion of irrigation from a low base. While buffalo population increased, cattle population decreased, balancing the density of population to some extent. In the humid (HR, LI) region, paddy production increased, but wheat area was stagnant. Due to faster growth in cattle, buffalo, and small ruminant population, the density of LSU also increased (Rao *et al* 2004).

For a sole feed source a good quality forage should possess certain attributes viz., (i) high feed digestibility (i.e. 65%); (ii) soluble protein in the diet producing about 30 kg N ha^{-1} digestible dry matter (DDM); (iii) by pass protein providing about same amount of protein as arises from microbes (i.e. 200g protein/kg DDM intake); (iv) a level off at in diet approaching that of temperate pastures (i.e. 4-8%).

Trees can be grown either in combinations with agriculture crops under agro forestry system or on separate lands usually not fit for common agriculture. Since, trees have the capacity of growing under different conditions and in areas where it may not be possible to grow conventional crops, such as saline and waterlogged soils, steep and rocky mountain slopes and areas with severe climatic conditions. The fodder production potential of some trees at Jhansi is presented below.

About 60% of the total feed available to sheep and goat comes form top feeds only. This is more in goat since they spend more than 90% time in browsing on top feeds and hardly 10% to graze on surface vegetation that too when sufficient top feeds are not available. Chemical composition and nutritive value of some important tree leaves used as top feed are given in Table 2.

Table 2. Important top feeds growing in different agroclimatic zones of India

Tree species	Region	Fodder availability period		Preference tolivestock
		Leaf fodder	Pod fodder	
Acacia albida	SAR	May-Feb	Mar-June	Goat, Sheep, Camel
Acacia nilotica	AR,SAR, IGP, CZ,COZ	May-Feb	Apr-June	Cattle, Goat, Sheep, Camel
Acacia senegal	AR, SAR	Apr-Dec	May-June	Goat, Sheep, Camel
Acacia tortilis	AR, SAR	May-Feb	May-June	Goat, Sheep, Camel
Ailanthus excelsa	AR, SAR	May-Mar		Goat, Sheep
Albizia lebbek	AR, SAR, IGP, CZ, NER	Apr-Nov	Aug-Noy	Goat, Sheep, Cattle
Albizia procera	SAR, CZ, NER	Jul-Mar	Jan-Mar	Cattle, Goat, Sheep
Anogeissus latifolia	SAR,AR	Apr-Oct	Jan-Mar	Cattle, Sheep, Goat
Anogeissus pendula	SAR,AR	Apr-Oct.	Jan-Mar	and Camel Cattle, Goats, Sheep
Artocarpus lakoocha	IGP, NER, STZ	Apr-Noy		and Camel Goat, Sheep, Cattle
Azadirachta indica	AR, SAR, IGP, CZ, NER	Through out the year	July-Aug	Goat, Sheep, Cattle
Bauhinia purpurea	STZ, CZ, NER, IGP	Apr-Nov		Goat, Sheep, Cattle
Bauhinia variegata	IGP, NER, CZ, STZ	Apr-Nov		Goat, Sheep, Cattle
Dalbergia sissoo	SAR, IGP	Feb-Dec	Apr-Sep	Goat, Sheep, Cattle
Dichyostachys cinerea	AR, SAR	July-April	Apr-Aug	Goat, Sheep, Cattle, Camel
Ficus reliogosa	SAR, CZ, IGP	July-Nov		Cattle, Sheep
Gliricidia sepium	SAR	Mar-Nov	-	Goat and Sheep
Gmelina arborea	SAR, CZ, IGP	July-Nov		Cattle, Sheep, Goat
Grewia optiva	STZ	May-Nov		Cattle
Hardwickia binata	SAR, CZ	Apr-June	Feb-Mar	Cattle, Sheep, Goat

Contd...

Table 2. Contd...

Leucaena leucocephala	SAR, IGP, CZ	Through out the year	Dee-May	Cattle, Goat, Sheep
Melia azedarach	!GP, SAR, STZ	July-Nov		Goat, Sheep, Cattle
Morus alba	!GP, CZ, STZ	Through out the year		Cattle, Goat, Sheep
Pithecellobium dulce	SAR	Through o ut the year		Cattle, Goat, Sheep
Prosopis cineraria	AR, SAR	Through out the year	May-June	Goat, Sheep, Camel, Cattle
Quercus incana	NER	Mar-Nov		Cattle, Sheep, Goat
Robinia pseudocassia	STZ	Mar-Nov		Cattle, Sheep, Goat
Sesbania grandijlora	IGP, CZ, NER	July-Nov		Cattle, Sheep, Goat
Sesbania sesban	SAR, IGP, CZ, NER	Through out the year	Dec-Mar	Cattle, Goat, Sheep
Tecomella undulata	AR, SAR	Mar-Nov.	Apr-June	Cattle,-Sheep, Goat, Camel
Terminalia belerica	COZ, STZ	July-Nov		Cattle, Sheep
Terminalia tomentosa	COZ, STZ	July-Nov	-	Cattle, Sheep
Ziziphus nummulasia	AR, SAR	Through out the year		Cattle, Goat, Sheep

(AR= Arid region, SAR= Semi arid region, IGP", Indogangetie plains, CZ=Central zone, COZ= Coastal zone, NER=North eastern region, STZ= Sub-temperate zone.) Cited from Rai et al. 2007

Table 3: Fodder production potential of some trees at Jhansi

Species	Year of Growth	Dry fodder yield (kg/tree)	Pod yield (kg/tree)
Acacia tortilis	7	3.2	-
Albizia amara	7	13.5	-
Albizia lebbek	6	5.4	4.5
Albizia procera	6	11.3	3.2
Dichrostachys cinerea	6	0.8	0.7
Hardwickia binata	10	3.7	-
Leucaena leucocephala	3	0.8	-
Leucaena leucocephala	4	1.9	-
Leucaena leucocephala (K8)	1.5	1.5	-
Sesbania grandiflora	3.5	0.3	-
Sesbania sesban	3.5	0.3	-

Source: Deb Roy *et al*, 1997, Gupta *et al*, 1983, Pathak and Patil, 1980, Rai *et al*, 1984, Roy *et al*, 1984, Roy, (1991) Cited from (Kundu *et al*, 2005).

Access of CPRs

In different ecosystems, CPRs can be major source of fodder for livestock. Maggs and Hoddinott (1997) noted that in developing countries, common property resources (CPRs) can be an important source of income for certain individuals within households. If a change in the management of CPRs imposes costs on these individuals, or causes a decline in the prices or productivities associated with goods produced from the CPRs, the intra household allocation of resources may alter in a manner detrimental to those individuals. The sustainability of CPRs is influenced by local people who are largely dependent on natural resources (fuel, fodder, minor forest products) for their livelihood. Qureshi and Kumar (1998) found that common lands, being a significant form of natural resource endowment in many developing countries, play a vital role in maintaining the ecological balance, and more particularly in supporting the people, especially the rural poor, in eking out their livelihood. The contributions of these lands in terms of variety and extent were many times higher in the households belonging to less developed than developed regions, and those households belonging to poor than wealthy classes.

Singh, (1994) concludes that a significant proportion of the total endowment of natural resources, comprising land, water, forests and fisheries, is used in common by people and are known as common pool resources (CPR). Lack of equitable access to and distribution of benefits from CPRs often lead to encroachment or appropriation of these resources, forcing an opportunistic and exploitative mode of resource use. In West Bengal, India, some of thc CPRs are part of a class conflict which is likely to increase in the future. Access to these resources is declining as CPRs are increasingly commercialized or becoming more scarce. The most important CPRs for the poor are not the grazing lands, but plants, fuel, fish and harvested grains (Beck, 1994). In the coastal areas of Orissa, India, the proportion of CPR land to the total area of the villages was 20%, and the persons per hectare of CPR land numbered 70. High density of population in these areas puts tremendous pressure on the natural resources. The area of the CPR lands in the villages has declined by 40% since 1950. The main cause of decline of CPRs is privatization. Transfer of CPR land to poor people through various social welfare programmes for their private use, and illegal occupation of the CPR land leading to subsequent legalization, were two important factors which resulted in large-scale privatization. The rural poor depend on CPRs for food, fodder, and fuel, in a variety of ways. Due to privatization, they have collectively lost a significant part of their source of livelihood in recent years (Sahoo and Misra, 1994). Sekar, 1999 observed that in Tamilnadu, about 17% of the area is under CPLRs in Tamil Nadu and in only three districts are CPLRs more than 10%. Per caput CPLR is very low (0.042 ha) in Tamil Nadu as against the national average (0.178 ha). Fluctuations in CPLRs are probably due to the poor attention paid to their conservation by individuals and public institutions, and non-existence of conceptual clarity and consistent techniques to estimate their magnitude.

The study was conducted to know the access of CPRs by different categories of farmers in three ecosystems of India i.e. arid-ecosystem, semi-arid ecosystem and sub-humid ecosystem. In arid ecosystem 56.79, 3.57 and 16.07 per cent farmers were using CPRs for fodder, other uses and rail and road side land respectively. The percentage of landless users was relatively less in comparison to other categories. In semi-arid ecosystem 48.60, 42.06 and 26.17 per cent farmers were using CPRs for fodder, other uses and rail and road side land respectively. In sub humid ecosystem 57.41, 3.70 and 7.41 per cent farmers were using CPRs for fodder, other uses and rail and road side land. In last two ecosystems, there was better access to landless farmers. In all ecosystems, the all categories of farmers were dependent on CPRs and relatively more percentage of resource rich farmers was accessing the CPRs compare to resource poor farmers. The rail and road side CPRs were better accessible to resource poor farmers in all ecosystems. It could be concluded that in arid ecosystem, the access of different CPRs for the resource poor farmers should be improved for fodder, timber, fuel wood and non-timber minor forest produce (Sharma *et al.*, 2008).

Meeting the Challenge

Opportunities for meeting the challenge of augmenting the feed and fodder situation lies in the wastelands and degraded lands, employing the technological interventions to increase the productivity of land besides others. The arid and semi arid regions need to be strategically developed to meet the current requirements of livestock together with making contingency plan to overcome the recurrent drought.

Forage from Boundary of Holding

The number of holding and the size of holding in different states falling under various arid and semi arid regions were worked out on the basis of the respective states. Taking the average size of the holding to be 2 ha, total number of holding was arrived at and the perimeter of all the holdings (800m per 2 ha) was calculated. Taking average spacing of 4 m per tree and av. Yield to be 5 kg dm / tree potential forage yield from the boundary of the holding was determined.

Forage from Holding Adopting Tree-grass Planting

Each holding of 2 ha is divided into 8 sub-plots of 0.25 ha each, on the boundary of which tree-grass is planted in 1m width with trees spaced at 4 m. This provided 600m of length inside the holding to be planted with tree-grass. Thus there would be 150 trees per holding (2 ha) producing 7.5 q dry matter and 6.0 q of dm obtained from grass per holding.

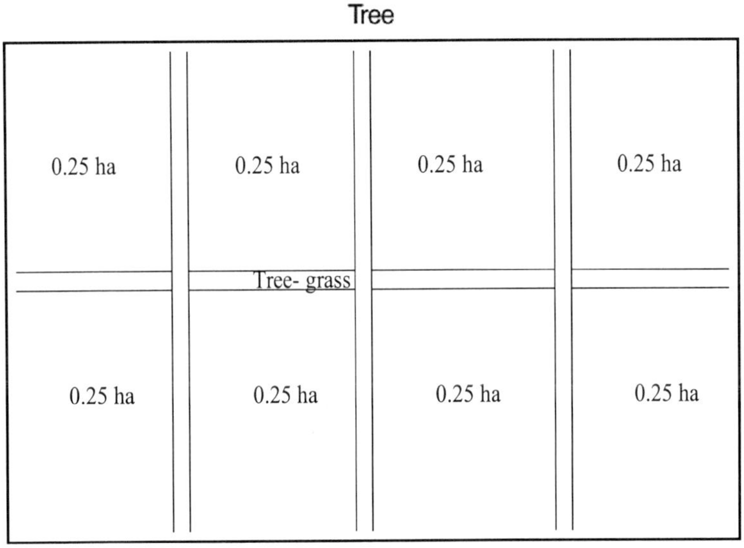

Tree

0.25 ha 0.25 ha 0.25 ha 0.25 ha

Tree- grass

0.25 ha 0.25 ha 0.25 ha 0.25 ha

Table 4: Potential production of forage from tree, tree-grass in arid region (million tones)

Region	Fodder from tree on boundary of holding	Fodder from tree inside holding	Grass inside holding
Arid Region	**8.01**	**6.67**	**5.34**
Cold Arid Region	0.16	0.13	0.11
Hot Western Plain Arid Region	6.98	5.82	4.65
Hot Deccan Plateau Arid Region	0.87	0.73	0.58
Semi Arid region	**19.37**	**16.14**	**12.91**
North central highland	5.39	4.49	3.59
Central Malwa highland	3.02	2.52	2.01
Deccan Plateau	5.46	4.55	3.64
Telangana Plateau & E. Ghat	3.02	2.52	2.02
E. Ghats & Daccan Karnataka Plt.	2.47	2.06	1.65
Grand Total	**27.38**	**22.82**	**18.25**

Adopting the above model in arid regions will yield about 20 million tones of dry matter which will not only help in overcoming the shortage of feed and fodder in the current year but also generate surplus to be used to tide over the drought years.

Proposed Land Use

The proposed changes in the land use pattern can fulfill the major part of DM requirement in dry zone. The 50 per cent area under forest, shrubs, grazing lands and wasteland if put to improved management practices and could yield DM of 5 t yr^{-1} in arid region and 8 t yr^{-1} in semi -arid regions, would satisfy the requirement. The proposed changes in land use and management practices are within the manageable range (Table 5).

Table 5 : Potential forage biomass production from silvopastoral system in dry zone (000 t)

Region	Forest	Shrubs	Grazing lands	Wastelands	Total
Arid Region	1646	724	2348	25206	29925
Cold Arid Region	34	41	16	5663	5754
Hot Western Plain Arid Region	1337	468	2282	18211	22297
Hot Deccan Plateau Arid Region	275	216	50	1332	1873
Semi arid region	25088	10348	4505	46654	86595
North central highland	4701	1123	2182	12796	20803
Central Malwa highland	3794	355	1042	7395	12587
Deccan Plateau	5397	1025	923	12595	19940
Telangana Plateau & E. Ghat	6649	6649	137	8550	21986
E. Ghats & Daccan Karnataka Plt.	4546	1196	220	5318	11279
Grand Total	26734	11072	6853	71860	116520

Meeting the Summer Fodder Need by *Stylosanthes* Leaf Meal

The significance of *Stylosanthes* species in grass intercropping and in rotation has been realised. *Stylosanthes* may nurse the companion crops by atmospheric nitrogen fixation and can serve as much to control weed and reduce evaporation due to its matting habit apart from providing nutritious fodder. The leaf meal can be good source of dry fodder during the scarcity period.

Strip Cropping in the Arid Region

In arid regions where the problem of shifting sand dunes prevail, strip cropping of grass (3 m) and crops (6 m) should be practiced and the strips area interchanged after three years will minimize the problems of sand dunes affecting crop production.

Fodder Conservation

To overcome the problems of preservation, storage and transport, hay Making feed pelleting, Densification/ Baling of Forages technology / equipment's are used for handling of forage and crop residues:

Conclusions

The analysis indicated that there is shortage of DM for the livestock of the region. However the situation can be managed by adopting the various technological options related to production, conservation, animal husbandry and land use pattern.

References

Beck, T. 1994. Common property resource access by poor and class conflict in West Bengal. *Economic-and-Political-Weekly, 29(4): 187-197.*

Kundu, S.S., Singh, Sultan, Mahanta, S.K. and Pailan, G.H. 2005. Feeding Farm animals. SSPH, New Delhi.

Maggs, P and Hoddinott, J .1997. The impact of changes in common property resource management on intrahousehold allocation. FCND Discussion Paper Food Consumption and Nutrition Division, International Food Policy Research Institute. (34): iv + 26

Pandey Deep Narayan .2007. Multifunctional agroforestry systems in India. *Current Science, 92(4): 455.*

Pathak, P.S. 2003. Prospects of feed crops in India: the role of CGPRT crops. CGPRT centre, UN/ESCAP.

Pathak, P.S. and Sunil Kumar, 2004. Forage and grazing resources in different agro-climatic Regions of India. In: Buffalo Production under Different Climatic Regions (S.S. Kundu, A. K. Mishra and P. S. Pathak, eds.), IGFRI, Jhansi. Pp 1-41.

Qureshi, M. H. and Kumar, M. Suresh.1998. Contributions of common lands to household economies in Haryana, India. *Environmental-Conservation 25(4): 342-353.*

Rao, P. Parthasarthy, Birthal, P.S., Kar, Dharmendra, Wikramratne, S.H.G., Shrestha, H.R. 2004. Increasing productivity in mixed crop -livestock systems in South Asia. ICRISAT, Patancheru 502324, AP, India.

Rai, P., Ajit and Samanta, A.K..2007. Tree Leaves, their Production and Nutritive Value for Ruminants: A Review. *Animal Nutrition and Feed Technology,* 7: 135-159

Sahoo, H.P and Misra, M.K. 1994. Village ecocomplex functioning with common property resources: a case-study on coastal Orissa, India. *Environmental Conservation, 21(1): 57-61.*

Sekar, C.1999. Common property land resources conceptualization, magnitude and issues: an experience from Tamil Nadu. *Artha-vikas, 35(1): 48-61.*

Sharma, Purushottam and Singh, Kiran. 2002. Shelter seeking behaviour of dairy cattle in various types of housing systems. *Indian Journal Animal Science, 72(1): 91-95.*

Sharma Purushottam, Pandey Sadhna, Meena B.S., Khan T.A., Biradar N., Radotra S., Ray , A.K. 2008. Access of CPRs for various categories of farmers in different ecosystems. National symposium on agroforestry knowledge for sustainability, climate moderation and challenges ahead , 15-17 December , Organized by Indian Society of Agroforestry and NRCAF, Jhansi, Pp 208.

Singh, Katar. 1994. Managing common pool resources: principles and case studies. Managing common pool resources:-principles and case studies. Oxford, UK: Oxford University Press, xx + 366.

Singh, Punjab; Pathak, P.S.; Roy, M.M. 1994. Agroforestry systems for land use. Oxford and IBH Publishing Co. PVT. LTD. New Delhi.

Yadav, R.L. and Subba Rao, A.V.M. 2001. Atlas of cropping systems in India. PDCSR, Modipuram, Meerut-250110, India.

Sources of data: Wasteland Atlas of India, NRSA, 1993; Forest Survey Report, 2002-03 and Wastelands in India-An untapped potential by Bagchi & Phillip, 1996, various Annual Reports which is duly acknowledged.

□□□

Agroforestry: Natural Resource Sustainability, Livelihood & Climate Moderation © 2009
Editors: OP Chaturvedi, A Venkatesh et. al., pp. 401-422
Satish Serial Publishing House, Delhi

Chapter 28

TOXIC CHEMICALS IN FEED AND FODDER AND THEIR IMPLICATION IN ANIMAL AND HUMAN HEALTH

Subir K. Nag
Indian Grassland and Fodder Research Institute, Jhansi-284 003 (U.P.)

Introduction

Animal feed stuffs like crop residues, agricultural by products (eg., oilseed cakes, chunnies), grains, green fodders, herbage etc. may often be contaminated with diverse type of toxic chemical substances like pesticides & POPs which include industrial pollutants such as dioxins, heavy metals, radionuclides, mycotoxins, nitrates and others.

Pesticides owing to their toxic potential attract much concern among the people world wide because residues are found present in all sorts of edible commodities as well as in each compartment of the environment. Pesticide residues have been reported to be present in wide range of animal feedstuffs as a result of direct application at the time of growing the crop in the field and during storage or indirectly, through unintentional spray, drift etc. Other persistent organic pollutants (POPs) like PCBs, PCDDs, PCDFs, etc. are some of the toxic industrial pollutants present in the environment which also contaminate the feeds particularly herbages.

Pollution of the environment with heavy metals like Pb, Cd, As, Hg, Se, Ni etc. result from various anthropogenic and industrial activities like burning of fossil fuel, mining, metallurgy industries, transport sectors, irrigation, fertilizers etc. which redistribute these toxic substances in to the environment. Plants grown in such environment accumulate these metals in their tissues and transfer them to man and animal.

Sometimes animal drugs and chemicals are also used as feed additive to promote growth, improve feed efficiency, breeding performance and enhance feed acceptability. Residues of these substances in food animals can have toxicological significance for consumers.

Presence of these varying types of toxic substances in feed material is undesirable because from there they enter into the animal body system and redistribute themselves in different tissues, organs and deposited there for a long time due to their persistence nature. They directly affect the growth of animals, incite many diseases, affect and reduce the immune system, reduce the reproductive efficiency which ultimately leads to loss of production efficiency. They generally combine with one or more bioactive ligands such as -OH, $-COO^-$, OPO_3H, >C=O, -SH, -S-S-, $-NH_2$ and –NH that is essential for normal physiological function and activation of several enzymes. Most importantly they contaminate human food chain like milk, meat etc. and thus enter human body and pose serious health problems.

Pesticide Residues

Most of the pesticide molecules and/or their metabolites are toxic xenobiotics and are reported to have toxicity related health problems like carcinogencity, teratogenicity, mutagenicity, genotoxicity, neurotoxicity and many other adverse effects. Over the years there has been a mounting fear and concern that indiscriminate and disproportionate use of pesticides may lead to harmful and ill effects in human beings and animals and may have deleterious effect on the environment. The term 'Pesticide Residues' has thus become much talked about environment related issue infamously popularized in the society and a matter of dispute in trade and commerce both nationally and internationally. The term 'Pesticide residue' may be defined as any specific substance in food, agricultural commodity, or animal feed resulting from the use of a pesticide. The term includes any derivative of a pesticide such as conversion product, metabolite, reaction product and impurities considered to be toxicological significance. Whenever pesticides are applied on a crop or commodity or on any substrate to control pests, diseases, weeds etc. they invariably leave toxic residues on the substrate on which they are applied. The amount of residues left depends on the nature of the chemical, dose and volume of application, nature of substrate and prevailing weather condition. The occurrence of pesticide residue in ecosystem may have a threefold impact on man, by reducing the reproductive capacity of his environment or by contamination of either his food or work environment. Man and animals are exposed to low level of pesticide residues in air, water and food chain for a prolonged period of time, which pose problem to all animals, birds and other organisms and also human being. Man is the ultimate consumer of pesticide. As we see in the pesticide cycle, residues of pesticides in the grains, pulses, oil and oilseeds, milk, ghee, coffee, fish, meat etc. is ultimately consumed by man. Through fodder, water, air and other foodstuffs pesticide residues reach animals and then through milk and/or meat products it is accumulated in the body of human being. This accumulation is five times higher in man in comparison to crops. The bioaccumulation and biomagnifications of certain pesticide residues in human food chain is an important aspect of pesticide induced environmental pollution, which needs attention.

The presence of pesticide residues has been detected in various items and in food chain. Among the different group of pesticides organochlorines are most persistent in nature. Being lipophilic they are deposited in body fat of man and animal. Such pesticides have an affinity for lipid material and are not easily biodegradable. So they cause contamination of animal of animal products like milk, meat, fish, egg, ghee, butter etc. Animal products contribute 70 - 90% of the total oragnochlorinated pesticides in the diet of an Indian, particularly milk and milk products (Kannan *et al.*, 1992). The residues of these pesticides in animal products also decline at a very slow rate even after the sources of contamination are eliminated. All the feed resources like cereal grains, oilcakes, brans, legume and nonlegume forage crops, crop residues are found to be contaminated with organochlorinated and other pesticides.

Pesticides can enter into the animal body in three possible ways i.e. (i) through contaminated feed and fodder (ii) Through contaminated water and (iii) through skin pores when sprayed or drenched to animal body surface for killing ecto parasites (Fig. 1).

The main source of entry is the contaminated feed and fodder offered to animals for consumption. The feed materials may be contaminated by pesticides either at the time of cultivation of the crops in the field, from which they have been prepared like oilseed crops (mustard, linseed, sesamum, cotton etc.) in case of oilseed cakes, cereals (paddy, wheat, maize, etc.) in case of straw, chuni and hay. During the storage of the feed materials, grains, seeds etc. in the warehouse, godowns and other places pesticides are used to keep away the storage pests and diseases. Green fodders and/or dual purpose crops like sorghum, maize, bajra, oat, berseem, cowpea, different grasses grown in the field are often sprayed with pesticides to save them from the onslaughts of diseases and pests. This may act as direct source of contamination because many a times fodder crops after being treated with pesticides are fed to animals straight way without waiting sufficient time for the residues to go down below maximum residue limit. Pesticides are also used in a large scale for their useful properties to kill ectoparasites like ticks, mites, insects on the animal body. As a result they may enter into the animal body through skin pores when applied by drenching. Water sometimes may also be a source of contamination although concentration may not be such as to worry about. After entry into the animal body pesticides are distributed in different tissues and organs. Since many of them particularly organochlorines and synthetic pyrethroids are lipophilic in nature, they have a tendency to accumulate in the adipose tissues and usually they do not get converted into water soluble metabolites. As the concentration in the fatty tissues exceeds a threshold level, the excess residues are translocated into the lactating glands of the animals from where they are secreted in the milk. Because of special position of milk in the diet of infants and children the presence of significant amount of residues in milk is undesirable. That is why the residue limit for milk and dairy products tends to be more severe than those for other products.

Pesticide Residues in Feed

Feed is main source for the contamination of animal body with pesticide residues. As discussed earlier, feeds coming as by products of agricultural crops such as straw, oilseed cakes, chunnies, brans are contaminated in the field in many ways directly or indirectly by pesticides. Apart from applying pesticide to the growing crops in the field, spraying of pesticide on feed concentrate bags in godowns for controlling storage pest is also a common practice which can also lead to contamination. The readymade feed concentrate sold in the market is actually a mixture of many grains and jaggery/molasses which many contain pesticide residues. Rice and wheat straw are one of the important feed materials used in India for cattle and buffaloes. Rice consumes second largest share of total pesticide (17%) consumed in India. Rice bran, a rich source of HCH and other residues is used as an ingredient in the readymade feed concentrate. Although wheat crop as such does not consume much pesticide but during the storage of wheat straw, pesticides are used to prevent the attack of insects, termites etc.

Numerous example of the contamination of animals through animal feed have been reported in the literature (Von Houweling *et al.*, 1977). But, public awareness of this did not become acute until 1974 when polybrominated biphenyls (PBD) were associated with an accident in Michigan in USA. The direct impact of this incident resulted in the destruction of 30,000 cattle and 6000 swine as well as sheep, poultry and dairy products. In USA screening of random samples of feed comprising of grains, cereals, by products and miscellaneous for organochlorine insecticides like lindane, â-BHC, aldrin, dieldrin, pp[1]-DDE, op[1]-DDT, pp[1]-DDD, pp[1]-DDT, heptachlor, and its epoxide revealed average contamination levels of less than 10ppb (Pierson *et al.*,1982).

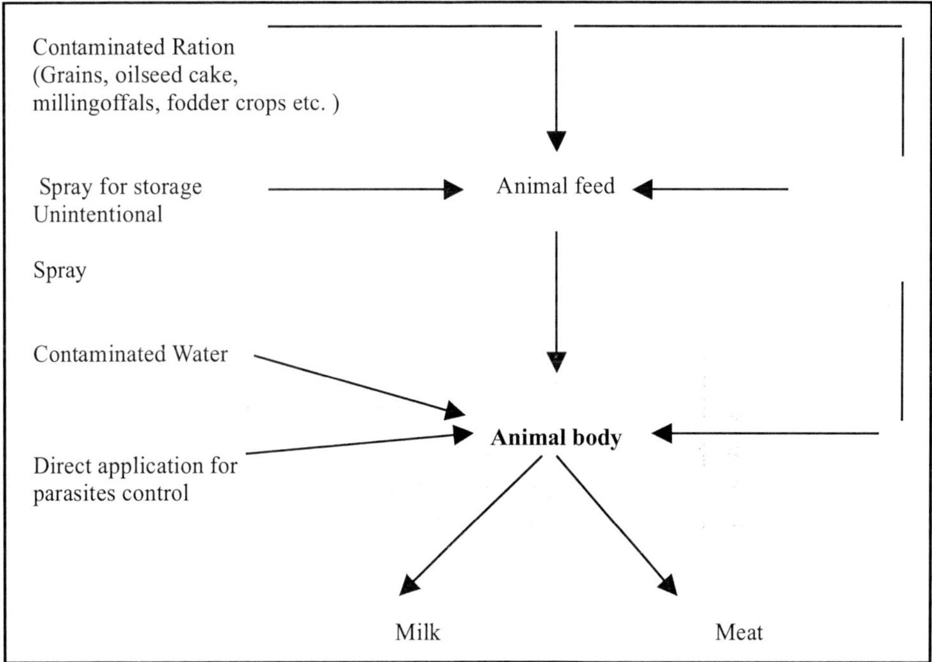

Fig. 1: Possible sources of contamination of animal body and animal products with pesticides.

Information regarding the status of pesticide residues in feed and fodder crops in India are very scanty unlike those for food and horticultural crops for which plenty of work has been done. Only little information is available about the status of residues in feed and fodder crops that too mostly concentrated on organochlorines like DDT and HCH, which are summarized below.

Contamination of cattle feeds by DDT and HCH in some species of fodder grass was reported by Kaphalia and Seth (1982). Shastry (1983) depicted results of monitoring of different categories of feed samples like poultry and cattle feed, different feed ingredients, hay, grains, cereals, pulses, oilseeds and some green fodders. Out of 1241 samples tested 747 (60.19%) were found to contain residues of two the then most important insecticides DDT and BHC (HCH), along with endrin and lindane in some samples.

Punjab as a leading state in the era of green revolution benefited maximum by implementing the technologies and boosting the production. But the state has also suffered maximum due to the negative impacts of green revolution like degradation of soil health, loss of soil fertility and productivity, lowering of ground water table, contamination of soil and water, and adverse effects of large scale use of pesticides on man and environment. Widespread and high level contamination of different food and agricultural commodities with pesticide residues has been reported many times from different sources. Singh *et al.*, (1988) determined insecticide residues in wheat grains and straw arising from their storage in premises treated with BHC and DDT under malaria control programme in Punjab. All the 120 samples of wheat grains collected from BHC as well as DDT sprayed districts contained residues of both HCH and DDT at varying concentration of 0.01-5.18mg kg^{-1} and 0.01-2.14 mg kg^{-1} respectively. Analysis of 40 wheat straw samples also showed presence of BHC (mean concentration 0.24 mg kg^{-1}) and DDT (1.35 mg kg^{-1}) residues in all the samples. Again, Battu *et al.*, (1996) on monitoring of 105 samples of different feeds collected form different places of Punjab found 100 and 80% contamination with residues of HCH and DDT respectively. Feed concentrates, straw and green fodder showed total HCH residues comprising α, β, γ and ä isomers up to 0.26, 2.28 and 0.69 mg kg^{-1} and DDT residues up to 0.28, 6.70 and 0.35 mg kg^{-1} respectively. The tolerable content of 0.02 mg kg^{-1} β-HCH and 0.10 mg kg^{-1} ÓDDT in the feed samples (dry weight basis) were exceeded in about 35% samples, while those of α, γ and δ HCH exceeded in about 3% samples. In addition to HCH and DDT, residues of endosulfan and malathion were also detected in a limited number of feed concentrate samples out of a total of 17 samples tested by Kang *et al.* (2002) at Ludhiana (Punjab). Dicofol, an acricide organochlorine group, was also detected in two feed concentrate samples. Limiting value of β, γ and δ HCH and total DDT were exceeded in about 13, 6.5, 65 and 9.7% of animal feed samples on dry weight basis. However these results showed substantial decline in pesticide residues in feed samples in Punjab with respect to β-HCH and total DDT as compared to earlier report of 35% samples exceeding the limiting values of two insecticides (Battu *et al.*, 1996).

Analysis of 32 cattle feed samples collected from local market of Lucknow revealed presence of α, β and γ isomer of HCH in all the samples and the mean concentration of total HCH was 0.393 ppm (Dikshit *et al.*, 1989b). Residues of pp^1 DDE, op^1 DDT, pp^1 DDD and pp^1 DDT were also present in all the samples and the total DDT content comprising of all these isomers was

0.236 ppm. The level of aldrin was found to be 0.156 ppm. Endrin, one of the most toxic organohlorines, was also present (0.02ppm) in all the feed samples along with α-endosulfan (0.029 pm). Dikshith *et al.* (1989a) in their attempt to quantify residues of different organochlorine pesticides in oilseeds used in the country found that mustard, groundnut and sesamum seeds obtained from Lucknow were contaminated with DDT to the extent of 130-3200 ppb and HCH to the extent of 91-250 ppb. From down South Unnikrishnan *et al.* (1998) also reported that BHC, DDT and endosulfan were detected in samples of cattle feed and feed ingredients collected from Bangalore. Total BHC content of the compounded feeds ranged from 0.0076-0.1197 mg kg^{-1}, DDT ranged from traces to 0.01 mg kg^{-1}. The active isomer γ–BHC constituted only a minor portion of total BHC content in most of these samples. In groundnut cake samples the predominant contaminant was DDT, while BHC was the major contaminant in compounded feed.

In a multilocational study conducted by All India Coordinated Research Project on Pesticide Residues (Agnihotri, 1999) in 15 locations on straw to determine HCH and DDT residues it was revealed that out of 377 samples analysed, most samples were found to contain residues of both HCH and DDT. The mean value was from <0.01-0.60 mg kg^{-1} for DDT and <0.01-2.27 mg kg^{-1} for HCH. The presence of residues of HCH and DDT in straw and feed in almost all samples of different locations in amounts above the prescribed limit proved that they were a potential source of contamination in milk (Agnihotri, 1999). HCH and DDT residues were also monitored in certain dry fodder (pearl millet, wheat and barley) and feed concentrates (sesamum, mustard, and cottonseed cake) collected from Jaipur in Rajasthan (Gupta *et al.*, 2000). Although no DDT residue was detected in any of the samples but HCH residues was present in the range of 0.3-1.73 mg kg^{-1} in feed concentrates. In case of dry fodder wheat and barley had no residues of HCH but pearl millet samples were contaminated with HCH in the range of 0.06-0.11 mg kg^{-1}.

Monitoring of HCH residues in feed samples from Hisar, Haryana showed that all four isomers of HCH i.e. α, β, γ and δ were present in mustard cake, cottonseed cake, churi and wheat straw. The mean concentration of total HCH varied 0.2 to 0.76 mg kg^{-1} (Singh *et al.*, 1997). In another study conducted at Karnal, Haryana (Prasad and Chhabra, 2001) it was found that contamination in feed concentrates and roughages was caused by BHC (73.3%), endosulfan (8%), heptachlor (8%) DDT (4.2%) and aldrin (2.6%). The content of BCH residues was highest in concentrate mixture followed by maize, groundnut cake, wheat bran, rice bran, cotton seed cake, barley, and mustard cake. Among the roughages it was highest in maize fodder followed by oat, berseem, wheat straw, lucerne and paddy straw in the decreasing order. Among the different isomers of HCH, α and γ-isomers were more than β and δ. The two isomers of DDT (op^1 and pp^1) were present in all the forages and in most of the concentrate feeds as compared to fodders. Again, Sharma *et al.* (2005) in a multiresidue monitoring programme of feed concentrates done at Karnal, found DDT contamination in oilseed cakes, groundnut husk, wheat straw and fodder to the extent of 0.107, 0.228, 0.37 and 0.05 mg kg^{-1} respectively.

Feed samples collected from Jorhat district of Assam shown presence of lindane or γ-BHC (0.025-0.041 mg kg^{-1}), β–endosulfan (0.0028 mg kg^{-1}) and endosulfan sulfate (0.045-0.049 mg kg^{-1}) residues in them (Deka *et al.* 2004).

Monitoring of feed and fodder samples for multiresidues of different groups of pesticides like organochlorines, organophosphates & carbamate and synthetic pyrethroids have been carried out in 533 feed samples (oilseed cakes, straw, chunnies. compound feed etc.) collected from different places of Bundelkhand region (Nag and Raikwar, 2006). Out of them 301 (56.47%) samples were found positive containing residues of different organochlorinated pesticides like HCH isomers, DDT complex, endosulfan. The mean total HCH concentration varied from 0.0099 to 0.3055 mg kg^{-1} in different samples. Among different isomers of HCH the concentration of b-HCH was highest. Endosulfan (a,b and sulfate) was detected in the range of 0.0095-0.237 mg kg^{-1}. Total DDT varied from 0.0161 to 0.1179mg/kg. Dicofol, an acaricide and DDT analogue, could be detected in very few samples. In samples positive with HCH isomers and DDT complex the concentration of residues was found to exceed the respective limiting values in very limited samples. Otherwise they were mostly within safe limits. Contamination of feed samples with organophosphate (OP) pesticide residues was lesser as out of 533 samples 274 (51.4%) were contaminated. Out of ten selected OPs, residues of mainly chlorpyriphos, malathion and quinalphos were detected. Others like dichlorvos, phorate and dimethoate were also found but less frequently. The concentration of different OP compounds in the contaminated samples was very less in most of the cases and the range was found to be from traces to 0.8 mg kg^{-1}. In case of synthetic pyrethroids (SP) contamination was still lower as 264 (49.53%) samples were contaminated with SP residues, mainly contributed by cypermethrin, fenpropathrin, cyhalothrin, fenvalerate and deltamethrin. The concentration of cypermethrin in some mustard and cotton seed cake samples was quite high (about 2 mg kg^{-1}). Otherwise, it was mostly low as envisaged from the range of concentration in the positive samples which varied from traces to 0.5 mg kg^{-1}. Among the different type of feed samples mustard, linseed, cottonseed cake, gram and lentil chunni samples were found to be more contaminated than others.

Green Fodder

Work on green fodder crops with respect to pesticide residues, their dissipation pattern and safety evaluation has been very limited. Endosulfan, the recommended and commonly used insecticide in forage crops, was evaluated for persistence of residues and safe-waiting period following application at the recommended dose in fodder crops like sorghum, cowpea, lucerne and maize. The residual half-life around 3-4 days and safe-waiting period around 9-14 days have been recommended (Nag and Shah, 1999, 2001, 2002, Raikwar *et al.*, 2003). Carbendezim, the systemic fungicide recommended and used in fodder crops was also evaluated for persistency in sorghum (Raikwar and Nag, 2004). Residual half-life of 6days was observed and safe-waiting period around 9-10 days was recommended.

The status of pesticide residues in green fodder can be assessed only from the very few monitoring reports available on the subject. Gupta *et al.* (2000) monitored HCH and DDT residues in certain green fodder crops like pearl millet, guinea grass, mustard, fenugreek, oat, chinese cabbage and lucerne collected from Jaipur (Rajasthan) and they concluded that all the samples contained HCH residues above MRL. However, DDT residue was not detected in any sample. Prasad and Chhabra, (2001) analysed few green fodders such as maize, oat, sorghum, berseem and lucerne collected Karnal, Haryana for organochlorine pesticide residues. According to their results, total HCH content was abnormally high ranging from 15-44 mg kg^{-1} DM basis, lowest in lucerne and highest in maize. Total DDT and endosulfan residues ranged from 1.3-2.8 mg kg^{-1} and 1.9-5.9 mg kg^{-1} respectively. Fourteen green fodder samples collected from Ludhiana district of Punjab were also evaluated for pesticide residues by Kang *et al.* (2002). All the 14 samples contained HCH residues (range 0.03-0.82 mg kg^{-1}) but in only two and three samples total DDT (BDL- 0.18 mg kg^{-1}) and endsulfan (BDL-1.58 mg kg^{-1}) residues were detected. Among the OPs, malathion was found in eight samples.

Sorghum, maize, bajra, oat, berseem and sugarcane top collected from different places and locations of Bundelkhand region were monitored for pesticide residues (Nag and Raikwar, 2006). About 50% samples were contaminated with different organochlorine pesticides like DDT, HCH and endosulfan ranging from Tr-0.8310 mg kg^{-1}, 0.0067-0.0290 mg kg^{-1} and Tr-0.6985 mg kg^{-1} respectively. In case of organophosphate group of insecticides, the contamination level was 45%. Malathion, chlorpyriphos and quinalphos were mainly detected. Synthetic pyrethroids particularly cypermethrin, cyfluthrin and fenvalerate were found present in some samples and overall contamination was 42%. However in grasses like setaria, guinea, dinanath, sirsa, paragrass, Bothriochloa, hybrid napier and fodder cowpea collected from Jorhat district of Assam no detectable residues of any pesticide was found (Deka *et al.* 2004).

POPs

Persistent organic pollutants (POPS) include OC pesticides, polybrominated diphenyl ethers (PBDEs), polychlorinated dibenzo-p-dioxins (PCDD), polychlorinated dibenzo furans (PCDF), polychlorinated biphenyls (PCB) etc. PCBs were introduced in 1929 and were first detected in milk of cow in 1970. They are extremely stable and undergo biomagnifications, i.e., their bioaccumulation increases at every trophic level. Transformer, capacitors, similar electrical appliances are the suspected sources of PCBs although there are few other sources like ship breaking activity, weapons, aircrafts, tanks and explosives. The presence of PCBs in milk has been traced back to cattle feed and other sources such as silos coated with sealant or paint, oils and grease containing PCBs, sewage and sludge application to agricultural pasture. PCBs are lipophilic in nature, which makes their elimination from milk and milch animals quite necessary. Dioxins and dibenzo furans are formed as a byproduct from many different industrial chemical processes where chlorinated products are used. These chlorinated hydrocarbons enter the milk presumably through the contamination of the grass with smoke particles. Both these compounds are lipophilic, chemically stable and of low volatility. The dioxins and furans preferentially accumulate in the liver of the organism as revealed by the liver adipose tissue ratio of different organisms in different studies. The most known and toxic in this group is 2,3,7,8 tetra chlorodibenzo-p-dioxin. POPs are transferred across the borders and are accumulated in animals such as polar bears and seals. POPs are causing contamination on global scale. To prevent or reduce global environmental contamination caused by these compounds through cooperation among countries around the world, the Stockholm convention on POPs was adopted in May, 2001 and came into effect in May, 2004. The occurrence, transport and distribution of dioxins and POP compounds depend on physico-chemical and biochemical properties. Soils in the wide open municipal dumping sites were found to have PCDDs and PCDFs at considerable concentration. PCB and dioxin pollution has been regarded as a problem for industrialized nations in Europe, North America and Japan. However, there is a growing realization that due to rapid industrialization PCBs and dioxins have been detected in tropical Asia.

Mycotoxins

Mycotoxins or fungal toxins are low molecular weight secondary metabolites produced by a wide range of fungus, particularly moulds. There are about 400 secondary metabolites of moulds but only a few of them are toxins causing pathological or physiological alterations in man and animals known as mycotoxicosis. There are over 100 known species of moulds that produce mycotoxins. But most of the mycotoxins of importance are produced mainly by the species of three genera viz., *Aspergillus, Penicillium and Fusarium*. The most studied mycotoxins are aflatoxins (AFB1, AFB2, AFG1, AFG2, AFM1 and AFM2), Ochratoxins, Fumonisin, Zearalenone, Deoxynivalenon (DON), Diacetoxyscirpenole (DAS), Nivalenone (NIV), T-2 and HT-2 toxins. Mycotoxin contamination of animal fodder occurs in the field owing to infestation by particular pathogenic fungi or with symbiotic endophytes and also during processing and storage of harvested products and feed in appropriate environmental condition. In general mould growth and mycotoxin production require a moisture content of greater than 14%, temperature of at least 25°C and some degree of aeration. When these conditions are met, mould infestation and mycotoxin production in target crops are likely to occur. During the process of host colonization fungi may produce toxins that are harmful for humans and animals. This occurs mainly under specific conditions of plant stress, such as those produced by adverse weather conditions. For instance, drought has been shown to prompt aflatoxin production in groundnuts. Differences in mycotoxin production are also related to plant genotype. Clinical manifestation of mycotoxicoses in animals depends on exposure, concentration and type of mycotoxin. Differences in tolerances to mycotoxins exist among poultry, pigs and cows.

Aflatoxins

The most important mycotoxin is aflatoxin (bisfurano coumarins) produced by *Aspergillus flavus, A. parasiticus* and *A. nomius*. The aflatoxins are potent carcinogenic, mutagenic, teratogenic, immuno suppressive agents and cause liver damage, decrease milk yield, egg production and suppression of immunity in animals. The importance of aflatoxins in animal health was first noticed in 1960 when in the UK 100,000 turkeys died from acute necrosis of the liver and hyperplasia of the bile duct which was called at that time as Turkey X disease. Following detail investigation it was revealed that the cause of the death was the mycotoxins which came after consumption of groundnut infected with *A. flavus* which was imported from Brazil for use as a protein supplement in animal diet. Aflatoxin producing strains of *Aspergillus* are widely distributed in soil and air. Under favourable environmental conditions colonization and mould growth can easily occur in substrate like a feed or seed. The nature and concentration of aflatoxin will vary greatly according to prevailing weather condition, substrate and the type of mould. *A. flavus* produces primarily B_1 and B_2 on corn while *A. parasiticus* on corn produces all four major aflatoxins B_1, B_2, G_1 and G_2. Three major feed stuffs viz., corn, cottonseed and peanuts have high potential for invasion by *Aspergillus sp.* during growth, harvest, transportation or storage. In case of soybean and small grains colonization generally occurs in storage. Poor storage condition, moisture, temperature, lack of aeration, kernel damage and spore dissemination caused by storage insects etc. aid in colonization and promote aflatoxin production.

Chronic poisoning or aflatoxocosis can result when low levels of toxins are ingested over a prolonged period. The toxic effects are not nearly as specific or clinically evident as in acute toxication. In general, affected livestocks exhibit decreased growth rate, lower productivity and immunosuppression. Among the biological effects, aflatoxins have effect on protein synthesis as they interact with a number of different cell components involved in protein synthesis. Aflatoxins have a high affinity for nucleic acids and polynucleotides (RNA and DNA) particularly at nucleophilic sites on the guanine base. Formation of these aflatoxin-nucleic acid covalently bonded adducts can lead to a number of different biological effects, including cancer. Aflatoxin is carcinogenic in several species including rats, ducks, mice, trout and subhuman primates. AFB1 is considered the most potent carcinogen known with AFG1, AFB2 and AFG2 in their order of decreasing potency.

Aflatoxin B_1 in feed is metabolized to aflatoxin M_1 in milk which is heptatoxic and hepatocarcinogen. Concentration of AFM1 in milk increases proportionally with amount of AFB1 in the diet. When ingestion is continuous milk concentration will increase until equilibrium with intake is established.

Zearalenones

The Zearalenones are a family of phenolic compounds produced by several species of *Fusarium* that cause estrogenic effects and infertility in animals. These compounds are one of the two major categories of *Fusarium* toxins, the other group is the trichothecenes. Zearalenone, sometimes called F-2 toxin, is 6-(10-hydroxy-6-oxo-trans-1-undecenyl)-β-resorcyclic acid lactone. Production of zearalenone by *Fusarium spp.* usually occurs in storage when moisture and temperature conditions are optimal. However, in the field infected corn ears may develop a rot of the crown or cob aptly named "Gibberalla rot". The primary biological effect induced by zearalenone in animals is hyperestrogenism which is a chronic toxicosis. Zearalenone is metabolized into a and â zearalenols which are eliminated quickly from the body and in lactating cow less than 1% of a dose is transmitted in to milk.

Trichothecenes

The trichothecenes are a family of tetracyclic sesquiterpenoid substances produced by several species of *Fusarium* and few other genera. Of the more than 40 naturally occurring trichothecenes that have been identified the most notable one with regard to animal agriculture are T-2 toxins, diacetoxyscirpenol (DAS) and deoxynivalenol (DON) also known as vomitoxin. Feed refusal and vomiting (vomitoxin) are associated with most trichothecenes.

Ochratoxins

The ochratoxins are a family of isocoumarin derivatives of the amino acid phenyl alanine. Of the nine ochrtoxins that have been identified only ochratoxins-A, B and C are the most commonly recognized members produced by several species of *Aspergillus* and *Penicillium*. The toxin was first isolated from *Aspergillus ochraceus* from which the name was derived. Ochratoxins-A (OTA) is the most toxic followed by OTB and OTC. OTA is primarily nephrotoxic, secondarily hepatotoxic, immunosuppressive, mutagenic, genotoxic, teratogenic and carcinogenic in nature. Interference with carbohydrate metabolism is considered to play a key role in ochratoxocosis. Its target organs are the kidneys and the developing nervous system. The principal source of ochratoxins is cereal grains including barley, rye, wheat, corn, sorghum and oats but these may also occur in other commodities such as cottonseed, nuts etc. Due to ubiquitous nature of OTA producing fungi which are distributed worldwide and common in soil, decaying plants, insects, stored grains and cereals, ochratoxicosis has been encountered in the form of naturally occurring syndrome in animals as well as in human being through diet containing contaminated food and feed.

Citrinin

Citrinin is a nephrotoxin produced by several species of the genus *Penicillium* and three species of the genus *Aspergillus*. It is a quinone methide first identified from *Penicillium citrinum*. The fungi producing citrinin are found in temperate regions. Feedstuffs known to be colonized by the toxigenic strains include most grains such as wheat, oat, barley, rye and corn.

Table: Mycotoxins, their source and effect on animals

Mycotoxin	Producing Fungus	Plants/hosts	Effect
Aflatoxins (B1,B2,G1,G2)	*Aspergillus flavus, A. parasiticus, A. nomius*	Corn, small grains, peanuts, cottonseeds, mixed feed, most nut crops, mustard etc.	Aflatoxicosis (carcinogenic, affect protein synthesis, DNA & RNA, immunosuppression)
Ochratoxins (A,B,C)	*Aspergillus ochraceus Penicillium viridicatum P. cyclopium, P. verrucosum*	Corn, wheat, oats, barley, beans, peanuts, mixed feed	Ochratoxocosis (Interference with carbohydrate metabolism, nephrotoxic, hepatotoxic, immunosuppressive, mutagenic, genotoxic, teratogenic & carcinogenic)
Trichothecenes like T-2 toxin, Diacetoxyscirpenol, Nivalenol, Deoxynivalenol or DON (Vomitoxin)	*Fusarium sporotrioides F. poa, F. trichinctum, F. roseum*	Corn, wheat, oats, barley, millet, rye, mixed feed	Feed refusal, vomiting
Fumonisins (B1,B2,B3) moniliformin, fusaric acid	*F. moniliforme*	Maize, mustard, mixed feed	Porcine pulmonary oedema, equine leucoencephalomalacia, human oesophageal cancer
Zearalenone	*F. roseum,*	Corn, wheat, oats, barley, sorghum	Hyperestrogenism
Ergotamine, Ergometrine alkaloids, Ergotoxin	*Claviceps purpurea*	Rye, wheat, barley, rice, millet, sorghum etc.	Persistent diarrohoea, convulsions etc.
Citrinin	*Penicillium viridicatum P. citrinin, Aspergillus*	Rye, wheat, barley, oat, corn, mustard	Nephrotoxin
Rubratoxin	*Penicillium rubrum* and *P. purpurogenum*	corn	Hepatitis, nephrosis and hemorage

Rubratoxin

Rubratoxins are bisanhydride metabolites produced by *Penicillium rubrum* and *P. purpurogenum*. There are two known toxins, Rubratoxin A (RA) and Rubratoxin B (RB). Comparatively RB is twice as toxic as RA. Fungi that produce RA and RB are commonly found in food and feedstuffs. These fungal toxins were originally found in moldy corn and caused a disease characterized by hepatitis, nephrosis and hemorage when fed to cattle and swine.

Approaches to eliminate mycotoxins from contaminated grains/feed are not always worthwhile in practical and economical terms. Strategies used to reduce mycotoxin levels in cereal grains include dilution of infected with healthier grains or the addition of adsorbents to the feed. Several compounds such as activated charcoal, aluminosilicates, special polymers and processed cell walls of *Saccharomyces sp.* (esterified glucomannan) have shown to be effective as mycotoxin adsorbents in feed. Though adsorbents have been effective in vitro and in vivo studies but it is very difficult to quantify their benefits in commercial operations.

Heavy Metals

Special attention has been laid from the last decade regarding heavy metal contamination in the environment and its impact on man and animals. The toxic metals of high environmental concern are lead (Pb), arsenic (As), mercury (Hg), cadmium (Cd), chromium (Cr), Nickel (Ni) and Selenium (Se). Industrial effluents, waste chemicals, municipal sewage and sludge and inorganic fertilizers contaminate water and soil with the toxic metal residues. Irrigation water if not properly treated carry with them all the pollutants which ultimately deposited in the soil. Soils actually act as the reservoir of the toxic metals. Combustion of fossil fuel and emission of industrial smoke pollute the atmosphere and surrounding environment and release heavy metals which ultimately get deposited nearby crops and soil. From soil many crop plant species uptake those heavy metals alongwith other essential nutrients and deposit in their tissues. Heavy metal contaminants enter animal body mainly through ingestion of contaminated forages, feed and water. Once in the animal system these metals can persist for several weeks even after discontinuation of the exposure. Elevated levels of lead could be detected in animal tissues up to 12-19 months after exposure has ceased (Hatch, 1988). Bioaccumulation of heavy metals in tissue and their secretion in milk is a matter of great concern just like presence of any other toxic matter in milk. Normally very low concentrations of these metals are secreted into milk because the carry over rate of metals from forage to milk is appx. 1:500 (Bluthgen *et al.*, 1997). The accumulation of heavy metals in dairy animals adversely affects health and milk production (Dey *et al.*, 1996, Dogra *et al.*, 1996). It has been found through experiment that cardiovascular and immune response affected before the appearance of clinical symptoms because of heavy metal poisoning (Dey *et al.*, 1993, Haneef *et al.*, 1995).

Lead

Lead is the most common heavy metal responsible for widespread toxicity and has no known beneficial biological function in the body. The important sources are industry, automobile exhaust and natural occurrence in the environment. It is absorbed from the GI tract and lungs and once it is in the blood stream, majority of it is bound with hemoglobin. In chronic poisoning different systems of body manifest signs of plumbism. It affects smooth muscle of GI tract producing metallic taste, anorexia and abdominal cramps. Lead poisoning causes reversible and irreversible nephropathy depending upon duration of exposure. Higher concentration of lead in milk and blood were reported in animals reared in polluted areas of urban India (Dey *et al.*, 1996, Bhatia and Chowdhury, 1996) and elsewhere in the world (Bars *et al.*, 1990). Milk lead concentration is exponentially related to blood lead. Because of its toxicity and passage into milk lead has been considered by Codex International specification in milk and milk products (0.02ppm in milk and 0.05ppm in butter).

Cadmium

Cadmium is toxic to virtually every system in the animal body. There are many sources from which environment and human body gets contaminated with Cd. Soil levels of Cd is increased by the Cd present in water, sewage contamination, air and phosphatic fertilizer. The root uptake by plants especially at lower pH is the main route of Cd entry into the food chain. The most likely source of contamination of milk is by the animal ingesting Cd from fodder and drinking water. High concentration of Cd in milk was reported due to polluted environment (Haneef *et al.*, 1995, Swarup *et al*, 1997). Absorption of Cd is more from respiratory system than from GI tract. Following absorption Cd binds with blood cells and albumin and then distributed to liver and kidneys as metallothionein-Cd complex. Cadmium causes chromosomal abnormalities, inhibition of repair of DNA damage and inhibition of plasma membrane Ca^{+2} channels and Ca^{+2}-ATPase. It inhibits -SH containing enzymes and oxidative phosphorylation. In acute Cd poisoning symptoms include vomiting, diarrhea, abdominal cramps, pneumonitis, dizziness etc. and finally death due to pulmonary edema. In chronic poisoning kidneys are first affected followed by lungs and GI tract. It is also reported to be carcinogenic.

Arsenic

Arsenic has wide distribution in the environment, naturally occurring in soil and also present in air and water. Among different forms of As organic arsenicals are least toxic followed by As^{5+} then As^{3+}, while arsine is most toxic agent. Arsenic produces toxic effect in all the systems. It causes mild to moderate vasodilation and increase capillary permeability that may lead to edema. In skin it causes hyperpigmentation and eventually skin cancer. It also induces mild abdominal cramps, gastroenteritis etc. In chronic poisoning garlic odour of the breath and presence of white tansverse line (Mee's lines) in the fingernails, scaling of skin and jaundice is manifested. In soil its concentration may vary from 1 to 40ppm. The principal means of contamination of milk appears to be from animals grazing in arsenic contaminated area. Contamination of As mainly occurs through ground water used for irrigation or drinking. An estimated 6 million people of West Bengal, India are drinking water contaminated with As at a level above the WHO permissible level (Chowdhury *et al.*, 2001). A survey conducted in two villages (declared to be arsenic contaminated) in two districts of West Bengal showed that feed, roughages given to animals contained highly toxic levels of As. According to a rough estimate lactating animals were consuming 44 mg day^{-1} of As from all feed resources including water which is far from tolerable levels. The average excretion of As in milk was found to be 77ppb (Singh *et al.*, 2002). In another report (Rana *et al.*, 2008) from high As contaminated area in Nadia district of West Bengal, the concentration of As in milk was found to be 0.156mg/kg. Permitted As level in milk should be less than 10 ng g^{-1} (IDF, 1986). Groundwater contamination with As was also reported from Vietnam, Cambodia, Bangladesh, Taiwan, Argentina, Japan, Thiland, Chile, Mongolia, Finland, Hungary and the likes (Rahman *et al.*, 2001).

Mercury

Mercury is considered a highly toxic contaminant and biologically non essential. It is released in the environment from both natural and anthropogenic sources. Man made sources including coal combustion, nonferrous metal production, waste disposal etc. release substantial quantity of mercury in the environment. Industrial wastes and sewage water from chloroalkali industry are also major source of mercury pollution. Toxicity depends on the chemical form. Elemental mercury is not toxic as it is seldom absorbed while organic mercurials are readily absorbed from GI tract and cross blood brain-barrier and placenta to produce neurological and teratogenic effect. Methyl mercury is most toxic and stable form of mercury and most avian and mammalian predators at the top of contaminated trophics suffer from this form. The last century witnessed several incidences of mercury poisoning due to pollution causing death of man and animals in Minamata Bay in Japan, Guatemala, Iraq and Pakistan. Minamata disease is a classic example of ecotoxicity due to mercury affecting various animal species and man in a food-web. Milk and milk products may be contaminated with mercury through fodder and drinking water. Value of mercury level in normal milk reported in literature range from 0.001 to 0.015 mg kg^{-1}. According to IDF Heavy Metals Group (1997) normal content of mercury in milk is around 1 mg kg^{-1}.

Chromium

Chromium toxicity depends on its chemical speciation, so associated health effects are influenced by the chemical form of the exposure. Of the various forms hexavalent chromium i.e. Cr^{6+} is most toxic and certain Cr^{6+} compounds are carcinogenic in human. Pollution of chromium is due to its release through industrial activity and contamination of aquatic ecosystem, thus producing toxicity to flora and fauna.

Selenium

It is an important trace element in animals and deficient feed (0.1ppm) produce deficiency disorders like white muscle disease, exudative diathesis, infertility etc. in animals. However, it becomes toxic also at a not very high concentration i.e. the difference between the deficiency and toxic level is very narrow. Consumption of vegetation by animals containing as low as 5ppm Se causes health hazards. Contamination of pasture from industries emitting fly ash particularly soft coal may cause Se poisoning in grazing animals. In 1980-89, poisoning characterized by alopecia and foot scaling was reported in pigs in China due to use of Se rich stone coal by farmers for fumigation of soil and as fertilizers (Mao *et al.*, 1996). Relationship between Se toxicity and plant Se level has been established (Ghosh *et al.*, 1999) and Se excess was considered as one of the factors causing Dengnala disease that affected thousands of cattle and buffalo in India and Pakistan (Dhillon *et al.*, 1990).

Normally contents of heavy metals in milk and its products are low unless they are contaminated by other sources like utensils, packaging materials, containers etc. The animal body acts as biological filter against transfer of heavy metals to milk. The carryover of heavy metals and mycotoxins to milk is very less as compared to lipophilic pesticides and other POPs.

Only a few reports are available for the removal of the heavy metals. Decontamination techniques available may not be economically feasible and commercially viable. The most important aspect is the regular and careful monitoring of soil, feed, fodder and water and taking of corrective measures accordingly.

Nitrates and Nitrites

Nitrates are naturally occurring form of nitrogen and an integral part of the Nitrogen cycle. Manures, nitrogenous fertilizers, decaying plants and animals, and other organic residues generate nitrates, which is also produced in the human and animal body. Due to increased use of fertilizers and manures in intensive crop production system level of nitrates is increasing in cereals, other field crops and water than that in the past. Forage crops may also be source of high level of nitrates. Sudan grass, oats, rape, wheat, barley and maize accumulate nitrate. Nitrate *per se* is relatively non toxic. But approximately 5% of all ingested nitrate is converted in saliva and GI tract to the toxic nitrite formed by the reduction by bacterial enzymes. Nitrite and N-nitroso compounds formed by binding nitrite to other substances like amines, amides etc. are toxic and can lead to severe pathologies in humans. N-nitroso compounds have also been reported to be carcinogenic in more than 40 animal species including mammals, birds, reptiles and fish. The best known effect of nitrite is its ability to react with haemoglobin (Oxyhb) to form methhaemoglobin (metHb) and nitrate. MetHb can not bind oxygen and as a result oxygen delivery to tissues is impaired. Once the proportion of metHb reaches 10% of normal Hb levels, clinical symptoms like cyanosis, asphyxia occur. Nitrate reduction in rumen competes with other metabolic reactions such as methanogenesis and in the formation of other end products leading to abortion, infertility and goiter. Only good agricultural practices can negotiate with this problem. Although milk and milk products are, either secretory or post secretory, contaminated with nitrates, nitrites or nitrosamines but the level does not pose any health risk due to these compounds. The nitrate level in milk is naturally relatively constant in the range of <1-12mg/kg raw milk, while nitrite is practically absent. Traces of volatile nitrosamines can be detected in milk.

Epilogue

Now there is a growing awareness about the food safety. Consumers worldwide expect protection from hazards occurring along the entire food chain. It is, therefore, the need of the hour to emphasis greatly on food safety issues. There is a direct link between the safety of feed and the safety of food of animal origin because feed and food are interrelated as one compliments the other with the supply of feed for production of milk, meat, egg etc. Feed production must, therefore, be subjected to the same way as food production, quality assurance including safety systems based on principal of HACCP (Hazard Analysis and Critical Control Point) endorsed by the Codex Alimentarius Commission (CAC), an inter governmental body, which has formulated international standards for a wide range of food products and specific requirements concerning pesticide residues, food additives, veterinary drug residues, hygiene, food contaminants etc. Access of countries to the food export markets depends on their capacity to meet the regulatory requirement (non tariff trade barriers) of the importing countries. Trade is an important stimulus to a country's economic development and in the current global economy it is not possible for any one country to remain isolated from changing demands of international requirements for food safety and quality regulations. The agreement on the application of Sanitary and Phytosanitary measures (SPS Agreement) under WTO (World Trade Organization) cover those aspects of food that are related to health i.e., food additives, pesticide residues, veterinary drugs, mycotoxins, heavy metals, codes of hygienic practices etc. This agreement recognizes CAC standards.

References

Agnihotri, N.P.1999. In: Pesticide safety evalution and monitoring. All India Coordinated Research Project on Pesticide Residues, Indian Agricultural Research Institute, New Delhi, Pp. 104-111.

Baars, A. J., Beek Huvan, Visser, I.T.R., Delft W Van, Fennema, G, Lieben, G.W., Lautenberg, K., Nieuwenhujis, J.H.M., Coulander, P., Pluimers, F.H., Haar G Van De, Jrna T., Tuinistra, L.G.M.T., Zandstra, P and Bruins, B. 1990. Lead poisoning among cattle in North Netherlands between October 1989 and January 1990. *Tijjjdschrift voor Diergeneeskunde*, 115: 882-90.

Battu, R.S., Singh, B., Chahal, K.K. and Kalra,R.L. 1996. Contamination of animal feed with residues of HCH and DDT. *Pesticide Research Journal*, 8: 172-175.

Bhatia, I and Chowdhury, G.N. 1996. Lead poisoning in milk – the basic need for foundation of human civilization. *Journal of Animal Health*, 40: 24-26.

Bluthgen, A., Burt, R and Heeschen, W.H. 1997. Heavy metals and other trace elements. Monograph on Residues and Contaminants in Milk and Milk Products, International Dairy Federation, Brussels, Belgium. Pp. 65-73.

Chowdhury, U.K., Rahman, M.M., Mandal, B.K., Paul, K., Lodh, D and Biswas, B.K. 2001. Groundwater arsenic contamination and human suffering in WB, India and Bangladesh. *Environmental Science*, 8: 393-415.

Deka, S.C., Barman, N. and Baruah, A.A.L.H. 2004. Monitoring of pesticide residues in feed, fodder and butter in Assam. *Pesticide Research Journal*, 16: 86-89.

Dey, S., Dwivedi, S.K. and Swarup, D. 1996. Lead concentration in blood, milk and feed of lactating buffaloes after acute poisoning. *Veterinary Record*, 138: 336.

Dey, S., Swarup, D. and Singh, G. 1993. Effect of experimental lead toxicity in cardiovascular function in calves. *Veterinary and Human Toxicolgy*, 35: 501-503.

Dhillon, K.S., Srivastava, A.K., Gill, B.S. and Singh, J. 1990. Experimental chronic selenosis in buffalo calves. *Indian Journal of Animal Science*, 60: 532-535.

Dikshit, T. S. S., Kumar, S. N., Tandon, G.S., Raizada, R. B. and Ray, P. K. 1989a. Pesticide residues in edible oils and oilseeds. *Bulletin of Environmental Contamination and Toxicology*, 42: 50-56.

Dikshit, T.S.S., Kumar, S.N., Raizada, R.B. and Srivastava, M.K. 1989b. Organochlorine insecticide residues in cattle feed. *Bulletin of Environmental Contamination and Toxicology*, 43: 691-696.

Dogra, R.K.S., Murthy, R.C., Srivastava, A.K., Gaur, J.S., Shukla, L.J. and Varmani, B.M.L. 1996. Cattle mortality in Thane district, India: A study of cause/effect relationship. *Archives of Environmental Contamination and Toxicology,* 30: 292-297.

Ghosh, A., Sarkar, S., Pramanik, A.K., Ghosh, S. and Basak, D.K. 1999. Selenium toxicosis of buffaloes with relation to plants. *Buffalo Journal,* 15: 91-95.

Gupta, A., Singh, B., Parihar, N.S. and Bhatnagar, A. 2000. Monitoring of HCH and DDT residues in certain animal feed and feed concentrates. *Pestology,* 24: 47-49.

Haneef, S.S., Swarup, D., Kalicharan and Dwivedi, S.K. 1995. The effect of concurrent lead and cadmium exposure on cell mediated immune response in goats. *Veterinary and Human Toxicolgy,* 37: 428-429.

Hatch, R.C. 1988. Poisons causing nervous stimulation or depression. In: Veterinary Pharmacology and Therapeutics, 6th edn., (Eds. N.H.Broth and LeMacdonald). Iowa State University, Amsterdam, Pp. 1083-1091.

IDF 9701. 1997. Monograph on residues and contaminants in milk and milk products.

Kang, B. K., Singh, B., Chahal, K.K. and Battu, R.S. 2002. Contamination of feed concentrates and green fodder with pesticide residues. *Pesticide Research Journal,* 4: 308-312.

Kannan, K., Tahabe, S., Romesh, A., Subramanian, A. and Tatsukawa, R. 1992. Persistent organochlorine residues in food stuff from India and their implication on human dietary exposure. *Journal of Agricultural and Food Chemistry,* 40: 518-524.

Kaphalia, B.S. and Seth, T.D. 1982. Organochlorine pesticides contamination in some species of fodder grasses. *Environmental Pollution Series* B, 3: 231-237.

Mao, D.J., Su, H.C., Hu, W.H. and Zheng, B.S. 1996. Survey of selenium poisoning in pig caused by soot contamination of soil. *Chinese Journal of Veterinary Science and Technology,* 26: 15-17.

Nag, S.K. and Raikwar, M.K. 2006. Final Report of Lal Bahadur Shastri Young Scientist Scheme "Monitoring of pesticide residues in animal feed, fodder and milk with special reference to Bundelkhand region" submitted to ICAR, New Delhi.

Nag, S.K. and Shah, N.K. 1999. Residual persistence of endosulfan in fodder sorghum. *Pestology,* 23: 52 – 55.

Nag, S.K. and Shah, N.K. 2001. Endosulfan residue and its dissipation in Lucerne (*Medicago sativa* L.). *Pesticide Research Journal,* 13: 109-111.

Nag, S.K. and Shah, N.K. 2002. Residues and efficacy of endosulfan against pest complex of fodder cowpea. *Journal of Entomological Resesrach,* 26: 133-139.

Pierson, D.A., Hoffman, J.S., Nord, P.J., Gebhart, J.E. and Frank, C.N. 1982. Distribution of chlorinated pesticides in animal feed components and finished feeds. *Journal of Agricultural and Food Chemistry,* 30: 187-189.

Prasad, K.S.N. and Chabara, A. 2001. Organochlorine pesticide residues in animal feeds and fodders. *Indian Journal of Animal Science,* 71: 1178-1180.

Rahman, M.M., Chowdhury, U.K., Mukherjee, S.C., Mandal, B.K., Paul, K., Lodh, D., Biswas, B.K., Chanda, C.R., Basu, G.K., Saha, K.C., Roy, S., Das, R., Palit, S.K., Quamruzzaman,

Q and Chakraborty, D. 2001. Chronic Arsenic toxicity in Bangladesh and West Bengal, India. A Review and Commentary. *Clinical Toxicology*, 39: 683-700.

Raikwar, M.K. and Nag, S.K. 2004. Dissipation of carbendazim residues in fodder sorghum. *Indian Journal of Agricultural Sciences*, 74: 227-229.

Raikwar, M.K., Nag, S.K., Banerjee, T. and Shah, N.K. 2003. Persistence behaviour of endosulfan in fodder maize. *Pesticide Research Journal*, 15:186-190.

Rana, T., Sarkar, S., Mandal, T.K., Bhattacharyya, K., Roy, A. and Kol, L. 2008. Contribution of arsenic from agricultural food chain in Nadia District of West Bengal in India. *The Internet Journal of Thoracic and Cardiovascular Surgery*, 4 (2).

Sharma, V., Wadhwa, B.K., Stan, H.J. 2005. Multiresidue analysis of pesticide in animal feed concentrate. *Bulletin of Environmental Contamination and Toxicology*, 72: 242-245.

Shastry, M. S. 1983. Monitoring of pesticide residues in animal feeds and animal products. *Pesticide*, December issue. Pp. 36-38.

Singh, P.P., Battu, R.S. and Kalra, R. L. 1988. Insecticide residues in wheat grain and straw arising from their storage in premises treated with BHC and DDT under malaria control programme. *Bulletin of Environmental Contamination and Toxicology*, 40: 696-702.

Singh, R., Kumari, B., Madan, V.K., Kumar, R. and Kathpal, T.S. 1997. Monitoring of HCH residues in animal feeds. *Indian Journal of Animal Science*, 67: 250-252.

Singh, R.B., Saha, R.C. and Misra, R.K. 2002. Arsenic profile in livestock feeds and products in Ghentuga and Gontra district of West Bengal. Project Report of the work done at Eastern Regional Station of NDRI, Kalyani-741235. In : Contaminants in milk and milk products and measures to minimize their levels (A. Chhabra). *Indian Dairyman*, 56: 115-125.

Swarup, D., Dwivedi, S.K. and Dey, S. 1997. Lead and cadmium levels in blood and milk of cows from Kanpur city. *Indian Journal of Animal Science*, 67: 222-223.

Unnikrishanan, V., Surendara Nath, B., Gayathri, V., Sampath. K.T. and Rama Murthy, M.K. 1998. Organochlorine pesticide residues content in feed and feedstuff. *Indian Journal of Dairy & Biosciences*, 9: 59-64.

Van Houweling, C.D., Bixler, W. B. and Mc Dowell, J.R. *Journal of American Veterinary Medical Association*, 171: 1153.

□□□

Agroforestry: Natural Resource Sustainability, Livelihood & Climate Moderation © 2009
Editors: OP Chaturvedi, A Venkatesh et. al., pp. 423-431
Satish Serial Publishing House, Delhi

Chapter 29

TOP FEEDS IN ALTERATION OF RUMEN FUNCTION AND ITS ENVIRONMENT FOR EFFICIENT NUTRIENTS UTILIZATION

Sultan Singh
Indian Grassland and Fodder Research Institute, Jhansi-284 003 (U.P.)

Introduction

Top feeds (multipurpose trees and shrubs) are of key value for economy of several arid and semiarid regions in the globe. They provide quality fodder, timbers, fuel wood, charcoal, gums and other products. Trees and shrubs also improve to soil stability and fertility through reduced soil erosion and N fixation, respectively. Multipurpose trees and shrubs also help in improving environment through protection against weather conditions like heat, cold, strong winds or solar radiation. (Baumer, 1991). Leaves of these trees and shrubs are rich in nitrogen, minerals and used as feed in dry season or conserved dry for later use. These browse plants are of particular importance for animals during periods when quality and quantity of cultivated green fodder is limited/poor. Supplementation with browses invariably alleviates nitrogen deficiency and also sulphur and other mineral deficiencies, thus increasing the intensity of rumen microbial activity. Leaves of these top feeds often contain variable amounts /levels (low to high) of secondary compounds like tnnins, cyonogens, sapponins non protein amino acids (mimosine, indospicine, canavanine), phytohaemagglutinins, alkaloids, oxalate etc which may have both positive and adverse effect on animal performance. These secondary compounds present in top feeds have been exploited in livestock diet to manipulate the rumen environment to benefit the host animals. This includes their use as protein protection, anti bloat, parasite control, defaunation, anti methanogeneis agent etc to provide more balanced nutrients to animals for their higher productivity

Source of Nutrients

Top feeds (tree leaves and shrubs) are rich source of protein and energy. Most of the top feeds contain more than 7 % CP which is sufficient to meet the maintenance requirement of the animals. Feeds are also good source of energy and better than crop residues and straws but comparable to energy sources. Top feeds are relatively low in (NDF, ADF and cellulose) but higher in lignin contents than the crop residues, straws and stovers. In spite of the higher lignin contents the digestibility of top feeds is better than grasses, straws and stovers. Higher CP and more digestibilities of top feeds make them a good source to supplement poor quality roughages to improve the animal productivity. The chemical constituents of some of the tree leaves and shrubs are given in Table 1.

Table 1: Chemical composition of top feeds (DM Basis)

Foliage	CP	OM	EE	NDF	ADF	Cellulose	Lignin
Leucaena leucocephala	22.50	87.50	5.2	28.50	17.50	10.00	7.20
Albizia lebbek	19.70	91.00	4.7	47.60	31.80	19.00	12.50
Hardwickia binata	7.80	90.75	2.5	38.60	28.00	16.20	9.75
Grewia optiva	16.00	86.83	4.2	35.30	18.60	13.70	4.40
Anogeissus pendulla	8.20	0.80	4.1	34.90	23.60	15.00	8.20
Azadirachta indica	17.60	91.10	2.4	45.9	35.50	23.60	11.80
Albizia amara	27.60	88.90	5.60	45.20	26.20	12.30	14.10
Ficus religiosa	19.80	87.30	3.8	48.20	33.80	22.90	14.60
Butea monosperma	14.00	89.50	3.8	65.40	48.30	28.10	20.00
Securengia virosa	11.30	90.30	1.9	27.50	17.20	13.40	4.10
Helictris isora	10.60	86.90	3.1	39.80	24.40	19.10	4.82
Dichrostachys cinerea	12.50	86.70	1.5	42.05	29.01	12.05	14.50
Acacia catechu	12.80	87.40	3.1	38.00	24.55	13.60	9.90
Zizyphus xylophyrus	11.50	91.30	2.4	32.30	19.20	8.50	8.70

Anti Nutritional Factors

The anti nutritional factors may be defined as those substances generated in natural feed stuffs by the normal metabolism of animal species and by different mechanisms e.g. inactivation of some nutrients, interferences with the digestive process or metabolic utilization of feed which exert effect contrary to optimum nutrition. The use of leaves, seed, pods and edible twigs from shrubs and tree leaves as animal feed is limited by the presence of these anti nutritional factors. Plants contain a number of compounds which depending upon their situations can have beneficial or harmful effects on livestock consuming them. Anti- nutritional factors usually are not lethal but reduce the animal productivity and may be toxic to animals when fed in large amounts usually during scarcity or when animal is in confinement. The commonly occurring anti nutritional compounds in tree leaves and shrubs are given in Table 2. Tree leaves and browse species differ in concentration of total phenolics and condensed tannins (Table 3) and accumulation of these compounds vary widely across the seasons of plant growth (Table 4).

Effect on Rumen Function and Environment (Rumen Ecosystem)

The rumen is a large pregastric fermentation chamber present in the digestive tracts of all ruminants. Rumen is the main site of ruminant's digestive tract where digestion and metabolism of fibrous feeds occur due to the presence of rumen microbes. The type and population of microbes influence the rate of digestion and production of metabolites from a particular feed/diet. Proliferation and growth or rumen microbes depend on the nutrients availability and rumen environment constituted by pH, temperature, osmotic pressure and concentration of rumen metabolites (total volatile fatty acids and nitrogen metabolites). Carbon dioxide and methane are the main gases produced during fermentation and the energy of a fed available to the animal varies with the amount of methane production. There are several factors such as composition of feed, the degree of physical processing and the presence of feed additives all affect the numbers, proportions and digestive activity of rumen micro organisms. These dietary factors not only affect the microbial activity but also affect the release /degradability of nutrients in rumen and rate of passage which ultimately influences the microbial protein synthesis and nutrients digestive efficiency. Addition of top feeds in ruminant diets has been reported to influence the rumen function and environment in respect of metabolites concentration, type and activity of rumen microbes and rumen dilution rate. All these factors govern the nutrients utilization and their availability to animal. Supplementation of top feeds in animal diets has a number of beneficial effects on rumen function and nutrients availability to the animals.

Table 2: Anti-nutritional factors in top feeds (tree leaves and shrubs

Anti nutritional compound	Plant species
Non protein amino acids	
Mimosine	*Leucaena leucocephala*
Indospicine	*Indigifera spicta*
Glycosides	
1. Cyanogens	*Acacia giraffae*
	Acacia cunninghamii
	Acacia sieberiana
	Bambusa bambos
	Manihot esculenta
2. Saponins	*Albizia stipulate*
	Bassia latifolia
	Sesbania sesban
Phytohhemagglutinins	
Ricin	*Bauhinia purpurea*
Robin	*Robinia pseudoacacia*
Ricin	*Ricinis communis*
Polyphenolic compounds	
1. Tannins	All vascular oplants
2. Lignins	All vascular plants
Alkaloids	
N -methyl -B -phenethylamine	*Acacia beriandieri*
Sesbanine	*Sesbania drummodii*
	Sesbania punicea
Triterpenes	
Azadirachtin	*Azadirachta indica*
Limonin	*Azadirachta indica*
	Azadirachta aneura
Oxalate	*Azadirachta aneura*

Source: Kumar and D' Mello (1977)

Table 3 : Total phenolics and condensed tannins in different seasons of some browse species

Browse species	Total phenolics				Condensed Tannins			
	Monsoon	Winter	Spring	Summer	Monsoon	Winter	Spring	Summer
H. isora	1.34	1.45	2.15	1.12	0.41	1.67	2.84	0.89
S. virosa	2.67	2.19	6.53	1.79	0.86	2.68	3.89	0.75
G. flavescenns	0.58	0.75	1.05	0.65	0.85	2.05	6.74	3.67
F. indica	1.34	2.55	3.87	2.75	1.03	2.14	4.56	0.48
A. catechu	11.70	7.76	14.60	13.50	6.97	10.30	13.60	10.60
Z. xylopyrus	6.40	5.01	8.31	7.34	1.35	1.11	2.80	1.17
Z. nummularia	1.82	1.28	2.13	1.87	2.02	1.63	3.47	2.19

Source: Negi *et al.*, 2003

Table 4. Polyphenolic compounds (%) of certain tree leaves on dry matter basis

Tree leaves/ Constituents	*Acacia auric-uliformis*	Cashewnut	*Gliricidia*	Jack	*Sesbania grandiflora*	Subabul	Yellow gold mohur
Total phenolics[1]	13.44	20.31	5.63	15.63	9.38	11.25	12.50
Non-tannin phenolics[1]	0.48	0.86	0.35	0.94	0.74	0.48	0.53
Total tannin phenolics[1]	12.96	19.45	5.28	14.69	8.64	10.78	11.98
Condensed tannins[2]	12.29	16.43	3.44	13.23	5.71	7.28	11.03
Calcium	0.90	0.68	1.83	1.10	2.42	2.24	1.08
Phosphorus	0.52	0.46	0.37	0.64	0.36	0.30	0.36

[1]*as tannic acid equivalent;* [2]

Source: Reddy and Elanchezhian, 2008

Supplementation of Critical Nutrients

Cereal straws, stovers, dry grasses are the main roughage source for ruminants in India. These are characterized by low nutrients (nitrogen, minerals, vitamins and available energy) and high fiber. Top feeds contain both micro and macro minerals in adequate concentration to meet the maintenance need of the livestock. These are better source of nutrients than the crop residues and other poor quality feeds.

Supplementation of these poor feed resources with critical nutrients will balance the ration for energy, protein and minerals to improve rumen fermentation (reducing CO_2 and methane production) and increase the efficiency of growth and milk production. Since the top feeds are rich source of nutrients and their addition in animal diets will provide the critical nutrients to microbes for their proliferation. This will change VFA production pattern resulting in more propionate production and reduced methane emission. Methane production reduced from 59.0 to 54.8 l kg^{-1} DDM and from 64.6 to 55.5 l kg^{-1} DDM on supplementation of urea molasses mineral block (UMMB) to untreated and treated straw, respectively. Supplementation of urea mineral to straw based ration can reduce the methane emission by 50 %.

Protein Protection

Ruminant animals have the ability to utilize low quality/high fibrous diets to produce high quality animal products. This is due to the symbiotic relationship between host animal and rumen microbes. They digest and ferment plant cell wall and cell contents (mainly protein) to produce volatile fatty acids and nitrogen metabolites and thus provide energy and protein to the animal. For the improve rumen function these metabolic end products should be released/produced in optimum concentration for their efficient utilization. The efficiency of rumen metabolism not only beneficial to the animal but also influences the output of environment polluting waste products (poor utilization of ammonia, more N excretion and methane production). Excess protein degradation in rumen results in inefficient utilization of ammonia (N retention) by rumen microbes leads to excretion of N rich waste. Tannins present in the top feeds inhibit the degradability by binding with dietary proteins and provide the optimum production of ammonia for efficient utilization of good quality protein at lower level of digestive system for higher animal productivity. Effect of different levels of tannins from Acacia catechu as protein binding agent was studied with different protein sources in terms of ammonia production. There was decrease in ammonia production with the increase in tannin level with all the protein sources in pellet feeds (Table 5).

Antimethanogenic Property

Top feeds contain many compounds (saponins, phenols etc) which inhibit the growth of rumen bacteria and fungi like protozoa. In absence of methanogens (methanogenic bacteria), the available hydrogen can be used by *Eubacterium limosum*, which forms acetate and butyrate from hydrogen and carbon dioxide. Hence studies are needed to document the rumen conditions under which the growth of *Eubacterium limosum* is favoured over that of methanogens and the utilization of hydrogen is diverted towards the synthesis of volatile fatty acids instead of methane.

Table 5: Effect of tannin on NH_3-N productions of protein tannin complex (pellet feeds)

Protein Source	Control			Tannin level		
	0%	2%	4%	6%	8%	Mean
Ground nut cake	31.35	17.50	11.66	9.77	8.87	15.83
Mustard cake	27.82	17.27	14.7	10.26	13.06	16.62
Soybean cake	63.17	46.20	38.03	28.47	30.57	41.29
Cotton seed cake	15.60	7.93	7.00	5.8	6.06	8.48
Linseed cake	14.65	18.43	16.80	13.77	13.76	15.48
Mean	30.52	21.46	17.64	13.61	14.46	

Source: Gupta, 2004

Decreased activity of methanogenic microbes particularly the methane-producing bacteria seem to be the other route to inhibit the CH_4 production due to fermentation of feeds. Methanogenic bacteria produce methane rapidly (about 500 times their own cell volume/minute), however they contribute very little (2-3%) of total bacterial niche of the rumen. Elimination of methanogenenic bacteria by dietary means would have no serious effect on animal fed balanced diet and it can be very well compensated for by increased synthesis of microbial mass of another species that will occupy the methanogenic niche. Specific inhibition of CH_4 may be a good economic proposition. Methane inhibition not only reduces the energy tax of animal rather increase the energy value of feed available to the host animal. In a reduced state of CH_4 production from feeds/fodder the supply of nutrients particularly the energy will be different for different feeds in relation to the existing feeding standards.

Defaunation

Elimination/removal of protozoa from the rumen is known as defaunation. It improves the wool growth, weight gain, feed utilization efficiency and performance of the animas Defaunation decreases the methanogensis, the CH4 production, because most of the methanogens remain attached to protozoa pellicle and exhibit an ecto-symbiotic relationship, utilizing hydrogen to convert into methane.. Saponin containing plants have been shown to be defaunating agent. There is decrease in protozoa numbers due to feeding of pure saponins and saponin containing plants. A number of plant and plant extracts have also been found to impact protozoal population, including hot water extract of soapnut, Yucca schidigera extracts, fruit of Sapindus saponaria, leaves of Entrobium timoba and tree leaves of Enterolobium ciclocarpum. Saponins from Quillaja saponaria, Acacia auriculoformis and foliage from Sesbania sesban have shown anti protozoal activity in vitro. In vivo studies with Yucca shidigera in heifers and Sesbania sesban in sheep have shown decrease in protozoa activity. The sensitivity of protozoa to saponins is due to the presence of sterols in protozoa. Feeding of Acacia salinga reduced more protozoa concentration than Sebania sesban in sheep.

Methane Inhibition

Methane is produced as a result of fermentation of feedstuffs in the rumen (Adam, 2000). Methane production is influenced by feed intake, animal digestive efficiency and dietary composition/quality. Amount of methane productiont depends on the efficiency of fermentation and conversion of feed into animal food products. Of the the volatile fatty acids (VFAs) produced during the fermentation, acetate and butyrate are methanogenic and spare hydrogen during formation while propionate is glucogenic in nature and utilizes the hydrogen. Hydrogen produced is then utilized by the methanogenic microbes (archaebacteria) in the production of methane and later belched out of the animal's mouth and nose, or it can convert the acetic acid into methane. The proportion of VFAs decides the extent of methane (CH4) production. The production of VFAs is basically a function of the dietary nature. The acetate to propionate ratio is generally lower for cereal grains than for forages. There is an inverse relationship between propionate and methane production. About 8 to 9 percent of gross energy (GE) is lost as a result of methane production.

Use of high tannin level cultivars of crops viz. trefoil, sulla, vetch, chicory, lotus in housed indoor sheep reduced CH_4 losses (g CH_4 Kg^{-1} DMI) between 20 and 55 % as compared to animals grazed on rye grass-white clover pasture (Ramirez-Restropp and Barry 2005). Similarly goats fed condensed tannin containing forage sericea lespedeza (*Lespedeza cuneata*) produced more than 30 % less methane compared with goats fed a mixture of crabgrass (*Digitaria ischaemum*) and tall fescue (*Festuca arundinacea*). Further the plants rich in saponins can be possible alternates to reduce methane emission from livestock because of saponins anti protozoal activities. Feeding tannin containing legumes viz. Sulla and Birds foot resulted in less methane production than ryegrass on feeding to sheep. Reduction in methane production was reported in sheep fed Lotus pendunculatus. This was due to the proanthocyanidins (PA) present in the Lotus pendunculatus. There was decrease by 16% by feeding 8 g kg^{-1} DM proanthocyanidins.

Future Research on Top Feeds

1. Exploitation of phenols and tannins and other anti-nutritional agents as potential methane inhibitor or possible rumen fermentation manipulator.
2. Chemicals of plant origin and plant extracts particularly from the unconventional resources should be tested for both for their defaunation and anti-methanogenesis properties
3. Development of total mixed rations to harness dual object of eco-friendly and higher animal productivity through varied proportions of crop residues, tree leaves, legume and non-leguminous crops based on their precise chemical evaluation for varied carbohydrate and protein fractions, their utilization and interaction with anti-nutritional constituents.

References

Adam, D. 2000. Tracking methane emission from cows just got a little easier. *url.http:://www.nature.com*

Baumer, M. 1991. Trees as browse to support animal production. Legume Trees and other Fodder Trees as Protein Sources to Livestock, Vol 102. FAO Animal Production and Health, Pp. 1-10.

Gupta, A. 2004. Characterization of protein and carbohydrate fractions of various feed resources and their utilization due to tannins in complete feeds. Ph D. Thesis, Bundelkhand University, Jhansi.

Ramirez-Restrepo, C. A. and Barry, T. N. 2005. Alternative temperate forages containing secondary compounds for improving sustainable productivity in grazing ruminants. *Animal feed science and technology,* 120: 179-201.

Kumar, R. and D' Mello, J. P. F. 1977. Anti-nutritional factor in Forage Legumes. Tropical Legumes in Tropical Nutrition. CAB Internacional, Pp. 95-133.

Reddy, D. V. and Elanchezhian, N. 2008. Evaluation of tropical tree leaves as ruminant feedstuff based on cell contents, cell wall fractions and polyphenolic compounds. *Livestock Research for Rural Development,* 20 (5) 2008.

❑❑❑

Agroforestry: Natural Resource Sustainability, Livelihood & Climate Moderation © 2009
Editors: OP Chaturvedi, A Venkatesh et. al., pp. 433-446
Satish Serial Publishing House, Delhi

Chapter 30

PROPAGATION OF TREE BORNE OIL SPECIES TOWARDS PRODUCTIVITY IMPROVEMENT

P. Ratha Krishnan
National Research Centre for Agroforestry, Jhansi-284 003 (U.P.)

Introduction

All life on this earth depends upon energy and today we are living in the era of advancement in science and technology. However the increase in population, industrialization and the modern lifestyle of human will overtake the advancements and exhaust the finite / fossil fuel reserves. In combination with the fast depletion of resources, the increase in crude oil price has made people to think about the alternative source of fuel. Environmental concern is alarming and the valuable thought like "We all have just one earth to live and life has to go on ....." is also evolving.

India consumes about 127 million tones of petroleum products annually (Biswas *et al.*, 2006) of which only 40 million tones are produced in the country. In order to save the huge amount of foreign exchange and self dependency of energy, the best opportunity in our hand is the bio-fuel. Where as bio-diesel is the name for alkyl ester of fatty acids made from vegetable oil or animal fats. Raw materials being exploited commercially by the bio-diesel producing countries constitute the edible fatty oils derived from rape seed, soybean, palm, sunflower, coconut, linseed, etc., (Reed, 1993). Use of such fatty oils to produce bio-diesel in India will not be possible in view of a large gap in demand and supply of edible oils. Further horizontal expansion of oil seed cultivation area is not possible due to the scarcity of cultivable land. In spite of the above demands and limitations, India has got 24.4 M ha of marginal / wastelands with degraded soils (Annual Report, 2000). And there is an urgent need to expand the area under forest cover to maintain the ecological balance.

As a progressive step in the direction of self sufficiency in oil and expansion of green cover is possible by planting suitable oil tree species which can be grown even under waste lands and unfavorable environmental conditions. Seeds of more than 100 species of forest plants - Tree Borne Oil seeds (TBOs) have been identified as source of fatty oils with an estimated potential of 11.3 Lakh tones (Dayal *et al.*, 1999). Standardization of propagation technique is the pre requisite for the improvement of such TBOs.

Propagation

Due to the long gestation period and unavailability of variety / improved seeds of trees, the production of genetically improved seedling have certain limitations. One such practice to overcome this limitation is to collect fresh and viable seeds from the good mother tree and use it for propagation. In case of short rotation species such as *Casuarina, Eucalypts,* Poplars and *Leucaena* the improved clones are available and they have to be used as mother tree for further Propagules production. So, it is clear that the long rotation species are mostly propagated sexually where the short rotation species is by asexually.

Sexual Vs Asexual propagation

Sexual propagation means propagation of species through seeds, which expresses segregation. Asexual propagation means the propagation of species by cuttings, grafting, budding, air layering or by micro propagation, which gives true to type progenies. Each method has its own advantages and limitations. True to type, early flowering and maturity, improved quality and any time production possibility and availability of planting material are the advantages of clonal propagation. Whereas the requirement of special propagation structures, skilled labor, higher initial costs are the limitations of it.

Seed Propagation

The main requirement of this method is the fresh, viable and quality seeds. Mother bed (raised / sunken), containers (Polybags / pots / dona / root trainers), nursery mixture (Forest soil, sand, FYM @ 2:1:1 ratio) are the other requirements. The following are some of the TBOs propagated by seeds, its number of seeds kg^{-1}, best seed collection period, viability, pre-sowing methods to break dormancy and germination %.

Species	No. of seeds kg^{-1}	Best seed collection time	Pre-sowing treatment	Viability	Germination %
Azadirachta indica	1750-4000	June - July	Fresh, Depulping and sowing	2 weeks	50
Jatropha curcas	1200 – 1800	Feb–Mar, Oct-Nov	Fresh, Water soaking 12–24 hrs	1 yr	>90
Pongamia pinnata	800 - 1500	March - May	Fresh, No treatment	12 months	80
Simarouba glauca	1000 -1500	Jan - April	Fresh, No treatment	1 yr	>80

(Banerjee, 1987, Ratha Krishnan, *et al.*, 2008)

Clonal Propagation

It requires the explants from the improved / selected mother tree. The required propagation structures for this method are: green house, mist chamber with cooling pad, polytunnels, shade / screen house, tissue culture room with temperature and light control. The required hardwares are root trainers (Composite / block / single cell), other containers, grafting machine, secateur, cutting knife, scissors and vermiculite. The following are the proven vegetative propagation methods for some TBOs.

Species	Type of propagation	Best month	Treatment	Reference
Azadirachta indica	Cutting, grafting, air layering	Feb - March	IBA	Ratha Krishnan *et al*, 2005, Gupta *et al*, 2008
Pongamia pinnata	Young shoot with heal cutting	April	4000ppm IBA	Thomas, 1988
Simarouba glauca	Grafting, air layering	Oct - Jan	4000ppm IBA	Ratha Krishnan, 2001

Improvement of TBOs

This chapter is an attempt to list the propagation activities carried by the author towards the improvement of TBOs namely *Simarouba glauca, Azadirachta indica* and *Jatropha curcas*. The following are the species wise attempts made by the author.

Simarouba Glauca Improvement

The growth performance of *Simarouba glauca* in the nine locations of Tamil Nadu, reported higher height, girth growth and seed yield in Mettupalayam location. The adaptability and suitability of this species was proved by the higher survival percent, good height and diameter growth in all the locations. This is due to the efficiency of this species to make successful colonization with the native VAM species (*Glomes mosseae*) (Kalai Selvi *et al.,* 2004; Ratha Krishnan *et al.,* 2008).

In the nursery experiment, the combined inoculation of *Azotobacer* (5g) and VAM (10g) registered the maximum root length, number of leaves, fresh and dry weight of root, root/shoot ratio, quality index, chlorophyll b content, nitrogen and potassium content and uptake, and VAM colonization per cent (Table 1). However, from the stand point of shoot length, collar diameter, volume index, chlorophyll a, total chlorophyll, total carbohydrate, total phenol and soluble protein content, the combined application of *Azotobactar* and *Phosphobacteria* performed superior to other biofertilizer treatments and uninoculated control. Considering the shoot dry weight, total dry matter production, relative growth rate and total phosphorus content the combined application of *Azospirillum*, *Phosphobacteria* and VAM registered the higher values (Ratha Krishnan *et al.,* 2004).

Table 1. Effect of biofertilizer treatments on growth and productivity of *Simarouba* seedlings - 180 DAS.

Treatments	SL (cm)	RL (cm)	TDMP (g)	Q.I
Azospirillum	21.50	31.80	6.49	1.2369
Azotobacter	21.52	31.74	6.24	1.2451
Phosphobacteria	21.50	30.87	6.26	1.3043
VAM	20.93	31.67	6.13	1.2517
Azo + Phospho	23.13	33.03	7.09	1.3301
Azo + VAM	22.83	39.07	6.75	1.3508
Azo+ Phospho+VAM	24.17	35.20	7.49	1.3287
Azoto + Phospho	24.60	33.30	6.95	1.2877
Azoto + VAM	22.90	38.47	6.96	1.3411
Azoto+ Phosp +VAM	23.97	36.23	6.85	1.2856
Phospho + VAM	22.03	36.10	6.94	1.3863
Control	19.27	25.03	5.79	1.1812
SEd	0.155	0.437	0.039	0.0012
CD (P=0.05)	0.380	0.861	0.075	0.0024

In case of asexual propagation, three methods were employed viz., air layering, cutting and grafting. Air-layers treated with 3000 ppm IBA registered more callus, whereas increases in rooting per cent, number of secondary roots, length of secondary roots and highest survival of translated layers were exhibited by IBA 4000 ppm. The higher dosage of 5000 ppm IBA produced more number of primary roots, increased length of primary roots and more height in translated layers (Table 2). Cuttings were not promising to rooting even with various hormonal applications in different rooting media with different types of cuttings.

Table 2. Effect of IBA on air-layers of *Simarouba glauca*

Treatments	Success %	Survival (%)	Ht of air layered material (cm)
1000 ppm IBA	44.25	43.75	17.25
2000 ppm IBA	78.92	55.00	19.25
3000 ppm IBA	76.75	63.75	21.25
4000 ppm IBA	65.08	65.00	24.75
5000 ppm IBA	42.00	63.75	23.75
Control	26.08	41.25	15.25
SEd	1.431	4.099	0.699
CD (P=0.05)	2.869	8.738	1.490

In case of grafting, the cleft grafting recorded the highest grafting success rate of 82.50 per cent with the diameter class of 8.60 to 10.00 mm roots stock (Table 3). The seasonal studies revealed that the grafting during October to January recorded a graft success of 82.50 to 87.50 per cent. April and May months were not found suitable for grafting which were recorded the minimum success rate of 45 to 50 per cent (Table 4) (Ratha Krishnan *et al.*, 2005 b).

The *Simarouba* seeds ripen in the month of March are eaten by mynas, parakeets, crows and monkey resulted for the spread of seeds during summer, which germinates in the subsequent rainy season by natural regeneration up to the distance of 500m from the source / mother trees. By this action, natural regeneration of Simarouba was noticed under *Santalum album, Logerstroemia, Acacia leucophloea, Syzygium cumini, Cassis hybrida* and *Millingtonia hortensis,* which are near to water source. The competition is very low. This natural regeneration potential of this species tends to prove its ecological adaptability to the new environment. (Ratha Krishnan *et al.,* 2005 c).

Table 3. Grafting method and root stock diameter class effect on grafting success in *Simarouba glauca* and *Azadirachta indica*

Methods	Grafting success %t for Simarouba				For neem
	Diameter class			Mean	Mean
	7.0-8.5mm	8.6-10.0mm	10.1-11.5mm		
Splice grafting	30.50	51.50	50.75	44.25	62.75
Cleft grafting	71.25	82.25	83.25	78.92	76.50
Whip & tongue grafting	69.00	80.50	80.75	76.75	48.50
Saddle grafting	59.25	68.25	67.75	65.08	54.50
Four flap grafting	39.75	46.50	39.75	42.00	34.00
Side veneer grafting	24.00	28.50	25.75	26.08	38.00
Mean	48.96	59.58	58.00		
	T	S	T x S	SEd	2.368
SEd	2.024	1.431	3.506	CD (P=0.005)	5.049
CD (P=0.05)	4.058	2.869	7.028		

Table 4. Seasonal effect of grafting in *Simarouba glauca*

Months	Grafting success % in Simarouba
January	82.5
February	77.5
March	65.0
April	50.0
May	45.0
June	55.0
July	50.0
August	52.5
September	70.0
October	87.5
November	85.0
December	85.0
SEd	2.041
CD (P=0.05)	4.213

Azadirachta indica Improvement

Screening for Neem CPTs were done by scoring methods for the traits viz., epicormic branches, branching, bole straightness, tapering, crown shape, fluting, pest and diseases. Selection of plus trees from the top scored candidate trees was done by regression analysis. Mean biometric data of the all selected plus trees were varied. Among the plus trees, plus tree from Tirunelveli registered highest basal girth (58.57%), GBH (72.93%) and volume (223.05%) increase over its group mean. (Ratha Krishnan and Paramathma, 2008a).

Semi-hardwood cuttings of neem treated with different concentrations (1000, 2000, 3000, 4000 and 5000 ppm) of IAA, IBA and NAA. Among the sixteen treatments, IBA 3000 ppm exhibited higher sprouting per cent of success followed by IBA 4000 ppm. The lowest sprouting was recorded in control followed by IAA 1000 ppm. In case of rooting IBA 3000 ppm induced higher success of rooting followed by IBA 2000 ppm. The shoot length and root length of the hormonal treated cuttings vary and the higher values recorded for the treatment IBA 3000 ppm whereas cuttings treated with IAA 2000 ppm produced higher number of roots (table 5). Seasonal study with cuttings revealed that the cuttings collected during February, prior to flowering produced higher sprouting and rooting than the cuttings collected during August. The growth of cuttings collected in August and February showed significant difference in growth parameters also. Cuttings collected during February recorded higher shoot length; root length and number of roots than the August. (Parthiban *et al.*, 2005; Ratha Krishnan & Paramathma, 2008 b).

Table 5. Effect of hormones and season on the sprouting and rooting of *Azadirachta indica* cuttings

Treatments	Percentage of sprouting			Percentage of rooting		
	August	February	Mean	August	February	Mean
Control	26.67	43.33	35.00	13.33	26.00	19.67
IAA 1000 ppm	30.33	43.33	36.83	13.33	34.33	23.83
IAA 2000 ppm	33.33	65.00	49.17	16.67	30.33	23.50
IAA 3000 ppm	33.33	66.67	50.00	20.00	41.00	30.50
IAA 4000 ppm	26.67	61.67	44.17	17.00	41.67	29.34
IAA 5000 ppm	30.00	56.00	43.00	13.67	35.67	24.67
IBA 1000 ppm	26.67	48.33	37.50	16.67	44.67	30.67
IBA 2000 ppm	33.33	70.00	51.67	23.33	59.66	41.50
IBA 3000 ppm	60.00	75.67	67.84	30.00	68.33	49.17
IBA 4000 ppm	43.33	81.67	62.50	20.00	49.00	34.50
IBA 5000 ppm	33.33	68.33	50.83	23.33	54.33	38.83
NAA 1000 ppm	26.67	61.67	44.17	13.33	42.67	28.00

Contd...

Table 5. Contd...

NAA 2000 ppm	30.00	53.67	41.84	16.67	52.00	34.34
NAA 3000 ppm	23.33	71.67	47.50	13.67	39.00	26.34
NAA 4000 ppm	33.33	65.00	49.17	23.67	53.00	38.34
NAA 5000 ppm	30.00	53.33	41.67	13.33	45.67	29.50
Mean	32.52	61.58	.	18.00	44.83	.
	T	S	TXS	T	S	TXS
SEd	4.627	1.636	6.544	4.882	1.726	6.904
CD (P=0.05)	9.245	3.268	10.073	9.753	3.448	13.793

Among the six grafting methods, cleft grafting recorded higher grafting success per cent followed by splice grafting. The lowest grafting success was recorded by four flab grafting (Table 2). Scions collected from the selected plus trees used to make cleft grafting in neem. The success per cent and growth of the grafts vary with source of scion. The highest grafting success per cent was obtained for the scions of plus tree from Madurai, whereas Ramanathapuram and Dharmapuri registered the low. In case of growth the higher shoot length and collar diameter were obtained for the scions from Tiruchendur. (Ratha Krishnan and Paramathma, 2007).

Air-layering in neem was done by applying 800 ppm IBA and using sphagnum mass as covering material. This 60 days observation showed 100% callus formation during February to August and October, where as 100 % rooting was achieved only during July. The calluses formed during June to September are vigorous and the roots are healthy and profuse. Rooting (during April) was increased from 40% with months and reaches 100 % during July. Again it starts declining in proceeding of months and reaches zero during November to February. The analysis of rooting success with climate data and phenology of neem revealed that July (starting of monsoon), just after fruiting gave higher rooting while the winter season (November to February) exhibited no rooting which may be due to low atmospheric humidity and temperature and also by the physiological dormancy. (Ratha Krishnan *et al.*, 2005 a; Gupta *et al.*, 2008).

Field evaluation of neem clones through biometrical, biochemical and physiological traits were done in the field. Among the clones, C_2 recorded superior height, basal diameter, number of leaves, chlorophyll a, chlorophyll b, total chlorophyll, net photosynthetic rate and transpiration rate which was followed by C_{20}, C_{11} and C_{14}. Among the five biometrics observed, volume index registered higher PV, GV, PCV, GCV, heritability and genetic advance as percentage of mean, whereas the lower heritability and genetic advance as percentage of mean displayed by number of leaves. In case of biochemical and physiological traits highest PCV and GCV recorded for chlorophyll b and highest heritability was recorded for the trait IAA oxidase. The lower heritability was recorded for specific leaf area followed by stomatal conductance. D^2 analysis on the 6 MAP data diversified the clones in to 10 clusters. In that clone C_1, C_8 and C_{12} comes under one cluster and C_6, C_{11} and C_{14} grouped in another cluster. The diversity indices revealed that volume index and total chlorophyll contribution much to the genetic divergence of clones (32.38% and 16.19% respectively).The NTSYSpc analysis recorded higher Euclidean distance among C_2 and C_{18} where the lower recorded between C_2 and C_{10}. The clones showed divergence at different coefficient levels. In a very narrow coefficient level of 1.34 the locations C_7 and C_{12} showed divergence. In RAPD analysis the amplified product varied in number and intensity among the locations. The higher number of amplified products observed in the profile of OPR 10 followed by OPR 02. The similarity index obtained from this study showed lowest similarity coefficient between C_{13} and C_{18}. The highest similarity coefficient value was recorded between C_6 and C_7. The geographical distribution of the clones was not expressed in the clustering pattern. [Ratha Krishnan and Paramathma, 2007; Ratha Krishnan and Paramathma, 2009 (in press)].

Table 6. Seasonal effect of air-layering in neem – IBA 800ppm

Months	Callus formation (%)	Rooting success (%)
January	100	15
February	20	0
March	0	0
April	30	0
May	100	0
June	100	60
July	100	40
August	100	60
September	100	90
October	100	100
November	100	80
December	80	40

Jatropha curcas Improvement

Preference analysis, popularization, and standardized cultivation practices of *Jatropha curcas* was done in the southern zones of Tamil Nadu. (Ratha Krishnan and Paramathma, 2006).

The combined application of Azospirillum (5g) + Phosphobacteria (5g) + VAM (10g) exhibited synergistic effect on growth, productivity and quality of *Jatropha curcas* seedlings. It was expressed by higher shoot length, root length, collar diameter, number of leaves, Shoot fresh and dry weight, root fresh and dry weight, root shoot ratio and quality index in both the stages The next best treatments suggested for the production of quality Jatropha seedlings were Azospirillum alone and Azospirillum + Phosphobacteria + Vermicompost (Table 7). The influence of treatments on VAM colonization % was not satisfactory and non significant too (Gajendra Singh *et al.,* 2008).

Table. 7. Effect of biofertilizers on growth and productivity of *Jatropha curcas* seedlings -90 DAS.

Treatments	SL (cm)	RL (cm)	TDMP (g)	Q.I
Urea	39.33	20.67	8.57	0.5003
Azospirillum	38.57	21.00	8.77	0.5747
Phosphobacteria	39.40	17.00	7.45	0.3458
VAM	40.47	19.33	7.86	0.5373
Vermicompost	41.40	21.17	8.03	0.4586
Azo + Phospho	39.93	18.00	7.32	0.4899
Azo + VAM	42.33	23.83	7.50	0.3778
Azo + Vermi	38.23	19.10	7.66	0.3940
Phospho + VAM	41.67	17.83	7.06	0.3296
Phospho + Vermi	38.90	21.83	7.91	0.3853
VAM + Vermi	38.87	19.50	6.25	0.5029
Azo + Phospho + VAM	46.23	25.67	9.32	0.6266
Azo+ Phospho + Vermi	41.43	21.83	8.47	0.5570
Azo+ VAM + Vermi	41.07	21.83	6.95	0.4084
Phospho+VAM + Vermi	38.00	23.43	7.09	0.4830
Azo+Phos+VAM+ Vermi	41.10	22.83	8.56	0.5824
Control	35.20	16.17	6.59	0.3875
SEd	2.78	2.11	0.64	0.0417
CD (P = 0.05%)	5.64	4.29	1.30	0.0849

The calculation on cost of Jatropha seedling production was showed Rs. 2872/ 1000 seedlings. In which the maximum (56 % of total) cost was incurred for the seedling maintenance (Rs. 1600.00). The cost and benefit analysis of *Jatropha curcas* seedlings production by the application of treatments showed the advantage of saving of Rs. 249.20 and Rs. 206.80 by the single application of Vermicompost and urea respectively, while the combination of vermicompost and biofertilizers increased the cost involved in the production of plantable Jatropha seedlings. When quality and economics were equally considered single application of Azospirillum or Phosphobacteria can produce quality seedlings with good savings. (Gajendra Singh *et al.,* 2008).

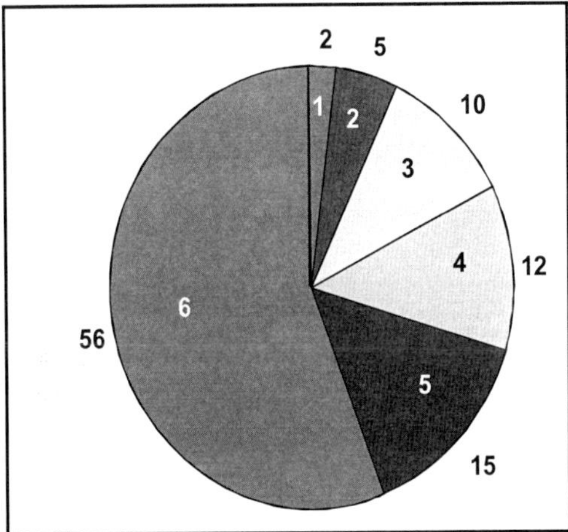

Fig. 1. Share (%) of costs involved in *Jatropha curcas* seedling production

The success of inter-specific grafting in *J. gossypiifolia* as root stock and *J. curcas* as scion was 80 % and union was noticed within a short period (21 days) whereas *J. curcas* as root stock and *J. gossypiifolia* as scion gave 80 % of success in 35 days. The grafting success by *J. gossypiifolia* as root stock and *J. curcas* as a scion is a potential technique which can be exploited to incorporate more hardiness and adaptability in *Jatropha curcas* for the Bundelkhand region. The grafting union success in the single stem / early stage is also more (80 %) than the matured / multi stem stage (40 %). The nursery maintenance (1year) of the grafted material and field planting doesn't show any grafting incompatibility symptoms. (Ratha Krishnan *et al.,* 2007).

Advantages Due to the above Techniques

- Increased growth and productivity of TBOs seedlings
- Cost of seedling production will be minimized
- True to type progeny production

- Identification of best genotype of TBOs
- Early flowering and fruiting which facilitates early assessment of generations.

Conclusion

Scientific research is the backbone for societal growth and prosperity. The investigations lead to be proven technologies, when they are standardized and exploited in better way. The above manuscript clearly explains the possible options to reduce the nursery period and the production on true to type progenies of TBOs namely *Simarouba glauca, Azadirachta indica* and *Jatropha curcas*. When our aim is to improve the productivity of TBOs and the cost of seedling production is mainly shared by the maintenance cost the above standardized techniques have to be used to tap the benefits. If the above standardized techniques are used for the production of TBOs seedling, it will give good quality planting material to meet our future energy needs with lesser inputs.

References

Annual Report, 2000. Ministry of Environment as Forest, New Delhi.

Banerjee, A.K. 1987. Nursery manual for species commonly used in social forestry in India, Rep: National Wasteland Development Board / 001 / 007 / 1987, Kapoor press, New Delhi, Pp. 64.

Biswas, S., Srita, G. and Narmala, K. 2006. Biodiesel: Fuel of the future? Biodiesel resources can provide energy security. The Hindu, August, 3, 2006.

Dayal., Rameshwar, Jain, P.P. and Soni, P.L. 1999. Non-wood Forest Products –A chemical Approach. Proc: International workshop on Forestry Research in Conservation of Natural Forests (Negi, R.S et al., eds)UNDP- ICFRE, Dehradun, Pp. 42 – 47.

Gajendra Singh, Ratha Krishnan, P., Gupta, V.K. and Anil Kumar. 2008. Biofertilizer technology for the production of quality Jatropha seedling production. Abstracts In: National symposium on Agroforestry knowledge for sustainability, climate moderation and challenges ahead. NRCAF, Jhansi, 15 – 17, Dec, 2008. Pp. 198.

Gupta, V.K., Ratha Krishnan, P., Ahlawat, S.P., Kumar, R.V. and Dutta, A. 2008. Influence of season on air-layering of *Azadirachta indica* by IBA. In: Abstracts of Golden Jubilee Conference on "Challenges and emerging strategies for improving plant productivity" held at New Delhi, 12 – 14, Dec, 2008, Pp. 243

Kalaiselvi, T., Gunasekaran, S. Sundersingh Rajapandian, J. Ratha Krishnan, P. Sekar, I. Thyageswari, S. and Neelakantan, K.S. 2004. Mycorrhizal dependency of *Simarouba glauca*. *Mycorrhiza News*, 16 (1): 14 – 15.

Ratha Krishna, P. and Paramathma, M. 2007. Standardization of neem grafting technique and its success variation with respect to CPTs of Tamil Nadu. In abstracts of "World Neem Conference" held at Coimbatore, 21 – 24, Nov, 2007, Pp. 76.

Ratha Krishnan, P. 2001. Studies on elite seedling production in *Simarouba glauca* (Linn.) by sexual and asexual propagation. M.Sc (Forestry) thesis, Tamil Nadu Agricultural University, Coimbatore, Pp.141.

Ratha Krishnan, P. and Paramathma, M. 2006. Studies on comparative assessment of tree borne oil species in Southern zone of Tamil Nadu, India. In: Proceedings of the "International Forestry and Environment Symposium" held at University of Sri Jayawardenapura, Sri Lanka, 22 – 23 Dec, 2006, Pp. 55.

Ratha Krishnan, P. and Paramathma, M. 2008a. Physiochemical traits and RAPD analysis of grafted neem progenies in Tamil Nadu – Preliminary evaluation. In: Abstracts of Golden Jubilee Conference on "Challenges and emerging strategies for improving plant productivity" held at New Delhi, 12 – 14, Dec, 08, Pp.196 – 197.

Ratha Krishnan, P. and Paramathma, M. 2008b. Seasonal variation in rooting of *Azadirachta indica* cuttings treated with auxins. In: Abstracts of Golden Jubilee Conference on "Challenges and emerging strategies for improving plant productivity" held at New Delhi, 12 – 14, Dec, 2008, Pp. 235.

Ratha Krishnan, P. and Paramathma, M. 2009. Preliminary evaluation of growth, physiological and biochemical traits of grafted neem progenies in Tamil Nadu. *Indian Journal of Agroforestry*, 11 (1). (In Press).

Ratha Krishnan, P. Paramathma, M. and Parthiban, K.T. 2005 a. Screening and multiplication of superior genotypes of neem for quality wood production. Proceedings of IUFRO International Conference on multipurpose trees in the tropics: Assessment, growth, and management, Jodhpur, Nov 22 – 25, Pp. 679 - 684.

Ratha Krishnan, P. Paramathma, M. and Sundersingh Rajapandian, J. 2004. Nursery studies on the influence of microbial biofertilizers on growth and biomass accumulation in pollypotted *Simarouba glauca* Linn. seedlings. *Journal of Non-timber Forest Products*, 11 (1): 57 – 60.

Ratha Krishnan, P. Sundersingh Rajapandian, J. and Neelakantan, K.S. 2005 b. Evaluation of growth and standardization of propagation methods of exotic *Simarouba glauca* in Tamil Nadu. Proceedings of IUFRO International Conference on multipurpose trees in the tropics: Assessment, growth, and management, Jodhpur, Nov 22 – 25, P 749 - 753.

Ratha Krishnan, P., Kumar, R.V., Handa, A.K. and Gupta, V.K. 2007. Grafting of *Jatropha curcas on Jatropha gossypifolia* to induct drought hardiness and adaptability. *Agroforestry Newsletter*, 19(4): 2-3.

Ratha Krishnan, P., Sundersingh Rajapandian, J. and Paramathma M. 2008. *Simarouba glauca* – Potential exotic for greening of wastelands of Tamil Nadu. In: Exotics in Indian Forestry, Eds: Sanjeev K. Chauhan, S.S. Gill, Rajni Chauhan and S.C. Sharma, Pub: Agrotech Publishing Academy, Udaipur, Pp. 278 – 282.

Ratha Krishnan, P., Sundersingh Rajapandian, J., Kumaran, K and Neelakantan K.S. 2005 c. First report on Natural regeneration of *Simarouba glauca*. *Indian Forester*, 131 (1) 129 – 130.

Reed,T.B. 1993. Proceeding of First Biomass Conference of the Americas, 22: 797.

Thomas, K.I. 1988. FRC Pub. (43) vii + 52 pp. 17 ref. Samdakan.Sabah, For.Res.Centre.

□□□

Agroforestry: Natural Resource Sustainability, Livelihood & Climate Moderation © 2009
Editors: OP Chaturvedi, A Venkatesh, pp. 447-460
Satish Serial Publishing House, Delhi

Chapter 31

IMPACT OF TREE SHELTERBELTS IN SUSTAINABLE MANAGEMENT OF ARID FARMING SYSTEM

Rajendra Prasad and R.S. Mertia[1]
National Research Centre for Agroforestry, Jhansi-284 003 (U.P.)
[1]Central Arid zone Research Institute, RRS, Jaisalmer-01 (Raj.)

Introduction

The hot arid region of India, covering an area of about 32 million ha, spread in states of Rajasthan (61%), Gujarat (20%), Haryana (4%), Punjab (5%) and Andhra Pradesh (7%). Low and erratic rainfall, extreme temperatures, very high wind velocity, low humidity and frequent droughts characterize the hot arid ecosystem. Low productivity is associated with inadequate length of growing period. The productivity remains restricted due to un-conducive environment, limited choice of crops and aberrant weather conditions. Shelterbelt plantations, which are vegetative barriers of trees/ shrubs/bushes, are used as protective measures to minimize hazardous effects of wind, particularly in hot arid region where wind erosion is a dominant phenomenon. They are designed to reduce wind velocity thus soil erosion, provide protection to roads, railways, water bodies, canals etc and create favorable environment for cropping, horticulture and livestock enterprises. Besides, shelterbelts are also viewed as mean to fulfill objectives of social forestry in supplementing local demands of fuel, fodder, shelter and timber. With the expectation of these results, the work on shelterbelt was started way back in 1950s. Initial research work was aimed to develop design of shelterbelt/windbreaks, screen suitable tree species for its composition and develop techniques for plantation establishment. In past 50 years, volume of data has been generated on design, composition, suitable tree species, planting technique etc for shelterbelts (Mertia, 1986). The current chapter discusses various impact created by shelterbelt technology in extremely hot arid ecosystem.

Design of Shelterbelt Plantations

The early research work (in late fifties and early sixties) on shelterbelt was aimed to develop its suitable design having optimum planting geometry, porosity, thickness (No. of rows) and shape. Simultaneously, the emphasis was also given on screening of most suitable trees and developing techniques for establishment of planted seedlings. For providing maximum shelter zone, the shelterbelts having 40 -50 % porosity with evenly distributed gaps was assessed to be the best (Nageli,1946). For arid condition in Thar Desert, Kaul (1969) suggested five-row shelterbelt with a pyramidal shape having one row of tall trees such as *Acacia tortlis, Tamarix articulata, Azadirachta indica* followed by two rows of smaller trees like *Acacia senegal, Prosopis juliflora, Parkisonia aculeata* etc. and then followed by two rows of shrubs such as *Aerua tomentosa, Zizyphus spinachristi* and *Calligonum polygonoides* at the edge in the flank rows.

In certain regions shelterbelts have mainly been designed and established to control soil erosion and help maintain soil fertility, improve microclimate and appropriate long-term yields. Single row belts spaced at 30 to 150 m, or even more have been used to control soil blowing. In intensively - cultivated areas, where all crop material including root is removed from the ground for food, fuel or shelter, rows are spaced at 6 to 20m in one direction and 30 to 150 m apart at right angles to the first (Chepil, 1949). In order to have minimum loss of water through extraction to tree-roots in a shelterbelt, trees with deep rooted systems have been preferred at the edges of the fields (Said, 1954).

Choices of species and planting techniques have always been crucial factors for establishment of shelterbelt plantations. Seedlings raised in sun dried bricks or polytubes and well adapted in nursery when planted in pits of 45 cm^3 or 60 cm^3 have proved very successful. Such deep planting facilitates root access to moisture from lower depths. Seedlings raised in galvanized iron containers and planted in duny areas have also given successful results. Tree species such as *Acacia tortilis, Albizia lebbeck, Dalbergia sissoo* and *Prosospis juliflora* have been raised successfully using above techniques. Pre-sprouted branch cutting of certain species like *Calligonum polygonoides* and *Tamarix articulata* have also been planted successfully. To keep livestock away from the trees, the technique adopted has been to surround a central planting pit with a 30 cm wide and 30 cm deep circular trench. However, in loose sandy areas tree guards are more suitable because trenches collapse. Nowadays, fencing with an iron angle post and barbed wire has been adopted to protect seedlings from cattle, sheep, goats and camel (Mertia, 1992).

With the passage of time the protective forestry has transformed into productive forestry and accordingly, the planting geometry and choice of species have changed. Earlier the protective forestry, mainly comprised of belt plantations for conservation of soil and water, anchoring sand dunes, reducing speedy wind and providing shelter to crops and cattle from wind and scorching heat was under practice. However, in productive forestry the objective of producing fodder (top feed), fuel and small timber was also coupled with the objectives defined for protective forestry. Hence, many multi-purpose tree species (MPTs) were screened and evaluated for the purpose of belt plantations. In recent years, single, double or triple row shelterbelts comprising mainly of *Acacia nilotica, Azadiracta indica, Prosopis cineraria, Albizzia lebbeck* etc have been planted along boundaries of agricultural fields, canal sides and roads. For different purposes, suitable species and design of shelterbelts have been given in Table 1.

Table 1. Design and species for shelterbelt plantations

Purpose	Design	Suitable species
Road side	3 to 5 staggered rows	*A tortilis, A. lebbeck, A. indica, D. Sissoo, P. aculeata T. juliflora* and *Tamarix articulata*
Railway lines	6 rows	*P. acculeata, P. juliflora, T. articulata*
Canal side	6 rows	*A. nilotica, Eucalyptus spp. T. undulata, A. tortilis P. juliflora, D. sissoo, P. cineraria A. nubica*
Farm boundary (rainfed)	1 / 2 / 3 rows	*Accacia tortilis, A. Lebbeck, A. indica, D. sissoo, P. aculeata, P. juliflora* (most common)
Farm boundary (irrigated)	2 rows	*A tortilis, A. lebbeck, Dicrostachys cineraria, P. juliflora*

Protection of roads and railways through side-belt plantations is a major requirement of plantation forestry in arid zone. The state forest department has undertaken roadside planting in eleven districts of Western Rajasthan. Figures in respect of Jaisalmer district are given in Table 2.

Table 2. Details of road side plantations in Jaisalmer

Time period	No. of sites	Area covered (Rkm)*	Area (ha)
1960-70	5	38	16
1970-80	7	115	56
1980-90	33	1128	564
1990-98	51	742	1484

* Rkm= Running kilometers; Source: Office of Dy Conservators of Forests, Jaisalmer

To check engulfing of railway line from blown sand in part of Bikaner, Sikar, Churu, Jaisalmer, Barmer and Jodhpur districts of Western Rajasthan, a program on railway line plantations was initiated in 1980s. In beginning a selection of 100 km along Sikar-Loharu, Sikar-Fatehpur and Palsana-Deshnoke lines was planted with a six row tree belt using *Parkinsonia aculeata, Prosopis juliflora* and *Tamarix articulata* (Mann and Muthana, 1984). With the advent of Indira Gandhi Nahar Pariyojna (IGNP) in Western Rajasthan, massive tree belt plantation work was undertaken to check silting up of canal through wind blown sand. Six row tree belts consisting of *Acacia nilotica, Eucalyptus spp, Tecomella undulata, Acacia tortilis, Prosopis cineraria, Dalberjia sisso, Tamarix aphylla* etc have been planted on either side of the canal (Table 3).

Table 3. Detail of canal side plantation in Jaisalmer

Year	No. of sites	Area covered	
		Rkm	ha
1991-92	14	1056	352
1992-93	35	2679	893
1993-94	61	3954	1318
1994-95	65	3744	1248
1995-96	60	3714	1238
1996-97	64	4680	1560
1997-98	102	7986	2662
Total	401	27813	9271

Source: office of Chief Conservator of Forests, Bikaner

Impact of Shelterbelt Plantations

Wind erosion: Tree shelterbelts have been found effective in reducing wind speed and thus soil erosion (Mertia, 1992). The wind speed reduction is higher during monsoon season than in summer due to thicker canopy growth (Table 4). Different shelterbelts are associated with different degrees of wind reduction in different seasons and over the surrounding area.

Table 4. Percent reduction of mean wind speed at different distance in the leeward side of different Shelterbelt (1977-79) at Jodhpur.

Species	Distance from shelterbelt					
	Summer season			Monsoon season		
	2 H	5 H	10 H	2 H	5 H	10 H
Prosopis juliflora	33	17	12	38	26	21
Cassia Siamea	36	17	13	46	36	24
Acacia tortlis	36	25	13	46	36	20

Due to reduction of wind speed the soil loss in shelterbelt area was reportedly reduced by 50 percent as compared to an unsheltered area (Gupta *et al.*, 1984). The shelterbelt based on *Cassia siamea* proved most effective in conserving soil. While assessing impact of shelterbelt in arid ecosystem, Mertia *et al.*, (2006) observed that different shelterbelts had varying effects on reduction of wind speed in leeward side of the belt. The maximum extent of reduction in wind speed of leeward side was observed at distance of 2H (H is the average height of the shelterbelt in meter) which slowly got nullified up to 20H (Table 5). They further observed that on an average, irrespective of the species, double row shelterbelt were more effective in reducing wind speed (Fig 1) as compared to single row belt. However, single row belt provided more effective area in comparison to double row belts.

Table 5. Effect of shelterbelt on wind speed in leeward side of the belt

Shelterbelt	Direction and number of rows	Mean height (m)	Per cent reduction in mean wind speed				
			2 H	5H	10H	15 H	20 H
Tube well command, Lathi series							
Acacia tortilis	S-W 2	8.9	29.6	24.5	10.7	3.3	0
Eucalyptus camaldulensis	S-W 1	15.5	24.8	18.6	12.5	9.4	7
Dalbergia sissoo	S-W 2	7.8	32.4	17.9	5.5	1.5	0
IGNP command, Mohangarh							
Dalbergia sissoo	S-W 1	6.2	21.5	14.2	6.2	3.2	0.2
Dalbergia sissoo	S-W 2	8.8	36	28.7	12.2	1.4	0
Tecomella undulata	SW 1	10.2	24	19.7	9.7	3.5	2

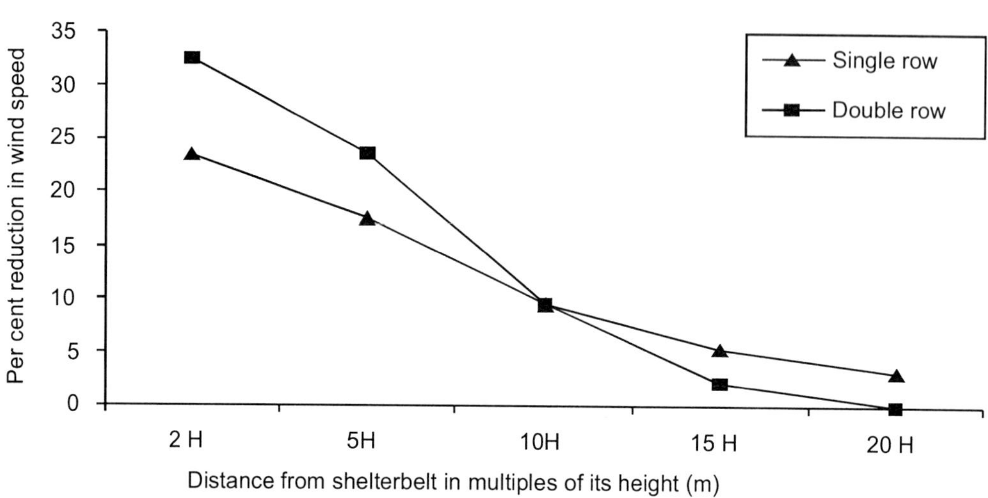

Fig 1. Effect of type of shelterbelt on reduction in wind speed in leeward side of the shelterbelt.

It appears that the structure of canopy, density (density refers to the proportion of solid material such as foliage, branches etc), age, height and direction of the shelterbelt plantations are major factors which decide its effectiveness in controlling the wind flow. Fairly tall and dense double row *D sissoo* reduced maximum wind speed in IGNP area but its effectiveness was only up to distance of 15H whereas, 20 years old single row *E. camaldulensis* provided sheltered area up to distance of 20H in tube well command area of Lathi series. In other shelterbelts the reduction in wind speed in leeward side was more pronounced up to distance of 10 H and limited to the distance of 15 H. For maximizing effectiveness of shelterbelt plantations for providing large sheltered area in leeward side, it is desirable to plant those species which, can grow fast and attain maximum height in short period. Belt of multiple rows apparently provide more resistance due to dense canopy resulting in decrease of sheltered area in the leeward side of the belt.

Blockage of Roads & Railways: The main Objective with which road side plantations or shelterbelts along roads are raised is to abate deposition of sand on the road and reduce the velocity of moving wind. Survey of selected road sides in Jaisalmer has indicated that road side plantations have been efficacious in reducing deposition of blown sand (Mertia *et al.*, 2006) which, otherwise deposited and blocked roads (Table 6).

Table 6. Road blockages due to blown sand in absence of afforestation.

Sr. No.	Name of Road	Location (Km)		Length (Mtrs)	Height (Mtrs)
1	Ramgarh -	11.20	11.50	300	3
	Longewala	14.00	14.35	350	3
		28.20	28.30	100	1
2	Ramgarh - Ashutar	37.10	37.50	400	3
	- Ghotaru	44.60	44.80	200	1

Source: Garison Reserve Engineers Force, Jaisalmer (2000).

Silting of Canals: Sand deposition in canals has reduced considerably after shelterbelt plantations along canal sides. It is inferred that canal side belt plantations have been able to achieve their main objective for which they were raised. The pre-plantation scenario (before 1991) in the area where there are now 8 years old tree belt exists was drastically different. Shifting dunes, in the absence of any check in the form of plantations, deposited huge quantities of blowing sand in the main canal and its distributaries thereby either completely choking or reducing the flow of the water. In order to de-silt the canals to restore flow of water, huge expenditures were incurred. The seriousness of the problem can be gauged from the fact that about 0.57 million cu m of blown sand had to be removed from the main canal (km 383.2 to km 444. 8) at an approximate cost of Rs. 72 lakhs in December 1986 (Bithu, 1989). A study of Upadhaya (1991) in IGNP stage II area in Jaisalmer have revealed that quantum of sand deposited in canal was reduced by 513 m^3 and 1023 m^3 per Km at one and two years old canal side tree-belt plantations , respectively as compared to unplanted sites (control). The corresponding savings on de-silting of deposited sand was estimated to be in the tune of Rs. 6,156 and 12,276 per km. The study further revealed that quantum of sand movement varied significantly with respect to various months of a year (Table 7).

Table 7. Effect of canal side plantation on sand deposition (M^3 / RM of canal) in canal

Month	Control	One year old plantation (M^3)	Two years old plantation (M^3)
Nov. 89	0.089	0.071	0.041
Jan.90	0.094	0.068	0.041
March 90	0.150	0.109	0.069
May 90	0.261	0.225	0.141
June 90	0.323	0.289	0.157
July 90	0.198	0.101	0.051
Sept. 90	0.068	0.041	0.021
Oct. 90	0.039	0.033	0.016

Source: Upadhyaya (1991).

Available information on removal of blown sand from canal beds has revealed that inverse relationship exists between the age of tree belt and the quantity of sand deposition. The older the plantations the lesser the quantity of sand deposited in canals. The data on removal of sand from canal bed in different years would appreciate this hypothesis (Table 8).

Table 8. Reduction in blown sand deposition in canal (M^3 per km) after canal side plantation

Branch/Distri/MInors	1994-95	1996-97	1997-98	1998-99
Sagarmalgopa Sakha	4125	2877	2079	1269
Mandau Distributry system	2296	777	373	718
Sankla minor	1949	1165	1009	388
Tibrewala MInor	1875	563	1250	1250

Source: Chief Engineer, IGNP, Jaisalmer.

Microclimate: Significant changes in microclimatic conditions have been assessed under shelterbelts as compared to unsheltered area. Favorable impact of canal side (Table 9) plantations on microclimate has also been noticed (Mertia *et al.*, 2006). Trees having good crown cover markedly affect the inside temperature and relative humidity levels irrespective of their age. Wherever, crown density is poor, very little or simply no difference in the inside and outside values is observed. Unlike canal side plantations, the roadside plantations did not show any significant variations in microclimatic conditions between inside and outside of the tree belt (Table 10). This could be due to the emission effects of running vehicles on road. It is inferred that canal side shelterbelt plantation and other afforestation activities have brought about significant change in microclimate at different places, which may have some impact on macroclimate of the district as a whole. Rise in decadal mean rainfall due to the effect of an increase in vegetation cover has been reported (Ramkrishna and Rao, 1991).

Mertia *et al.* (2006) revealed that shelterbelts brought significant changes in microclimatic conditions on farmer's fields. Considerable reduction in air temperature in sheltered area of leeward side was noticed, which varied with the type of shelterbelt plantations. Irrespective of type of shelterbelt plantation, maximum reduction in air temperature was observed beneath the shelterbelt (Table 11). The reduction in temperature decreases water loss from the soil through evaporation. Muthana *et al.* (1984) reported that the evapotranspiration (PE) is largely influenced by the wind in arid region and shelterbelt had reduced 5 -14% pan evaporation values on either side of the belt (Table 12). The microclimatic variations within shelterbelts and their modification with respect to seasons have significant impact on associated agricultural crops (Table 13). Positive effects of shelterbelt on conservation of soil moisture on its leeward side have been reported (Mertia, 1992).

Table 9. Impact of canal side plantation on micro climate

Sr. No.	Site	Time	Temperature (°C) In Side	Temperature (°C) Out Side	RH % In Side	RH % Out Side
1	40-50 RDSBS	2.20 PM	33.0	36.0	47	42
2	14-20 RD	5.00 PM	38	38	28	28
3	30-40 RD SMGS	12.10 PM	36.5	40.0	42	27
4	0-10 RD Sadha disty R/S	10.35 AM	32.5	24.0	52	51
5	70-80 RD Sultana Disty	9.30 AM	37.0	39.0	31	29
6	12-20 RD Sadha Disty R/S	5.00 PM	39.5	39.5	23	22
7	165-175 RD Sadha B/S	6.30 PM	32.0	32.0	58	58

Table 10. Impact of road ride plantation on microclimate

Sr No.	Site	Time	Temperature (°C) Inside	Temperature (°C) Outside	Relative Humidity (%) Inside	Relative Humidity (%) Outside
1	54RDBS to Digha	8.00 AM	28.0	28.0	81	70
2	Sultana Kabir Basti	5.00 PM	40.0	40.0	27	27
3	86-104 SBS R/S	1.30 PM	38.0	38.0	38	38
4	190-212 RD	10.30 AM	33.0	33.0	47	47

Table 11. Effect of different tree shelterbelts on daily air temperature (°C)

Shelterbelt	Direction and no. of rows	Daily air temperature 0c 0H	1H	2H	5H	10 H
Tube well command Lathi series						
A. tortilis	S-W 2	22.5	23.4	24.2	24.8	26.5
E. camaldulensis	S-W 1	22	23.5	24.6	25	26.5
D. sissoo	S-W 2	23	24.5	25.8	26.4	26.5
IGNP command Mohangarh						
D. sissoo	S-W1	41	43.2	44.1	44.5	44.5
D. sissoo	S-W2	36.2	42.8	43.8	44	44.1
T. undulata	SW 1	40.4	40	42.5	44	44

Table 12. Effect of *Cassia siamea* shelterbelt on pan evaporation

Location	Pan evaporation (mm day^{-1})			
	April	May	June	July
Windward side	11.0	14.5	13.4	6.0
Leeward side	9.5	13.4	12.5	5.7
Percent decrease	14	8	8	5

Table 13. Effect of shelterbelt on yield of pearl millet

Distance from shelterbelt	Pearl millet yield (q ha^{-1})	Control	Consumptive use (mm)	Water use efficiencies (kg ha^{-1} mm^{-1})
5 H	3.7	-	174	1.5
10 H	6.1	-	223	1.51
15 H	6.3	-	216	3.09
20 H	8.8	-	216	3.79
25 H	9.8	-	227	5.62
30 H	5.9	-	194	2.65
---	---	4.8	---	---

Soil quality: Presence of shelterbelt plantations on farmer's field have significant influence on soil characteristics of sheltered area in leeward side. Mertia *et al.* (2006) found that presence of shelterbelt increased soil organic carbon (Table 14) and electrical conductivity (Table 15), and reduced soil reaction (Table 16) near the shelterbelt plantations. These effects were more pronounced up to the distance of 2H and nullified after distance of 5H. On an average, maximum accumulation of OC was found in sheltered area of *A. tortilis* in tube well command, Lathi series followed by *T. undulata* and *D. sissoo* in IGNP command, Mohangarh. The accumulation of organic carbon decreased with increase in distance from the belt. In general, organic carbon (OC) decreased with depth of soil in sheltered area of all the shelterbelt. The built up of OC in sheltered area appears to be due to decomposition of leaf litter.

Crop Production: With the creation of irrigation facility due to IGNP, the cropping pattern of the region has changed completely. In irrigated areas of desert region, orchards of pomegranate, guava, grapes, *Citrus spp.* and *Cordia spp.* are maintained. Today more than two million ha land in Thar Desert grow brinjal, chillies, tomatoes and potatoes. Ber fruit especially a fleshy grafted variety is an important source of income. The above advancements are directly related to advent of canal irrigation, however, indirectly the shelterbelt plantation carried out along field boundary for protection has helped in sustaining changed cropping pattern. In addition, tree-belt plantation along canals to protect the channels from sand deposition has immensely helped in maintaining the flow and volume of water in the channels. In the absence of shelterbelts, the much needed water for irrigating agricultural fields would not have reached the farmers of Jaisalmer.

Table 14. Effect of different tree shelterbelt on organic carbon content of surface soil in sheltered area of leeward side on farmer's field.

Shelterbelt	Location	Depth (cm)	Organic Carbon content (%) at distance multiples of shelterbelt height				
			0H	1H	2H	5H	10 H
D. sissoo	IGNP Mohan garh	15	0.11	0.08	0.05	0.04	0.04
		30	0.07	0.06	0.07	0.05	0.05
		60	0.07	0.05	0.04	0.04	0.03
		Mean	0.08	0.06	0.05	0.04	0.04
T. undulata	IGNP Mohan garh	15	0.37	0.21	0.05	0.08	0.07
		30	0.12	0.11	0.03	0.06	0.06
		60	0.09	0.06	0.04	0.03	0.03
		Mean	0.19	0.13	0.04	0.06	0.05
A. tortilis	Tube well command Lathi series	15	0.46	0.21	0.16	0.14	0.15
		30	0.22	0.17	0.12	0.11	0.07
		60	0.17	0.12	0.12	0.12	0.13
		Mean	0.28	0.17	0.13	0.12	0.12

Table 15. Effect of different tree shelterbelts on reaction (pH) of soil in sheltered area of leeward side on farmers' field.

Shelterbelt	Location	Depth (cm)	Surface soil pH at distance multiples of shelterbelt height				
			0H	1H	2H	5H	10 H
D. sissoo	IGNP Mohan garh	15	7.84	7.84	7.93	7.9	8.05
		30	7.82	7.88	8	7.94	7.99
		60	7.87	7.92	8.03	8.09	8.1
		Mean	7.84	7.88	7.99	7.98	8.04
T. undulata	IGNP Mohan garh	15	7.6	7.76	7.98	7.95	8.05
		30	7.81	7.76	7.92	7.95	8.08
		60	7.84	7.88	7.99	7.99	8.01
		Mean	7.75	7.8	7.96	7.96	8.05
A. tortilis	Tube well command Lathi series	15	7.5	8.5	8.92	8.99	8.93
		30	8.53	8.59	8.92	8.89	8.73
		60	8.66	8.57	8.94	8.97	8.86
		Mean	8.23	8.55	8.93	8.95	8.84

Table 16. Effect of different tree shelterbelts on Soil electrical conductivity (dSm^{-1}) in sheltered area of leeward side on farmer's field.

Shelterbelt	Location	Depth (cm)	Soil EC at distance multiples of shelterbelt height				
			0H	1H	2H	5H	10 H
D. sissoo	IGNP Mohan garh	15	0.40	0.38	0.28	0.27	0.28
		30	0.39	0.34	0.23	0.27	0.27
		60	0.36	0.33	0.22	0.24	0.25
T. undulata	IGNP Mohan garh	15	0.88	0.42	0.22	0.21	0.21
		30	0.45	0.41	0.22	0.24	0.25
		60	0.43	0.36	0.21	0.21	0.21
A. tortilis	Tube well command Lathi series	15	0.67	0.34	0.34	0.36	0.34
		30	0.41	0.38	0.34	0.36	0.39
		60	0.4	0.37	0.36	0.36	0.42

Biotic Pressure on Natural Forest: Frequent drought and famine are common phenomena in arid region. Natural grasslands and forests are often subjected to excessive biotic interference in the form of overgrazing and removal of fuel woods. Introduction of multipurpose trees (MPTs) like *Accacia tortilis, Albizia lebbek, Azadirachta indica, Acacia nilotica* etc for tree-belt along roads, canals, field boundary of farm land and rangelands is aimed to supplement local demand of fuel and fodder along with its main objectives of reducing wind, drifting of sand and providing shelter. Trees on rangeland provide nutritious top feed in the form of leaves and pods to grazing animal (Ahuja, 1977). Quantified data on supply of fodder from belt plantation is not available but top feed is viewed as alternative source to supplement decreasing fodder productivity of grasslands in traditionally pastoral economy of arid region.

Excessive extraction and uprooting of indigenous trees/bushes from the woodlands to meet ever increasing demand of fuel wood and household timber is another threat to natural vegetation resource of arid region. Some of the species like *Calligonum polygonoides,* which used to be the main vegetation in sandy desert of Jaisalmer, Barmer and Bikaner districts of Rajasthan, is now facing serious threat of its extinction. Supply of fuel wood and small timber from shelterbelt would certainly put a break on such devastating processes. It is assumed that seven to eight years old belt plantations can provide sufficient quantity of fuel wood and timber to meet some of the needs of local people. Due to improvement in agricultural production and availability of fodder for livestock, there has been a change in migration pattern of nomads. The migration of sheep to neighboring areas has reduced to some extent (Table 17).

Table 17. Year wise migration of sheep from Jaisalmer district

Years	No. of sheep migrated	Percentage of total sheep population in district
1988-89	65320	13.48
1990-91	53140	10.97
1992-93	24610	5.08

Source: District sheep and wool development office, Jaisalmer

Employment Generation: Massive afforestation work taken by State Forest Department along canals, roads and railways has created employment opportunities for both men and women. For raising plantations and their maintenance, skilled as well as unskilled work force has been utilized. According to an estimate, employment of about 2.4 million man-days has been generated under JBIC project during 1990-2002 in Jaisalmer district of Rajasthan.

Summary

Tree-belt plantations, in arid region, were concieved designed and raised with the primary objectives of providing protection against harsh effects of high velocity winds. Selection of proper design and suitable species holds the key for efficacious impact of tree-belts on various components of arid ecosystem. In past five decades, after independence, massive afforestation work on tree-belt plantation has been done in Indian Thar Desert particularly in eleven district of Western Rajasthan. Major thrust has been on belt plantation along roads, canals and field boundaries of agricultural fields and grasslands. In Jaisalmer district alone, about 1305 Rkm roads-side and 27812 Rkm canal-side, tree-belts have been raised. In addition, about 14245 ha area has been covered under sand dunes, range lands, farmlands etc.

Impact assessment studies have revealed that tree-belt plantation has by and large been effective and fulfilled its prime objective of controlling hazards resulting from high speed winds. Significant reduction in sand deposition on roads and canals have been noticed thus resulting in savings of millions of rupees which otherwise, spent on de-silting operations. Favorable changes in microclimate have been noticed in plantation area. Work on tree belt has generated employment of about 2.4 million-man days and caused reduction in migration of nomads.

References

Ahuja, L.D. 1977. CAZRI monograph, 24: 161-166.

Bithu, B.D. 1989. Problem of soil erosion and its control in western Rajasthan. In: Proceedings of International Symposium on Managing Sandy Soils, Part II, Pp. 551 - 555.

Chepil, W.S. 1949. *Agronomy Journal*, 41: 127-129.

Kaul, R.N. 1969. Shelterbelts to stop creep of the desert. *Science and culture*, 24:406-409.

Mann, H.S. and Muthana, K.D. 1984. Arid Zone Forestry (with special reference to Indian arid zone), CAZRI Monograph, 23: 1- 48.

Mertia, R.S. 1986. Shelterbelt studies in India's Thar Desert. M.Sc. Thesis. University of Wales, U.K.

Mertia, R.S. 1992. Shelterbelt research in arid zone. *Journal of Tropical Forestry*, 8 (3): 196-200.

Mertia, R.S., Rajendra Prasad, Gajja, B.L., Narain, P and Samra, J.S. 2006. Impact of shelterbelts in arid region of western Rajasthan. Central Arid Zone Research Institute, Jodhpur, Pp. 96.

Ramakrishna, Y. S., Rao, A. S., Singh, R. S., Kar, A. and Singh S. 1990. Moisture, Thermal and Wind measurements over selected stable and unstable sand dunes in the Indian desert. *Journal of Arid Environments*, 19: 25-38.

Said, M. 1954. *Pakistan Journal of Forestry*, 4: 175-179.

Upadhyaya, A.K. 1991. Shelterbelt plantations effectively check sand deposition in Indira Gandhi canal. *Indian Forester* , 117 (7): 511-514.

❑❑❑

Agroforestry: Natural Resource Sustainability, Livelihood & Climate Moderation © 2009
Editors: OP Chaturvedi, A Venkatesh et. al., pp. 461-473
Satish Serial Publishing House, Delhi

Chapter 32

ALLELOPATHY AND ITS ROLE IN AGROFORESTRY

S.K. Das and A.K. Rai
Indian Grassland and Fodder Research Institute, Jhansi-284 003 (U.P.)

Definition of Allelopathy

According to International Allelopathy Society "any process involving secondary metabolites produced by plants, algae, bacteria and fungi that influence the growth and development of agricultural and biological systems" is called Allelopathy (Anonymous, 1996). It refers to the beneficial or harmful effects of one plant on another plant, both crop and weed species, by the release of chemicals from plant parts by leaching, root exudation, volatilization, residue decomposition and other processes in both natural and agricultural systems. Allelopathic inhibition is complex and can involve the interaction of different classes of chemicals like phenolic compounds, flavonoids, terpenoids, alkaloids, steroids, carbohydrates, and amino acids, with mixtures of different compounds sometimes having a greater allelopathic effect than individual compounds alone.

There is convincing evidence that allelopathic interactions between plants play a crucial role in natural as well as in manipulated ecosystems:

1. The credit for a specific vegetation pattern has mostly been given to the competition. However, in recent times evidence is accumulating that all types of plants, viz. herbs, shrubs and trees, allelopathically affect the patterning of vegetation, largely in their immediate vicinity.
2. One of the most worked out aspects of allelopathy in manipulated ecosystems is the role of allelopathy in agriculture. In this, the effects of weeds on crops, crops on weeds and crops on crops have been invariably emphasised. In addition, the possibility of using allelochemicals as growth regulators and natural pesticides (number of them are either commercially available or in the process of large-scale manufacture) promotes sustainable agriculture.
3. Allelopathic interactions have been demonstrated to play a crucial role in natural as well as man-made forests. Such interactions are pivotal in determining the composition of the vegetation growing as understorey and in understanding the forest regeneration problems. Results obtained so far have shown that almost all types of plants (viz. angiosperms, gymnosperms, lower plants like ferns and micro-organisms, including mycorrhizae) present in forests indulge in allelopathic interactions.
4. Some of the recent fundings have demonstrated that tree-crop interactions may have significant bearings on the total productivity of an agroforestry system (simultaneously or sequentially combined production of crops and forest plants). Therefore, it seems essential that the allelopathic compatibility of crops with trees should be checked before being introduced to an agroforestry system.

To meet the ever-increasing demand for food, fodder, vegetables, fuel wood, timber, medicines etc., acceleration in human and livestock population necessitated acquisition of more and more land under cultivation continued. Demographic pressure has further forced man to seek unconventional methods of agriculture to utilize land to the maximum extent. National Forest Policy (1988) has set a goal that 33% of the country's geographical area should be under forest and tree cover. Under Tenth Five Year Plan, the Government of India has set a target to increase the country's forest cover up to 25%. According to the latest report of Forest Survey of India (2001), the forest cover in the country is 675,538 km2, constituting 20.55% of its total geographical area. Out of this, dense forest constitutes 12.68% and open forest 7.87%. The forest cover in the hills districts is only 38.34% compared with the desired 66% area.

The practice of introducing trees in any form to the farming world has played a significant role in enhancing productivity on per unit area basis. Agro-forestry has shown that beside sustainable agriculture it can also help promote a better environment. Appropriate definition of agroforestry can be that "it is a collective name for land-use system and technologies in which woody perennials, trees, shrubs, bamboo etc. are deliberately combined on the same land management unit with herbaceous crops and animals in some form of spatial arrangement of temporal sequence".

Allelochemicals

Allelochemicals are defined as "biocommunicators". Allelopathic chemicals can also persist in soil, affecting both neighboring plants as well as those planted in succession. Although derived from plants, allelochemicals may be more biodegradable than traditional herbicides but may also have undesirable effects on non-target species, necessitating ecological studies before widespread use. Allelochemical concentrations in the producer plant may also vary over time and in the plant tissue produced.

Integrated Pest Management

Contemporary plant production systems rely heavily on inputs of chemicals to provide protection against insects, diseases and weeds. This extensive use of chemicals in plant protection has given rise to concerns about pesticide residues in the environment and to the development of pesticide resistance by many organisms. Alternative approaches to plant protection include the continued use of pesticides as components of integrated pest management programmes and more radical development, such as enhancing and harnessing the array of defensive chemicals produced naturally by plants. An increasing body of evidence points to their ability to contribute to plant protection (Lovett, 1990). Allelopathic chemicals have already been used to defend crop plants against insects, nematodes and diseases, however, parallel approaches in control of weeds may be possible by finding compounds that inhibit seed germination, inhibit plant growth or prevent propagule formation. There is a greater emphasis on the control of weeds in sustainable agricultural systems by reduced or non-chemical means that are compatible with reduced tillage (Regnier and Janke, 1989).

Natural Phytotoxic Compounds

Green plants produce hundreds or thousands of compounds that are not involved in the primary metabolism of the plants, i.e., those are secondary products. The compounds involved in interspecific chemical interactions (allelopathy) with higher plants are often phytotoxic or herbicidal to other species or even to the species producing them (Duke, 1986). Secondary compounds in plants may act as attractants or repellants, phagostimulants or antifeedants, nutrients or toxins to phytophagous insects. In initial efforts to develop botanical herbicides from allelochemicals, active principles of allelopathic chemicals have been isolated and identified in Asia and the U.S.A. (Kim, 1995). Examples of some of the allelochemicals are given in table 1.

Table 1. Allelochemicals produced by plants.

Plant	Chemicals
Datura stramonium	Tropane alkaloids, Scopolamine, Hyoscyamine
Secale cereale L. (Rye)	2,4-dihydroxy-1,4(2H)-benzoxazin-3-one (DIBOA), 2-(3H)-bezoxazolinone (BOA)
Avena sativa L., *Sorghum bicolor* L. (Sorghum), *Triticum aestivum* L. (wheat)	Ferulic Acid, *p*- Cumaric Acid
Triticum aestivum L.	2,4-Dihydroxy-7-methoxy-1,4-benzoxazin-3-one (DIMBOA)
Melilotus alba Desr. (Sweet clover)	Coumarin
Avena sativa L. (Oats)	Scopoletin
Sorghum halepense L. Pers. (Johnson grass)	Dhurrin (a cyanogenic glycoside) and its epimer taxiphyllin
Sorghum bicolor	*p*-benzoquinone, dimethoxy substituted *p*-bezoquinone, 2,6-dimethoxy benzo-quinone (most active primary haustorical inducer)
Sorghum bicolor	Sargoleone
Trianthema portulacastrum L. (Horse purslane)	Caffeic, chlorogenic, *p*-hydroxybenzoic, *p*-coumaric and ferulic acids
Pistia stratiotes L.	Linoleic acid, gamma-linoleic acid, alpha-asarone
Barley	Hordenine
Wheat	Hydroxamic acid
Mungbean	Glucosyl flavonoids (vitexin and isovitexin)
Catharanthus rosea L. (periwinkle)	Vinblastine, Vincristine
Medicago sativa L.	Saponins
Chenopodium ambrosioides L.	Ascaridole, alpha-terpinene, gamma-terpinene, Limonene

Source: Khalid *et al.*, 2002

Plants evolved different defence mechanisms, which can be described as mechanical, phenological and chemical (Coley 1983; Aide 1993; Rosenthal and Berembaum 1991). All defences are used in synergistic action, and provide an excellent adaptation capacity for plants in many ecological situations. Chemical defence is based on secondary metabolites (SM), which, in addition to their defence role, are responsible for the interaction of plants with their surroundings, by means of plant-plant, plant-insect and plant-microorganism interactions. These metabolites can be synthesized by the plant from the beginning of its growth and development (constitutive defence), as in the case of intraspecific plant-plant interaction, or after exogenous stimulation (herbivore action). This last process is mediated by a metabolic alarm which starts or catalyses the secondary metabolites biosynthetic routes (induced defence) (Agrawal *et al.*, 1999; Karban and Baldwin 1997; Tallamy and Raupp 1991). It is not surprising that a wide variety of bioactivities, measured in vitro or in vivo, are associated with these SM (Singh *et al.*, 2003). Studies of the dynamics of these chemical defences, from an ecological point of view, include two main aspects: first, the observation of the natural phenomenon, and after this, (due to the great variety of conditions found in nature), imitation in the laboratory of the observed effect, by controlling various parameters; and second, to describe the phenomenon in an accurate manner. These studies are a part of allelopathy research, and the mediators of these interactions are allelochemicals (Rice 1984). Allelopathic science applies the fundamentals of ecological studies: observation of the natural phenomenon, and the simulation of them under controlled bioassay conditions. These phenomena can be described by an ecological component (the evidence of their existence in nature), a chemical component (to isolate, identify and characterise allelochemicals), and a physiological component (the mechanism of interference of chemicals with organisms, at both cell, and molecular level).

The concentration of the allelochemical inside the plant does not define exclusively its allelopathic potential, measured as the amounts released to the environment. These chemicals have been quantified in shoots, but the amounts released correlate better with the concentrations recorded in roots. Thus, roots are transit organs for the allelochemicals before their exudation. Starting allelochemicals, in addition to partial and complete degradation products are present in the root exudates, and all of them contribute to the measured bioactivities of the donor plants. The allelochemical DIBOA is transformed into benzoxazolinone (BOA) and aminophenoxazine (APO) also under sterile conditions. Thus, the measurement of the biological (allelopathic) activity is masked by the biotransformation products. This fact can be assumed in the case of any allelochemical. In the bioassays carried out with plantula, the target plant is never chemically analysed in the search for the allelochemicals released by the donor, but only the macroscopic effects (germination and growth), together with some enzymological properties are recorded. This kind of analysis can help to determine which chemicals truly act in an allelopathic phenomenon. Target plants can absorb the released allelochemical, in addition to its degradation products, and this fact must be considered in biological activity studies. This is especially important when dealing with mode of action determination, since each chemical in these compound mixtures could have a different mode. It is also important to note that the activities recorded in a bioassay depend on the different biologically and non-biologically induced transformations of the studied allelochemical. All these factors complicate the study of the allelopathic phenomenon, but their control can offer a more accurate description of what really occurs in the environment. None of the different parameters is static, but allelopathic phenomena are instead dynamic and time-dependent. The bioassay conditions themselves induce variations, though some of them can be controlled. On the other hand, the chemical nature of the allelopathic agent can not be controlled, this determining its persistence and integrity in the environment.

The allelopathic effects are selective (Stowe, 1979; Melkania, 1983) and vary with different trees since these plants will vary in the amount of indigenous secondary metabolites and would release different amounts of the phytotoxins. Generally leaves are the most potent source of allelochemicals, however, the toxic metabolites are also distributed in all other plant parts in various concentration. The secondary compound released from litter or formed by microbes decomposing the litter will be influenced by the type of crop being leached or decomposed (Putnum and Duke, 1978). In fact, litter leaching and decay are the major pathways of the release of allelochemes form plants. Harborne (1977) proved that higher plants (tree crops) release some phytotoxins into soil, which adversely affect the germination and yield of crops. Such type of tree crop interactions called phytochemical ecology/ ecological biochemistry. In the traditional agroforestry system of country people are growing several tree species in or around the agricultural fields. However, although recent attempts have been made to use available land more efficiently, agricultural losses are being experienced by marginal and sub-marginal farmers, who are concerned about the adverse effect of farm trees on cultivated land and standing crops (Bhatt *et al.*, 1993). Due to the adverse effect of trees most of the farmer are now neglecting to grow the tree species in agricultural field because, to grow food crop is essential for subsistence. Therefore the planting of tree crops has not been practiced on large scale. In agroforestry importance of multipurpose tree species cannot be overlooked as they provide food, fodder, fuelwood and social security to the growers. Some species improve the soil but at the same time some species may cause adverse effect on long-term basis (Gill, 1992; Mughal, 2000). The component plant species in agroforestry system depends on the same reserve of growth resources such as light, water and nutrients and hence there will be influence of one component of a system on the performance of the other components as well as system as a whole. These are referred to as tree-crop interactions. These interactions may be positive or negative (Basavaraju and Gururaju, 2000). The balance between these positive and negative effects determines the overall effects of the interactions in a given agroforestry combination.

Baker (1966) reported the *Eucalyptus globules* produces volatile emanations that inhibit root growth of Cucumis species seedlings and also the growth of hypocotyls, but not the roots of Eucalyptus seedlings. Singh and Bawa (1982) found leaf leachates of Eucalyptus globulus to be inhibitory to seed germination of *Glaucium flavus*. Many other species also reported for allelopathic to plant growth are Celtis laevigata, Rhododendron albiflorum, *Grevillea robusta, Quercus falcata, Quercus alba* (Rice, 1974, 1979), *Pinus roxburghii, Cedrus deodara, Quercus leucotrichophora, Myrica esculenta* (Melkania, 1983). Various workers have been reported allelopathic influences on certain tree crops (Saxena and Singh, 1987; Melkania, 1984; Suresh and Vinaya Rai, 1987; Bhatt and Todaria, 1990) for the different parts of the country. Kaletha *et al.* (1996) also done the similar study for aqueous extract of leaves and bark of *Grewia oppositifolia, Ficus roxburghii, Bauhinia variegata* and *Kydia calycina* on test crops *Echinochloa frumentacea, Eleusine coracana, Zea mays, Vigna unguiculata,* and *Glycine max* and found that the bark and leaf aqueous extracts of tree species were most toxic to food crops. Similarly Bhatt and Chauhan (2000) found allelopathic influenced of Quercus species on *Triticum aestivum, Brassica campestris* and *Lens culinaria* and found leaf and bark extract suppressed the germination, plumule and radicle length of all food crops.

A very important point concerning allelopathy is that its effect depends on a chemical compound being added to the environment. It is thus separated from competition, which involves the removal or reduction of some factor from the environment that is required by some other plant or microorganism sharing the habitat" (Rice 1987; Rizvi and Rizvi, 1992). To reach organisms within the rhizosphere, especially where signalling allelopathic agents rather than the more general biocides are involved, would require mobility of the chemical. However, it should not be assumed that the only mechanism by which soil mobility can be achieved is through water solubility. A number of soil-active nematicides and herbicides are indeed highly water soluble, e.g., the nematicidal carbamates aldicarb and carbofuran. However, some compounds, active as pesticides in the soil, such as the volatile pyrethroid, tefluthrin (Tomlin, 1997), operate by having relatively high vapour pressures, which allow transmission within and between the aerial interstices of the soil. All main aspects of the secondary metabolic biosynthetic pathways (Edwards and Gatehouse, 1999) are represented in the generation of allelopathic agents. Furthermore, interactions with almost all of the possible biochemical targets within the responding organism are represented (Putnam, 1983). However, more recent developments reviewed here involve signaling mechanisms by which the allelopathic agents effect changes indirectly by means of receptor systems (as yet, largely undefined) in the responding species.

There are allelopathic effects, which can readily be seen beneath coniferous tree stands, and these are largely caused by release of large amounts of low-molecular weight biocidal isoprenoid products. The isoprenoid pathway has recently been elaborated (Lichtenthaler *et. al.*, 1997; Rohmer, 1999) in work originally on eubacteria and green algae, and then on higher plants. A novel pathway, additional to that from mevalonate, involving glyceraldehyde 3-phosphate and pyruvate, was found to generate isopentenyl diphosphate via 1-deoxy-D-xylulose 5-phosphate and 2-C-methyl-D-erythritol 4-phosphate. The grand fir *(Abies grandis)* has been the model leading to a greater understanding of this biosynthetic pathway and its associated molecular genetics (Bohlmann *et. al.*, 1998, 1999). This tree generates low-molecular weight isoprenoids which are superficially expressed in the bark of the tree and which contribute to the allelopathic competition against other plants within its stands. In angiosperm trees, similar types of allelopathy can occur, although other biosynthetic pathways are often represented. For example, the notorious allelopathy of the walnut tree *(Juglans regia,* Juglandaceae) is largely accounted for by the release of juglone, which is an extremely toxic naturally occurring quinone with high potency against both plants and animals. Well studied higly 'apparent' plants such as the oak trees in the Quercus genus (Fagaceae) (Feeny, 1976) employ biosynthesis based on the phenylalanine ammonia-lyase pathway to release large amounts of allelopathic phenolics in their defence (Harborne and Baxter, 1999).

A series of relatively small molecular weight allelochemicals have been identified in the gooseweed, or 'pak pawd' as it is known in Thailand (*Sphenoclea zeylanica*, originally in the Campanu1aceae, now in Sphenocleaceae) (Macfas *et. al.*, 1999, Hirai *et. al.*, 1999). Such compounds can provide interesting leads for the development of synthetic herbicides.

In Himalaya, majority of medicinal herbs are collected unscientifically from the wild, which has resulted in the local disappearance of favoured and effective resources and reduced species diversity (CAMP 2003). The cultivation of medicinal plants requires in-depth knowledge of many variables (e.g. biological interactions between different plant species), which may aid in producing high quality medicinal crops (Chiej 1984), One of the interactions between different species is allelopathy. Almost all the traditional food crops were found sensitive (in terms of germination and plumule and radicle growth) to the 2% leaf/rhizome aqueous extracts of *Saussurea lappa* and *Potentilla fulgens*. In general, *Amaranthus caudatus* and *Eleusine coracana* (except for its germination) were found most sensitive whereas, Phaseolus mungo was most resistant to the 2% leaf/rhizome aqueous extracts of all three medicinal species. *Rheum emodi* was found allelopathically less effective as compared to *Saussurea lappa* and *Potentilla fulgens* (Table 2 and 3). These findings also confirm that extracts of many plant species including medicinal herbs contain allelochemicals which have been reported to affect enzyme responsible for plant hormone synthesis, in addition to inhibition of nutrient and ion absorption by affecting plasma membrane permeability (Fujii *et al.*, 1991). Effect of a given compound or plant metabolites might be inhibitory or stimulatory depending on their concentration in the surrounding medium (Khailov, 1974).

Table 2. Effect of 2% leaf/rhizome aqueous extract of medicinal plant on germination of traditional agriculture crops after 7 days of sowing (values in parenthesis indicate reduction percentage in germination as compared to control) [Source: Nazir *et. al.*, 2007].

Species	*Triticum aestivum*	*Eleusine coracana*	*Fagopyrum esculantum*	*Amaranthus caudatus*	*Phaseolus mungo*	*Phaseolus vulgaris*
Rheum emodi	92.0^{ab} (8.0)	92.0^{a} (3.2)	94.0^{a} (-)	76.0^{ab} (24.0)	90.0^{ab} (10.0)	96.0^{ab} (4.0)
Saussurea lappa	80.0^{c} (20.0)	88.0^{a} (7.4)	76.0^{b} (17.4)	56.0^{b} (44.0)	76.0^{b} (24.0)	74.0^{b} (26.0)
Potentilla fulgens	86.0^{bc} (14.0)	94.0^{a} (1.1)	84.0^{ab} (8.7)	89.0^{a} (11.0)	77.0^{b} (23.0)	78.0^{b} (22.0)
Control	100.0^{a}	95.0^{a}	92.0^{a}	100.0^{a}	100.0^{a}	100.0^{a}

Source: Nazir *et. al.*, 2007

All the above-mentioned results suggested that (i) *R. emodi* can be grown in mixed cropping system consisting of T*riticum aestivum, F. esculantum, P. mungo* and *P. vulgaris* and *S. lappa* can be mixed with *P. mungo.*

Table 3. Effect of 2% leaf/rhizome aqueous extract of medicinal plant on radicle and plumule growth of traditional agriculture crops after 7 days of sowing (values in parenthesis indicate reduction percentage in radicle and plumule growth as compared to control) [Source: Nazir *et. al.*, 2007].

Species	*Triticum aestivum*		*Eleusine coracana*		*Fagopyrum esculantum*		*Amaranthus caudatus*		*Phaseolus mungo*		*Phaseolus vulgaris*	
	R	P	R	P	R	P	R	P	R	P	R	P
Rheum	13.63^{ab}	10.52^{ab}	6.3^{ab}	5.30^{ab}	20.5^{ab}	12.3^{ab}	6.32^{a}	5.3^{a}	9.63^{ab}	7.53^{ab}	16.2^{ab}	8.52^{ab}
emodi	(10.5)	(16.0)	(23.5)	(26.6)	(19.7)	(13.6)	(23.2)	(14.9)	(05.9)	(11.6)	(16.2)	(17.5)
Saussurea	11.11^{bc}	9.51^{b}	5.33^{b}	4.33^{b}	16.33^{b}	10.3^{bc}	2.52^{b}	2.32^{b}	8.53^{bc}	7.00^{b}	13.63^{b}	7.52^{b}
lappa	(27.1)	(24.0)	(35.2)	(40.0)	(36.0)	(27.6)	(69.4)	(62.8)	(16.6)	(17.8)	(29.5)	(27.2)
Potentilla	10.23^{c}	9.5^{b}	5.55^{b}	4.3^{b}	15.33^{b}	8.55^{c}	5.43^{ab}	4.5^{a}	7.5^{c}	6.63^{b}	12.33^{b}	7.55^{b}
fulgens	(32.8)	(24.1)	(32.6)	(40.4)	(39.9)	(39.9)	(34.0)	(27.8)	(26.7)	(22.2)	(36.2)	(26.9)
Control	15.23^{a}	12.52^{a}	8.23^{a}	7.22^{a}	25.52^{a}	14.23^{a}	8.23^{a}	6.23^{a}	10.23^{a}	8.52^{a}	19.33^{a}	10.33^{a}

Source: Nazir *et. al.*, 2007

To demonstrate allelopathy, plant origin, production, and identification of allelochemicals must be established as well as persistence in the environment over time in concentrations sufficient to affect plant species. Allelopathy can involve more sophisticated activities in addition to large inundation of the rhizosphere with biocidal material. These should be the focus of studies in the future, since there would be an economy in terms of plant energetics and because toxicologically benign natural products are involved. Allelochemistry may provide basic structures or templates for developing new synthetic herbicides.The characterisation of signalling systems and the biosynthetic pathways involved could be exploited more extensively using the tools of molecular genetics.

References

Agrawal A.A., Tuzun, S. and Bent, E. 1999. Induced Plant Defences Against Pathogens and Herbivores. APS Press, St. Paul.

Aide, T.M. 1993. Patterns of leaf development and herbivore in a tropical understudy community. *Ecology,* 74: 455-466.

Baker, H.G. 1966. Volatile growth inhibitory produced by Eucalyptus globules. *Madrona S. Francisco,* 18: 207-210

Bhatt, B.P. and Chauhan, D.S. 2000. Allelopathic effects of Quercus species on crops of Garhwal Himalaya. *Allelopathy Journal,* 7(2): 265-272

Bhatt, B.P. and Todaria, N.P. 1990. Studies on the allelopathic effects of some agroforestry tree crops of Garhwal Himalaya. *Agroforestry Systems,* 12: 251-255.

Bhatt, B.P., Chauhan, D.S. and Todaria N.P. 1993. Phytotoxic effects of tree crops on germination and radicle extension of some food crops. *Tropical Science,* 33: 69-73.

Bohlmann, J., Crock, J., Jetter, R. and Croteau, R. 1998. Terpenoid-based defenses in conifers: cDNA cloning, characterisation and functional expression of wound-inducible (E)-a-bisabolene synthase from grand fir: (*Abies grandis*). *Proc. Natl. Acad. Sci. USA*, 95(12): 6756-6761.

Bohlmann, J., Phillips, M., Ramachandiran, V., Katoh, S. and Croteau, R. 1999. cDNA cloning, characterisation and functional expression of four new monoterpene synthase members of the Tpsd gene family from grand fir (*Abies grandis*). *Arch. Biochem. Biophys*, 368 (2): 232-243.

Conservation Assessment and Management Priority Workshop (2003). Threat assessment and management prioritization of selected medicinal plant in western Himalaya, Simla, May 22-26.

Chiej, R. 1984. The Macdonald encyclopedia of medicinal plants. London, U.K.:Macdonald Co. (Publishers) Ltd., Pp. 447.

Coley, P.D. 1983. Herbivory and defensive characteristics of tree species in a lowland tropical forest. *Ecological monographs*, 53: 209-233.

Feeny, P. 1976. Plant apparency and chemical defense. In Recent Advances in Phytochemistry: Biochemical Interaction between Plants and Insects. Eds. J W Wallace and R L Mansell. pp. 1-40. Plenum Press, New York.

Fujii Y, Furukawa M, Hayakawa Y, Sugahara K, Shibuya T 1991. Survey of Japanies medicinal plants for the detection of allelopathic properties. *Weed Res Tokyo*, 36: 36–42

Gill, A.S. 1992. Allelopathy in agrofoestry "Allelopathy in Agroecosystem" Tauro P & SS Narwal (Eds), Indian Society of Allelopathy, Haryana Agriculture University, Hissar, India, 9

Harborne, J.B. 1977. Introduction to Ecological Biochemistry. Academic Press, New York.

Harborne, J.B. and Baxter, H., 1999. The Handbook of Natural Flavonoids. Wiley, London.

Hirai, N., Sakashita, S-I, Sano, T., Ohigashi, H., Premasthira, C-U, Asakawa, Y., Harada, J. and Fujii, Y. 1999. Potent allelochemicals from a tropic weed *Sphenoclea zevlanica*. International Workshop on Bioactive Chemicals in Communication, 8-10 November, 1999. Natl. Inst. of Agro-environmental Sciences, 3-1-1, Kannondai, Tsukuba, Ibaraki, Japan.

Kaletha, M.S., Bhatt, B.P. and Todaria, N.P. 1996. Tree-crop interaction in traditional agroforestry systems of Garhwal Himalaya. 1. Phytotoxic effects of farm trees on food crops. *Allelopathy Journal*, **3**(2): 247-250.

Karban, R. and Baldwin, I. 1997. Induced Responses to Herbivory. The University of Chicago Press, Chicago.

Khailov, K.M. 1974. Biochemical traphodynamics in marine coastal ecosystems. Kiev, USSR, Naukova Dumka

Khalid, Shahida, Ahmad, Tahira and Shad, R.A. 2002. Use of Allelopathy in Agriculture. *Asian Journal of Plant Sciences*, **1**(3), 292-297.

Kim, K.U. 1995. Current status of biological control and integrated management of paddy weeds in Korea, with emphasis on Allelopathy of weeds. In: Proceedings of the International Symposium on Biological Control and Integrated Management of Paddy and Aquatic weeds in Asia, Tsukuba, Japan, 285-303.

Licillemhalcr H K, Rohmer, M. and Schwender, J. 1997 Two independent biochemical parhways for isopentenyl diphosphate and isoprenoid biosymhesis in higher plants. *Physiologia Plantarum*, 101: 643-652.

Lovett, J.V., 1990. Chemicals in plant protection: is there a natural alternatives. Bassett, C., Whitehouse, L.J., Zabkiewicz, J.A. (ed.). Alternatives to the chemical control of weeds. Proceeding of an International Conference. Rotorua (New Zealand), Ministry of Forestry, Forest Res. Institute, 57-67.

Macfas, F.A., Castellano, D., Oliva, R.M., Cross, P. and Torres, A. 1997. Potential use of allelopathic agents as natural agrochemicals. Proceedings Brighton Crop Protection Conference - Weeds. Vol. 2. Pp. 33-38.

Melkania, N.P. 1983. Influence of certain selected tree species on ground flora, Ph.D. Thesis, Kumaun Unviersity, Nainital. Pp. 442.

Melkania, N.P. 1984. Influence of leaf leachates of certain woody species on agricultural crops. *Indian Journal of Ecology*, 1182-86

Mughal, A.H. 2000. Allelopathic effect of leaf extract of *Morus alba* L. on germination and seedling growth of some pulses. *Range Management and Agroforestry*, 21(2): 164-169

Nazir, Tahir Uniyal, A.K. Todaria, N.P. 2007. Allelopathic behaviour of three medicinal plant species on traditional agriculture crops of Garhwal Himalaya, India. *Agroforest System*, 69: 183–187

Putnum, A.R. and Duke, W.B. 1978. Allelopathy in agroecosystems. *Ann. Rev. Phytopathol.*, 16, 431-451

Regnier, E.E. and Janke, R.R., 1989. Evolving strategies for managing weeds. In: C.A. Edwards, R.Lal, P.Miller and G.House (Eds.). Sustainable agricultural systems. Soil and Water Conservation Society, Ankeny, Iowa, USA: 174-202.

Rice, E.L. 1974. Some roles of allelopathic compounds in plant communities. *Biochemical Systematic and Ecology*, **5**: 201-206

Rice, E.L. 1979. Alleopathy-An update. *Botanical Review*, 45: 15-109

Rice E.L. 1984. Allelopathy. Academic Press, Orlando/FL.

Rice E.L. 1987 Allelopathy - An Overview. *ACS Symp. Ser.* 330: 8-22.

Rizvi S J H. and Rizvi V 1992 Exploitation of allelochemicals in improving crop productivity. In Allelopathy: Basic and Applied Aspects. Eds. S J H Rizvi and V Rizvi. pp. 443-472. Chapman and Hall, London.

Rohmer, M. 1999. The discovery of a mevalonate-independent pathway for isoprenoid biosynthesis in bacteria, algae and higher plants. *Nat. Prod. Rep.,* 16: 565-574.

Rosenthal, G.A. and Berembaum, M.R. 1991. Herbivores, their interactions with secondary plant metabolites. Academic Press: San Diego/CA.

Saxena, S. and Singh, J.S. 1987. Influence of leave leachates from *Eucalyptus globulus* and *Aesculus indica* on the growth of *Vigna radiata* and *Lolium perenne. Indian Journal of Ecology,* 5: 148-158.

Singh, H.P., Batish, D.R., and Kohli, R.K. (2003). Allelopathic interactions and allelochemicals: New possibilities for sustainable weed management. *Critical Reviews in Plant Science,* 22: 239-311.

Singh, R. and R. Bawa, R. 1982. Effect of leaf leachates from Eucalyptus globulus Labill. And Aesculus indica Colebr. On seed germination Glaucium flavum Crantz. *Indian Journal of Ecology,* 9: 21-28

Suresh, K.K. and Vinaya Rai R.S. 1987. Studies on the allelopathic effects of some agroforestry tree crops. *The International Tree Crops Journal,* 4: 109-115.

Stowe, L.G. 1979. Allelopathy and its influence on the distribution of plants in an Illinois old field. *Journal of Ecology,* 67: 1065-1085.

Tallamy, D.W. and Raupp, M.J. 1991. Phytochemicals Induction by Herbivores. John Wiley& Sons, New York.

□□□

Agroforestry: Natural Resource Sustainability, Livelihood & Climate Moderation © 2009
Editors: OP Chaturvedi, A Venkatesh et. al., pp. 475-483
Satish Serial Publishing House, Delhi

Chapter 33

BELOW GROUND DIVERSITY IN AGROFORESTRY AND ITS ROLE IN SOIL MANAGEMENT

Sharmila Roy
Indian Grassland and Fodder Research Institute, Jhansi-284 003 (U.P.)

Introduction

Soil organisms are integral part of natural ecosystem on account of their role in maintaining soil health, ecosystem functions and production. However, below ground biodiversity is less researched part of agro-ecosystems. In order to sustain land productivity we need to focus our attention on the rhizosphere - hot spot of below ground biodiversity (CBD, 2001). Soil biota is highly sensitive to changes and is indicator of soil health/quality. The shape and size of biological diversity is as complex as that of above ground but more resilient of anthropogenic perturbations. The ongoing degradation of the biological base of soil productivity in agro ecosystem has led to realization that efforts are required for understanding below ground biodiversity - their community structure, characterization, dynamics and functionality (NAAS 2006). By understanding and manipulating these biological processes, the agroforestry systems can be made more sustainable in yield.

Belowground Biodiversity

Biodiversity is the richness of life indicated by the variety of biota and inter-related chemical processes in a habitat. Bio diversity of populations is correlated with the stability of an ecosystem. Soil is a living heterogeneous system with characteristic biotic and abiotic factors. The soil organisms participate in the genesis of the habitat wherein they live. They constitute one of the five interactive factors in soil formation; the other four are; climate, topography, parent material and time.

Biological activity in soils is mainly concentrated in the top soils. The biological component occupy a small fraction (<0.5%) of the total soil volume and make less than 10% of total organic matter in soil. This living component of the soil organic matter consists of 5-15% of plant roots, 85-90% soil organisms. The variation in soil profile, resource availability, microclimate, chemical and physical structure influence the soil biota size, composition and distribution (Buscot and Verma, 2005).

All living things can be classified into one of the five fundamental kingdoms of life namely, Monera, Protista, Fungi, Plantae, Animalia and are well represented in soil ecosystem. The soil flora and fauna comprise thousands of species with a wide range of ecological strategies. They may be classified either on the basis of functional groups viz., mycophagous/herbivores, omnivores and predators or on the basis of body width viz., micro (<1mm), meso (<2mm) and macro-organisms (>2mm).The free-living components of soil biota are bacteria, fungi, algae and the fauna.

Soil Flora

Bacteria are the most abundant group of microorganism in soil, probably half of the total microbial biomass in the soil. Their population ranges from one hundred thousand to several hundred million per gram of soil, depending on the condition of the soil. The most common soil bacteria come under the genera *Pseudomonas, Arthrobacter, Clostridium, Achromobacter, Bacillus, Micrococcus, Flavobacterium, Chromobacterium, Sarcina, Aerobacter* etc. Majority of soil bacteria are heterotrophs, utilizes atmospheric nitrogen *e.g. Rhizobium, Azotobacter, Beijerinckia, Derxia, Azospirillum.*

Fungi are eucaryotes and are primarily responsible for the decomposition of organic residues, though in plate count bacteria outnumbered them. Some important soil fungal genera are *Dictylostelium, Physarum, Pythium, Plasmopara, Phytophthora, Saprolegnia, Allomyces, Rhizophydium, Mucor, Mycotypha, Rhizopus, Zygorhyncus, Endothia, Ceratocystus, Claviceps, Saccharomyces,*

Agaricus, Poria, Boletus, Fomes, Aspergillus, Trichoderma, Penicillium, Helminthosporium, Fusarium, Athrobotrys, Rhizoctonia etc.

Actinomycetes are commonly regarded as an intermediate group between the bacteria and fungi. *Micromonospora, Nocardia, Streptomyces, Streptosporangium and Thermoactinomyces* are some of the important genera of soil actinomycetes.

Soil Fauna

Free-living protozoa in soil feed on dissolved organic substances and on other organisms. Many feed wholly by grazing and predation. The soil ciliates depend primarily on bacteria as food. Amoebae feed on bacteria, other protozoa, and fungal spores. Protozoa occurring in soil are presented by Small flagellates (Oikomonas, Bodo), amoebae (*Vahlkampfia, Naegleria, Hartmannella, Nuclearia, Dictyoste*), Ciliates (Colpoda, Leptopharynx, *Chilodenella, Cyrtolophosis, Uroleptus, Keronopsis, Gonostomum Vorticela striata Spathidium, Bresslana, Tetrahymena rostrata, Podophrya*) and Testacea (*Cyclopyxix, Phryganella, Trinema, Corythion, Centropyrix, Nebela, Difflugia, Arceil, Microchlamys*). The micro fauna can directly affect the structure of microbial communities.

Following the protozoa, nematodes are the second most dominant group of soil fauna in numbers and biomass. Most of the terrestrial nematodes are <2mm long and 0.05mm wide. Nematodes feed, move and reproduce like protozoa in water films around soil particles.

The most abundant soil arthropods, in terms of number of individuals and species, are the acarina (mites) and collembolans (springtails). The predominant macro arthropods populations were coleopteran and hymenoptera. The density of gastropods, isopods diplopods and chilopods were relatively low in arable soil because of their sensitivity to agricultural practices. Earthworms, ants, and termites do fragmentation and transportation of organic matter in deeper soil layers. Many small soil dwelling invertebrates, such as millipedes, nematodes, centipedes, mites, annelids, spiders, insects etc have their effect on the physical and chemical properties of soil and detritus and on the structure of microbial communities.

Biological Processes and Soil Fertility

The microorganisms and invertebrate animals play key role in the primary productivity of the ecosystem they inhabit. Their diverse role can be categories into (i) facilitating nutrient acquisition by the vegetation through the mycorrhiza and N- fixing organisms, (ii) Regulating the flow of nutrients through decomposition, mineralization and immobilization, (iii) mediating and breakdown of organic matter, (iv) modification of soil structure which influence water availability to the plants, (v) Modifying the health of the plant by parasitism and pathogenesis.

Soil flora is the primary consumer in the detritus food web. They act on the organic wastes and convert them either into useful or into innocuous and less harmful substance. In nature fungi, bacteria, and invertebrates interact and influence decomposition and functioning of rhizosphere.

Soil fauna governs the distribution, abundance and activity of soil micro flora in rhizosphere. They also graze on root surface. In saprophytic succession, six mechanisms of interaction are important. Two control micro-flora distribution and abundance (I) selective grazing of micro-flora, (II) dispersal of micro-flora inoculums. Four mechanisms stimulate microbial activity; (I) direct supply of mineral nutrients in urine and faces, (II) stimulation of microbial activity by faunal activity, (III) compensatory fungal growth due to periodic grazing, (IV) release of micro-flora from competitive stasis.

Biodiversity in Agroforestry

Environmental factors and anthropogenic actions, in particular, agricultural and forestry management practices, affect soil biological communities and their functions. Agroforestry systems contain enormous numbers of diverse living organisms, ranging from the myriad of invisible microbes, bacteria and fungi to the more familiar macro-fauna such as earthworms, termites and plant roots. However, the observed diversity and the mathematical theories guide in exploring the below ground diversity to some extent Roy et al., 2008). These organisms interact with one another and with the various plants and animals in the ecosystem forming a complex web of biological activity.

The land management practices alter soil conditions and the soil community of micro-, meso- and macro-organisms. The changes in their organization influence the soil fertility and productivity. However, the relationship between specific practices and soil functions is less clear. In general, the structure of soil communities is largely determined by ecosystem characteristics and land use systems. The activity of different species depends on specific management practices as these affect the micro-environment conditions, including temperature, moisture, aeration, pH, pore size, and type of food sources e.g. in dry land situations where earthworms are few, but termites, ants and other invertebrates are abundant, perform similar functions.

The soil biodiversity was also affected by vegetation pattern. In a study by Roy *et al.,* (2006a), it has been found highest number of species in natural grassland (118 sp.) > silvipasture system (114 sp.) > sown pasture (113 sp.) > tree grove (106 sp.). The annual cultivation systems characterized by high dominance of only a few species (71 sp.) matched with the degraded land situation (57 sp.). The superiority of perennial systems in terms of high species richness and medium abundance is a reflection of the high level of maturity attained, allowing other species to flourish and thus leading towards a sound ecosystem (Fig. 1).

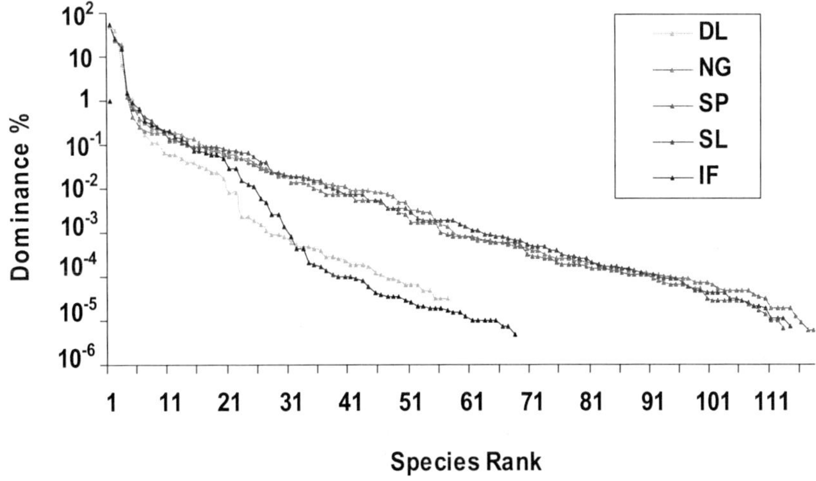

Fig.1. Species Diversity of soil biota in different land use systems (DL = Degraded Lands; SP = Sown Pasture; NG = Natural Grassland; SL = Silvipasture; IF = Intensive fodder crop cultivation)

Detailed analysis of soil biodiversity and inter species associations in various systems through use of Shannon Diversity Index supported the conclusion that perennial systems, especially natural grassland, provide better microclimate to support soil biodiversity (Table 1).

Table 1. Shannon Wiener Diversity indices for various groups of arthropods in different systems

Systems		Soil Biota	
	Collembolla	Acari	Other Arthropods
Bare Plots	1.59	1.45	1.81
Natural Grassland	1.85	2.40	2.99
Sown Pasture	1.64	2.44	2.76
Fodder Tree Plantation	1.69	2.11	2.71
Silvipasture	1.76	2.41	3.09

Another study conducted to evaluate the four silvipasture systems (three were single tree species based while fourth was multiple tree species based) for the various soil fertility aspects, revealed that besides good fodder and fuel wood yield they support high biodiversity and density of the soil biota (Table 2). It seems that the diversified plant components maintain higher belowground biodiversity (Roy *et al.*, 2006b).

Table 2. Species Diversity of soil biota in different silvipasture systems

Soil Biota (no. of species)	Silvipasture Systems			
	SL-1	SL-2	SL-3	SL-4
Collembola	10	9	7	13
Cryptostigmata	20	13	19	27
Prostigmata	5	5	3	7
Mesostigmata	5	9	9	10
Astigmata	1	1	-	1
Other arthropods (including pest sp.)	20	22	20	48
Micro-flora (including pathogens)	20	19	20	22

(SL 1= *Accacia tortilis*; SL 2= *Albiziz amara*; SL 3= *Hardwikia binata*; SL 4= Multispecies Silvipasture)

Soil biota may increase or reduce productivity depending on its composition and the effects of its different activities. *Vice versa,* farming practices modify soil life including the total number of organisms, the diversity of species and the activity of the individual organisms and the aggregate functions of soil biota. These changes can be beneficial or detrimental to the soil biota and its functions and its regenerative capacity (Table 3).

Table 3. Density of soil biota in different land use systems

Soil Biota (Abundance)	Degraded Land	Natural Grassland	Sown Pasture	Silvipasture
Actinomycetes (x 10^5 cfu/g)	99.10	181.4	204.1	281.9
Bacteria (x 10^5 cfu/g)	36.3	105.9	133.1	114.3
Fungi (x 10^5 cfu/g)	6.81	11.0	10.36	19.75
Azotobacter (x 10^4 cfu/g)	3.5	----	-----	10.2
Microfauna (x 10^4 /m^2)	8.60	42.76	39.33	36.75
Mesofauna(x 10^4 /m^2)	2.09	20.25	14.89	19.34
Macrofauna (/m^2)	29.06	52.21	41.06	59.90
Productivity (t/ha)	------	0.10 - 1.50	2.10 - 5.80	3.00 – 7.80

In low input production ecosystems, like grassland and agroforestry systems, where litter production and its decay strongly regulates the nutrient availability for uptake by plants, the population of soil meso fauna has a profound influence on the course of litter decay and played a major role in diversity, abundance and dynamics of decomposer flora and the decomposing litter nutrients (Table 4).

Table 4. Correlation between micro arthropod populations, litter mass loss, decomposing flora and litter nutrients

	Coll	Mite	ML	TN	TP	TK	PPh	Fungi	Bact	Actin
Coll	1.00									
Mite	0.42*	1.00								
ML	0.85**	0.59**	1.00							
TN	-0.19	-0.04	-0.13	1.00						
TP	-0.09	0.12	-0.12	-0.20	1.00					
TK	0.23	-0.04	0.27	0.35*	-0.37*	1.00				
PPh	-0.37*	-0.36*	-0.51**	0.18	-0.13	-0.07	1.00			
Fungi	-0.45**	-0.35*	-0.55**	0.41*	-0.35*	-0.01	0.28	1.00		
Bact	0.46**	0.40*	0.22	-0.16	0.35*	-0.19	0.10	-0.06	1.00	
Actin	-0.16	0.04	-0.15	-0.27	0.39*	-0.46*	0.03	-0.10	0.14	1.00

Coll = Collembola; Bact = Bacteria; Actin = Actinomycetes; ML = Mass Loss; TN = Total Nitrogen; TP = Total phosphorus; TK = Total potassium; PPh = Tannin equivalent of poly phenols

The goal of efficient pastures and silvipasture systems is to develop a sustainable ecosystem with minimal dependence on agrochemical and energy inputs. These systems have found developed ecological interactions and synergy among biological components which made the systems to sponsor their own soil fertility and production. These soil organisms in the soil also partially determine soil resilience. Bhatt *et al.*, (2006) evaluated four silvipasture systems along with three pastures for their carbon restoration capacity. The study indicated that the soil biota abundance is closely linked with the carbon sequestration capacity of particular systems (Table 5).

Table 5. Carbon Sequestration in grasses and silvipasture systems

Systems	Micro flora population (10^5 cfu/g)			Micro fauna population ($10^2/m^2$)			CSV (mgC/ha)	CSS (mgC/ha)	TCS (mgC/ha)
	Fu	Am	Ba	Mt	Cl	Nm			
Pm	6.73	75.40	47.93	48.86	40.72	1994	6.59	3.61	10.20
Cc	5.13	87.53	38.47	44.45	29.52	1819	4.27	1.13	5.40
Sh	5.00	99.40	74.67	60.74	21.72	1628	2.93	3.38	6.31
Pm+Sh+Ll	7.87	92.20	66.20	63.79	23.41	1896	8.63	2.70	11.33
Pm+Sh+Go	6.20	111.3	71.93	54.63	44.45	2047	8.48	2.93	11.41
Cc+S.h+Ds	2.87	95.93	46.80	82.80	26.81	1711	5.70	2.26	7.96
Cc+ Sh +At	4.47	95.93	38.13	59.72	34.95	1839	6.68	2.03	8.71
CD (5%)	1.74	37.87	29.76	17.15	10.34	106.4	-	-	-

(Am = Actinomycetes; Ba = Bacteria; Fu + Fungi; Cl = Collembola; Mt = Mites; Nm = Nematode; CSV = Carbon sequestration in vegetation; CSS = Carbon sequestration in soil; TCS = Total Carbon sequestration; Pm = *Panicum maximum*; Cc = *Cinchrus ciliaris*; Sh = *Stylosanthus hamata*; Ll = *Leucaena leucocephala*; Ds = *Dalbergia sisoo*; At = *Acacia tortilis*; Go = *Grevia optiva)*

Conclusion

Soil degradation is hardly ever ascribed to the decline of soil fauna and reduction in their activities. However, recent Research results have shown that the process which eliminate the beneficial soil flora and fauna communities have resulted in low yield and have degraded soil fertility especially in low input systems. In man made production systems, the emphasis about potential manipulation of beneficial soil biota in design of management practices is extremely rare. Holistic approaches for assessing soil quality and health are required to sustain finite soil resources for future generations. Thus an understanding of soil biological process is critical if on farm materials are to be used effectively.

The declining land productivity and inadequate response of inputs in terms of the outputs on account of deterioration in soil health is recognized as a major problem of recent times. So the interest in maintenance of soil ecosystems and monitoring of soil health is growing. As the biological activities of soil are intimately linked with the maintenance of soil structure and productivity, sound understanding of various aspects of soil flora and fauna in maintenance of soil fertility in agroforestry systems is vital.

Therefore, designing of agroforestry systems that either prevent destruction of beneficial soil biota or provide favorable environmental condition to facilitate their activities should receive adequate emphasis. Finally, careful integration of local knowledge of soil organism in relation to fertility should also be integrated in designing production systems with a view to improve its technical feasibility, economic profitability and adoptability for a particular ecological zone.

References

Bhatt, R. K., Baig, M. J., Yadav, R. B. and Roy, S. 2006. Effect of high level CO_2 on biomass production and carbon sequestrationin silvipastoral systems under semiarid tropics: AP Cess report. ICAR N. Delhi. Pp. 50.

Buscot, F. and Verma, A. 2005. Microorganisms in Soils: Roles in genesis and functios. Springer-Verlag, Berlin Heidelberg, Germany. Pp. 419.

CBD. 2001. Soil Biodiversity and sustainable agriculture. Food and Agriculture Organization, UN. Pp. 28.

NAAS. 2006. Belowground Diversity in Relation to Cropping Systems. Policy Paper No. 36, National Academy of Agricultural Sciences, New Delhi. Pp. 8.

Roy, S., Nag, S. K. and Bhatt, R. K. 2006a. Biodiversity and Dynamics of soil fauna affecting sustainability of rainfed forage farming systems: AP Cess report. ICAR N. Delhi. Pp. 40.

Roy, S., Bano, R. Nag, S. K., Bhatt, R. K. and Roy, M. M. 2006b. Soil arthropods: Species diversity and richness in perennial forage based systems in Bundelkhand Region. In: abstract: National Symposium on Agroforestry for livelihood security, environment protection and biofuel production. Pp. 153.

Roy, S., Saxena, P and Roy, M. M. 2008. *Soil biodiversity under forage production systems*. IGFRI, Jhansi (India), Pp. 50.

□□□

Agroforestry: Natural Resource Sustainability, Livelihood & Climate Moderation © 2009
Editors: OP Chaturvedi, A Venkatesh et. al., pp. 485-490
Satish Serial Publishing House, Delhi

Chapter 34

CLIMATIC VARIABILITY AND CHANGE: IMPLICATIONS FOR INDIAN AGRICULTURE

J. B. Singh and Pradeep Behari
Indian Grassland and Fodder Research Insitute, Jhansi-284 003 (U.P.)

Introduction

India is a large developing country with nearly 700 million rural populations directly depending on climate-sensitive sectors (agriculture, forest) and natural resources (such as water, biodiversity, coastal zones, grasslands) for their subsistence and livelihoods. Climate change is one of the most important global environmental challenges facing humanity with implication for food production and natural ecosystem. The past two decades have witnessed globally a rapid increase in the awareness about climate change and triggered widespread apprehension among scientists and government about their global and regional implications (Gadgil, 1996). The global climate is certainly changing; it is no more considered uncertain but a reality. There is greater consensus that in future climatic variability will increase leading to more frequent extremes of weather. The Inter-Governmental Panel on climatic change in its recent assessment (IPPC, 2007a) has projected an increase in earth mean temperature by 0.74 ^{0}C between 1906 and 2005. Further, the temperature increase by the end of 2100 is likely to be in the range of 1.5 to 4.0 ^{0}C. There is clear evidence that climatic changes in recent times is the increase in the atmospheric temperatures which may be attributed to increased levels of green house gases such as CO_2, CH_4, O_3 and CFCs. The CO_2 concentration which was 280 ppm in the pre industrial period (1850) is rising at the rate of 1.5 to 1.8 ppm per year. The concentration of CO_2 is likely to be doubled by the end of 21st century. Such global climatic change will affect agriculture through their direct and indirect effects on crop, livestock threatening food security of several developing countries including India.

Climatic Variability Over India

Climatic variability is the major factor influencing the agricultural productivity. One of the most important variables describing the climate is the surface air temperature which is strongly linked to other atmospheric processes such as evaporation and precipitation. Changes in monthly and annual temperatures are of interest also in different field of human economic activity such as agriculture, hydrology and water management. In India, studies by several workers regarding long term variations in the surface air temperature over the past century for Indian region have demonstrated an increasing trend in surface temperature (Rupa kumar *et al.*, 1994, Hingane *et al.*, 1985). The main results that emerge from these studies is that the mean annual surface temperature over India has warmed at the rate of about 0.4 ^{0}C per century. This warming trend is comparable to global mean trend of 0.5 ^{0}C in last 100 year. There have been large regional, seasonal and diurnal variations in this warming. Predominant warming has been in the post monsoon and winter season rather than in summer, and large increase has occurred in maximum rather than in minimum temperature. The warming trend over the period 1878-1993 in the post monsoon season (0.52 ^{0}C/100yr) and in the winter is more marked. The monsoon temperatures do not show a significant trend in any major part of the country. This indicates that there are some regions of India where more substantial warming has occurred than others. The significant warming trends are seen most clearly over northern and southern parts of west coasts and in the northeast regions of the country. Recent climate model suggest that India may experience a further rise in temperature of 1^{0}C by 2050, which is about 4 times the rate of warming over past 100 years. Regardless of any change in precipitation amount, this temperature rise is likely to decrease soil moisture availability as the evaporation increases. Anticipated climatic changes may have adverse implications for agriculture in India.

Because of importance of water supply for Indian agriculture, a considerable amount of work on observed trends and periodicities in Indian rainfall has been undertaken. The all-India summer monsoon rainfall series, obtained by the weighted rainfall of 306 stations spread over Indian Plains has a mean value of 852.6 mm with c.v. of 9.9 %. Years of excess rainfall during 1871-2000 were 20, out of which 17 were La Nina years and deficit rainfall occurred in 22 years, of which 12 were associated with El Nino years (Kulkarni, 2003). Recent work concerning the analysis of homogeneous Indian monsoon (June-September) rainfall record clearly demonstrates that there is no systematic increasing or decreasing trend in Indian rainfall over the last century, although large interannual variability and some decadal variations are seen. The rainfall variability has been largely random over the century. The increasing trend in the seasonal rainfall have been found along the west coast, North Andhra Pradesh and North West India, whereas, decreasing trend has been observed over east Madhya Pradesh and northeast India. Further, there is a significant trend during 20th century in the frequency of monsoonal cyclonic disturbances in northern Indian Ocean.

Climatic Projections for India

The recent state-of-the art global climate results (Lal *et al.*, 1995) suggest that the surface air temperature, over the Indian subcontinent is likely to rise from 1.0 ^{0}C (during monsoon) to 2.0 ^{0}C (during the winter) by the middle of next century. The rise in surface temperature could be quite significant across the semi-arid regions of N-W India. The UKMOGCM model predicts (Bhaskaran *et al.*, 1995) a total precipitation increase of approximately 20% and increase in winter or *rabi* crop season temperature by 1-4 ^{0}C with increased CO_2 concentration. Specific humidity increases by 19%, indicating that the increased monsoon rainfall is largely due to the water content of the atmosphere. The climatic element which affect plant growth and development, hence agriculture in a broader sense, viz. temperature, CO_2 concentration, precipitation and radiation are likely to be altered with the increased build up of green house gases in the atmosphere(Sinha, 1993). In a recent report(Lal *et al.*, 2001), it was estimated that CO_2 level will increase to 379-416 ppm by 2010s from the present CO_2 level of 371 ppm and this would further increase by 605-755 by 2070s. They projected between 1 to 1.4 ^{0}C and 2.23 to 2.87 ^{0}C area averaged annual mean warming by 2020 & 2050, respectively. The increase in temperature is projected to be more in *rabi* season than in *kharif* crop growing season. However, a large uncertainty is associated with projected *rabi* rainfall than *kharif* rainfall in 2050s. Recent assessment of IPCC has projected 0.5-1.2 ^{0}C rise in temperature by 2020 and 0.88-3.16 ^{0}C by 2050 depending on the recent SERES Scenario of future development (IPCC, 2007 b).

Implications for Agriculture

Despite technological advances such as improved crop varieties and irrigation system, weather and climate are still key factors in agricultural productivity. In India, the food production potential today is already quite vulnerable to climatic variability. The impact of future climatic change may be felt more severely in developing countries such as India whose economy largely depends on agriculture and is already under stress due to the current population increase and associated demand for energy, fresh water and food. Studies have also revealed that drought is the area in which climate change poses the greatest risk for agriculture and consequently arid and semi-arid regions are likely to be most vulnerable to climate change.

Future changes in climate could as well provoke serious reductions in agricultural productivity during the years to come. It is estimated that by 2020, food grains requirement would be almost 30-50% more than the current demand (Paroda and Kumar, 2000). The IPCC indicate that by 2100 AD, India may lose crop production between 10-40% despite beneficial effects of higher CO_2 on several crops. A number of studies also confirms the declining trend of agricultural production with climate change. The work carried out at Indian Agricultural Research Institute predicts the possibility of wheat production loss of 4-5 million tones with every rise of 1 ^{0}C temperature throughout the growing period (Aggarwal, 2008).

The global change will also affect the water resources of the country which include enhanced evaporation due to warming, geographical changes in precipitation intensity, duration and frequency, average runoff, and frequency and severity of droughts and floods. Relatively small climate changes can cause large water resource problems in many areas especially arid and semi-arid regions such as north-west India. If water availability decreases for this region, it could have significant implications for agriculture. With still largely unregulated river system, India is particularly vulnerable to hydro meteorological changes.

The changing scenario of climate will affect the crop, soils and pests. The increase in temperature can reduce crop duration, increase crop respiration rates and evapotranspiration. There may be considerable effects (Aggarwal, 2003) on land use due to snow melt, spatial and temporal rainfall variability, availability of irrigation, frequency and intensity of inter and intra seasonal droughts and floods, soil organic matter transformations, soil erosion, change in pest profiles, decline in arable areas due to submergence of coastal lands. All together this will lead to serious impact on agricultural production and food security of the region.

In spite of the uncertainties about the precise magnitude of climate change on regional scale, an assessment of the possible impacts on Indian agriculture is important for formulating response strategies. The identification of suitable response strategies is key to sustainable agriculture. The IPCC report suggests several adaptation strategies to deal with projected climatic changes which include changing varieties; more efficient water use; altering the timing or location of cropping activities; breeding of plants that are more resilient to variability of climate and making better use of seasonal climatic forecasts to reduce production risks. These adoption strategies may have substantial potential to offset negative impacts from climate change and take advantages of positive impacts.

References

Aggarwal, P. K. 2003. Impact of climate change on Indian agriculture. *Journal of Plant Biology,* 30(2):189-198.

Aggarwal, P. K. 2008. Global climate change and Indian agriculture: impacts, adaptation and mitigation. *Indian Journal of Agricultural Sciences,* 78(10): 911-919.

Bhaskaran, B., Mitchell, J. F. B., Lavery, J. R. and Lal, M. 1995. Climatic response of Indian subcontinent to doubled CO_2 concentrations. *International Journal of Climatology,* 15: 873-892.

Gadgil, S. 1996. Climate change and agriculture - An Indian perspective. In *Climate variability and Agriculture* (eds Abrol, Y.P. *et al.*) Narosa Publ. House, New Delhi, Pp.1-18.

Hingane, L. S., Rupa Kumar, K. and Raman Murty, B.V. 1985. Long term trends of surface air temperature in India. *Journal of Climatology,* 5: 521-528.

IPCC. 2007a. *Climate Change 2007: The Physical Science Basis.* Summary for Policymakers. Inter- Governmental Panel on Climate Change.

IPCC. 2007b. *Climate Change 2007: The Physical Science Basis.* Summary for Policymakers. Inter- Governmental Panel on Climate Change.

Kulkarni, B.D. 2003. *Assessment of climate change in India and mitigation policies* (Eds. Dash, S.K. and Rao, P.), World Wide Fund for Nature-India, New Delhi, Pp. 62-71.

Lal, M., Nozawa, T., Emori, S., Harasaw, A., Takahashi, K., Kimoto, M., Abe-Ouchi, A., Nakajima, T., Takemeura, T., and Numaguti, A. 2001. Future climate change:Implications for Indian summer monsoon and its variability. *Current Science,* 81: 1196-1207.

Lal, M., Cubasch, U., Voss, R. and Waszkewitz, J. 1995. Effect of transient increases in greenhouse gases and sulphate aerosols on monsoon climate. *Current Science,* 69(9): 752-763.

Paroda, R.S. and Kumar, P. 2000. Food production and demand in South Asia. *Agricultural Economics Research Reviews,* 13(1): 1-24.

Rupa Kumar, K., Krishna Kumar, K., and Pant, G. B. 1994. Diurnal asymmetry of surface temperature trends over India. *Geophysical Research Lettrs*, 21: 677-680.

Sinha, S. K. 1993. Response of tropical ecosystem to climate change, *International Crop Science*, Crop science society of America, Madison, WI 53771, U.S.A., Pp.281-289.

❑❑❑

Agroforestry: Natural Resource Sustainability, Livelihood & Climate Moderation © 2009
Editors: OP Chaturvedi, A Venkatesh et. al., pp. 491-505
Satish Serial Publishing House, Delhi

Chapter 35

CLIMATE CHANGE WITH REFERENCE TO RESPONSE OF FODDER CROPS TO ELEVATED CO_2

R.K. Bhatt
Indian Grassland and Fodder Research Institute, Jhansi-284 003 (U.P.)

Introduction

Since in the mid 1800's, the increasing human activities have contributed to atmospheric CO_2 which has increased from roughly 250 to present day 380 ppm and further doubling of the global CO_2 within the next century. Human activities such as fossil fuel combustion, industry, land clearing and agriculture are increasing atmospheric concentrations of greenhouse gases such as carbon dioxide, methane and nitrous oxides. There is growing evidence that these changes are impacting on physical processes, species and ecosystems globally. This increasing CO_2 level has caused us to re evaluate our knowledge of plant responses to increasing CO_2 research regarding the potential impact of elevated CO_2 on photosynthesis and on the potential for these enhanced rates to increased crop and forest productivity. The increasing level of CO_2 conc. is accepted to have substantial effects on the world's ecosystems by directly affecting plant growth and development.

As CO_2 is the sole source of carbon for growth in plants, the response of photosynthesis and growth to increasing CO_2 has been the subject of a number of studies. In recent years researchers have been predicting that the green house effect will dramatically change global weather pattern in the next century, i.e. air temperature and CO_2 level will rise in the coming years. Therefore, it is very essential at this stage to assess the effect of this CO_2 and take necessary measures to avoid any adverse impact that this change may have.

Enhanced levels of CO_2 in the atmosphere affect the carbon balance in the rhizosphere and photosynthetic carbon assimilation in plants, thereby affecting the agricultural productivity. Predicting climate change and evaluating the effects of changing levels of carbon dioxide in the atmosphere on agro ecosystems, requires the use of models. However, these models must be developed and evaluated with data obtained under realistic field condition.

It has been assumed that the rates of C sequestration will sufficiently high which suggests that rangelands may be a substantial global sink for atmospheric carbon. The significant increases in soil organic carbon sequestration, forage production and forage quality demonstrate the potential of this practice to improve rangeland health and assist in the mitigation of elevated atmospheric carbon dioxide levels.

A massive research effort has been undertaken by International Geosphere - Biosphere programme to determine the direct (e.g. photosynthesis, growth) and indirect (climatic change responses) effects of elevated CO_2 and other trace gasses on agricultural and natural ecosystems. Although there is still uncertainty concerning the degree of climatic change that may result from increases in atmospheric trace gasses, a consensus is emerging that significant global warming will occur along with altered precipitation patterns and increased storm intensity.

Long-term Effect of Elevated CO_2

At present atmospheric general circulation model (GCM) predicts an increase in mean surface temperature of several degrees with a doubling of current CO_2 concentration (Watson *et al.*, 1990). Increase in CO_2 concentration and/or temperature may alter the plant's ability to compete with individuals with in a given environment. Two major photosynthetic path ways are C_3 (95% of all known plant species) and C_4, it is generally agreed that the growth of plant with the C_3 pathway should be stimulated more by CO_2 enrichment than the growth of C_4 plants (Imai and Murata, 1979). Kimball (1983) reported biomass increase in response to a doubling of ambient CO_2, which range from 10 to 43% in C_3 crop plants and 24% in C_4 plants.

Global temperature will have important impacts upon plant species production, distribution and plant community composition. In general, relatively large increases in temperature should favor warm season plants. (Alward *et al.*, 1999). However, there may be interactions with CO_2 that modify or even cancel temperature effects on vegetation. For instance, while warming may favor warm season C_4 species, some of the effects of CO_2 enrichment on C_3 photosynthesis will tend to counter that response, rendering C_3 /C_4 distribution changes relatively insensitive to increases in temperature (Polly *et. al.*, 2000).

A large body of knowledge exist describing different plant species responses to elevated CO_2, with some progress towards identifying different functional plant groups in terms of their potential response to CO_2. Plants processing the C3 photosynthetic capacity are generally expected to exhibit a greater growth enhancement from CO_2 enrichment since their photosynthetic apparatus is limited more by present ambient CO_2 concentrations compared with C_4 plants. Evidence from numerous studies indicates that this expectation holds (Poorter, 1993, Wand *et al.*, 1999), although it can be complicated by other aspects of plant/soil responses to CO_2, as well as interactions of plant CO_2 responses with other environmental parameters (Hunt *et al.*, 1996. Owensby *et al.*, 1999).

The responses of grasses to CO_2 were recently reviewed by Wand *et al.*, (1999). They report strong growth enhancements of both C_3 and C_4 grasses from elevated CO_2, although greater responses were with C_3 grasses. While photosynthesis in C_4 plants can respond directly to increases in CO_2 above present atmospheric concentrations (e.g. Le Cain and Morgan, 1998), the response is considerably more limited compared to that of C_3 species.

Average photosynthetic rates of the youngest fully expanded leaves of *L. perenne* increased by 30 to 80% under elevated CO_2 as compared to ambient CO_2 (Rogers *et. al.*, 1998, Isopp, 2000). Total symbiotic N2 fixation increased by 66% in the grass legume mixture under elevated CO_2 (Zanetti *et al.* 1997, Lüscher *et al.*, 2000)

Paspalum dilatatum was stimulated by elevated CO_2. The mechanism for this has not been identified, but Ghannoum *et. al.*, (2000) reported that increased intercellular CO_2 partial pressure, changes in diurnal CO_2 fixation patterns, improvements in shoot water relations and increased leaf temperatures as possible ways in which C_4 species might respond to elevated

CO_2. Recruitment from seed is an important mechanism of species compositional change in the enriched CO_2 environment and is perhaps an underestimated mechanism in other systems.

Chem *et al.,* (1994) observed that the maximum carboxylation rate of Rubisco, potential, electron transport rate and quantum efficiency was increased by CO_2 enrichment, while the maximum phosphoenol pyruvate carboxylase activity was decreased. They also predict that the light saturated leaf photosynthesis will increase in atmospheric CO2 and increasing atmospheric CO_2 significantly decreased the stomatal conductance. High CO_2 strongly reduced the stomatal conductance of both species, while it affected the Rubisco content (30% decreases) of high light C_3 species only (Oula Ghannoun *et. al.,* 1997). The increase in the net rate of CO_2 (NCE) in various C_3 crop species grown under elevated CO_2 is also reported (Campbell *et al.,* 1988). However, the response of crop plants in terms of photosynthetic enhancement varies from species to species.

At very high atmospheric concentration of CO_2 (1200 ppm), the CO_2 out burst is completely eliminated in tissues that show the outburst in air and such condition inhibits the photorespiration due to high levels of CO_2. Such inhibition of photorespiration by high levels of CO_2 was found in bean & sunflower leaves (Egle and Fock, 1967). Thus the high conc. of CO_2 inhibits photorespiration as reflected by the CO_2 out burst after illumination probably because synthesis of the substrate for photorespiration is inhibited.

Gas exchange and water balance research in Kansas (Ham *et. al.,* 1995; Nie *et. al.,* 1992, Owensby *et. al.,* 1999) and Colorado (Lapitan *et. al.,* 2000; Morgan *et al.,* 1998, 2001) indicate that the primary effect of CO_2 on production of the tall grass prairie and short grass steppe is through improved water relations that result in higher water use efficiency. CO_2 induced growth enhancements of C_3 and C_4 grasses in the short grass steppe occurred because they were driven primarily by the effects of CO_2 on water relations (Morgan *et al.,* 1998, 2001), which benefited C_3 and C_4 grasses alike.

The partitioning of biomass between above-and below - ground organs in response to environmental perturbations like CO_2 is another important plant trait that has been studied extensively. Several researchers (Hunt et al 1998) have suggested that such CO_2 -induced changes in plant biomass partitioning among organs may be best understood through an examination of the effects on labile C and N pools and the functional balance hypothesis. Production increases resulting from growth at elevated CO_2 are often accompanied by decreases in plant N concentration, and there is some evidence that N may limit plant response to rising CO_2 (Ghannoum and Conroy, 1998,

Poorter *et. al.*, 1996). Similarly, there is some indication that plants capable of fixing their own N may have a greater capability to respond to rising levels of atmospheric CO_2 (Poorter, 1993). In addition to predicted productivity increases under elevated CO_2, species composition of plant communities are likely to change, although the exact nature of these changes are difficult to predict (Polley *et al.*, 2000). But it can be assumed that it will bring serious implications for land use and land cover in India in future.

Effect of Elevated CO_2 on Photosynthesis and Biomass Production of Fodder Crops

Forage species particularly perennial grasses cover large land area and these multicuts and perennial species can serve the major sink of increased atmospheric carbon dioxide. Therefore, it is important to under stand the impacts of increased CO_2 on productivity and nutritive quality of these pastures species. Finding low-cost methods to sequester carbon is emerging as a major international policy goal in the context of increasing concerns about global climate change. Berseem (*Trifolium alexandrinum*) and Oat (*Avena sativa*) are the dominant cultivated fodder species with their multicut in nature and the range grasses (*Cenchrus ciliaris* and *Panicum maximum*) and legumes (*Stylosanthes hamata*) are perennial in nature and cover the maximum area under thier production in the tropical region has been studied as a response to elevated CO_2 on their growth and productivity. The crop species studied in this investigation are of wide economic significance with large sink of carbon as per their productivity.

Cenchrus ciliaris+ Stylosanthes hamata

The *Cenchrus ciliaris and Stylosanthes hamata* were grown in intercropping system with or without elevated CO_2 for long period in open field and under open top chambers (OTC). The leaf area index, which indicates the leaf area production/m^2, increased significantly in *C. ciliaris* and *S. hamata* under elevated CO_2 over the open field grown crops in intercropping system. A significant increase in LAI was observed when total leaf areas of both the crops under intercropping systems. The total leaf area index of *C. ciliaris* and *S. hamata* increased by 55.55% over the control (8.46). The average leaf area index of three years increased by 1.66 fold over the LAI of open field grown crops (Figure 1). The increase in leaf area index under elevated CO_2 may be due to increase in tiller/branch and leaf production per plant and also increase in leaf size.

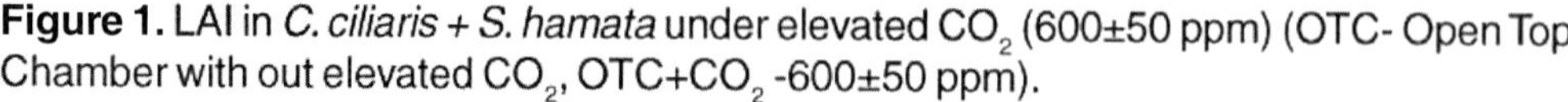

Figure 1. LAI in *C. ciliaris + S. hamata* under elevated CO_2 (600±50 ppm) (OTC- Open Top Chamber with out elevated CO_2, OTC+CO_2 -600±50 ppm).

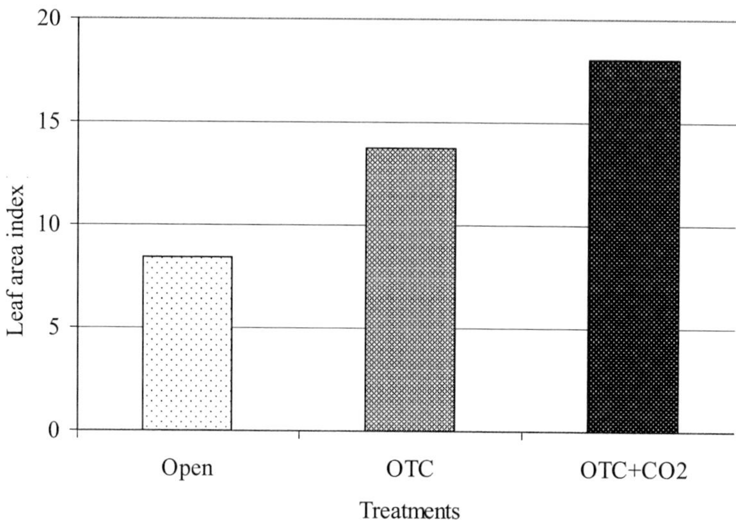

The canopy photosynthesis, which is calculated by multiplying the rate of photosynthesis with leaf area index (PN x LAI) in *C. ciliaris* and *S. hamata* increased significantly in OTC with elevated CO_2 (Figure 2). The canopy photosynthesis increased by 2.5 folds in *C. ciliaris* and 1.55 folds in *S. hamata* under elevated CO_2 (600±50 ppm) over the open field grown crops. The higher canopy photosynthesis has maximized the growth and biomass production in these crop species under elevated CO_2 and therefore, recommended as major sink of the increasing level of CO_2 in the atmosphere.

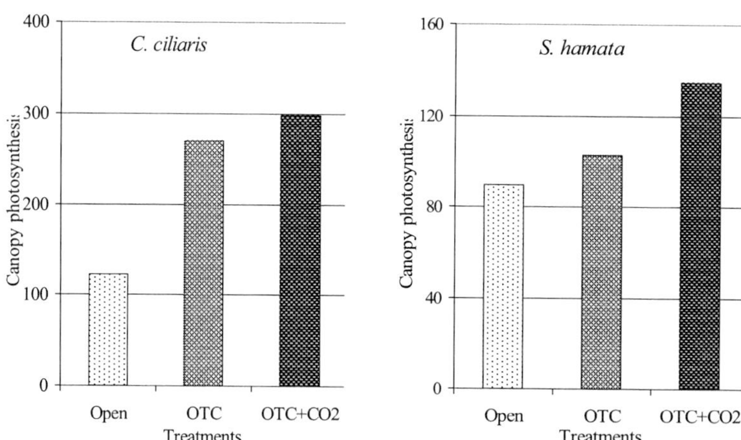

Figure 2. Effect of elevated CO_2 on canopy photosynthesis (PN x LAI) of *P. maximum* and *S. hamata* under intercropping system.

The cumulative green biomass production in *C. ciliaris* and *S. hamata* was recorded 2.05 and 1.65 fold higher under elevated CO_2 respectively. *Cenchrus ciliaris + Stylosanthes hamata* produced 106% higher fresh biomass under elevated CO_2 which over control. (Table 1). Similarly dry biomass production in all the plant parts of *C. ciliaris* and *S. hamata* increased significantly under elevated CO_2 (600±50 ppm) over the control. In *C. ciliaris* and *S. hamata* dry matter accumulation increased by 2.52 folds and 1.76 folds respectively under elevated CO_2 over the open field grown crops in intercropping system. On average of all the three years total dry matter accumulation increased by 141% over the crops grown in open field (Table 2).

Table 1. Fresh biomass production (g/m^2) in *Cenchrus ciliris+ S. hamata* under different environmental conditions during different years of growth.

Treatments	*Cenchrus ciliris+ S. hamata*				
	1st year	2nd year	3rd year	Total	Average
Open	783.80	2490.00	1607.82	4881.62	1627.20
OTC	2269.38	3146.66	2569.45	7985.49	2661.83
OTC + CO_2	3375.01	3990.00	3307.07	10672.08	3357.36

Table 2. Dry biomass production(g/m^2) in *Cenchrus ciliris+ S. hamata* under different environmental conditions during different years of growth.

Treatments	*Cenchrus ciliris+ S. hamata*				
	1st year	2nd year	3rd year	Total	Average
Open	266.15	716.57	430.77	1413.49	471.16
OTC	779.20	937.94	662.18	2379.32	793.10
OTC + CO_2	1142.28	1315.32	952.27	3409.87	1136.62

Panicum maximum + Stylosanthes hamata

The effect of elevated CO_2 (600±50 ppm) was also observed in *Panicum maximum* and *Stylosanthes hamata* in intercropping system with or without elevated CO_2 for long period in open field and under open top chambers (OTC). The leaf area index was also significantly increased under elevated CO_2 (600±50 ppm) in both the crops in intercropping system. The cumulative leaf area index of *P. maximum* and *S. hamata* increased by 1.5 fold as the crop grown under elevated CO_2 (600±50 ppm) over the control. The increase in leaf area index may be due to increase in leaf length and leaf width and the total number of tillers/branches and leaf production per plant (Figure 3).

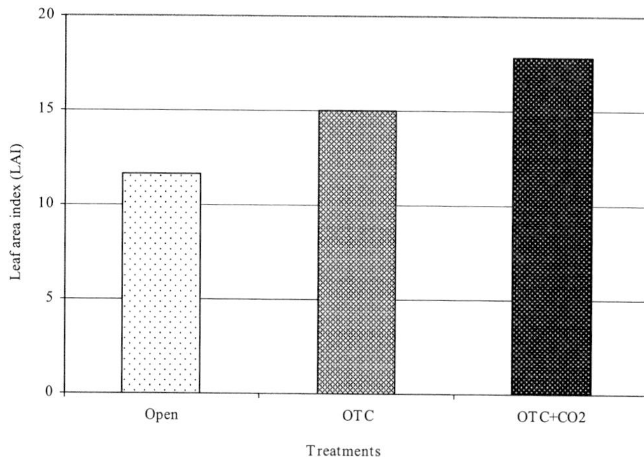

Figure 3. Average Canopy photosynthesis (PN x LAI) of *P. maximum* and *S. hamata* under OTC with and without elevated CO_2 and in open field.

Canopy photosynthesis in both the crops increased significantly in OTC with elevated CO_2 (600±50 ppm). The canopy photosynthesis increased by 152% in *P. maximum* and 1128% in *S. hamata* under elevated CO_2 in intercropping system over the open grown crops (figure 4). The increase in canopy photosynthesis in these crop species may be due to increase in the rate of photosynthesis and leaf area index. The increase in the canopy photosynthesis under elevated CO_2 indicated that there is more assimilation of CO_2 which intern higher biomass production in these two crop species nder intercropping systems.

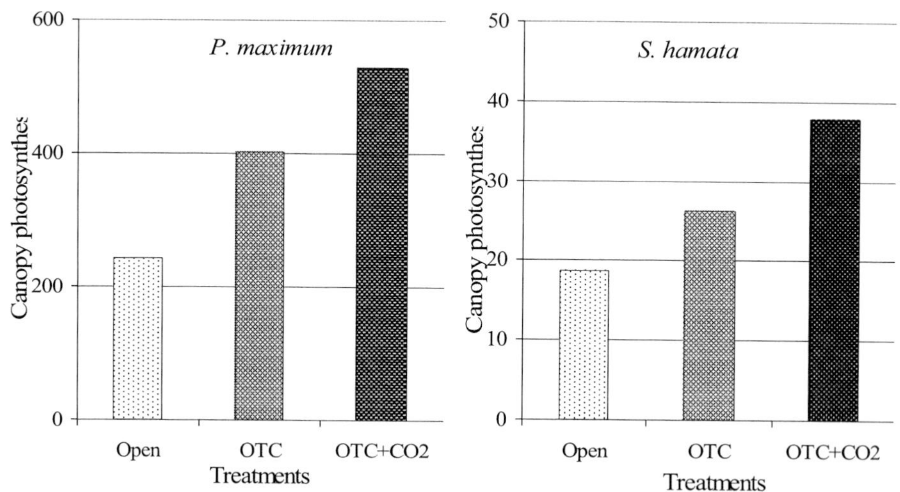

Figure 4. Effect of elevated CO_2 on canopy photosynthesis (PN x LAI) of *P. maximum* and *S. hamata* under intercropping system.

The biomass production (fresh and dry) in both the crop species under intercropping systems increased significantly in OTC with elevated CO_2. *P. maximum* + *S. hamata* produced 4159.07 g m^{-2} fresh biomass under elevated CO_2 which was 67% higher over the control (2491.46 g m^{-2}) (Table 3). Similarly dry biomass production increased by 85% under elevated CO_2 (600±50 ppm) over the crops grown in open field (684.70 g m^{-2}). The increase in biomass production may be due to enhanced rate of photosynthesis and higher fixation of carbon and its allocation to the plant components. Elevated CO_2 stimulated total dry biomass accumulation which is a common physiological response to high CO_2 (Atkinson *et al.* 1997). Elevated CO_2 increased leaf size, plant height and dry mass of shoot in *Trifolium alexendrinum* (Pal *et al.* 2004).

Total green biomass production of all the three years was observed under the OTC with elevated CO_2 (600±50 ppm) (Table 3). The fresh biomass increased by 67% under elevated CO_2 over the control. Similarly the total dry biomass has also increased by 1.84 fold (85%) under elevated CO_2 over the control and 67% over OTC with out elevated CO_2 (Table 4).

Table 3. Fresh biomass (g m^{-2}) production in *P. maximum + S. hamata* under intercropping system under various environments during different years of growth.

Treatments	*P. maximum + S. hamata*				
	1st year	2nd year	3rd year	Total	Average
Open	1479.49	2875.32	3119.59	7474.40	2491.46
OTC	2875.63	3386.66	3395.76	9858.05	3219.35
OTC + CO_2	3457.43	4194.98	4824.81	12477.22	4159.07

Table 4. Dry biomass (g/m^2) production in *P. maximum + S. hamata* under inter-cropping system under various environments during different years of growth.

Treatments	*P. maximum + S. hamata*				
	1st year	2nd year	3rd year	Total	Average
Open	453.98	802.22	797.91	2054.11	684.70
OTC	917.84	983.49	887.38	2788.71	929.57
OTC + CO_2	1172.95	1326.07	1303.52	3802.54	1267.51

Long term exposure of crop *P. maximum* + *S. hamata* to elevated CO_2 (600±50 ppm) in open top chambers resulted in a significant enhancement in the growth and biomass production. This increase in growth was due to the production of more photosynthates and their partitioning to different plant parts which ultimately increased the total biomass production.

Oat (*Avena sativa*)

A significant change was recorded in leaf area index (LAI) in the crops grown under elevated CO_2 over the open grown crops in both the cutting stages of the crop. The LAI was observed 13.18 and 10.41 during the first and second cut respectively under elevated CO_2 whereas it was only 7.96 and 5.91 under first and second cut respectively in the open grown crop. The canopy photosynthesis (P_NXLAI), indicting the over all assimilation of CO_2 of the crop increased significantly under elevated CO_2 over the open grown crops and crop grown in OTC at ambient CO_2 (Figure 5).

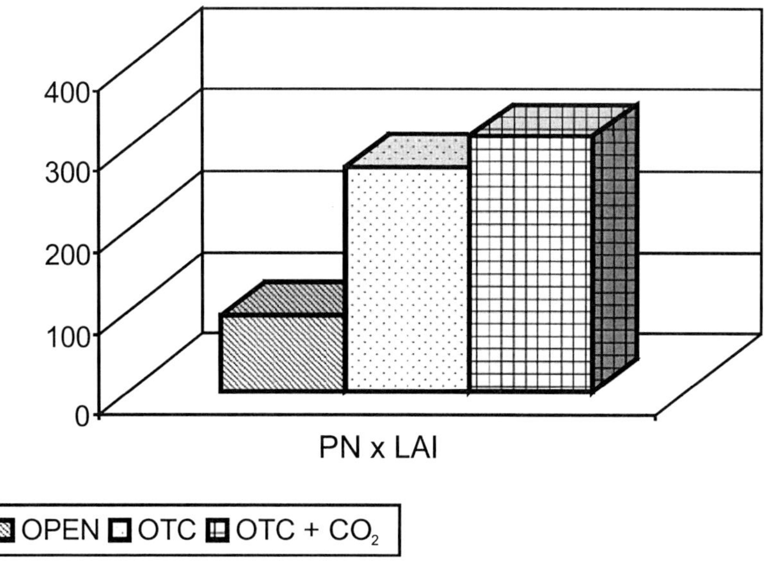

Figure 5. Effect of elevated CO_2 on canopy photosynthesis of Oat cv. JHO 822.

The fresh biomass production was increase to double fold under elevated CO_2 in both the cuttings. The dry matter accumulation was observed to 1130.5 g m^{-2} in the first cut and 1258.5 g m^{-2} in second cut of the crop under elevated CO_2 where as it was 746.8 g m^{-2} and 1071.7 g m^{-2} in first and second cut of crop respectively in open grown crop. The cumulative biomass production in terms of fresh and dry matter accumulation of two successive cuttings showed an increase of 196.22% (fresh) and 140.97% (dry) under elevated CO_2 over open condition and the increase in the biomass production in C_{OTC} was recorded to a tune of 148.39% (fresh) and 125.88 (dry) over Ca (Table 5).

Table 5. Biomass production in Oat cv. JHO 822 under different environmental conditions.

Treatments	1st cut				2nd cut		Cumulative	
	FW gm^{-2}	DW gm^{-2}	Ht cm	Tillers m^{-2}	FW gm^{-2}	DW gm^{-2}	FW gm^{-2}	DW gm^{-2}
OPEN	3197.3	746.8	110.3	215.5	3177.8	1071.7	6375.1	1818.5
OTC	4621.9	1061.5	134.0	283.3	4811.1	947.9	9433.0	2009.4
OTC +CO2	6387.3	1130.5	143.1	494.4	6122.2	1258.5	12509.5	2389.0
CDat 0.05 p	547.18	247.31	17.78	75.45	1061.7	268.2	1000.4	349.44

Berseem (*Trifolium alexandrinum*)

The leaf area index (LAI) increased significantly under the elevated CO_2 over open grown crops under all cutting stage of crop growth. The expansion of the leaf was more under elevated CO_2 through which the assimilatory surface was increased and in tern maximize of biomass production. The canopy photosynthesis which is calculated by multiplying the rate of photosynthesis with leaf area (P_N X LAI) increased significantly in OTC with elevated CO_2 in all the cuttings. The increase in canopy photosynthesis was 37.06%, 29.20% and 59.01% in 1st, 2nd and 3rd cut respectively over the open condition. The increase in canopy photosynthesis has indicated the fact that there was higher assimilation of CO_2, which has intern the maximum biomass production (Table 6)

Table 6. P_N and P_N x LAI in berseem cv. Wardan in successive three cuttings.

Treatments	1st CUT		2nd CUT		3rd CUT	
	P_N	P_N x LAI	P_N	P_N x LAI	P_N	P_N x LAI
OPEN	25.15	55.63	28.31	45.08	24.21	74.23
OTC	26.58 (5.37)	64.48 (13.73)	31.42 ((9.89)	94.06 (52.08)	26.34 (8.09)	86.06 (13.75)
OTC+CO2	29.82 (15.66)	88.39 (37.06)	34.21 (17.25)	104.85 (29.20)	30.12 (19.62)	109.97 (59.01)
CDat 0.05p	8.423		6.24			

Data in parentheses indicates percent increase over open

The biomass production (fresh and dry) under all the three cuttings increased significantly in OTCs with or without elevated CO_2 but the maximum biomass production was recorded under elevated CO_2. The fresh and dry matter accumulation increase to double fold in OTC with elevated CO_2 over Ca. The percent increase in the cumulative biomass of three cuts showed that 40.18% (fresh) and 31.43 % (dry) increase in biomass production when the crop grown under C_{600} over open field grown crops, however, the percent increase in C_{OTC} was recorded as 29.52 (fresh) and 24.54 (dry) over open field condition (Figure 6).

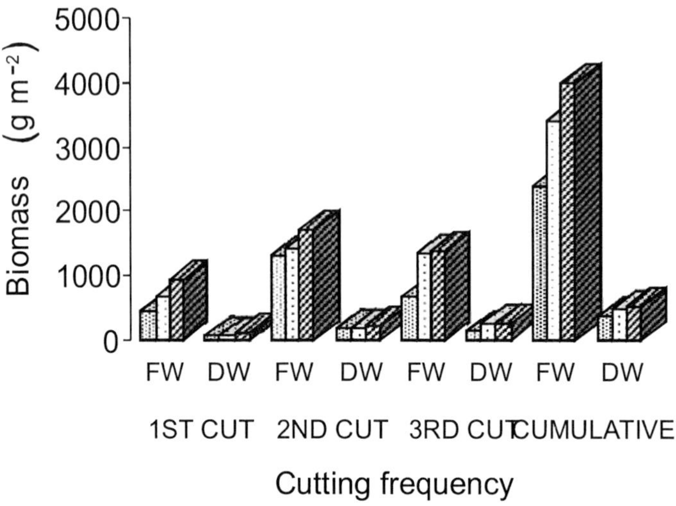

CD at 0.05p 1st cut: FW = 221.23, DW = 16.09, 2nd cut: FW = 388.16 DW = 43.17 3rd cut: FW = 334.23, DW = 61.23

Figure 6. Biomass production (fresh & dry) influenced by elevated co_2 in berseem cv. wardan

The increase in canopy photosynthesis, fresh and dry matter accumulation under elevated CO_2 indicating that these crop species should be promoted for higher biomass production and carbon sequestration in the semi arid tropical environmental conditions. These results revealed that these fodder crop species having the potential to assimilate atmospheric CO_2 even under increasing level of CO_2 (600±50 ppm) and therefore will help in understanding the relationships between productivity and CO_2 sequestration in different agro-ecosystems in monoculture and intercropping systems. The resulting data can be incorporated in to predictive models of the response of these agro-ecosystems on CO_2 change.

References

Alward, R.D., Detling J.K. and Milchunas D.G. 1999. Grassland vegetation changes and nocturnal global warming. *Science*, 283: 229-231.

Atkinson, C.J., Taylor, J.M., Wilkins, D., Besford, R.T. 1997. Effect of elevated CO_2 on chloroplast components. Gas exchange and growth of oak and cherry. *Tree Physiology*, 17: 319-325.

Campbell, W.J., Allen Jr, L.H. and Bowes, G. 1988. Effect of CO_2 concentration on RuBiscoactivity, amount of photosynthesis in soybean leaves. *Plant physiology*, 88: 1310-1316.

Chen, D.X., Coughenour, M.B., Knapp, A.K. and owensby, C.E. 1994. Mathemathetical simulation of C_4 grass photosynthesis in ambient and elevated CO_2. *Ecological modeling*, 73 (1-2): 63-80.

Egle, K., and Fock, H., 1967. Light respiration correlations between CO_2 fixation, O_2 pressure & glycollate concentration. In the "*Biochemistry of Chloroplasts*". (T.W. Goodwin, ed.) Vol.2, Pp 79-87. Academic Press, New York.

Ghannoum, O. and Conroy J.P. (1998). Nitrogen deficiency precludes a growth response to CO_2 enrichment in C_3 and C_4 *Panicum* grasses. *Australian Journal of Plant Physiology*, 25: 627-636.

Ghannoum, O., Von Caemmerer S., Ziska L.H. and Conroy J.P. 2000. The growth response of C_4 plants to rising atmospheric CO_2 partial pressure: a reassessment. *Plant, Cell and Environment*, 23: 931-942.

Ham, J.M., Owensby C.E., Coyne P.I. and Bremer D.J. 1995. Fluxes of CO_2 and water vapor from a prairie ecosystem exposed to ambient and elevated atmospheric CO_2. *Agriculture & Forest Meteorology*, 77: 73-93.

Hunt, H.W., Elliott E.T., Detling J.K., Morgan J.A. and Chen D.-X. 1996. Responses of a C_3 and C_4 perennial grass to elevated CO_2 and climate change. *Global Change Biology*, 2: 35-47.

Hunt, H.W., Morgan J.A. and Read J.J. 1998. Simulating growth and root-shoot partitioning in prairie grasses under elevated atmospheric CO_2 and water stress. *Annals of Botany*, 81: 489-501.

Imai, K. and Murata, Y. 1979. Effect of carbondioxide concentration on growth and dry matter production on crop plants. VII. Influence of light intensity and temperature on the effect of carbondioxide environment in some C_3 and C_4 species. *Japanese Journal of Crop Science*, 48: 409-417.

Isopp, H., Frehner M., Almeida J.P.F., Blum H., Daepp M., Hartwig U.A., Lüscher A., Suter D. and Nösberger J. 2000. Nitrogen plays a major role in leaves when source-sink relations change: C and N metabolism in *Lolium perenne* growing under free air CO_2 enrichment. ***Australian*** *Journal of Plant Physiology*, 27: 851-858.

Kimball, B.A. 1983. Carbondioxide and agricultural yield: an assemblage and analysis of 430 prior observations. *Agronomy Journal*, 75:779-788.

Lapitan, R.L., Mosier A.R. and Morgan J.A. 2000. Fluxes of CO_2, N_2O and water in a short grass steppe under ambient and CO_2-enriched atmospheres. *American Society of Agronomy Abstracts*, Minneapolis, MN. (in press)

Lecain, D.R. and Morgan J.A. 1998. Growth, photosynthesis, leaf nitrogen and carbohydrate concentrations in NAD-ME and NAD-ME C_4 grasses grown in elevated CO_2. *Physiologia Plantarum*, 102: 297-306.

Lüscher, A., Hartwig U.A., Suter D. and Nösberger J. 2000. Direct evidence that symbiotic N_2 fixation in fertile grassland is an important trait for a strong response of plants to elevated atmospheric CO_2. *Global Change Biology*, 6: 655-662.

Morgan, J.A., Lecain D.R., Mosier A.R. and Milchunas D.G. 2001. Elevated CO_2 enhances water relations and productivity and affects gas exchange in C_3 and C_4 grasses of the Colorado Shortgrass Steppe. *Global Change Biology* (in press).

Morgan, J.A., Lecain D.R., Read J.J., Hunt H.W. and Knight W.Gg 1998. Photosynthetic pathway and ontogeny affect water relations and the impact of CO_2 on *Bouteloua gracilis* (C4) and *Pascopyrum smithii* (C_3). *Oecologia*, 114: 483-493.

Nie, D., He H., Mo G., Kirkham M.B. and Kanemasu E.T. 1992. Canopy photosynthesis and evapotranspiration of rangeland plants under doubled carbon dioxide in closed-top chambers. *Agricultural and Forest Meteorology,* 61: 205-217.

Oula Ghannoum, Susanne von Caemmerer, Edward W.R. Barlow and Jann.P. Conroy 1997. The effect of CO_2 enrichment and irradiance on the growth, morphology and gas exchange of C_3 (*Panicum laxum*) and a C4 (*Panicum antidotale*) grass. *Australian Journal Plant Physiology,* 24: 227-237.

Owensby, C.E., Ham J.M., Knapp, A.K. and Auen L.M.(1999. Biomass production and species composition change in a tallgrass prairie ecosystem after long-term exposure to elevated atmospheric CO_2. *Global Change Biology,* 5: 497-506.

Pal, Madan, Karthikeyapandiam, V., Jain, V., Srivastava, A.C., Anupam, R., Sengupta, U.K. 2004. Biomass production and nutritional levels of berseem (*Trifolium alexandrinum)* grown under elevated CO_2. *Agricultural Ecosystem Environment*, 101: 31-38.

Polley, H.W., Morgan J.A., Campbell B.D. and Stafford Smith M. 2000. Crop ecosystem responses to climatic change: Rangelands. In: Reddy, K.R. and Hodges, H.F. (eds) *Climate Change and Global Crop Productivity*, CABI Publishing Co., Wallingford, Oxon, UK, Pp. 293-314.

Poorter, H. 1993. Interspecific variation in the growth response of plants to an elevated ambient CO_2 concentration. *Vegetatio,* 104/105: 77-97.

Poorter, H., Roumet C. and B.D. Campbell. 1996. Interspecific variation in the growth response of plants to elevated CO_2: A search for functional types. In: Korner, C. and Bazzaz, F.A. (eds), *Carbon dioxide, populations and Communities*, Academic Press, New York, Pp. 375-412.

Rogers, A., Fischer B.U., Bryant J., Frehner M., Blum H., Raines C.A. and Long S.P. 1998. Acclimation of photosynthesis to elevated CO_2 under low-nitrogen nutrition is affected by the capacity for assimilate utilization. Perennial ryegrass under Free-Air CO_2 enrichment. *Plant Physiology*, 118: 683-689.

Wand, S.J.E., Midgley G.Ff, Jones M.H. and Curtis P.S. 1999. Responses of wild C4 and C_3 grass (Poaceae) species to elevated atmospheric CO_2 concentration: a meta-analytic test of current theories and perceptions. *Global Change Biology*, 5: 723-741.

Watson, R.T.,Rodhe, H., Oescheger, H. and Siegenthaler, U. 1990. Green house gases and aerosol. In climate change: *The IPCC Scientific Assessment*. (Eds. J.T. Houghton, G.D. Jenkins and J.J. Ephraums), Cambridge University Press, Cambridge, Pp. 1-40.

Zanetti, S., Hartwig U.A., Van Kessel C., Lüscher A., Hebeisen T., Frehner M., Fischer B.U., Hendrey G, Blum H. and Nosberger J. 1997. Does nitrogen nutrition restrict the CO_2 response of fertile grassland lacking legumes. *Oecologia*, 112: 17-25.

❑❑❑

Agroforestry: Natural Resource Sustainability, Livelihood & Climate Moderation © 2009
Editors: OP Chaturvedi, A Venkatesh et. al., pp. 507-519
Satish Serial Publishing House, Delhi

Chapter 36

ASSESSMENT OF CLIMATE CHANGE THROUGH SATELLITE REMOTE SENSING

J.P. Singh, Neeraj Kushwaha, R.S. Chaurasia and Kamlesh Kumar
Indian Grassland and Fodder Research Institute, Jhansi-284 003 (U.P.)

Introduction

The atmosphere of the earth has proper depth and chemical composition. About 30% of incoming solar energy is reflected back to space while the rest reaches the earth, warming the air, oceans, land and maintaining an average surface temperature of about 15 °C. The Earth is the only planet in our solar system that supports life. The complex process of evolution occurred on earth only because of some unique environmental conditions viz., water (an oxygen-rich atmosphere), and a suitable surface temperature. If all the incoming energy were to be absorbed completely, the earth would gradually become hotter and hotter. But actually, the earth both absorbs and, simultaneously releases it in the form of infra red waves. All this rising heat is not lost to space, but is partly absorbed by some gases present in very small (or trace) quantities in the atmosphere, called GHGs (greenhouse gases). Greenhouse gases (for example, carbon dioxide, methane, nitrous oxide, water vapour, ozone), re-emit some of this heat to the earth's surface. If they did not perform this useful function, most of the heat energy would escape, leaving the earth cold (about -18 °C) and unfit to support life. However, ever since the Industrial Revolution began about 150 years ago, man-made activities have added significant quantities of GHGs to the atmosphere. The atmospheric concentrations of carbon dioxide, methane and nitrous oxide have grown by about 31%, 151% and 17%, respectively, between 1750 and 2000 (IPCC 2001). The Earth's climate system constantly adjusts so as to maintain a balance between the energy that reaches it from the sun and the energy that goes from Earth back to space. This means that even a small rise in temperature could mean accompanying changes in cloud cover and wind patterns. Some of these changes may enhance the warming (positive feedback), while others may counteract it (negative feedback). Negative feedback (causing a cooling effect) may result from an increase in the levels of aerosols (small particles of matter or liquid that can be produced by natural or man-made activities). Positive feedback may result from an increase in water vapour (because of greater evaporation with temperature rise), which itself is a GHG and can further add to the warming effect. The threat of climate change that led to the Framework Convention on Climate Change (FCCC) at Rio, is perceived differently by different countries. This fact has delayed any effective international agreement on how to deal with the problem. In the case of the Montreal Protocol covering ozone-depleting substances, there was a wide consensus and effective action was mobilized quickly. Thus, an understanding of perceptions and positions of different countries makes it easier to explore possibilities of effective action.

In this regard Parikh and Parikh (2002) have documented the problem of climate change and sustainable development; the kind of negotiating positions that follow from these perceptions; the policies India has undertaken so far and finally India's possibilities for action that can help contain the threat of climate change. All the factors described above complicate the work of scientists who try to predict the fallout of climate change. Despite these uncertainties, the Third Assessment Report published by the IPCC states, 'there is new and stronger evidence that most of the warming observed over the last 50 years is attributable to human activities' (IPCC 2001). Convincing evidence is available now on the worming of the earth's atmosphere due to higher carbon emissions from the human activity. The discussion on climate change therefore, touches upon some emerging area of concerns. This section of the survey looks at adaptation to climate change (Hindu, 2008).

Evidence of Climatic Change

Climate change refers to a statistically significant variation in either the mean state of the climate or in its variability, persisting for an extended period (typically decades or longer). In the recent past several scientific investigations and models were carried out on climate change at global or regional scales. It has been noticed on earth's surface temperature, rainfall variability, rise in sea level, surface water balance, Change in biodiversity and productivity, emission of green house gasses and resultant phenomena.

Earth's Surface Temperature

Over both the last 140 years and 100 years, the best estimate is that the global average surface temperature has increased by 0.6 ± 0.2 °C. Although climate change has been a salient issue for scientific communities for several decades, it exploded upon the world stage in full force only in 2007. The change in the annual average variation of the temperature of the earth surface is presented in table 1 (Goddard, 2005). The best scientific consensus as reflected by the forth assessment of the IPCC (Inter governmental Panel on Climate change) predicts a rise in global mean temperature of anywhere from 1.4°C to 4°C (within a wider "most likely" range of 1.1°C to 6.4°C depend on our assumption about the further course of GNG emission) by the end of 21st century. This will be accompanied by much greater variability in both temperature (more frequent heat waves and cold snaps) and precipitation (more frequent drought). Over the Indian subcontinent the projected increase in temperature (in terms of both magnitude variability) is greater for the North and then for the South, in part because of the buffering influence of the ocean on the peninsula. The most recent example of unexpected weather is the south-the most since data has been collected for this month-that has ruined range of crops from mango to paddy following this, the heavy and unexpected rains during the first week of April in the north have damaged the standing wheat crop that was almost ready for harvest (Hindu, 2008).

Rainfall Variability

Monsoon variability may be defined as the tendency of seasonal rainfall to fluctuate about its long-term climatic normal value. When year-wise sample observations are available, one can find the normal value as the time average of the data series. The standard deviation of the sample provides a simple characterization of the variability. It is the usual practice to represent this deviation, as a percentage of the normal or mean value, and refer to it as the coefficient of variation. For example, the all-India (southwest) monsoon rainfall has a mean value of 85.24 cm with a standard deviation of 8.47 cm. This gives the coefficient of variation over a period of 130 years (1871–1990) to be nearly 10%. This would be higher for a smaller region such as a district or a State. It would be of interest to know whether this variability remains constant over time in a statistical sense. If this parameter were to be stable, it would hint that floods and droughts have been part of this normal climatic variability (Iyengar, 2004).

The bigger challenge, however, has been to change in pattern of rainfall. The Indian monsoon is a complex phenomenon that most global model fail to replicate satisfactorily though they are getting much better at doing so therefore model projection for change in rainfall over India have to be viewed with considerable caution until more reliable ones are developed. Bundelkhand region is characterized as hot semi arid eco region along with growing period of 90–150 days. The south-west monsoon starts by 27th Standard Meteorological Week (July 2-8) and ends by 37th Std. Met. Week (Sept.10-6). About 90 per cent of total precipitation is received between July to September with occasional showers during winter months. The annual rainfall ranges from 769 to 1251 mm over the region with annual mean of 906.5 mm specifically for Jhansi. The average annual climatic data (rainfall and temperature) from 1971 to 2000 of Bundelkhand region is presented in figure 1. It shows a gradual declining trend in mean rainfall during past 30 years.

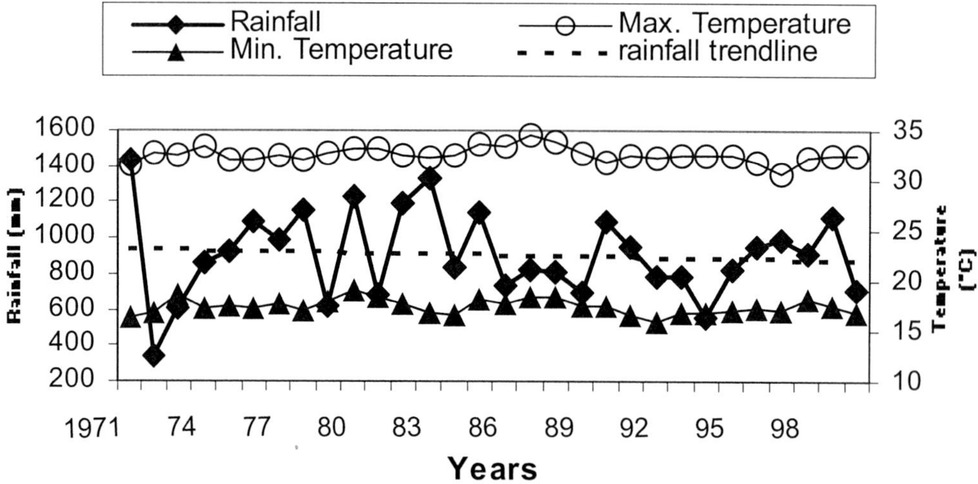

Fig.1. Trend of rainfall and temperature in Bundelkhand Region during 1971-2000.

Table 1. Average Global Temperature (1881 - 2004)

Year	(^{0}C)	Year	(^{0}C)	Year	(^{0}C)	Year	(^{0}C)	Year	(^{0}C)	Year	(^{0}C)	Year	(^{0}C)
1881	13.88	1901	13.95	1921	13.95	1941	14.11	1961	14.1	1981	14.4	2001	14.56
1882	14	1902	13.7	1922	13.91	1942	14.1	1962	14.05	1982	14.1	2002	14.7
1883	13.96	1903	13.64	1923	13.84	1943	14.06	1963	14.03	1983	14.34	2003	14.64
1884	13.59	1904	13.58	1924	13.89	1944	14.11	1964	13.65	1984	14.16	2004	14.6
1885	13.77	1905	13.75	1925	13.85	1945	13.99	1965	13.75	1985	14.13		
1886	13.75	1906	13.85	1926	14.04	1946	14.01	1966	13.93	1986	14.19		
1887	13.55	1907	13.6	1927	13.95	1947	14.12	1967	13.98	1987	14.35		
1888	13.77	1908	13.7	1928	14	1948	13.97	1968	13.91	1988	14.42		
1889	14.04	1909	13.69	1929	13.78	1949	13.91	1969	14	1989	14.28		
1890	13.78	1910	13.79	1930	13.97	1950	13.83	1970	14.04	1990	14.49		
1891	13.44	1911	13.74	1931	14.03	1951	13.98	1971	13.9	1991	14.44		
1892	13.6	1912	13.67	1932	14.04	1952	14.03	1972	13.95	1992	14.16		
1893	13.61	1913	13.72	1933	13.89	1953	14.12	1973	14.18	1993	14.18		
1894	13.68	1914	13.98	1934	14.05	1954	13.91	1974	13.94	1994	14.31		
1895	13.68	1915	14.06	1935	13.92	1955	13.91	1975	13.98	1995	14.47		
1896	13.73	1916	13.8	1936	14.01	1956	13.82	1976	13.79	1996	14.36		
1897	13.85	1917	13.54	1937	14.12	1957	14.08	1977	14.16	1997	14.4		
1898	13.79	1918	13.67	1938	14.15	1958	14.1	1978	14.07	1998	14.71		
1899	13.76	1919	13.91	1939	13.98	1959	14.05	1979	14.13	1999	14.44		
1900	13.95	1920	13.85	1940	14.14	1960	13.98	1980	14.27	2000	14.41		

Source:Goddard Institute for Space Studies, NASA Goddard Space Flight Center, Earth Sciences Directorate, **"Global Temperature Anomalies in .01 C,"** <http://www.giss.nasa.gov/data>, updated January 2005.

The distribution of rainfall is often erratic and even wet months (July & August) many times experience long dry spells. Once the rainy season sets in, mid-season droughts are likely to occur during 33-35th Std. Met Week, i.e., mid of August. The chances of occurrence of deficient rain in July and August are only 3 out of 20 while in June and September chances are 7 out of 20 cases. In July and August drought would occur after every 7 years. The recurrence interval of deficient rains for two consecutive years in September is 11 years. The moisture index varies from –40 to –60. Diurnal variation in temperature is quite high. May and June are the hottest months. The maximum-recorded temperature on a particular day often touches 47-48°C during summer. The annual potential Evapotranspiration varies from 1400-1600 mm.

Rise in Sea Level

Sea level rise may have a physical impact on coastal areas and island states. People in low-lying areas like Bangladesh, live in the fear of severe flooding. Islands like the Maldives face the threat of being swallowed by the sea. India's long coastline of over 6000 km is densely populated. Several million people may be impacted in India alone if the sea level rises by a metre. By the end of this century, there will be a sea level rise of half a metre, according to the IPCC. Even small changes can have significant impact in local communities. Take for example, the large mangroves of the Sunderbans. Nearly six hundred thousand people who work in industries that use raw materials from the area; and climate change that destroys the mangroves will affect their livelihood directly.

Climate change is predicted to result in increases in mean sea level, as well as possible increases in the frequency and intensity of coastal surges and cyclones that already cause significant damage to coastal populations. These predicted changes threaten life, livelihoods and infrastructure, and policy making therefore needs to take account of both physical measures to reduce impacts (such as sea walls) as well as policy measures such as disaster preparedness efforts to reduce vulnerability. One quarter of the Indian population live along the country's coasts, and are largely dependent on coastal livelihoods. Climate change effects on sea level can impact coastal areas in two ways - through increase in mean sea level, and through increased frequency and intensity of coastal surges and storms. Climate change is of concern to India in view of the damages that occur along the east coast of India from the cyclones that form in the Bay of Bengal. Any increase in the frequency or intensity of tropical disturbances due to climate change in the future could cause increased damages to life and property in the coastal regions. The National Institute of Oceanography (NIO) conducted a study on the impacts of climate change on sea level to assess the degree to which mean sea level and the occurrence of extreme events may change.

Surface Water Resources and Glaciers

A much-quoted the study of 466 glaciers using satellite images by the Indian Space Research Organization (ISRO), found that between 1962 and 2001 there was an average reduction of over 2007 in the surface area of these glacier. The Himalayan glaciers that feed the major river of the North, such as the Ganges and the Brahmaputra, have been retreating. Himalayan geology has been making precise ground measurements of the extent of some glacier, and found that the chorabari glacier has retreated at the rate of 9 meters a year in recent years while the Dokriana has lost 207 of its volume over the past three decades. The famous Gangotri glacier has also been in retreat - at an astonishing rate of 17 meters a year between 1971and 2004. The Himalayas possess one of the largest resources of snow and ice, which act as a huge freshwater reservoir. Monitoring the glaciers is important to assess the overall reservoir health. In this investigation, glacial retreat was estimated for 466 glaciers in Chenab, Parbati and Baspa basins from 1962. Expeditions to Chhota Shigri, Patsio and Samudra Tapu glaciers in Chenab basin, Parbati glacier in Parbati basin and Shaune Garang glacier in Baspa basin were organized to identify and map the glacial terminus. The investigation has shown an overall reduction in glacier area from 2077 sq. km in 1962 to 1628 sq. km at present, an overall deglaciation of 21%. However, the number of glaciers has increased due to fragmentation. Mean area of glacial extent has reduced from 1.4 to 0.32 sq. km between the 1962 and 2001. In addition, the number of glaciers with higher areal extent has reduced and lower areal extent has increased during the period. Small glaciarates and ice fields have shown extensive deglaciation. For example, 127 glaciarates and ice fields less than 1 sq. km have shown retreat of 38% from 1962, possibly due to small response time. This indicates that a combination of glacial fragmentation, higher retreat of small glaciers and climate change are influencing the sustainability of Himalayan glaciers (Kulkarni *et al*, 2007).

In the Bundelkhand region, the basic constraint for crop production is the inadequate water supply. Uncontrolled runoff from slops, apart from loss of water, causes land degradation through soil erosion. It is noteworthy that rain water, the vital source of crop production, becomes a dominant factor (as runoff) for land degradation, to the extent that the once fertile lands are converted into degraded wastelands/ ravines. Hence, it is essential that rain water assessment should be taken up in an integrated manner. The major source of surface water resources in Bundelkhand region is monsoonal rainfall but occasionally little amount of precipitation is also received during winter season due to westerly disturbances. The average annual rainfall in the region is 950 mm but due to landforms / topography/ physiographic characteristics and its strategic location the rainfall amount, nature and spatial distribution is found uneven. The south and southeastern part of the region receives about 1200 mm rainfall where as northeastern parts get hardly 850 mm. Since the general slope of the region is from south to north thus the available rain water flows through drainage network after soil moisture utilization (SMU/ SMR), actual Evapotranspiration (ET), evaporation through plants and ground water recharge through percolation as well as storage of water in man made / natural reservoirs which is being used for irrigation and drinking water supply. On the basis of number of rainy-days and amount of rainfall broadly the region may be divided into 3 zones viz., high rain (1200 mm and above), medium rain (1000-1199 mm) and low rain (<1000 mm). Based on average annual precipitation in different zones, the amount of total surface water may be estimated as $65.1 \times 10^9 m^3$. Further, based on bio-physical characteristics, a balance sheet of surface water on annual basis has been calculated as:

Estimated surface Water	=	$65.1 \times 10^9 m^3$.
Actual Evapotranspiration (potential $91.14 \times 10^9 m^3$)	=	$28.15 \times 10^9 m^3$
Soil Moisture Utilization/ Recharge	=	$12.69 \times 10^9 m^3$
Ground Water Recharge (Through percolation)	=	$11.72 \times 10^9 m^3$
Storage of Water (In man made/ natural reservoirs/ dam/ check dams etc.)	=	$2.08 \times 10^9 m^3$
Other use	=	$0.167 \times 10^9 m^3$
Surface runoff (Through drainage networks)	=	$10.29 \times 10^9 m^3$

Surface runoff or unused / uncontrolled valuable water resources; which has the potential to (i) irrigate the entire rabi cropped area, (ii) provide drinking water and (iii) restore the degraded land through Forestry, Agroforestry, Silvipastoral, Hortipasture and Range-grasses etc. creating havoc like situation especially flash floods during heavy rain in middle reaches and heavy soil loss (ravines) in lower reaches. Researchers have the view that the surface runoff can be managed through using following two broad techniques considering the economic viability, eco-friendly and wide range of acceptability.

Change in Biodiversity and Productivity

If the projected climate change materializes, natural or artificial regeneration of forest with local seed sources will become increasingly difficult. Insect and diseases can be expected to cause significant losses to forest, and in addition fire severity may also increase. It is also excepted that the climate will change forest then tree species are able to respond (IPCC,1990). Climate change may also lead to forest dieback. Thus socioeconomic impact of climate change could be significant in regions dependent on forest. Climate change could have positive as negative impacts on woody biomass production for energy. Climate change of the predicted magnitude are expected to affect the net primary production (IPCC 1990). The impact of climate change on biodiversity and biomass production is clearly visible in the remotely sensed satellite image of north-western part of Bundelkhand region (Fig. 2).

Satellite Remote Sensing

The application of remote sensing technology in the study of the climate change or natural resources has resulted in the development of methodologies for mapping and monitoring of degraded lands in a coast effective manner. This is due to the fact that remote sensing satellites have synoptic view and covers same area at regular intervals. Remotely sensed data from space borne sensors are being regularly used for mapping and monitoring of various degraded lands like ravines, salt affected soils, eroded areas, shifting cultivated areas, water logged areas and bio-physical resources etc. in a coast effective manner.

Available Satellites and Sensors

Monitoring for climate change requirements places particular emphasis on the long-term temporal consistency, standard archiving and calibration of satellite data. To support the aims of the UN Framework Convention on Climate Change, the Committee on Earth Observing Satellites (CEOS) has developed a strategic plan to help rehabilitate past satellite records and to ensure that future space missions for climate change purposes can achieved the required level of homogeneity. Groups like CEOS and the Coordinating Group on Meteorological Satellites (CGMS) provide effective mechanisms for international cooperation to optimize the overall benefits of remote sensing from space. Satellite data are critical to weather prediction in the southern hemisphere owing to the large fraction of ocean compared with the northern hemisphere. Data assimilation provides the mechanism for the optimal use of satellite data in numerical weather prediction models. Tropical cyclones are a natural hazard for both India and Australia, and numerical modeling provides an important tool for predicting their path and intensity. GPS technology and hyper-spectral data are increasingly used to estimate the vertical profiles of humidity and air temperature.

Hyper-spectral satellite data from the NASA demonstration mission EO-1 are evaluated in India to support future natural resource management applications. For example, crop classification can be carried out at fine scales using hyper-spectral data combined with neural network algorithms. Remote sensing applications are vitally dependent on the quality assurance of the data streams fed into the algorithms, and there is a substantial effort in India to ensure the quality of the data at all levels in the processing cycle. India has developed a range of microwave sensors in space, and its future plans include the launches of Oceansat-II in early 2008 and Megha-Tropiques in 2009. There could be opportunities for Australia and India to cooperate on calibration, validation and application of the data from these two spacecraft. In particular, the establishment of a number of *in situ* moorings with precision radiometers and water-quality sensors, as part of the new Australian Integrated Marine Observing System (IMOS), may be of use in some of these collaborative projects (IARS, 2007). The Indian Remote Sensing (IRS) Satellites viz., IRS 1A, 1B, 1C, 1D, P4, P6, Cartosat I and Cartosat II etc. is being widely used.

Assessment of Climate Change

Several studies have been conducted in the past using remote sensing technique. Remotely sensed satellite data and GIS have the capability to assess and map the ravines affected area for grazing lands and its developmental planning. The area under ravines has increased during the recent past and the trend is upward. It is mainly due to variability in the temperature and rainfall. The general phenomena in the study region is deteriorated forest cover became grazing lands or wastelands and finally being converted into ravines. Precise information on the extent, conditions and forage availability from grazing lands is essential for developmental planning. There are evidences of excessive degradation of tropical grazing lands. Excessive grazing and deforestation further converted them into heavily degraded ravines. The current data shows that about 158 million ha area in India has turned into wastelands. Grazing lands remain an important land use in India and supply about 40% of the total forage need. Since ancient times cattle breeding and milk production has been the second most important profession in India, next only to agriculture. Grazing lands are distinct biophysical entities of nature having a broad three dimensional feature made up of the range plants, the animal eating them and the soil supporting them. This 3-dimensional view has different shades according to the environmental setting, the driving variables or climate, and finally, the human factor through its utilization by way of hay harvest or grazing livestock.

Based on IRS P6 data the assessment and monitoring of grasslands of Bundelkhand region was carried out. The result shows that owing to change in climate in past 100 years, more particularly between 1980-2006, the soil and associated vegetation cover have deteriorated as it is presented in figure 2.

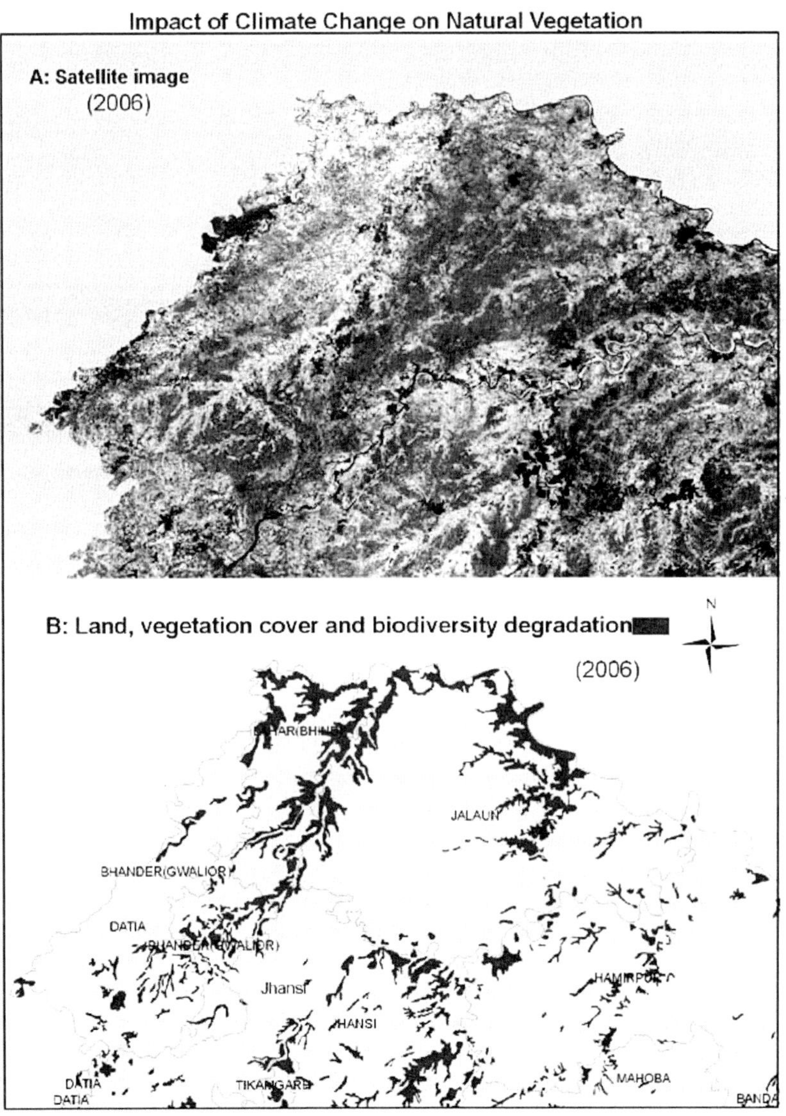

Fig. 2. Assessment of soil and vegetation cover using IRS P6 Image

The high resolution QB image showing deterioration of soil and land cover is presented in figure 3.

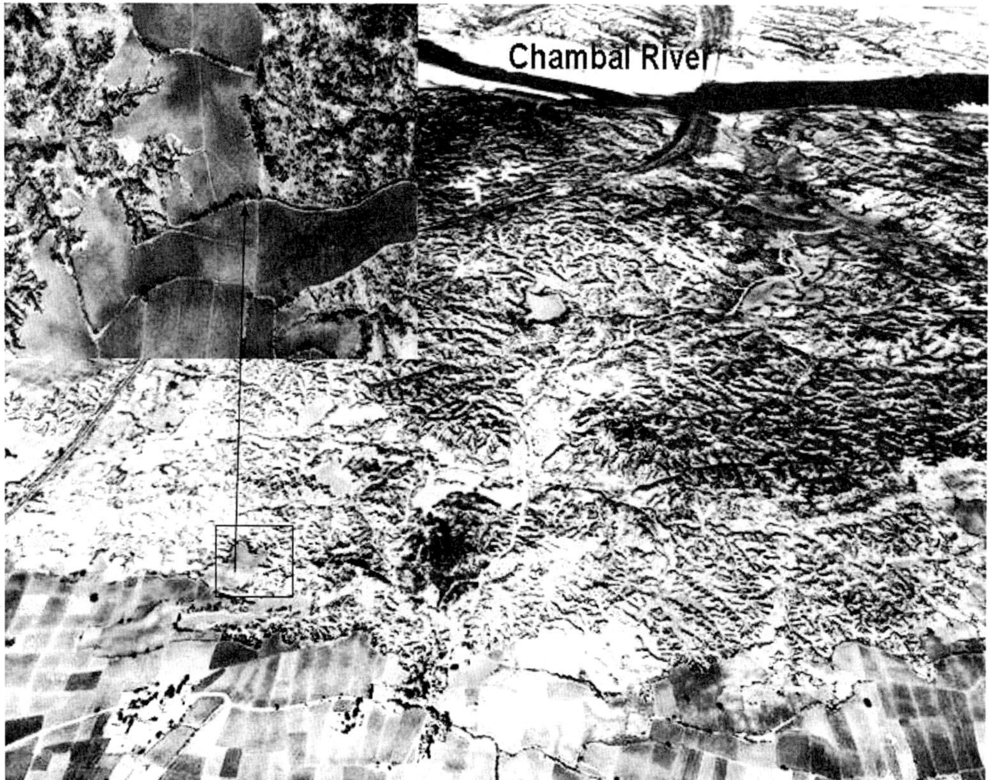

Fig. 3. High Resolution QB Image: Land and Vegetation cover degradation.

Impact of Climate Change

Climate change in India represents an additional stress on ecological and socioeconomic systems that are already facing tremendous pressure due to rapid urbanization, industrialization and economic development. With its large and growing population, and an economy that is closely tied to its natural resource base, India's population is vulnerable to the impacts of climate change such as changes in forest and water resources and sea level rise. Understanding climate change scenarios for India, and their impacts, is therefore critically important for policymakers to protect lives, as well as the assets upon which India's economy is dependent. Equally important for policymakers is to understand the impact of industrialization on climate change itself. As the country industrialises, India's production of greenhouse gases (GHGs) that contribute to climate change will increase.

Awareness by policymakers about climate change and its possible effects on industry, people, and livelihoods will be important as they seek to balance economic growth and development with a changing resource base. The Indian Institute of Tropical Meteorology (IITM) in collaboration with the Hadley Centre for Climate Prediction and Research, UK carried out an analysis of climate change scenarios for India. IITM used the Hadley Centre Regional Climate Models (RCMs) for the Indian subcontinent to model the potential impacts of climate change.

References

Goddard. 2005. Goddard Institute for Space Studies, NASA Goddard Space Flight Center, Earth Sciences Directorate, "Global Temperature Anomalies in .01 C," <http://www.giss.nasa.gov/data>, updated January 2005.

Hindu. 2008. Survey of environment- climate change.29-39.

IARS,2007:India-Australia Remote Sensing Workshop Report Michael Manton Chair ATSE International Strategy Group. 2 -5 December 2007

IPCC.2001. Climate Change. Cambridge: Cambridge University Press

Iyengar, R. N. 2004. Description of rainfall variability in Br_hat - samhita of Varâha-mihira, *Current Science*, 87(4): 531.

Kulkarni, Anil V. Bahuguna, I. M. Rathore, B. P. Singh, S. K. Randhawa S. S., Sood R. K. and Dhar Sunil. 2007. Glacial retreat in Himalaya using IndianRemote Sensing satellite data. *Current Science*, 92 (1): 10

Parikh, K. and Parikh, Kirit. 2002. Climate Change: India's Perceptions, Positions Policies and Possibilities. OECD. 2002.

George C. Marshall Institute *Marshall Institute Policy Outlook* 1625 K Street, NW, Suite 1050. Washington, DC 20006. info@marshall.org. http://www.marshall.org

IARS. 2007. Technical Report of India-Australia Remote Sensing Workshop. December 2 -5.

❑❑❑

Agroforestry: Natural Resource Sustainability, Livelihood & Climate Moderation © 2009
Editors: OP Chaturvedi, A Venkatesh et. al., pp. 521-529
Satish Serial Publishing House, Delhi

Chapter 37

APPLICATION OF GIS TOOLS FOR ALTERNATE LAND USE OPTIONS

D.V.K. Nageswara Rao
Indian Grassland and Fodder Research Institute, Jhansi - 284 003 (U.P.)

Concept of Agroforestry

Agroforestry is the intentional combination of agriculture and forestry technologies to create integrated, diverse, productive, profitable and suitable land use systems (Rietveld, 1995). Examples of agroforestry practices that are focused on meeting the environmental, social and economic needs of people on private lands include alley cropping, forest farming, riparian forest buffers, silvopature and windbreaks. Environmentally, agroforestry practices can reduce erosion, improve water infiltration and quality, moderate microclimates, enhances nutrient cycling and provide wildlife habitat (Sanchez, 1995). Socially, communities may be revitalised through partnerships that implement publicly owned agroforestry projects like community shelterbelts or ecobelts (Josiah *et al.*, 1999, Bentrup *et al.* 2001). Lal (2000) cited agroforestry as one of the future technological innovations that will be needed to meet the food demands for the global increase in population. Economically, agroforestry practices can reduce production costs by lowering the needs for chemical, water, energy and/or labour inputs while increasing overall agricultural output (Lassoie and Buck, 2000; Olson *et al.* 2000). Another financial benefit for landowners is the potential to grow and sell speciality products (Bentrup and Kellerman, 2003). Through careful management, agroforestry systems can be harvested substantially for several products like fruits, nuts, wood, fodder, gums and medicinal products etc. When these agroforestry species or products are combined, a synergistic system can be created that maximises landowner benefits while providing environmental services to all (Bentrup and Kellerman, 2003).

The goals and objectives that agroforestry systems can achieve partially depend on the spatial scale at which they are planned and designed. Landowner goals, like minimising soil erosion and providing crop production, as stated by Bentrup and Kellerman (2003), can be often accomplished by just focussing on the site conditions within the property boundaries. However, a large scale perspective is often required for community -driven goals. To integrate these different concerns and goals within the agroforestry systems, a multi-scale planning process needs to used (Rietveld and Francis, 2000).

Alternate Land Use Options

Agroforestry, as an approach to integrated land management or alternate land use system, is widely thought to have enormous potential as a source of technological solutions to problems of rural development. Although, agroforestry is an ancient form of land use practice by many generations of rural people in many parts of India, it is relatively a new field of organised scientific activity. Well designed agroforestry systems can contribute to the improvement of rural welfare through a variety of direct production roles of 7F's (food, fodder, fuel, fibre, furniture, fruits and fencing) as well as through whole range of indirect service roles (soil and water conservation, fertility improvement, microclimate amelioration etc.). Farmers of the country are not familiar with improved food, fodder and fuel production techniques due to various reasons. The situation, therefore, need to be tackled by producing most sustainable food-forage-fuel-finance from marginal, sub-marginal and less productive and problematic sites in rural areas by development of suitable agroforestry systems like agrisilviculture, alley cropping, agrihorticultute etc. (Subrahmanyam, 2009).

Geoinformation and its Critical Importance

State- of the- art strategic planning relies on three major pillars of content handling-remote sensing based imaging, satellite based positioning and satellite communication and its intensity or capacity determines the success over the antagonist. Geoinformation by virtue of its integrative ability inherent in maps, rather dynamic as opposed to the conventional static ones, creates a baseline to plug in data on range of phenomenon at differing details/resolutions. Continuum of information built in these systems with logical connectivity would bring in the convergence of information into a new virtual reality, helping to analyse, understand, question, store the spatial data for decision making. In the context of assessing change happening across geographical scale and predicting it, geoinformation provides mammoth scope to realise situations.

Geographical Information Science (GIS)

Use of geographical information system (GIS), which is of late renamed as geographical information science, a collection of computer hardware and software used to analyse and display geographically referenced information, can facilitate this planning process. Spatial data layers like soil type, slope and land cover can be used to develop suitability assessments that can identify optimal locations for agroforestry practices to solve landowner and community concerns (Bentrup and Kellerman, 2003). By selecting data with appropriate spatial data with appropriate spatial resolution, this assessment process can be used at any scale for planning agroforestry practices. The most significant benefit of using GIS-guided suitability assessments is the ability to combine different assessments to determine locations where multiple objectives can be achieved. Although suitability assessments have been used for several decades to identify locations for different land uses, this technology has yet to be used extensively in agroforestry (Ellie *et al*., 2000).

Applications of Geoinformatics to the Cause of Agroforestry

D & D survey (diagnosis and design survey) is the preliminary step in planning an agroforestry project that encompasses all aspects of a given situation including socio-economic issues. Once the problems and potentials are understood, it would be easy to plan for alternate land uses. In general, ALU (alternate land use) systems are appropriate for rehabilitating degraded or marginal or waste lands or saving arable lands from further degradation. Agroforestry is an appropriate ALU system to all classes of land aiming at generation of assured income with minimum risk through efficient utilisation of available resources. ALU options must match the LCC (land capability class) to be effective and profitable. Obviously, we need to generate the theme maps to establish the decision rules to help decision making process and the GIS tools can help creating these maps and analysis to decide upon the suitable agroforestry system for a given geographical area.

The Scheme of GIS Operations

The following theme maps help to document the features of a given geographical area at a suitable scale.

1. *Soil map:* The soil map of the given area can be obtained from the relevant agencies like the Soil Survey Department of the respective State or from the available reports of the maps prepared by the NBSS & LUP or from the other sources like libraries archiving the theses of student research or even can be prepared, depending on the condition. It was a general observation that although the coarse fragment content in volumetric terms was given in the soil survey reports, the idea of ESV (effective soil volume as introduced by Rao and Jessy, 2007) is not applied. With the available results of study conducted by Rao and Jessy (2007) it is understood that ESV had a direct impact on growth and yield (when studied using rubber as a test crop). This ESV map certainly helps taking the decision on the species to be included in the ALU system.
2. *Slope map:* The slope map can be obtained after digitisation of the contours indicated in the toposheets or using the modern tools like geographical positioning system (GPS) and interpolation. It was seen that use of freely available digital elevation model (DEM) created from the data collected during the SRTM (Shuttle Radar Topography Mission) was of considerable use in the generating the contour maps (Rao *et al.* 2006).
3. *Present land use/Land cover map:* With the access to the technology and customised software inputs it is a very easy task to prepare land use/ land cover maps at any level of resolution using satellite imagery. It is now possible to generate a high quality image with even a less than 1 meter resolution using Ikonos or Quickbird kind of sensors.
4. *Soil erosion and displacement map:* Soil erosion can be estimated by the Universal Soil Loss Equation (USLE) developed by Wischemier and Smith (1978). This equation can be incorporated in the GIS. Soil erodibility factor (K-map), topographic factor (LS-factor) and crop management (CP-map) and supporting practice factor map were derived from the soil map, slope map and land use/land cover map, respectively, by assigning suitable values for each polygon in the respective maps. These maps can be integrated and an item soil loss = RKLSCP can be created in the integrated map. The rainfall erodibility factor (R) can be taken as a constant as the amount of rainfall is uniform over an area, for example a watershed.
5. *Soil and water conservation map:* The erosion map and displacement map, soil map and slope map can be integrated and based on the allowable

erosion rate for the corresponding soil depth in each polygon, decision rules for soil and water conservation measures can be incorporated in the integrated map.

6. *Hydrology map:* This map shows the distribution of areas irrigated by surface and groundwater. These are the areas, on the areas, on the downstream of lakes that get water for irrigation.
7. *Land Capability map:* It can be obtained from the published reports or can be made with the available information. In all, eight land capability classes have been recognised (Soil Survey Manual, 1970). Soils in classes I to V1 are considered to be suitable for agriculture, whereas those in Classes V to VII are not suitable for cultivation and are recommended for silvipasture and other uses. Soils of Class VIII are not suitable for any cultural use. Land capability classes are further subdivided as per the predominant hazard and limitation of ground such as 1. Erosion and ruf off - e, 2. Excess of water - w, 3. Root zone limitations - s and 4. Climate limitations-c. The land capability map is a derived map, prepared from the soil map, based on above criteria.
8. *Alternate Land Use Options Map:* Combining the above set of maps one can create the Alternate Land Use Options Map considering all the relevant requirement and realties like crop specifications and responses etc, land suitability, market access and socio-economic situations etc. By making appropriate decision rules considering the existing relevant issues, suitable ALU options can be derived to achieve the goals.

There are several reports on the application of GIS techniques to watershed development plans, identifying agroforesty options etc both in India as well as elsewhere with and ultimate aim of achieving the very objects of sustainable growth defined by the term 'Agroforestry'.

Additional Dimensionality, the Visualisation Technique

Visualisation of a terrain is an important idea to understand the given terrain in three dimensions. One can easily understand the land attributes that determine the soil properties with further information on slope, elevation, aspect, plan curvature etc. which have significant agricultural importance. This can be easily achieved with even freely available software. However, the requirement is the resolution of the DEM that is planned to be visualised. In the first instance one can use freely available DEMs acquired by the SRTM , which is in 90 meter resolution to visualise a given terrain. This will be help in the initial stratification of the area under study. Depending on the requirement, high resolution DEMs can be created for locating for example sites for constructing soil conservation structures etc.

A Case of Visualisation

Consider the image shown in Fig. 1 where the soil map was described based on the measured soil properties. There were very shallow to deep soils with other properties in permutations and combinations. However, one can not know the terrain from this 2D map. Similarly, the satellite image shown in Fig. 2, which very clearly showed the CRF area even the field boundaries with some kind of vegetation present at the time of imagery acquisition. In this case also slope and elevation information can not be perceived. However, the images in fig. 3 and 4 add more to the understanding of the terrain both in terms of vegetation and other surface features as well as soils, which the other images could not do. This was the advantage of the visualisation where the information about slope, elevation, aspect etc could be derived and visualised which have got agricultural significance (Rao and Tripathi, 2008).

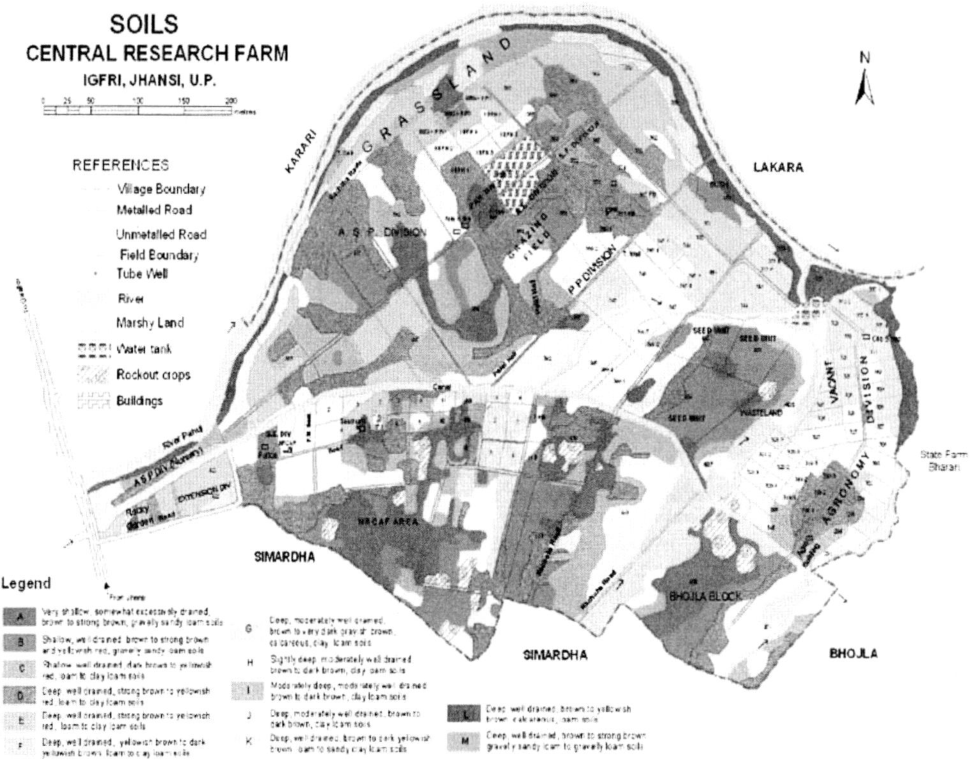

Fig 1. Soil map of the Central Research Farm, IGFRI, Jhansi

Fig. 2. High resolution satellite image pertaining to the CRF, IGFRI, Jhansi.

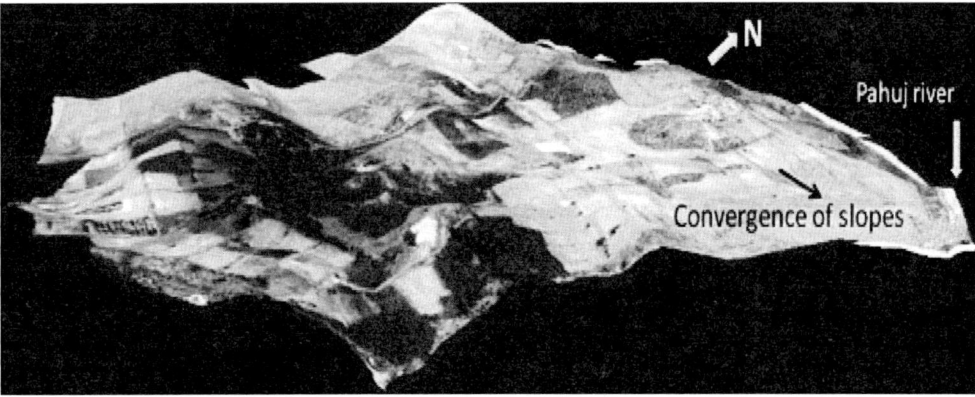

Fig 3. Draped satellite image on DEM.

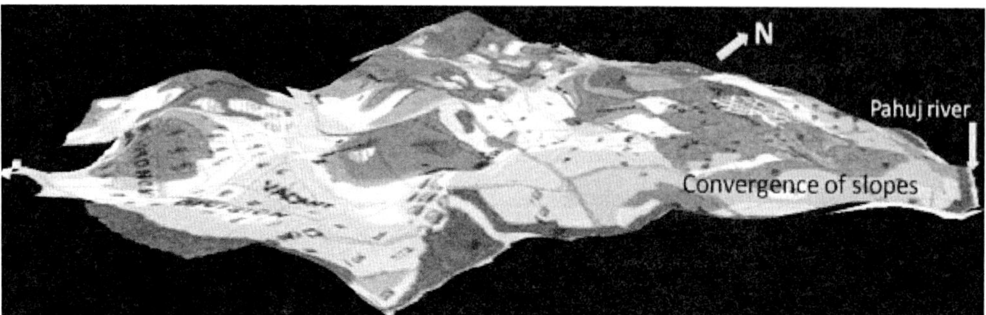

Fig 4. Soil map draped on DEM.

Concluding Remarks

Application of GIS tools to the cause of agroforestry as an alternate land use option/system can certainly ne successful because of the inherent potential of the GIS tools. Besides that visualisation technique coupled with expression of EVS add more to the understanding of the soil resources based on which the ALU is developed. Hence, it is wisdom of the user to include these two aspects in the D&D survey for proper understanding of the problems in the first instance.

References

All Indian Soil and Land Use Survey. 1970. Soil Survey Manual, Ministry of Agriculture, Government of India. Pp.1-123.

Bentrup, G. and Kellerman, T. 2003. Agroforestry and GIS: Achieving land productivity and environmental protection. In "Proceedings of 8th North American Agroforestry Conference, 23-25 June 2003, Corvallis, OR. Pp.15-25.

Bentrup, G., Schoeneberger, M.,M., Josiah, S. and Francis, C. 2001. Ecobelts: reconnecting agriculture and communities – case studies. In: Steward, W.C. and Lisec, A. (eds), *Proceedings of the Ecospheres Conference*. University of Nebraska, Lincoln, NE. June 10-12.

Ellis, E.A., Nair, P.K., Linehan, P.E., Beck, H.W. and Blanche, C.A. 2000. A GIS-based database management application for agroforestry planning and tree selection. *Computers and Electronics in Agriculture*, 27, 41-55.

Josiah, S.J., Gordon, L.J., Streed, E. and Joannides, J. 1999. Agroforestry in Minnesota: A Guide to Resources and Demonstration Sites. University of Minnesota Extension Service, St. Paul, MN.

Lal, R. 2000. A modest proposal for the year 2001: We can control greenhouse gases and feed the world... with proper soil management. *Journal of Soil and Water Conservation* 55(4), 429-433.

Lassoie, J.P. and Buck, L.E. 2000. Development of agroforestry as an integrated land use management strategy. In: Garett, H.E., Rietveld, W.J. and Fisher, R.F. (eds). *North American Agroforestry: An Integrated Science and Practice.* American Society of Agronomy, Madison, WI. Pp. 1-29.

Olson, R.K., Schoeneberger, M.M. and Aschmann, S.G. 2000. An ecological foundation for temperate agroforestry. In: Garett, H.E., Rietveld, W.J. and Fisher, R.F. (eds) North American Agroforestry: An Integrated Science and Practice. *American Society of Agronomy*, Madison, WI, Pp. 31-61.

Rao, D.V.K.N. and Jessy, M.D. 2007. Impact of effective soil volume on growth and yield of rubber (*Hevea brasiliensis*). *Geoderma* 141, 332-340.

Rao, D.V.K.N. and Tripathi, S.B. 2008. Statistical and visualisation techniques to understand the soils. *73rd Annual Convention of Indian Society of Soil Science*, 27-30 November 2008, Bangalore, India.

Rao, D.V.K.N., Nair, P.V., Prem, E. and Nazeer, M.A. 2006. Information Technology Tools in Rubber Cultivation. In: AFITA 2006-The *Fifth International Conference of the Asian Federation for Information Technology in Agriculture* (eds. Patil V.C. and S. Ninomiya). MacMillan India Ltd, New Delhi. Pp. 507-524.

Rietveld, W.J. 1995. Agroforestry: a maverick science and practice. In: Reitveld, W.J. (ed) Proceedings of Agroforestry and Sustainable Systems Symposium, August 7-10, 1994, Fort Collins,CO. General Technical Report RM-GTR-261. USDA, Forest Service, Rocky Mountain Forest and Range Experiment Station.

Rietveld, W.J. and Francis, C.A. 2000. The future of agroforestry in the USA. In: Garett, H.E., Rietveld, W.J. and Fisher, R.F. (eds) *North American Agroforestry: An Integrated Science and Practice. American Society of Agronomy*, Madison, WI, Pp. 387-402.

Sanchez, P.A. 1995. Science in agroforestry. *Agroforestry Systems,* 30, 5-55.

Subrahmanyam, M.V.R. 2009. Agroforestry – An Alternate Land Use Option for Sustainable Dryland Agriculture. In: Winter School on Alternate Land Use Systems for Resource Conservation, emerging market needs and mitigation of climate change in rainfed regions- A compendium of lectures, 16th January to 5th February, 2009, Central Research Institute for Dryland Agriculture, Hyderabad, India, Pp. 99-102.

Wischemier, W.H. and Smith, D.D.1978. Predicting rainfall erosion losses, a guide to conservation planning. *Agricultural Handbook,* 537, USDA.

□□□

Agroforestry: Natural Resource Sustainability, Livelihood & Climate Moderation © 2009
Editors: OP Chaturvedi, A Venkatesh et. al., pp. 531-547
Satish Serial Publishing House, Delhi

Chapter 38

APPLICATIONS OF GIS AND REMOTE SENSING TECHNOLOGIES IN NATURAL RESOURCE ANALYSES WITH EMPHASIS ON AGROFORESTRY

R.H. Rizvi, I.A. Khan and K.D. Kaushik
National Research Centre for Agroforestry, Jhansi-284 003 (U.P.)

Introduction

Agroforestry is an ancient practice that farmers have used through out the world. There are innumerable examples of traditional land use practices involving combined production of trees and agriculture species on the same piece of land in many parts of the world (Nair, 1993). In India, the origin of agroforestry is believed to be since the Vedic period, while organized agroforestry research began in 1979 at Imphal (Puri and Panwar, 2007). In 1983, All India Co-ordinated Research Project on Agroforestry (AICRPAF) a large network project with 20 centers was initiated and later on more centers were added. At present there are 36 AICRPAF centers covering all the major agro-ecological zones of India with headquarters at National Research Centre for Agroforestry, Jhansi. During the last twenty five years significant research work has been done on different aspects of agroforestry. The major research thrusts under this initiative have been on diagnostic survey and appraisal of existing agroforestry practices, collection and evolution of MPTs for agroforestry, studies on management practices of agroforestry systems, environment, watershed and community land development and integration of different components in agroforestry (NRCAF, 2007).

The growing awareness of the importance and potential of agroforestry has resulted in an invaluable proliferation of site specific case studies (Unruh and Lefebvre, 1995). In India the diagnostic survey and appraisal of agroforestry practices in the country revealed that there are enumerable practices in different agro-ecological zones (Pathak *et al*, 2000). These systems/practices occupy sizeable areas. Though an effort has been made by Dhyani *et al.*, (2006) to estimate the agroforestry area in the country, however, these estimates are not the true reflection as they are not based on ground truthing. The integrated uses of spatial technologies like, Geographical Information System (GIS), Remote Sensing (RS) and Geographical Positioning System (GPS) have the potential to over come the above constraints. The integration of satellite remote sensing data into GIS is one of those great ideas which have made valuable contribution in other fields but need to be utilized in this area. Furthermore, remote sensing is often the most cost effective source of information for updating a GIS and it is a valuable source of current land use/land cover data. Remote sensing techniques has been utilized successfully in certain areas of application, including forestry, watershed management, agriculture and related fields, especially in developed countries were agriculture patterns are well defined and methodologies developed. In agroforestry, however, these technologies have yet to be used extensively (Ellis *et al.*, 2000). The areas of application in agroforestry have been the identification/classification of MPTs and associated crops, inventory of agroforestry system acreage, assessment of tree biomass, crop yield and soil survey, etc.

Spatial Technologies

Manual (traditional) methods of mapping take a relatively long time and high cost. GIS enables the storage, management and analysis of large quantities of spatially distributed data (De Mers, 1997). GIS has gained considerable use in land use planning and natural resource management, providing a spatial framework to aid in the decision process (Zeiler, 1999). Remote sensing, in the broadest sense, may be defined as a technique of acquiring information about some object or phenomenon by recording devices such as cameras, laser, radio frequency receivers, radar system, magnetometers etc. without being in physical or intimate contact with the object or phenomenon under study (Chaturvedi and Khanna, 1982). The data analysis process involves: examining the data by various viewing instruments to analyze pictorial data which is called the 'visual image interpretation techniques and by computers to analyze digital data through process known as 'digital image processing' (Figure 1). GPS technology has provided an indispensable tool for management of agriculture and natural resources. GPS is a satellite and ground based radio navigation and locational system that enables the user to determine very accurate location on the surface of earth. By using the GPS traverse one can directly gets the nodes of the polygon and in large traverses the time is saved by downloading the data into computer.

Applications of Spatial Technologies in Different Fields

Spatial technologies play an important and integral role for information, management and decision-making in all disciplines related to natural resource management. Constant accretion of data and information on agriculture, forestry, agroforestry and natural resource management has created the need to synthesize, organize and maintain this growing knowledge base and facilitate its accessibility and use for education, research and decision-making (Schmoldt and Rauscher, 1996).

Forestry

GIS and RS have now become the potential tools for the forest managers with rapid and economical methods of acquiring plantation information related to forest productivity. The forest cover assessment was started in 1987 in India and is the first country in the world to start an operational system for monitoring countries forest cover. Brazil has started the same in the year 1989 for monitoring the forest cover of Amazon region only (50% of Brazil's area). Other countries such as Bolivia, Indonesia and Peru have initiated similar monitoring systems using satellite remote sensing imagery on a project basis (FSI, 2005). The forest regions of Kolli hills in Western Ghats, India are prone to degradation. Using GIS and RS technologies about 25 ha loss has been observed in dry evergreen forests and about 35 and 1360 ha loss was observed in semi-evergreen and dry deciduous forest, respectively during 1990-1999. (Jayakumar *et al.*, 2002). The GIS overlay techniques used to provide a quantitative historic documentation of deforestation and land-use dynamics in the middle mountains of Nepal between 1947 and 1990 showed that 1960 was the most critical deforestation year and 86% of the recently afforested land is now under *Pinus roxiburghii* (Schreier *et al.*, 1994). In the area of eastern Nepal, based on a comparison of remote sensing data over a 20 year interval, multispectral analysis and supervised classification of Landsat TM (1992) and Landsat MSS (1972) data estimate approximately 7000 ha of low elevation riparian forest (Zomer *et al.*, 2001). The utility of satellite imagery (Landsat ETM+) was evaluated to predict canopy attributes of black wattle (*Acacia*

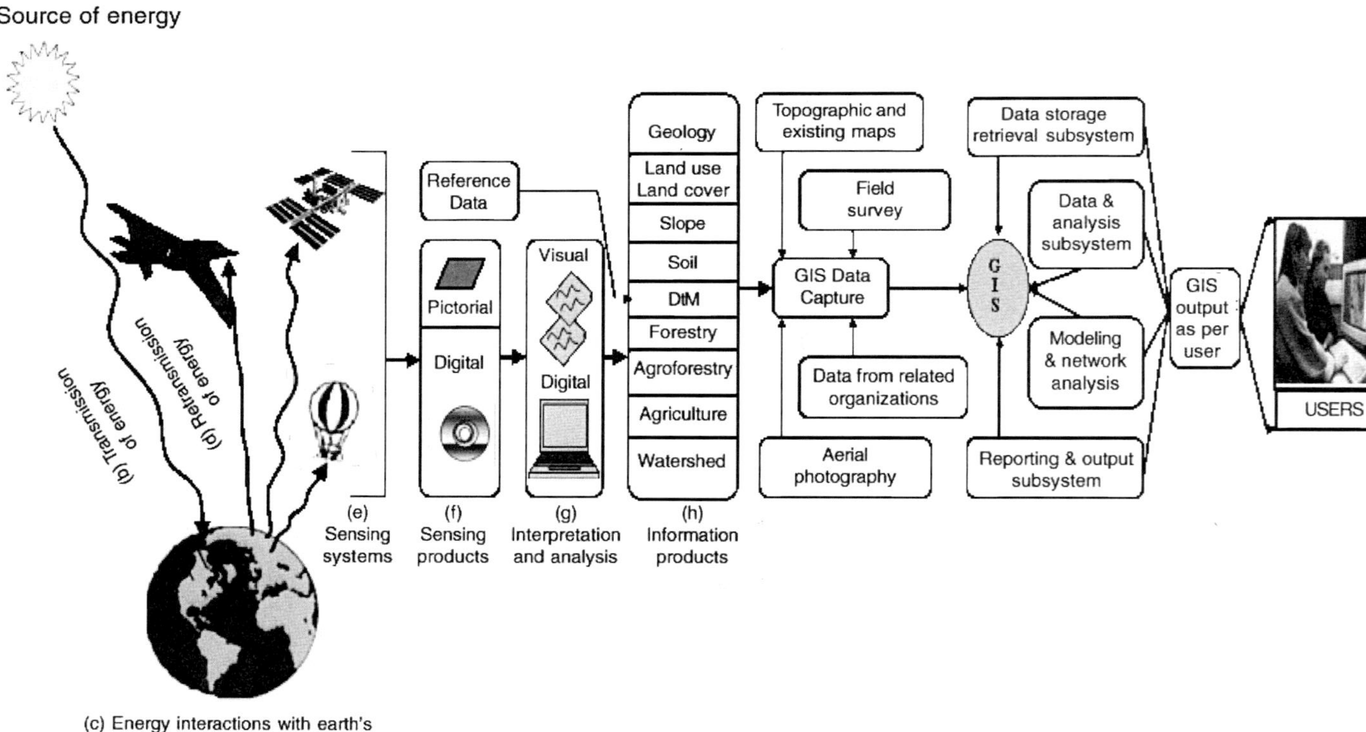

Figure 1. Electromagnetic Remote Sensing Process with overview on GIS.

mearnsii), an economically important commercial plantation species grown in the summer rainfall area of South Africa (Ghebremicael *et al.*, 2003). Forest Survey of India has undertaken a project on mapping of forest cover and types of India based on Champion & Seth classification. It is expected that with application of GIS and GPS tools, generation of such maps would be convenient and geo-referenced forest maps for the whole country showing the latest boundaries of recorded forest area would be available on 1:50,000 scales (FSI, 2005). Using temporal IRS-IC Wide Field Sensors (WiFS) data for monitoring the phonological fluxes of forested landscape of northeast India, the Normalized Difference Vegetation Index (NDVI) was evaluated for monitoring seasonal changes in vegetation (Roy and Joshi, 2002). Roy *et al.*, (2001) use GIS and remote sensing for mapping of popular medicinal shrub *Hippophae rhamnoides* in the adjoining areas of Kaza in Kinour district of Himachal Pradesh. An automated computer based system was proposed for individual tree identification in high resolution remote sensed imagery for open forest cork oak stands (montados) which serves as a substitute for a human interpreter (Surovy *et al.*, 2004). The extensive fire which savaged extensive forest area during summer of 1995 in the Western and Central Himalyas of India drew wide attention of the forest managers and environmentalist. The digital satellite remote sensing data was used to provide the estimates of forest fire damaged areas in Tehri district of Garhwal Himalayas (Pant *et al.*, 1996). The Indian Remote Sensing Satellite 1B, LISS-II (IRS-B, LISS-II) data of pre-fire and post fire period (1993-95) were also studied and analyzed digitally in the IBM RS/6000, EASI/PACE computer system (Roy, 2000). The unique potential of RS and GIS for forest cover assessment and monitoring was also demonstrated by Shrivastava *et al.*, (2002). They highlighted the deforestation and encroachment in the moist deciduous and other forest areas in Sonitpur district of Assam, India. The time series analysis of satellite imageries (1994, 1999 and 2001) showed that there was an alarming rate of conversion of well-stocked forests into cultivable land. The spatial distribution of forests showed progressive decline from 1994 to 2001 and overall loss of 232.19 km^2 of forests was noticed in the district.

Agriculture

Agro-ecological zoning is one of the most important approach for agricultural development planning because survival or failure of particular land use or farming system in a given region heavily relies on careful assessment of agro-climatic resource, and modern tools such as satellite remote sensing and GIS have been providing newer dimensions to effectively monitor and manage land resource in an integrated manner for agro-ecological characterization (Patel, 2004). Aggarwal *et al.*, (1998) studied the agricultural land use option for the sate of Haryana using symphonic use of expert knowledge, simulation modeling, GIS and optimization techniques. The study area was divided into agro-ecological land units by overlaying maps of soil, soil organic carbon and climatic normal rainfall in raster GIS IDRISI. The original soil mapping units based on 198 soil properties were reclassified based on soil texture, level and extent of salinity and sodicity, slope and ground water depth. The organic carbon and normal rainfall maps were generated by inverse square interpolation of observed data points followed by segmentation. With the help of GIS based techniques and land-use/land-cover map prepared with the help of satellite remote sensing tools, it is estimated that the Tehri reservoir in the upper catchment of Ganga river would directly affect 2687 ha of agricultural land and another 3347 ha around the reservoir rim would be rendered unfit for cultivation (Rautela, *et al.*, 2002). Prediction and estimation of yield is closely related to the capability of identifying crop species and certain agronomic variables such as maturity, density, vigour and disease which can be used as yield indicators (Nualchawee, 1984). Sehgal et al.,(2002) demonstrated a technique for estimating date of sowing using RS-derived spectral-temporal crop growth profile and Crop Growth Monitoring System (CGMS) simulation capability and evaluated the capability of CGMS for spatial yield mapping and district level yield prediction for Haryana State during 2000-01 crop season. Rastogi *et al.*, (2000) tested Price model on farmers field during 1996-97 season in Karnal (Haryana, India) and 1997-98 in Delhi using IRS LISS-III data and estimated wheat attenuation coefficients. Conventional agriculture is practiced for uniform application of fertilizer, herbicide, insecticide, fungicide and irrigation, without considering spatial variability. To alleviate the ill effect of over and under usage of inputs, the new concept of precision farming has emerged (Mandal and Ghosh, 2000). GIS-GPS-RS technologies are used in combination for precision farming and site-specific crop management and to evaluate the variable parameters that can affect agriculture production. A GIS based expert system for Gansu Pastoral Agriculture Development (GPAD) in China can show users the knowledge, experience and techniques accumulated by pastoral agriculture specialists, both in theory and practice over many years (Liang *et al.*, 2004). The use of remote sensed data in crop acreage estimation has been demonstrated by various researchers in different parts of the world (Saha and Jonna, 1994; Moulin *et al.*, 1998; Dadhwal, 1999; Hartkamp *et al.*, 1999; Dadhwal and Ray, 2000; Maracchi *et al.*, 2000; Dadhwal *et al.*, 2003; Prachi *et al*, 2005).

Soil and Watershed

Soil reflectance properties depend on numerous soil characteristics such as mineral composition, texture, structure, percentage of organic matter and moisture content (Lillesand and Kiefer, 1987). GIS provided a fast and efficient means of generating the input data required for the model and allowed for easy assessment of the relative erosion hazards over the watersheds under different land management options (Cox and Chandra, 1998). The Indo-Gangatic alluvial plain is subjected to large scale soil alkalization. In order to map and characterize these soils and applying management techniques, high resolution IRS-LISS II, geocoded FCC images on 1:50 000 scales were used by Sharma *et al.*, (2000). For precise inventory of watershed (Gondkhari, India), components like drainage density and pattern, slope percentage and direction, physiography, soils, land use and land cover, were visually interpreted using the geocoded false color composites of IRS-1B LISS II and IRS-1C LISS III at 1:50 000 scale (Saxena *et al.*, 2000). The use of GIS proved effective in estimating and evaluating potential erosion and phosphorus transport in the Beaver Lake watershed, USA (McKimmey *et al.*, 1998), and the GIS database were used to develop a map of the small watershed in Nevada, USA which delineates areas of low potential infiltration from areas of high conceptualization in computer modeling of runoff (Sullivan *et al.*, 1996). The GIS and RS technologies were also used for the prioritization of watersheds using morphometric parameters and assessment of surface water potential (Suresh *et al.* 2004; Chopra *et al.*, 2005), development of management options for agricultural watershed (Pandey *et al.*, 2004) and application for monitoring and management of watershed (Katpatal and Dube, 2004).

In the recent years there are many research finding of efficient use of GIS and RS in watershed management in many parts of the world. Using the latest satellite Quick Bird and GIS technology, the volume of soil erosion in the Quingshuigo watershed in China was evaluated (Pan JingHu and Dong XiaoFeng, 2006), where as in the Yangou watershed, China, on the basis of field survey and laboratory analysis, by using multi-linear regression and GIS, organic C density and its relationship with land use, landform was studied (Wang XiaoLi *et al.*, 2007). The GIS based methodologies supports a programme for monitoring the territorial dynamics that are occurring in Estoraos rural watershed (NW Portugal) (Alonsa *et al.*, 2007). Different land uses/land covers in Nagini watershed of Uttrakhand, India were digitized through GIS using visual interpretation of digital satellite data (Singh *et al.*, 2007). Landsat TM images using a GIS-RS approach, land use/cover changes were analyzed in tropical and biodiverse mountainous area of central-eastern Mexico (Munaz Villers and Lopez Blanco, 2008).

Biomass and Carbon Sequestration

Traditionally biomass estimation involves harvesting of the trees. As the forest cover decreased, it is imperative to use non-destructive methods for volume/biomass estimation. Methods were developed to relate the biomass with girth, height etc., component-wise biomass equations were developed, which were used to estimate biomass at the plot level. In last couple of year's satellite remote sensing has been successfully used for biomass and productivity estimation (Kale *et al.*, 2002). Increasing atmospheric CO_2 also necessitates assessment of carbon storage in tree biomass because above ground biomass of forests is an important component of the global carbon cycle (Zheng *et al.*, 2007). The carbon produce due to anthropogenic activities is sequestered by the terrestrial plant biomass by the process of photosynthesis (Roy and Shirish, 1996). The importance and contribution of various variables in predicting forest biomass have been examined using remote sensing technology in Western Ghats Region of Karnataka, India. Various variables derived from remote sensing methodology were used to predict the forest growing stock. There was a weak linear correlation among the predictor variables where as Vegetation Index, Normalized Difference Vegetation Index and rainfalls have higher linear correlation with the biomass (Benchalli and Prajapati, 2004).

A temporal analysis of urban forest carbon storage using remote sensing (Myeong *et al.*, 2006), combining remote sensing imagery and forest age inventory for biomass mapping (Zheng *et al.*, 2007), a GIS based model to improve estimation of above ground biomass of secondary forests in the Philippines (Magcale *et al.*, 2006), biomass mapping using forest type and structure derived from Landsat TM imagery (Luther *et al.*, 2006), mapping of carbon storages in GIS (Sivrikaya *et al.*, 2007), a linear mixed-effects model of biomass and volume of trees using Landsat ETM+images (Meng *et al.*, 2007) and development of Geographical Information System database and spatial analysis for a forest biomass resource assessment in USA (Hughes and Mackes, 2006) are some of the recent reports of efficient use of GIS and RS technologies for estimation of biomass and carbon sequestration.

Agroforestry

An initial effort to use computers to manage agroforestry data began in the late 1980s with Agroforestry System Inventory Database (AFSI) developed by the International Centre for Research in Agroforestry (ICRAF), Nairobi, Kenya, now the World Agroforestry Centre (Ellis *et al.*, 2004). Another early agroforestry database was the Multi-Purpose Tree and Shrub Database (MPTs), also developed by ICRAF in 1991 (von Carlowitz *et al.*, 1991). In India the agroforestry database development (agroforestryBASE) was initiated at

National Research Centre for Agroforestry, Jhansi, in 2003, containing information on various aspects of agroforestry under four independent module/database namely MPTs, research projects, economic analysis and agroforestry intervention/innovations (Ajit *et al.*, 2003). Recently the information on biofuel species like Jatropha and Pongamia has also been added in the database. The database is widely accessible and available on the Internet at http://www.mirror.iasri.res.in/net.

An early GIS application compiled information on 173 species including their description, soil and climatic preferences and management characteristics for Africa (Booth *et al.*, 1989). On regional scale similar applications were fashioned for Zimbabwe, identifying areas within a country climatically suitable for particular tree species (Booth *et al.*, 1990). In a spatial database approach suitable areas for agroforestry were estimated in sub-Saharan Africa (Unruh and Lefebvre, 1995) and suitable areas of *Annona cherimola* agroforestry system were determined in Southern Ecuador (Bydekerke *et al.*, 1998). The role of GIS in the characterization and monitoring of agoroforestry parks was also highlighted by Bernard and Depommier (1997). In temperate alley cropping spatial analysis using ground penetrating radar (GPR) to evaluate root biomass and distribution and soil nutrient crop-tree interactions was done by Jose *et al.*, (2001). An assessment of the current status of the West African agroforestry parklands was launched in 2002 by ICRAF. The GIS tool was used to find out links between tree biodiversity and land use among peasant farmers in three adjacent village territories in these parklands (Rouxel *et al.*, 2005).

The applications of spatial technologies enable the storage, management and analysis of large quantities of spatially distributed data. These data are associated with their respective geographic features. For example in agroforestry the type of tree species and associated crops would be related with a sampling site, represented by a point. Data on existing agroforestry systems and area dwell in might be associated with fields or experimental plots, represented on a map by polygons. The power of GIS lies in its ability to analyze relationship between features and associated data (Samson, 1995). The new generation of GIS that integrated satellite images with maps data means that this technology can be successfully used for remotely monitoring agroforestry systems. When image analysis and GIS are combined into one package, it can offer a very efficient and cost-effective solution (Buchan, 1997). Satellite images are used to identify what is growing, while GIS component is used to assess area, categorize it and locate its position on earth's surface to provide complete record of the site.

Initiative for Agroforestry Systems Mapping in India

In India the use of GIS-RS-GPS has also widened from natural resource management to habitat analysis. Though the agroforestry land use occupy large areas but use of this technology to estimate area has been recently initiated. At National Research Centre for Agroforestry, Jhansi, a project entitled '*Spatial and Temporal analysis of agroforestry intervention in North-western India using GIS and Remote Sensing*' has been initiated in October, 2007.

In the north western India the agroforestry systems are very prominent land use systems, especially the adoption of Poplar and Eucalyptus based agroforestry systems. About 70 million trees of poplar producing 10.40 million m^3 industrial wood are standing in N-W India alone (FSI, 2005). With this view the two districts one from Haryana i.e. Yammuna Nagar and another from U.P. i.e. Saharanpur were selected for the study. Reason for selecting these two districts is that both the districts are characterized by intensive commercial agroforestry. Through this study, area under agroforestry systems in general and Poplar and Eucalyptus based systems in particular using GIS and Remote Sensing technologies will be estimated/mapped. Any changes occurred in area under agroforestry over a period of 10 years as a result of declining wood prices will also be assessed. Further, the methodology developed as an outcome of this study may be replicated for agroforestry systems existing in other parts of India, for example Eucalyptus and *Leuceana* based agroforestry systems in Andhra Pradesh.

This study has been proposed with the objectives: (1) To generate maps of districts for different land use and land cover, (2) To develop methodology for identification of areas under agroforestry, (3) To delineate areas of Poplar and Eucalyptus based agroforestry systems and estimate areas under two agroforestry systems, and (4) To assess temporal change in area under agroforestry in the two districts. The methodology to be adopted under this study is depicted in Figure 2.

Six blocks of Yamunanagar district namely Radaur, Mustafabad Chhachrauli, Salura, Bilaspur and Jagadhri were recently surveyed for collecting primary data on different agroforestry systems prevalent in these blocks. The prominent Agroforestry systems found on the farmers fields were Poplar based Agrisilviculture system, Poplar based block plantation, Poplar based boundary plantation, *Eucalyptus* based block plantation and Eucalyptus based boundary plantation. Some plots of Agroforestry and pure crops were tracked through GPS and saved. These tracks will serve as training sites for signature analysis of different vegetation cover. Toposheets required for study districts viz., Saharanpur and Yamunanagar at 1:50000 scales have been procured from Survey of India, Dehradun. These Toposheets have been scanned, georefrenced and joined to get the actual boundary and area of the two districts. The GIS and RS lab has been established for spatial analysis and remote sensing data processing. Remote Sensing data of IRS P-6 LISS III and IRS 1D LISS III has been procured from NRSA, Hyderabad.

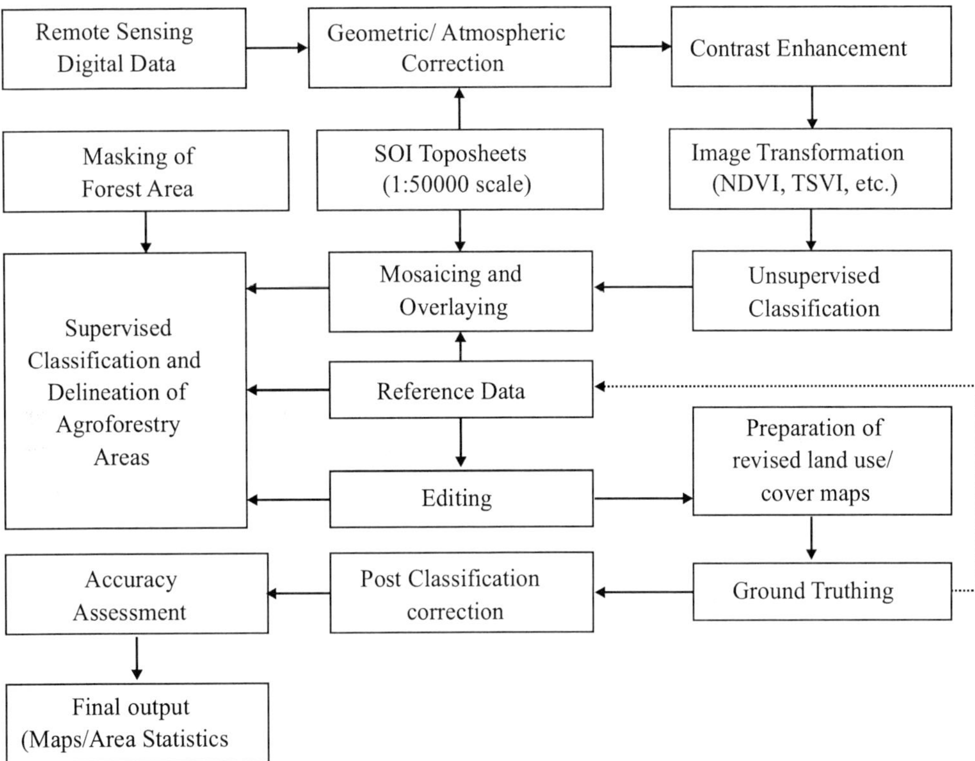

Figure 2. Flow chart showing methodology for agroforestry mapping.

Conclusion

Application of GIS and RS technologies in agroforestry is very limited in India. Although these technologies have great potential in the field agroforestry field and may be used for estimating system production (biomass/yield), assessment of carbon sequestration, identification of areas suitable for agroforestry intervention etc. However, there are some constraints while applying these spatial technologies in this field. There may be the limitations on account of spatial, spectral, temporal, and radiometric resolutions. Besides the satellite limitations, errors in interpretation and classification may be caused due to cloud or shadow effects. Other constraints may be due to identification of boundary plantation with corps and distinct spectral signature between young plantation and crops like sugarcane. There may be the wrong assessment of areas under agroforestry system due to road side and canal side plantations. Hence, a systematic planning and judicious use of these technologies is essential in agroforestry research. However, with the launch and functioning of high resolution and hyper spectral remote sensing satellites, these constraints would not be the major limitations.

References

Aggarwal,P.K., Kalra, N., Bandhyopadhyay, S.K., Pathak, H., Sehgal, V.K., Kour, R., Rajput, T.B.S., Joshi, H.C., Choudhary, R. and Roetter, R. 1998. Exploring agricultural land use options for the State of Haryana: Biophysical modeling, pages 59-65. In: Exchange of methodologies in land use planning (Toetter, R., Hoanh, C.T., Luat, N.V., van Lttersum., M.K and Laar, H.H. eds.), Sys Net Research Paper Series No. 1, IRRI, Los Banos, Philippines.

Ajit, Rai, P., Handa, A.K., Sudhakar Chaudhary., Krishna Prasad, Y.V. and Ajay Babu Pilli. 2003. Application of Database Management tools for on-line dissemination of information on agroforestry in India through World Wide Web. *Indian Journal of Agroforestry*, 5(1&2): 109-114.

Alonsa, J., Rey, N., Castro, P. and Guerra, C. 2007. GIS based land use planning and watershed monitoring as tools for sustainable development. *Ecosystem and Sustainable Development*, VI: 205-214.

Benchalli, S.S. and Prajapati, R.C. 2004. Biomass prediction through remote sensing. *Indian Forester*, 30(6): 599-604.

Bernard, G. and Depommier, D. 1997. The systematic approach and the role of GIS in the characterization and monitoring of agroforestry parks. XI World Forestry Congress, Antalya, Turkey, 13 to 22 October 1997. Vol. 1: Pp. 87.

Booth, T.H., Stein, J.A., Hutchinson, M.F. and Nix, H.A. 1990. Identifying areas within a country climatically suitable for particular tree species: an example using Zimbabwe. *International Tree Crop Journal*, 6: 1-16.

Booth, T.H., Stein, J.A., Nix, H.A., and Hutchinson, M.F. 1989. Mapping regions climatically suitable for particular species: an example using Africa. *Forest Ecology and Management*, 28: 19-31.

Buchan, P. 1997. Satellite imagery for regulatory control. Modern Agriculture: Journal for Site-Specific Crop Management, 1(2): 20-23.

Bydekerke, L., Van Ranst, E., Vanmechelen, L. and Groenemans, R. 1998. Land suitability assessment for cherimoya in southern Ecuador using expert knowledge and GIS. *Agriculture Ecosystem Environment*, 69: 89-98.

Chaturvedi, A.N. and Khanna, L.S. 1982. Forest Mensuration. International Book Distributors, Dehra Dun, India. Pp. 406.

Chopra, R.; Dhiman, R. D. and Sharma, P. K. 2005. Morphometric analysis of sub-watersheds in Gurdaspur district , Punjab using remote sensing and G.I.S. techniques. *Journal of the Indian Society of Remote Sensing*, 33 (4): 531-539.

Cox, C. and Chandra, M. 1998. Application of geographic information systems in watershed management planning in St. Lucia. *Computers and Electronics in Agriculture*, 20(3): 229-250.

Dadhwal, V.K. 1999. Remote Sensing and GIS for agricultural crop acreage and yield estimation. International Arch. Photogramm & Remote Sensing, XXXII, 7-W9, 58-67.

Dadhwal, V.K. and Ray, S.S. 2000. Crop Assessment using remote sensing-Part II: Crop conditions and yield assessment. *Indian Journal Agriculture Economics*, 55(2): 55-67.

Dadhwal, V.K.Sehgal, V.K., Singh, R.P. and Rajak, D.R. 2003. Wheat yield modeling using satellite emote sensing with weather data: Recent Indian experience. *Mausum*, 54 (1): 253-262.

De Mers, M.N. 1997. Fundamental of Geographic Information Systems, Wily New York, Pp 486.

Dhyani, S.K., Kareemulla, K. and Handa, A.K. 2006. Agroforestry potential and returns across agroclimatic zones, Special Report, NRCAF, Jhansi.

Ellis, E.A., Bentrup, B. and Schoeneberger, M.M. 2004. Computer-based tools for decision support in agroforestry: Current state and future needs. *Agroforestry Systems*, 61: 401-421.

Ellis, E.A., Nair, P.K.R., Linehan, P.E. Beck, H.W. and Blanche, C.A. 2000. A GIS-based database management application for agroforestry panning and tree selection. *Computers and Electronics in Agriculture*, 27: 41-55.

FSI, 2005. State of Forest Report, Forest Survey of India (Ministry of Environment & Forests), Dehradun, Pp. 171.

Ghebremicael, S.T., Ahmad, F.B. and Smith, C.W. 2003. Estimating the leaf area index of black wattle (*Acacia mearnsii*) from Landsat ETM+satellite imagery. ICRAF-Bulletin Series. (07/2003):23.

Hartkamp, D.A., White, J.W and Hoogenboom, G. 1999. Interfacing geographic information system with agronomic modeling: A review. *Agronomy Journal*, 91: 761-772.

Hughes, E.M. and Mackes, K.H. 2006. Developing a Geographical Information System database and spatial analysis for a forest biomass resource assessment. *Journal of Testing and Evaluation*, 34(3): 153-157.

Jayakumar, S., Arockiasamy, D. I. and John Britoo, S. 2002. Conserving forests in the EasternGhats through remote sensing and GIS - A case study in Kolli hills. *Current Science*, 82(10):1259-1267.

Jose, S., Gillespie, A.R., Seifert, J.R. and Pop, P.E. 2001.Comparision of mini rhizotron and soil core methods for quantifying root biomass in a temperate alley cropping systems. *Agroforestry Systems*, 52: 161-168.

Kale, M.P., Sarnam Singh and Roy, P.S. 2002. Biomass and productivity estimation using aerospace data and Geographic Information System. *Tropical Ecology*, 43(1): 123-136.

Katpatal, Y. B. and Dube, Y. A. 2004. Monitoring and management of Pioli watershed near Nagpur urban area through remote sensing and G.I.S. In: International conference on Advanced Modelling Technique for Sustainable Management of water Resources, 28-30 Jan 2004: Pp. 359-362.

Liang, T.G., Chen, Q.G., Ren, JiZhou., Wang, Y.S. 2004. A GIS-based expert system for pastoral agriculture development in Gansu Province, PR China. *New Zealand Journal of Agricultural Research*, 2004. 47(3): 313-325.

Lillesand, T.M., and R.W. Kiefer. 1987. Remote Sensing and Image Interpretation. 2nd edition. John Wiley & Sons.

Luney, P.R. and Dill, H.W. 1970. Uses, potentialities and needs in Agriculture and Forestry. *In* Remote Sensing with Special Reference to agriculture and Forestry. National Academy of Sciences, Pp. 1-34.

Luther, J.E., Fournier, R.A., Piercey, D.E., Guindon, L. and Hall, R.J. 2006. Biomass mapping using forest type and structure derived from Landsat TM imagery. *International Journal of Applied Earth Observation and Geoinformation*, 8(3): 173-187.

Magcale, Macandog, D.B., Delgado, M.E.M. and Ty, E., Villarin, J.R.T. 2006. A GIS based model to improve estimation of aboveground biomass of secondary forests in Philippines. *Journal of Tropical Forest Science*, 18(1): 8-12.

Mandal, D. and Ghosh, S. K. 2000. Precision farming – The emerging concept of agriculture for today and tomorrow. *Current Science*, 79(12): 1644-1647.

Maracchi, G., Perarnaud, V. and Kleschenko, A.D. 2000. Application of geographical information systems and remote sensing in agro- meteorology. *Agriculture and Forest Meteorology*, 103: 119-136.

McKimmy, J.M., Scott, H.D. and Keisling, T.C. 1998. Identification of critical source areas for control of sediment and phosphorus transport within a large watershed. Special Report Arkansas Agriculture Experiment Station, No. 186: 70-78.

Meng, Q.M., Cieszewski, C.J., Madden, M. and Borders, B. 2007. A linear mixed-effect model of biomass and volume of trees using Landsat ETM+images. *Forest Ecology and Management*, 244(1/3): 93-101.

Meyong, S.J., Nowak, D.J. and Duggin, M.J. 2006. Atemporal analysis of urban forest carbon storage using remote sensing. *Remote Sensing of Environment*, 101(2): 277-282.

Moulin, S., Bondeau, A. and Delecolle, R. 1998. Combining agricultural crop reflectance estimates: a regional case study on wheat. *International Journal of Remote Sensing*, 20(1): 213-218.

Munoz Villers, L.E. and Lopez Blanco, J. 2008. Land use/cover changes using Landsat TM/ETM images in a tropical and biodiverse mountainous area of central eastern Mexico. *International Journal of Remote Sensing*, 29(1/2): 71-93.

Nair, P.K.R. 1993. An introduction to Agroforestry. Kluwer Academy Publishers, Dordrecht, The Netherlands, Pp. 499.

NRCAF. 2007. NRCAF-Perspective Plan Vision 2025. National Research Centre for Agroforestry, Jhansi, Pp 46.

Nualchawee, K. 1984. Remote sensing in agriculture and role of groundtruth as supporting data. Proceeding of Symposium. Third Asian Agricultural Remote Sensing Symposium, Feberuary 6-11, 1984. Chiangmai, Thailand, Pp. 269-285.

Pan JingHu and Dong XiaoFeng. 2006. GIS-and-Quick Bird-based quantitative assessment of soil erosion in small watershed. *Journal of Ecology and Rural Environment*, 22 (2): 1-5.

Pandey, Vinay K.; Panda, S. N. and Raghuwanshi, N. S. 2004. Development of management options for agricultural watershed using G.I.S. and remote sensing In: International conference on Advanced Modelling Technique for Sustainable Management of water Resources, 28-30 Jan 2004, Pp 353-358

Pant, D.N., Das, K.K. and Roy, P.S. 1996. Digital mapping of forest fires in Garhwal Himalaya using IRS. *Indian Forester*, 22(5): 390-395.

Patel, N.R. 2004. Remote Sensing and GIS application in Agro-ecological zoning. Proceedings of the Training Workshop, 7-11 July, 2003, Dehra Dun, India, Pp. 263-289.

Pathak, P.S., Pateri, H.M. and Solanki, K.R. 2000. Agroforestry Systems in India: A diagnosis & Design Approach. NRCAF (ICAR), New Delhi, Pp. 223.

Prachi, M.S., Anil, R., Singh, R., Handique, B.K. and Rao, C.S. 2005. Integrated approach based on remote sensing and GIS for estimation of area under paddy crop in north-eastern hill region. *Journal of Indian Society of Agricultural Statistics*, 59(2): 151-160.

Puri, S., and Panwar, P. 2007. Agroforestry Systems and practices. New India Publishing Agency, New Delhi, Pp. 643.

Rastogi, A., Kalra, N., Agarwal, P.K. Sharma, S.K., Harit, R.C., Navalgund, R.R. and Dadhwal, V.K. 2000. Estimation of wheat leaf area index from satellite data using Price model. *International Journal of Remote Sensing*, 21(15): 2943-2949.

Rautela, P., Rakshit, R., Jha V. K., Gupta, R.K. and Munshi, A. 2002. GIS and remote sensing-based study of the reservoir-induced land-use/land-cover changes in the catchment of Tehri dam in Garhwal Himalaya, Uttaranchal (India). *Current Science*, 83(3): 308-311.

Rouxel, C., Barbier, J., Niang, A., Kaya, B. and Sibelet, N. 2005. Relationship between village territories and the biodiversity of tree species. Case study on three villages in the Segou Region (Mali). *Bio et Forest des Tropiques*, 283: 33-49.

Roy, P.S and Joshi, P.K. 2002. Forest cover assessment in north-east India- the potential of temporal wide swath satellite sensor data (IRS-IC WiFS). *International Journal of Remote Sensing*, 23(22): 4881-4896.

Roy, P.S. 2000. Assessment of forest fires in India through remote sensing. In: Natural Disaster and their Mitigation- Remote Sensing and Geographic Information System Perspectives, Edited by P.S. Roy, C.J. van Westen, V.K. Jha, R.C. Lakhera & P.K. Champati Ray. Indian Institute of Remote Sensing, Dehra Dun, Pp. 110-129.

Roy, P.S. and Shirish, A. Ravan. 1996. Biomass estimation using satellite remote sensing data-An investigation on possible approaches for natural forest. *Journal of Bioscience*, 21(4): 535-561.

Roy, P.S., Porwal, M.C. and Lalit Sharma. 2001. Mapping of Hippophae rhamnoides Linn. in the adjoining areas of Kaza in Lahul and Spiti using remote sensing and GIS. *Current Science*, 80(9): 1107-1111.

Saha, S.K., and S. Jonna. 1994. Paddy acreage and yield estimation and irrigated cropland inventory using satellite and agrometeorological data. *Asian Pacific Remote Sensing Journal*, 6(2):79-87.

Samson, S.A. 1995. Determining need for a Geographic Information System (GIS). Journal of extension, 33(5). (on-line) http://www.joe.org/joe/1995october/tt2.html.

Saxena, R.K., Verma, K.S., Chary, G.R., Rajeev Srivastava. And Barthwal, A.K. 2000. IRC-1C data application in watershed characterization and management. *International Journal of Remote Sensing*, 21(17): 3197-3208.

Schmoldt, D.L. and Rauscher, H.M. 1996. Building Knowledge Based Systems for Natural Resource Management. Chapman & Hall, New York. Pp 386.

Schreier, H., Brown, S., Schmidt, M., Shah, P., Shrestha, B., Nakarmi, G., Subba, K., Wymann, S. 1994. Gaining forests but losing ground: a GIS evaluation in a Himalayan watershed. *Environmental management*, 18 (1): 139-150.

Sehgal, V.K., Rajak, D.R., Chaudhary, K.N. and Dadhwal, V.K. 2002. Improved regional yield prediction by crop growth monitoring system using remote sensing derived crop phenology. *Remote Sensing and Spatial Information Science*, 34(7): 329-334.

Sharma, R.C., Saxena, R.K. and Verma, K.S. 2000. Reconnaissance mapping and management of salt affected soils using satellite images. *International Journal of Remote Sensing*, 21(17): 3209-3218.

Singh, R., Sharma, H.C. and Dwivedi, R.P. 2007. Analysis of land use in Nagini watershed of Uttrakhand using GIS and Remote Sensing. *Range Management and Agroforestry*, 28(2A): 69-70.

Sivrikaya, F., Keles, S., Cakir, G. 2007. Spatial distribution and temporal changes of carbon storage in timber biomass of two different forest management units. *Environmental Monitoring and assessment*, 132(1/3): 429-438.

Srivastava, S., Singh, T. P., Harnam Singh, Kushwaha, S. P. S. and Roy, P. S. 2002. Assessment of large-scale deforestation in Sonitpur district of Assam. *Current Science*, 82 (12): 1479-1484.

Sullivan, M., Warwick, J.J. and Tyler, S.W. 1996. Quantifying and delineating spatial variation of surface infiltration in a small watershed. *Journal of Hydrology Amsterdam*, 181 (1-4): 149-168.

Suresh, M.; Sudhakar, S.; Tiwari, K. N. and Chowdary, V. M. 2004. Prioritization of watersheds using morphometric parameters and assessment of surface water potential using remote sensing. *Journal of the Indian Society of Remote Sensing*, 32 (3): 249-259.

Surovy, P., Ribeiro, N.A., Oliveira, A.C., Scheer, L. 2004. Automated aerial imagery analysis system for individual tree identification in cork oak stands. *Advances in Geoecology*, 37: 287-296.

Unruh, J.D., and Lefebvre, P.A. 1995. A spatial database for estimating areas for agroforestry in SubSaharan Africa: aggregation and use of agroforestry case studies. *Agroforestry Systems*, 32(1): 81-96.

Von Carlowitz, P.G., Wolf, G.V. and Kemperman, R.E.M. 1991. Multipurpose Trees and Shrubs Database: An Information and Decision Support System. ICRAF, Nairobi, Kenya & GTZ, Eschborn, Germany, Pp. 104.

Wang XiaoLi ., Duan JianJun and Guo ShengLi. 2007. oraganic carbon density of topsoil and its spatial distribution of small watershed in hilly region of loess plateau. *Journal of Northwest A and F University Natural Science*, 35(10): 98-102.

Zeiler, M. 1999. Modeling our World: the ESRI guide to geodatabase design. ESRI Press, Redland, CA. pp 19.ghebremicael, S.T., Ahmad, F.B., Smith, C.W. 2003. Estimating the leaf area index of black wattle (*Acacia mearnsii*) from Landsat ETM+ satellite imagery. ICFR-Bulletin-Series. (07/2003):23.

Zheng, G., Chen, J.M., Tian, Q.J., Ju, W.M., Xia, X.A. 2007. Combining remote sensing imagery and forest age inventory for biomass mapping. *Journal of Environmental Management*, 85(3):616-623.

Zomer, R.J., Ustin, S.L., and Carpenter, C.C. 2001. Land cover change along tropical and subtropical riparian corridors within the Kakalu Barun National Park and Conservation Area, Nepal. *Mountain Research and Development*, 21(2): 175-183.

❑❑❑

Agroforestry: Natural Resource Sustainability, Livelihood & Climate Moderation © 2009
Editors: OP Chaturvedi, A Venkatesh et. al., pp. 549-562
Satish Serial Publishing House, Delhi

Chapter 39

INDICATORS FOR ASSESSMENT OF SOIL QUALITY IN AGROFORESTRY

Rajendra Prasad, S.K. Dhyani and O.P. Chaturvedi
National Research Centre for Agroforestry, Jhansi-284 003 (U.P.)

Introduction

Assessment of soil quality refers to measurement of relative changes in soil characteristics over time brought by human management under different land use systems. It may be based either on simple visual observation or involve complex laboratory analyses of soil tests. Assessment of soil quality is a holistic approach to view soil in its landscape setting and how it functions within the defined ecosystem. This chapter discusses concept of soil quality and its link with agricultural sustainability, indicator for soil quality assessment and development of a soil quality index with especial reference to agroforestry land use.

Soil Quality and Agricultural Sustainability

Soil quality may give different meanings to different persons according to their perspective. To land managers and crop advisors, it may mean sustaining or enhancing productivity and maximizing profits, while maintaining soil resources for future generations. For conservationists, soil quality may mean protecting the environment and sustaining soil resource for future generations. For the environmentalist, it may mean a holistic, interconnected view of soil in an ecosystem with respect to maintenance or enhancement of biodiversity, water quality, nutrient cycling and biomass production. For consumers of agricultural products, it means plentiful, healthful and inexpensive food for present and future generations. The view on soil quality also changes with the scale of concerns. For a farmer, his field may be important while a community will have concern of soil quality within their watershed area. For national policy makers, national assessment of soil quality and its trends are important. Need of assessment of soil quality has significant bearings from the misuse and mismanagement of non-renewable and finite soil resources. Today, soil degradation has become a major global issue. As per one approximation, the degraded area increased from 10% of the world's arable land in early 1970s (Biswas and Biswas, 1974) to 40% in early 1990 (Oldeman, 1994). In India, nearly 188 m ha land area suffers from one or other kind of land degradation. Various forms of soil / land degradation, which reduce soil's capacity to perform, have been depicted in Fig 1. Reduction in the functional capacity of soil results in decline of productivity. The declining trends in food grain productivity threaten agricultural sustainability.

The concept of agricultural sustainability expressed during last two decades of twentieth century is being magnified in twenty first century because of limited arable land resources, rapid increase in population, conversion of agricultural land to other uses and persistence of hunger and malnutrition in several regions of the world (Lal, 1998). Therefore, the question of agricultural sustainability became extremely relevant and led to considerable interest in productivity trends of long-term experiments (Lal and Stewart, 1995). It is also emphasised that maintaining the quality of natural resource base would ensure sustainability of agricultural production.

Soil quality is considered a key element of sustainable agriculture (Warkentin, 1995) and refers to its capacity to perform specific function (SSSA 1987). Soil function represents a composite function of soil's physical, chemical, hydrological and biological properties that provide a medium for plant health, regulate water in the environment and serves as buffer in the formation and degradation of hazardous compound (Larson and Pierce, 1991). In terms of agricultural production, therefore, soil quality refers to its ability to sustain productivity. Strong link exists between soil quality and agricultural sustainability. If an agricultural system is unsustainable, it may partly be due to fact that soil quality is declining over time. The decline in soil quality may be caused by reduction in top soil depth by accelerated erosion, increase in soil salinity due to water logging, drop in groundwater table due to excessive irrigation, depletion of soil organic matter and fertility due to residue removal/ burning or inadequate application of fertilizer and manures. Threats to agricultural sustainability, concerns for environmental quality and economic progress for well being of mankind have sparked the interest in concept of soil quality and its assessment.

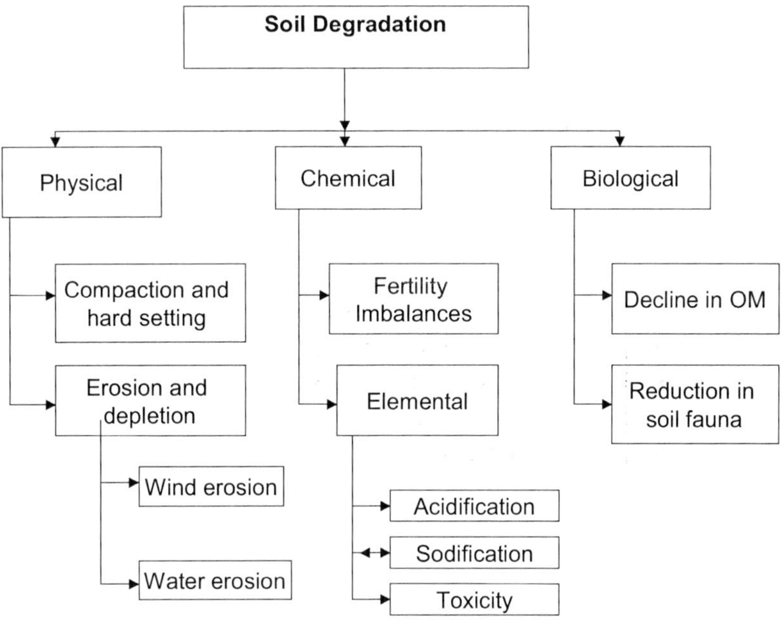

Fig 1. Graphical depiction of various forms of soil degradation

Need to Assess Soil Quality of Agroforestry Systems

The overriding objective of agroforestry, since its inception with establishment of ICRAF in 1977 has been to develop integrated land management systems involving trees, crops and/ or animals, which would contribute substantially to decrease deforestation, increase food production, enhance biodiversity and protect environment (Nair, 1993). It is argued that presence of woody perennials in agroforestry system affects several bio-physical and bio- chemical processes that determine the health of soil substrate. The most obvious effects of trees on soil include amelioration of erosion primarily through surface litter cover and under story vegetation; maintenance or increase of organic matter and diversity through continuous degeneration of roots and decomposition of litter; N fixation; enhancement of physical properties such as soil structure, porosity and moisture retention due to extensive root system and the canopy cover; and absorb and recycle nutrients in the soil that would otherwise be lost through leaching (Sanchez, 1987). Although soil conservation for restoring quality and health of degraded soil has been advocated as primary goal of agroforestry (Young, 1989), the empirical estimates of accrued benefits are lacking due to temporal and spatial complexity of agroforestry systems and soil resource dynamics. It is only in recent times that the scientific literature has been focussing on soil quality concept, its definition as well as its assessment.

Assessment of soil quality or health in agroforestry acquires more relevance due to expansion of area under agroforests in tropics and sub tropics, and their enhanced contribution toward sustainability of agricultural production. In India, as per one estimate, about 25.72 million ha area has been covered under various types of agroforestry plantations. The task force on Greening India (Anonymous, 2001) has identified a potential of 10 million ha in irrigated land and 18 million ha in rainfed areas that could be developed through agroforestry on a watershed basis. The main driving force and objective of covering such large area under agroforests is to keep the soil resource productive on sustainable basis. It is hypothesized that various agroforestry systems being advocated are productive, remunerative and environment friendly. To prove this hypothesis, it is essential to assess impact of various agroforestry management practices on soil quality.

Selection of Indicators and Reference Value

Assessment of soil quality depends on indicators that relate to various soil functions. Depending on approaches viz. descriptive or qualitative and quantitative soil quality assessment, the indicators can be observational features of soil or plants growing on it, and measure of physical, chemical and biological soil attributes. The quantitative approach offers greater precision but requires more resources for doing assessment.

Before selecting soil quality indicators for assessment programme, the basic fact regarding soil to be given due consideration is that the soil quality embraces two general elements viz. i) inherent properties of soil or inherent capacity and ii) dynamic nature of soil. The inherent capacity of soil is the result of factors of soil formation- climate, topography, vegetation, parent material and time. Attributes of inherent soil quality such as mineralogy and particle size distribution are almost static and show little change over time. The dynamic nature of soil encompasses those soil attributes that can change over relatively short time in response to human management. These soil attributes are referred as indicators of soil health. Hence, assessment of soil quality refers to assessment of dynamic soil attributes (Doran and Parkin, 1994). The selected indicators of soil quality should have greatest sensitivity to changes in soil function (Andrews *et al.*, 2004). It is emphasised that soil quality indicators should correlate well with ecosystem processes; integrate soil processes, accessible to many users and sensitive to management and climate (Daran and Parkin, 1996). Further, the group of indicators selected to measure soil functions must be sufficiently diverse to represent the chemical, biological and physical properties and processes of complex system.

For indicators to be useful in assessing soil quality, standards or reference values must be established as baseline from which current state of soil quality is assessed. The current state of soil must be compared to some standards to assess degree of excellence of soil quality. Reference values can be given specific limits or ranges for each indicator for a particular soil or group of similar soils. These ranges or limits are based on the values for indicators, which define a condition representative of soil functioning at full potential (Seybold *et al.*, 1996). Full potential is defined as the maximum capacity of a soil to function under a particular use. In this form a soil is either healthy (with the acceptable range) or at risk (out side the acceptable range). Another way to express reference value is to develop scoring functions or equations (Karlen *et al.*, 1994). Scoring functions show gradations from very poor quality to very good or excellent soil quality for a particular indicator. These functions are usually normalized to a scale of 0 to 1, with 1 being excellent quality and 0 being very poor quality. There can be three conditions for developing indicator reference values viz. i) the natural condition, ii) specific human use, and iii) altered soils. For soils in their natural condition, reference value represents the inherent ability of a soil to function as defined by the soil forming factors and processes in its native or natural state. Reference values for soil that are intensively managed, as production agriculture, should be based on what is sustainable under a specific land use as defined by best management practices. For the third case, reference values on altered soils are based on present condition of soil in conjunction with best management practices.

Minimum Data Sets (MDS)

Since it is impractical to measure every soil property, many researchers have proposed a minimum data set, which is smallest set of soil properties or indicators needed to measure or characterise soil quality. Identifying key soil properties or attributes that are sensitive to change in soil functions establish a minimum data set. The concept of how indicators are related to soil function and quality is illustrated in Fig 2. In most cases, one or two indicators can sufficiently represent each function; however, indicators may be related to many soil functions.

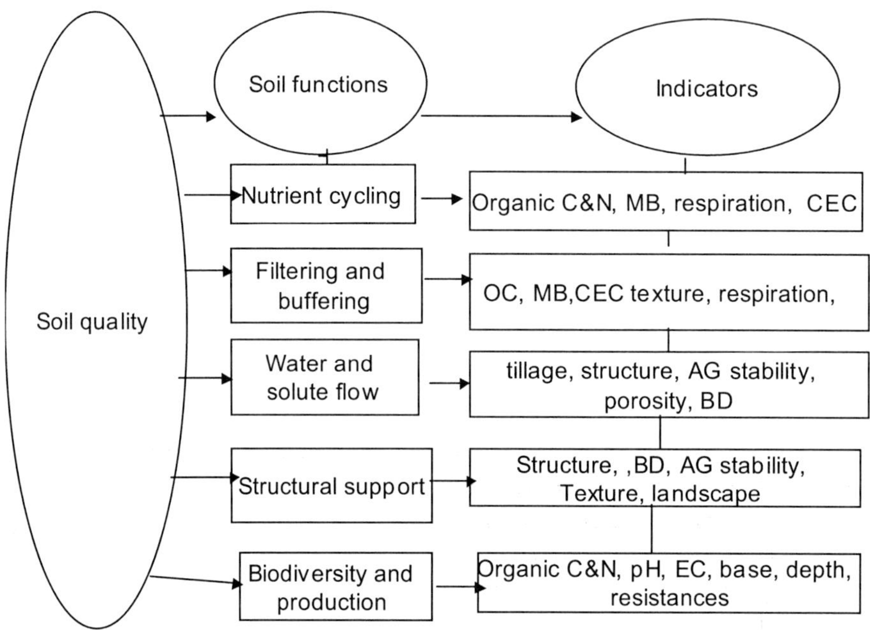

Fig. 2. Graphical representation of concept of soil quality using soil function and indicators (Seybold *et al.*, 1996). OC: Organic carbon, N: Nitrogen, MB: Microbial biomass, CEC: Cation exchange capacity, AG: Aggregate, BD: Bulk density.

A minimum data set does not usually encompass all relevant properties for a region or farming system. It is an example of a minimum set of indicators required to obtain a comprehensive understanding of the soil being evaluated. Each minimum data set is tailored to a particular region or a soil map unit and includes only those properties which are relevant to soil types, farming system and land uses of area being evaluated. Several minimum data sets of indicators have been proposed (Larson and Pierce, 1991, Doran and Prakin, 1994, Seybold *et al.*, 1996; Gregorich et *al.*, 1994). Some of important soil health indicators include aggregation, water holding capacity, pH, EC, NPK reserves, organic carbon, microbial biomass, and community composition. In specific situations some additional properties like exchangeable sodium percentage (ESP), micronutrients status, pesticide residues and heavy metal pollutants may also be selected as indicators of soil quality. An example of minimum data set is given in Table 1 (Seybold *et al,* 1996).

Table 1. Minimum data set of physical, chemical and Biological indicators for screening quality and health of soil.

Indicator	Relationship to soil condition and function
Physical	
Texture	Retention and transport of water and chemicals
Depth of soil, top soil and rooting	Estimate of productivity potential and erosion
Infiltration and bulk density	Potential for leaching, productivity and erosivity
Water holding capacity	Related to water retention, transport and erosivity
Chemical	
Soil organic matter (SOM)	Defines soil fertility, stability and erosion extent
pH	Defines biological and chemical activity thresholds
Electrical conductivity	Defines plant and microbial activity thresholds
Extractable N, P, and K	Plant available nutrients and potential for N loss
Biological	
Microbial biomass C and N	Microbial catalytic potential and repository for C and N
Potentially mineralizable N	Soil productivity and N supplying potential
Soil respiration	Microbial activity measure

Selection of Soil Health Indicators for Agroforestry Systems

Minimum data set (MDS) discussed in previous section is comprehensive and reflects various soil function, and therefore can be used for determining optimal functioning, stability and sustainability of agroforestry systems. However, drawing on the fact that success of any agroforestry system is guided by the complex array of positive and negative interactions and strategies to manage them for maximum benefits; assessment of soil quality may require different data set. The temporal and spatial complexity of agroforestry systems and resource dynamics have to be given due consideration in assessing soil health benefits based on selected indicators. The management goal remains the driving principle behind a particular soil attribute to be chosen as a soil health indicator. The main goal of soil management through agroforestry is to make best use of its resource-conserving and resource-sharing potential. Hence, the assessment of soil quality in agroforestry should no longer be limited to mere productivity aspects, but include broader environmental effects (Karlen *et al.*, 1997). Management goals pertaining to productivity and environmental conservation may differ from region to region and own much from its physiographic settings, thus affecting choice of soil quality indicator. In context to agroforestry systems/ practices followed in India, the examples of region specific challenges and management goals, which can play key role in deciding indicators for quality of soil and sustainability of the ecosystem, are given in Table 2 (Pandey, 2007).

Table 2. Examples of region specific challenges and management goals of agroforestry practices in India.

Region	Challenge/ management goal	Observed changes due to agroforestry
Himalaya (Kurukshetra)	Improvement of sodic soils	Increase in microbial, tree biomass and soil carbon; enhanced N availability
Himalayas	Restoration of abandoned agricultural sites	Improvement in soil physico-chemical properties, carbon sequestration, and biomass production
Western Himalayas	Reducing soil and water loss agro-ecosystem in steep slopes	Reduced runoff and soil loss
Sikkim Himalayas	Enhancing litter production and soil nutrients dynamics	Increased N fixation by N fixing trees, enhanced litter production and greater release of N and P
Indo-Gangetic plains	Biomass production and nutrient dynamics in nutrient deficient and toxic soils	Biomass production
Meghalaya	Enhancing tree survival and crop yield	Crop yield did not decrease in proximity of *Albizia* trees.
Western India (Rajasthan)	Compatibility of trees and crops	Density of 417 tree/ha was found ideal for cropping with pulses
Western India (Karnal)	Improvement of soil fertility of moderately alkaline soils	Increase in soil microbial biomass, soil carbon
Central India (Raipur)	Biomass production in N and P stressed soils	*Azadirachta indica* tres were found to produce biomass in depleted soils
Central India	Soil Improvement	Decline proportion of soil sand particles, increase in soil Organic C, N, P and Mineral N
Southern India (Kerala)	Growing commercial crops and trees	*Ailanthus triphysa* helps ginger grown in interspaces in developing better rhizome in comparison to solo farming.

Numerous physical, chemical and biological properties of soil can be measured as part of soil quality or soil health assessment programme depending on their sensitivity and control over specific function. The set of soil attributes that has been used as MDS to assess soil quality and develop a soil quality index for different agroforestry systems at National Research Centre for

Agroforestry (NRCAF), Jhansi include i) physical attributes: bulk density, porosity, soil depth/ rooting depth, water holding capacity, infiltration rate; ii) chemical attributes: soil organic carbon, pH, EC, CEC, extractable NPK; and iii) biological attributes: microbial biomass and soil respiration.

Development of Soil Quality Index (SQI) for Agroforestry Systems

Measurement of changes in selected indicator value in relation to established base line or reference value over time is the way of quantifying soil quality. Changes in indicator reflect the combined effect of land use and climate. However, unless a particular indicator is strongly related to a specific soil function, the individual properties are of little value for assessing comparative soil health of different ecosystem. Also, due to predominance of site specific agroforestry approaches and inter correlation of many soil properties with each other, using a single dependent variable in isolation may not provide a complete and accurate assessment of agroforest soil health. To overcome this problem in assessing potential impact of changes in soil properties on soil health of agroforesry ecosystems, it is desirable to develop a soil quality index (SQI) that integrates measured soil physical, chemical and biological properties into a single parameter that could be used as an indicator of overall soil quality of any ecosystem.

Different methods have been suggested to assess soil quality index of any ecosystem integrating available actual data on various physical, chemical and biological characteristics of soil. The approach of Roming *et al.,* (1995) is very useful as qualitative measure of soil quality. Their approach involves i) interviewing farmers on their concept, perception and measure of healthy soil; ii) summarizing and ranking of the results of the interviews; iii) ranking measures of soil health into three classes: healthy, impaired, unhealthy; and iv) developing a score card with criteria for each class of soil health. The scores of each indicator are totalled and used as baseline for assessing soil quality. Using a quantitative approach for calculating soil quality index, Karlen and Stott (1994) suggested that for each indicator, a scoring function and realistic baseline and threshold values need to be established. After these values are established, weights can be assigned based on importance of each soil function and indicator as determine by the land use. For assignment of weights to soil functions and their respective soil quality indicators, a nested scheme has been described by Karlen *et al.,* (1994). Thereafter, the scores of each indicator are multiplied by appropriate weight. An overall soil quality index is calculated by summing the weighted score for each soil function. Barbiroli *et al.,* (2004) proposed tree-structured approach to formulate an overall index by going through successive levels of aggregation. The procedure consisted of selecting parameters used in the soil characterization; defining variability ranges for the chosen parameters; identifying suitable normalization functions to transform values of each parameter in an index within a 0-10 range (where 0 means poor quality and 10 high quality); weighting parameters, if needed; and aggregating parameters into intermediate and final indices through the use of an aggregation function.

The use of index value based on summarized soil properties have already been suggested and used to assess soil quality of agricultural systems (Doran and Parkin, 1996, Kang, *et al.*, 2005). While evaluating sustainability of rice-wheat cropping system on a Vertisol in India, Mohanty *et al.*, (2005) developed a soil quality index (SQI) based on bulk density (BD), penetration resistance (PR), water stable aggregates (WSA) and soil organic matter (OM) and concluded that SQI values of 0.84–0.92, 0.88–0.93 and 0.86–0.92 were optimum for rice, wheat and the combined system of rice + wheat, respectively. The maximum yields for rice and wheat were 5806 and 1825 kg ha^{-1} obtained at SQI values of 0.85 and 0.99, respectively. Using zero tillage (ZT) for wheat had a positive effect on soil quality regardless of the treatments used for rice. Regression analyses to predict sustainability of the various tillage and crop residue treatments showed that as puddling intensity for rice increased, sustainability without returning crop residues decreased from 6 to 1 year. When residue was returned, the time for sustainable productivity increased from 6 to 15 years for direct seeded rice, 5 to 11 years with low-intensity puddling and 1 to 8 years for high-intensity puddling.

Similar to agricultural systems, SQI has also been used to assess forest soil health. To assess trends in forest soil health under forest inventory analysis (FIA), a single value SQI integrating 19 physical and chemical properties of soil has been developed by Amacher *et al.*, (2007) to monitor changes in soil properties of forest soil with time. They assigned an index value to each soil property and all individual index values were summed up to obtain a total SQI. The SQI values were developed for six FIA plots of the United States. The average SQI values varied from 43% to 67% for different FIA plots.

Use of SQI value based on summarized soil properties to assess soil quality of agricultural and forest ecosystems is well established, however, the systematic information on use of a unified value of SQI to assess impact of various agroforestry systems on soil quality is lacking. Development and application of a soil quality index to national ongoing agroforestry programme such as All India Coordinated Research Project on Agroforestry (AICRPAF) would provide a mechanism for evaluation of changes in soil quality across the regions.

Example of Using SQI in Agroforestry: Out of different methods prescribed for calculating SQI, the method used by Amacher *et al.*, (2007) is the simplest and easy to understand. An example calculation of soil quality index using hypothetical data of three agroforestry systems viz. A: *Acacia nilotica* based agri-silvi; B: *Albizia lebbek* based agri-silvi; and C: *Ailanthus excelsa* based silvi-pasture system is demonstrated. The soil property threshold levels and associated soil index values are listed in Table 3. The observed soil parameter values measured at the start of experiment and after 10 years, and assigned quality index values along with total soil quality indices have been given in Table 4 (all the values are hypothetical & not specific to any soil or land use). The values are shown for demonstration purpose only and it is urged to readers to use their own data and assign index values as per their existing conditions).

Table 3. Soil property threshold levels and associated soil index values

Indicator	Level	Index
Soil Organic Carbon (%)	>2.0	5
	1.5-2.0	4
	1.0-1.5	3
	0.5-1.0	2
	<0.5	0
Soil pH	<5.5	0
	5.5-6.5	2
	6.5-7.5	5
	7.5-8.5	2
	>8.5	0
Microbial biomass (mg C kg^{-1} soil)	>650	5
	450-650	4
	250-400	3
	100-250	2
	<100	0
Total N (%)	>0.75	5
	0.5-075	4
	0.2-0.5	3
	0.1-0.2	2
	<0.1	0
Bulk density (g cm^{-3})	>2.0	0
	1.6-2.0	1
	1.4-1.6	3
	1.3-1.5	4
	<1.3	5
Porosity (%)	>60	0
	50-60	2
	40-50	5
	20-40	2
	<20	0

Table 4. Effect of different agroforestry systems on Soil quality index (SQI).

Soil properties/ Indicator	Measured Soil properties				Quality index score			
	At start	After 10 years						
	R	A	B	C	R	A	B	C
SOC (%)	0.40	0.56	0.68	0.65	0	2	2	2
Soil pH	8.1	7.8	7.4	7.5	2	2	5	5
Microbial biomass (mg C kg^{-1} soil)	121	212	256	275	2	2	3	3
Total N (%)	0.10	0.18	0.28	0.56	2	2	3	4
Bulk density (g cm^{-3})	1.7	1.52	1.58	1.4	1	3	3	4
Porosity (%)	36	42	40	45	2	5	5	5
Total Soil quality index					9	16	21	23
Soil Quality index (%)					30.0	53.3	70.0	76.7

R: Reference soil data at the start of experiment, A: *Acacia nilotica* based agri-silvi; B: *Albizia lebbek* based agri-silvi; and C: *Ailanthus excelsa* based silvi-pasture system

The individual index values for all soil properties are summed up to give total SQI of particular agroforestry system. The maximum possible value of SQI based on six soil indicators under reference is 30. The observed total soil quality index is also expressed as percentage of maximum possible value of total SQI. From this model exercise, simple and general inferences drawn are:

i. In comparison to reference soil quality (soil at the start of experiment), integration of agroforestry practices have improved the soil quality.

ii. After 10 years of agroforestry practices, maximum soil improvement was brought by *Ailanthus excelsa* based silvi-pasture followed by *Albizia lebbek* based agri-silvi and *Acacia nilotica* based agri-silvi system.

iii. The functional ability of soil improved from 30% of its maximum potential at the start of experiment, to 77%, 70%, and 53% in 10 year with integration of *Ailanthus excelsa* based silvi-pasture, *Albizia lebbek* based agri-silvi and *Acacia nilotica* based agri-silvi system; respectively.

Conclusion

Soil quality refers to functional ability of soil and is a key factor of ecosystem's sustainability. Monitoring and evaluation of soil health is the key process to assess soil conservation benefits of agroforestry. Assessment of soil quality depends on selection of indicators and relating them to specific soil function. Indicators can be qualitative (ocular observation) or quantitative such as various physical, chemical and biological properties of soil. Reference or standard values of soil indicators of identified minimum data set are to be established to assess current status of soil quality. Due to complex array of positive and negative interactions; and temporal and spatial complexity of agroforestry systems; selection of indicators for assessing soil quality has to be done carefully. The management goals should guide selection of soil health indicators and development of a unified soil quality index of different agroforestry systems.

References

Amacher, M.C., O'Neil, K.P. and Perry, C.H. 2007. Soil vital signs: Quality index for assessing forest soil health. *USDA Forest Service Res Paper*. RMRS-RP-65.

Andrews, S.S., Karlen D.L. and Cambardella, C.A. 2004. The Soil Management Assessment Framework: A quantitative Soil Quality Evaluation Method. *Soil Science Society America Journal*, 68: 1945-1962.

Anonymous. 2001. Report on the Task Force on Greening India for Livelihood Security and Sustainalble Development. Planning Commision. Govt of India, Pp. 231.

Barbiroli, G., Casalicchio, G. and Raggi, A. 2004. A new approach to evaluate multifunctional soil quality index. *Journal Soils and Sediments*, 4 (3): 201-204.

Biswas, A.K. andBiswas, M.R. 1974. Environmental Considerations of Increasing World Foood Production. UNEP, Nairobi, Kenya.

Doran, J.W. and Parkin, T.B. 1994. Defining and Assessing Soil Quality, in *Defining Soil Quality for Sustainable Environment*. J.W. Doran, D.C. Coleman, D.F. Benzdick, and B.A. Stewart, Eds. SSSA Spec. Pub. No. 35, Soil Science Society America, American Society of Agronomy, Madison, WI, 1994, Pp. 3-21.

Doran, J.W. and Parkin, T.B. 1996. Quantiative indicators of soil quality: A minimum data set. P25-37. In J.W. Doran and A.J. Jones (ed) 1996. *Methods for assessing soil quality*. SSSA Spec. publ. No. 49. SSSA, Madison, WI

Oldeman, L.R. 1994.The Global Extent of Soil Degradation, in D.J Greenland and I. Szabolcs, Eds., *Soil Resilience and Sustainable land*, CAB, Innternational, Wallingford, U.K., Pp. 99-118.

Gregorich, E.G., M.R. Carter, D.A. Angers, C.M. Monreal, and B.H. Ellert. 1994. Towards a minimum data set to assess soil organic matter quality in agricultural soils. Can. J. Soil Sci. 74:367–385Lal R. 1998. *Soil Quality and Agricultural Sustainability* (eds). Ann Arbor Press, Chelsea Michigan

Karlen, D.L. Wollenhaupt, N.C., Erbach, D.C, Berry, E.C. and Swan, J.B. 1994. Long-Term Tillage Effects on Soil Quality. *Soil Tillage Research*, 32: 313-327

Karlen, D.L., M.J. Mausbach, J.W. Doran, R.G. Cline, R.F. Harris and G.E. Schuman. 1997. Soil quality: A concept, definition, and framework for evaluation. *Soil Science Society America Journal*, 61: 4–10.

Karlen , D.L. and Stott, D.E. 1994. A framework for evaluating physical and chemical indicators of soil quality, in Defining soil quality for a sustainable environment. J.W. Doran, D.C. Coleman, D.F. Benzdick, and B.A. Stewart, Eds. SSSA Spec. Pub. No. 35, Soil Sci. Soc. Am., Am. Soc. Agron., Madison, WI, 1994, Pp. 53-72.

Lal, R and Stewart, B.A. 1995. *Soil Management: Experimmental basis for sustainability and environmental quality (eds)*. CRC/Lewis Publishers, Boca Raton, FL

Lal R. 1998. *Soil Quality and Agricultural Sustainability* (eds). Ann Arbor Press, Chelsea Michigan

Larson Larson, W.E., and Pierce, F.J. 1991. Conservation and enhancement of soil quality. *In Evaluation* for sustainable land management in the developing world. Vol. 2: Bangkok, Thailand. International Board for Research and Management 1991. IBSRAM Proccedings No. 12(2): 1991, Pp. 175-203.

Larson, W.E. and Pierce, F.J. 1991. Conservation and Enhancement of Soil Quality, in *Evaluation for Sustainable Land Management in the Developing World*, Vol 2: Technical papers. Bangkok, Thailand; International Board for Research and Management, 1991. IBSRAM Proceedings No 12 (2) 1991, Pp. 175-203.

Mohanty, M., Painuly, D.K., Mishra, A.K. and Ghosh, P.K. 2005. Soil quality effects of tillage and residue under rice-wheat cropping on Vertisol in India. *Communication Soil Science Plant Analysis*, 36 (9): 1162-1176

Nair, P.K.R. 1993. *An Introduction to Agroforestry*. Kluver, Boston

Pandey, D. N. 2007. Multifunctional agroforestry systems in India. *Current Science*, 2 (4): 455-463

Roming, D.E., Garlynd, M.J, Harris, R.F. and McSweeney, K. 1995. How farmers assess soil health and quality. *Journal of Soil Water Conservation*, 50: 229-236

Sanchez, P.A., 1987. Soil productivity and sustainability in agroforestry sytems. In H.A. Seppler and P.K.R. Nair(eds): Agroforestry: A Dcade of Development. International Council for Research in Agroforestry, Nairabi, Kenya, Pp. 205-223.

Seybold, C.A., Halvorson, JJ., Karlen, D.L., and Rogers, H.H.1996. *Quantification of Soil Quality . Advances in Soil Science*, 1996.

SSSA (soil Science Society of America). 1987. *Glossary Soil Science Terms*. SSSA, Madison, WI. 1987

Warkentin, B.P. 1995. The changing concept of soil quality. *Journal of Soil Water Conservation*, 50: 226-228

Young, A. 1989. Agroforestry for Soil Conservation. CAB International, Oxford, U.K.

❑❑❑

Agroforestry: Natural Resource Sustainability, Livelihood & Climate Moderation © 2009
Editors: OP Chaturvedi, A Venkatesh et. al., pp. 563-577
Satish Serial Publishing House, Delhi

Chapter 40

MARKETING ISSUES AND CONSTRAINTS IN AGROFORESTRY

D.R. Palsaniya, S.K. Dhyani, R.K.Tewari, Ramesh Singh and R.S. Yadav
National Research Centre for Agroforestry, Jhansi-284 003 (U.P.)

Introduction

Agroforestry is an age old practice where farmers have been developing and managing agroforestry systems by nurturing trees on their farms, pasture lands and homesteads. However, on scientific lines, agroforestry in India is being promoted since the last 50 years, though the efforts are more concerted and focused since last two and half decades. Agroforestry is a dynamic, ecologically based, natural resources management system that, through the integration of trees on farms and in the agricultural landscape, diversifies and sustains production for increased social, economic and environmental benefits for land users at all levels. Agroforestry have many direct and indirect benefits. In case of annual crops the private benefits are high as compared to social benefits while in agro-forestry, social benefits are relatively high as compared to private benefits. Traditionally, agroforestry produces a wide variety of products such as timber, fuel wood, fruits, fodder, fibers, vegetables, spices, resins, and medicines, primarily to meet household needs and also to generate some income through sale in local markets. The supply of timber from natural forests has declined considerably. At present, imports of forest raw materials, promoted by industrial and trade liberalization, are filling the gap. The wood based industry is growing rapidly with increasing demand for furniture, housing, construction material, packaging, agricultural implements, sports goods, plywood, veneer, matches, and so on. The total wood based imports for various uses today is already over US$ 1 billion. Depending on the regeneration of wood and the changing demand profile, India is likely to face a wood deficit of 20 to 70 million cubic metres by 2020. Certainly, the case of agroforestry is obvious. Non timber forest products, also lends itself to medicinal and herbal applications, and is a very important source of sustenance for forest dwellers. It is estimated that over 75 % of forest export revenue comes from these sources. Approximately 40% of the forest products are supplied from outside forest areas. Similarly, the fuel wood supply from agroforestry is over 95 per cent. Most of the timber requirement is being meet from trees outside the forests. Furthermore, it provides employment to millions of people. Taken together, the wood based value chain, including plantations, has the potential to create sustainable livelihood for nearly 100 million people.

Agroforestry Based Industries

A number of industries like paper and pulp, plywood, particle board, fibre board, furniture, housing and construction material, agricultural implements, sports goods, packaging and matches box and value addition industries are based on wood which is contributed from forestry or agroforestry. Owing to rapid depletion of our forests a need was felt to conserve the natural forest resources by encouraging reconstituted wood products such as Plywood, Hardboard, Particleboard and Medium Density Fibre board (MDF) to meet the rising demand of wood from consumers including individuals, Railways, Defence, Furniture and Laminate manufacturers, builders etc. This led to a greater scope for agroforestry.

Pulp and Paper Industry

India is the major consumer of paper in the world. In India, paper is manufactured from a variety of raw materials like wood, recovered paper, bagasse, wheat straw, rice husk, etc. Wood is the main raw material for paper in India and the paper industry's per annum demand for wood is expected to grow from 5.2 million tones in 2000 to 13.2 million tones by 2020. The government source accounts for 20 percent, mostly from bamboo while non-government source covers 80 percent of the total supply. At present, India is deficit in paper production and nearly 50 percent of the industry's requirements are met through imports, which are increasing. Current imports amount to 2 million tones, worth US$500 million. A major deterrent to competitiveness and growth is the lack of raw material in the form of wood fibre. Unavailability of sufficient raw material prevents mills from growing in size to achieve economies of scale and attain global benchmarks. As an alternative, many major mills have begun sourcing their raw material from agro-forestry sources. This may boost agroforestry adoption in India.

Plywood Industry

Due to its multiple and widespread utility, plywood industry is one of the fastest growing wood based industry in India. The plywood industry in India is more than 75 years old. Industrial Licensing is governed by the Industries (Developmental and Regulation) Act, 1951 under which the wood based industry were covered. There was no restriction of putting up plywood Mill up to 1970. However, during the protective era, the wood based industry was under Compulsory Industrial Licensing from 1988 under which a small scale unit or a Large/Medium unit, new undertaking or substantial expansion would require Industrial License. After switching over to liberalization and privatization regimes, Government of India totally delicensed wood based industries in 1997 resulting in establishment of more than 2000 small-scale units.

Particle Board Industry

Particle board is a reconstituted constructional panel particularly developed as a substitute for natural constructional wood and is made from low grade waste woods or from certain agricultural ligneous wastes. It is widely being used for wall paneling and internal decoration of houses and offices. In India the first Particle board plant was set up in late 1950's at Sitapur in Uttar Pradesh. At present, there are about twelve wood based and agriculture ligneous waste based Particle board units in India.

Fibre Board Industry

Fibre board is constituted of sheet materials of widely varying diversities manufactured from refined or partially refined wood fibres or other vegetable fibres. Bonding agents and other materials may be incorporated in it to increase strength, resistance to moisture, fire or decay. The first Fibre board industry was established in India during late 1950's.

Furniture, Housing, Construction Material and Agricultural Implements

Given the major environmental advantage of wood as a carbon sink, greater usage of wood in Furniture, housing and construction material is a better alternative than the use of cement and steel which have much larger ecological footprints. By 2010-12, more than 500 million sq.ft. of space is expected to be built in India, mostly in urban areas. Even a limited use of wood for door and window frames, flooring, wooden cabinets etc. would mean a demand for wood products valued at US$ 3 billion. With greater usage of wood as a predominant material for housing and construction material, in rural and urban areas, this potential can increase exponentially. Today, the Indian furniture market is estimated at US$ 8 billion and employs around 3,00,000 people. Wooden home furniture is the largest segment. Imports of furniture are growing at a fast clip with a rate as high as 60 percent.

Bio Chemical Industries

Trees are also a valuable source of bio chemicals. The cellulose part of the structure, used primarily in pulp and paper industry, represents only 20 to 30% of the total, and the balance can be a valuable resource for extraction of bio chemicals with wide applications from food to medicine. These non timber forest products are also very important source of sustenance for forest dwellers and contribute over 75 % of total forest export revenue.

Sports Goods, Packaging and Matches Box

Most of the sports goods, packaging materials and matches boxes are made from wood and their demand in future is likely to increase indicating greater potential of agroforestry to produce required raw material. They also provide employment to a huge population.

Marketing Issues and Constraints in Agroforestry

Marketing of surplus produce is an important issue before the farmers to get adequate price. For crop products, there are enough marketing options available but for tree based products such options are either absent or not enough. Markets for tree products are important to small, resource-poor farmers in their integration into a market economy. In these markets, however, farmers encounter competition, and policy constraints make it difficult for them to compete. The cumbersome and costly bureaucratic procedures lead the farmers to depend on intermediaries to market their produce. Relatively low returns to agro-forestry largely result from inappropriate policies, complicated transit rules, unfavorable institutional environment and bureaucratic hindrances. In most Indian states, farmers can only harvest and transport wood products after a long and cumbersome process to obtain permits. Further, there is a ban on timber export and restrictions on interstate movement of high quality wood.

So far there is no adequate government marketing policy for tree based products. Wood marketing is governed by rigid and age old forest laws and legislations. Government of India announced Vishesh Krishi Upaj Yojana for enhancing employment and boosting export of flowers, fruits, vegetables, minor forest produce and their value-added products, thus, implicitly recognizing agroforestry potential (Foreign Trade Policy, 2004-09). Smallholder agroforestry practices are of increasing importance in both sustainable food production and safeguarding environmental services such as enhancing tree cover, biodiversity conservation and carbon sequestration. There is rising demand of wood from consumers including individuals, railways, defense, furniture and laminate manufacturers, builders etc. Marketing is an important issue which affects the decision making and ultimately determines the adoption of agroforestry.

Unique and Distinctive Features of Wood Marketing in India

In agroforestry, marketing is unique for several reasons and we can identify several distinctive features of wood marketing *vis-à-vis* food grain marketing in India.

Price Policy

Government of India has a well defined price policy for food grain crops and other cash crops. Government announces minimum support price (MSP) for different crops at the start of the growing season every year. In case of bumper production and when the market prices falls below MSP, farmers can sell their produce to government agencies like food corporation of India (FCI) at MSP and thus, they can save themselves from market vagaries. Government also announces procurement price at the time of procurement which may be even higher than the MSP. However, such price control mechanism is absent in case of tree based products particularly wood. In India, wood marketing is governed by rigid and age old forest and environmental laws and policies wherein the farmer's rights to trees on their own lands are vague and ill defined.

Marketing Information

Farmers in India can obtain information regarding prevailing market prices of food crops and cash crops at markets or Mandis across the country or state from multiple sources like news paper, radio and television on daily basis. Accordingly, they can take marketing decision for their produce. As per the Farmers Commissions recommendations, Government of India is also planning to establish a village resource centre in every village of country where farmers can have access to such informations at their doorstep. However, such marketing information is not available for tree based products mainly the wood. Farmers sell their wood to local merchants at the price they offer.

Market Access

In case of grain crops, vegetables and fruits good market avenues are available at village, local, regional, state, national and international levels. Farmer can sell their produce at any of the above levels as per his capacity. However, in case of wood marketing, there is poor market access and the farmers contact is limited to village buyers only. Further, they are exposed to large number of middlemen and their high profit margins.

Credit Availability

Credit for agriculture is available both from institutional and non institutional sources. Moneylenders are the main player in the informal sector in India who are as old as its villages. In formal sector, the various institutions are the commercial banks, regional rural banks and the cooperative credit structure (CCS). The short term rural credit needs are fulfilled by the State Cooperative Banks (SCBs), District Central Cooperative Banks (DCCBs) and the large network of the Primary Agricultural Credit Societies (PACS) in the villages. The CCS is refinanced by the National Bank of Agricultural and Rural Development (NABARD). It is easy to get loans and insurance cover for raising food crops and cash crops but these facilities are not available on the basis of wood resource which is a major hindrance for adoption of agroforestry.

Rigid Forest and Environmental Laws

Wood marketing in India is governed by rigid forest and environmental laws wherein the farmer's rights to trees on their own lands are vague and ill defined. Farmers can plant trees on their lands but they can only harvest and transport wood products after a long and cumbersome process of obtaining permits. However, in the states of Haryana and Punjab there is no such regulation and some other States have also exempted agroforestry species. Further, there is a ban on timber export and restrictions on interstate movement of high quality wood. However, other agricultural products are not bound by such rigid forest and environmental laws and other restrictions making their cultivation hassle free. There is an urgent need of separate laws and policies for agroforestry.

Value Addition

Value-adding is the process of increasing the value of a raw product anytime between harvesting and sales of the final product. Value-addition is an important activity in agriculture which offers farmers a higher return, open new market avenues, create brand recognition and add variety to farming processes. Value-addition has the potential to capture a larger share of the consumer market and to increase the profit margin of the farmers besides increasing employment opportunities. The typical post harvest and value-addition includes processing in some way such as cleaning, sorting, grading, cutting, packaging, smoking, drying, freezing, extracting, preserving, etc. Post harvest and value-addition infrastructure is more common and widely used in agriculture while it is less common and poor in case of wood and other tree based products.

Infrastructural Facilities

Lack of adequate basic infrastructure, both social and physical, continues to remain a major obstacle in the development of numerous villages and their habitants. Both public and private investment in agriculture is not only but also decreasing year by year. The private investors are reluctant to invest in agriculture and other rural infrastructure projects due to fear of relatively low returns on their investment and uncertainties in realizing revenue from the poverty-stricken rural users. Further, if we compare infrastructural facilities available for agricultural crops and forest based products, the former is in better position. It is easy to find grain storage facilities but we can not find wood storage facilities. Woods are stored in open and are exposed to various degradation forces like environmental factors and insects.

Grading and Quality Standards

There are enough national and international grading or quality standards are available for grain crops, cash crops, vegetables and fruits. However, these are poor and not well defined in case of wood and other tree based products.

Understanding Marketing Forces and Market Analysis

Market analysis and understanding of market forces is very much essential for designing and making effective marketing strategies. We should carefully study, analyse and understand the five market forces namely, potential entrants, potential suppliers, potential buyers, potential substitutes, and potential competitors. By understanding these unique and distinctive features of wood marketing in India and competitive forces within agroforestry, market strengths, weaknesses, opportunities and threats can be identified and successful strategies can be developed. The following seven secrets of successful marketing should be kept in mind while making agroforestry marketing strategies.

The Seven Secrets of Successful Marketing

1. Know your customer in terms of age, sex, caste, religion, food habits, taste, preference, size, etc.
2. Know your product in terms of quality, quantity, perishability, substitution, etc.
3. Know the process of production, manufacturing and distribution.
4. Know your costs in terms of input, production, processing, value addition, transportation, storage, etc. so as to minimize it.
5. Know your competitive edge over others.
6. Know the communication process for effective advertisement and publicity.
7. Know yourself to develop self confidence.

With the careful market analysis, we can say that enough commercial opportunities exist for farm communities to transform their traditional agroforestry systems towards market orientation. Most farmers produce small quantities of many products primarily for household consumption with limited market sales, which they often sale at the local markets. Forest product producers and gatherers often cannot sell what they have produced to the processors directly. Thus, large processors are unable to get sufficient quantities of raw materials. This is an indication of a frequent mis-match between demand and supply and is believed to be due to lack of sufficient production of quality products and unavailability of timely good market information with both suppliers and buyers for correct decision-making. Smallholder farmers must develop intensive agroforestry management systems designed to yield quality products from priority species that meet market specifications. Further, the market informations based on primary survey data should be disseminated among the private and public agents including gatherers, growers, processors, traders, buyers, and government decision-makers to help them in reducing, locally and nationally, the mis-match between demand and supply. The following factors help in development of market oriented smallholder agroforestry systems:

i. Security of land tenure/use conditions.
ii. Supportive policy framework/conditions.
iii. Access to and knowledge regarding the management of quality planting material.
iv. Tree management skills and information and
v. Adequate market information and linkages.

Sound Marketing Strategies in Agroforestry

Market forces, as they stand today, do not adequately reward long term objectives of building a sustainable economy. A more enduring solution will emerge only when policy makers and market participants assign a larger priority to promoting economic activity in a way that does not compromise the interests of the future for short term gains of the present. By understanding the above scenario of marketing problems and prospects in agroforestry, successful marketing strategies can be developed. Economic viability is an important aspect of agroforestry development. The marketer must be able to identify how different sectors of the market vary from each other, and how the different requirements of each sector may be targeted. The successful marketing stems from sound knowledge, quality products, innovative distribution channels and successful publicity and communication processes. Following suggestions can be adopted to develop a sound agroforestry marketing strategy.

Promotion of Co-operatives in Agroforestry

Cooperative marketing is viewed as an important solution of marketing problems being faced in agroforestry. It is appropriate for farmer groups to assume a marketing role through proactive and cooperative involvement with willing market agents. The farmers cooperative societies may be formed which may focus on one or more of the following marketing activities:

- Production activities like field preparation, sowing and harvesting.
- Transportation, storage, wholesaling and other mid-channel activities.
- Processing and value addition in raw materials and manufacturing finished goods.
- Organizing agro-forestry extension campaigns.
- Transferring relevant technology and impart necessary training.
- Organizing supplies of propagation material.
- Making necessary institutional and fiscal arrangements.

The Nashik District *Eucalyptus* Growers' Co-operative Society is a famous example of involvement of cooperatives in agroforestry production and marketing in India.

Improving Rural Credit System

The rural credit delivery system should focus on strategies that are required for tackling issues such as sustainability and viability, operational efficiency, recovery performance, small farmer coverage and balanced sectoral development. Agroforestry should receive adequate credit from all sources like commercial banks, regional rural banks and the cooperative credit societies (CCS) on priority basis. Further, SHGS should be formed in agroforestry sector also as they show lower transaction costs and higher recovery performance. These groups are also reported to have favourably impacted the social and economic status of their members. Insurance cover should also available on the basis of wood resource to minimize risk.

Providing Policy Support through Institutional Reforms

Many provisions of the forest-related legislations in India have acted as serious disincentives to tree farming on private lands. The spirit of existing legislations in this sector is geared primarily to curtail the felling of trees, as opposed to facilitating the growing of renewable forests. Wood marketing in India is governed by rigid forest and environmental laws wherein the farmer's rights to trees on their own lands are vague and ill defined. Under the 1981 Rural Development and Rural Employment Guarantee Programmes, tree planting incentives include subsidies, exemption from sales tax and direct planting on small farms and marginal lands. The National Forest Policy of 1988 states that as far as possible, the forest-based industry should work with individual suppliers supported with credit, technical advice and final harvesting and transport services for procuring raw materials. Under the 1999 National Forestry Action Programme the annual afforestation/ rehabilitation of two million hectares were planned, 10 percent of which would be "farm forestry". Farmers can plant trees on their lands but they can only harvest and transport wood products after a long and cumbersome process of obtaining permits. Though, the import policy for wood and wood products has been liberalized since the mid-1990s, yet, there is a ban on timber export and restrictions on interstate movement of high quality wood. The import of timber under Open General License (OGL) and the inconsistencies in inter-state timber trade/transit rules have upset the wood production by farmers. This calls for evolving appropriate policy packages to popularize agroforestry, covering aspects such as harvesting, processing, marketing, and utilization of farm-grown wood. Government of India announced Vishesh Krishi Upaj Yojana for enhancing employment and boosting export of flowers, fruits, vegetables, minor forest produce and their value-added products, thus, implicitly recognizing agroforestry potential (Foreign Trade Policy, 2004-09). There is an urgent need of separate laws and policies explicitly for agroforestry in this regard. Adequate policy protection is much needed in agroforestry related investments because of long gestation periods involved in yielding adequate returns on investment.

Promotion of Wood Based Rural Enterprises

Promotion of wood based rural enterprises will not only encourage agroforestry adoption but also to total income and employment in the rural areas. Farmer-based enterprise encompasses any activity that contributes to farmers' livelihoods or incomes. Possibilities for a variety of rural enterprises based on agroforestry are not only growing and harvesting, but also value-added processing, packaging, and transportation. Many small-scale ventures may be initiated with adequate information and planning and can enter markets selling. Potential agroforestry based entrepreneurs should complete four evaluations before investing in a new venture. These include a personal evaluation, a resource evaluation, a market evaluation, and a project feasibility evaluation.

Providing Infrastructural Facilities

Development of infrastructural facilities in rural areas can promote the development of industries which in turn encourage taking up of agroforestry based ventures. The basic requirements of water, power, transportation, communication, roads, storage, value addition, etc. can be provided to promote agroforestry based ventures. Further, the forces of globalization have created opportunities for the integration of rural populations in a larger marketplace than was ever available to them before. Similarly, revolutions in the Information and Communications Technologies (ICT) have the potential to remove the barriers to information asymmetries that were impeding the working of markets that are critical for economic growth.

Agroforestry Management

Sound agroforestry management is required to produce a raw material of desired quality and quantity. Knowledge of project goals, commercial requirements, planting site conditions, and the range of potentially suitable species is essential for such sound strategies. The productivity of such products in agroforestry systems can be improved by adopting suitable agrotechniques and enhancing management skills of the farmers which can be achieved through following strategies.

i. Selecting suitable tree species.
ii. Supply of quality saplings.
iii. Selection of suitable agroforestry systems that match farmer's land, labour, capital and socioeconomic limitations - including annual crops, tree crops, intercropping and under storey cropping options.
iv. Sound tree management options to produce high quality products.
v. Proper nutrient, water, weed, pest and disease management.
vi. Participatory training of farmers for capacity building and management skills regarding nursery development, improved agrotechniques, awareness and leadership.
vii. Use of modern techniques like biotechnology for a promising opportunity in plantation forestry. It possesses the same productivity advantages of tree improvement such as higher yields, better wood quality, lower risk associated with pests and disease management.

Value Addition

Value-addition offers higher return, open new markets, and create brand recognition and higher employment opportunities. Agroforestry offers ample opportunities of value addition at any time between production and harvesting to sale of the final product through a variety of services or processes like cleaning, cutting, sorting, grading, packaging, smoking, drying, freezing, extracting, or preserving. There is need of increase in value addition infrastructure for this purpose.

Extension

India has an established scientific and extension machinery to support agroforestry adoption; however extension agents have yet to master the socio-economic tools to analyze communities and enhance local social structures. Extension helps to empower and motivate farmers to enhance and diversify the productivity/ profitability of their agroforestry systems and also in strengthening their capacity to seize market opportunities, both existing and developing. Extension agents can help in greater adoption of agroforestry by educating the farmers in techniques like seedling propagation and nursery management, tree and agroforestry system management, farmer-market linkages, and farmer-operated commercial enterprises.

Simple Carbon Credit Trading Policy for Agroforestry

Agroforestry is being seen as one of the most important instrument of climate change mitigation. The Prime Ministers council on Climate Change recently unveiled its National Action Plan. As part of 8 missions announced, the National Mission for a Green India envisages the afforestation of 6 million hectares to enhance forest and tree cover to 33%. Agroforestry will be an important component of this. The plan has further suggested in its Mission for Enhanced Energy Efficiency, a market based mechanism that would certify energy savings that could be eventually traded. In a similar fashion, to promote biomass energy from agroforestry, the differential in cost of generation between the biomass energy and the least cost generation from fossil fuel based energy sources could also be reimbursed in the form of energy credits, which can be traded. Needless to state that an appropriate regulatory frame work would be required to mandate purchase of energy credits by polluting industries.

Promoting Contract Farming in Agroforestry

It is obvious that innovative public-private-people partnership will have to be forged to make a meaningful difference in respect to common issues of inclusive growth, sustainability, livelihood security, energy security, food security and environmental security. Corporates, on their part, are uniquely placed to play a constructive role in this process. Positioned, as they are, in the frontline of engagement with civil society, corporates have the wherewithal, including the more crucial managerial resources, to deliver social projects efficiently at a lower incremental cost. Corporates engaged in processing and marketing of agricultural products are entering into contract with the farmers wherein the farmers are provided inputs and required expertise to plant the contractor's crop on his land on predefined terms and conditions. They buy back the product at a pre decided price. The terms and conditions of the contract are well defined and spelt out in advance and a consent letter is obtained from both the parties. Contract farming has manifold advantages for every stakeholder- farmer, contractor, consumer and the society as a whole in terms of income, security, productivity, quality, employment, etc.

Conclusion

There is an urgent need for evolving appropriate policy packages to popularize agroforestry, covering aspects such as harvesting, processing, and utilization of farm-grown wood, as well as ensuring credit and extension services to smallholder producers. By understanding these competitive forces within an industry, market strengths, weaknesses, opportunities and threats can be identified and successful strategies can be developed. It is appropriate for farmer groups to assume a marketing role through proactive and cooperative involvement with willing market agents. They may focus on one or more of the following activities: i) transportation, wholesaling and other mid-channel activities ii) processing raw materials and manufacturing finished goods. A variety of wood based rural enterprises involving not only growing and harvesting, but also value-added processing, packaging, and transport are available and can contribute to farmers' livelihoods or incomes. Development of infrastructural facilities in rural areas can promote the development of wood based industries which in turn will encourage taking up of agroforestry tree crops. Public private partnership should be promoted through appropriate incentives. The forces of globalization and liberalization have created opportunities for the integration of rural populations in a larger marketplace than was ever available to them before. Government and NGOs can play a crucial role in imparting training to farmers in cultivation, value addition and trade of agroforestry tree produce through extension related activities. However, we have to keep patience as agroforestry is more a livelihood sustaining and strengthening system whose potential on commercial scale is yet to develop or establish in case of India.

Concerted efforts to promote wood plantations in farmer's field in India could potentially change the fortunes of small and marginal farmers and consequently the opportunities in rural India. It would be desirable to extend to agroforestry all the benefits applicable to agriculture, so that farmers can be encouraged to cultivate their private lands including wastelands with equivalent benefits like price support, unrestricted movement of such produce, long term credit support on priority basis, and so on. To promote public private partnership and enable larger corporate involvement in renewable agroforestry, and to enhance linkages for commercial application, industry must be given appropriate incentives. Simple carbon credit trading policy should be developed for the agroforestry sector separately to enhance area under it.

References

Anonymous, 2008. The Hindu, August, 2, 2008.

Anonymous, 2004-2009. The Foreign trade policy 2004-2009, GOI, New Delhi

Anonymous, 2007. Current affairs, competition wizard, Career Classics, New Delhi.

Anonymous, 2008. India 2008, Publication Division, GOI, New Delhi.

Gupta, S.K. 2002. Contract farming for rural development. *Kurukshetra*, February 2002: 13-19.

Palsaniya, D.R., Yadav, R.S. and Tewari, R.K. 2008. Marketing issues and constraints in agroforestry. Abstract: National symposium on "Agroforestry knowledge for sustainability, climate moderation and challenges ahead, December 15-17, 2008, NRCAF, Jhansi, Pp. 197.

❑❑❑

Agroforestry: Natural Resource Sustainability, Livelihood & Climate Moderation © 2009
Editors: OP Chaturvedi, A Venkatesh et. al., pp. 579-588
Satish Serial Publishing House, Delhi

Chapter 41

ROLE OF MARKET INTELLIGENCE AND COMMODITY FUTURES IN RISK MANAGEMENT AND PRICE DISCOVERY

A.K. Vasisht
Indian Agricultural Statistical Research Institute, New Delhi-110 012

Introduction

Formalized future trading in agricultural commodities in India has been in place since 1918-19. But the futures trading in commodities through commodity exchanges, especially in agricultural commodities, has started only very recently. After India embarked on economic liberalization policies and signed the GATT agreement in the early nineties, the government realized the need for futures trading to strengthen the competitiveness of Indian agriculture and the commodity trade and industry. Futures trading in several commodities was revived by the government to effectively and efficiently serve its twin economic functions of price discovery and price risk management and to provide farmers alternative markets for enabling them to take right decision for sowing, harvesting, and selling at the right market at the right time. There are three main commodity exchanges, namely, the Multi Commodity Exchange, the National Commodity and Derivatives Exchange and the National Multi Commodity Exchange. These exchanges are permitted to trade in as many commodities as they chose. However, it is mandatory for them to establish electronic trading platforms for anonymous automatic order matching, which lead to transparency in futures trading organized by these exchanges.

System of Regulation of Forward Trading in Commodities

There are three tiers of regulation of forward trading in commodities viz. the Central Government, Forward Markets Commission and the Recognized Commodity Exchanges / Associations. The Central Government broadly determines the policy as to the commodities in which forward trading is to be permitted as well as the territorial area in which the same is to be permitted and the recognition of the Exchange/Association through which such trading is to be permitted. The Forward Markets Commission performs the role of approving the Rules and Regulations of the Exchange, subject to which trading is to be conducted, accord permission for commencement of trading in different contracts, monitor market conditions continuously and take remedial measures whenever required.

Role of Commodity Exchanges

The markets for various agricultural commodities are widely dispersed and are located at great geographical distances. The communication channels between markets participants are also poor and thus the prices are determined by a handful of traders with a partial knowledge about the market fundamentals covering a limited geographical area. In the traditional marketing system, particularly in respect of products which were produced in few pockets of the country and whose consumers are spread throughout the country, there was a poor price relationship among the markets as in the case of wheat, Chana, and Urad. The value of the correlation coefficients between some of the markets was 'nil' meaning that over-supply in a producers' market often had no or little effect on the consumer's market or vice-versa. Thus, the middlemen used to remain as the "price setters" in the spot markets of the past. They used to take advantage of their partial knowledge about market fundamentals, and thus ignore other stakeholders in the market.

Economics of Commodity Futures Market

Forward contracts have historically been a feature of commodity markets, partly because while the demands of actual users of many commodities is continuous across the year, supplies come into the market at specified points in time such as harvest time. Producers of commodities have to make production decisions based on expectations of the price that they would receive when the output arrives, and purchasers of commodities as inputs or final goods make judgments of the availability and cost of the commodity at different points of time during the year. To guard against price volatility and uncertainty in production and availability, sellers and buyers often enter into forward contracts, specifying the quantity, quality and price of the commodity they would deliver for sale or acquire for purchase at a pre-decided date in the future.

Thus forward contracts are means of hedging against the risk of price volatility or uncertainty of supply. Futures contracts differ from forward contracts in the sense that futures contracts are standardized contracts to buy or sell a standard quantity of a standard quality of a commodity. These are traded in exchanges, through brokers, with no need for the buyer and seller to meet and negotiate. An important feature is that a contract need not be settled by actual delivery. It can be matched by an offsetting contract taken by the buyer or seller, and the two can be squared at any point at some gain or loss. The administration of the exchange guarantees that contracts would be settled, and requires traders to pay up margins to cover ongoing losses, if any, to secure the viability of the exchange. To avoid paying margins, traders can buy an option to offer or acquire a contract at some specified future date. If the option is not exercised, because price movements are contrary to expectations, the loss is restricted to the premium paid to hold the option and the transaction costs of acquiring it.

Futures contracts are purchased when the investor expects the price of the underlying security to rise. This is known as going long. Because the holder has purchased the obligation to buy goods at the current price, he will profit if the price goes up, allowing him to sell his futures contract for a profit or take delivery of the goods on the future date at the lower price. The opposite of going long is going short. In this case, the holder acquires the obligation to sell the underlying commodity at the current price. He will profit if the price declines before the future date.

Futures trading — taking place on a transparent, electronic exchange with nationwide access — brings in a host of new players. In fact, there are hundreds of people in India who are watching world markets, processing information, and putting trades onto overseas commodity futures markets. Physical proximity to the market becomes a non-issue once an electronic exchange is in operation. Someone in Maharashtra can be trading guar seed, even though he may have never been to Bikaner which is the traditional trading centre for guar.

Numerous studies have investigated how farmers should use forward pricing markets, but only limited research exists on how farmers actually use these markets. Less government intervention in commodity markets has heightened farmers' exposure to price risk and has created a stronger desire among farmers to attempt to manage this risk by using forward pricing markets. Many researchers have derived optimal hedging rules, wherein the hedge ratio has been derived based on the ratio of the covariance between cash and futures prices to the variance of futures prices. As such, it is easily estimated from a simple regression where cash prices are regressed on futures prices.

Role of Futures Trading

Futures trading in commodities plays two important roles of price discovery and price risk management. Future contracts are often relied upon for price discovery as well as for hedging. It gives an advance price signal of what is expected price based on which one can decide the course of action to be taken by him. The prices of the commodities are governed by the supply and demand factors and also by the number of market players participating in buying and selling activity due to the existence of diverse domestic and global information on the price of the traded commodity. These factors result in an efficient price discovery because of large number of buyers and sellers transacting in the futures market. In many physical commodities (especially agricultural commodities), cash market participants base spot and forward prices on the futures prices that are 'discovered' in the competitive, open auction market of a futures exchange. Hedgers who minimize their risks get attracted in futures market. Forward and futures trading is promoted on the ground that it helps traders deal with market uncertainty by hedging their transactions, and stabilise prices for the final producers. There may be other traders who have market information and price judgement. This, in turn, encourages competition. The traders hold an immediate knowledge of the market, whereas hedgers hold a long term view. Futures market helps in efficient price discovery of the respective commodities and does not impair long-run equilibrium price of commodities.

Need for Price Risk Management

Commodity exchanges also provide a specialized risk management tool for the farmers, producers, traders and the consumers encompassing the entire ecosystem. There are price risks in most commodities and hence there is a need of 'hedging'. The larger, the more frequent and the more unforeseen is the price variability in a commodity, the greater is the price risk in it.

The insurance companies offer suitable policies to cover the risks of physical commodity losses due to fire, pilferage, transport mishaps, etc., but they do not cover similarly the risks of value losses resulting from adverse price variations. The reason for this is obvious. The value losses emerging from price risks are much larger and the probability of the recurrence is far more frequent than the physical losses in both the quantity and quality of goods caused by accidental fires and mishaps, or occasional thefts.

Commodity producers, merchants, stockists and importers face the risks of large value losses on their production, purchases, stocks and imports from the fall in prices. Likewise, the processors, manufacturers, exporters and other market functionaries, entering into forward sale commitments in either the domestic or export markets, are exposed to heavy risks from adverse price changes.

The price variability may also lead to high level of gains, when prices move favorably. In the long run, such gains may even offset the losses from adverse price movements. The greater the exposure to commodity price risks, the greater is the share of the commodity in the total earnings or production costs. Hence, the need for price risks management or hedging through the use of futures contracts. Hedging involves buying or selling of a standardized futures contract against the corresponding sale or purchase respectively of the equivalent physical commodity. It performs the economic function of helping to reduce significantly, if not eliminate altogether, the losses emanating from the price risks in commodities. The benefits of hedging flow from the relationship between the prices of contracts (either ready or forward) for physical delivery and those of futures contracts.

Commodity Profile

During the year 2007-08 foodgrains production is estimated at 227.32 million tonnes during 2007-08 including the rice production of 95.68 MT and 76.68 MT of wheat, according to the Third Advance Estimates of production of major crops grown in the country, released by the department of agriculture and cooperation (Table 1). The Coarse cereal production is estimated at 39.67 MT and of pulses at 15.19 million tonnes. The Oilseeds production during 2007-08 is estimated at 28.21 million tonnes with groundnut estimated at 8.87 million tonnes, soyabean at 9.43 million tonnes and rapeseed and mustard at 6.43 million tonnes. Production of cotton is estimated at 23.19 million bales of 170 kg each. Production of sugarcane is estimated at 344.23 million tonnes during 2007-08.

Compared to 2006-07, rice production is estimated to increase by about 2 million tonnes, wheat by about 1 million tonne, coarse cereals by about 6 million tonnes (mainly contributed by maize) and pulses by about 1 million tonne during 2007-08. The oilseeds production is estimated to increase by about 4 million tonnes (mainly contributed by groundnut) and cotton by about 6 lakh bales during 2007-08 over 2006-07.

Table 1. Food grains Production (Million tonnes)

Crop/Year	2003-04	2004-05	2005-06	2006-07				2007-08	
	Final estimates		Target	Second	Final Estimates	Target Estimates		Second	3rd Advance Estimates
Estimates									
Rice	88.5	83.1	91.8	92.8	90.1	93.4	93.0	94.1	95.7
Wheat	72.2	68.6	69.4	75.5	72.5	75.8	75.5	74.8	76.8
Coarse Cereals	37.6	33.5	34.1	36.5	32.0	33.9	37.5	36.1	39.7
Pulses	14.9	13.1	13.4	15.2	14.5	14.2	15.5	14.3	15.2
Total Food grains	213.2	198.4	208.6	220.0	209.2	217.3	221.5	219.3	227.3

Source: Ministry of Agriculture; Govt. of India

There has generally been a shortfall in the achievement of target of food grains, pulses and oilseeds during 2000-01 to 2006-07. The actual production of food grains on an average was 93 per cent of the target. Actual production, however, was only 87.7 per cent of target for pulses and 85.3 per cent of target for oilseeds. Production of sugarcane and cotton, however, over achieved their respective targets in 2005-06 and 2006-07 (Table 2).

Some of the major agricultural commodities traded in the Commodity Futures Market are rice, wheat, jute, gur, cotton, coffee, major pulses like urd, arhar, chana, edible oilseeds like mustard seed, coconut oil, groundnut oil, sunflower, and spices (pepper, chillies, cumin seeds and turmeric) among others. By the end of December 2006, 94 commodities were traded, as compared to 59 in January 2005. Guar seed accounted for 11 per cent and chana 10 per cent of the trade in terms of value. The prices of most of the commodities continued to be under pressure for most part of the year. Recently futures trading in pulses (urad and tur) had been banned to ensure orderly market conditions, yet prices remained high. Historical trend reveals that the prices are supply sensitive. The consumer cost expands in large part due to the multiple levels of intermediation and inefficiencies in marketing, transportation, and storage. It is difficult to point to any particular causal agent for the current increase in the prices of pulses.

Table 2. Actual production relative to targets (per cent)

Crop	2001-02	2002-03	2003-04	2004-05	2005-06	2006-07	Average (2001-07)
Rice	101.5	77.2	95.2	88.9	104.5	100.6	94.7
Wheat	93.3	84.3	92.5	86.3	91.8	100.4	91.4
Coarse cereals	101.1	79.0	110.6	90.9	93.3	92.9	94.6
Pulses	89.1	69.6	99.4	85.8	88.4	93.7	87.7
Foodgrains	97.6	79.4	96.9	88.1	97.0	98.8	93.0
Oilseeds	37.8	55.0	102.0	93.0	105.2	82.6	85.3
Sugarcane	91.4	89.8	73.1	87.8	118.4	131.7	98.7
Cotton	68.9	57.5	91.5	109.5	112.1	122.3	93.7
Jute & mesta	106.2	94.0	93.1	87.1	96.1	99.9	96.1

*Economic Survey 2007-08

But there seems to be a deteriorating demand – supply mismatch for most of the pulses during the last decade. The demand of pulses has been growing steadily with an increasing population and with higher income levels of the people; its consumption across all sections of the society is going to rise. The availability of pulses has fallen to half the level of the 1960s, though pulses remain the primary protein source for a large section of the Indian population that is vegetarian. Thus, there could be an absolute increase in demand pulling up the prices. The twin factors of supply push inflation and demand pull inflation are putting an extra pressure on the prices of all commodities in general. The domestic consumption of pulses, which is estimated to be around 15-16 million tonnes per year, is steadily increasing probably due to rising incomes. The price of pulses tends to be highly sensitive to supply shortfalls as it is having limited price elasticity. The import of pulses has been steadily increasing over the years. It was 1.3 MT in 2004-05, 1.6 MT in 2005-06 and up to January 2007, India imported 1.7 million tonnes of pulses.

Future Trading Vs Price and Inflation

It has been the stand of the policy makers that futures trading lead to speculation and introduced price instability in the spot market, which would go to the disadvantage of the farmers and the consumers in the country. Domestic markets in essential agricultural commodities, especially the food grains, therefore were kept insulated from the influence of the international / border prices and the role of futures market in risk management and price discovery as a reference price in the spot market was almost totally denied.

There was a temporary ban on futures trading of agricultural commodities in two phases from the start of the year 2007. During January the Government banned futures trading in urad and tur dal, while it was banned on wheat and rice at the end of February. Recently, the Government banned futures trading in four more commodities namely, chana, soy oil, potato and rubber. In spite of the Government's ban on futures trading in dal to curb inflation, the prices had not fallen. Several studies and opinions are available on positive and negative influences of trading in future derivatives and options in agricultural commodities, on the efficacy of price discovery, price stability and risk management.

The Forward Market Commission admits that the futures trading of a commodity would not really impact their prices. The futures market provides a platform for price discovery and risk management. It is nothing but a forecast of likely demand and supply and prices at a future point of time. It does not impact prices. The ban was affected because of perceptions that markets and speculations were leading to the rise in prices. The high prices, in all probability, may be due to a mismatch in supply and demand of these two commodities. In fact, official figures also support that. The production of pulses has stagnated in the last few years.

It is reverse in the case of wheat and rice ban as the wheat prices have started falling. The downslide in wheat prices continued unabated on persistent offerings by stockists against slowdown in buying by rolling flour mills after Forward Markets Commission announced the ban. There is not much activity in future trading in rice and thus the price of this commodity has remained almost steady. Thus, it cannot be concluded that futures trade is one of the reasons for the hike in inflation. An efficient and well-organised commodities futures market encourages competition by attracting traders who hedge their bets and minimise risks on the basis of their own market information and price judgment. The utility of commodity exchanges can be enhanced by regulating trading activities through them and by ensuring transparency as well as timely flow of information to the market players. This will ensure efficient price discovery and minimise price shocks triggered by unanticipated supply- demand mismatches.

The derivatives trading is almost completely backed by the estimated physical availabilities. But futures multiplier is normally only a small fraction of the total availabilities in case of agricultural commodities. This is mainly because of three reasons: i) a very small portion of the total production is available for trading, ii) future trading is further confined to a very small proportion of marketed surplus, and iii) periods of transactions in futures is generally between six to nine months.

Role of Market Intelligence

Economic and Market Intelligence is an important element in the dynamics of commodity marketing. There are cells for this purpose in national exchanges. Commodity prices are determined by the factors affecting the demand for and supply of commodities. These factors are many, and they vary from commodity to commodity and from day to day. These include the weather conditions, the availability of inputs and cost, the conditions of crop growth, the pace of market arrivals, the stocks at different stages of marketing and consumption, the pattern of demand for investment and end uses, the import and export requirement, the demand for substitutes, the prices prevailing in other domestic and overseas markets, the rate of inflation, the interest and foreign exchange rates etc. The price and cross-price elasticity of supply and demand as well as the income elasticity of demand also affect the commodity prices.

In order to perform the price discovery function efficiently, a commodity market must receive the requisite data and information on all the factors affecting supply and demand promptly. The swift collection and dissemination of all economic and market intelligence is a pre-requisite for all dynamic commodity markets so that the market players can formulate their own judgements on the present and prospective prices. This will enable the markets to discover an equilibrium price of a commodity for the present and future dates, which can serve as a reference price for all other entering the physical contracts for distant deliveries in the domestic and overseas markets.

Proposed Mechanism of Trading and Farmer's Participation

The procurement operations of Food Corporation of India were meant to serve two purposes; i) to provide fair price to the farmers through the "minimum support price" mechanism and to create a buffer for food security. After the Green Revolution in the 1960s, there was surplus agricultural production and hence farmers needed to be protected against a sharp fall in prices. Today, India is moving towards a situation of shortage, and the prices are firm. Farmers could get fair prices by participating in the futures market that is enabled by the commodity exchanges. By using the exchanges, FCI could avoid buying only at the harvest time and holding the stock in its godowns if it buys sufficient stocks for, say, one quarter. Farmers will benefit because they do not have to sell all their produce soon after the harvest, when the prices are typically the lowest. FCI will benefit by elimination of transportation and holding costs. While there was no certainty that the farmer would deliver, which would happen if the spot prices were higher than the contracted prices, the exchange would protect itself by collecting market-to-market margins from the farmers, through the banking system. If there was a delivery failure, the exchange would ask the buyer to buy from the open market and reimburse the difference between the contracted price and the spot price.

The farmers can derive benefits from the futures market either through direct participation or by using the price signals emanating from the Commodity Exchanges in planning their acreage and marketing decisions. The farmers generally do not directly participate in the futures market, even in the developed countries. The farmers, however, can derive benefits from the price discovery functions of the futures markets.

There is a need to creating awareness of the benefits of futures trading and to educate the farmers on the use of commodity exchanges for hedging price risks. The farmer's associations should act as an intermediary in marketing chain. Co-operative societies and state level marketing agencies can play the role of aggregators on behalf of farmers. In this connection, there is a need to build necessary institutional capacity for warehouses. The commodity exchanges can act as a market place for trade in warehouse receipts, which would help in the development of a sound warehousing and warehouse receipt financing system.

In order to imbibe more transparency, efficiency and reliability in the marketing system, it is essential to disseminate market information uniformly across all the market players. The government must take some steps towards creating a productive and sustainable market driven support system.

References

Johl, S.S. 2007. Trading in Commodity Futures in India. Inaugural address delivered at the seminar on "Futures Trading in agricultural commodities - its impact on Agricultural Production and Prices", organized by ISAE and MCX, Mumbai, Sept. 2007.

Sahadevan, K.G. 2002. Sagging agricultural commodity exchanges: growth constraints and revival policy options. Economic and Political Weekly, vol. XXXVII No. 30, July 27-aug.02, Pp. 3153-60.

Sahadevan, K.G. 2004. Commodity Derivatives and Futures Trading: A Study of the Sources of Market Failure and the Policy Options for its Reviva. Forward Markets Commission, Mumbai.

Shah, J. 2007. "Dynamics of Commodity Markets", *Commodity Vision*, Vol. 1:1.

Vasisht, A.K. and Bhardwaj, S.P. 2007. Futures Trading in Agricultural Commodities— an Economic view-point. Paper presented in a seminar organized by ISAE and MCX, Mumbai, Sept. 2007.

❑❑❑

Agroforestry: Natural Resource Sustainability, Livelihood & Climate Moderation © 2009
Editors: OP Chaturvedi, A Venkatesh et. al., pp. 589-603
Satish Serial Publishing House, Delhi

Chapter 42

BIOTECHNOLOGICAL APPROACHES IN AGROFORESTRY RESEARCH

Suresh Kumar and A. Chandra
Indian Grassland and Fodder Research Institute, Jhansi-284 003 (U.P.)

Introduction

For centuries, trees have been essential component of several crop- and livestock-production systems. But during the past few decades, the ancient traditions of harmony and balance between man and nature have been seriously disturbed. More than 1000 million people and 500 million livestock have exerted increasing pressure on forests, tree cover and land use in general. Of the total geographical area of 329 million ha, forests constitute 67 million ha, of which approximately 50% is in degraded condition. It is estimated that about 60 million ha of agricultural land falls under eroded or degraded category. Tree planting and agroforestry interventions have been suggested as viable strategies for rehabilitating these lands. As a consequence, agroforestry and social forestry programs have received considerable emphasis in the country and are promoted by governmental, non-governmental and academic institutions.

Agroforestry systems have developed over the years based on the people's needs and site-specific characteristics. But systematic agroforestry research was initiated in the country only about three decades ago, since then considerable progress has been made through conventional breeding methods. The interactions between and among the tree, crop, grass, and animal components have been studied, and several agroforestry technologies have been developed and tried on farmers' lands. In stead, productivity and plant performance are not making headway for a number of reasons. This is where biotechnological tools and techniques have become important and relevant part of the agroforestry research. Wise management of agricultural and forest lands using biotechnological approaches has become essential for the growth and prosperity of the developing countries.

Biotechnology uses living organisms or part thereof to make or modify products, to improve plants, animals or to develop microbes for specific uses. It has created unprecedented opportunities, not only for the manipulation of biological systems for the benefits of mankind, but also for undertaking studies for understanding fundamental processes of life. Today, scientists can isolate any gene of interest from any living organism and can incorporate that gene into any other living organism. Scientist can cut apart the gene and paste them together almost at will regardless of the source of the gene. It has empowered scientists to tailor make genes and have them expressed in any functional cell. All of these have become possible due to the revolutionary developments in the field of molecular biology and biotechnology. The potential of biotechnology lies in its ability to manipulate deoxyribonucleic acid (DNA.), which carry genetic information required for the development of any organism. Due to advances in the molecular biology of DNA and the availability of technologies (genetic engineering) it has become possible to engineer certain characters of living organisms at our will. In today's perspective, biotechnology has become one of the fastest growing and most rapidly changing technologies of the world.

In recent years, techniques for manipulation of prokaryotic as well as eukaryotic genome have witnessed a remarkable development, which allow breakage of a DNA molecule at two desired places to isolate a specific DNA segment or gene and then insert it in another DNA molecule at a desired position. The product thus obtained is called recombinant DNA (rDNA) and the technique often known as genetic engineering. Genetic engineering has now become an essential component of today's biological science and it is believed that in future it may become a major force for human existence. Of all living organisms, microbes can be genetically manipulated the most easily, and they have great potential to provide useful products such as enzymes, other organic molecules and metabolic machinery.

Plant cells exhibit a variety of characteristics that distinguish them from animal cells. A striking feature of plant cells is their totipotency, i.e. most plant cells retain the ability to give rise to mature plants. Plants have capacity to regenerate from modified plant cells called protoplasts, the plant cells whose cell wall has been removed by enzymatic digestion, which can be easily fused together (somatic hybridization) or transformed with a foreign gene. The ability of a plant cell to regenerate into a whole plant enables scientists for genetic manipulation of the cell and let the cell develop into a transgenic plant. It allows to examine the whole spectrum of physical and growth effects of the genetic manipulation within a relatively short period of time. Such a process in plant cell is far more straightforward than the parallel process in animal cell, which cannot regenerate a fully-grown adult animal. Therefore, the effects of any genetic manipulation are usually easier to be examined in plant system than in animal system.

A transgenic plant is that which has been genetically engineered using recombinant DNA technology either to promote or to prevent the production of a particular protein, with the objective of introducing or enhancing a desirable characteristic in the plant. Currently used transgenic plants provide resistance to herbicides, diseases, and insect pests. Herbicide-tolerant crops simplify weed control by allowing growers to apply broad-spectrum herbicides without harming the crop plant. Herbicide-tolerant plants make it more convenient to adopt minimum tillage or no-tillage practices that reduce soil erosion. Disease and insect-resistant plants reduce pesticide use and therefore are safer for consumers, field workers and the environment. Transgenic plants for improved nutritional qualities are under development. In the year 2007, more than 114 million hectares of transgenic crop plants (Soybean, Maize, Cotton, Canola and Rice) were grown by approximately 12 million farmers in 23 countries world over (Figure 1).

Biotechnology has already shown potential for accelerating technological developments in agriculture. Current studies in biotechnology can provide significant advances toward improving agroforestry practices. Molecular biology can play important role in identification and development of plant genotypes suitable for agroforestry. Research areas such as micropropagation, somatic embryogenesis, genetic engineering, marker-aided selection, and molecular diagnostics can help producing better trees for agroforestry plantation. A combination of classical and molecular biology research can be used to improve insect-pest resistance and stress tolerant genotypes. Tree improvement through sexual means is a slow process as the breeding cycle of tree species is very long. Marker assisted selection and tissue culture of elite germplasm can play very important role in the selection and rapid propagation of genotypes for successful establishment. Biotechnological tools and techniques such as micropropagation, genetic transformation, marker-aided selection, and molecular diagnostics are merging with traditional forest biological studies to help identify and produce better-suited trees for agroforestry. Combination of classical research and molecular biology could be used to improve forest trees and its physiological functioning with varying success rates.

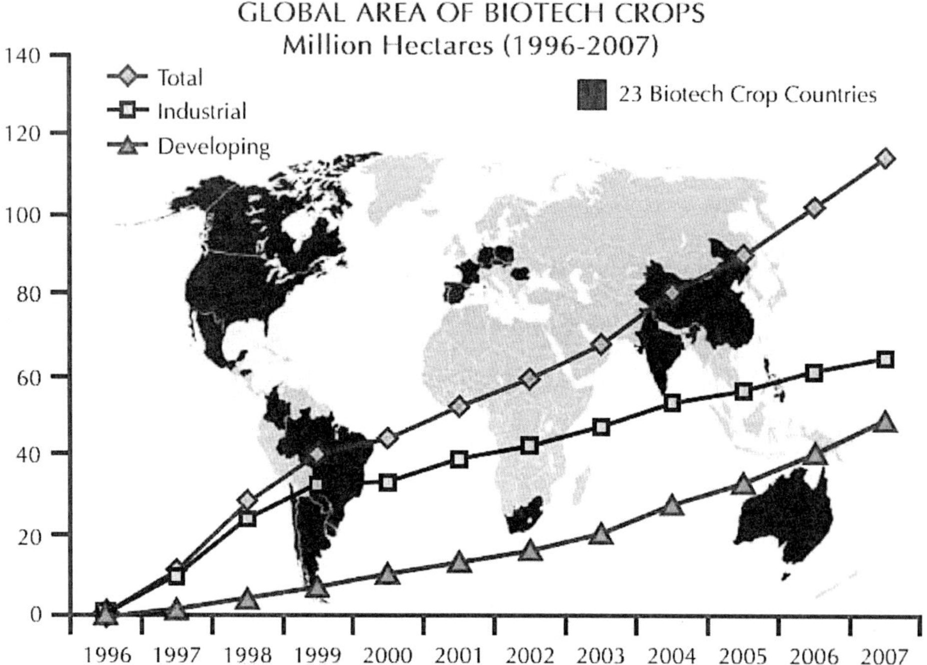

Fig. 1. Growing area of transgenic crops being grown world over in the developing and developed countries for the last decade. *Source:* Clive James, 2007

Tree improvement research as a whole can be categorized as (i) supportive research, i.e. collection of data on reproductive biology and genetics necessary to support effective breeding; and (ii) strategic research, which aims at the development of better breeding methods. Certainly, careful prioritization of research objectives is important and biotechnology should be used where there is an intimate basic knowledge of the species being experimented. Nonetheless, if basic information and knowledge are available in addition to the sound tree improvement programmes, biotechnology can prove to be a powerful tool. The most important and primary step in all agroforestry practices is the need to provide right tree for the right purpose on a given site. Full benefits of agroforestry are sacrificed when mal-adapted trees, used for the plantation, encounter biotic and abiotic stresses. Mal-adapted genotypes also fail to function adequately as windbreak, soil stabilizer, shade provider, riparian buffer, living snowfence, etc. Establishment of trees in agroforestry is both time consuming and expensive. Tree sites are typically of low quality and tree maintenance is often neglected when maintenance priorities are given to the herbaceous crops. To survive and grow over their optimal life span, trees must tolerate changing environmental stresses and resist continual threats from changing insect and pathogen populations. Hence, biotechnology and molecular biology are being adopted to select and multiply right kind of planting material. Advances in biotechnology and genomics can help in selection and development of tree varieties with desired traits such as fast growth, drought tolerance, resistance to insect-pests and diseases that can be grown on poor-quality, marginal farmlands. Thus, the possible benefits of biotechnology in forestry are perhaps even greater than that in agriculture because of the potential gain of time in certain tree improvement processes. The challenges facing foresters regarding production or outturn, whether of wood or other products, are no less urgent than those faced by agriculturists. Biotechnological approaches can be used in the agroforestry research for the various applications as discussed below.

Micropropagation

For many of the tree species conventional vegetative propagation methods are not very reliable and successful. Generally, rooting ability of cuttings is lost with the increasing age of mother tree and rooted cuttings tend to grow in horizontally, which is also known as plagiotropism. Until their terminal buds revert their growth vertically, it shows plagiotropism which may introduce some unwanted inter and intra clonal variability. Because of this reason, evaluation of genotypes becomes difficult in a selection experiment. AS a remedy, tissue culture technology can be used in tree improvement programmes. Micropropagation refers to *in vitro* plant propagation method. The attraction towards micropropagation lies in its ability to multiply elite material very rapidly. More than 1000 plant species have been micropropagated, including more than 100 forest tree species (Bajaj, 1991; Thorpe *et al.*, 1991). Successful experimental practices probably could be developed for most tree species. Micropropagation has an immediate application, however, in integrated clonal propagation systems featuring the commercial planting of cuttings harvested from rapidly multiplied, micropropagated stool plants of the selected clones.

There is also possibility of encapsulating somatic embryos to be used as artificial seeds which can then be handled like conventional seeds. Somatic embryogenesis, or production of embryos from somatic cells, is in fact a cloning technique as opposed to the zygotic embryogenesis in which germinal cells give rise to seedlings that are all genetically different. The process of somatic embryogenesis derives usually from callus formation induced by exogenous application of growth regulators, e.g cytokinin or auxin, to juvenile plant tissues. Under the favourable conditions, some undifferentiated cells of the calli may develop into somatic embryos characterized by a shoot–root bipolar structure, similarly to the zygotic embryos. This basically differentiates somatic embryogenesis from microcutting consisting of shoot formation followed by adventitious root development. In the case of conifers, somatic embryogenesis, especially when derived from a single cell, seems the most suitable regeneration and propagation technique.

The development of large-scale vegetative propagation techniques, essentially based on micropropagation or somatic embryogenesis, makes it theoretically possible to deploy superior planting materials rapidly and effectively. This is already the case for some coniferous and broad-leaved tree species including *Eucalyptus* and teak (*Tectona grandis*). Micropropagation techniques are also necessary to regenerate GM plants. The majority of GM trees is likely to be used in the form of clonal materials and will need to be vegetatively propagated. In broad-leaved species, vegetative propagation of GM materials is likely to use a combination of micropropagation and rooted stem cuttings, at least in the beginning.

One of the major disadvantages of micropropagation is that this method is expensive as compared to the conventional methods. Another concern of micrpropagation is that *in vitro* culture may cause anatomical abnormalities in the leaves, stems and roots of micropropagated plants. It may also affect major physiological processes, reducing the frequency of survival of micropropagated trees at the acclimatization stages. Two major strategies are adopted for the successful acclimatization of trees. The first involves encouraging micropropagated plants to become autotrophic and the second reducing water stress by changing culture conditions. Both the strategies may be applied at the preparatory (*in vitro*) step as well as in the basic stage (*ex vitro*) of acclimatization. Several modifications can be made at the preparatory stage, including decrease in the sugar concentration in culture medium, increasing light intensity and modifications in gas exchange. Aerated vessels with bacteriological filters have successfully been used in case of several hardwood and conifer species. The beneficial effects of aeration have also been reported in the initial stages of micropropagation.

Water and mineral uptake, root system development and the prevention of infection by soil pathogens, are markedly improved by the addition of mycorrhizae. Trees that form ectomycorrhizae may be inoculated in the preparatory stage. However, the *in vitro* technique is found unsuitable for trees that form mycorrhizae with endomycorrhizal fungi. For such species, mycorrhiza development is generally stimulated *ex vitro* by transferring the plants to inoculated soil. Mycorrhiza development has been reported to increase the survival rate of almost all tree species (conifers, orchard or broad-leaved forest trees) tested. These techniques if used in combination may facilitate successful acclimatization of micropropagated trees. However, the techniques must be chosen on a case-by-case basis.

Cryopreservation and *in vitro* Storage

Cryopreservation comprises of maintaining cells, tissues or organs under the condition where growth is slowed (by reduction of light, temperature or nutrients) or suspended (by storage at ultralow temperature in liquid nitrogen). Although, technical difficulties are involved, particularly in the subsequent regeneration of plants from the cultures, yet results are encouraging. Regeneration from cryopreserved tissues has been induced for more than 70 plant species, including coconut, rubber, cocoa and coffee, and for several forest tree species. These results have led to the hope that the technology may have number of applications in tree improvement.

In vitro storage and cryopreservation have also been used as a technique for gene conservation, particularly for the storage of threatened germplasm of agricultural species (Engelmann, 1991). Presently the technique has little to offer for this purpose with regard to forest trees. Gene pools of most of the established industrial species are reasonably well preserved in stands, both *in situ* and *ex situ,* and in seed storage. Undoubtedly, the gene pool of many tree species is threatened, particularly among the tropical hardwoods and non-industrial species. Major impediments to the preservation of forest tree germplasm are: the inadequacy of resources for the survey and collection work that would be required before any germplasm could be stored. Even for the recalcitrant (hard-to-store) species, priority would be better directed to the establishment of *ex situ* plantings, which should facilitate urgently needed evaluations of the material. In the longer term, cryopreservation and *in vitro* storage may have some application as a backup conservation strategy, but only for populations of well-surveyed, recalcitrant species.

Molecular Markers

In the early period of studies, morphology, anatomy, physiology and embryology were used for genetic analysis to determine variability in the population. These methods have the advantage of being readily available, do not require sophisticated equipments and are the most direct measure of phenotype, thus they are available for immediate use, an important attribute. However, morphological determinations need to be taken by an expert in the species, they are subjected to changes due to environmental factors, may vary at different developmental stages and their number is limited. To overcome the limitations of such morphological traits, molecular markers have been developed at both protein and DNA level. The use of molecular markers involves examination, using sophisticated biochemical techniques, of variations in cellular molecules such as DNA and proteins. As an alternative to traditionally measured features such as vigour, stem quality and various morphological aspects, molecular markers offer the advantages of being unaffected by the environment or the developmental stage of the plant while also being very numerous. The DNA-based molecular markers have acted as efficient and versatile tools, and complemented the classical strategies for genetic analysis in various fields like taxonomy, physiology, embryology, genetic engineering, etc. Hence, DNA-based molecular markers are more suitable and ubiquitous to most of the genetic studies. These characteristics have led to a number of potential applications of molecular markers in tree improvement.

Genetic fingerprinting: The inherent characteristics of molecular markers render them much more useful than morphological traits in establishing the identity of a particular tree or tracing its genetic relationship to other trees. For example, using molecular markers, it was possible to identify each of 39 peach cultivars individually (Ballard *et al.*, 1992). Molecular markers have important immediate applications in supportive research for advanced breeding programmes with industrial species mainly for quality control, e.g. checking clonal identification, orchard contamination and within-orchard mating patterns by DNA fingerprinting. They also have important immediate applications in research on tropical hardwoods and non-industrial species, in particular for essential taxonomic studies and investigations of mating systems.

Quantification of genetic diversity: Molecular markers are potentially more useful for quantifying genetic diversity than the morphological markers which are influenced by environmental variations (i.e. it is not clear if traits are produced genetically or by external factors). Markers have been used to compare within- and inter-population variation in several tree species (Muller-Starck *et al.*, 1992). The quantification of genetic variation to aid in sampling strategies for the development of gene conservation and breeding populations of new industrial and non-industrial species is a potentially useful application of molecular markers. They have provided a way of estimating genetic diversity and allowed comparison between individuals in different growth and development conditions. They have also been used in profiling to tag and then single out individual trees. These markers may, however, provide underestimates of genetic variation with respect to traits (e.g. vigour and stem quality) that are more subject to evolutionary pressures and, therefore, will need to be used with caution.

Marker-assisted selection: This refers to indirect selection on the basis of markers shown to be associated with gene of interest. Unaffected by the environment or developmental stage, molecular markers offer the possibility of highly effective and early selection of forest tree (e.g. selection for wood quality at the young seedling stage). Marker-assisted selection has given further impetus to tree breeding and selection. Molecular markers are genetically linked to a given allele on a given locus and can therefore be used to predict the presence of the allele with great accuracy. Co-segregation of RAPD and AFLP type markers has allowed the construction of genetic maps. Genetic maps can be used to assess the degree of evolutionary relatedness of a number of species and identify the zones where gene variations are statistically linked to the variability of quantitative traits. These zones are called QTLs (quantitative trait loci). QTL research focuses on the genetic architecture of traits of interest. However, QTL research results are not easily transferred to other tree populations. Therefore, the use of candidate genes became necessary for applying MAS to forest trees. The candidate gene approach was made possible by the development of gene search methods (EST and cDNA banks), transcriptomics and proteomics. In order to use the candidate gene approach in selection, their molecular variability must be linked to the variation of the targeted trait of interest. This requires associating appropriate field experiments with genetic mapping.

Although the possibilities are very attractive, there are certain limitations with using marker assisted selection: (i) marker analysis is currently too expensive to permit the screening of large populations of seedlings, (ii) associations between markers and economically important traits have to be established separately for different species, thus, even when markers are available, marker-assisted selection will apply mainly to advanced and sophisticated breeding programmes.

Genetic Engineering

Genetic engineering refers to a set of technologies that is being used to change the genetic makeup of cells and moving genes across species boundaries to produce novel organisms. Genetically modified organisms (GMOs) are called transgenic organisms, indicating that a gene has been transferred into the organism and the transferred gene is referred to as a transgene. Transformation technology is adopted when a desired trait, governed by one or a few genes, is not available in the plant species or related species of interest. Therefore, gene can be isolated from any other organism and used for genetic transformation of the plant of interest. Several methods of introducing foreign DNA/gene into protoplasts and intact cells are available. Following methods have been repeatedly successful: (i) polyethyleneglycol (PEG)-mediated and (ii) electroporation-mediated introduction of foreign DNA into protoplasts, (iii) biolistic and (iv) *Agrobacterium*-mediated methods of gene transfer into intact cells, and (v) microinjection, a most common method for introduction of foreign genes into animal cells. In the absence of *Agrobacterium*-mediated T-DNA delivery system, protoplast transformation was the only method of choice for plant transformation. However, development of fertile transgenic plants from protoplasts has been a laborious, time-consuming and genotype dependent process. In the past decade, methods for transformation of intact cells have been developed, and these include biolistic and *Agrobacterium*-mediated methods of transformation. Biolistic or gene-gun method of gene transfer has emerged as a simple and promising alternative for monocotyledonary plant transformation and has been successfully used for transformation of many plant species.

Genetic modification has been applied to forest trees mainly during the last two decades, at the experimental stage in the laboratory, to alter various traits such as herbicide, metal, salt, or insect resistance or cold tolerance. The rationale for producing trees with reduced lignin content for the pulp and paper industry is generally based on the expected environmental and economic benefits (Christensen *et al.*, 2000). Genetic modification is often considered as a tool to improve knowledge of tree biology and functioning, with significant potential applications for studying cell wall properties and wood formation. Since genes regulating secondary tissue and lignin formation in the annual plant *Arabidopsis thaliana* were identified (Goujon and Jouanin, 2003), research on the genetic basis of wood quality in forest trees has significantly increased, mainly in *Eucalyptus, Pinus* and *Populus*. A tree crop into which genes have been inserted is the poplar, and many projects are under way, e.g. the reduction of lignin biosynthesis.

Herbicide Resistance

The introduction of herbicide-tolerant gene in trees is of some interest but the advantage of practicing unguarded herbicide applications may not be sufficient to pay for the research programme. The first approach of developing herbicide resistant plant has been introducing a mutated version of the gene encoding the enzyme target for various herbicides such as glyphosate (the herbicide that blocks synthesis of amino acids tryptophan, tyrosine and phenylalanine) and chlorosulfuron (the herbicide that blocks synthesis of leucine, isoleucine and valine). A second and more frequently used strategy consists of introducing a microbial gene encoding an enzyme for the detoxification of the herbicide, and has been applied to *Populus, Eucalyptus, Pinus* and *Picea.*

Insect Resistance

Insect resistance is of potential value, for example in poplars and some tropical hardwoods. However, the work involves introducing several different resistance genes expressed at higher level sufficient enough to ensure that insects do not develop resistance. For the development of insect resistant trees, the approach has been introducing *Bt* gene encoding an endotoxin which binds to the receptors in the intestine of Lepidopteran, Coleopteran and Dipteran insects, lysing the organ and killing them. This was successfully achieved in *Populus* and *Picea* and GM *Populus* and *Pinus* trees expressing *Bt* endotoxin have been obtained.

Second approach of induced insect resistance is based on the introduction of a gene coding for a protease inhibitor that modifies insect digestion, causing the death of the pest. Studies used potato gene *pin*2, a protease inhibitor introduced into *P. alba and P. grandidentata* through *Agrobacterium*-mediated transformation method (Klopfenstein *et al.*, 1991), and a gene of rice protease inhibitor introduced into *P. tremula* and *P. tremuloides* (Heuchelin *et al.*, 1997). Third approach focuses on simultaneous introduction of two genes for enhanced and sustainable resistance to insects. This was achieved in *Liquidambar styraciflua*, combining a peroxidase anionic enzyme gene involved in cell wall growth and development with a *Bt* gene (Sullivan and Lagrimini, 1993).

Quantitative and Qualitative Modification of Lignin

Alteration in lignin biosynthesis is a very valuable objective for the pulp species, therefore modification in lignin composition or content is being actively pursued because of the expected financial gains from pulp processing improvements. Lignins, which enhance cell wall mechanical properties and hardness, are difficult to process and are a significant limitation in processing wood into paper pulp by chemical treatment. Genetic transformation to modify lignin characteristics is a key research feature on species used in the paper industry. The aim is to regulate the activity of key enzymes involved in the lignin biosynthesis pathway (Lapierre *et al.,* 1999; Jouanin *et al.,* 2000; Li *et al.,* 2003). Active on-going research targets the effects of lignin biosynthesis in *Populus*.

Flowering Modification

In order to produce sterile trees and prevent possible dispersal of transgenic pollen in the environment, an approach based on genetic ablation has been tested on poplar (Skinner *et al.,* 2000). This technique, which consists of expressing a cytotoxic gene under the control of a very specific poplar floral promoter, resulted in more than 90 percent of transformed lines lacking floral structures. Other approaches, based on the suppression of key flowering genes, are being tested.

Other Traits

Other traits that have been considered for genetic modification include: resistance to diseases in *Populus*, *Betula pendula*, *Alnus* and *Juglans.*; bioremediation in *Populus*; enhanced growth through higher efficiency of nitrogen assimilation in *Picea* and *Populus* (Gallardo *et al.,* 1999), and modification of gibberellin synthesis in *Populus*. Cold-tolerance is likely to be of some commercial value for many species, in particular the eucalypts. Much remains to be done, however, to establish that sufficient tolerance can be conferred using antifreeze proteins and to extend the work to tree species. The major factor limiting application of genetic engineering in forest trees is the state of knowledge of molecular control of the traits that are of most interest - those relating to growth, adaptation and stem and wood quality. Genetic engineering of these traits remains a distant prospect. Research projects of this type are necessarily intensive and must be regarded as long-term with only a modest expectation of success.

The interest induced by the first genetic modifications of tree species at the end of the 1990s has since declined. Commercial deployment of new biotechnology-based forest tree varieties of a limited number of taxa (*Eucalyptus, Pinus taeda, P. radiata, P. pinaster, Populus*) can be expected in the near future. The impact of biotechnology applications on global wood supply may take longer to materialize. In this context, genetic modification appears to be the most significant and controversial sector of forest biotechnology. An often overlooked research component is the extensive testing that would be required before a responsible recommendation for the large-scale plantation of transgenic trees could be made. Regulatory frameworks have been established in many countries for GMOs field experiment or commercial deployment, with increased testing costs. In a number of countries, hostile reaction from the public or/and environmental groups has also limited deployment of GM trees. Prevention of the escape of genes into wild populations is likely to become an important concern, and sterility should be an early target of genetic engineering work with forest tree species.

Conclusions

With the exception of micropropagation, for which some applications are reported, most of the biotechnological approaches relates to research activities. Field applications appear extremely limited. Nevertheless, in spite of the unclear economic rationale, use of biotechnology in forestry research has increased and provided new knowledge on tree biology and functioning. It also appears that forest biotechnology is increasingly funded by the public sector in academic research, while the private sector seems to focus investments more on specific traits of commercial interest such as wood quality.

In addition to the time-tested methods of plant improvement, new tools of molecular biology and biotechnology have now become available for raising the yield ceiling. Molecular genetics provides unusual scope for tailoring plants to suit different growing conditions and consumer needs but research in this area must be based on sound principles of biosafety and bioethics. Since, advances in biotechnology need highly trained personnel and state-of-the-art equipment and facilities, the product of biotechnology may become tool for stakeholders to gain more profits keeping the poor remain on the periphery. Therefore, it must be ensured that the technology is developed and extended in a way which will benefit the small poor farmers. It has now become very clear that transgenics will be accepted by people only when the doubts related with general risks and hazards, long term safety to human health and nutrition, environment and sustainability of agriculture are properly dispelled. Thus, we need to organize mass awareness campaign to educate our people about the advantages and potential dangers of genetic modifications, to enable them to judge at their own the vital role the biotechnology can play.

References

Bajaj, Y.P.S. 1991. Automated micropropagation for *en masse* production of plants. *In* Y.P.S. Bajaj, ed. *Biotechnology in agriculture and forestry 17 High-tech and micropropagation 1,* Berlin, Springer Verlag, Pp. 3- 16.

Ballard, R.E., He, G., Abott, A.G., Mink, G. and Belthoff, L.E. 1992. Molecular biology of forest trees. DNA fingerprinting of *Prunus* varieties using low copy sequence probes. *Proc. IUFRO Working Party S2.04.06 Workshop.* Carcans-Maubuisson France. INRA.

Christensen, J.H., Baucher, M., O'Connell, A., van Montagu, M. and Boerjan, W. 2000. Control of lignin biosynthesis. *In* S.M. Jain & S.C. Minocha, eds. *Molecular biology of woody plants,* Vol. 1, pp. 227–267. Dordrecht, Netherlands, Kluwer.

Engelmann, F. 1991. *In vitro* conservation of tropical plant germplasm - a review. *Euphytica,* 57: 227-243.

Gallardo, F., Fu, J., Cantón, F.R., García-Gutiérrez, A., Cánovas, F.M. and Kirby, E.G. 1999. Expression of a conifer glutamine synthetase gene in transgenic poplar. *Planta,* 210: 19–26.

Goujon, T. and Jouanin, L. 2003. Genes involved in the biosynthesis of lignin precursors in Arabidopsis. *Plant Physiology Biochemistry,* 41: 677–687.

Heuchelin, S.A., Jouanin, L., Klopenstein, N.B. and McNabb, H.S. 1997. Potential of proteinase inhibitors for enhanced resistance to *Populus* arthropod and pathogen pests. *In* Klopfenstein, N.B., Chun, Y.W., Kim, M.S. & Ahuja, M.R., eds. *Micropropagation, genetic engineering, and molecular biology of* Populus. Gen. Tech. Rep. RM-GTR-297, pp. 173–177. Fort Collins, CO, US Department of Agriculture, Forest Service, Rocky Mountain Research Station.

James, C. (2007) Global status of commercialized biotech/GM crops, ISAAA Brief No. 37, ISAAA, Ithaca, New York.

Jouanin, L., Goujon, T., de Nadaï,V., Martin, M.T., Mila, I., Vallet, C., Pollet, B., Yoshinaga, A., Chabbert, B., Petit-Conil, M. and Lapierre, C. 2000. Lignification in transgenic poplars with extremely reduced caffeic acid-*O*-methyltransferase activity. *Plant Physiology,* 123: 1363–1373.

Klopfenstein, N.B., Shi, N.Q., Kernan, A., McNabb, H.S., Hall, R.B., Hart, E.R. and Thornburg, R.W. 1991. Transgenic *Populus* hybrid expresses a wound-inducible potato proteinase inhibitor II-CAT gene fusion. *Canadian Journal of Forest Research,* 21: 1321–1328.

Lapierre, C., Pollet, B., Petit-Conil, M., Toval, G., Romero, J., Pilate, G., Leplé, J.C., Boerjan, W., Ferret, V., de Nadaï, V. and Jouanin, L. 1999. Structural alterations of lignins in transgenic poplars with depressed cinnamyl alcohol dehydrogenase or caffeic acid *O*-methyltransferase activity has an opposite impact on the efficiency of industrial kraft pulping. *Plant Physiology,* 119: 153–163.

Li, L., Zhou, Y., Cheng, X., Sun, J., Marita, J.M., Ralph, J. and Chiang, V.L. 2003. Combinatorial modification of multiple lignin traits in trees through multigene co-transformation. *PNAS*, 100: 4939–4944.

Muller-Starck, G., Baradat, P. and Bergmann, F. 1992. Genetic variation within European tree species. *New Forests*, 6: 23-47.

Skinner, J.S., Meilan, R., Brunner, A.M. and Strauss, S.H. 2000. Options for genetic engineering of floral sterility in forest trees. *In* S.M. Jain & S.C. Minocha, eds. *Molecular biology of woody plants*, Vol. 1, Dordrecht, Netherlands, Kluwer, Pp. 135–153.

Sullivan, J. and Lagrimini, L.M. 1993. Transformation of *Liquidambar styraciflua* using *Agrobacterium tumefaciens*. *Plant Cell Reproduction*, 12: 303–306.

Thorpe, T.A., Harry, I.S. and Kumar, P.P. 1991. Application of micropropagation to forestry. *In* P.C. Debergh & P.H. Zimmerman, eds. *Micropropagation: technology and application.* Dordrecht, the Netherlands Kluwer Academic.

❑❑❑

Agroforestry: Natural Resource Sustainability, Livelihood & Climate Moderation © 2009
Editors: OP Chaturvedi, A Venkatesh et. al., pp. 605-616
Satish Serial Publishing House, Delhi

Chapter 43

PROSPECTS OF BIOCHEMICAL AND MOLECULAR MARKERS IN CROP IMPROVEMENT: A CONCISE OVERVIEW

A. Chandra and Suresh Kumar
Indian Grassland and Fodder Research Institute, Jhansi-284 003 (U.P.)

Introduction

Amino acid is the building block of all proteins which precisely govern the physiology and biochemistry of all living organisms. Any change in amino acid, especially at crucial positions of the polypeptide chains, may sometimes lead to the genetic disorder which can be observed or phenotyped easily. Contrary to this, change in nucleotides of DNA sequences, i.e. wobble sequence, yields different form of the gene which ultimately generates different forms of proteins having similar properties. Such polymorphic protein is called isozyme. As most of the genes are found in multi-copies in genome, proteins (enzymes) are also present in different forms. Such variations in protein profiles have been used in characterizing germplasm at biochemical level. Though such variations are directly associated with expressing macromolecules, the number of such markers is less and also get influenced by the environmental factors. On the other hand, molecular markers like DNA-based are numerous and not affected by environmental factors. Such markers are largely developed based on the coding and non-coding regions of the genome. Both biochemical and molecular markers have created unprecedented opportunities for studying the natural variations that exist in nature. Such variations have been studied in many food and forage crops. Using these markers it is possible to tag and isolate genes of interest from any living organism for their incorporation into other systems of interest.

Polygenic characters which were previously considered to be very difficult to analyze using conventional plant breeding methods, would now be easy to be tagged with molecular markers. Molecular markers which are particularly useful in selection of desirable characters include random amplified polymorphic DNA (RAPD), restriction fragment length polymorphism (RFLP), polymerase chain reaction (PCR) based markers such as sequence characterized amplified region (SCAR), sequence tagged sites (STS), amplified fragment length polymorphism (AFLP) and amplicon length polymorphism (ALP) using various genetic stocks like F_2, BC populations, near isogenic lines (NIL) and recombinant inbred line (RIL).

Though biochemical markers are co-dominant in nature, their number and time required to perform limits their applicability in plant breeding. The effective use of molecular markers in plant improvement programmes require that it should be highly polymorphic and co-dominant in nature. Among the suite of DNA markers available, microsatellites are preferred for plant genome analysis. Microsatellite marker, also known as SSR (simple sequence repeat) or STR (short tandem repeat), is developed on the basis of flanking sequences present on both sides of the nucleotide repeats. Sequence tagged microsatellite (STMS) or SSR markers reveals polymorphism due to variation in the length of microsatellite at specific locus. Being polyallelic and co-dominant in nature, it has been proved to be a very useful marker. SSRs occur at a high frequency and exhibit high mutation rates, preferential association with non-repetitive regions of the genome, co-dominant inheritance, inter-specific transportability, a high throughput assay and cost efficiency. The difficulty in generating SSRs from genomic DNA is mitigated to some extent by *in-silico* SSR discovery using expressed sequence tag (EST) or other sequence data bases (Fig. 1). EST-derived SSRs occur at a high frequency in the genome and have greater transferability than genomic SSRs, with the difference becoming more pronounced with increasing phylogenetic distance. EST-SSRs also exhibit high levels of polymorphism despite their association with sequences experiencing direct selection pressure.

The advantages of using gene-associated sequences as genetic markers include the ability to utilize bioinformatics and whole genome sequencing initiatives for comparative genome analysis, to construct function oriented maps and to identify candidate genes for direct investigation of sequence-function relationships. The existing genetic maps of diploid and tetraploid *Medicago sativa*, complete sequence information of *M. truncatula* and markers from white clover provide comprehensive markers to be utilized for genome analysis of other species of the same tribe Trifolieae, which contains, among others, the genera *Trifolium* and *Medicago.*

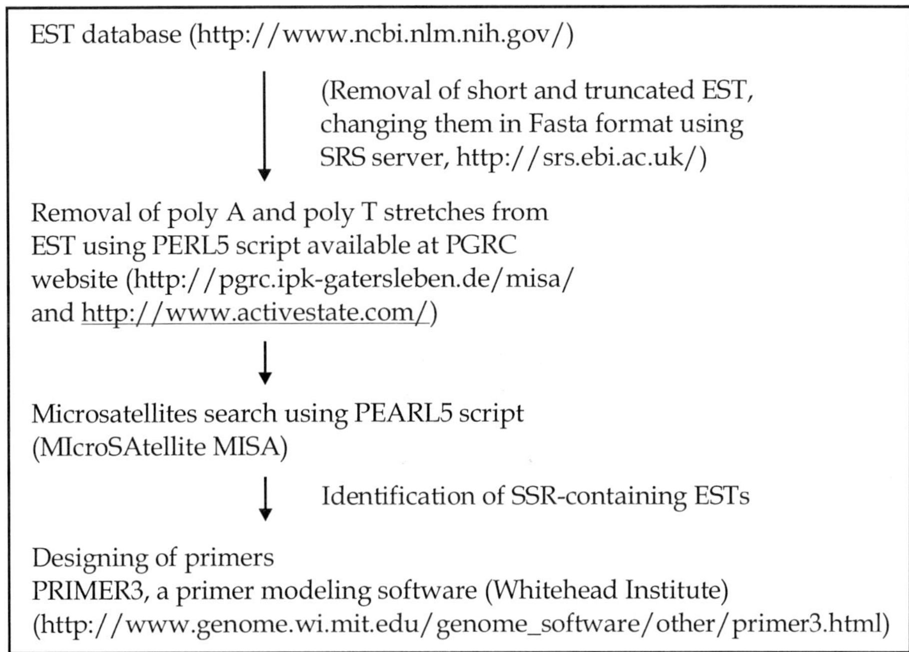

Fig. 1. Steps involve in development of EST-derived microsatellite markers. Removal of poly A and poly T and microsatellites search are achieved through Perl module (thick border) whereas operation of EST database like changing them in Fasta format and designing of primers were achieved through additional programs (lighter box).

Assessment of Genetic Diversity

Genetic diversity has been created at inter- and intra-specific levels in plants by evolutionary forces and is an important parameter utilized for plant improvement either by selection or applying various breeding methodologies. Information on genetic diversity, relationship within and between crop species and their wild relatives is essential for efficient utilization of plant genetic resources and management of germplasm collections and also for evolving conservation strategies. Different types of marker systems have been used for genetic analysis and characterization studies (Table 1). These include morphological, cytological, biochemical and DNA-based marker systems (Fig. 2). However, variation caused by environmental factors and occurrence of narrow genetic base as well as low resolution power of isozyme limit the use of morphological and biochemical markers in identification of genotypes (Lallemand *et al.*, 1994). Nevertheless, isozyme markers have been extensively used in construction of zymogram for better understanding of the variability and to pin point exactly the position of isoform which gets activated or expressed as a new isoform (Fig. 3). Usefulness and reliability of any genetic marker is dependent on its heritability and the level of polymorphism, it exhibits. The more polymorphic and heritable the trait, the greater is its potential value for germplasm characterization (Porter and Smith, 1982). DNA-based markers are considered to be the best for determining genetic relationships/diversity, as they are unlimited in number, highly polymorphic and independent of environmental interaction i.e., highly heritable. Several molecular markers namely RFLP, AFLP (Vos *et al.*, 1995), variable number of tandem repeats (VNTR) (Nakamura *et al.*, 1987) and SSR (Jacob *et al.*, 1991) have been deployed to assess genetic diversity and relationship, but RAPD (Williams *et al.*, 1990) data can be generated faster and with less labour compared to other molecular markers. RFLP analysis is labour intensive and time consuming but highly expensive. For the development of VNTR and SSR markers prior information about the genome is necessary and thus poses problems where large germplasm samples are to be analyzed.

Table 1. Utility of markers in plants, animals and microorganism.

Plants	Animals	Microorganism
- Genetic Diversity Analysis - Variety Identification - Acceleration of Inbred Conversions - Phylogenetic Analysis - Linked Marker Assay Development - MAS - Seed and Plant Quality Assay Development - Genetic Map Construction - Microsatellite Discovery	- Genetic Diversity Analysis - Line Identification - Linked Marker Assay Development - Genetic Map Construction - Microsatellite Discovery	- Genetic Diversity Analysis - Strain Identification - Phylogenetic Analysis - Epidemiology - Detection Test Development

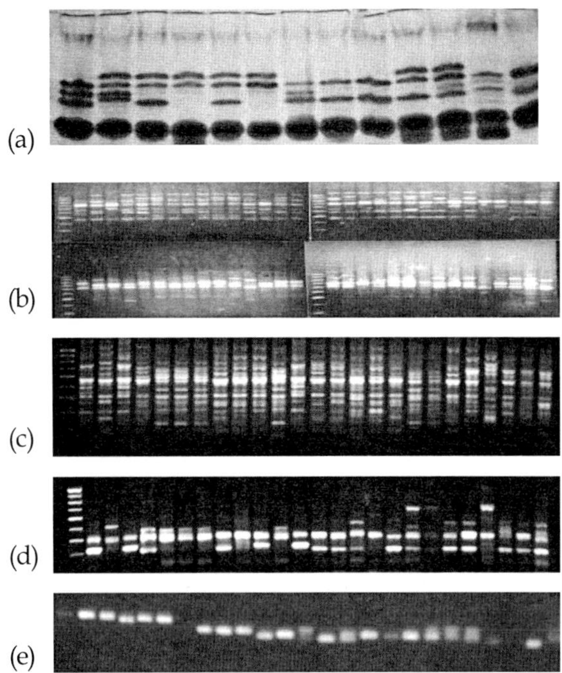

Fig. 2. Representative gel photographs of different sets of markers can be utilized in genetic diversity estimate. (A) Esterase biochemical marker, (B) RAPD, (C) ISSR, (D) STS and (E) SSR.

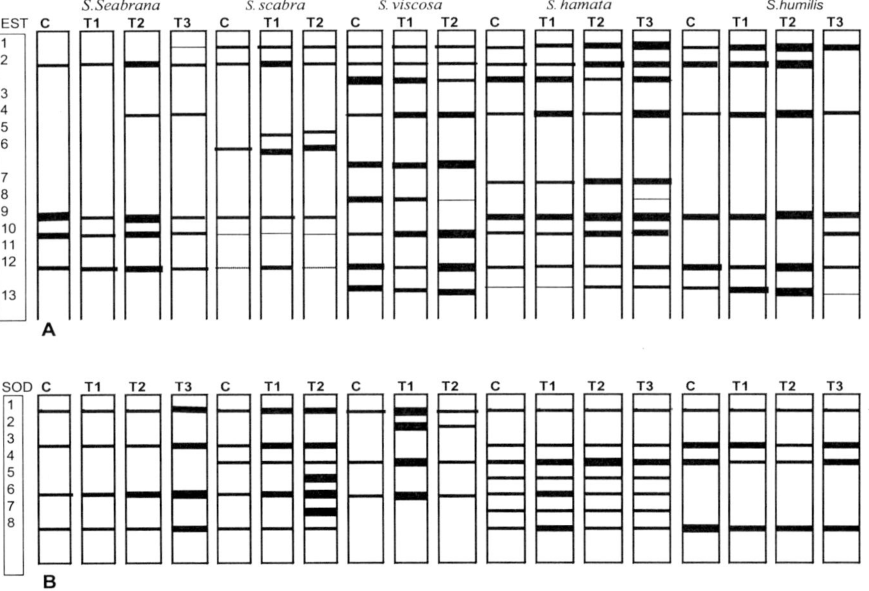

Fig. 3. Zymogram developed using biochemical markers data.

RAPD employs single short primer (10 mer) with an arbitrary sequence to generate genome specific 'fingerprint' of multiple amplification products. Polymorphism found between RAPD profiles can serve as a genetic marker (Williams *et al.*, 1990). It is being used extensively to search polymorphism in plants (Hu and Quiros, 1991; Yang and Quiros, 1993; Iqbal and Rayburn, 1994; Khandka *et al.*, 1996; Chandra *et al.*, 2004). RAPD has been used to identify asparagus and *Dichanthium annulatum* as well as their diagnostic potential was demonstrated on individual plant basis (Chandra *et al.*, 2006; Khandka *et al.*, 1996). One hundred decamer arbitrary primers have been utilized in identifying polymorphic and genotype-specific markers in an apomictic grass *Dichanthium annulatum* (Chandra *et al.*, 2004). Genetic similarities based on RAPD profiles were estimated via coefficient of DICE and then the data were processed by cluster analysis (UPGMA) (Fig 4). Like other tropical grasses, *Cenchrus* is also apomictic in nature hence characterization of accessions and varieties is often difficult due to lack of reliable and distinguishable morphological traits. The situation becomes more complicated when no publicly accessible molecular markers are available. At present, using 11 agronomic attributes 322 accessions of *C. ciliaris* and *C. setigerus* have been grouped in six categories largely on the basis of rhizome development, plant maturity and yield (Pengelly *et al.*, 1992). Nevertheless in recent past we have genetically assessed eight species of *Cenchrus* using RAPD markers (Chandra and Dubey, 2008), however repeatability of such markers is always questionable. STS markers have been developed and used in genetic relationship and progenitor analysis in *Stylosanthes*, a tropical fodder legume (Liu *et al.*, 1997; Vander Stappen *et al.*, 1999; Ma *et al.*, 2004). Low levels of genome specificity of such markers have been reported in *Stylosanthes* (Liu *et al.*, 1996). Since tropical fodder grasses have been less intensively studied than other members of the *Poaceae* and limited DNA sequences are available in the databases, the transferability of STS markers developed largely on the basis of functional gene sequences have been attempted for genetic analysis of tropical grasses. The STS markers based on sequence of related species of *Gramineae* have been used in studying genetic relationship in ryegrass (*Lolium perenne* L.) (Lem and Lallemand, 2000). When eight species of *Cenchrus* were analyzed with 11 STS markers, developed on the basis of cDNA sequences of functional genes of stylo, 10 have generated scoarble bands whereas only four reacted to the six STS developed based on the sequences of clones of *Pst1* library of *Stylosanthes*. In total 195 bands were scored, of which 83 % were polymorphic (Chandra and Dubey, 2007). STS markers separated annual species from perennial, however *C. myosuroides* being perennial clustered with annual species, possibly a bridging species among two growth habits. Results possibly suggest the use of trans-genus STS markers in analysis of polymorphism preferably at species level (Chandra and Dubey, 2007). Using STS genome specific markers, ten *S. seabrana* lines have been identified from the existing germplasm stock of *S. scabra*. All of them displayed STS phenotypes of typical diploid species. With primer pairs SHST3F3/R3 all putative *S. seabrana* yielded single band of ~550 bp and *S. viscosa* of ~870 bp whereas both of these bands were observed in allotetraploid *S. scabra*. The newly identified lines were characterized along with the available *S. seabrana* lines using RAPD and STS markers indicated a high level of similarities (> 92% and 80% respectively) among them. Since this primer pairs is known to amplify single or no band with diploid and two bands with tetraploid species, the amplification patterns corroborated that all newly identified *S. seabrana* lines were diploid in nature. The information obtained through DNA fingerprinting using RAPD and RFLP not only gives a comprehensive picture on diversity and relatedness but also determines efficacy of each marker to be used in diversity studies (Fukuoka *et al.*, 1992; Heun *et al.*, 1994; Castagan *et al.*, 1997; Landry *et al.*, 1997; Suh *et al.*, 1997). Molecular markers have been also used to characterize exotic and endemic population of pathogen corresponding to different diseases. In case of *Stylosanthes,* ninety isolates from several countries have been analyzed and have shown extensive genetic diversity in *C. gloeosporioides* population causing anthracnose disease (Chakraborty *et al.*, 1998).

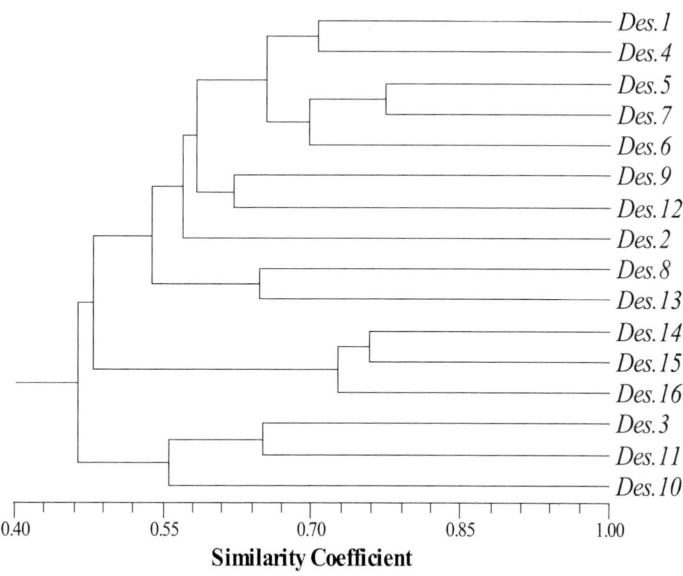

Fig. 4. UPGMA based dendrogram generated using EST based SSR markers of a forage legume accessions. Bootstrap values indicating significance of clustering (>30) showed on various nodes.

Construction of Dendrogram and Cluster Analysis

For both biochemical and molecular markers, presence and absence of bands is transformed to binary mode using scores 1 and 0 for presence and absence of bands/allele, respectively. The binary data is then used to derive Dice coefficients (as indicator of genetic similarity) followed by phenetic clustering using the Sequential Agglomerative Hierarchical and Nested (SAHN) method. The analytical routines can be carried out using NTSYSpc ver 2.02 (http://www.ExeterSoftware.com). The reliability or the significance of the generated dendrogram can be tested by bootstrap analysis using WinBoot software employing more than 500 iteration values. Usually when bootstrap values are more than 40 the generated clusters/nodes are said to be reliable and observed clusters are significant in terms of closeness among accessions (Fig. 4). On the basis of gel patterns, zymograms of enzymes can be developed. Isozyme bands are usually numbered from top of the gel (sample loading side) in both acrylamide and starch gels.

Polymorphic information content (PIC) values are calculated for individual primer to evaluate discriminatory power of the markers. The PIC value is calculated applying the formula $PIC_i = 2f_i(1\text{-}f_i)$, where f_i is the frequency of the amplified allele (band present) and $(1\text{-}f_i)$ is the frequency of the null allele (band absent) of marker *i* in a given accession. Marker index (MI) can be determined as the product of PIC and the number of polymorphic bands per assay unit.

Markers and Stress Tolerance

Stylosanthes scabra is a widely sown pasture legume in the semi-arid tropics of northern Australia where water stress is a major problem. Studies have shown that transpiration efficiency (TE) is one of the traits contributing to drought resistance in *S. scabra* (Thumma *et al.,* 1998) and a significant negative relationship between TE and carbon isotope discrimination (CID) exists. The study has also been executed in eight species of *Cenchrus* and results demonstrated a significant correlation between these traits (Chandra and Dubey, 2008). Thus, a negative relationship between TE and specific leaf area (SLA) has been established in both grass and legumes indicated their utilization in screening of lines for such (drought) traits. It was suggested that CID and/or SLA may be used as indirect measures of TE. DNA based markers such as RFLP, RAPD, AFLP and microsatellites provide opportunities to study quantitative traits such as drought resistance through QTL analysis. A QTL is a region on chromosome linked to a marker which has a significant effect on the quantitative trait (Tanksley, 1993). QTL analysis not only provides information on the number of loci involved in the expression of a trait, but also indicates the possible location of the loci. Recently QTL analysis has been used to test the relationships between physiological traits (Lebreton *et al.,* 1995; Simko *et al.,* 1997). For two traits to be causally related, the significant QTL effect of one trait should have a measurable effect on the other trait (Lebreton *et al.,* 1995). Coincidence of QTL for two traits, with allelic differences corresponding to the expected causal relationship between the traits, is strong evidence that the two traits are related (Quarrie *et al.,* 1997). Generation of linkage map and hence the QTL analysis performed for drought indicated a causal relationship between traits related to drought resistance in *S. scabra* (Thumma *et al.,* 2003).

Breeding Implications of Traits Associated with Agronomical Traits

For any trait to be used for selection in breeding programs, its measurement should be easy and rapid. Such a trait should have high genetic correlation with the trait of interest and it should have high heritability. Carbon isotope discrimination (CID) and SLA may fall into this category. Many studies have shown high heritability and low genotype C environment interaction for CID (Farquhar *et al.,* 1989). A drawback in its use is the cost of measurement although when compared to using molecular markers it is not very high. Therefore using CID itself rather than markers associated with it may be sufficient in breeding programs to improve TE. However, in some studies CID was shown to be positively related to yield, indicating selection for low CID would improve TE but it also could lead to lower yield (Condon *et al.,* 1987; Ray *et al.,* 1999). Therefore, it has been suggested that selection should be applied for both traits simultaneously. In this context molecular markers are very valuable in pyramiding the traits such as low CID, SLA and high MHP accumulating capacity.

References

Castagan, R. S., G. M. Perenzin and M. Heun. 1997. Genetic variability of the wild diploid wheat *Triticum urartu* revealed by RFLP and RAPD markers. *Theoretical and Applied Genetics,* 94: 465-471.

Chakraborty, S., C. J. Liu, D. F. Cameron, A. Chandra, C. R. Ramesh, C. D. Fernandes, M. J. Charchar, L. Guodao and S. Kelemu. 1998. Application of molecular markers to breed and select anthracnose resistant *Stylosanthes* and to characterize endemic and exotic populations of *Coletotrichum gloeosporioides*. In: Proc.4th Asia–Pacific Conference on Agricultural Biotechnology, Darwin, Australia, Pp. 53-56.

Chandra, A. 2006. A partial genetic map of *Stylosanthes* developed using inter-specific F_2 cross. *Cytologia,* 71: 353-358.

Chandra, A. and A. Dubey. 2007. Transferabilty of STS markers for studying genetic diversity within the genus Cenchrus (Poaceae). *Current Science,* 72 : 961-967.

Chandra, A. and A. Dubey. 2008. Evaluation of genus *Cenchrus* based on malondialdehyde, proline content, specific leaf area and carbon isotope discrimination for drought tolerance and divergence of species at DNA level. *Acta Physiologia Plantrum,* 30: 53-61.

Chandra, A. and R. K. Bhatt. 2008. Carbon isotope discrimination function analysis and drought tolerance of stylo species grown under rain-fed environment. *Acta Physiologia Plantrum,* 30: 63-69.

Chandra, A., R. Saxena and A. K. Roy. 2006. Polymorphism and genotype-specific markers for *Dichanthium* identified by random amplified polymorphic DNA. *Genetic Resources and Crop Evolution,* 53 : 1521-1529.

Chandra, A., R. Saxena, A. K. Roy and P. S. Pathak. 2004. Estimation of genetic variation in *Dichanthium annulatum* genotypes by RAPD technique. *Tropical Grasslands,* 38: 245-252.

Condon, A. G., R. A. Richards and G. D. Farquhar. 1987. Carbon isotope discrimination is positively correlated with grain yield and dry matter production in field grown wheat. *Crop Science,* 27: 996-1001.

Farquhar, G. D., J. R. Ehleringer and K. T. Hubick. 1989. Carbon isotope discrimination and photosynthesis. *Annual Review of Plant Physiology and Molecular Biology,* 40: 503-537.

Fukuoka, S., K. Hosaka and O. Kamijima. 1992. Use of random amplified polymorphic DNA (RAPDs) for identification of rice accessions. *Japanese Journal of Genetics,* 67: 243-252.

Heun, M., J. P. Murohy and T. D. Phillip. 1994. A comparison of RAPD and isozyme analysis for determining the genetic relationships among *Avena sterilis* L. accessions. *Theoretical and Applied Genetics,* 87: 689-696.

Hu, J. and C. F. Quiros. 1991. Identification of broccoli and cauliflower cultivars with RAPD markers. *Plant Cell Report,* 10: 505-511.

Iqbal, M. J. and A. L. Rayburn. 1994. Stability of RAPD markers for determining cultivar specific DNA profiles in rye (*Secale cereale* L.). *Euphytica,* 75: 215-220.

Khandka, D. K., A. Nejidat and A. Golan-Goldhirsh. 1996. Polymorphism and DNA markers for asparagus cultivars identified by random amplified polymorphic DNA. *Euphytica,* 87: 39-44.

Lallemand, J., F. Briand, F. Breuils, D. Denoue and C. Rameau. 1994. Identification of asparagus varieties by isozyme patterns. *Euphytica,* 79: 1-4.

Landry, B. S., R. Keseli, H. Leung and R. W. Michelmore. 1997. Comparison of restriction endonucleases and source of probes for their efficiency in detecting restriction fragment length polymorphism in lettuce (*Latuca sativa* L.). *Theoretical and Applied Genetics,* 74 : 649-653.

Lebreton, C., V. Lazic-jancic, A. Steed, S. Pekic and S. A. Quarrie. 1995. Identification of QTL for drought response in maize and their use in testing causal relationships between traits. *Journal of Experimental Botany,* 46: 853-865.

Lem, P. and J. Lallemand. 2000. Grass consensus STS markers: an efficient approach for detecting polymorphism in *Lolium. Theoretical and Applied Genetics,* 107: 1113-1122.

Liu, C. J. and J. M. Musial. 1997. *Stylosanthes* sp. Aff. *S. scabra : a* putative diploid progenitor of *Stylosanthes scabra* (Fabaceae). *Plant Systematics Evolution,* 208: 99–105.

Liu, C. J., J. M. Musial and F. W. Smith. 1996. Evidence for a low level of genomic specificity of sequence tagged sites in *Stylosanthes. Theoretical and Applied Genetics,* 93 : 864-868.

Ma, Z.Y., A. Chandra, J. M. Musial and C. J. Liu. 2004. Molecular evidence that *Stylosanthes angustifolia* is the third putative diploid progenitor of the hexaploid *S. erecta* (Fabaceae). *Plant Systematics Evolution,* 248 : 171-176.

Nakamura, Y., M. Leppert, P. O. Connel, R. Wolgg, T. Holm, M. Culver, C. Martic, E. Fijimoto, M. Hoff, E. Kumlin and R. White. 1987. Variable number of tandem repeat (VNTR) markers for human gene mapping. *Science,* 235: 1616-1622.

Pengelly, B. C., J. B. Hacker and D. A. Eagler. 1992. The classification of a collection of buffel grasses and related species. *Tropical Grasslands,* 76: 6-11.

Porter, W.M. and D. H. Smith Jr. 1982. Detection of identification errors in germplasm collections. *Crop Science,* 22: 701-703.

Quarrie, S. A., D. A. Laurie, J. Zhu, C. Lebreton, A. Semikhodsky, R. Tuberosa and C. Calestani C. 1997. QTL analysis to study the association between leaf size and abscisic acid accumulation in droughted rice leaves and comparisons across cereals. *Plant Molecular Biology,* 35 : 155-165.

Ray, I. M., M. S. Townsend, C. M. Muncy and J. A. Henning. 1999. Heritabilities of water-use efficiency traits and correlations with agronomic traits is water-stressed alfalfa. *Crop Science,* 39: 494-498.

Simko, I., S. McMurry, H. M. Yang, A. Manschot, P. J. Davies, E. E. Ewing. 1997. Evidence from polygene mapping for a causal relationship between potato tuber dormancy and abscisic acid content. *Plant Physiology,* 115: 1453-1459.

Suh, H.S., Y. I. Sati and H. Morishima. 1997. Genetic characterization of weedy rice (*Oryza sativa* L.) based on morpho-physiology, isozymes and RAPD markers. *Theoretical and Applied Genetics,* 94: 316-321.

Tanksley, S. D. 1993. Mapping polygenes. *Annual Review Genetics,* 27: 205-233.

Tewari, S. and A. Chandra. 2008. Genetical assessment of diploid progenitors of *S. scabra* by Isozyme, RAPD and STS markers: a possible strategy for improvement of drought tolerant allo-tetraploid *S. scabra* species. *Euphytica* 162: 39-50.

Thumma, B. R., B. P. Naidu, A. Chandra, D. F. Cameron, L. M. Bahnisch and C. Liu. 2001. Identification of causal relationship among traits related to drought resistance in *Stylosanthes Scabra* using QTL analysis. *Journal Experimental Botany,* 52: 203-214.

Thumma, B. R., B. P. Naidu, D. F. Cameron and L. M. Bahnisch. 1998. Transpiration efficiency and its relationship with carbon isotope discrimination under well-watered and water-stressed conditions in *Stylosanthes scabra. Australian Journal of Agricultural Research,* 49: 1039-1045.

Vander Stappen, J., I. Weltjens, S. V. Campenhout and G. Volckaert. 1999. Genetic relationships among *Stylosanthes* species as revealed by sequence-tagged site markers. *Theoretical and Applied Genetics,* 98: 1054-1062.

Vos, P., R. Hogers, M. Bleeker, M. Reijans, T. V. D. Lee, M. Hornes, A. Frijters, J. Pot, J. Peleman, M. Kuiper and M. Zabeau. 1995. AFLP: a new technique for DNA fingerprinting. *Nucleic Acids Research,* 23: 4407-4414.

Williams, J. G. K., A. R. Kubelic, J. Livak, J. A. Rafalski and S. V. Tingey. 1990. DNA polymorphisms amplified by arbitrary primers are useful as genetic markers. *Nucleic Acids Research,* 18: 6531-6535.

Yang, X. and C. Quiros. 1993. Identification and classification of celery cultivars with RAPD markers. *Theoretical and Applied Genetics,* 86: 205-212.

□□□

Agroforestry: Natural Resource Sustainability, Livelihood & Climate Moderation © 2009
Editors: OP Chaturvedi, A Venkatesh et. al., pp. 617-627
Satish Serial Publishing House, Delhi

Chapter 44

DECISION SUPPORT SYSTEM AND DATA BASES IN AGROFORESTRY

P. Ratha Krishnan and R. Indumathy[1]
National Research Centre for Agroforestry, Jhansi-284 003 (U.P.)
[1]NRCAF Colony, National Research Centre for Agroforestry
Jhansi-284 003 (U.P.)

Introduction

Indian soils are flooded with culture and civilization. Our agriculture began by 9000 BC as a result of early cultivation of plants, and domestication of crops and animals. The settled life soon followed with implements and techniques being developed for agriculture made it a prime source of livelihood. The progress made by agriculture in the last four decades has been one of the biggest success stories of free India. Agriculture and its allied activities are the means of livelihood of about 2/3 of the work force in the country. But, the present climatic uncertainties, increased cost of cultivation and scarcity of labor made the farmers to opt for the alternate land use technologies.

Application of recent technologies like computer / information technology in this ancient originated field is the needed strategy of this hour. Computer made the calculations easier, predicts accurately and recently used for the information management and support decisions. Using computer / information technology by the way of Decision Support System (DSS) and databases for the dissemination and extension of technology in agriculture/ agroforestry can optimize the production while conserve the resources.

What is DSS ?

DSS is an interactive flexible computer based information system, which helps decision makers to utilize data & models to solve unstructured problems like complex nutrient, irrigation, protection and marketing problems faced by the farmers. Hence it provides valid representation of complex real world problems. DSS is composed of four subsystems viz., data, model, communication and knowledge management subsystem. The data management subsystem consists of DSS data base, database management system, data dictionary and query, while the model management subsystem can be of model based (static or dynamic), model based management system and modeling language. From the above, DSS data base can be classified as relational data base, object oriented database and relational object oriented database.

Decision making is a process of choosing among alternative course of action for the purpose of attaining goal. It can be done by three levels viz., operational, managerial and strategic. Decision making is being done in four phases i) Intelligence phase (Organizational goals are set and objectives are identified. Hence the problems will be evaluated and predict the outcomes), ii) Design phase (Models formulated and criteria will be chosen), iii) Choice phase (Models worked after sensitive analysis model), and iv) Implementation phase (implement the solution). The farmer's query will be collected and stored in the database and maintained in the proper RDBMS with corresponding data dictionaries. According to the queries answers will be provided by researchers, expert in that field like gyandoot, echoupal, ikisan, n-logue, etc. (Indumathy and Ratha Krishnan, 2008).

Principle of DSS

The following two graphical pictures (Fig. 1 & 2) explains the principle of DSS.

What is Database?

It is the collection of information or data which is possible to access in any mode and answerable for any query. It may be of personal system supported files or available in the internet. It covers the information from many sources. It organizes and facilitates the management and querying of large quantities of data and information.

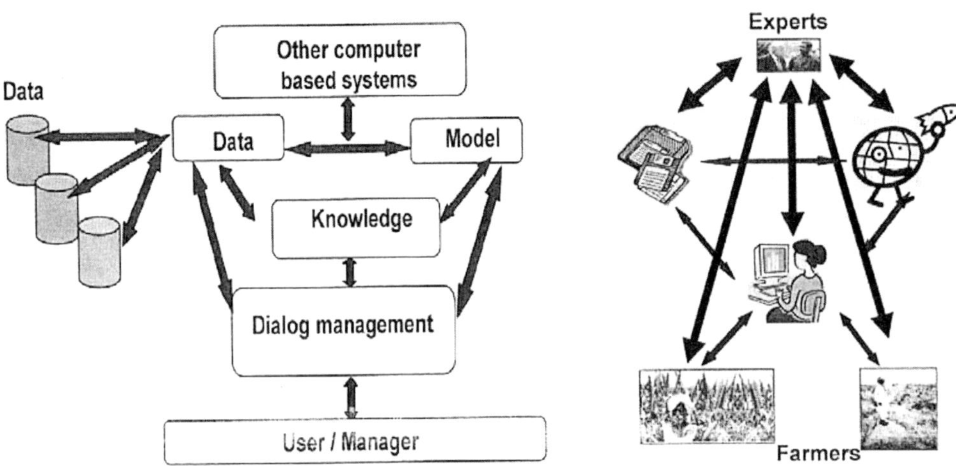

Fig 1. Collection of information / data from all possible sources

Fig 2. Collection and dissemination of information / data

Why DSS and Database are Important in Agroforestry?

Three decades of agroforestry research in our country has led to the development of several agroforestry systems that are based on people's need and site specific characteristics. Presently, the expectations from agroforestry are quite high in this country both in rural and urban areas. In recent years sizeable increase in tree cover outside the forest areas has taken place in many states (FSI, 2005). However, it is realized that there exists tremendous scope of dissemination of knowledge to land owners, farmers and extension agents about not only the availability and suitability of various agroforestry options but also their management in such a complex environment. In order to achieve this an appropriate Decision Support Systems (DSS) may be useful. Such a DSS has to cover a wide array of systems, tools and technologies in an interactive manner with the objective of helping managers and other clients to make decision. It should provide a platform to its clients for retrieving, summarizing and analyzing data relevant for decision making.

Agroforestry Databases

Agroforestry Systems Inventory Databases (AFSI) has been developed by the World Agroforestry Centre (WAC, Nairobi). It was done by a global collection of data / information on agroforestry systems using a questionnaire as the survey instrument. Data and information collected and entered into the database included general description, geographical location, biophysical characteristics, socioeconomic aspects, system evaluation, components of the system and their uses and identification of research gaps. With AFSI, the user is able to query the database, extracting information such as geographical locations of different agroforestry systems and the species found within these systems in different locations (Nair, 1987). AFSI was apparently developed as a research and information tool for researchers. But, AFSI was failed to be maintained and upgraded and therefore, fade with time and presently no more.

Another database viz., Multipurpose Tree and Shrub Database (MPTS) was developed by ICRAF in 1991 (Von Carlowitz *et al.,* 1991). It was meant for the researchers and extension agents as an aid to select the right tree or shrub species for agroforestry practices. This database was primarily meant for the tropics and subtropics containing information on 1093 species on the aspects related to site-specificity, morphological and phenological descriptions, management characteristics and environmental responses (Schroder and Jaenicke, 1994). A simple climate model was included to predict climatic conditions based on the input of geographical coordinates. Tree and shrub species were selected via a database search that matched the descriptors that the user selected. The descriptors included 19 different criteria covering aspects of location, climate and soil conditions, products, environmental services, management, cultivation and references on selected agroforestry species (Schroder and Jaenicke, 1994). The current and revised version of MPTS is now the Agroforestry Tree Database (AFTD). It is a database management system intended for use by researchers and fieldworkers to select agroforestry trees that are being deliberately grown and managed for more than one output and expected to make significant economic and/or ecological impacts (Word Agroforestry Centre, 2003a).

In India, the Web based application "agroforestry BASE" was designed and developed by NRCAF, Jhansi (Ajit *et al.,* 2005). It includes six databases *viz.,* Research Project Database, Database on Economic Analysis, Database on MPTs, Agroforestry Interventions and Poplar and Eucalyptus based Bibliographic Systems. This data base is upgraded and serves the need of the researchers till date.

Computer Based Technologies in Agroforestry

In recent past the use of computer-based technologies in agroforestry and natural resource management has been impressive. Like in most natural resource management options, agroforestry is characterized by high complexity of which we have limited understanding the data (Sanchez, 1995; Nair, 1998). However, the advanced cartographic science and its application in agroforestry can be greatly enhanced through the use of GIS, GPS and SRS. Important categories of computer-based spatial information technologies are presented in Table 1.

Table 1. Major categories of useful computer based spatial information technologies for agroforestry.

Category	Description
Databases	Organizes and facilitates the management and querying of large quantities of data and information.
Geographical Information Systems (GIS)	Brings in a geographic or spatial component to a database; manages, manipulates and analyzes spatial data.
Computer-Based Models	Mathematical computer models that represent real world processes and predict outcomes based on input scenarios
Knowledge-Based or Expert Systems (KBS)	Adopts 'Artificial Intelligence' in the form of organizing, manipulating and obtaining solutions using knowledge in the form of qualitative statements, expert rules and a computer language representation system for storing and manipulating knowledge.
Hybrid Systems	Integrates two or more of the above computer-based technologies (e.g. GIS, KBS and Models) for more versatile, efficient and comprehensive DSTs.

Geographic Information System (GIS) can be defined as a data management system designed to input, store, retrieve, manipulate, analyze and display spatial data for the purpose of research and decision-making (De Mers, 1997). Use of GIS can facilitate the planning process of agroforestry development. Spatial data layers like soil type, slope, and land cover can be used to develop suitability assessments that can identify optimal locations for agroforestry practices to solve landowner and community concerns.

The Land Cover Classification System (LCCS) has already been developed by the FAO of the United Nations Environment Programme (UNEP) to meet the need for improved access to reliable and standardized information. By selecting data with the appropriate spatial resolution, this assessment process can be used at any scale for design and management of different agroforestry options. The most significant benefit of using GIS-guided DSS is the ability to combine different assessments to determine locations where multiple objectives can be achieved. In a GIS, a database is associate with map features, and data values are geographically referenced, so resource managers can spatially represent information such as soil types or plant communities.

As the GIS technology is widely available and affordable today and agroforestry is directly dependent upon spatial characteristics; it is logical to expect to have several agroforestry-specific GIS based DSSs. Ellis *et al.* (2004) have reviewed many DSS tools used in agroforestry research and development. Several such systems are depicted in Table 2. GIS guided assessments are being used to evaluate four key issues of the Western Corn belt : biodiversity, soil protection, water quality, and agroforestry products. By combining these assessments, information is generated for use in identifying opportunities and constraints on the landscape where multiple benefits from conservation buffers, especially agroforestry plantings, can be achieved (Bentrup *et al*,. 2000).

Table 2. GIS based decision support tools used in agroforestry.

GIS based decision support tool	Type	Important Features
AFSI (Agroforestry Systems Inventory Databased) (Oduol *et al.* 1988)	Database	Agroforestry system inventory database describing geographic location and biophysical, socio- economic and species characteristics.
MPTS (Multipurpose Tree and Shrub Database) (Schroder and Jaenicke, 1994)	Database	Multipurpose tree and shrub database used for tree selection and species information
Agroforestry Tree Database (Salim *et al.* 1998)	Database	Internet and CD-ROM application for reference and selection guide of agroforestry trees.
Forestry Compendium (CABI 1998; Tree CDs)	Database	Compilation of knowledge on forestry, Agroforestry and plantations and information on trees for management decision-making and species selection.
Agroforestry Parklands in Burkina Faso (Bernard and Depommier, 1997)	GIS	Spatial analysis of dynamics of agroforestry parklands and species distribution due to human impacts
Historical Transformation of Agroforestry Landscape in Canada (Paquette and Domon, 1997)	GIS	Spatial analysis of census and geomorphologic data to explore dynamics of agroforestry in 19th century Canadian landscape
Field-level spatial analysis of temperate agroforestry system (Jose *et al.*, 2001)	GIS	Spatial analysis using ground penetrating radar (GPR) to evaluate root biomass and distribution and soil nutrient crop-tree interactions in temperate alley cropping.
AME (Agroforestry Modeling Environment) (Muetzelfeldt and Taylor ,1997)	Modeling Tool	Object oriented tool to graphically visualize, construct, integrate and exchange agroforestry models

contd...

Table contd...

FALLOW (World Agroforestry Centre, 2003b)	GIS and Model	Model to evaluate impacts of shifting cultivation and fallow rotations at a landscape-scale evaluating transitions in soil fertility, crop productivity, biodiversity and carbon stocks World Agroforestry.
Agroforestry Planning Tool in China (Liu *et al.*, 1999)	Hybrid GIS, Models and KBS	Hybrid DST integrating GIS data, regression models plus expert knowledge to assess biophysical, social and economic suitability of Paulownia.
Plantation & Agroforestry Species Selection Tool (PLANTGRO) (Hackett and Vanclay, 2003)	Hybrid GIS/KBS	Plantation and agroforestry species selection tool integrates GIS and expert system on plant growth.
Southeastern Agroforestry Decision Support System (SEADSS) (Ellis *et al.*, 2003)	Hybrid Database, GIS, KBS	Landscape and site-scale AF planning and species selection, DST for landowners and extension agents of Southeast US that integrates GIS, tree and shrub database and expert knowledge.
Agroforestry BASE (Agroforestry Information System) (Ajit*et al.*, 2005)	Database	An internet based agroforestry information quarry and data retrieval system.

Status and Response to DSS

Agricultural decision support systems (DSS) may be argued to have passed sequentially through phases of unbelief, euphoria and disappointment and to be passing into either a phase of maturity with realistic expectations of the technology or to abandonment.

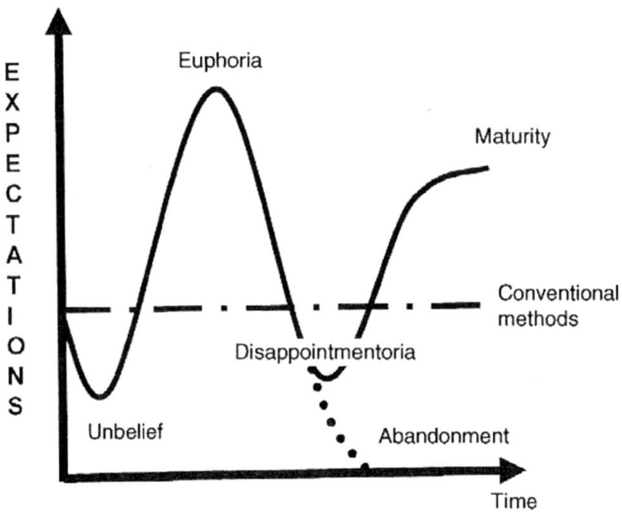

Figure 3. Expectation phases for technologies -modified (Biethahn and Nissen, 1995)

In a survey by Mathew *et al.* (2005) the following results were obtained. To some extent this met with expectations - high use of personal judgement (95% - though numbers using only judgement were not recorded) and low use of DSS (3%). The use of computer based tools was significant with spreadsheets and GIS represented at 44 and 34% (table 3). This would seem to back up McCown (2002) observation that software that is controlled by the decision maker who is more popular than software that requires any degree of ceding of control. The use of external services at 34% was interesting since in nearly all cases the services were from specialist consultants, from which it was concluded that there was a market for the advice but that it was a niche already filled.

Table 3. Tools and resource being used – Survey

Tool/Resource	Resp	%
Personal judgement	58	95
Crop yield information	27	44
Livestock yield information	23	37
Other information	25	41
Spread sheets	27	44
GIS	21	34
DSS	3	5
Other software	10	16
External services	21	34
None of the above	1	2
Others	7	12

Table 4. Likely response to DSS

Response to DSS	Resp	%
Find out more	22	36
Want to see relevant success stories	18	29
Sceptical	19	31
Tools already developing with way	3	5
Not relevant	13	21
Already adopted	4	7
Tried but abandoned	0	0
Others	2	3

(Matthews *et al.*, 2005)

From the table 4 it is clear that while the DSS capabilities are attractive the reality was that decision makers while interested to find out more (36%) were skeptical (31%) and would need previous success stories to convince them (29%). This would seem to chime with Carberry *et al.* (2002) experience in FARMSCAPE with interest but an initial credibility gap to be overcome. Those interested were predominantly from the agri-businesses and land-agents categories, perhaps seeking a competitive advantage. There was a significant proportion of respondents for which the DSS was not considered relevant but since there was no opportunity to say why it is difficult to interpret this result. Lastly it was interesting to note that there were a small number of respondents that had adopted DSS, and of those none had abandoned then.

Conclusion

Just in time the above (DSS and Data Bases) can resolves farmer's queries. Assures farmers to collect information globally & act locally. The raised conflict / debate can be resolves or at least overcomes by the above applications. DSS is based on observations on the real system but less ambitious to PAR. PAR based approaches are advised to address the key issues of engagement & credibility. Socio-political, Institutional and Geographic dependency is a major bottleneck of DSS. So, DSS can assist to select compatible tree & crop combinations towards increased productivity and is a recent facilitating mechanism to nourish the roots of agricultural practices.

References

Ajit, Rai, P., Handa, A.K. 2005. Structure Design and Utility of Agroforestry Information System "agroforestry BASE". National Research Centre for Agroforestry, Jhansi.

Bentrup G., Dosskey M., Schoeneberger M., Wells M., Leininger T. and Klenke K. 2000. Planning of multipurpose riparian management. In: Riparian Ecology and Management in Multi-land Use Watersheds (Eds. Wigington P.J. and Beschta R.L.). American Water Resources Association, Middleburg, VA, Pp. 423-426.

Bernard C. and Depommier D. 1997. The systematic approach and the role of GIS in the characterization and monitoring of agroforestry parks. XI World Forestry Congress, Antalya, Turkey, 13 to 22 October 1997, 1: 87.

Biethahn, J. and Nissen, V. 1995. Evolutionary algorithms inmanagement applications. Springer, Berlin, Pp. 195.

CABI (Commonwealth Agricultural Bureau International). 1998. The Forestry Compendium : A Silvicultural Reference. CABI, Oxford, U. K. CD-ROM.

Carberry, P., Hochman, Z., McCown, R., Dalgliesh, N., Foale, M., Poulton, J.H.P.L., Hargreaves, D., Cawthray, S., Hillcoat, N. and Robertson, M. 2002. The FARMSCAPE approach to decision support: farmers', advisers, researchers' monitoring, simulation, communication and performance evaluation. *Agricultural Systems*, 74:141–177.

De Mers M.N. 1997. Fundamental of Geographic Information Systems. Wiley, New York, Pp. 486.

Ellis E.A., Nair P.K.R. and Jeswani S.D. 2003. The southeastern agroforestry decision support system (SEADSS): a Web-based application for agroforestry planning and tree selection. In : Decision Support for Multiple Purpose Forestry. (Vacik H., Lexer M.J., Rauscher M.H., Reynolds, K.M. and Brooks R.T. (eds). Pp.1-12.

Ellis, E.A., Bentrup, G. and Schoeneberger, M.M. 2004. Computer-based tools for decision support in agroforestry : Current state and future needs. *Agroforestry Systems*, 61: 401-421.

FSI, 2005. State of Forest report- 2005. Forest Survey of India, Dehradun, Pp. 171.

Hackett C. and Vanclay J.K. 2003. Mobilizing expert knowledge of tree growth with the PLANTGRO and INFER systems. Plantsoft Services, Stirling Australia and CIFOR, Jakarta, Indonesia, Pp. 18.

Indumathy, R. and Ratha Krishnan, P. 2008. Decision support system in agriculture and agroforestry. Abstracts In: National symposium on Agroforestry knowledge for sustainability, climate moderation and challenges ahead. NRCAF, Jhansi, 15 - 17, Dec, 2008. Pp. 112.

Jose S., Gillespie A.R., Seifert J.R. and Pop P.E. 2001. Comparison of minirhizotron and soil core methods for quantifying root biomass in a temperate alley cropping system. *Agroforestry Systems*, 52: 161-168.

Liu J., Shao G. and Li W. 1999. Expert knowledge-based model for regional agroforestry planning. Proceedings of Geoinformatics'99 Conference. Ann Arbor, Michigan, 19-21 June. University of California, Berkely, CA. Pp.1-8.

Matthews, K., Buchan, K., Sibbald, A. and Craw, S. 2005. Combining deliberative and computer-based methods for multi-objective land-use planning. *Agricultural Systems*, 77: 148 - 160.

McCown, R. 2002. Changing systems for supporting farmers' decisions: problems, paradigms and prospects. *Agricultural Systems*, 74:179–220.

Muetzelfeldt R.B. and Taylor J. 1997. The Agroforestry Modeling Environment. In : Annual Report : Agroforestry Modeling and Research Coordination. Institute of Terrestrial Ecology. Natural Environment Research Council. Edinburgh, UK. Pp.10-20.

Nair P.K.R. 1987. Agroforestry System Inventory. *Agroforestry Systems*, 5 : 319-338.

Nair P.K.R. 1998. Directions in tropical agroforestry research : Past, Present and Future. *Agroforestry Systems*, 38: 223-245.

Oduol, P.A., Mauraya, P., Fernandes, E.C.M. and Nair, P.K.R. 1988. The agroforestry systems database at ICRAF. *Agroforestry Systems*, 6: 253-270. 420.

Paquette S. and Domon G. 1997. The transformation of the agroforestry landscape in the nineteenth century: a case study in southern Quebec, Canada. Landscape Urban Plan, 37: 197-209.

Salim A.S., Simons A.J., Waruhiu a., Orwa C. and Anyango C. 1998. Agroforestry Database: A Tree Reference and Selection Guide. Version 1.0. ICRAF, Nairobi, Kenya. CD-ROM.

Sanchez P.A. 1995. Science in agroforestry. *Agroforestry Systems*, 30: 5-55.

Schroder J.M. and Jaenicke H. 1994. A computerized database as decision support tool for the selection of agroforestry tree species. *Agroforestry Systems*, 26: 65-70.

Von Carlowitz P.G., Wolf G.V. and Kemperman R.E.M. 1991. Multipurpose Tree and Shrub Database : An Information and Decision Support System. ICRAF, Nairobi, Kenya & GTZ, Eschborn, Germany. Pp. 104.

World Agroforestry Centre, 2003a. Agroforestry Database. http://www. world-agroforestrycentre.org/Sites/TreeDBS/ AFT/AFT.htm

World Agroforestry Centre, 2003b. Forest, Agroforest, Low value Landscape or Wasteland? http://www.worldagroforestrycentre.org/sea/agromodels/FALLOW/ fallow. htm Accessed September, 12, 2006.

❑❑❑

Agroforestry: Natural Resource Sustainability, Livelihood & Climate Moderation © 2009
Editors: OP Chaturvedi, A Venkatesh et. al., pp. 629-637
Satish Serial Publishing House, Delhi

Chapter 45

STATISTICAL SOFTWARE TOOLS FOR DATA ANALYSIS

Ajit, A.K.Handa and Uma
National Research Centre for Agroforestry, Jhansi-284 003 (U.P.)

Introduction

The role of statistical analysis in scientific research is quite clear, and specifically it is more predominant in agroforestry research. Recommendation of specific agroforestry system to be adopted, choice of tree/crop species, tree spacing and other components of a agroforestry system cannot be determined without certain knowledge of statistics. This leads to the problem of choosing a tool for data analysis and subsequent activities that require statistical content. Excel is a much-used tool and there are add-ins for Excel that can greatly improve its value in statistical analysis. Yet, Excel is not in its essence a statistical tool. The other approach is to use languages and packages that are able to perform statistical functions. However, these languages require extensive learning and training. The user has to learn programming with them well before they can be put into good use. Of course, there are the professional commercial statistics packages like SAS, SPSS, SYSTAT, STATA, MINITAB etc. that allow command line input and programming as well as use of graphical user interface (GUI) analysis. However, these are often expensive and in many cases beyond the means of the individual users. There are quite a number of different packages and programs that have statistical functionality, however before deciding upon a statistical software one should be clear about the requirements of statistical software.

The first question to be answered is what requirements should a statistical software package fulfill to make data analysis possible? There are many software packages with different characteristics and abilities. The software packages that have established themselves within the statistical community are SPSS, SAS, SYSTAT, STATA, MINITAB etc. These are indeed professional tools with great capabilities. Consequently, they are very expensive tools, sometimes not easy to use, and most of the time beyond the financial capabilities of most research and teaching departments. In addition, unless statistics and data analysis is their primary activity, most departments will not have these packages. The fact is that most statistical functions needed in agroforestry research are in the areas of statistical designs of experiments, statistical inference, multivariate analysis, sampling theory etc. The amount of data to be analyzed is not extensively high. There are some free statistical tools available to meet the requirements of statistical analysis.

The second requirement is ease of use. The learning curve should not take too much time. If a software package is complicated to use and learn, it will not be adopted by majority of users. Ease of use for statistical software generally means that reading the data is simple, at least several formats of input data are available, the analysis is simple with a number of prewritten functions, the package has a GUI which is simple and easy to understand, accompanying documentation is easy to read, and software has good capabilities in producing graphs and plots.

The third requirement is tutorial and teaching components of the software. If there are such functions incorporated into the package, for example explanations of some functions, probability distributions, statistical plots and various exercises, learning is much easier and consequently simpler for analysis and interpretation.

Of course, a certain number of statistical functions and procedures should be supported so that the most common tasks can be performed without extensive additional programming effort. Linear regression, univariate and multivariate analysis, box-plots, normal, Chi-square and t distribution generation, scatter plots and histograms, frequency tables and ANOVA analysis should be the minimum functions available within the software.

Finally, since most of the users are used to Windows operating systems, we have concentrated on packages that can run under 11Windows. Needless to say, there are many excellent tools for Mac OS or Linux operating systems and these might be a topic of another discussion, but a vast majority of users are running Windows on their computers.

Here, we will describe important characteristics and capabilities of several software packages.

ViSta

ViSta is a very nice graphical tool that can work as standalone application or with Excel, and has nice plotting capabilities. Its GUI could be more appealing. There are some bugs, and to perform multivariate analysis, lognormal analysis, homogeneity analysis, multilevel analysis and application development, one has to download plug-ins. It can be downloaded from www.visualstats.org. ViSta runs on Windows, UNIX and Mac OS. It is open and free. ViSta allows users to define new functions and software, to build web applets and to write scripts. It comes with teaching functions that actually explain what is going on. It can execute LISP, C++ and FORTRAN programs. ViSta can read SAS data and Excel spreadsheets and free-form text data files. However, more functionality in the applications' input data files would be welcome.

AM

AM is a statistical package from American Institute for Research. The software can be used freely for noncommercial purposes. It can be downloaded from http://am.air.org/. AM has the following features:

1. AM offers sophisticated statistics with an easy-to-use drag and drop interface, and an integrated help system that explains the statistics as well as how to use the system
2. AM includes statistical procedures for analyzing assessment and survey data, including models estimated via marginal maximum likelihood (MML). MML procedures recognize that tests can't pinpoint proficiency, but instead define a probability distribution over the proficiency scale. AM will also analyze the "plausible values" used in programs like NAEP.
3. AM includes procedures for analyzing non-assessment survey data as well.
4. All of AM's procedures automatically provide appropriate standard errors for complex samples. All procedures are available with a Taylor-series approximation for the standard errors, and a few offer an option to use a jackknife or other replication technique.

5. AM is completely free and may be shared and copied. It was developed, in part, with funding from the National Center for Education Statistics.

AM can import data from about 100 different file formats, including SAS and SPSS. The manual that comes with the software is well documented and the GUI is easy to use. The manual gives a mathematical description of the functions and procedures. AM is primarily aimed at survey data analysis. It allows univariate and multivariate analysis. The graphing capabilities are limited to bar chart, line chart and density plots. Command line programming, script writing and random number generation are not supported.

WinIdams

WinIdams is a software package for data analysis developed by UNESCO. It allows users to define their own programs, but it is also capable of interactive data analysis through WinIdams GUI. Plotting capabilities are good. The manual in HTML form is easy to read and also comes with an explanation of plots and functions. WinIdams can perform univariate and multivariate analysis, linear regression, ANOVA, frequency tables and the other most common statistical procedures. The results are output in the form and table form. WinIdams is free and can be downloaded from http://www.unesco.org/idams.

Irristat

IRRISTAT can perform univariate and multivariate analysis, regression, ANOVA and other usual statistical procedures. It comes with examples and a tutorial. However, it does not allow the user to make programs and new functions. Everything is done via a GUI that is well developed and easy to use. The graphical capabilities are not very good although scatter and biplots can be generated. IRRISTAT has been developed primarily for the analysis of data from agricultural field trials and that might account for its somewhat limited capabilities in general data analysis. The software can be downloaded from http://www.irri.org/science/software/irristat.asp.

Openstat

OpenStat3 can perform univariate and multivariate analysis and has a set of statistical process control tools. The drawback is that it does not allow programming and defining new functions. Also, only OS3, OS2, comma separated, tab separated and space separated files can be imported. Otherwise, everything is performed through a GUI and pop-up windows. The software is free and can be downloaded from http://www.statpages.org.

Instat

Instat+ can be downloaded from http://www.reading.ac.uk/ssc. It has somewhat questionable plotting capabilities, but it is very well suited for teaching statistics. In fact, that was one of the main intentions of the authors of the software. All the most common statistics functions are supported, including multivariate analysis but not nonlinear regression. In fact, most free packages do not support nonlinear techniques. Instat+ is free for single noncommercial users, but it requires a one-time department site license fee. It is very much based on a worksheet approach and it does not allow programming, although macros can be used.

SAS

1. A very complete package for statistical analysis.
2. Can link to Access Database via ODBC.
3. Data can be put in several different locations including mainframes.
4. Can be run on several different platforms i.e. large mainframe computers down to PC's.
5. SAS programming is used as a means to access the data for several other purposes besides statistical analysis. SAS is an integrated suite of software tools used for purposes such as data warehousing, executive information systems, data visualization, application development, etc. besides statistical analysis.
6. There are a number of highly experienced, dedicated SAS programmers around the world.
7. There is probably nothing in statistical analysis, which cannot be done with SAS but getting it done is always very difficult. There is so much literature available in it that manuals.
8. It is expensive (unless one get an academic discount). Base SAS is quite limited, but for SAS/STAT and other modules the prices add up.
9. SAS requires a lot of support, by someone who knows computers very well and knows SAS very well. It is a pain to learn. This is partially mitigated by the fact that SAS is very modular. Basically, there is the data step, where SAS data sets are made, manipulated and saved. Then there is a long list of procedures (called PROC's), which can be learned one at a time, for the most part.
10. The graphics in SAS are difficult to a certain extent. They came out with SAS/GRAPH, which has (theoretically) vast capability, but is almost impossible to use.

However, SAS appears to be more appropriate for an enterprise solution, where the data may reside in many different formats and SAS is the tool used to get at the data and perform Statistical analysis. However, if one is looking only for a single user PC some what limited use solution the other packages appear more appropriate.

SPSS

1. SPSS is owned by the same company, which owns some other statistical package copyright such as SYSTAT. This experience of statistical programming makes it more advanced and user friendly.
2. SPSS owes a lot to the days when the manual was one of the best introductory books around on statistics.
3. Non programmers find it easier to use than some other statistical packages; more menu driven versus programming
4. Slower performance
5. Tends to be used more in the professional field.
6. Training available everywhere.
7. It seems that SPSS is not the software of choice in industries that need data analyst. But SPSS has its audience too. It depends on our area of expertise.

SPSS and SAS

SPSS is a nice program for doing social research. SAS is a wonderful tool for doing data mining. They both have their place. It has been observed that in smaller databases, such as market research, SPSS is a far better tool for some quick and painless analysis. When you need to get some numbers flying around and are doing transformations like a maniac you had better know how to do some SAS. In terms of creating easily exportable, attractive output SPSS blows SAS away. It is not even close. SAS/Graph is well, not so good. The reason SAS is more valuable is because SPSS cannot really handle large databases very well. If there is a need to manipulate thousands and thousands of cases, SAS has the power and reliability to get you there. It also exists in a UNIX/mainframe environment, SPSS does not.

STATA

1. Interactive, very fast performance. Analysis can be done iteratively or through programming (called ado files). STATA loads the entire data set into RAM (which makes it much faster than SPSS in the normal case) which means the performance degrades considerably when data

set size exceeds available memory. Whether this matters depends on the exact size of the data sets, and availability of RAM within the system.

2. Fast performance requires large memory. Model could be constrained by lack of memory.
3. Offers "NetCourses"; less expensive training accomplished via E-mail.
4. Analysis tools and support is available, both from the STATA company and from other users on the STATALIST.
5. STATA does not play as nicely with other programs as SAS does. Also, STATA does not have many of the modules available in SAS.
6. STATA can handle extremely large datasets, both in terms of number of cases and variables.
7. Updates are easy to install, and there are many very good user-written plug-ins available.
8. STATA's documentation is superior to those available from SAS.
9. STATA is much more competitively priced for the entire package rather than getting pieces of it.

STATA and SPSS

STATA has many advantages from a statistics point of view. They have excellent support for complex samples (cluster, stratified, weighted) and offer lots of statistics especially for limited dependent variables. SPSS is ahead of STATA in terms of the friendly interface and it is easier to do data management with SPSS. If one is looking for a single program to handle big data sets and do relatively simple analyses, SPSS is certainly up to the job.

MINITAB

MINITAB is very quick to learn. More than 400 textbooks supplements reference makes it easy to use MINITAB in academic courses. There are a lot of macro's written for MINITAB. MINITAB Statistical Software is available for PC and Macintosh microcomputer systems. MINITAB is also available for mainframes, minicomputers, and workstations including VAX and other DEC platforms, Sun, IBM, Prime, Data General, Hewlett-Packard, and others.

MINITAB, SPSS, SAS

MINITAB is much more suited to statistical analysis than Excel, but definitely has some significant disadvantages compared to SAS or SPSS. It has a good interface and decent graphics, a reasonable macro facility, and decent support.

SYSTAT

SYSTAT offers robust algorithms from leading statisticians to give meaningful results even with extreme data. Fit regression models in nested, 2-level data when assumptions of independent observations are violated and the estimates from standard least squares general linear model (GLM) are suspect. Create missing value estimates using regression-based point estimation or an EM algorithm, then save the data or the triangular matrix as input to GLM, ANOVA or other models. Analyze geostatistical data with deep spatial statistics capabilities including 2-D and 3-D Variogram and Kriging. Analyze industrial data and take your process to higher sigma levels. Obtain complete distributions and standard errors using SYSTAT's bootstrapping capability implemented globally across 21 statistical procedures even when normality assumptions are violated and no model is available. Fit suitable models to your data using SYSTAT's wide choice of standard probability distributions. Use SYSTAT's fine documentation and training material-printed and online-in educational environments.

SYSTAT offers a large number of scientific and technical graphing options and a great deal of interactivity for a desktop statistics package. Compare subgroups, overlay charts, transform coordinates, add geographic projections, change colors, symbols and more to create insightful presentations. Change graph locations, point-and-click to alter axis labels, scales, colors and symbols. Create unique graphs that bring out the true meaning in your data with advanced chart options including normal and kernel densities, multiplots, maps, Voronoi tessellations, function plots, contours, scatterplot matrices with 20 diagonal density choices and 126 non-parametric smoothing options, just to name a few. Speed up your analysis by rotating your 3-D graphs to visually determine the perfect power or log transformation to normalize your data using the Dynamic Explorer. Create insightful presentations with advanced chart types such as maps, multiplots and kernel densities. Speed up your analysis with SYSTAT's interactive graphic tools. Simply point-and-click to perfect your graph's appearance.

Conclusion

Several software packages have been described along with their capabilities, compatibilities, usefulness and limitations. Many statisticians use several of the statistical packages at the same time. The core statistical capabilities exist in each of the packages. Each of the packages has its own strengths and ease of use features for the different types of analysis. While all of them have comparable GUI capabilities, it appears that ViSta and WinIdams offer the most functionality amongst the open/free access software's. The ability of defining user programs offered by both these packages is important for more advanced uses. The ease of use by GUI helps in research/teaching statistics so that the users can concentrate on the learning statistical principles and analysis, not spending a lot of time and effort in learning the software. Graphic abilities of both software packages are quite good. ViSta might have a small advantage in easy access to help functions and tutorial and educational contents. On the other hand, WinIdams is an effort of UNESCO and it might in the future be developed faster or better maintained than ViSta.The other open access downloadable software packages have some nice capabilities, but it is questionable how they can be used in more advanced procedures if the user cannot write user-defined programs and scripts. All the Statistical Packages yields almost same results (number of decimal places or some decimal points may change at best) although it follows different algorithm in each cases. In addition, some more information is being provided in each package compared to others. Thus, according to need, users should choose their packages carefully.

References

http://am.air.org/, retrieved 13th January, 2009

http://www.irri.org/science/software/irristat.asp, retrieved 13th January, 2009

http://www.reading.ac.uk/ssc, retrieved 13th January, 2009

http://www.statpages.org , retrieved 13th January, 2009

http://www.unesco.org/idams, retrieved 13th January, 2009

http://www.visualstats.org/, retrieved 13th January, 2009

❑❑❑

Agroforestry: Natural Resource Sustainability, Livelihood & Climate Moderation © 2009
Editors: OP Chaturvedi, A Venkatesh et. al., pp. 639-659
Satish Serial Publishing House, Delhi

Chapter 46

ECONOMIC ANALYSIS OF AGROFORESTRY SYSTEMS METHODS AND ILLUSTRATIONS

K. Kareemulla
Central Research Institute for Dryland Agriculture, Hyderabad (A.P.)

Introduction

Since the resources are scarce, economic analysis of agricultural systems attains prominence. This helps in assessing the returns from alternate systems or projects, which need to be prioritized or evaluated from different points of view. This evaluation can be before the systems/ projects are developed or after the development. The economic analysis is normally done from the viewpoint of impact of the project/ system on beneficiaries, environmental impact and externalities to the natural resources and the society. The economic analysis of agricultural and agroforestry developments are almost similar with a difference that the agroforestry systems have more than one component with complementary and supplementary roles. Economic analysis of agroforestry in the past three decades focused on comparative analysis of agroforestry systems with that of farmers traditional cropping/ farming systems in financial terms with tools like net returns and benefit cost analysis.

Another issue that was mostly studied was on the factors for adoption of agroforestry practices. The pace and depth of analysis lagged the trend set by bio-physical sciences. However, the wide challenges set by of Scherr, Sanchez, and Mercer and Miller, who have worked on various aspects of agroforestry economics, have been addressed to a large extent by many economists (Janaki and Mercer, 2004).

In agroforestry trees play a key role in agriculture as they not only add to diversification but also contribute to nutrition and energy security of the people. An estimated 25 per cent of rural people in Asia use agroforestry products (Leakey and Sanchez, 1997). This meant an estimated 566 million consumers, which is a formidable market by any means. The rising incomes in the developing economies are diversifying the food basket of the middle-income groups, which are creating demand for value added and nutrition foods including fruits and the processed foods. Therefore the scope for fruit tree cultivation is increasing. Owing to pressure for land demand for various uses, intensification of land use by introducing agroforestry systems that integrate horticulture species like fruits is an ideal way. The major socio-economic advantage of such a system is that the farmer will have risk bearing capacity in the event of one component, say seasonal crop component, performing below par. Similarly, the avenues for distribution of employment beyond the traditional crop cycles in tree related activities and scope for non-farm enterprising in the processing and marketing of tree products. Further the scope for linkages like nursery and other activities is another distinct opportunity linked to integration of fruit trees into agriculture i.e. through agrihorticulture system. The forward and backward linkages possible with agroforestry systems create huge employment potential.

The How and What of Economic Analysis

Adoption of agroforestry by the farmer, community, region and the country as a whole need an assessment from different points of view like technical, physical, economic and others. This is an attempt to describe and demonstrate the economic implications of agrihorticulture. Broadly based on the level of analysis, the criteria for economic analysis of agriculture and allied activities, including agrihorticulture, could be at three levels, viz. Farm level analysis, Regional level and Country level/ macro level analysis (Sullivan *et al*, 1992). Besides looking at the returns, the economic analysis is also concerned with adoption behaviour, marketing of agroforestry products and different management regimes.

Data Needs

General data for economic analysis of agroforestry must be of the following heads:

Location - study country/region, community, farm etc.

Physical and biological aspects- population, land tenure, social services, community groups, infrastructure, literacy, industrialization etc.

Institutional aspects - operational area of the project, incentives if any, land tenure rights, tree harvest laws, agencies for input and extension

Market aspects- market location, prices for tree, crop and animal products

Systems of production- type of system, management, tree species, crop species, animal species.

Economic Tools

Some of the economic tools for agroforestry economic analysis are :

Analysis of Farm Management Efficiency: Three types of farm management efficiency analyses are farm income/ comparative analysis, farm budgeting and gross margin analysis. Farm management analysis evaluates the production revenues and costs, utilization of labour, land, and capital, constraints of production etc.

Analyzing Returns to Investment: To evaluate agroforestry activities involving high initial investment or projects for development funding, a return on investment is a critical criterion. This involves two modes of analysis viz. cost benefit analysis and payback analysis. Cost benefit analysis (CBA) assesses the resource inputs, costs, and benefits over the lifetime of the proposed (ex-ante) or existing (ex-post) agroforestry technology. Pay back analysis deals with the investment, rate and spread of returns in a project mode that would give a scope for comparing alternative projects.

As the scope for analysis extends from a farm to community and regional level, one should be careful in identifying the costs and returns and in arriving at arriving at the monitory value of the services. In other words, the costs and returns must be considered in economic sense, rather than in financial sense.

Economics of Agrisilviculture Systems

The system wise economic analysis of agroforestry considering the rotation period, cropping pattern and the actual number of farmers who have harvested the produce in the past three years was done. The farmers in general practice bund planting with Poplar as well as Eucalyptus. For the sake of arriving at the comparative economics of exclusive Poplar (*P. deltoides*) and Eucalyptus based systems with the average population maintained per ha, the B:C was worked out. The cost benefit analysis of various systems is given in Table-1. Among the four systems the farmers preferred the bund system due to its less detrimental impact on the grain/commercial crop pattern (Kareemulla *et al.*, 2003).

Table 1. Cost- benefit analysis of agroforestry systems in Saharanpur

System	Rotation (years)	Tree density ha^{-1}	B:C ratio (10% Discounted)	Net returns (Rs. yr^{-1})
Poplar based bund system	8	146	2.84	17282
Poplar based agrisilviculture	7	481	3.00	28879
Eucalyptus based bund system	8	146	2.68	16124
Filed crops		Paddy –Wheat system		11734

A study was conducted with the aim of evaluating the performance of poplar-based agroforestry in terms of income, employment and environmental impact from the farmers' perspective. Poplar-based agroforestry was found to be economically viable and more profitable than many of the crop rotations followed in the study area. This land-use system was also found to be capable of providing employment opportunities on farms. Sensitivity analyses indicate that this system is not highly risky. However, costs charged by the promoting industry for technical advice substantially reduced the income from poplar plantation. With agroforestry experience, farmers can expect high dividends in subsequent rotations. Life- time matrices developed through group interviews were useful for thorough economic analyses of agroforestry projects, particularly in cases where data over a period of time are not available (Jain and Singh, 2000).

Economics of Silvipasture Systems

Kareemulla *et al.* (2002 a) had estimated the economics of introduction of silvipasture system in the degraded lands both for conservation and livelihood. The results (Table -2) indicate that in an eight year rotation on one ha with *Albizia amara+D. cinerea+ L. leucocephala* with *Chrysopogon fulvus* + *S.hamata* +*S. scabra* the B:C ratio was 1.52 at a 12 per cent discount factor.

Table 2. Economics of a Silvipasture system in one ha on degraded lands

Parameter	Indicator or value
Gross returns (Rs.)	56936
Cost of cultivation (Rs.)	36835
Net returns (Rs.)	20101
Annual average net return (Rs.)	2512
B:C ratio at current prices	1.55
B:C ratio at 12 % discount prices	1.52

Economics of Agrihorticultural Systems

Similarly, agrihorticulture based on *Emblica officinalis* (Aonla) with Blackgram under on-station conditions gave the following financial indicators (Kareemulla, 2003).

Table 3. Financial indicators of Aonla based agroforestry

Indicator	Aonla + Blackgram	Aonla	Blackgram
Discounted B:C (at 12%)	3.62	5.94	1.10
IRR (%)	59	50	-
Payback period (years)	7	8	-

Nurseries exclusively dealing with aonla have come up in Pratapgarh district. Pratapgarh district of Uttar Pradesh is the hub of aonla production with over 12000 ha of aonla plantation (Pathak, 2003). Nurseries as a viable enterprise have the potential to sustain the aonla industry besides providing employment. The present study was taken up to assess the present status of the nurseries and to analyse the costs and benefits of the enterprise.

Cost of cultivation of the sample nurseries was worked out considering the average production capacity of 30,000 saplings per annum,. The annual cost of cultivation of an average aonla nursery worked out to be Rs. 1,24,399 costing Rs.4/- per sapling (Table-4). Almost 92 percent of the cost of cultivation was accounted for by the operating (variable) costs and the rest 8 percent was on account of fixed costs. Among the variable cost, the highest was the imputed salary for the nursery entrepreneur followed by opportunity cost of investment and seed. The depreciation on tube well, pump set, pipeline and buildings were taken into account under the fixed costs.

Table 4. Cost of cultivation for an average production of 30000 seedlings per year

Item/Particular	Cost (Rs)
Seed	11900
Electricity	3000
Labour	24470
Salary of supervisor (imputed wages for nurseryman)	36000
FYM, fertilizer, irrigation etc	7300
Plant protection	500
Interest on working capital @ (8%)	6654
Opportunity cost of investment@ (8%)	24600
Total variable costs	114424 (92%)
Total fixed costs (installation of tube well, pipeline etc.)	9975 (8%)
Total cost per annum	124399 (100%)

The financial analysis of nurseries was done at an estimated investment of Rs. 3.25 lakh per nursery being the imputed value of land and capital investment on irrigation. The returns were estimated at an average sale price of Rs. 11.2 per sapling. The net annual returns worked out to be Rs. 1,27,186. The pay back period was 3 years. The discounted B:C ratio worked out to 1.14 and the IRR was 23.69 per cent. These financial ratios indicate the soundness and economic viability of the nurseries.

Economics of Hortipasture

Mango, guava, citrus, pomegranate, ber, aonla and bael are most adapted fruit species in the lower Shiwaliks. Aonla has proved to be most successful as it thrives best on adverse agro-climatic conditions even on degraded lands. Aonla started bearing from third year onwards and produced much higher yield (108 kg/plant) in aonla + chrysopogon as compared to pure stand (104 kg./plant) and lowest (92 kg./plant) in aonla + napier combination (Prasad et al, 2005). Thus, study suggested that horti-pastoral system involving aonla+chrysopogon was most promising for productive and economic utilization and rehabilitation of degraded Shiwalik foothills.

Economics of Woodlots

Based on a survey of 100 villages in northern Ethiopia, the nature of community management of woodlots was examined and the determinants of collective action and its effectiveness in managing woodlots was investigated, Despite limited current benefits received by community members, the woodlots contribute substantially to community wealth, increasing members' willingness to provide collective effort to manage the woodlots. It was noted that benefits are greater and problems less on woodlots managed at the village level than those managed at a higher municipality level, and that the average intensity of management is greater on village-managed woodlots. The factors that do significantly affect collective action include population density (higher collective labor input and lower planting density at intermediate than at low or high density), market access (less labor input, planting density and tree survival where market access is better), and presence of external organizations promoting the woodlot (reduces local effort to protect the woodlot and tree survival (Gebremedhin *et al*, 2003).

The economics of woodlots for afforestation of wastelands in India is given in Tables 5 and 6. It may be noted that with similar plant population, eucalyptus gave better returns between the two species.

Table 5. Model project on Eucalyptus for marginal lands

Particulars	Value
Tree density	4444 (1.5x1.5m)
Investment (unit cost)	Rs. 50,800/=
Rotation	7 years
Mortality	20 %
Sale price of pole	Rs. 80
Fuel wood price t^{-1}	Rs. 500
Average net income yr^{-1}	32666
BCR	2.85
IRR	36%

Source: NABARD (www.nabard.org)

Table 6. Model project on Subabul cultivation in marginal lands

Particulars	Value
Tree density	4444
Investment (unit cost)	29489
Rotation	4 years
Mortality	20%
Average Net Income yr^{-1}	Rs. 12810
BCR (at 15 % DF)	2.54
IRR	50%

Source: NABARD (www.nabard.org)

Comparative Returns from Agroforestry and other Land Management / Cropping Systems

Financial appraisal of agroforestry is similar to appraisals applied to pure forestry. Existing techniques for valuing non-market effects could also be applied. Agroforestry offers a quantifiable value in short-term sustainability, and might meet recent economic definitions of long-term sustainable development, without encountering the full dangers to future generations embodied in conversion of natural to human capital (Price, 1995).

The expected rate of return from agroforestry ventures in Chattisgarh was estimated by adopting tools like discounting, net present value, benefit cost ratio and internal rate of return by Marothia (2003). The results of the same are as follows:

Table 7. Comparative returns on investment in agroforestry and agriculture activities

S.No.	Activities	Expected life (years)	NPV (Rs'00000)			BCR at			IRR
			Discount Rates (%)						
			5	10	12	5	10	12	
1	Silvi-Pastoral	15	8	3	2	3	2	2	23
a.	Subabul	15	14	7	5	8	5	4	29
b.	Acacia spp.	20	21	11	9	21	13	11	67
2	Agri-horticulture	56	107	48	38	28	17	14	115
a.	Lime orchard	30	15	8	6	45	34	30	115
3	Land treatment	50	3	1	1	2	1	1	21
4	Water harvesting tank	50	24	11	9	2	2	2	47
5	Remodeling of old tanks	50	11	4	3	2	1	1	20
6	Tube wells	20	17	10	8	2	2	2	34
7	Pasture Land	7	0	0	0	1	1	1	18
8	Overall	50	172	79	61	4	3	3	45

(Source: Marothia, 2003)

The comparative returns on investment were much higher (above 100%) in the case of agri-horticulture and pure orchard compared to silvipasture and and silviculture as seen above. This goes to prove that the rate of return is highly lucrative in agroforestry compared to other conventional projects of land development.

Marketing of Agroforestry Produce

The market surplus and the channels of marketing for agroforestry are discussed below:

Marketed Surplus

The production and consumption of various tree products in Jhansi district were estimated. Out of the sample 60 farmers surveyed, 53 had harvested some tree product or the other during the reference year. The average figures have been worked out for the total 60 farmers. It was observed that among the tree products extracted by farmers, fuel wood production was the highest followed by fruit, timber and flower. Average production of fuel wood for the sample farmers was 6.18 q per year. Fruit, timber and flower production per farmer stood at 2.17, 1.62 and 0.63 q, respectively. Most of the sample farmers do not collect tree produce like Mahua seed and Neem fruit on their own. On the other hand the landless and children collect such produce for sale and meet their petty expenses (Kareemulla *et al*, 2004). It is reflected from the figure-1 that the farmers consume on an average about 55.4 percent of total fruit production in their home. The consumption of fuel wood, timber and flower was found to be 29.9, 23.4 and 12.3 percent, respectively.

Fig. 1. Consumption and sale of farm tree produce

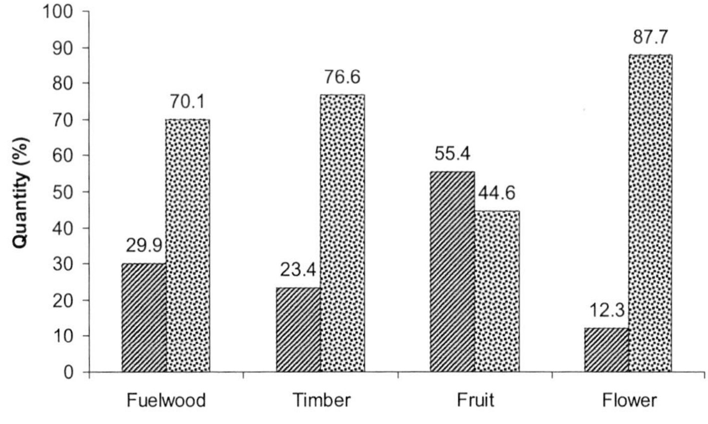

Studies on the demand and supply for agroforestry products and the price analysis can be done. Similarly estimating the marketable and marketed surplus of agroforestry produce is also possible. The usual market channels and the price spread with estimates of marketing efficiency also fall in the ambit of economic analysis.

The market channels identified for the two major tree components of agroforestry in Saharanpur district are as follows (Kareemulla *et al*, 2003):

Poplar

Product line- 1 (Trees to logs to plywood)

i. Farmers - Itinerant traders - Commission agent - Peeling/ plywood mills

ii. Farmers – Commission agent – Peeling/ plywood mills
Product line- 2 (Trees to logs to paper)

i. Farmers – Itinerant traders – Paper mills
Product line- 3 (Trees to logs to chat spoons/ match sticks)

i. Farmers – Itinerant traders – factory
Product line- 4 (Tree stumps, roots, branches to households etc.)

i. Farmers-Itinerant traders-Fuel wood depots-Hotels/Households/ Factories.

Eucalyptus

Product line – 1 (Trees to logs – Paper mills)

i. Farmers- Itinerant traders – Paper mills ii) Farmers – Paper mills

Product line – 2 (Trees to logs/poles)

ii. Farmers – Itinerant traders – Commission agent – Pole wood traders – Construction trader/timber yards.

Marketing Channels for Farm Tree Produce

The most common marketing channels observed in the district for the farm tree produce are given below

Fuel wood:
1. Farmers– Wholesaler – Urban households
2. Farmers – landless (consumer)

Fruit:
1. Farmers - Village trader - Wholesaler - Retailer/ hawker–households (consumer)
2. Farmers – Households (consumer)

Timber: 1. Farmers – Middlemen/Village Traders – Sawmill - Carpenter/ Trader - Consumer
2. Farmers – carpenter – Household

Mahua Flower: Farmers - Village Traders - Wholesaler - Retailer (liquor makers) - Households (consumer)

Jayasankar and Muraleedharan (2000) studied the bamboo marketing in India and had identified four channels in the process.

Case Study of Agrihorticulture System in Pratapgarh, U.P.

Aonla (*E. officinalis*) is a medicinal fruit species which is extensively grown in Pratapgarh district of Uttar Pradesh. Aonla area in Pratapgarh district has increased from about 8252 ha in 1994-95 to 12724 ha by 2004-05. The growth rate (CAGR) of area was 4.02 per cent. The increase in aonla based farming system could be attributed to two factors viz. assured market and promotion scheme from the department of Horticulture. The area occupied by aonla based farming system in 1994-95 accounted for about 3.81 percent of the net sown area of the district, which enhanced to 6.02 percent by 2004-05. This indicates greater acceptance of the system by the farmers. The production from the aonla plantations increased from 47329 to 82690 t in the reference period at a growth rate of 5.2 percent. The average productivity in ten years period increased from 5.7 to 6.5 t ha^{-1}. The specific decline in production during the period 1999 to 2001-02 was due to replacement of some old seedling route plantations with improved grafted varieties like NA 7 and Chakaiya on a rapid pace, as reported by the Horticulture department. As the grafted plants start yielding in the fourth year, the production picked up since 2002-02, onwards. The impact of introduction of improved grafted varieties might be seen in terms of higher productivity too from 2002-03 onwards. The trends in the area and production of aonla are also depicted in Fig. 2.

Marketing of Agroforestry Produce

In the early 1980s farmers in northwest India planted Eucalyptus on a massive scale for sale as poles and pulpwood. Later the farmers in this region stopped growing Eucalyptus, as their experience with its marketing was not a happy one. The pole market got saturated, paper mills did not pay a remunerative price, and fuel wood prices were low and uneconomical. More important, because of legal restrictions on the transport and sale of wood, and other institutional factors, the gap in the farm gate price and the consumer price remained very wide. Wood markets generally have, on the whole, exploited the farmers, rather than helped them (Saxena, 1991).

Fig. 2. Aonla area(ha) and production(t) in Pratapgarh District

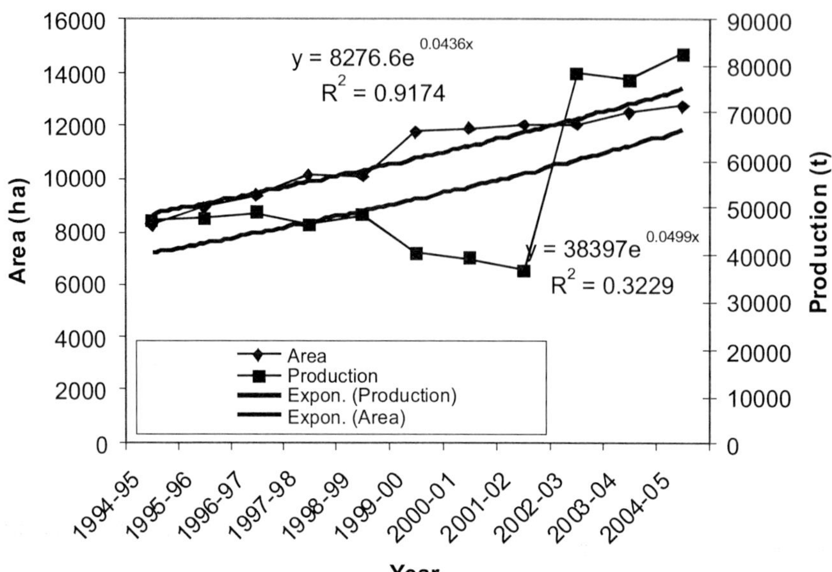

Analysis of Aonla fruit arrival in the regulated APMC, Pratapgarh during the period 1995-2004 indicate that the CAGR was 21.31 per cent. There was a wide variation in the arrivals over the years as indicated by a Coefficient of Variation of 76.21 per cent. This was probably due to direct off take of the fruits from the farm gate by the traders, bypassing the APMC for sale in consumption centres. The mandi arrivals grew by almost seven fold in a decade (Fig. 3). This indicates that productive plantation has increased in the past 10 years. A correlation analysis between the aonla production in Pratapgarh district and the arrivals at the APMC gave a correlation coefficient of 0.75. This clearly indicates that as the production increased higher arrivals are attracted to the mandi. The aonla arrives in the APMC during October-January. The aonla fruits trading at APMC take place in the evening hours during this season.

Reliable data on aonla prices at APMC, Pratapgarh were available only for the period 2001 to 2004. During this period, the current prices fell from Rs. 603 to Rs. 396 per q i.e. by 34 per cent. The price fall at constant prices was even steeper at 44 per cent. This fall obviously seems to be influenced by higher arrivals. A correlation analysis between the arrivals and prices indicated that the correlation coefficient was significant (-0.85 and -0.79) at current and constant prices, respectively.

Fig. 3. Trend of aonla arrivals at APMC, Pratapgarh

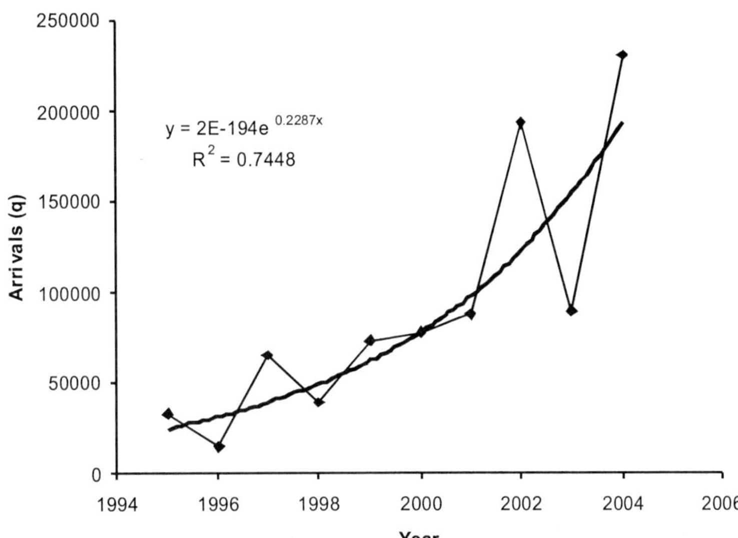

Modelling Agroforestry Adoption

A regression model was developed to predict the tree planting levels by the farmers in a semi arid district of central India (Kareemulla *et al*, 2002). The model descriptives and coefficients are given in Table-8.

Table 8. Model descriptives and coefficients in adoption of agroforestry

Variables	Mean	SD	Coefficient value	SE
No. of trees planted (Y)	10.00	13.90	-0.229	0.660
Cultivated land in ha (X_1)	6.35	4.16	0.550**	0.117
Uncultivated land (X_2)	1.35	1.86	0.090	0.454
Family size in no. (X_3)	10.00	5.60	-0.152	0.905
No. of literates in family (X_4)	4.00	2.80	0.295	0.578
No. of livestock (X_5)	6.00	4.08	0.208	0.093
No. of existing trees (X_6)	23.00	25.65	-0.229	0.660
Constant			(2.331)	4.747
R^2			0.48	

** Significance at 0.05 level

From the model it may be inferred that for adopting agroforestry, land-holding size as indicated by the extent of cultivated land is the significant variable that determines the level of adoption.

Similarly, the factors influencing the agroforestry adoption were ascertained from the sample farmers interviewed in Saharanpur district. Out of the total sample farmers', majority (71.3 %) have indicated that agroforestry is a source of additional income. The other determinants of agroforestry as pointed out by the farmers are source for emergency cash (17.5%), supplemental employment (4.4%), soil conservation (4.4%) and fuel wood (2.5%).

Willingness to Pay for a Resource

The willingness of the resource users can be modelled using the decision variables influencing such behaviour (Marothia, 2003). Factors like total area owned by the families, followed by time spent in collection of the produce, family size and distance to the resource influence significantly on the willingness to pay for a resource (forestry/ agroforestry produce) by the people.

Table 9. Modelling willingness to pay for a resource by people

Independent variable	Linear		Semi - log	
Age	0.909	(0.852)	0.006	(0.621)
Sex	-14.125	(0.341)	-0.995	(2.674)***
Family size	-4.897	(2.015)**	-0.002	(0.112)
Household income	-0.0001	(0.333)	-2.530	(0.693)
Education level (0/1)	-39.767	(1.176)	-0.194	(0.638)
Total area	36.022	(2.802)***	0.184	(1.598)*
Total livestock	3.209	(0.793)	0.046	(1.253)
Interest in maintenance (0/1)	0.826	(0.005)	-0.089	(0.053)
Distance to resource	-65.851	(1.954)**	-0.145	(0.480)
Perception of resource degradation (1/0)	104.365	(0.611)	-1.148	(0.748)
Time spent in collection	55.794	(7.730)***	0.099	(1.400)
Constant	19.661		3.630	
R^2	0.458		0.176	

Source: Marothia, 2003; Significance at * = 0.10, **= 0.05 and *** = 0.01 level

Janaki *et al* (2003) postulate that if non-market goods and services can be internalized to the benefit of landowners, the adoption of AFS will increase. Using the theory of externality as a conceptual framework their study provided an environmental economic logic for developing incentive policies to internalize environmental services especially in the industrialized countries. It has been found that the profitability of silvopasture would increase, relative to conventional ranching, if environmental services are included. Estimates of public willingness to pay for environmental services associated with silvopasture and estimates of ranchers' willingness to accept for the adoption of silvopasture will provide a scientific basis for policy development.

Optimization and Simulation Models: Mathematical programming models and production function analysis allows the analyst to identify those technical options that provide the best returns. Mathematical programming entails optimization under specific resource constraints and objective function (eg. maximizing income from alley cropping by optimum allocation of leafy biomass between mulch and fodder, on a specified land area with specified available labour).

Environmental Resource Economics

The role of forestry and agroforestry systems in the flux and long-term storage of carbon (C) in the terrestrial biosphere has increased global interest in these land-use options to stabilize greenhouse gas (GHG) emissions. Early assessments suggest that some agroforestry systems (e.g., agrosilvicultural) can be CO_2 sinks and temporarily store C, while other systems (e.g., ruminant-based silvopastoral systems) are probably sources of GHG (e.g., CH_4). Under tropical conditions, it is estimated that one ha of sustainable agroforestry can provide goods and services which potentially offset 5–20 ha of deforestation. At a global level, agroforestry systems could potentially be established on 585–1275 million ha of technically suitable land, and these systems could store 12–228 Mg C ha^{-1} under current climate and edaphic conditions (Dixon, 1995).

The environmental impact of adopting biomass based power generation using *P. juliflora* was done. The total CO_2 emission reduction potential (as a consequence of not using the coal in thermal plants) of the biomass-based power plants (on account of 16000 MW potential/yr) is 35.3million t/ year. Although many plants have been commissioned, they are yet to be registered with the United Nations Framework Convention on Climate Change (http://unfccc.int). As on March 6, 2006 only two plants from A.P.

(18 MW) have registered with UNFCC for a saving of 39670 t of CO_2 per annum. At a moderate rate of $4/ t CO_2 the potential environmental impact through CO_2 reduction will be $ 140 million / year.

Total Economic Value

In order to capture the direct and indirect benefits as also to take in to account the intangible benefits, the only method is to work out the total economic value (Cavatassi.2004), which is estimated as follows:

TEV = *{Use Values}* + *{Option Values}* + *{Non-use Values}*

Which can be rewritten as

TEV = *{Direct Use V. + Indirect Use V.}* + *{Option V.}* + *{Existence V. + Bequest V.}*

Total economic value provides a convenient framework for pooling the different classes of value. The indirect market values may be estimated using surrogate values, replacement or avoided value, opportunity cost, travel cost, hedonistic methods etc. Use value refers to the actual use of a resource. For example, the cost of fertilizer when used in some production is a use value. Option value is the value attached to the future use of a resource. Non-use values are independent of present use of a resource. They will be called existence values in the case of an endangered species and are called bequest values in case certain produces are preserved for future generations.

Employment Generation

Besides the entrepreneurial opportunity for the nursery promoter, the aonla nursery provides ample scope for both skilled and unskilled labour. The budding activity created opportunity for gardeners who specialize in budding skills with formal/informal training. In studied nurseries there were 1-2 such gardeners employed on a regular basis depending on the production capacity of the nurseries. Besides this, an average nursery provides continuous employment of 762 mandays per year, of which the skilled labour employment worked out to be about 19 percent. Weeding was the major operation that generated maximum employment in nursery management.

Impact Assessment of Community Forestry

A community forestry project was implemented by Indian Farmer's Fertilisers Cooperative (IFFCO) and its subsidiary Indian Farm Forestry Development Cooperative (IFFDC) in three states viz. Uttar Pradesh, Madhya Pradesh and Rajasthan adopting the cooperative model. The project promoted 145 such grass root level institutions with a membership of with about 28000 members. The members belonged to both landed and landless communities. In order to encourage community leadership and micro enterprising, Self Help Groups (SHGs) were also promoted as part of the project. The project was implemented in two phases with the first (pilot) phase initiated in 1986, while the second phase was implemented with funding support from India-Canada Environment Facility (ICEF). Besides other impact indicators, economic indicators were worked out for the project (Kareemulla *et al*, 2006).

The revenue generated by the pilot societies from agroforestry and micro enterprises during the project period worked out to Rs. 30491 per society/ year (Table-10). Across the states this figure was the highest in Rajasthan at Rs. 39660 probably due to higher tree harvests. Lower revenues in Madhya Pradesh is obviously due to lower plantation area, which is about 46 ha/ society compared to 155 and 184 ha in UP and Rajasthan. Thus on a per ha basis, the average revenue works out to be Rs. 212 on a overall basis with the highest of Rs. 410/yr in MP followed by Rs. 216 and Rs. 179 in Rajasthan and U.P., respectively. Nurseries, fertiliser marketing and some micro-enterprises were the major income sources for the PFFCS.

Table 10. Revenue generated by Pilot PFFCS from plantation and micro enterprises.

State	Revenue earned/ society (Rs.) during 1986-2003	Average per year (Rs.)	Average Rs. yr^{-1} ha^{-1}
UP	470765	27692	179
MP	320333	18843	410
Rajasthan	674222	39660	216
Overall	518345	30491	212

Viability of grass root level institutions

Based on the actual revenues and the estimates of standing and projected tree stock worth, the economic evaluation was done for the pilot (phase-I) plantations (Tables 11-12). Similarly, by accounting the environmental contributions of the agroforestry plantations, the environmental cum economic evaluation was also done (Tables 13-14).

Table 11. Economic evaluation of pilot PFFCS at 17 years age (amount in Rs.)

State	Gross returns ha^{-1} yr^{-1}	Net returns ha^{-1} yr^{-1}	Gross returns $member^{-1}$ yr^{-1}	IRR (%)
UP	5139	4229	4177	17.0
MP	2931	2022	2382	11. 7
Rajasthan	2785	909	3193	11.5

Table 12. Economic evaluation of pilot PFFCS at 30 years (amount in Rs.)

State	Gross returns ha^{-1} yr^{-1}	Net returns ha^{-1} yr^{-1}	Gross returns $member^{-1}$ yr^{-1}	IRR (%)
UP	9822	9309	7984	33.4
MP	5216	4701	4239	27.0
Rajasthan	5168	4653	5924	28.9

Table 13. Environmental cum economic evaluation of pilot PFFCS at 17 years (amount in Rs.)

State	Gross returns ha^{-1} yr^{-1}	Net returns ha^{-1} yr^{-1}	Gross returns $member^{-1}$ yr^{-1}	IRR (%)
UP	5471	3611	4447	20.39
MP	3138	1278	2550	13.94
Rajasthan	2996	1136	3435	13.96

Table 14. Environmental cum economic evaluation of pilot PFFCS at 30 years of age (amount in Rs.)

State	Gross returns ha^{-1} yr^{-1}	Net returns ha^{-1} yr^{-1}	Gross returns $member^{-1}$ yr^{-1}	IRR (%)
UP	10269	8409	8347	38.88
MP	5463	3603	4440	30.09
Rajasthan	9564	7704	10964	32.73

Livelihood Impact

The effect of a particular agroforestry system on the livelihoods of the farmers and other stakeholders specifically in terms of income and employment will give an understanding of the efficacy of the agroforestry system. This could be done by way of comparing the present income and employment levels from the existing land use systems and then comparing with the proposed or introduced agroforestry systems. Agroforestry provides the scope for distribution of income and employment across seasons, unlike crop production. This is an important dimension that infuses the stability to the farming systems and the livelihoods of the farmers. New initiatives in agroforestry are seeking to integrate into tropical farming systems indigenous trees whose products have traditionally been gathered from natural forests. This is being done in order to provide marketable products from farms that will generate cash for resource-poor rural and peri-urban households. This poverty-alleviating agroforestry strategy is at the same time linked to one in which perennial, biologically diverse and complex mature-stage agroecosystems are developed as sustainable alternatives to slash-and-burn agriculture (Leakey and Simons, 1997).

The livelihood impact due to agroforestry and rural development under the IFFCO/IFFDC project was measured in terms of employment, self help groups and micro-enterprises. Employment for the members and non-members of the society was generated on account of the agroforestry operations in the project areas. In pilot phase societies of Uttar Pradesh, the total employment generation was 429/ ha person days in a period of 16 years @ 27 person days/ha/year. On the other hand in phase II it was about 23 person days/ ha/yr. The share of female employment was 47 percent with higher share in Rajasthan.

In order to actively involve the weaker sections and women in rural development, besides the wastelands development, Self Help Groups were formed by the PFFCS across the three states. Three types of SHGs viz. women, men and mixed groups were formed as part of the IFFDC project, initially in phase II societies which were gradually spread to phase I societies. The groups initially encouraged the habit of thrift and internal lending through such savings. Subsequently, from the IFFDC project funds, a revolving fund facility was extended to the SHGs through the societies. While the SHGs' own funds were meant for both productive and consumptive purposes, the assistance from revolving fund is meant only for productive purposes. The number of SHGs in phase I were only 76, while it was almost ten folds at 800 in phase II societies. Among the three states, U.P. had the maximum number of SHGs in both phase II & I. The total membership of these SHGs in phase II stood at 8084.

Conclusions

Agroforestry with its multiple impacts needs to be assessed for its economic and environmental contributions. This could be possible by assessing the rates of return on investment for various alternative agroforestry practices. Similarly the impact on livelihoods through income and employment by adopting a particular agroforestry model can be assessed. The extent of value addition and the scope for non-farm sector development in the process can also be evaluated. Further, there is need to assess the effect of agroforestry development on the society, region and the environment. There are many methods and tools for economic analysis, but their selection and application depends on the context and the objective of assessment.

References

Cavatassi, R.2004. Valuation Methods for Environmental Benefits in Forestry and Watershed Investment Projects ESA Working Paper No. 04-01, FAO, Rome

Dixon, R.K. 1995. Agroforestry systems: sources or sinks of greenhouse, gases? *Agroforestry Systems,* 31:99-116.

Gebremedhin, B, John Pender [a2] and Tesfay, Girmay. 2003.Community natural resource management: the case of woodlots in Northern Ethiopia Environment and Development Economics, 8 (1):129-148

Janaki R.R.A and Mercer, D. E. 2004. Valuing Agroforestry Systems Methods and Applications. Kluwer Academic Publishers, New York. Pp. 314

Janaki .R.R.A, Shrestha, R.K., Stainback, G.A. and Matta, J.R. 2003. Agroforestry development: An environmental economic perspective *Agroforestry Systems,* 61-62:299-310

Jain, S. K. and Singh, P. 2000. Economic analysis of industrial agroforestry:poplar (*Populus deltoides*) in Uttar Pradesh (India). *Agroforestry Systems,* 49: 255–273

Kareemulla.K., Dwivedi, R.P and Rizvi, R.H. 2004. An exploratory study of farm tree produce marketing in Jhansi district. Annual report-2003-04 NRCAF. Jhansi.

Kareemulla.K., Dwivedi, R.P., Singh, Ramesh and Rizvi, R.H. 2003. An economic diagnosis of agroforestry systems in western Uttar Pradesh, Completion report of APCF scheme. NRCAF. Jhansi.

Kareemulla.K., Ram Newaj, Yadav, R.S. and Solanki, K.R. 2003. An economic analysis of aonla based agroforestry system under rainfed conditions. *Agricultural Economic Research Review,* 16 (1): 55-59

Kareemulla.K, Rai, P. Rao, G.R. and Solanki, K.R. 2002. Economic analysis of a silvipastoral system for degraded lands under rainfed condition. *Indian Forester,* 128 (12): 1346-1350

Kareemulla.K., Rizvi, R.H., Singh, Ramesh and Dwivedi, R.P. 2002. Trees in rainfed agro-ecosystem- a socio-economic investigation in Bundelkhand region. *Indian Journal of Agroforestry,* 4(1):53-56

Kareemulla. Rizvi, R.H., Munna Ram and Yadav, R.S. 2006. Evaluation of Community/ farm Forestry from Natural Resources and Livelihood Management perspective . Final Report of the AP Cess Fund Scheme, Pp. 75

Leakey, R.R.B. and Sanchez, P.A., 1997. How many people use agroforestry products? Agroforestry Today 9 (3), 4–5 Mahapatra, Richard, 2003. The multibillion Dollar Fuel wood Trade is the last resort for India's poor. Info change Features, www.infochange.org.

Leakey, R. R. B and Simons, A. J. 1997. The domestication and commercialization of indigenous trees in agroforestry for the alleviation of poverty. *Agroforestry Systems,* 38: 165-176.

Marothia, D. 2003. Valuing agroforestry systems from natural resource economics perspective. Paper presented in Symposium on Agroforestry and Sustainable Production.7-9, Nov. 2003. Jhansi.

Pathak, R. K. 2003. Status report on genetic resources of Indian goose berry-Aonla (*Emblica officinalis* Gaertn.) in South and South East Ashia. IPGRI, office for South Asia, NASC, New Delhi, Pp. 105.

Prasad, R., Aggarwal, R.K., Agnihotri, Y., Yadav, R.P. and Samra, J.S. 2005. Fruit Based Hortipastoral Land Use Systems for The Rehabilitation of Degraded Shiwalik Foothills of North-West India. *Acta Horticulture*, 696: 597-600.

Price, C. 1995. Economic evaluation of financial and non-financial costs and benefits in agroforestry development and the value of sustainability *Agroforestry Systems,* 30: 75-86, 1995.

Saxena N.C.1991. Marketing constraints for Eucalyptus from farm lands in India. *Agroforestry Systems,* 13: 73-85

Sullivan, G.M., Huke, S.M. and Fox, J.M.1992. Financial and economic analysis of agroforestry systems. Proceedings of a workshop held in Honolulu, Hawaii, USA, July 1991. Paia. HI: Nitrogen Fixing Tree Association.

UNFCCC.2006. http://unfccc.int. United Nations Framework Convention on Climate Change, Registered CDM projects.

❑❑❑

Agroforestry: Natural Resource Sustainability, Livelihood & Climate Moderation © 2009
Editors: OP Chaturvedi, A Venkatesh et. al., pp. 661-674
Satish Serial Publishing House, Delhi

Chapter 47

BIOMETRICS AND ECONOMICS OF POPLAR BASED AGROFORESTRY SYSTEMS IN NORTHERN INDIA

A. Venkatesh, R.K. Tewari, Rajendra Singh, K. Kareemulla A. Kumar, O.P. Chaturvedi and S.K. Dhyani
National Research Centre for Agroforestry, Jhansi-284 003 (U.P.)

Introduction

Timber, fuel wood, fodder, fruits and medicine are the major products from forestry and agroforestry. Among these, timber is of great commercial significance as it has greater relevance to trade situation. India has been importing timber and other forest produce for industrial and domestic purposes. The gap of 20 million m^3 of industrial timber between demand and production in 1997 is increasing rapidly and is projected to be around 110 million m^3 by 2090 in India (NRCAF, 2007; Kareemulla et al. 2005; Singh and Marzoti, 1996). More than one thousand crore rupees worth of raw material is imported every year for meeting the demand of forest based industries in 1999 and it is likely to increase to twelve thousand crore rupees by 2010 (Singh and Kumar, 1999). Fuel wood is another important contribution of forests and vegetation on farmlands which is supplemented by agroforestry in the past few decades. Still the domestic cooking energy needs are met by using about 120 million tonnes dry cow dung in the country (Deb Roy, 1999). Thus, there is a vast potential to propagate large biomass yielding trees in agricultural lands under agroforestry mode.

To meet the commercial/industrial and domestic demand, *Populus deltoids* Bartr. ex. Marsh (poplar), a fast growing species was introduced in north western India (above 28^0) from North America during 1950s (Mathur and Sharma, 1983). The main purpose was to overcome poor adoptability and slow growth of indigenous poplars viz., *Populus ciliata* Wall. ex Royale, *P. gamblei* Dode, *P. jacquenmontana* Dode var *glauca* (Haines), *P. alba* Linn. *P. nigra* Linn. and *P. euphratica* Oliv. (Tiwari, 1993). Poplar based agroforestry is one of the most successful and widely adopted systems/ and found in north-western parts of India covering Western Uttar Pradesh, Uttarakahnd, Haryana and Punjab (Kareemulla et al. 2003; Singh et al. 1999). However, poplar cultivation is now being successfully extended to central and eastern Uttar Pradesh, Bihar, parts of Madhya Pradesh, Maharashtra and West Bengal (Ballal et al. 2001; Puri et al. 2001; Chandra, 2001; Singh et al. 2001c; Singh and Jha, 2001).

Poplar wood is used for various purposes, match wood, ply wood, particle boards, cardboards, crates, pulp, packing material, pallets, furniture, plank, fuel wood etc. In north India, poplar is the principal species for making ply wood (Singh et al. 2000; Singh and Jha, 2001; Singh and Kumar, 1999). In addition, it contributes nitrogen to the soil as much as 43 kg in 3rd year rising to 102 kg in 11th year (Kaul et al. 1983). Improvement in soil structure and chemical properties under poplar are also reported (Singh et al. 1989).

Poplar in Agroforestry – Indian Scenario

Out of 100 species of *Populus* in the world, seven species are reported in India, but none is suitable for commercial cultivation in northern plains of the country. Around 500 clones/varieties of *Populus* species have been introduced and evaluated since 1950s (Singh and Kumar, 1999) and clones viz., G-3, G-48, D-121 were found most productive (Chaturvedi, 1982) and their cultivation started on large scale.

From 1980 to1990 *P. deltoids* has been planted commercially under agroforestry by farmers of north western India (Chandra, 2001) owing to its fast growth [short rotation of 5-8 years (Chaturvedi and Rawat, 1994)], high productivity [> 30 $m^3 ha^{-1} year^{-1}$ (Chandra, 2001)], deciduous nature in winter (FAO, 1979) and intercropping with wide range of arable, vegetables, spices and fodder crops (Newman, 1997).

The average productivity of agroforestry plantations with poplar is 20-25 $m^3 ha^{-1} yr^{-1}$ in comparison to 4.5 $m^3 ha^{-1} yr^{-1}$ from forest plantation. Some plantation in Punjab have yielded mean annual increment of 49 m^3 ha^{-1} yr^{-1} under intensive management using genetically superior clones and this is probably the highest productivity ever achieved in India by agroforestry trees (Singh et al. 2001 a, b). The spectacular success of poplar based agroforestry in northern parts of India has halted further deforestation and/or deterioration of poorly stocked forests and as a result it has brought substantial relief to wood-based industries (Singh and Kumar, 1999).

Extent of Practice

About 70 million trees of poplar, producing 10.40 million m^3 industrial wood annually, are standing in north western India alone (FSI, 2005). This is equivalent to 0.14 million ha pure plantation of poplar at a national density of 500 trees ha^{-1}. About 25, 000 ha equivalent plantations of poplar are established every year in association with agricultural and horticultural crops under a rotation of 6 to 8 years. It is estimated that currently the requirement of ETPs (entire transplants) would be around 15 million yr^{-1} while, 13 million ETPs are grown annually in nurseries by agencies, farmers and NGOs. In 8th five year plan, the agricultural area brought under poplar based agroforestry in Uttar Pradesh, Punjab and Haryana was equal to the extent of forest plantations raised by Forest Departments of these states (Chandra, 2001).

Given the key role played by the poplar-based agroforestry a study was conducted to document the performance of the system. A transect survey was made in three states viz., Uttarakhand, Haryana and Punjab states between 2003 and 2005 to asses the present scenario of poplar based agroforestry in terms of growth, productivity, cropping sequence, marketing, economics etc.

Methodology

A transect survey was made starting from Pant Nagar (Uttarakhand) to Muzzafarnagar (Uttar Pradesh), Yamuna Nagar (Haryana), Ludhiana, and Amritsar (Punjab). Five farms in each age category in every state were selected to compare the growth pattern in these states. Thus, a total of 40 farmers from Uttarakhand, 35 from Punjab and 30 from Haryana were selected (n=105). To know the impact of management option (pruning interval) on the growth of trees was done in Uttarakhand state, and five farms of each category were identified (n=25). For comparison of tree growth and productivity of different crops in block and boundary plantations 2 farms each of age (1-7 years), crop combination and type of plantation were selected (n=84) from Uttarakhand. For taking biometric observations, in each environment, about 15-20 trees were measured for tree height, girth at breast and the quantum of light available to the intercrops. To derive the economics of block and boundary plantations ten farmers in each state were interviewed using a pre-tested questionnaire (n=30). However, in case of poplar with aquaculture, only one farmer's field was investigated in Punjab since the occurrence of the system is very rare (n=1). In this investigation, main emphasis was given to have an idea about crop combinations with poplar, growth of poplars in different states, response of poplars to frequency of pruning, productivity and economic returns of poplar based agroforestry system. In each environment, 20 trees were measured for tree height, girth at breast and the quantum of light available to the intercrops.

Results and Discussion

Prominent Agroforestry Systems

Mostly poplar trees are grown in block plantations with spacing of 4m x 5m or 4.5m x 4.5m and on the farm boundaries with line spacing of 2.5m to 3.5 m were adopted in the poplar based agroforestry system. By and large, poplar plantation is raised by planting ETP (Entire Tree Planting), which are prepared, by planting cuttings. Among the clones, G-3 and G-48 were mainly grown in all the three regions, however some clones viz., WSL-38, S7/C15 and White were also found in Yamuna Nagar and other areas in Haryana. Farmers opined that G-48 is comparatively resistant to pest and disease over G-3.

Cropping Systems

The following main cropping systems were observed in the study states. Where ever the water logging condition exists, the farmers tend to grow poplar on the bunds of boundary in Uttarakhand.

Poplar Plantation Type	Crop sequence along with Poplars		
	Uttarakhand	**Haryana**	**Punjab**
Boundary	Paddy + Wheat	Paddy + Wheat	Paddy + Wheat
Boundary	Sugarcane	-	Fishery
Boundary	Paddy (Kharif) + Paddy (summer) + Pea (Rabi)	-	-
Block	Sugarcane (I&II year) + Wheat (III year onwards)	Sugarcane (I&II year) + Wheat (III year onwards)	Sugarcane (I&II year) + Wheat (III year onwards)
Block	Forage Berseem	Forage Berseem	Potato
Block	Ginger / Turmeric	Mango / Litchi + Ginger / Turmeric	Ginger / Turmeric
Block	Wheat / Potato	Wheat / Potato	Fodder crops, Wheat / Potato
Block	-	Vegetables (I & II year) + Wheat (III year onwards)	Flower plants (Rose / Marigold)

It was observed that most of the farmers grow sugarcane in first two years of poplar plantation and then for remaining rotation period wheat is grown. Occasionally, crops like potato /and turmeric are also grown as under under-storey crop. Wheat is sown in the last week of October when the poplar sheds leaves on account of its deciduous nature. Thus the wheat crop escapes shade during maturity stage and yield will be higher. If early sowing is done, the fallen leaves must be swept away from the field. In case of late sowing, seed rate has to be increased by 20% and no sweeping is required. But at the time of grain filling, the tree-crop competition will be more due to emergence of new shoots from poplar that results in low crop yield.

No *kharif* crops were grown in block plantation, however, in boundary plantation paddy is grown during *kharif* season. This is so mainly because of two reasons (i) during *kharif*, poplar is lush green as such shade is high (ii) paddy cultivation requires standing water in field which often causes uprooting of trees as poplar is shallow rooted. No severe insect pest/disease interaction was found between poplar and associated crops except hairy caterpillar.

Growth Pattern of Poplar in Different Edapho-climatic Conditions

In order to have a thorough understanding of growth of poplar along age series in different poplar growing regions, growth data was collected from Punjab, Haryana and Uttarakhand. The growth of poplar in terms of height and girth at breast height (GBH) was more in Haryana than that of Punjab and Uttarakhand. Minimum growth rate was recorded in Uttarakhand. Mean annual increment in height was 4.48, 3.41 and 3.16 m in Haryana, Punjab and Uttarakhand, respectively while GBH was 13.1.7, 11.5 cm and 10.1 cm, respectively. This is attributed to edapho-climatic conditions of respective growing regions. Further, in Haryana, poplar cultivation is common along Yamuna river where soils are light and soil moisture is adequate round the year. These ETPs are uniform in growth and have 2.8-3.0 m height and 10-12 cm GBH at the time of planting. In six years, poplars attain a height of 22.3 to 26.9 m and GBH of 70.6 to 82.1 cm (Fig 1 & 2). The growth of poplars in six years is comparable with the results showed by Singh et al. (1993) in Punjab.

Management Options

Pruning is the most important silvicultural practice in poplar cultivation. Pruning affects clean bole development and ultimate production. Farmers generally prune 50 % of the canopy each time and practice four types of pruning viz., a) pruning in the first year and repeated every year, b) pruning in the first year and repeated in alternate years, c) pruning in the second year and repeated every year, and d) pruning in the second year and repeated in alternate years depending upon their resources. Trees pruned in the first year and continued in every year resulted in maximum height and GBH with narrow crown diameter. However, alternate pruning starting from 1st year or 2nd year resulted in reduced growth in terms of height and GBH and wider canopy diameter. Unpruned trees recorded 17 m (height) and 60 cm (GBH), whereas pruned trees showed height between 19.5 and 22.5 m; and GBH between 65 and 72 cm (Table 1). On enquiry, it was learnt that sale of pruned biomass largely compensates pruning costs. Currently, the farmer pays Rs 2-4 per plant for pruning depending upon the age of trees. Pruning from 2nd year and subsequently every year was most common with the farmers growing sugarcane as inter crop in the first two years of the rotation. Other pruning systems were observed with paddy - wheat or other crops as companion crop.

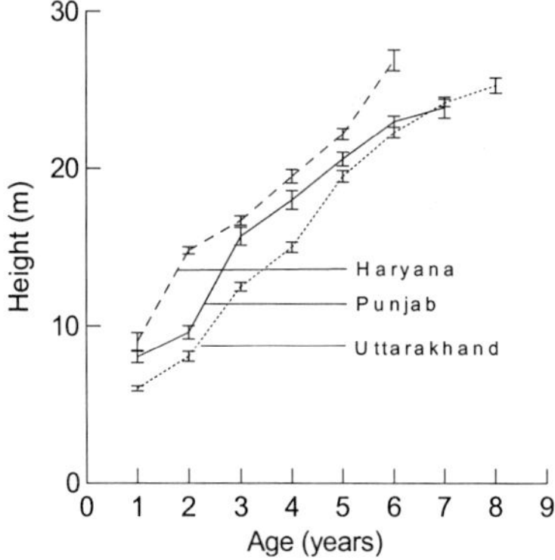

Figure 1. Variation in height growth of poplars in three states (mean ± SE)

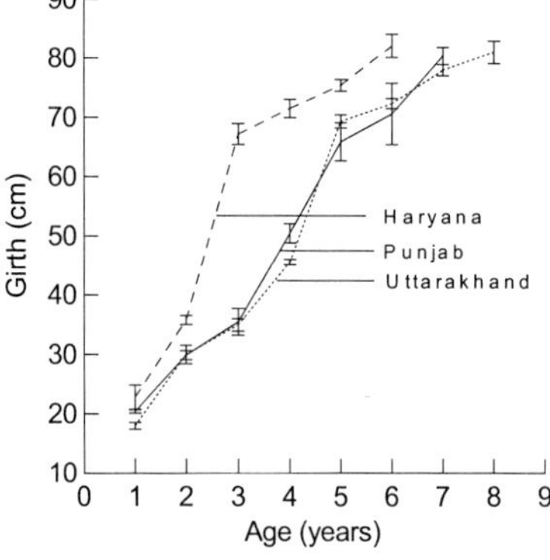

Figure 2. Variation in GBH of poplars in three states (mean ± SE)

Table 1. Effect of pruning on growth of poplar (6 years old, Uttarakhand)

Pruning system	Height (m)	Girth (cm)	Canopy diameter (m)
Pruned from 1st year every year	22.5±0.29	72±0.083	4.5±0.14
Pruned from 1st year then alternate year	20.5±0.44	68±0.71	5.0±0.14
Pruned from 2nd year every year	21.5±0.35	70±1.41	4.7±0.10
Pruned from 2nd year alternate year	19.5±0.27	65±1.00	5.2±0.11
With out pruning	17.0±0.58	60±1.14	7.0±0.19

Tree-Crop Interactions and Effect on Crop Productivity

Block plantation (spacing 5mx4 m)

Growth data obtained for block plantations at 5mx4 m spacing revealed that the light interception decreased with increase in age of tree due to positive increment in tree height and GBH (Fig 3). The study revealed that sugarcane/ wheat, wheat and paddy produced maximum yield (59.0, 4.25 and 4.75 t ha^{-1} respectively) in the first year and decreasing trend was observed with the increase in age of poplars. Crop yield for the first year under poplar is considered as same in the open condition. Hence, the poplars were planted amidst the crops in the auger hole with out disturbance to sole crop. The yield reduction ranged between 42 % (sugarcane & wheat, and wheat) to 45 % (paddy) in the final rotation period *i.e.,* at 7th year (Table 2). It might be due to the above and below ground competition between the trees and under story crops for nutrient, minerals, moisture, light etc. Ralhan et al. (1992) have mentioned that a wheat variety (HD-2329) showed a different yield reduction under different levels of stresses caused by poplar of different ages (1-6 years). The reduction in yield of three wheat varieties under six year old poplar (Singh et al. 1993) and along the age series of poplar plantations (Singhal and Panwar, 1992) has already been reported.

Table 2. Poplar growth and crop yield reduction along age series in block plantation (spacing 5x 4 m)

Age (yrs)	Height (m)	GBH (cm)	Light reduction (%)	Crop yield (t ha^{-1})		
				Sugarcane & wheat	Wheat	Paddy
1	6.0±0.15	18.0±0.52	21.3	59.1[a] (00%)	4.25 (00%)	4.75 (00%)
2	8.0±0.30	30.0±1.39	32.3	39.4[a] (33.3%)	3.93 (7.5%)	4.35 (8.5 %)
3	12.5±0.25	35.0±0.93	37.9	3.72 (12.5%)	3.72(12.5%)	4.19 (14 %)
4	15.0±0.30	45.6±0.42	46.7	3.40 (20%)	3.40(20%)	3.66 (23%)
5	22.8±0.36	69.3±1.98	50.8	2.98 (30%)	2.98 (30%)	3.11(34.5%)
6	26.3±0.19	79.0±0.97	53.5	2.55 (40%)	2.55 (40%)	2.73(42.5%)
7	26.8±0.27	84.0±1.59	53.5	2.47 (42%)	2.47(42%)	2.61 (45%)

(Value in parentheses indicate % yield reduction over first year)

[a] Yield of sugarcane, since for the initial two years sugarcane only is grown and from third year onwards wheat is grown

Boundary Plantation (spacing 2 m)

Poplars in boundary plantation at 2 m spacing attain a height of 27.3 m with GBH of 99.3 cm in 7 years. The growth of poplars in the boundary was superior over block plantation. As progress in the age of poplars, the reduction in crop yield ranged from 6 -7.5 % in 4th year and almost double in 7th year (13-15%). On the other hand poplar recorded maximum current annual increment for height (7 m) and GBH (27 cm) in 5th year. After 6th year the CAI for height was decreasing while GBH increasing. The mean annual increment of poplar was 3.9 m (height) and 14.2 cm (GBH). The reduction was more in the crop combinations of paddy + poplar (15 %) followed by wheat + poplar (14 %), while it was less in sugarcane + wheat + poplar (13 %) as compared to first year yield (Table 3). Chaturvedi and Pandey (2001) also recorded drastic reduction in yield of inter crops viz., wheat, maize, turmeric, pigeon pea and turmeric with increase in age of poplar plantations. A progressive reduction in agricultural production (6-25%) under different ages of *Eucalyptus* boundary planting has been reported by Saxena (1990). Similarly, the effect on wheat yield of *Eucalyptus* planted on boundaries has been studied by Dhillon et al. (1982).

Table 3. Poplar growth and crop yield reduction along age series in boundary plantation (spacing 2 m)

Age (yrs)	Height (m)	GBH (cm)	Crop yield (t ha^{-1})		
			Sugarcane	Wheat	Paddy
1	6.1±0.23	18.0±0.39	59.05 (0.0%)	4.25 (0.0%)	4.75 (0.0%)
2	8.4±0.38	32.1±1.39	57.87 (2.0%)	4.18 (1.8%)	4.67 (2.0%)
3	13.2±0.27	37.3±1.71	56.98 (3.5%)	4.08 (4%)	4.55 (4.5%)
4	16.1±0.42	48.0±2.05	55.51 (6.0%)	3.97 (6.5%)	4.39 (7.5%)
5	23.1±0.31	75.0±0.82	54.33 (8.0%)	3.86 (9.0%)	4.28 (10.0%)
6	26.8±0.24	86.2±0.60	52.55 (11.0%)	3.74 (12.0%)	4.11 (13.5%)
7	27.3±0.33	99.3±2.29	51.37 (13.0%)	3.66 (14.0%)	4.04 (15.0%)

(Values in parentheses indicate % yield reduction over first year)

Instability of Market Price for Poplar Wood

Stable and remunerative prices are an indication of the good performance of a crop in the market context. Poplar had unstable market in the past couple of years. Table 4 indicates that there is a decreasing trend in market prices of poplar wood from 1997-98 to 2003-04, there after price increased. The price fluctuations were more in case of higher girth class compared to lower one. This certainly is a cause of worry for the farmers. In order to avoid fluctuations in market price and encourage farmers for poplar plantations, the Government should announce a support price for farm wood produce, as it is done for agricultural commodities.

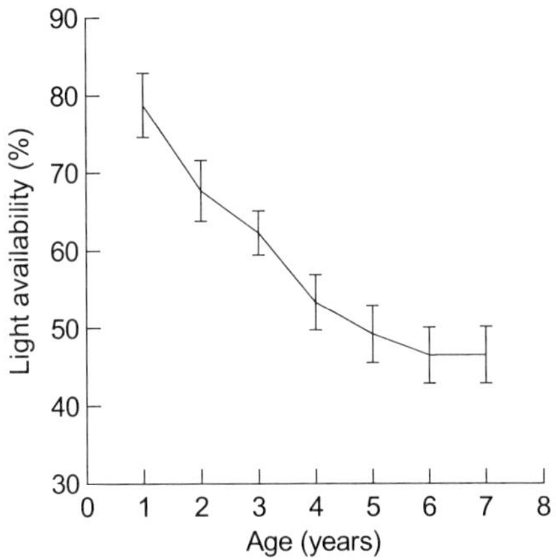

Figure 3. Availability of light along with age (mean ± SE)

Economics of Different Systems

Poplar + Sugarcane & Wheat (6 years rotation)

The study revealed that poplar + sugarcane for first two years and wheat for remaining four years of a total rotation of 6 years fetched on an average Rs. 3,42,180 per ha (1 US$ = 47.31 Rs. November 2008) which works out to Rs. 57,030 ha^{-1} yr^{-1} as net income. Major contribution to the gross income was from poplar (73.7%) followed by sugarcane (14.9%) and wheat (11.4%). The net return from poplar based agroforestry system (sugarcane + wheat) when compared with pure cropping system seems to be more remunerative, although the market prices of poplar have gone down sharply from Rs. 350 q^{-1} (1995) to Rs. 175 q^{-1} (2005) for the timber having >25 inches of girth (Kumar, 2007). Still, this land use is more remunerative than cropping alone. This system is most common in Punjab, Haryana, parts of Uttar Pradesh and Uttarakhand. Dhillon et al.(2001) reported that the average annual net returns for poplar in Haryana state were Rs. 54,753 ha^{-1} yr^{-1} (as pure crop) and Rs. 72,480 ha^{-1} yr^{-1} (with intercropping) at 12 % discount rate over seven year rotation. The benefit: cost ratios in the above two planting situations were 1.92:1 and 2.13:1, respectively.

Poplar Plantation on Boundary

Boundary plantation of poplar is quite common in Uttarakhand tarai. In this system, poplar is grown on the field boundaries. Economic analysis of this system indicated a net return of Rs. 3,21,795 ha^{-1} on 7 years rotation. Thus, annual average return worked out to be Rs. 45,971 ha^{-1} $year^{-1}$. In this system, contribution of returns from poplar worked out to be 31.3% and from sugarcane 22% while from paddy 18.5% and wheat 28.0%.

Poplar + Fish

Another land use system involving poplar plantation was observed in Punjab where in poplar was planted on the bunds of fishpond. The system appeared to be economically viable and highly remunerative. However, the frequency of occurrence of this system is very less. It was observed that farmers earn about Rs. 5,52,872 ha^{-1} in a 6 year rotation of plantation. In this system major contribution of net returns (81.4 %) comes from fisheries while 19.6 % returns come from poplar plantation. Growth of poplar under this system was excellent whereas input cost was minimal as the plants on pond bund were totally rain fed. The coarse texture of soil facilitates flow of sufficient seepage water from pond to meet the moisture requirement of trees.

Singhal and Panwar (1992) reported benefit : cost ratio of 4.45 under a crop rotation of sugarcane, main crop (1st year), sugarcane coppiced (2nd year), wheat-jowar fodder (3rd and 6th years) and wheat fallow (7th and 8th year) along with poplar plantation in north-western Uttar Pradesh, India. Similarly Chaturvedi and Pandey (2001) recorded benefit : cost ratio of 5.01 under maize-wheat-turmeric and 6.69 under pigeon pea-turmeric with 10 year rotation of polar.

Public Private Partnership

Traditionally state forest departments have been playing the role of technological support to a limited extent to reach out to the private forestry or agroforestry. The initiative by corporate like WIMCO (Western Indian Match Company Limited) for Poplars and ITC (Indian Tobacco Company Limited) for *Eucalyptus* and Subabul to meet the farmers need for saplings and for tie-ups with banks for credit for taking up agroforestry on an extensive scale were started in the mid-1980s which was intensified in the 1990s. This was in the aftermath of the National Forest Policy, 1988. Institutions like NABARD also played a key role in drawing up special schemes for financing to poplar cultivation in *terai* belt. Shrinking supply of cheap raw material from forests due to policy changes, and increased competition drove the industry to seek alternate supplies. Hence companies encouraged tree planting by farmers through free or subsidized seedlings. However, experience indicates that farmers do not take adequate care of such seedlings. Many companies have attempted direct partnerships with farmers. Thus there were two types of schemes viz. bank loan scheme and leasing of farmers' land or crop sharing schemes. The companies are currently focusing on tree improvement activities through research and development (R&D) in order to identify suitable clones and its mass multiplication. The most significant contributions by WIMCO in the development of poplar clones which is appreciable. WIMCO Limited is the leading safety match manufacturing company in India. The impact of the WIMCO initiative as on today is that, a large number of poplar nurseries have sprung up in Uttar Pradesh, Punjab and Haryana. Although the WIMCO scheme promoted large-scale poplar cultivation, as the area increased, the capacity of WIMCO to buy back could not match the production levels. Further, there was considerable confusion over the buy-back guarantee offered by WIMCO and several farmers failed to understand the concept of 'harvestable tree'. They believed that WIMCO would purchase *all* their trees at the predetermined rates after eight years. This led to much litigation (Saigal et al. 2002). However, the growth of alternate users viz. plywood mills for poplars kept the momentum going. In order to make the public private partnership more resilient, it is desirable that mechanisms to ensure minimum support price, quality control and favorable trade policies must be in place.

Deveshwar (2008) emphasized that to enable larger corporate involvement in renewable agro-forestry, and to promote linkages for commercial application, Industry must be permitted to use degraded forest land for afforestation on long lease at nominal terms. Concessional lease could be linked to commitments from industry relating to community involvement, employment generation over a time period, and ensuring availability of fuel wood. This would require enabling amendments to be made in the Forest Conservation Act, 1980 and the related Rules. Policies would also need to be devised to promote industrial plantations in the economic vicinity of mills through ownership, lease, contract farming or sustainable models of social farm forestry. This would require amendments to certain State laws relating to land use. It would be desirable to extend to agro-forestry all the benefits applicable to agriculture, so that farmers can be encouraged to cultivate their private wastelands with equivalent benefits like concessional power, unrestricted movement of such produce, and so on. In order to support farmer investment, institutional credit needs to be provided in the form of long term loans, depending on the species being planted. Such loans should qualify as priority sector lending by banks and financial institutions.

Future Course of Action

i. A mechanism for identifying and certifying planting material of poplar should be devised and only certified material should be made available to the growers to ensure genetic purity.

ii. To regulate the price fluctuations in the market, the import of wood, veneer and plywood should be cautiously regulated in order to safeguard the interest of local farmers.

iii. Suitable shade tolerant crop varieties and poplar clones should be identified for cultivation in order to minimize the losses in crop yield.

Conclusions

Among the existing two predominant poplar based agroforestry systems poplar + sugarcane and /or wheat were found to be most remunerative. Better growth of the poplars was found in Haryana compared to Punjab and Uttarakhand. The coverage of poplar based agroforestry was more from the agri-silviculture system having 5x4 m spacing than boundary plantation and mono cropping (pure poplar plantation). Of late the price for poplar is slowly increasing; thereby new plantations of poplars are expected in these regions. From the farmers' economics view point, the loss of returns from crop component is compensated by the tree component. However, only in case of farmers with concerns of food security, this would be an issue. Thus, it may be concluded that poplar based agroforestry systems are a better option for farmers in the favorable agro-climatic conditions.

Table 4. Fluctuation in price of poplar wood (Rs/q) in Yamunanagar

Year	Girth classes (inches)		
	Over (>24)	Under (18-24)	Sokhta (10-18)
1997-98	350-450	225-275	170-180
1998-99	350-425	225-260	150-175
1999-00	310-400	180-230	125-160
2000-01	275-350	150-200	110-140
2001-02	250-275	150-175	110-120
2002-03	225-250	125-140	80-100
2003-04	155-200	85-130	70-100
2004-05	175-280	125-225	90-180
2005-06	275-500	175-300	125-220

Kumar (2007)

References

Ballal, A.N., Gogate, S.D., Dakshindas and Karunakaran, M. 2001. Introduction of poplar species in Maharashtra state – an assessment. *Indian Forester,* 127 (1): 306-843

Chandra, J.P. 2001. Scope of poplar cultivation. *Indian Forester,* 127(1):51-60

Chaturvedi, A.N., Rawat, B.S. 1994. Poplar tree improvement program. *Indian Forester,* 120: 110-118.

Chaturvedi, A.N. 1982. Poplar farming in Utter Pradesh. U.P. Forest Bulletin 45, Nainital

Chaturvedi, O.P., Pandey, I.B. 2001. Yield and economics of *Populus deltoids* G_3 Marsh based inter-cropping system in eastern India. *Forests, Trees and Livelihood,* 11: 207-216

Deb Roy, R. 1999. Sustainable development of degraded land through agroforestry in India. In: Compendium of Short Course on Sustainable Rehabilitation of Degraded Lands through Agroforestry (Solanki KR, Bisaria AK and Handa AK). NRC for Agroforestry, Jhansi, India, 21-30 June, 1999. Pp. 18-22

Deveshwar, Y.C. 2008. Creating a sustainable and secure future - The Agro-Forestry potential at the 97th Annual General Meeting on July 30, 2008, http://www.itcportal.com/chairman_speaks/chairman-2008.html

Dhillon, A., Sangwan, V., Malik, D.P. and Luhach, M.S. 2001. An economic analysis of poplar cultivation. *Indian Forester,* 127 (1): 86-90

Dhillon, G.S., Surlit, S., Dhillon, M.S. and Atwal, A.S. 1982. Developing agrisilviculrual practices: studies on the shading effect of *Eucalyptus* on the yield of adjoining crops. *Indian Journal of Ecology,* 9(2):228-236

FAO, 1979. Poplars and Willows. FAO Forestry Series 10.

Forest Survey of India, 2005. State Forest Report. Ministry of Environment and Forests, Dehradun, Pp. 177.

Kareemulla, K., Samra, J.S. and Rai, P. 2005. Globalization of Industry Oriented Indian Agroforestry: A Case Study of Poplar, Eucalyptus, Teak and Subabul Based Agroforestry Sectors. In Proceedings of the South Asia Regional Conference on Globalization of Agriculture in South Asia: Has it Made a Difference to Rural Livelihoods. Organized by CESS, ANGRAU, IFPRI and IAAE, 23-25, March 2005 at Hyderabad, Pp.36.

Kareemulla, K., Dwivedi. R.P., Ramesh Singh and Rizvi, R.H. 2003. An economic diagnosis of agroforestry systems in western Uttar Pradesh, Completion report of APCF scheme. NRCAF. Jhansi

Kaul, O.N., Sharma, D.C. and Tandon, V.N. 1983. Biomass distribution and productivity in a poplar plantation. *Indian Forester,* 109(11): 822-828

Kumar, R, 2007. Poplar plantation in agroforestry system. *Indian Farming,* 57 (3): 21-25 & 39

Mathur, R.S. and Sharma, K.K. 1983. Poplars in India. *Indian Forester,* 109 (9):13-17.

NRCAF, 2007. NRCAF Perspective Plan – Vision 2025, National Research Centre for Agroforestry, Jhansi, Pp. 46.

Newman, S.M. 1997. Poplar agroforestry in India. *Forest Ecology and Management,* 90: 13-17.

Puri, S., Swamy, S.L. and Jaiswal, A.K. 2001. The potential of poplar (*P. deltoids*) in the sub-humid tropics of central India : survival, growth and productivity. *Indian Forester*, 127 (2) : 173-186.

Ralhan, P.K., Singh, A. and Dhanda, R.S. 1992. Performance of wheat as intercrop under poplar (*Populus deltoids* Bartr.) plantations in Punjab (India). *Agroforestry Systems*, 19: 217-222

Saigal, S., Arora, H. and Rizvi, S.S. 2002. The new foresters: the role of private enterprise in the Indian forestry sector. Instruments for sustainable private sector forestry series. Ecotech Services, New Delhi and International Institute for Environment and Development, London.

Saxena, N.C. 1990. Farm forestry in North-West India. Ford Foundation, New Delhi, Pp. 40.

Singh, N.B. and Jha, R.K. 2001. Poplar based Agroforestry in eastern Uttar Pradesh. Paper presented in agroforestry workshop. Forestry Training Institute, Kanpur. May, 2001

Singh, A., Dhanda, R.S. and Ralhan, P.K. 1993. Performance of wheat varieties under poplar (*Populus deltoids* Bartr.) plantations in Punjab (India). *Agroforestry Systems*, 22: 83-86

Singh, K., Chauhan, H.S., Rajput, D.K. and Singh, D.V. 1989. Report of a 60 month study on litter production, changes in soil chemical properties and productivity under poplar and eucalyptus inter planted with aromatic grasses. *Agroforestry Systems*, 9 (1): 37-45

Singh, K.D. and Marzoti, A. 1996. Forest Resources Assessment Project. FAO, Rome.

Singh, N.B. and Kumar, D. 1999. Status of poplar improvement in India. In : Proc. National Seminar on Poplar, 25-27 Nov. 1999, Forest Research Institute, Dehradun, Pp.31.

Singh, N.B., Gupta, R.K. and Singh, K. 2001a. Poplar in Agroforestry : Present status and future thrust. *Advances in Horticulture and Forestry*, 8: 247-264

Singh, N.B., Kumar, D. and Gupta, R.K. 2000. Future of Poplars in India. In: International Symposium on Tropical Forestry Research: Challenges in the New Millennium. Aug. 2000, KFRI, Peechi, Kerala

Singh, N.B., Kumar, D., Gupta, R. and Singh, K. 2001b. All India Coordinate Research Project on Poplar Improvement: need, implementation and achievement. *Advances in Horticulture and Forestry*, 8: 205-237

Singh, N.B., Kumar, D., Rawat, G.S., Gupta, R.K., Singh, K. and Negi, S.S. 2001c. Clonal evaluation of poplar (*P. deltoides*) in eastern Uttar Pradesh. II Estimate of genetic parameters in field testing. *Indian Forester*, 127 (2): 163-172.

Singhal, R.M. and Panwar, B.P.S. 1992. Commercial approach to poplar (*Populus deltoids* M) based agroforestry system. North western Uttar Pradesh. *Van Vigyan*, 30 (1): 29-38

Tiwari, D.N. 1993. Poplar. Surya Publications, Dehra Dun, India, Pp. 321.

❑❑❑

Agroforestry: Natural Resource Sustainability, Livelihood & Climate Moderation © 2009
Editors: OP Chaturvedi, A Venkatesh et. al., pp. 675-697
Satish Serial Publishing House, Delhi

Chapter 48

ECONOMIC APPRAISAL OF AGROFORESTRY PROJECTS

R.S. Tripathi
Central Soil Salinity Research Institute, Karnal-01 (Haryana)

Introduction

Traditionally, the agro-biological and social scientists have tended to view their roles as the generator and the evaluator of new technology, respectively. Now, the research policies are emphasizing the social acceptability and economic profitability of technological innovations, as well as their biological or technical soundness. Therefore, socio-economic analysis is encouraged, sometimes even mandated or taken for granted, as an integral component of the process of technology design, testing and delivery. It is important to demarcate the potential end users of a given technology before it is designed or evaluated. Indeed, this is what determines the criteria for socio-economic evaluation.

Agroforestry practices encompass an entire spectrum of land use systems in which woody perennials are deliberately combined with agricultural crops and animals in some spatial or temporal arrangement (Lundgren and Raintree,

1982). Advocates have contended that soil conservation is one of its primary benefits (Young, 1989). The presence of woody perennials in agroforestry systems may affect several bio-physical and bio-chemical processes that determine the health of the soil substrate (Nair, 1993). The less disputed of the effects of trees on soil include: amelioration of erosion, primarily through surface litter cover and under story vegetation; maintenance or increase of organic matter and diversity, through continuous degeneration of roots and decomposition of litter; nitrogen fixation; enhancement of physical properties such as soil structure, porosity, and moisture retention due to the extensive root system and the canopy cover; and enhanced efficiency of nutrient use because the tree-root system can intercept, absorb and recycle nutrients in the soil that would otherwise be lost through leaching (Sanchez, 1987).

The economists argue that the theoretical basis for rigorous economic analysis of agroforestry practices is woefully lacking and empirical analyses are rare (Scherr, 1992). For example, after reviewing 108 agroforestry project evaluations, Scherr and Muller (1991) concluded that only a few even attempted to assess economic impacts. At the same time, while there has been considerable economic research on the causes of change in agricultural productivity in general, conventional economics has rarely been used to understand the role of natural resources stocks such as soil in providing flows of non-market services such as soil conservation (Pagiola, 1993). The paucity of economic valuations of agroforestry can be explained by several factors including: the spatial and temporal complexity of agroforestry systems; heterogeneous farm conditions at the inter and intra household level; multiple inputs and outputs; and the existence of several non-markets costs and benefits. All these factors make rigorous statistical analysis very difficult. Understanding the long run impact of a change in soil quality on agricultural productivity in particular and societal welfare in general, requires extensive economic and agro-ecological panel data with significant length and width (time series and cross sectional variation).

Three-stage analyses for natural resource based enterprises such as agroforestry have been proposed by Freeman (1993) for examining the relationship between the economic concept of value and the bio-physical dimensions of the natural resources or resource systems being valued. Soil conservation and agroforestry benefits can be evaluated by the three stage analysis. Stage 1, quantifies the relationship between soil conservation, through agroforestry (A), and soil quality, (S). In Stage 2, the effects of changes in soil quality on individual household agricultural production (Y) are estimated. Finally, in Stage 3, these production changes are valued at net market prices. Stages 1 and 2 represent bio-physical relationships which are evaluated in economic terms in stage 3. These three stage analysis can be described with the following three equations Pattanayak, 1996.):

Stage I - Soil quality index (S) = f (Z, A).. (1)

Soil quality, represented by an index (S), is a function of management practice (A), including agroforestry practices, and a vector of environmental variables (z) composed of geologic material, topography, climate, time and biota.

Stage II- Agricultural yield (Y) = f (S, x).. (2)

Agricultural production (Y) is a function of soil quality (S) and a vector of other human and nonhuman physical and financial inputs (x).

Stage III- Economic value of agricultural production (I) = f (Y, x P, w)................ (3)

The economic value of the agricultural production (I) is a function of production (Y), a vector of the inputs used (x), and a vector of output and input prices (P and w respectively). Each of the above equations is expressed in a simplified form to highlight the significance of the variables of interest, A, S, Y and I. Moreover, to the extent possible, it is important to isolate the effects of the key variables, A, S, and Y, in the functional forms in each of the three stages.

This three stage framework can be used to isolate and estimate the soil conservation benefits of agroforestry practices. In Stage 1, the relationship between agroforestry (A), the land use change, and soil quality (S) is estimated. This requires the formulation of an aggregate soil quality index, S, composed of several key physical, chemical and biological soil characteristics. Using these characteristics as independent variables would likely create multi-colinearity problems. Previous studies used a weighted average of water holding capacity, aeration, bulk density, pH, and electrical conductivity as a "Productivity Index" to aggregate soil quality (Larson et al., 1983). Following the Universal Soil Loss Equation which links vegetative cover and erosion control practices to the volume of soil eroded (Wischmeier and Smith, 1978), our A value (agroforestry practice variable) is related to the aggregate index, S. Partial equilibrium analysis is used to isolate the effect of A on S.

The objective in Stage 2 is to relate a change in S to the agricultural production profile which includes yield changes as well as allocation of labor, capital and other inputs. Although some agronomy research relates individual soil characteristics to agricultural yields, aggregate soil quality indices are rarely. In general, agronomic analyses of soil productivity use one of the following three basic approaches for estimating yield as a function of soil properties:

i. Systematization of observed yield levels;

ii. Statistical analysis of observed yield levels; and

iii. Bio-physical simulations.

These three general approaches differ in degrees of accuracy, complexity, flexibility, and above all, data requirements. In practice, the method (or combination of methods) chosen is likely to be driven by the availability of data (Pagiola, 1993). Irrespective of the method chosen, one must identify the impact of soil quality change on productivity.

In Stage 3, the objective is to estimate the market or net economic value of the change in agricultural productivity which results from the change in soil quality, which in turn originates from the management decision to conserve the soil. Although environmental valuation is a fairly well charted field (Freeman, 1993), three caveats are necessary. First, the value of the conservation practice should be measured net of costs. Soil conservation practices require different amounts of physical and labor inputs directly and also directly and indirectly (through its impact on S). This implies different costs to the farmer. Therefore, the gains in agricultural productivity need to be measured over and above these costs. Second, market prices adjust to changes in demand and supply decisions made by economic agents as they realize that the values of the conservation practice and of the soil quality are changing. Assumptions regarding the market structure should be explicitly stated. Third, soil improvement or degradation is a temporal process; that is S is a time dependent "stock" variable. Hence, a time series of Y values, the yield stream is associated with the stream of changes in S. The I value calculated in Stage 3 is, therefore, the sum of the discounted yield stream valued net of costs.

Differences between Financial and Economic Efficiency

There are two fundamental differences between financial analysis and economic efficiency analysis. The differences relate to (1) what costs and benefits (or positive and negative impacts) are included in the assessment; and (2) how those costs and benefits (impacts) are valued (Table 1). The term financial analysis is used to describe the type of analysis that is concerned only with actual monetary flows from (cost) and to (return) specific individuals or groups of individuals within society - farmers, private firms, public corporations, and others. In this sense, financial analysis deals only with those goods and services for which people pay or are paid. It deals with the actual monetary payments involved, e.g., for labor, capital, land. Financial analyses always have to be done from a specific interested party's point of view - government agency, private firm or individual, cooperative, etc.

Economic efficiency analysis, on the other hand, is concerned with the costs and benefits to society as a whole, regardless of who pays and who gains. It deals with benefits measured in terms of what society actually is willing to pay for goods and services, and costs in terms of the opportunity costs involved, e.g., the values of the opportunities foregone when a resource is used for one purpose rather than its next best use that actually would have occurred. This concept is valid regardless of whether or not money actually is paid for a good or service.

Table 1. Relationship between steps in a financial and an economic efficiency assessment

Financial Analysis	Economic Efficiency Analysis
1. Identifying and quantifying inputs and outputs:	
Direct inputs provided by the financial entity and outputs for which the entity is paid are included.	In addition to direct inputs and outputs, indirect effects are included, i.e., effects which are not included in the financial analysis since they are not bought or sold within the project context. These are effects on others in society.
2. Valuing inputs and outputs:	
Market prices are used. For inputs and outputs which occur in the future, future market prices are estimated.	Consumer willingness to pay (w.t.p.) is used as the basic measure of value. In cases where market prices adequately reflect w.t.p., such prices are used. In other cases, "shadow prices" are estimated to provide best measure of w.t.p.
Inputs and outputs are multiplied by market prices to arrive at total costs and returns which are then entered in the cash flow table. Transfer payments (taxes, subsidies, loan transactions, etc.) are added to the cash flow table.	Inputs and outputs are multiplied by unit economic values to arrive at total economic costs and benefits which are then entered in a total value flow table. Transfer payments are not treated separately, but included as part of economic costs or benefits as appropriate.
3. Comparing costs with benefits:	
Using cash flow table, calculate chosen measures of project worth or commercial profitability.	Calculate chosen measures of economic efficiency or economic worth, using the information in the total value flow table.
4. Dealing with uncertainty: Sensitivity analysis:	
Test results for uncertainty by varying values of key parameters in a sensitivity analysis.	Test results for uncertainty by varying values of key relationships/parameters in a sensitivity analysis.

The economic efficiency analysis, like the financial analysis, is concerned with profitability, but it is profitability from society's point of view, which is related to the return society as a whole can obtain with a given use of its limited resources. In most cases, the nation is taken as the unit of society. But it could just as well be a state or smaller unit (part of a nation). In economic analysis, market prices often are adjusted to more accurately reflect social or economic values. These prices are referred to as accounting or shadow prices. For example, if there is significant structural unemployment, and a project employs otherwise unemployed laborers, then in the economic efficiency analysis we might apply a labor value lower than the ongoing wage to reflect a lower opportunity cost for such labor that otherwise would be unemployed.

If both financial and economic efficiency questions are being asked, they generally are carried out together since both have much in common in terms of information requirements and procedure. The steps in a financial analysis are more straightforward to carry out and clearer in concept. Therefore, assessments generally start by answering the financial questions related to financial profitability and then the results of this step are used as a starting point for the parallel step in the economic efficiency analysis.

When proposing that an economic impact assessment be conducted, decision makers sometimes neglect to consider the perspectives or questions of all the interested parties involved. The interested parties control resources and must decide whether to commit these resources to a given project and how much to commit. The assessor has to keep this in mind when defining the full array of questions to be dealt with. For example, a project can be extremely attractive from a national economic efficiency point of view, but if it is not also financially attractive to all private entities which have to commit resources to it, then it will not be undertaken as planned. A financially unattractive project can be made financially attractive if the government provides subsidies in terms if capital and inputs (incentives).

Whether or not such subsidies are considered justifiable in a social economic context depends directly on their required magnitude in relation to the economic surplus associated with the project (economic benefits minus economic costs, appropriately adjusted to take time into account). Similarly, analyses which show that a project appears to be more attractive financially than economically may provide some indication of the desirability to tax the interested parties involved. Questions also need to be answered for developing financially and economically feasible and viable agroforestry projects. Some of the differences are indicated in Table 2.

Table 2. Questions relevance at different stages in the process of project evaluation.

Relevant Questions for Project Assessment	Project Identification	Project Preparation	Project Appraisal	Project Implementation	Project Sustainability (ex post evaluation)
Financial efficiency (overall cash flow) questions	What is a rough estimate of the monetary returns of the project compared to costs involved? Does the project agency have capability to pay recurrent costs?	What is the detailed relationship between monetary returns and outlays for interested parties involved in the project and what is the cash flow likely to be (timing of costs and returns)?	Does the appraisal team agree with the estimates provided by the project planner? If not, why not? How do the estimates differ? Does the net cash flow and financial returns meet acceptable criteria for acceptability?	How are actual costs and returns progressing in relation to planned costs and returns? If they differ, why and what can be done about it?	Were costs and returns as initially expected? (Since they seldom are, the real question is: why did actual costs and returns in the project differ from planned?) What is the ex-post financial rate of return compared with ex-ante ROR and to be sustained?
Questions related to distribution of costs and returns of the project (who pays and who gains? what are the foreign exchange impacts)?	What major groups are likely to gain and which ones are likely to lose?	Who specifically will pay for and who will get benefits of the project? What will be the path of costs and benefits over time?	Are distributional implications of the project likely to be as envisioned by project planners? Are they acceptable in terms of given equity & income distribution?	Who is actually gaining and paying? Are the flows of benefits and costs differing from those envisioned in the plan? If so, why? Should anything be done, and how?	Who is better off because of the project? Did the project actually reach the target groups? Can the benefits be sustained after the project?
Economic efficiency questions	What are the likely market and non-market benefits and costs involved?	What are the specific market and non-market benefits and costs involved in the project and what is the likely flow of such benefits and cists?	Does the appraisal team agree with the estimates of costs and benefits and the analysis provided by the project planners? If not, how do they differ?	How are actual economic costs and benefits progressing in relation to planned ones? Included here is monitoring and assessment of project externalities.	Were the actual benefits sustainable over time and cost of project in line with planned ones? If not, why did they differ? What is the ex-post economic rate of return in relation to ex-ante rate provided by the project plan?

Efficiency Criteria for Investment Analysis in Agroforestry

The economic evaluation of agroforestry projects are based on the financial as well as economic parameters such as net present value (NPV) or net present worth (NPW), financial internal rate of return (FIRR), weighted average cost of capital (WACC) and the benefit-cost ratio (BCR). The FIRR and the NPV are computed based on the validated submissions of the project proponents of benefit and cost. A project is financially viable in the "all capital" approach if the resulting FIRR is greater than the WACC and the NPV is greater than zero using the WACC as the discount rate.

A project is financially viable under the "equity capital" approach when the resulting FIRR exceeds the cost of equity contribution of the proponent while NPV should be greater than zero using the cost of equity capital as discount rate. The FIRR is that discount rate which equates the present values of the project's benefits and costs so that the NPV is zero, and the BCR is one. Sensitivity analyses are conducted to determine whether or not the project will remain feasible if changes in the assumptions used in the calculations / projections were to take place. Three sensitivity analysis scenarios are used in financial evaluation of the project.

Forest economist and managers plan for investments using capital budgeting techniques. Capital budgeting is that branch of applied economics that involves decisions with respect to how money is invested in forestry projects, which usually require investment capital, followed by further production costs in order to receive incomes in later year. Capital budgeting deals with ranking alternative projects in order to find investment alternatives for the same asset, which can maximize its value. Five criteria are mainly used for ranking projects: Net Present Value, Land Expectation Value, Benefit Cost Ratio, Payback Period, and Internal Rate of Return.

The Net Present Value, also known as Net Discounted Value (NDV) or Net Present Worth (NPW), of an investment is the present value of its revenues minus the present value of its costs where T is the horizon of the investment. NPV considers the benefits and costs over the entire span of the project. According to the decision rule of the NPV, an investment is acceptable if the NPV is equal or more than 0. A remark about this criterion is that it is difficult to interpret which one is more profitable for investments with different scales.

The Land Expectation Value (LEV) or Soil Expectation Value is the maximum an investor can invest in the land asset and still earn the MAR on the invested capital. The decision criterion is to accept the investment if the LEV equal or more than the land market price or the calculated LEV for an alternative land-use investment on the same piece of land. This criterion is used for land base investments; however, it could be adapted for other types of assets. Thus the Soil Expectation Value refers to the cash amount one may be willing to pay for bare land given the pattern of costs, revenues that one can expect from the plantations raised on these lands. The algebraic form of the Soil Expectation Value (Se) can be expressed as below:

$$Se = \frac{\sum_{t=0}^{n} Bt\,(1=i)^{n-t} - \sum_{t=0}^{n} Ct\,(1+i)^{n-t}}{(1+i)^{n}}$$

where;	Se	=	Soil expectation value
	Bt	=	Revenue received at time t
	Ct	=	Cost incurred at time t
	N	=	Age of rotation in years
	t	=	Time with present difference as O
	i	=	Interest rate

Expected Value of Rotation, also known as first man's formula, refers to the expected value of a tree rotation raised on the forest land. This provides information to the forester that whether the rotation is viable economically or not. The algebraic form of the expected value of rotation is as fellows:

$$EVR = \frac{Y^{r} + \Sigma\ ta\,(1+i)^{r-a} - C\,(1+i)^{r}}{(1+i)^{r} - 1} - E$$

where;	EVR	=	Expected value of rotation at the end of the rotation
	E	=	e/i
	e	=	sum of all annual expenses (maintenance cost)
	y^{r}	=	Not value of final felling at the end, of rotation, i.e., r years
	Ta	=	Net value of several thinnings
	C	=	Initial cost
	i	=	Interest rate.

The Break even size of the enterprise or the unit is one which covers total costs of production below which there is a loss to the producer. In other words, it is a point or size of in unit at which the producer neither incurs loss nor gains profit in producing a particular produce.

$$\text{Break even point} = \frac{\text{Cost of production (Rs /Farm)}}{\text{Selling price of produce (Rs/ Unit)}}$$

It can be expressed in the form of Liner equation as follows:

$$TC = TR$$

or

$$a_c + bS_c = a_r + bS_r$$

$$TC = a+bS$$

$$TR = a+bS$$

where;

TC = Total cost.

TR = Gross Return.

S = Size of the unit or enterprise.

The Benefit/Cost Ratio (B/C), also called the profitability index, is the present value of revenues divided by the present value of costs using the investment's MAR. The decision rule states that an investment is accepted if the B/C equal or more than 1.

The Payback Period is a method that identifies the period in which the capital invested in the project is recovered directly from the benefits produced within this period. The project with the shortest payback period is selected. However, the main critic of this method is that benefits and cost occurring after the payback period are ignored. The payback period is not an investment efficiency criterion, but a financial one because it does not consider the value in the present but how to solve cash availability in the shortest time. This criterion should be used in relation to the NPV and the LEV criteria.

The Internal Rate of Return (IRR) is the interest rate at which the present value of revenues equals the present value of costs. The IRR decision rule accepts a project if the IRR is equal to the investment's MAR. There are several technical and theoretical problems with the use of the IRR. The first one is that the IRR is more difficult to calculate, the second is the multiple roots problem and the assumption with respect to reinvestment and additional investment. Moreover, the IRR differ from the NPV or the LEV decision criteria in that the IRR is an average rate while the others are marginal rates in respect to time.

Financial Analysis

The financial appraisal of agroforestry projects is aimed to find out whether the project is economically reasonable and able to provide justified return on the investment made on it. The financial feasibility involves detail analysis of cost-benefit analysis. For benefit-cost analysis, we have to estimate capital requirement for plantation, annual operational and maintenance cost and benefits generated by the project. In the agroforestry projects, the initial investment is made once for plantation whereas the returns obtained from the project is spread over several years in future. As a matter of fact, the cost-benefit analysis is a decision-making tool for investment choice with respect of total costs and total benefits of the technology. It helps in comparing the cost and benefit of alternative technologies and also profitability of the enterprises. Nevertheless, all the costs and benefits are difficult to quantify in financial terms in case of the forestry projects because a new technology may have positive and/ or negative side effects on the life quality of society and accordingly required to be accounted for the assessment. When side effects of a technology are accounted both direct and indirect benefits and costs, it is called social cost-benefit analysis. Generally, for the financial feasibility we consider only quantifiable direct benefits and costs for the simplicity of understanding.

There are many tools and measures to evaluate the feasibility of the agroforestry projects. Few of the important cost-benefit analysis measures are discussed here which are used widely to find out the financial and commercial viability of these projects.

1. Pay Back Period
2. Simple Rate of Return
3. Net Present Worth
4. Present Worth of Benefit-Cost Ratio
5. Internal Rate of Return.

The last three measures are most important and used widely to assess financial feasibility of agroforestry projects. In the successive paragraphs, these measures are discussed in detail to assess financial feasibility of the agroforestry projects and also to take decisions for investment choice with respect of total costs and total benefits of these projects.

Pay Back Period

The Pay Back Period measures the numbers of years it will take for the net undiscounted benefits to repay the investment of a project. If the pay out period is longer than some arbitrary limit, the project is rejected. If shorter, it is accepted. Thus, the pay back is the time period for an investment to generate sufficient incremental cash to recover its initial capital outlay in full. The following formula is used to calculate the pay back period, if the cash flows are uniform.

$$P = \frac{I}{E}$$

where P = Number of years required for pay back the investment,
I = Initial capital investment, and
E = Annual net benefits.

If the cash flow is not uniform per year, the payback period is determined by calculating the cumulative proceeds in successive years until the total is equal to the original outlay. It is computed to supplement the other measures used to judge the desirability of the projects. The shorter pay back period provides the greater profitability of the project.

The pay out period criterion is justified for the individual's point of view in the short run but for aggregate purposes such as a nation as a whole, some times it misleads the results. The crucial draw back of this measure is that it rejects all projects whose benefits take long time to materialize and favours only good short-term prospects. There is no reason to believe that all quick yielding projects are superior projects.

Simple Rate of Return

The simple rate of return of a project is used to compare with the rate that has been determined to cut off criterion or the minimum rate of return. The following formula is used to calculate the SRR of a project.

$$r = \frac{E - D}{I}$$

where r = Simple rate of return on the investment made for the project,
E = Annual net benefits expected from the project,
D = Annual average depreciation on the fixed assets, and
I = Initial capital investment on the project.

Net Present Worth

Investment theory uses banking and financial mathematics of interest as instruments for analyzing the intertemporal behavior of the investor. Therefore, it is important to review the Net Present Value (NPV) concept as a tool in forest economics, which deals with the problems of choice in intertemporal production of forest. The NPV is the present value of positive payments minus the present value of negative payments made at different points in time. All these future values are discounted to the present by using the investor's interest rate. The NPV expresses a value in terms of the tradeoff between present and delayed consumption. This means the individual by borrowing and lending (saving) reallocates the consumption over time within the limitations of current and future income. The theory indicates that the rational man would choose the investment yielding the largest present value because it is equivalent to a larger choice. The NPV is the keystone model for finding the investor's value of any asset yielding a future cash flow. The NPV is a powerful tool for valuing forest resources (trees and land) based on the future monetary benefits and costs, all discounted to the present using the investor's minimum acceptable rate of return.

If some one offers a choice between receiving Rs 100 today or receiving Rs 100 after five years, the decision would be clear to get the money immediately and deposit in a bank account and earn 12% interest which would give Rs 176 after five years. If you are willing to take risks and invest it in a business, you may earn much more. This is an illustration of time value of money or the cost of waiting. Money received today is always worth much more than the same money received in the future. The same is true for the economy also. Resources available to the country this year can be invested to produce goods and services over the next several years. If these resources were available five years later, the economy would lose the benefit of that output for the next five years. Thus in project appraisals, it is necessary to account for the cost of waiting. Benefits and costs accruing earlier are valued more than the benefits and costs accruing later. Just as one can not meaningfully add kilograms of sugar and meters of cloth one cannot meaningfully add costs or benefits of 1991 to costs or benefits of 2001. The process of weighing cash flows according to the year in which they occur is called discounting. The percentage difference between the value of one hundred rupees now and its value a year later from now is called the discount rate. Further costs and benefits are expressed and compared in terms of present value or present worth. Let us assume that we invest Rs 200 at an interest rate of 10% per year. After 2 years it will yield Rs 200 $(1+1/10)^2 = 242$. Now looking from the other angle, if we want to have Rs 242 in hand two years from now, how much must be deposited today? The answer is Rs $242 / (1+1/10)^2 =$ Rs 200. Thus Rs 200 is the present value of Rs 242 which will be available after 2 years at 10% discount rate. If we assume that Rs 242 is only available after 3 years, then the present value at 10% discount rate will be 242 / $(1+0.10)^3$ = Rs 181.81, which is less than the present value Rs 200 in 2 years time. Let us now assume that the discount rate is 5%, then the present value of Rs 242 in two years would be 242 / $(1.05)^2$ = Rs 220 which is greater than the present value at 10% discount rate.

These simple calculations prove the three important features of present value.

i. The present value is always less than the nominal value that occurs in the future.

ii. The longer the delay, the less is the present value.

iii. The higher the interest rate, the lower the present values.

This means that waiting has a cost and the longer you wait the larger the cost. The third point proves that if money and other assets are more productive, as reflected in the higher discount rate, the waiting is costlier. This raises the question as to which interest rate should be used in project appraisal. Generally, the discount rate should reflect the cost of capital to the investor. In the estimation of NPW, the return achieved at different future dates is made commensurable by assigning to them equivalent present values. This is an expression of net revenues from the crop production discounted to a common time point for ensuring costs and returns comparability, which occur at different periods of time.

The NPW can be calculated by taking the difference between present worth of benefits and present worth of cost. The positive values of NPW reflect viability of the project whereas negative NPW indicates economic loss in the project. Once future benefits and costs have been expressed in terms of present values, we add them to find out the NPW of the project. The general formula used for estimation of NWP is:

$$NPW = \sum_{t=1}^{n} \frac{Bt - Ct}{(1+i)^t}$$

where Bt = Benefit received each year,

C = Cost incurred each year,

t = Time in years (with present difference as 0),

n = Number in years of the project duration, and

i = Rate of interest for discounting the cost or benefit.

Present Worth of Benefit-Cost Ratio

The basic purpose of benefit-cost analysis (BCA) is to assess the economic attractiveness of an action or project. A single project may be assessed to see whether it is profitable or meets a minimum profitability standard. BCA is also commonly used to rank alternate projects and select the most attractive option. Project assessment is based on a comparison of the costs of the project against the expected benefits. There are several ways in which the costs and benefits can be compared, but all of these measures share certain key characteristics. In forestry projects, the benefits and costs are often incurred over a long period of time and they have to be discounted to allow for a consistent comparison.

The benefit-cost ratio is the most popular criterion in social project appraisals. It is the ratio of present worth of benefit and present worth of cost expected at different points of time for a particular project. It is calculated by dividing the total discounted benefits by total discounted costs. Thus, the benefit-cost ratio is a ratio of present value of benefit to present value of cost of a project under assessment. In other words, it is return to one rupee invested on the project. Therefore, at the economic feasibility level of the project the BCR should be more than unity. The project is accepted if the BCR is 1 or more. The projects with the highest ratios are given higher ranking. The ratio more than 1 reflect economic viability of the project whereas less than 1 indicating loss in taking up the project. The BCR can be calculated with the help of the following formula:

$$BCR = \frac{\sum_{t=1}^{n} \frac{Bt}{(1+i)^t}}{\sum_{t=1}^{n} \frac{Ct}{(1+i)^t}}$$

Internal Rate of Return

The Internal Rate of Return (IRR) is the discount rate that generates a zero net present value for a series of future cash flows. This essentially means that IRR is the rate of return that makes the sum of present value of future cash flows and the final market value of a project (or an investment) equal its current market value. Internal Rate of Return provides a simple 'hurdle rate', whereby any project should be avoided if the cost of capital exceeds this rate. Usually a financial calculator has to be used to calculate this IRR, though it can also be mathematically calculated using the following formula:

$$CF_0 + \frac{CF_1}{(1+r)^1} + \frac{CF_2}{(1+r)^2} + \frac{CF_3}{(1+r)^3} + \frac{CF_n}{(1+r)^n} = 0$$

In the above formula, CF is the Cash Flow generated in the specific period (the last period being 'n'). IRR, denoted by 'r' is to be calculated by employing trial and error method.

Internal Rate of Return is the flip side of Net Present Value, where NPV is the discounted value of a stream of cash flows, generated from an investment. IRR thus computes the break-even rate of return showing the discount rate, below which an investment results in a positive NPV. A simple decision-making criteria can be stated to accept a project if its Internal Rate of Return exceeds the cost of capital and rejected if this IRR is less than the cost of capital. However, it should be kept in mind that the use of IRR may result in a number of complexities such as a project with multiple IRRs or no IRR. Moreover, IRR neglects the size of the project and assumes that cash flows are reinvested at a constant rate.

The internal rate of return is used to find out the rate of return, which a project is likely to earn over its useful life. This measure is practically used for all economic and financial analyses of projects by the international financing agencies. When the internal rate of return is used in economic analysis, it is called internal economic rate of return (ERR) whereas on the financial analysis it is called internal rate of return (IRR). The IRR is the discount rate at which the NPW is equal to zero. In calculating the NPW, we independently chose a discount rate based on the opportunity cost of capital and then found the differences between discounted benefits and costs. The IRR calculation reverses the procedure as we use for NPW. Instead of selecting the discount rate, we set the NPW at zero and try to solve for the discount rate which finally gives results. Since higher discount rates reduce the present value of future cash flows, the higher the discount rate, the lower the NPW. The process of finding the IRR involves trial and error method. An arbitrary discount rate is used to find NPW. If the result is positive a higher rate is used to find the NPW, if negative a lower rate is used and the process is repeated until the NPW is reduced to zero. At this discount rate, BCR is equal to one. The IRR is compared with the minimum acceptable rate of return and if it is either higher than or equal to the minimum acceptable rate of return, then the technology is assessed to be superior/ desirable. Suppose, the IRR is 18% this means that a discount rate of 18% the project just breaks even, i.e., it will earn back all the capital and operating costs extended upon it and pay 18% for the use of money in the mean time.

The internal rate of return or discounted cash flow rate of return is the marginal efficiency of capital or discounted cash flow of the investment on a project. It is the rate at which the discounted cash flows are equal to the investment outlay of the enterprises. So, the IRR is that rate of interest which applies to the expenditures incurred at different times for finding compounded sums equal to revenues compounded at the same time. The rate of discount, which makes net present worth of the investment exactly equal to zero, is known as internal rate of return of a project. Thus, IRR is that rate of discount which makes present value of benefits zero.

The IRR is a trial and error solution in which we choose a discount rate at random. The investment is considered to be desirable if the IRR is higher than the cost of capital in a project. If NPW >0, we choose a higher discount rate (accept the project viability) and NPW<0, we choose lower discount rate (reject the proposal). The rate of discount at which the NPW is equal to 0 is the actual IRR and at this stage the procedure is completed. The IRR can be expressed in algebraic form as:

$$\text{IRR} = \sum_{t=1}^{n} \frac{Bt - Ct}{(1+i)^t} = 0$$

In estimation of the IRR, the first step is to discount the cash flow at the cost of capital. If the NPW is negative, we know the project cannot pay such a high rate of interest. It means that we have chosen a high discount rate. Now, choose a discount rate (lower rate) which will give a positive NPW. If in the first step, NPW is positive we should choose a new discount rate (higher), which will decrease the NPW and make it negative. The real IRR lies between these two rates, and we can successively narrow down the limits. The easier and widely adopted method employed for estimation of true IRR is the interpolation formula. The interpolation formula is as follows:

$$\text{IRR} = \text{Lower discount rate} + \begin{matrix}\text{Difference between} \\ \text{the two discount rates}\end{matrix} \times \frac{\text{NPW at lower discount rate}}{\begin{matrix}\text{Difference between NPW's} \\ \text{at the two discount rates}\end{matrix}}$$

It is very important to note that interpolation should not be carried out between a wider spread of discount rates (not more than five per cent). Since interpolation is a higher linear algebraic technique and the changes in IRR, NPW do not follow this pattern. In reality, the IRR rectified by actual verification and by narrowing down the limits between the two discount rates.

Computer Software and Calculation Procedure for NPW and IRR

Computer modeling is necessary to complement field research and extend research results to soils, farming and agroforestry practices, and long-term climatic regimes that would be economically unfeasible to study in the field. The economic feasibility assessment models provide quick, easy and accurate solutions of the agroforestry problems. These simulations will allow researchers to evaluate the probability of success of various agroforestry practices for minimizing environmental impacts and optimizing use of natural resources like soil and water.

The economic feasibility measures such as Net Present Worth and Internal Rate of Return can be calculated quickly, easily and accurately with the help of computer based financial functions. In manual estimation of these parameters, lot of time and energy are required for the job while in the case of computer models these can be calculate most efficiently. Numbers of software and programs are available for estimation of NPW and IRR parameters. The NPW and IRR of agroforestry projects can be estimated simply by using financial functions NPV and IRR available in Excel of Windows 95/98/2000/Xp. The brief description of the calculation procedure is as follows.

Calculation of NPV (Net Present Value)

In the financial function, using a discount rate and a series of future payments (negative values) and income (positive values) calculate the net present value of an investment.

Syntax

NPV (rate, value 1, value 2, ...)

Rate is the rate of discount over the length of one period.

Value 1, value 2, ... are 1 to N arguments representing the payments and income.

- Value 1, value 2, ... must be equally spaced in time and occur at the end of each period.
- NPV uses the order of value 1, value 2, ... to interpret the order of cash flows. Be sure to enter your payment and income values in the correct sequence.
- Arguments that are numbers, empty cells, logical values, or text representations of numbers are counted; arguments that are error values or text that cannot be translated into numbers are ignored.
- If an argument is an array or reference, only numbers in that array or reference are counted. Empty cells, logical values, text, or error values in the array or reference are ignored.

Remarks

- The NPV investment begins one period before the date of the value1 cash flow and ends with the last cash flow in the list. The NPV calculation is based on future cash flows. If your first cash flow occurs at the beginning of the first period, the first value must be added to the NPV result, not included in the values arguments. For more information, see the examples below.
- NPV is similar to the PV function (present value). The primary difference between PV and NPV is that PV allows cash flows to begin either at the end or at the beginning of the period. Unlike the variable NPV cash flow values, PV cash flows must be constant throughout the investment.
- NPV is also related to the IRR function (internal rate of return). IRR is the rate for which NPV equals zero.

Examples

Suppose you are considering an investment in which you pay Rs 10,000 one year from today and receive an annual income of Rs 3,000, Rs 4,200, and Rs 6,800 in the three years that follow. Assuming an annual discount rate of 10 percent, the net present value of this investment is:

NPV (10%, -10000, 3000, 4200, 6800) equals Rs 1,188.44

In the preceding example, you include the initial Rs 10,000 cost as one of the values, because the payment occurs at the end of the first period. Consider an investment that starts at the beginning of the first period. Suppose you're interested in buying a shoe store. The cost of the business is Rs 40,000, and you expect to receive the following income for the first five years of operation: Rs 8,000, Rs 9,200, Rs 10,000, Rs 12,000, and Rs 14,500. The annual discount rate is 8 percent. This might represent the rate of inflation or the interest rate of a competing investment.

If the cost and income figures from the field store are entered in B1 through B6 respectively, then net present value of the shoe store investment is given by:

NPV (8%, B2: B6)+B1 equals Rs 1,922.06.

In the preceding example, you don't include the initial Rs 40,000 cost as one of the values, because the payment occurs at the beginning of the first period.

Suppose your field store's roof collapses during the sixth year and you assume a loss of Rs 9000 for that year. The net present value of the field store investment after six years is given by:

NPV (8%, B2: B6, -9000)+B1 equals -Rs 3,749.47

Calculation of Internal Rate of Return

Return is the internal rate of return for a series of cash flows represented by the numbers in **values.** These cash flows do not have to be even, as they would be for an annuity. However, the cash flows must occur at regular intervals, such as monthly or annually. The internal rate of return is the interest rate received for an investment consisting of payments (negative values) and income (positive values) that occur at regular periods.

Syntax

IRR (values, guess)

Values are an array or a reference to cells that contain numbers for which you want to calculate the internal rate of return.

- Values must contain at least one positive value and one negative value to calculate the internal rate of return.
- IRR uses the order of values to interpret the order of cash flows. Be sure to enter your payment and income values in the sequence you want.
- If an array or reference argument contains text, logical values, or empty cells, those values are ignored.
- **Guess** is a number that you guess is close to the result of IRR.
- Microsoft Excel uses an interactive technique for calculating IRR. Starting with guess, IRR cycles through the calculation until the result is accurate within 0.00001 percent. If IRR can't find a result that works after 20 tries, the #NUM! error value is returned.
- In most cases you do not need to provide guess for the IRR calculation. If guess is omitted, it is assumed to be 0.1 (10 percent).
- If IRR gives the # NUM! error value, or if the result is not close to what you expected, try again with a different value for guess.

Remarks

IRR is closely related to NPV, the net present value function. The rate of return calculated by IRR is the interest rate corresponding to a 0 (zero) net present value.

The following formula demonstrates how NPV and IRR are related:

NPV (IRR (B1: B6), B1: B6) equals 3.60E-08 [Within the accuracy of the IRR calculation, the value 3.60E-08 is effectively 0 (zero)].

Examples

Suppose you want to start a retail sale business for forestry products. You estimate it will cost Rs 70,000 to start the business and expect to net the following income in the first five years: Rs 12,000, Rs 15,000, Rs 18,000, Rs 21,000, and Rs 26,000. B1: B6 contain the following values: Rs -70,000, Rs 12,000, Rs 15,000, Rs 18,000, Rs 21,000 and Rs 26,000, respectively.

To calculate the investment's internal rate of return after four years:

IRR (B1: B5) equals -2.12 percent

To calculate the internal rate of return after five years:

IRR (B1: B6) equals 8.66 percent

To calculate the internal rate of return after two years, you need to include a guess:

IRR (B1: B3, -10%) equals -44.35 percent.

Constraints in Assessment of Forestry Projects

The assessment should not only be effective, that is, provide appropriate answers to the questions being asked but also be realistic in its approach, in recognizing the practical constraints under which the assessment is being conducted. Consideration of these constraints will help the analyst plan and organize the assessment. Figure 1 indicates the considerations that should be taken into account.

1. *Technical considerations:* Technical considerations include data requirements and data availability, cost, time requirements, and reliability, reproducibility, and acceptance in the scientific community. Different assessment questions require different kinds of data. For example, an assessment that includes an analysis of the distribution of costs and benefits would require considerably more detailed data and information than one that considers only aggregate impacts. Such a distributional analysis also is likely to require considerably more time to obtain and process the required input data and to actually carry out the analysis, than would be required for an aggregate analysis. Some methods of analysis may be more reliable than others, and be more easily reproduced by other analysts. This may or may not lead to greater acceptance in the scientific community. For example, assessments based on observed and reported market values may be more easily reproduced by other analysts than assessments which involve unique estimates to account for non-market values, and thus be more readily accepted by the scientific community. Yet, some segments of the scientific community

may give greater credence to attempts to account for non-market values that are known to exist, even though the methods of accounting for those values are not as widely accepted.

2. *Decision maker considerations:* The actual purposes and importance of the assessment in decision making are primary considerations which should dictate how an assessment is designed and organized to give it maximum usefulness (make it timely and relevant for the intended use and purpose). Time and skills are important constraints in designing assessments in most developing countries. It takes time and appropriate skills to plan an assessment, to collect, process, analyze, and synthesize data and information used in an assessment, and to write up the report on the analysis, have it reviewed, and disseminate it to potential users. The amount of time available in which to conduct an assessment must be carefully considered in deciding which questions can be answered, and in the choice of methods by which the questions are to be answered. If the time necessary to complete all of the tasks on the assessment is longer than the time framework initially set by the decision maker, some adjustments will have to be made. Perhaps the time frame can be lengthened or more personnel obtained, or some of the questions modified, or assessment methods changed. Again, this must be an interactive and iterative decision process between decisionmakers and analysts.

Figure 1. Constraints affecting the assessment approach chosen for agroforestry

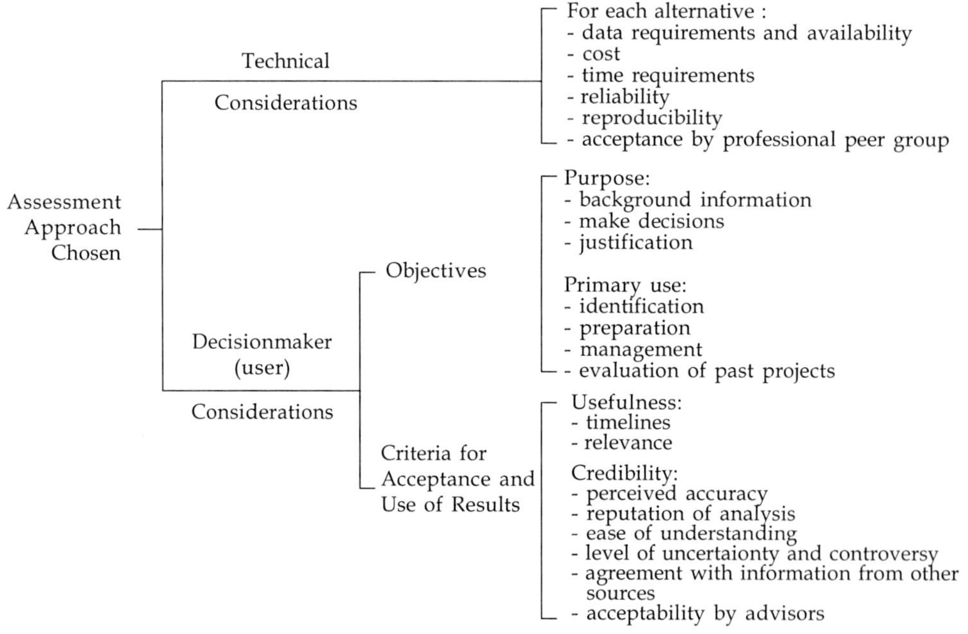

References

Freeman, A., 1993. The Measurement of Environment and Resource Values: Theory and Methods. Resources for the Future, Washington, D.C.

Larson, W., Pierce, F. and Dowdy, R. 1983. Loss in long term productivity from soil erosion in the United States. In: S.A. El-Swaify, W.C. Moldenhauer, and Lo (eds), Soil Erosion and Conservation, lowa.

Lundgren, B.O. and Raitree, J.B. 1982. Sustained Agroforestry. In B. Nestel (eds) Agricultural Research for Development: Potential and Challenges in Asia. ISNAEm The Hague, Netherlands. Pp. 37 - 49.

Nair, P.K.R., 1993. An Introduction to Agroforestry. Kluwer, Bosto.

Pagiola, S. 1993. Cost Benefit Analysis of Soil Conservation. In: Economic and Institutional Analyses of Soil Conservation Projects in Central America and the Caribbean. World Bank Environment paper No. 8. World Bank, Washington, D.C.

Pattanayak, Subhrendu and Evan Mercer, D. 1996. Valuing soil conservation benefits of agroforestry practices. Southeastern Center for Forest Economics Research, Research Triangle Park, NC. FPEI Working Paper No. 59, Pp. 21.

Sanchez, P.A. 1987. Sokl Productivity and sustainability in agroforestry systems. In H.A.

Scherr, S. J. 1992. Not out of the Woods Yet: challenges for Economics Research Agroforestry. *American Journal of Agricultural Economics*, 74 (3): 802-808.

Scherr, S. J. and Muller, E.U. 1991. Technology impact evaluation agroforestry projects. *Agroforestry Systems*, 13: 235 - 258.

Wischmeier, W. and Smith, D. 1978. Predicting rainfall erosion losses. Agriculture Handbook No. 537, USDA, Washington, D.C.

Young, A., 1989. Agroforestry for Soil Conservation. CAB International, Oxford, U.K.

□□□

Agroforestry: Natural Resource Sustainability, Livelihood & Climate Moderation © 2009
Editors: OP Chaturvedi, A Venkatesh et. al., pp. 699-714
Satish Serial Publishing House, Delhi

Chapter 49

EXTENSION TOOLS/APPROACHES FOR ENHANCING LIVELIHOOD SECURITY THROUGH TREE BASED SYSTEMS

R.P. Dwivedi
National Research Centre for Agroforestry, Jhansi-284 003 (U.P.)

Introduction

The concept of sustainable rural livelihoods is central to the debate about rural development, poverty reduction and environmental management; where the emphasis is on the goals of rural poverty reduction, empowerment of people and the promotion of increased security of livelihoods of rural people. A livelihood comprises the capabilities, assets and activities required for a means of living. Livelihoods are secure when households have secure ownership of or access to resources (both tangible and intangible) and income earning activities to offset risks, ease shocks and meet contingencies. Understanding the livelihoods with an economic, ecological and social perspective is imperative to get information for strategic planning to improve the livelihoods of rural people.

Small and marginal farmers constitute 78 per cent of the farming community in India (The Hindu, 2006). These farmers can be spared of the drudgery they are suffering from generations if science and technology helps them in increasing crop productivity. In addition, unemployment has emerged as a major problem plaguing youth in our country, both in urban and rural areas. But for rural youth, the problem assumes greater proportion since land–man ratio has gradually declined over the past years due to ever increasing population and rapid industrialisation. Continuous fragmentation due to ownership multiplication and division of joint families into nuclear families has also resulted in low per capita availability of land. Agricultural labourers constitute one of the vital inputs in the agricultural production process. Farm women and youth agricultural labourers perform most of the farm operations and it remained as a cause if drudgery faced by them.

Agroforestry is a production system in which crop, livestock, fodder, fisheries, bee keeping and trees are grown together in a unit of land. Due to modernization and urbanization, the agricultural land is decreasing in a fast rate. The negative consequence of deforestation is the reduction of tree cover in the country. As per the Forest Policy of Govt. of India, 1988 there should be 66 % and 33 % forest cover in the hills and plains respectively. The shrinking natural resource has resulted in the poor supply of fuel wood, fodder, fruit and food to the majority of Indian farmers.

It is important to examine the social dimensions particularly with reference to gender and social equity and ethics very carefully, while designing agroforestry research and development programmes. Only such integration can help us to impart a pro-nature, pro-poor and pro-women orientation to technology development and dissemination. Agroforestry being a multi-disciplinary and multi-institutional subject, synergies are required at various levels among scientists, planners, development agencies and all other stakeholders to harness maximum benefits of inputs. Linkages between research, education, extension and farmers have been established but their interactions need to be strengthened for appropriate technology generation and its effective transfer.

Since Independence, the diversion of forest land for non-forest land uses has continued unabated though with a reduced pace off late. Between 1951 to 1980, about 4.3 million hectares of forest land has been diverted for non-forest uses. This represents an average loss of forest land @ nearly 0.15 million hectares per year. The causes of forest degradation are not purely physical or biological but often sociological. This process of diversion of forest land to non-forest land use and their degradation has caused floods, water logging, increased erosion of fertile lands and the silting of rivers and dams. It is estimated that about 0.7 million hectare of forest lands have been encroached upon so for, by people living in the vicinity of forests or forest dwellers. As per the assessment of the State of Forest Report (2003), the forest cover is 678,333 km^2 constituting 20.64% of the geographic area of the country. The rate at which degradation of forest land is taking place is alarming. The Indian Forest Policy of 1988 gave conservation orientation and a human face to forestry. The policy emphasized the protective role of forests in maintaining ecological balance and environmental stability.

Forests and Agroforestry in Livelihood Support

The major products for which the country depends overwhelmingly on forest lands are fuel wood and green fodder. A recent survey has indicated the dependence of people in fuel wood to meet the energy requirement to the tune of 78% in rural areas and 26% in urban settlements. It is an accepted fact that most of the fuelwood is in the form of dead or dry wood and is extracted from forests on head loads by the villagers living near forests. This extraction of fuel wood goes unrecorded. But even after allowing for this unrecorded production, the total production from these forests, within silviculturally permissible limits would not exceed 40 million cubic meter or 25 million tonnes. Fuel wood production from various sources was around 169 million tonnes in 1990, which increased to 201 million tonnes by 2001. Forests account for over 50% of this supply, the rest comes obviously from the farmlands and community lands. About 10 million rural people are estimated to be engaged in collection of fuel wood. Besides this many more are engaged in the marketing of fuel wood in rural and urban areas.

It is estimated that the demand for timber (wood other than fuel wood) is likely to grow from 82 million cubic meters in 2006 to 116 million cubic meters in 2025. The supply of wood is projected to increase from 65 million cubic meters (12 from forests and 53 from plantation & agroforests) in 2006 to 116 million cubic meters (12 from forests and 104 from plantation & agroforests) in 2025. The deficit of 17 million cubic meters (in 2006) could be met by promoting agroforestry. The productivity of timber in India is only 0.7 cu. m ha^{-1} yr^{-1} whereas the world average is 2.1 cu. m ha^{-1} yr^{-1}. Increased production of timber and forest produce and their downstream processing could bring about accelerated employment and socio-economic development in rural areas.

The agroforestry system yields almost everything which people need. Sustainable land management, conservation and increase in productivity are important aims of agroforestry. Through agroforestry we can continue to improve productivity and profitability of small farms on an environmentally and economically sustainable basis. The Research and Development (R&D) efforts made so far has clearly indicated that agroforestry is sustainable, environmentally safe and economically attractive land use option which is already making major contributions towards the supply of timber to support local value addition by industries, meet the local demands of fuelwood, fodder and bio-energy; conserving biodiversity; environmental ameliorations; diversification away from agriculture and large employment generation specially for the socially disadvantaged sections of the population and poverty alleviation in rural areas. Besides, promotion of agroforestry is only viable option to achieve the National Forest Policy target of 33 % tree cover.

With the promotion of agroforestry in rural areas, small holders have emerged as the timber suppliers of the 21st century. In 2006, out of a total demand of 82 million m^3 of timber and other wood, 53 million m^3 i.e. 65 % was supplied from plantations out side forests and agroforestry alone. Agroforestry is also meeting almost half of the total demand of 201 million ton of fuelwood in the country. The contribution of agroforestry will increase substantially as more and more improved planting material is made available to farmers.

Possible Reasons for Non-adoption of Agroforestry

Some of the major factors influencing the rate of adoption of agroforestry are: Complexity of innovation, Divisibility of innovation, Divisibility or what Rogers (1983) terms 'trialability, Degree of congruence with farm and personal objectives, Effects on farmers' flexibility, Likely economic benefits, Implementation cost, Capital outlay, Implementation cost, Intellectual outlay, Risk and uncertainty, Conflicting information, Perceptions of the environment, Physical infrastructure, Social infrastructure, Farming subcultures and styles of farming, and Effectiveness of various media and methods.

Linkages

Though extension has to maintain effective linkages with several systems, only the Research- Extension linkages capabilities of the KVK and the quality of linkage have been emphasised. However, several measures to improve this linkage, though adopted, didn't yield positive results. "Information flow has been mostly top-down" (Macklin, 1992) and "the weak feedback has not resulted in any fundamental change in the way research priorities are set at the research stations" (Jha and Kandaswamy, 1994). Apart from linkages for obtaining technology, the capabilities of the extension agency to assess and refine them for its integration in their knowledge base has been very weak, mainly due to lack of qualified staff. The contribution of KVKs to do this important activity would eventually depend upon the it has with the line departments. In many cases, KVKs were found to be wanting in both. With its main emphasis on transfer of technology and implementation of schemes which are input distribution related, the extension system never did take the other systems seriously though its performance did depend upon all these factors. To facilitate both research and extension, there should be close collaboration between researchers, extension agents and landholders in the design, conduct and application of research on agroforestry.

Communication Strategy

Realization of sustainable development in a developing society like ours, needs forming of a communication strategy which takes into consideration the characteristics of the people in terms of information. Our communication strategies adapted, plan after plan, could not deliver the desired goods to the society. This may be essentially due to the reason that our planners were not having total perspective of various subsystems, their interaction and comprehension of people in terms of information strata. Therefore, there is a need to examine the entire developmental process in the above-mentioned perspective.

Till today developmental programmes have by passed the category of Aphonic people in the society. The first two categories of the information system (Information holders and seekers) are privileged one to derive the benefits of development programmes but the third category namely Aphonic persons being least privileged need special care. The characteristic of Aphonic people in different strata of information system forms the basis of Demand-Drive Model. The information provided to the people will generate demand which will ultimately drive them to action.

Demand - Drive Model of Communication

i) *Demand:* It connotes as right in relation to social economic, political, ecological perspectives.

ii) *Drive:* As a propallent i.e. thrust for taking appropriate actions for meeting out demands.

Features of the Demand - Drive Model

1. To build capability of the group to analyse their living realities and to generate fear from the existing consequences. Since, generations after generations they are forced to live in an exploited condition, hence, they should be involved in the process of analyzing their own situation and

made to realize the consequences which will ultimately arouse fear from the existing living conditions. This psychological instinct will lead to dissatisfaction with the present situation. Realized level of dissatisfaction will accelerate the process of change.

2. To identify the alternatives available in the social system for overcoming the severity of consequences.
3. To establish that the benefits of the selected alternatives exceed the adverse consequences of living in exploited situation.
4. Selected course of actions should be within the easy reach and acceptable to the group.

Competency

Competency refers to the capability/ability and liking of a person to perform the (assigned) task. To be an effective change agent he should have the following competencies namely – (i) technical; (ii) economic; (iii) scientific; (iv) occupational; (v) communication; and (vi) social competencies.

Technical Competency is the level of knowledge and understanding relevant to the activity in which the client group is engaged and includes the ability to diagnose correctly the problem and abnormalities related with the activity.

Economic Competency is the ability to weigh alternatives of production inputs and marketing strategies and to make recommendations in the light of economic relationship with in which client operates.

Scientific Competency is a basic understanding of a philosophy of science and the ability to conduct simple replicated field experiments for testing its validity for sustainable development in the client's own conditions.

Occupational Competency is the willingness and skills to perform the range of physical tasks involved in execution of a specific activity.

Communication Competency is the ability to plan, prepare and present adequate message for relevant audiences so as to increase the clients' knowledge of know-how and education with all ramifications of the activity for improving the quality of life.

Social Competency is the ability to mobilize and sustain a group for achieving the desired goals.

Extension Approaches

Extension agent works with the farmers differing in age, educational status, levels of living, cultural background and value systems, etc. These differences demand a wide range of approaches and a great variety of methods. The purpose of extension approach is to motivate the farmers to act towards some specified goal.

The objectives of the extension program and the availability of resources largely dictate the extension approaches and methodologies. Within this overall constraint there are many approaches and numerous methods of extension that can broadly be categorized as "top-down" (dictated from the outside) versus "down-top" (planned and managed by villagers). Overall, the extension approach in India is a "top-down" model. But in many cases, there have been some changes in extension approaches and methodologies along the lines of participatory development.

Where to Approach?

The attack on rural poverty is an attack on the way of life of the farmers or villagers. The best approach to the poor villagers is by indirect and continuous educational process of developing, motivating and by creating new incentives and building up social pressure to induce the people to change. A number of ways of approach, suggested below, can help overcome villagers' opposition and enhance the chance of acceptance and cooperation.

The approach to the whole family, rather than to individuals, usually brings good results. In general, from food supply, shelter and clothing, to the bringing up of children and educating the youth, the responsibilities have been more or less predetermined traditions and give little margin of deviation. The extension worker should recognize this specific structure and depend on it. Farming is a co-partnership in which women share as much responsibility as men, and home-making is also as important as farming. Until the interest of the house wife is aroused and her assistance secured to serve the programme of change, all plans for raising the standard of living of farmers are likely to fail.

Rural youth in India are much behind in education and professional training than the urban youth. If they have to adopt farming as their vocation they get the training from their own farms only. Proper care and efforts are required to organize the rural youth to provide necessary training.

Integrated Approach

Integrated rural development refers to the method or strategy for achieving rural development. Four basic requirements for such an integrated plan of action are:

a. Provision to the rural poor of more productive, remunerative and socially satisfying employment.
b. A more equitable distribution of the material and non material benefits of production and more equal access to services and to social security.
c. Fuller participation of the rural people in development activities and in the relevant decision-making processes.
d. Better or more rational management of renewable and particularly non-renewable resources.

Extension Tools

Mass Media

Public education and publicity are carried out through all known methods of mass media, such as radio, television, loudspeaker and public speeches, posters and booklets. The most common, effective and cheap way of public education in the rural India are the posters, which is simple message in slogan form posted or painted on the walls of the houses, on the terraces, on the river banks and hung up on lamp posts or trees along the roadside. The main slogans are "Lakadi chara phal aur ann, Krishivaniki hai jeevan"; "Plant trees!" and "Plant Trees for the benefit of Future Generations!", and "Soil Conversion for the benefit of Future Generation!" and so on. Slogans, posters and loud speakers on tree-planting are the most common and wide spread public propaganda contents across India. Loudspeakers are useful for education messages in rural areas, especially where many villagers are illiterate. All these mass media enable farmers or villagers being aware of the importance of tree planting all the time. All kind of books and teaching materials and various periodical have also been playing an important role in agroforestry public education to people in all levels. This general awareness had only been possible through the integrated nature of the training programs and public education, but also through the interdependent nature of agriculture, agroforestry, forestry and animal husbandry which, in fact, forms the basis of the Indian economy and the way of life.

Group Media

Some methods with local color such as plays, films, puppet shows, mimes and entertainments are often used at local level. Using locally known comedians and popular language, the play presents a parody of local situations. These methods have proved popular and effective, introducing or convincing people of the relevance of a particular subject.

Training and Field Visit

Apart from the above formal training or education, training by demonstration is also important. Once a technique, as for instance budding and grafting techniques or a new cultivation method, shows promise, then the local leadership, scientists and extension officers collaborate on the publicity of the new knowledge by all possible means such as publications, TV, radio and posters. One of the most effective and widely used ways is field visits. The candidates for field visits are usually extension officers, technicians and representatives of farmers or farm workers, who are selected by local government. Then these people set up their own demonstration sites for training the farmers.

Individual Contact

When necessary, extension personnel and farmers establish individual contacts to deal with the specific issues and problems through consultation and technical instruction. The roles of enabling policies, scientific research, various public education methods, field demonstration and professional and short term training have become more and more important in extension. What method to be chosen is usually dependent on the human and financial resources of a program or project as well as its goals? Generally, a combination of methods is used.

Channels of Communication

In the communication process extension teaching methods or channels of communication are the tools in the hands of extension workers in transferring new ideas. An extension worker has to deal with many rural people. One of the methods to deal with them is personal visits by the extension worker to the farms or homes of the farmers. A personal visit represents a best learning situation but it may not be always possible to visit the farmers. The extension worker has therefore, to use other teaching methods for reaching the maximum number of people effectively. Extension education is mainly concerned with the dissemination of useful and practical information relating to rural life and helping the rural people in the practical application of such knowledge to rural situations. The extension teaching methods are the means through which the extension worker can achieve this objective. It has been shown by research that extension workers using only one extension teaching method were able to influence a little over one-third of the village families to improve practices in various fields.

Extension Strategies

The extension in the agroforestry implementation has five main strategies:

- Raise the awareness of the public on the significance of the Agroforestry
- Increase the understanding and participation willingness of farmers
- Establish cooperation with those who influence the participation of farmers
- Include the farmers into implementation process of the Agroforestry
- Technology transfer.

Raise the Awareness of the Public on the Significance of the Agroforestry

This strategy targets the public whose participation is sought, raising their awareness through informing them about the consequences of soil losses and environmental degradation, the usefulness of the ecological service functions of agroforests to the rural society, the objective and significance of the agroforestry, and the experiences and results of the pilot work carried out elsewhere. This not only creates an enabling environment in the society but also helps produce new ideas.

All available tools of communication are used including both electronic media based instruments such as television, videos, radios, films and filmstrips etc. and print media based instruments such as posters, brochures, books, teaching materials, journals, newspapers and so on. Disseminating the achievements and educating the young people in school and college is also helpful in achieving the objective.

Increase the Understanding and Participation Willingness of Farmers

Farmers are the true implementers of the agroforestry and the real risk-takers of the program policies who undertake the task in the hope of being the beneficiaries in the near future. They are impacted directly by the agroforestry. The focus is on the village communities and farmers, especially the eligible farmers, increasing their understanding and willingness to participate in the agroforestry, through intensive education on the program policies, their assigned tasks, and services offered by the government and compensation that they can expect for their contribution to the program.

Establish Cooperation with those Who Influence the Participation of Farmers

This strategy focuses on those who influence the participation of farmers. They are mainly the government agencies at all levels. It is important that they are aware of the agroforestry and what they can do to further the program.

At the central level, the key players are the various commissions, boards and the State Development departments. Governments at Central and State levels are answerable to the higher levels for the implementation of the agroforestry within their jurisdictions. They are responsible for identifying priority areas for soil conversion, creating necessary conditions for implementation such as timely provision of matching funds, setting up lead groups to direct the work of relevant agencies, and examining and approving the proposed plans of relevant agencies. At the county-level Ministry of Agriculture, ICAR set up a agroforestry management to be responsible for formulation of relevant plans, implementation, monitoring and evaluation, management of program funds, formulation of specific management rules and provision of technical support. Cooperation amongst government's research institutes, SAUs, line departments, NGOs and voluntary organizations are desired for effective and meaningful participation of the farmers.

Include the Farmers into Implementation Process of the Agroforestry

The focus of this strategy is on the eligible farmers. Millions of farmers are the implementers of the agroforestry on the ground. They are typically encouraged to participate in the planning and design of the agroforestry at the field level. The participatory planning and design of the agroforestry should begin with awareness-raising about the policy and regulation in eligible village communities and farmers. The planning team, consisting of technical experts from the ICAR, SAUs, line departments' officials and local technical personnel, go to the villages to introduce the program and participatory planning process to farmers. They conduct field surveys regarding current land use and prepare maps accordingly. The planning team then decides on land use plans and the types of trees for planting together with farmers. A plan regarding the land and density of trees should be prepared in participatory mode.

Technology Transfer

This focuses on delivering techniques and services necessary to farmers, which involves mainly researchers, extension and technical personnel, and farmers. Major approach is informal education, including using appropriate methods such as posters, publications, demonstrations or short training course on relevant knowledge and technique, and individual problem- solving on site. Much research has already been conducted in various places on the technical aspects of vegetation restoration. Based on substantive field investigation and survey by the teams of researcher and technical personnel, the technical standard for the agroforestry in light of different geographical and climatic conditions are formulated and different cropland conversion models are designed. In addition, ICAR, SAUs and the State government, extension and technical personnel conduct field investigations in village or farmers field on the appropriate types of trees, the demands and sources of sapling in order to provide the service of supplying sapling to the farmers.

Information is power. We must try and ensure that the actual cultivator who tills the soil is himself encouraged to collect and maintain key information relating to his own arable land. We should enlist the support of Panchayati Raj institutions and some of our agricultural Universities in this task. It is essential to actively involve farmers in the development of new technologies for it is only with their active participation that we can achieve our goal of sustainable development. It would be equally important to bear in mind that our extension strategies should not only be pro-poor and pro-nature but also pro-women. The pioneering role played by women in movements undertaken by them for preservation of forests and topsoil is well recognised. The big leap in our country's milk production can be credited to the toil and labour put in by nearly 75 million women who are involved in the care of the cow and the buffalo. We must therefore endeavour to mainstream gender considerations in all areas of agroforestry research, education and extension. Women farm labour about 60 % is also playing an important role in our agroforestry operations.

Round the Year Livelihood Opportunities

Agroforestry provides livelihood opportunities round the year to the small and marginal farming community (Table 1). The operations involving labour input are: July-October: Management of nursery, seedling plantation, budding in Aonla and Ber, management of budded trees etc.; November-March: Intercultural activities of agroforestry systems, weeding, spray of insecticide and pesticides on trees and crops, harvesting of fruits, packaging and marketing of agroforestry produce; April-June: Digging of pits, layout of plantations, pruning in Ber, preparation for nursery, putting farmyard manure and compost in the pits and nursery sites, collection of fuelwood, fodder and tree fodder, bark, gum, resin, honey bee, seeds of multipurpose tree species, drying of seeds and tree leaves for medicinal purpose, making *Dona-Pattal* (Paper plate & cup) and selling them in the market. Youth of agricultural labourers frequently involve in formation of ridges and furrows, fertilizer application, earthing up and seed planting, weed management, harvesting and post harvest operations of agroforestry.

Forestry, agroforestry and related industries are a major source of employment in the primary, secondary and tertiary sectors all over the country. Forestry activities alone generate employment of approximately 240 million person days per day. At least 10 percent of the present tribal population of 50 millions derive their livelihood directly from forests. Agroforestry intervention in Himalayan region can provide employment opportunity to almost 6 million mandays which will have grate bearing on the migration of landless and marginalised communities from rural to urban settlements. Silvipasture generates 120 mandays/ha/yr employment directly. In southern Tamil Nadu, *Prosopis juliflora* provides employment to the tune of 6.34 million mandays and 7.03 million womendays per annum for charcoal making and bioenergy production. Biofuel plantations are likely to create employment for 0.44 million mandays only for seed collection. Bamboo sector generates employment for 48-60 million mandays for harvesting and 60-70 to million mandays for loading, unloading, handling etc., annually. These are only few examples which indicate the vast scope of employment generation and livelihood security from these sectors.

Table 1: Round the year management work for agroforestry

Sl. No.	Months	Management work for agroforestry
1.	April to June	Digging of pits, making rows for tree plantation, pruning in indigenous Ber trees, nursery preparation, application of Farm Yard Manure (FYM) and compost in pits and nursery, collection of fuel wood, gum, bark, tree leaves for medicinal purpose, collection, drying and packaging of multipurpose tree seeds, etc.
2.	July to October	Care of nursery, transplanting, budding and gootee in fruit trees, removal of unwanted tillers from budded branches, etc.
3.	November to March	Inter-cultural activities in trees, weeding, spraying of insecticide and pesticides, plucking of fruits, packaging of fruits, post harvest value addition, marketing, etc.

(Source: Solanki and Dwivedi 1999)

Reforms

To be effective and to remain relevant in the years to come, the state extension departments should initiate the following structural reforms in the organisation.

- Strengthen its understanding on matters with respect to technology, markets, prices, demand and policies. Departments have to either recruit specialists or have to hire the services of professionals in these areas.
- Recruit better qualified staff - States have to initiate (as Punjab has done) measures to ensure that, the minimum essential qualification for an extension staff should be a graduation in agriculture. In total the states employ some 1, 10,000 extension staff of whom presently only around 20 percent are graduates.
- Improve social science skills of extension personnel. Apart from technical skills, "extension personnel needs several social science skills with respect to need assessment, group formation, negotiation, conflict resolution, mobilisation, management of common property resources (CPRs), use of IT, data collection, analysis and documentation" (Suresh Pal, et al, 1998).
- Increase the allocation for operational expenditure. "Allocation of operating expenses in State Departments of Agriculture is around 15% whereas a fully functional extension system should have 30-35% of its total expenses as operational expenses"(Swanson, 1996).
- Decentralise the operations of the department and provide flexibility to field level officers to decide appropriate extension programmes. Initiate activities for developing Strategic Research and Extension Plans in all the districts to be followed up with Block level plans.
- Improve the capabilities of extension managers-Extension managers need skills to operate effectively in the pluralistic extension environment. They need to know, how their organization can do better or cheaper than other organisations?; how can it co-operate with other actors in this system to provide all farmers better knowledge to survive and succeed in a competitive society?; and how to create the social climate for a successful learning organisation? The main challenge for an extension manager would be managing the process of change in extension. A series of training programmes have to be initiated to provide these skills.

Action Plan for Future

- Government regulations on felling of trees from agroforestry lands and transport of agroforestry produce particularly of fuelwood, fodder and timber has to be carefully deregulated to benefit various strata of farming community (Dwivedi, 2003).
- National Commission on Farmers (NCF) 2005, in its report, adopted the approach, "before advising farmers, listen to them", since they know both the problems and solutions better than others. Therefore, the Joint agroforestry management by participatory approach at all levels should be practiced between the planners and executors to plant, maintain, protect and to draw the benefits across out of the plantations which may also pave the way for developing and incorporating other enterprises as one of the components for getting higher income per unit area of land.
- Strengthening of agroforestry extension needs considerable attention. Effective training should be given to agroforestry extension personnel to improve their technical skill.
- The eco-system of the country is so fragile that it cannot bear any more pressure. Therefore the need to review the situation and to evolve for the future, a new strategy of conservation of land and its sustainable management has become imperative.
- Cooperatives such as tree growers' cooperatives may be formed. This will facilitate farmers to market their products remuneratively and safeguard their interests from middlemen and price fluctuation.

Conclusions

The most important challenge for the future extension managers would be the Management of Knowledge. The success of a farmer in the years to come is going to be primarily dependent upon his level of knowledge. The permanent solution for land problems (degraded and waste lands) lies in sustainable land management i.e. changing the land use of the entire country for agriculture, forestry and open grazing. Then how will people in the area survive? The answer lies in Agroforestry. There is a basic difference between tree farming for commercial purposes and for survival. The basis of tree farming for trade is the demand of the market. Contrary to this, when tree farming becomes the basis of survival, its objective is to give priority to land & water conservation, environment protection and meet the needs of food, fodder, fuelwood, timber, fibre and various other products. Once the needs of the people in respect of forest products are satisfied, dependence of the people on forests will reduce. This will be of great help for enhancing livelihood security through tree based systems.

References

Cassata, M.B. and Asante, K.A. 1979. Mass Communication Principles and Practices, Macmillan Publishing Co. Inc., New York.

Dwivedi, R.P., Chauhan, J. and Meena, B.S. 2003. Farmers' perception about agroforestry in Bundelkhand region. *Indian Research Journal of Extension Education*, 3(1): 41-43.

Jha, D. and Kandaswamy, A. 1994. Decentralising Agricultural Research and Technology Transfer in India, ICAR. New Delhi and IFPRI. Washington. D.C.

Macklin. M.1992. Agricultural Extension in India, World Bank technical paper 190, World Bank, Washington, D.C.

National Commission on Farmers. 2005. Government of India. New Delhi.

Rogers, E.M. 1983. Diffusion of Innovations. The Free Press, New York.

State of Forest Report. 2003. Forest Survey of India. Dehradun.

Solanki, K.R. and Dwivedi, R.P. 1999. Krishivaniki me gramin yuvaon ki bhagidari. *Unnat Krishi*, 38(5): 45-47.

Suresh Pal, Farrington , J. and Rasheed Sulaiman V.1998. Improving the Effectiveness of Agricultural Research and Extension in India, Policy Paper 8, NCAP. New Delhi.

Swanson, B. 1996. Innovations in Technology Dissemination Component of NATP, Prepared for Ministry of Agriculture, Delhi.

The Hindu, 2006. India's national newspaper, March 27, 2006.

Van den Ban. A.W.1998. Supporting Farmers' Decision making Process by Agricultural Extension. *Journal of Extension Systems*, 14: 55-64.

❑❑❑

Agroforestry: Natural Resource Sustainability, Livelihood & Climate Moderation © 2009
Editors: OP Chaturvedi, A Venkatesh et. al., pp. 715-722
Satish Serial Publishing House, Delhi

Chapter 50

EXTENSION MANAGEMENT PERSPECTIVES IN AGROFORESTRY

R.P. Dwivedi
National Research Centre for Agroforestry, Jhansi-284 003 (U.P.)

Introduction

It is important to examine the social dimensions particularly with reference to gender and social equity and ethics very carefully, while designing agroforestry research and development programmes. Only such integration can help us to impart a pro-nature, pro-poor and pro-women orientation to technology development and dissemination. Agroforestry being a multi-disciplinary and multi-institutional subject, synergies are required at various levels among scientists, planners, development agencies and all other stakeholders to harness maximum benefits of inputs. Linkages between research, education, extension and farmers have been established but their interactions need to be strengthened for appropriate technology generation and its effective transfer.

The status of agroforestry management in India is determined by the very character of the problems to be solved by land-use technologies. Often, our technical problems are expressed in biophysical symptoms, but the primary causes are social not biophysical. Consequently, solutions must be social (livelihood oriented) in character as well technically sound. In the process of designing solutions, agroforestry management can direct technical research towards the farmer's problems with the highest payoffs. Also, agroforestry management can help clarify the values that define pay-off in productivity, sustainability and in equity terms from a broad social as well as client-defined perspective.

Extension has the greater responsibility in order to facilitate community awareness and participation for management and judicious use of natural resources, to promote agroforestry development and to strengthen social capital in rural systems through institution and capacity building programmes. Government of India and many State Governments have tried to bring all-sided development of the rural society. To fill up the gap between what was and what ought to be, in 1952 the Community Development (CD) Programme was launched and the next year (1953) National Extension Service (NES) was added. In 1959 Panchayati Raj also known as democratic decentralization, was added with the Community Development as a result of recommendation of the B.R. Mehta Committee on plan projects (1957) in order to make the community development movement self propelling by direct involvement of people. The grim food situation in subsequent years triggered off a search for directly production oriented programmes. This led to the formulation of a number of programmes in early 60's, such as Intensive Agricultural District Programme, Intensive Agricultural Area Programmes, High Yielding Varieties Programme and Multiple Cropping schemes. But all these production oriented programmes generated a great socio- economic gap in the rural areas due to the reason that maximum benefits were exploited by those who were potentially viable and responsive towards the activities of these programmes.

Globalization has triggered fundamental changes in social, economic and political systems of many nations of the world. Agricultural sector worldwide too is witnessing change in production and marketing spheres. Emerging opportunities and challenges pose task of a different perspective, wherein extension needs to reorient and revitalize itself to strengthen the nation's initiative for its prosperity. Besides the fulfillment of avowed objectives to wipe the tears of the poorest of poor, it has to prepare the farmers to withstand the pressures of WTO and globalize economy. Educated and rural youth need motivation and cognitive conditioning for self-employment through entrepreneurial development. Institution building at grassroots level needs augmentation for emphatic participation of poor in development process. At this juncture extension has to play a crucial and responsible role in propelling the society towards sustainable development with stability and equality.

In recent years agroforestry has been developed as a science that promises to help farmers increase the sustainability, productivity and profitability of production on their land. About 50 per cent of the total agricultural land in the country, is estimated to be marginal land and could be more productively used by revegetating with suitable species of trees, fruit trees and grasses in association with crops. In this context social science considerations on sustainable agroforestry in the light of shrinking natural resources such as land and water and need to maintain the ecological balance will be the important dimension. Researchers other than social sciences do not possess skills in dealing with farmers. When they do go to farmers, they find them indifferent, suspicious or violently protesting. Therefore, agroforestry extension has the potential to transform the agroforestry scenario by motivating and providing the farmers with the right kind of package of practices in the complex climatic and market system of the country. Five types of studies: assessment, tenure, institutional, community and adoption studies, illustrate the range of empirical studies that can help agroforestry plan and prepare for action, as well as providing a basis for project progress. Often, our technical problems are expressed in biophysical symptoms, but the primary causes are social not biophysical. Unless the social factors are changed, applying improved technology will have limited, if not negative results. Consequently, solutions must be social in character as well as technically sound. In the process of designing solutions, agroforestry extension can direct technical research toward the farmers' problems with the highest pay-offs. Also social science can help clarify the values that define pay-off from a broad social as well as client-defined perspective.

Agroforestry is an age old land use that has been practised for thousands of years by farmers the world over. In recent years it has been developed as a Science that promises to help farmers increase the productivity, profitability and sustainability of production on their land. Agroforestry is a land use that involves deliberate retention, introduction, or mixture of trees or other woody perennials in crop/animal production fields to benefit from the resultant ecological and economic interactions (Nair, 1985).

In India, the annual consumption of wood is reckoned to be about 150 million tonnes of fuelwood, five million tonnes of pulpwood and 22 million tonnes of timber, whereas the current production from the forest is abysmally low at 30 million tonnes, with the rest of the demand being met by imports, private forestry, agrowastes and agroforestry outputs. Thus, the scenario is very bleak and dismal. Of an estimated 143 million hectares of agricultural land in the country, 83 million hectares are estimated to be marginal land and could be more productively used by revegetating with suitable species of trees, grass and legumes in association with dryland agriculture. In this context, extension management for sustainable agroforestry in the light of shrinking natural resources such as land and water will be the important dimension of the agroforestry extension management perspectives in agroforestry research in India. Agroforestry extension has the potential to transform the agroforestry scenario by providing farmers with the right kind of package of practices in the complex climatic and marked system of the country.

Theoretical Considerations

The theoretical perspectives of the contributions that social sciences can make to agroforestry have been superbly reviewed by Burch (1991). He concluded that five types of studies illustrate the range of empirical studies that can help agroforestry plan and prepare for action, as well as providing a basis for project progress. These are:

1. *Assessment Studies:* Social benefits and costs are measured and assigned to government project, village and subvillage levels;
2. *Tenure Studies:* The theory and methods for examining the influence of agroforestry programmes;
3. *Institutional Studies:* The use of knowledge about the structure, function and evolution of social institutions and their effect on agroforestry projects;
4. *Community Studies:* Factors affecting adjustment and response by the community to different kind of innovations and
5. *Adoption Studies:* Elements that influence the adoption process of new or improved technologies by the clientele group.

The resource professional can begin to see how certain silvic or other treatments may affect distributional or socially sustainable factors in the dependent farmers community. When systematic extension management knowledge is applied to agroforestry activities, it is to aid basic problem solution.

Three areas in agroforestry where agroforestry extension management may be most useful are:

i. Participatory practices that administrative and development institutions rely on,
ii. Sociocultural, political, institutional and administrative factors of wasteland restoration and
iii. The potential role of agroforestry extension research in sustaining the utility of biophysical research for agroforestry.

Wasteland Development Considerations

One of the concerns of agroforestry is the wasteland development. The major problems in wasteland development are policy issues, economic issues, educational issues, sociocultural issues, technological issues and critical issues (Shah, 1992). A credibility gap exists between people and the various government agencies. The primary target groups of wasteland development programmes, the poor and landless, often only perceive the long-term economic, social and cultural benefits of trees to them and their communities. The target people must be educated about the need for wasteland development and how they can profit by participation. The research and training programmes must be adjusted to fit today's needs for rehabilitation of wastelands for enhanced productivity through MPTs which products are vital to the rural economy in general and to the sustenance of the poor in particular.

Sociocultural Considerations

Every agroforestry extension research activity should begin with a study of sociocultural environment to answer basic questions:

1. What are the social and cultural perceptions of the products and benefits from trees, shrubs and herbs?
2. What species of trees, shrubs, and herbs can satisfy these values to the maximum extent?
3. What are the sociocultural factors that have contributed to the destruction of forests? and
4. What changes in attitudes are the result of ill-conceived development, breakdown of social relationship, and the like?

The rural poor are commonly considered as the primary beneficiaries of agroforestry; consequently, agroforestry technologies are expected to be especially relevant and applicable to small-scale-land owners. Therefore, social acceptability is a much more important measure of success for agroforestry technologies than for commercially-oriented, high input agricultural and forestry technologies.

Technology Development and Transfer Considerations

Improving the transfer of technologies is the traditional reason for incorporating social sciences, most particularly agroforestry extension, into research programmes on agroforestry. In simple form, the working hypothesis is that, by understanding farmers, better information about technologies can be packaged in a more acceptable form to the farmers. Farming systems concepts were built on specific crop or animal agricultural technologies, and again the agricultural extension was added to help transfer technology. Because resource-poor farmers often are at least partially subsistence producers, production, transformation, and consumption operate at the household level. Consequently, a broad way array of social sciences became useful in developing the "new household economics".

The introduction of a stronger social science component into farming system research shifted attention from the technology transfer to identification of the critical needs of farmers as clients. Diagnosis of causes rather than simple identification of symptoms leads to focused research and more effectively designed solutions. Better solutions, defined in terms of the farmers problems, are much easier to transfer than marginal improvements, let alone changes that have negative consequences from the farmers' perspective.

Policy and Planning Considerations

The role of extension management is more obvious in policy and planning than in technology development, but politicians and other policy makers have rarely been enthusiastic about applying science to their activities. One factor that has helped in the incorporation of social sciences in more policy and planning process is to sell the application of social science ideas to prospective beneficiaries. Also, many problems that technology can help solve at the farm or small watershed levels are created more by social factors than by technological forces. Unless the social factors are changed, applying improved technology will have limited, if not negative, results.

Diagnostic Considerations

This is the process by which new hypotheses are generated. The simplest understanding of diagnostics is the method of asking questions that compare and contrast the evidence.After several iterations of questions, the social scientist has uncovered most, if not all, the objective information known about particular set of symptoms. At this point, the extension scientists may pose some hypotheses that explain the observed symptoms and fit the answers to the initial questions. The two cases namely soil erosion in the lower Shivalik hills (Seckler and Joshi,1982) and agroforestry design in the mid-hills (Raintree,1986) are the classical examples where the diagnostic step led to testable hypotheses about the role of social and technical factors in solving some problems that include agroforestry.

Social dimensions

It is important to examine the social dimensions particularly with reference to gender and social equity and ethics very carefully, while designing agroforestry research and development programmes. Only such integration can help us to impart a pro-nature, pro-poor and pro-women orientation to technology development and dissemination. Agroforestry being a multi-disciplinary and multi-institutional subject, synergies are required at various levels among scientists, planners, development agencies and all other stakeholders to harness maximum benefits of inputs. Linkages between research, education, extension and farmers have been established but their interactions need to be strengthened for appropriate technology generation and its effective transfer.

Action Plan for Future

Government regulations on felling of trees from agroforestry lands and transport of agroforestry produce particularly of fuelwood, fodder and timber has to be carefully deregulated to benefit various strata of farming community.

The Joint agroforestry management by participatory approach at all levels should be practised between the planners and executors to plant, maintain, protect and to draw the benefits across out of the plantations which may also pave the way for developing and incorporating other enterprises as one of the components for getting higher income per unit area of land.

Strengthening of agroforestry extension needs considerable attention. Effective training should be given to agroforestry extension personnel to improve their technical skill. Trainers' training is an important area which needs to be focused. We have now 561 KVKs, they can play crucial role in agroforestry dissemination once they are given training.

Private land plantation for industrial purposes and proper tie up arrangements between the tree growers and industries consuming forest products should be encouraged.

Cooperatives such as tree growers cooperatives as exist in many places may be formed. This will facilitate farmers to market their products remuneratively and safeguard their interests from middlemen and price fluctuation.

Conclusions

The status of extension management perspectives in agroforestry in India is determined by the very character of the problems to be solved by land-use technologies. Often, our technical problems are expressed in biophysical symptoms, but the primary causes are social not biophysical. Consequently, solutions must be social in character as well technically sound. In the process of designing solutions, agroforestry extension management can direct technical research towards the farmers' problems with the highest payoffs. Also, extension management can help clarify the values that define pay-off in productivity, sustainability and in equity terms from a broad social as well as client-defined perspective.

References

Burch,W.R.,Jr. 1992. Thinking social scientifically about agroforestry. In: Social Science Application in Asian Agroforestry (Burch, W.R.,Jr.and Parker, J.K. Eds), Winrock International, USA.

Nair, P.K.R. 1985. Classification of agroforestry systems. *Agroforestry Systems*, 2(2): 97-128.

Pathak, P.S. and Dagar, J.C. 1998. Agroforestry: a strategy for rehabilitation of degraded lands. In: Ecology Today: An Anthology of Contemporary Ecological Research (Eds. Gopal, B., Pathak, P.S. and Saxena, K.G.), International Scientific Publications, New Delhi, Pp. 333-371.

Raintree, J.B. 1986. D&D user's manual: an introduction to agroforestry diagnosis and design. ICRAF, Nairobi.

Seckler, D. and Joshi, D. 1982. Sukhomajri: water management in India. *The Bulletin of Atomic Scientist,* Pp. 38.

Shah, S.A. 1992. Wasteland Development in India: political, institutional and sociocultural imperatives. In: Social Science applications in Asian Agroforestry (Burch,W.r., Jr., and Parker, J.K. Eds.), Winrock International, USA.

❑❑❑

Agroforestry: Natural Resource Sustainability, Livelihood & Climate Moderation © 2009
Editors: OP Chaturvedi, A Venkatesh et. al., pp. 723-731
Satish Serial Publishing House, Delhi

Chapter 51

SIGNIFICANCE AND DEVELOPMENT OF INDIGENOUS KNOWLEDGE A FRAMEWORK FOR ACTION

S. N. Pandey
Development Alternatives, Jhansi-284 003 (U.P.)

Indigenous Knowledge (IK)

Herbal medicine is a good example of indigenous knowledge (IK) which has affected the lives of people around the globe. The literature on indigenous knowledge does not provide a single definition of the concept. This is in part due to the differences in background and perspectives of the authors, researcher, ranging from social anthropology to agricultural engineering. Nevertheless, the various definitions also have some common traits.

Indigenous Knowledge (IK) can be broadly defined as the knowledge that an indigenous (local) community accumulates over generations of living in a particular environment. This definition encompasses all forms of knowledge - technologies, know-how skills, practices and beliefs - that enable the community to achieve stable livelihoods in their environment. A number of terms are used interchangeably to refer to the concept of Indigenous Knowledge including Traditional Knowledge (TK), Indigenous Technical

Knowledge (ITK), Local Knowledge (LK) and Indigenous Knowledge System (IKS).

Indigenous Knowledge is unique to every culture and society and it is embedded in community practices, institutions, relationships and rituals. It is considered a part of the local knowledge in the sense that it is rooted in a particular community and situated within broader cultural traditions. It is a set of experiences generated by people living in those communities.

Indigenous knowledge is based on, and is deeply embedded in local experience and historic reality, and is therefore unique to that specific culture; it also plays an important role in defining the identity of the community. It has developed over the centuries of experimentation on how to adapt to local conditions. It therefore represents all the skills and innovations of a people and embodies the collective wisdom and resourcefulness of the community. Such knowledge systems are cumulative, representing generations of experiences, careful observations, and trial-and-error experiments.

When a knowledgeable old person dies, a whole library disappears. (A proverb)

Significance of Indigenous Knowledge

According to the 1998/99 World Development Report, knowledge, not capital, is the key to sustainable social and economic development. Building on local knowledge, the basic component of any country's knowledge system, is the first step to mobilize such capital.

Moreover, there is a growing consensus that knowledge exchange must be a two way street. A vision of knowledge transfer as a sort of conveyor belt moving in one direction from the rich, industrialized countries to poor, developing ones is likely to lead to failure and resentment.

Sharing knowledge with the poor is most effective when we also solicit knowledge from them about their needs and circumstances. Therefore, development activities, especially those that aim to benefit the poor directly, need to consider Indigenous knowledge in the design and implementation stages of the process.

The challenge for the development community is to find better ways to learn about indigenous institutions and practices and where necessary adapt modern techniques (i.e., "global best practices") to the local practices. Only then will global knowledge be rendered relevant to the local community needs.

Indigenous knowledge provides the basis for problem-solving strategies for local communities, especially the poor. It represents an important component of global knowledge on development issues. Indigenous knowledge is an underutilized resource in the development process. Learning from Indigenous knowledge by investigating first what local communities know and have, can improve understanding of local conditions and provide a productive context for activities designed to help the communities. Understanding local knowledge can increase responsiveness to clients. Adapting technological practices to the local setting can help and improve the impact and sustainability of development assistance. Sharing indigenous knowledge within and across communities can help and enhance cross-cultural understanding and promote the cultural dimension of development.

Application: *Preserving traditional cultivators in local communities.*

The international community is establishing gene banks to preserve genetic information of local varieties or indigenous species. Genetic traits of these species and the knowledge of cultivators may prove instrumental in future breeding programs to introduce resistance against pests or diseases or endurance for harsh climatic conditions. However, preserving genetic traits without preserving the knowledge of their husbandry may prove futile as the seeds and clones stored in seed banks do not carry the instructions on how to grow them. Hence, gene banks cooperate with farmers and communities who still cultivate local varieties to preserve such essential knowledge and skills *in situ*.

Lesson: *Local knowledge is vital for preserving bio-diversity.*

Exchange of Indigenous Knowledge

The integration of indigenous knowledge into the development process is essentially a process of exchange of information from one community to another. The process of exchange of IK within and between institution and community involved mainly the following way:

- The process typically begins with recognition and identification of knowledge as expressed in a technology or a problem solving strategy. However, identification of IK can at times prove difficult. For example, some IK may be embedded in a mix of technologies or in cultural values, rendering them unrecognizable at first glance to the external observer. Others may have become part of every day life of a community to an extent that makes it difficult to isolate such practices even by individuals or communities applying them. In such cases, technical and social analyses of certain practices may be needed to identify IK.
- The next typical step is to validate IK in terms of its significance and relevance (to solving one or several specific problems), reliability (not being an incidental occurrence), functionality (how well does it work), effectiveness and transferability. This is because most IK is stored in

tacit form, which in certain circumstances may make it transferable only through direct practice and apprenticeship. Proof of an efficient process at the point of origin does not necessarily ascertain its efficacy under seemingly similar conditions in other locations. Consequently, it is important to carry out pilots to test the new technology with the recipient. Nevertheless, in some cases it should be possible to undertake a general assessment of transferability, subject to confirmation with follow up pilots.

- The next step recording and documenting, is another major challenge again because of the tacit nature of indigenous knowledge. The scope of recording/documentation is largely determined by the intended use of the information. Thus, while scholars would want to understand and capture a more comprehensive view of knowledge with all its ramifications, a practitioner might be satisfied with an answer to the question "How did they do that?" The recording may require audio-visual technology, taped narration, drawings, or other forms of codifiable information. In case the tacit nature of a practice does not lend itself to such recording, information about locations, individuals or organizations that can demonstrate or teach a practice could be used as a pointer to the source of IK.
- Storage in retrievable repositories is the next typical step in the process. This involves categorization, indexing, relating it to other information, making it accessible and conserving, preserving and maintaining it for later retrieval. Meta-information needs to be produced to make retrieval more user-friendly. This could include electronically stored and indexed abstracts, directories of experts or applications. Storage should not, however, be restricted to only text documents. It should also include other retrievable types of repositories of information such as tapes, films, databases and IK practitioners.
- The transfer of IK goes beyond conveying it to the potential recipients. An important element of the transfer is to test the knowledge in the new environment. Economic and technical feasibility, social and environmental impact and other criteria as deemed necessary by the recipients need to be examined. Some local practices can only be transferred directly, from practitioner to practitioner. Only few people in a community will have the risk bearing capacity to accept substantial failure of an imported technology. Careful selection of cooperating partners and potential beneficiaries in a participatory process is a prerequisite for a successful transfer. The risk of failure is reduced if the new technology builds upon existing local knowledge.

> The World Bank-supported Sodic Lands Reclamation Project in U.P India was a farmer-driven effort to increase household incomes. The major constraints were sodic soils, a result of inappropriate irrigation management and brown plant hoppers, which often destroyed up to 50 percent of paddy and wheat crops. By combining local and modern knowledge, farmers applied gypsum, built contour bunds, leached the soil, started multi cropping, green manuring, crop rotation and composting, and reclaimed over 68,000 hectares of land belonging to 247,000 families. They controlled brown plant hoppers with Neem* extract, rice husk and green manure. After five years, paddy and wheat yields and incomes had reportedly risen by 60 percent. With World Bank support, farmers created a local farmers school to incorporate these practices into curriculum and outreach work. Today, farmers train and advise fellow farmers, reaching over 7,200 households in 65 villages. Recognizing and incorporating IK has not only produced technical and economic results, but has helped to create a farmer-owned training institution with tremendous credibility and outreach.
>
> *The Neem tree (Azadirachta indica) has been used in India for centuries in a variety of applications related to human and animal treatments and as a means to control insect pest.*

Once the transfers and adaptation process has been carried out successfully through a pilot, the dissemination of IK to a wider community adds the developmental dimension to the exchange of knowledge and could bring about a wider and deeper impact of the knowledge transfer. Depending on content and context, dissemination activities could include public awareness campaigns, public broadcasting, advertisements, seminars, workshops, distribution of information material, publications and the incorporation of IK in extension programs or curricula. Dissemination activities could be either targeted to specific groups or address the general public. Governments could encourage the process by creating a favorable political, economical and legal framework.

The exchange of indigenous knowledge is the ideal outcome of a successful transfer of technology. This is essentially a learning process whereby the community, where an IK practice originates, the agent that transmits the practice, and the community that adopts and adapts the practice all learn during the process.

Framework for Action

The framework to respond to the challenge of better integrating Indigenous knowledge into the development process requires,

Disseminating information. Key Actions Include

- Developing a database of IK practices, lessons learned sources, partners, etc.

- Identifying and testing instruments for capture and dissemination of IK.
- Publishing selected cases in print and electronic format.

Facilitating exchange of IK among developing communities. Key actions include

- Helping build local capacity to share IK, especially among the local Indigenous knowledge centers.
- Identifying appropriate methods of capturing, disseminating IK among communities.
- Facilitating a global network to exchange IK.

Applying indigenous knowledge in the development process. Key actions include

- Raising awareness of the importance of IK among development partners.
- Helping countries to prepare national policies in support of indigenous practices.
- Integrating indigenous practices in programs/projects supported by partners.

Building partnerships. Key actions include

- Learning from local communities and NGOs.
- Leveraging limited resources of partners to obtain greater development impact.
- Addressing the intellectual property rights issue of indigenous knowledge.

The "Typology and Sectoral Features of Indigenous Knowledge at Community Level" assist in determining which approach to chose in support of IK in a development program. The Matrix describes sector specific areas of indigenous knowledge; typical practices which are relevant for transfer; types of bearers of knowledge; manifestation of IK in hardware (artifacts) or software (laws, rituals, organization) which could assist in determining the relative ease of accessibility of IK to outsiders. The Matrix given below explain the process to the features of Indeginous knowledge at community level.

Typology and Selected Features of Indigenous Knowledge at Community Level (Example)

Category Sector	Area of Knowledge (examples)	Bearers of Knowledge	Manifestations of Knowledge (examples)	Accessibility of Knowledge for Outsiders
Agriculture	• soil and land classification • cultivation, plant protection • plant sociology • characteristics of crops under stress	• all farmers (according to gender or ethnic distribution)	• local varieties, • cultivation techniques, • plant protection techniques	good, if manifested in discernible practices and methods
Animal husbandry and ethnic veterinary medicine	• Breeding • animal treatments • plant sociology drought resistant species • treatment of ekto- and endoparasites	• nomads, • herds people, • animal keepers	• breeds • watering places • livestock routes	difficult, sometimes restricted rituals
Post harvest technologies and nutrition	• protection against vermin • treatment of seed and products • food preparation	• most farmers • specifically women	• storage systems • food products • cosmetic products	usually good, unless related to rituals
Use and management of natural resources and environment protection	• sources and qualities of raw materials • plant sociology, underutilized plants / products • erosion control measures • long term cycles of climate and environment	• most farmers • women • crafts people	• forestry produ cts • contour bunds, • regulation of commons, usufruct, village by-laws • supply of raw materials and products	fair
Saving and lending	• risk assessment • saving in kind	• savings groups • money-lenders	• group collateral • conditions • risk sharing	difficult
Community development	• power sharing • networking • lobbying	• elders • local leaders	• rituals • by-laws, regulations • advocacy	moderately difficult
Poverty alleviation	• risk aversion strategies, • neighborhood assistance	• elders • local leaders	• rituals • dependencies, • village by-laws, • integration of marginalized, kinship	fair

World Bank, (1999) World Development Report

Indigenous Knowledge Limitations

All knowledge systems have their limitations and weaknesses, and IK is no exception. Neither IK nor international science will be appropriate and accurate in all circumstances. In the same way that it has proven unwise to uncritically accept international science (for example, green-revolution technology), it would be unwise to accept all traditional knowledge as good practice or as sustainable practice. Indigenous peoples have at times mismanaged resources.

The nomadic hunters and gatherers who are not tied to any specific resource base may not have a conservation ethic. Some IK practices are less wise and unwise environmental practices and valid and invalid environmental beliefs coexist in many cultures. To assume differently is to assume that with respect to natural resource management indigenous peoples are either inherently superior or inherently inferior to the cultures of the developed world.

Challenges Ahead

The fact that some IK is lost naturally as techniques and tools are modified or fall out of use, the recent and current rate of loss is accelerating because of rapid population growth, growth of international markets, educational systems, environmental degradation, and development processes , pressures related to rapid modernization and cultural homogenization.

Considerable progress has been made in promoting recognition of indigenous knowledge is increasingly becoming part of the development agenda. The national initiatives and policies are emerging; civil society groups are forming a broad base of support; local initiatives are multiplying; and the number of development projects and programs integrating IK is increasing. Yet, some substantial challenges remain.

Conclusion

Indigenous knowledge is a critical factor for sustainable development. Empowerment of local communities is a prerequisite for the integration of IK in the development process. The integration of appropriate IK systems into development programs has already contributed to efficiency, effectiveness, and sustainable development impact. IK, like any other knowledge, needs to be constantly used, challenged, and further adapted to the evolving local contexts. Supporting local and regional networks of traditional practitioners and community exchanges can help to disseminate useful and relevant IK and to enable communities to participate more actively in the development process. While innovative mechanisms for the protection of IK need to be developed, many indigenous knowledge practices can at the same time be integrated into local, national, regional, or even global development efforts. However, experience has shown that this cannot be done by one institution alone. Therefore, partnerships are needed to support this process at all levels.

References

Biodiversity Conservation Strategy Update, 2006. Biological Resources and Institutions Program, World Resources Institute, 1709 New York Ave. NW, Washington, DC, USA.

Srivastava, J., Smith, N. J. H. and Forno, D. 1996. Biodiversity and agriculture : implications for conservation and development, World Bank Technical Paper WTP321

Technology and Social Change No. 15. Ames, Iowa: Technology and Social Change Program, Iowa State University.

Warren, D. M. 1989. Linking Scientific and Indigenous Agricultural Systems. In : The Transformation of International Agricultural Research and Development (Eds. J. L. Compton), Boulder: Lynne Rienner Publishers, Pp. 153- 170.

Warren, D.M., Slikkerveer, L.J. and Brokensha, D. 1995. The cultural dimension of development: Indigenous knowledge systems. London: Intermediate Technology Publications.

World Bank, 1997. Knowledge and Skills for the Information Age, The First Meeting of the Mediterranean Development Forum.

World Bank, 1999. World Development Report 1998/1999. Knowledge for Development.

❑❑❑

Agroforestry: Natural Resource Sustainability, Livelihood & Climate Moderation © 2009
Editors: OP Chaturvedi, A Venkatesh et. al., pp. 733-743
Satish Serial Publishing House, Delhi

Chapter 52

ROLE OF K.V.K. IN WOMEN EMPOWERMENT

Nishi Roy
KVK, Bharari, Jhansi-284 003 (U.P.)

Introduction

Women have been regarded as the nuclei of a nation. Just as the body is made up of millions of families depending on the women -housewife and mother for their progress and well-being. The health and the growth of the nation depend on the vitality of its women. Understandably, therefore, as the perennial strength of the nation, the women must appreciate their axial position as also their obligation towards nation building along with the know-how and do-how to fulfill this obligation (Kapoor, 1986). Her attention needs to centred on children, who are in the dawn of life, youth in the sunny phase, old who are in twilight and the disabled who are in the shadow of life, not to overlook the duties, she has to perform for her counterpart. Her paramount duty is considered to be service to her husband and family by thought, word and deed. She is to manage the resources, time, money and energy in an excellent way so that the demands in her family are met with. Women are thus nuclei of family life and pace setters in the family resource development but are considered as 'invisible workers' in the home (Laxmidevi, 1989).

Women constitute half of the world. They are not only the carriers of human race but civilization does seem to rest on them. 'Women' is the mother of the race and is the liaison between generations. Harmony is ingrained in their genes, as they are the least destructive. They are the best upholders of environmental, ecological and social balances, even at the cost of personal loss and sacrifice.

Women are the first of those to rise and the last to eat and sleep. They perform upto two-third of the manual labour in rural areas, not only in India but also across the developing world. Rural women in our country share abounded responsibilities and perform a wide spectrum of duties. They work as housewife, economic providers, food producers and dynamic community leaders. Even they own less than one per cent of physical assets. As their work does not involve money, it is undervalued.

There is no denying the fact that women have suffered utter neglect and deprivation over centuries. Though they have made significant contribution, they are not regarded as equal partners in the development process. Participation of women in agriculture in developing countries has been silently appreciated but without much recognition and recording of their contributions. They have not been prepared for active involvement in the development process. By and large, they have remained as 'invisible workers'.

Women have formed an organic component of the working force. At least 60 per cent of the rural women are involved in agricultural works viz. sowing, transplanting, weeding, manuring, harvesting, winnowing, shelling and storing of crops and also help in family enterprises such as bidi-making, rope making, poultry etc. They do all this in addition to their normal domestic chores as a housewife.

Genesis of Krishi Vigyan Kendra

Krishi Vigyan Kendra is a farm science centre. Dr. Mohan Sinha Mehta Committee (1974) specially appointed by the Indian Council of Agricultural Research for formulating the K.V.K. Scheme, enunciated the following basic concepts:

1. The Krishi Vigyan Kendra will impart learning through work-experience and hence will be concerned with technical literacy.
2. The K.V.K. will impart training to only those extension workers, who are already employed or to the practicing farmers and fisherman.

3. There will be no uniform syllabus for a K.V.K. The syllabus and programme of each Kendra will be tailored to the felt needs, natural resources and the potential for agricultural growth is that particular area.

The K.V.K. is the grass-root level vocation training institutions designed for bridging the gap between the available technologies at the one end their application for increased production at the other. The K.V.K. provides a strong training support for bringing about production break through in agriculture. The first K.V.K. was established in 1974 at Pondicherry under Tamil Nadu Agricultural University. Till now 562 K.V.K have been established in the country. The main objectives are as follows:

1. Planning and conducting survey of the operational area in order to prepare the resources inventory with special reference to identifying the training needs of the farming community.
2. Developing and organizing non-formal educational programmes by way of field days, farm visits, farmers fair, radio talk, farm science clubs etc. as the follow-up information supporting to training courses.
3. Developing and maintaining the campus farm and demonstration units an scientific lines as the facilities for providing work experience to the trainees as also disseminating the latest technical know-how.
4. Providing added training facilities in the areas for home-making and nutrition education for rural community.
5. Gradually enlarging the training facilities to encompass other important areas such as home crafts, cottage industries etc. consistent to requirements of the Integrated Rural Development in collaboration with concerned organizations.
6. Planning and conducting production oriented, need based short and long duration training courses both on campus as well as in the villages for various target groups with priority on the weaker and the poor.
7. Implementing all such schemes of the ICAR and other related organizations, which intend to strengthen the training programmes of the Kendra.

The specific objectives, nevertheless, may be many and varying depending upon the agro-eco-system of the area and the needs of the client, the resources available and economic status of the targeted groups.

Mandates of the Krishi Vigyan Kendra

1. To train farmers, farm women and school dropouts in agricultural and allied subjects relating to their needs, identified based on their farming situations and farming systems.
2. To undertake On-Farm Testing of the technologies developed by the SAU's and ICAR institutes in agriculture and allied subjects for their local suitability and identifying the constraints through further research or adoptive research and application.
3. To demonstrate the technologies tested and recommended for location specific situations and to find out their potentialities in maximizing the fields/income per unit area.
4. To train grass root level extension functionaries of line departments, connected with developmental activities.
5. To organize farm work including Lab-to-Land programmes and other extension activities.

Empowerment of Women

Empowerment of women is construed as providing them with a sufficient degree of control, to give them decision-making powers and enable them to raise the level of consciousness of their class and enhance their gender status and rightful entitlement. This can be done by securing greater access to knowledge as well as ideological shifts that bring about changes in addition to the community.

Empowerment in such a context calls modalities other than more economic betterment, because much of women's powerlessness stems from non-quantifiable rather than monetary impositions or handicaps. There is need of combating powerlessness/subordinate status, which are the crucial issues. Women must be empowered by enhancing their awareness, knowledge skills and technology use efficiency so that agricultural production multiplies at a faster rate, environmental degradation is reduced and resources are conserved as well as utilized properly, leading to overall development of the society.

The constitution of India grants equality to women in various fields of life. Yet a large number of women are either ill equipped or not in a position to propel themselves out of their traditionally unsatisfactory socio-economic conditions. Patriarchy continues to be embedded in the social system in many parts of India, denying a majority of women the choice to decide on how they live and have a say in community issues.

The promotion of gender equality and the empowerment of women was one of the central concerns of the tenth plan (2002-07), which spelled out a three-pronged strategy for empowering women:

- *Social Empowerment:* Create an enabling environment through adopting various policies and programmes for development of women besides providing them easy and equal access to all the basic minimum services so as to enable them to realize their full potential.
- *Economic Empowerment:* ensure provision of training, employment and income generation activities with both forward and backward linkages with the ultimate objectives of making all women economically independent and self-reliant.
- *Gender Justice:* Eliminate all forms of gender discrimination and thus enable women to enjoy not only *de jure* but also *de facto* rights and fundamental freedom on par with men in all spheres, viz, political, economic, social, civil, cultural etc.

Agricultural Production and Women Empowerment

Women's role is enormous in the activities related to food production, processing and marketing; especially women contribute in the area of: grain cultivation, plantation and horticultural crops, post harvesting and marketing, livestock rearing, fisheries, sericulture and forestry.

Grain Cultivation

Women play a pivotal role in producing staple crops like rice, wheat and maize. Women generally participate in most of the rice production activities. They dominate in planting, weeding, harvesting, and transporting the produce and all the post harvest activities. Except for ploughing, leveling and irrigating the field, all the other activities such as sowing, weeding, transplanting, harvesting, stocking of straw, husking, drying and storing are female dominated tasks. In the rice system, women supply on an average 70-80% of the labour for transplanting, 70 to 85% for weeding and over 60% for harvesting (Agarwal, 1983). They also have an important role in seed selection, storage and post-harvest procedures.

In India, women carry out as much as 80% of the rice production. They are involved right from seeding, transplanting, weeding, harvesting and processing to marketing of the produce and buying the necessities of life. Also, women labour contributes a higher proportion of total labour input in paddy production activities on all small size groups. Women labour is exclusively used for transplanting, harvesting, stubble collection, threshing and winnowing. Mishra and Awasthi (1988) concluded that women labour use was higher on high yielding paddy farms over local paddy farms. Singh *et al.*, (1993) explained that the maximum number of women participating in the operations of storage, harvesting, threshing, watching birds, sowing and fertilizer application belong to the age group of 30 years, while the majority of the women below 30 years participated in irrigation, plant protection measures, weeding and land preparation.

Plantation and Horticultural Crops

Plantation crops, e.g. tea, coffee, rubber, cashew nut, coconut, oil palm, areca nut, cocoa, cardamom, black pepper and spices are cultivated in India over an area of three million hectares. The share of women workers in the total production sector is within a range of 30 to 54%. It varies, between 20 to 80% in the processing sector. Women workers are engaged in the marketing sector, but their roles are marginal as compared to men. Some of the specific areas of operation performed predominantly by women are weeding and manuring in all plantation crops, mulching and pot watering in coconut and area nut gardens; plucking/picking and sorting of tea; transport of coconut and arecanut after harvest; splitting, sun drying, husking and grading of ripe arecanut, plucking of cashew apples and nuts, shelling, drying, peeling and grading of cashew kernels; picking and cleaning of cardamom and drying, cleaning and packing of black pepper. In the coir sector, women are involved in beating of husk and spinning of coir yarn. Of the 8.8-lakh plantation workers in South India, 3.9 lakh are women. They are 56% of the workforce on tea plantations, 46% on coffee and 30 to 35% on cardamom and rubber plantation. Sixty per cent of the women are involved in broom making and 45 per cent in oil extraction. Horticultural crops are important for their nutritive as well as economic value. There is a great export potential for vegetables, fruits, cut flowers, bulbs and foliage plants. Women play a prominent role in natural horticultural productivity. Women farmers are involved in planting and care and management of orchards. In vegetable growing and ornamental gardening they are involved in seed production, sowing, controlled pollination, harvesting and seed processing. All these operations are delicate and require great patience. The extent of participation of women in different horticultural production system depicts interesting regional differences in India. In North India, women mostly participate in fruits and vegetables processing, flower gardening, ornamental nursery, kitchen gardening and same extent in vegetable growing. In South India, beside these operations, women also participate in vegetable and flower marketing and nursery technology.

One of the most prominent operations was done by farm women is irrigation of nursery plants by watering can. In this operation, the amount of drudgery involved is great. Study on the role and participation of women in agricultural production system in the villages of coastal west-Bengal indicated that women were fully responsible for fruit and vegetable production in homestead farming (Chakravarthy, 1990). The strawberry farms in Mexico city prefer to employ women as they are considered to be better at labour intensive work of harvesting. They are also cheaper in wages than men. (FAO, 1984).

Post Harvesting Activities and Marketing

Across the developing world, women perform most of the food processing activities. It is almost exclusively women who process, preserve and store food as well as sell it in the market. In Africa as a whole 100% of the food processing activities are done by women (Anonymous, 1980). Women do many post-harvest activities include manual threshing, sun drying, traditional parboiling and milling using traditional devices. Women's participation in post harvest operations ranges between 89 and 99 per cent. It may be due to the fact that these activities are generally performed at the household level. Study on percentage share of women labour force in processing of plantation crops indicated that women's participation was 80% in coffee, 72% in areca nut, 84% in cashew, 70% in black pepper. Women constitute 90% of labour force in cashew processing industry, which is considerable.

In most of the developing countries women play a major role in marketing of food and agricultural products. Most of the women often transport and sell vegetables, fruits and other products needed for local consumption. Women are mostly involved in fish-trade, vegetable and flower vending, collection and selling of grass, sale of cow dung cakes, firewood and dairy products such as ghee and milk.

In Asia, women's post-harvest tasks of preparing food for table or the market stall include, transforming paddy into rice by milling in Indonesia, making puffed rice in Nepal, chopping bamboo shoots for picking and canning in Japan, processing oil-seeds in China, drying fish in Srilanka and canning Tuna in Thailand and cleaning and slicing fruit and vegetables for canning, picking and preparing puffed and beaten rice in India. Post-harvest activities in Bangladesh are traditional and labour-incentive. Women do many of the post-harvest activities in the rural areas. These activities include manual threshing, sun drying, traditional par-boiling and milling in traditional devices (Akhtar *et* al; 1996).

Livestock Rearing

Women play an important role in livestock production, care and management and in the processing and sale of livestock products. Women are responsible for dairy production and for processing of cow dung for fuel and manure. Small animal husbandry is one of the few sources of income that are fully under their control (FAO, 1998). Women's work includes collection of fodder and water, preparation of feed for animals, cattle shed cleaning and bathing of animals, taking animals for roadside grazing, milking, preparation of milk products, management and marketing of milk, collection of cow dung, preparation of cow-dung cakes and their storage and preparation of manure for the farm. The degree of participation of women vis-à-vis that of men in the hills especially, is significantly higher in livestock rearing and related activities. In livestock rearing women's participation is to the tune of 71% as against 21% for men (Singh and Sharma, 1990). Small livestock such as goats, sheep and poultry are usually predominantly the concern of women and are much more frequently owned by the women.

Sericulture

Women contribution in sericulture is very much prominent and has been playing a pivotal role for the development of the silk industry. Women carry out almost 60 to 65% of the activities of the sericulture industry. Nearly 60% of the labour requirements of sericulture industry is provided by women (Durgadevi, 1992). Women are involved in feeding of mulberry leaves to silk worms in addition to their normal household chores. According to Charslay (1993) silkworm rearing represents suitable domestic activities for the middle class sericulture families, especially for women folk who would not be expected to work outside the home. Women do hand spinning of pierced cocoons and silk waste reeling but their involvement is usually restricted to less skillful labour intensive activities both in cocoon and reeling sectors. Women are also involved in weaving of silk, dress designing and garment production.

Forestry

In developing world, a significant portion of forest industry work force is made up of women (FAO, 1998). Forests are a storehouse of many raw materials and foods such as medicinal plants, basketry material and fuel wood, value added items from the forests like fruits, nuts; mushroom and oils are sold or consumed in the home. Besides, providing rich nutritional supplements forest products can help in warding of famine. In the developing world a significant portion of forest industry workforce is of women. Much of what they gather from the forest is processed and marketed to bring in supplementary income to the family. Women's knowledge of forest resources is extensive e.g. tribal women in India have medicinal uses for almost 300 forests species. Fuel scarcity has negative impact on household nutrition as it may influence the amount of food cooked, resulting in damaging effects as child nutrition and incidence of diseases.

Fish Cultivation

Women play a leading role in fisheries development. They engage in fishing in shallow waters in coastal regions or protected total areas, harvesting molluscs or seaweeds and fish with traps or headlines from canoes. In addition to the important primary fishing activities, women living in fishing communities are very activities, women living in fishing communities are very active from the moment the boats are landed in sorting and grading the catch, drying, smoking or curing the fish, marketing fish paste and taking the product to the local market. In addition, women are also responsible for performing the skilled time consuming jobs that take place on store, for example net making and mending.

Women process the catch by sun drying, salting, smoking and preparing fish and fish derived foods and they are the ones who subsequently sell these products. Gopal Krishnan (1996) stated that the role of women in traditional Indian Coastal shrimp farming is significant. Where in women have the responsibility of broad stock and seed cultivation, pond construction, pond maintenance, collection and preparation of feeds, and harvesting and post harvest handling. The business of dry fish is the most widely practiced activity among the womenfolk of marine fisheries sector. Selling of fish in local markets is yet another activity in which women are involved. A few of the women folk are also involved is supplying semi-processed prawns to the local exporters. In prawn-processing units, women contribute around 77% of the labour carder (Sharma and Thakur, 1991).

Krishi Vigyan Kendra plays a vital role in the process of transfer of technology. It works for the development and enrichment of physical and human resources in its operational area. It provides good educational facilities to the farming community by establishing various instructional units in agriculture and allied areas. Through it's vocational training, on farm testing and front line demonstrations, trains the farming community farming which ultimately increases the productivity, and also trains farmers, farmwomen and rural youth in income generating activities, which enhance the self-employment opportunities in the rural area. It updates the technical know how of the grass root level extension workers. It extends its activities throughout the district through various extension activities.

It has been observed that with in a time span interaction of K.V.K. with the farming community it brought about changes in adoption of latest agricultural technologies leading to increased productivity and enhanced self employment opportunities. Thus, the Krishi Vigyan Kendra provides both food security and livelihood security to the rural women work force on ecologically sustainable basis.

References

Aggarwal, B.1983.Mechanization in Indian Agriculture. An Analytical study based on Punjab, Allied Publishers.

Akhtar, A., Mazed, M.A. and Ahmed, A.1996. Improving the rice post-harvest technology in Bangladesh, Bangladesh Rice Research Institute, Gazipur, Bangladesh.

Anonymous.1982. Report of survey of Coir workers; Directorate of Economics and Statistics, Government of Kerala, Trivandrum.

Chakravarthy,S.1990. Look into the role and participation of women in agricultural production- An analysis of the experience in coastal West Bengal in NIRD, Workshop on women in agriculture, Background paper, Oct, 29-Nov1, 1990.

Charsley, S.R.1993. Sericulture-the way of the life. *Indian Silk*, 27(1): 3.

Durgadevi, G.V.L. 1992. Women in Agriculture and Rural Development, Discovery Publishing House. New Delhi, Pp. 149.

FAO,1984. The status of food and agriculture world review: The situation in sub-Sahara Africa, Women in Agriculture, Rome, Pp. 126-136.

FAO,1991. Women and population in agricultural and rural development in Sub-Sahara Africa, Women in Agricultural development, No.5, Rome.

Kapoor, M.1986. Women and family life education in India, Print well Publisher, Jaipur, Pp. 1.

Laxmidevi, A.1989. Rural women management in farm and home, Northern Book Centre New Delhi, Pp. 8.

Mishra,A and Awasthi, P.K.1988. Technical Impact on female Labour participation in Agricultural Sector, In Abstract. International Conference on Appropriate Agricultural Technologies for Farmwomen, New Delhi.

Padmaja, M.1999. Role of Krishi Vigyan Kendra in rural development- A case study. Guntur (A.P.)

Pandey, H.2001. Understanding farm Women, NRCWA, ICAR, Bhubneshwar, Pp. 34-40.

Sharma, B.K. and Thakur, N.K. Women in Fisheries activities. In: Women in Agriculture, (Ed. R.K. Punia), Northern Book Centre, New-Delhi, Pp. 227.

Singh, M.P. and Sharma, N.1990. Women in rice based Farming system in Uttar-Pradesh. In: Women in Agriculture, their status and role (Ed. R.K. Punia), Northern Book Centre, New-Delhi, Pp. 99.

Singh, R.R.2008. Third National Conference on Krishi Vigyan Kendra, Pantnagar (Uttarakhand). Rural prosperity: Role of women empowerment, Pp 6-7, 8-13.

Swaminathan, M.S.1990. Importing a rural women users perspective to agricultural research and development, Women in Agriculture-Technological Perspective, Edited by Prasad and Shriram, International Federation for Women in Agriculture, Pp. 2.

❑❑❑

Agroforestry: Natural Resource Sustainability, Livelihood & Climate Moderation © 2009
Editors: OP Chaturvedi, A Venkatesh et. al., pp. 745-756
Satish Serial Publishing House, Delhi

Chapter 53

EFFECT OF TREE INTRODUCTION ON DISEASE INCIDENCE IN COMPONENT CROPS OF AGROFORESTRY SYSTEMS

Anil Kumar and Ashok Shukla
National Research Centre for Agroforestry, Jhansi-284 003 (U.P.)

Introduction

Agroforestry can be defined as an intensive land use management system that integrates trees, shrubs, crops, etc. on a landscape level to achieve optimum benefits. The system maximizes production, gives supplement food and fodder, improves soil, utilizes waste and degraded lands, provides employment opportunities, minimizes adverse effect of climatic factors, aids industrial growth and improves environment. Of late, researchers are becoming increasingly aware of the importance of disease problems in agroforestry systems. Limited amount of information is available on the issues that are likely to influence disease incidence, etiology and epidemiology in agroforestry situations (Schroth *et al.*, 2000). There is a need to identify main diseases of crops and tree species in various systems and different interactions, which occur between them. The interactions have been divided into primary, where component crops share diseases and secondary, where the influence of one component crop on the survival and growth of diseases on the other is considered (Huxley and Greenland, 1990).

Most of the information included in this paper is based on a study was carried out at National Research Centre for Agroforestry (NRCAF), Jhansi, Uttar Pradesh, India (24^0 11′ N latitude, 78^0 17′ E longitude and 271 m above msl) during 1998-2006. On agro-ecological zone map of India, Jhansi lies in agro-ecoregion 4 [Northern Plain and Central Highlands, Hot Semi Arid Ecoregion with Alluvium–derived Soil (Sehgal *et al.*, 1990)]. To study the disease interactions in the agroforestry systems, the data was mainly collected from five fields. Observations were made on disease incidence on various crops. Comparative observations on the field patterns of disease incidence in open and under tree shade were made. While selecting sites for recording data, care was taken that the observations on treatments were from the same field, same seed lot of single variety was used for sowing in the field, disease incidence in initial stages was uniform throughout the field and no fungicidal sprays were taken up. Observations on temperature, relative humidity (RH), radiation intensity, and presence of free water on leaves in various fields were also recorded under and outside canopy of windbreak at hourly intervals, when the diseases were at peak.

Primary Interactions

Alterations in disease severity of under-story crops in agroforestry systems (as compared to sole crop) are possible due to sharing of diseases among its component crops or modification of the microclimate by the tree species. Data on disease incidence for multipurpose trees, horticultural plants and field crops in various agro-forestry systems showed that pathogens common to component crops of a given agroforestry system were rare. Only certain closely related plant species recorded common pathogens. Among field crops, alternaria leaf blight was present on mustard and Taramira (*Eruca sativa*), yellow mosaic virus recorded on soybean, green gram and black gram, and cercospora leaf spot was shared by green gram and black gram. Among trees, citrus canker was recorded on lemon and Kinnow (*Citrus deliciosa*) while ganoderma fruiting bodies were observed on *Leucaena leucocephala, Acacia nilotica, Dichrostachys cineria* and *Dalbergia sissoo* (Table 1). Sharing of the diseases was either restricted to field crops or tree species. Pathogens, common to field crops and tree species, were not recorded, except cucumber mosaic virus (Table 2).

Table 1. Disease incidence in multipurpose trees, fruit plants and field crops recorded at NRCAF, Jhansi.

Crop	Name of Disease	Pathogen
Multipurpose trees		
Acacia sp.	Root rot	*Ganoderma lucidum*
Dalbergia sissoo	Powdery mildew	*Phyllactinia dalbergiae*
	Leaf rust	*Uredo sissoo*
	Root rot	*Ganoderma lucidum*
	Seedling blight	*Fusarium sp.*
Neem	Leaf spot	*Cercospora meliae*
Kala Siris	Leaf blight	*Alternaria sp.*
	Pod spot	*Collectotrichum sp.*
Deshi Siris	Leaf spot	*Cercospora albiziae*
	Pod spot	*Collectotrichum sp.*
Nutan	Root rot	*Ganoderma lucidum*
Hardwikia	-	-
Fruit trees		
Kinnow	Citrus canker	*Xanthomonas citri*
	Leaf blight	*Colletotrichum sp.*
Aonla	Fruit rot	*Aspergillus sp.*
Ber	Powdery mildew	*Oidium erysiphoides* f. *ziziphi*
	Sooty mould	*Isariopsis indica*
	Leaf spot	-
Mango	Powdery mildew	*Oidium mangiferae*
	Anthracnose	*Collectotrichum gloeosporiodes*
	Mango malformation	-
Guava	-	-
Rainy season crops		
Sesamum	Angular leaf spot	*Cercospora sesamicola*
	Leaf blight	*Phytophthora parasitica var. sesami*
	Phyllody	MLOs
Soybean	Yollow mosaic	Yellow mosaic virus
	Leaf blight	-
Groundnut	Tikka	*Cercospora arachidicola* *Cercospora personata*
Black gram	Powdery mildew	*Erysiphe polygoni*
	Yollow mosaic	Yellow mosaic virus
	Cercospora leaf spot	*Cercospora cruenta*
Green gram	Yollow mosaic	Yellow mosaic virus
	Cercospora leaf spot	*Cercospora cruenta*

contd...

Table contd...

Winter crops		
Wheat	Leaf blight	*Alternaria triticina*
	Loose smut	*Ustilago tritici*
Mustard	Downy mildew	*Pernospora parasitica*
	White rust	*Albugo candida*
	Leaf blight	*Alternaria brassicola*
		Alternaria brassicae
Gram	Fusarium wilt	*Fusarium sp.*
Taramira	Leaf blight	*Alternaria brassicola*
		Alternaria brassicae

Table 2. Summary of sharing of diseases among crops recorded at NRCAF, Jhansi.

Disease	Pathogen	Recorded hosts
Sharing among crops		
Alternaria leaf blight	*Alternaria brassicola, A. brassiae*	Mustard, Taramira
Yellow mosaic	Yellow mosaic virus	Soybean, green gram, black gram
Cercospora leaf spot	*Cercospora cruenta*	Green gram, black gram
Sharing among tree crops		
Citrus canker	*Xanthomonas citri*	Lemon, Kinnow
Ganoderma root rot	*Ganoderma lucidum*	*Leucaena leucocephala, Acacia nilotica, Dichrostachys cineria, Dalbergia sissoo*
Sharing among companion crops		
Mosaic	Cucumber mosaic virus	Papaya, cucurbits

Secondary Interactions

The data on secondary disease interactions, recorded at NRCAF, have been summarized in Table 3. Observations on mustard diseases were recorded in a field with windbreaks on three sides. The direction of the windbreaks was such that there was heavy shade under windbreaks 2 and 3, and it was sunny under windbreak 1 due to southern inclination of sun during winter. Incidence of white rust and downy mildew of mustard in terms of number of diseased panicles was more under tree shade as compared to open during 1997-98, which was a wet year (Figure 1). Diseased panicles were not noticed either under shade or in open for next two seasons, which were relatively dry. A cumulative rainfall of 298 mm, 8 mm and 16 mm was recorded from

October to February during 1997-98, 1998-99 and 1999-2000, respectively (Table 4). Thus, the results indicated that wet conditions favoured the disease development and it was more pronounced in shade as compared to open. Under dry conditions, the said difference was not noticed (Kumar *et al*, 1998). Damp, cool, and low light intensity conditions have been reported to increase severity of white rust and downy mildew of mustard (Kolte, 1985, Bains and Jhooty, 1979). In addition to white rust and downy mildew, alternaria leaf blight was also recorded on mustard, but no such variation was observed in disease levels. Observations recorded during 1997-98 and 1998-99 on incidence of wheat diseases indicated that the severity of alternaria leaf blight and loose smut were at par when grown under agri-horticultural system and a sole crop (control) plot (Table 5). In another observation, cercospora leaf spot caused by *Cercospora cruenta* and yellow mosaic virus were recorded on black gram and green gram, whereas powdery mildew caused by *Erysiphe polygoni* was observed only on black gram. Among black gram diseases, powdery mildew was significantly more under neem as compared to open area and no change was recorded for cercospora leaf spot and yellow mosaic (Table 7). Observations on green gram diseases at different distances from a wind-break, indicated that disease severity of cercospora leaf spot was more under shade, while yellow mosaic registered a reverse pattern (Kumar *et al*, 2004).

Figure 1. Field pattern of white rust and downy mildew systemic infection of mustard at immature pod stage during January 1998.

													North ↑
	Windbreak 1												
Windbreak 2	11	0	1	0	1	0	0	0	1	1	0	0	7
	21	3	5	9	5	1	0	0	1	0	1	0	0
	71	0	2	0	1	0	0	0	0	0	0	0	0
	19	0	1	2	1	0	0	0	0	1	0	1	1
	16	3	2	2	1	3	0	1	0	1	2	0	3
	17	10	6	2	0	6	4	4	2	0	0	0	1
	30	0	6	6	6	4	13	9	4	14	3	1	3
	7	23	11	3	9	14	15	14	9	32	5	14	4
	8	7	6	6	3	1	5	3	2	10	14	2	6
	5	13	4	2	5	5	3	4	7	6	9	20	15
	6	7	8	21	6	18	9	17	49	5	20	62	43
	Windbreak 3												

Table 3. Summary of secondary interactions recorded at National Research Centre for Agroforestry, Jhansi.

Crop	Type of plantation	Associated crop species	Disease	Remarks
Black gram	Block plantation	Neem	Powdery mildew	Disease severity increases under shade
			Cercospora leaf spot	No change in disease severity
			Yellow mosaic	No change in disease severity
Green gram	Windbreak	*Acacia sp.*	Cercospora leaf spot	Disease severity increases under shade
			Yellow mosaic	Masking of disease symptoms in tree shade
Mustard	Windbreak	*Dalbergia sissoo*	Downy mildew	Disease severity increases under windbreak in wet climate
			White rust	Disease severity increases under windbreak in wet climate
			Alternaria leaf blight	No change in disease severity
Wheat	Windbreak and Block plantation	*Dalbergia sissoo,* Leucaena, Cassia simea, *kinnow, ber, guava, anar*	Alternaria leaf blight	No change in disease severity
			Loose smut	No change in disease severity
Ber *(Ziziphus mauritiana)*	Block plantation	Wheat	Powdery mildew	Irrigated crops like wheat increases disease severity

Table 4. Rain fall (mm) received in different months from 1997 to 2001, at Jhansi.

Year	Jan	Feb	Mar	Apr	May	Jun	Jul	Aug	Sep	Oct	Nov	Dec
1997	0	0	1	21	17	50	172	281	148	108	29	159
1998	0	2	23	5	17	73	423	282	80	0	0	0
1999	8	0	0	0	22	19	250	323	479	16	0	0
2000	0	0	0	2	50	193	348	91	21	0	0	0
2001	0	0	0	19	0	378	408	290	28	17	0	0

Table 5. Effect of windbreak and block plantation of some horticultural tree species on incidence of wheat diseases.

Treatments	Associated tree species	Disease index of Alternaria leaf spot	Disease index of Loose smut
1997-98			
Wind break	Subabool, Shisham	22.5[a]	0.38[a]
Block plantation (spacing: 10 x 10m)	Anar, Ber,	24.5	0.08
Block plantation (spacing: 10 x 5m)	Guava, Kinnow	20.0	0.25
Block plantation (spacing: 5 x 5m)		19.4	0.13
Control (sole crop)		23.9	0.29
S.Em. (+)		1.53	0.13
C. D. (0.05)		NS	NS
1998-99			
Wind break	Subabool, Shisham	20.8	0.13
Block plantation (spacing: 10 x 10m)	Anar, Ber,	23.7	0.13
Block plantation (spacing: 10 x 5m)	Guava, Kinnow	26.0	0.54
Block plantation (spacing: 5 x 5m)		21.1	0.08
Control (sole crop)		23.1	0.50
S.Em. (+)		1.49	0.16
C. D. (0.05)		NS	NS

[a] Average of 24 observations

Table 6. Effect of tree shade on black gram diseases.

Treatment	Powdery mildew Mean	Powdery mildew Range	Cercospora leaf spot Mean	Cercospora leaf spot Range	Yellow mosaic Mean	Yellow mosaic Range
	Disease index of					
1998						
Under tree shade	84.8[a]	64.0-98.7	8.6	4.4-11.1	1.8	0.7-6.7
Outside tree shade	222	9.3-40.0	8.2	4.4-11.1	1.4	0.7-2.2
CD (0.05%)		-	NS	-	NS	-
1999						
Under tree shade	77.9	49.3-90.7	11.3	4.4-23.7	1.4	0.7-3.0
Outside tree shade	4.8	0.0-18.7	14.8	4.4-20.7	1.0	0.7-1.5
CD (0.05%)		-	NS	-	NS	-

[a] Average of 24 observations

Table 7. Severity of green gram diseases at different distances from wind-break

Distance from wind-break (m)	Disease index of Cercospora leaf spot		Yellow mosaic	
	Mean	Range	Mean	Range
I	45.6[a]	3.11-68.8	7.9[a]	0.-33.3
II	29.0	22.2-44.4	29.6	0-55.6
III	26.8	12.2-57.8	36.0	0-77.6
CD (0.05%)	5.7	-	10.5	-

[a] Average of 21 observations

Cultivation of an inter-crop in an agroforestry system also affected the disease status of the tree species in some cases. The data on build-up of powdery mildew of *Ziziphus mauritiana* with and without wheat as an inter-crop showed that the disease intensity was more in the former (Kumar *et al,* 1999). The disease started 10 days earlier, the infection rate was higher and the final amount of the disease was 7-8 times more in the inter-cropped field as compared to control ber plants (Table 8). The extent of damage in terms of per cent cracked fruit surface area was 52.8% in the inter-cropped field and 10.6% without the inter-crop. The difference in disease severity, in above mentioned plots, was more during dry winters as compared to wet winters. Analysis of microclimate data recorded in above mentioned situation indicated that drop in daytime temperature (2-3°C) and radiation intensity (2300-8500 lux) coupled with increase in RH (2-3%) and availability of free water for longer duration during morning hours in the wheat field might have favoured the faster development of the disease. The increase can also be due to larger canopy size of the ber plants, higher soil moisture, greater soil fertility, and application of fertilizers in the inter-crop. Cultivation of inter-crops also increased the disease severity of alternaria leaf blight of *Albizia procera* and cercospora leaf spot of neem at seedling stage in respective agroforestry systems (Kumar *et al,* 2007).

Table 8. Build up of powdery mildew of ber with and without wheat as an inter-crop during 1998.

Date	Disease index*	
	Without inter-crop	With inter-crop
23.10.98	0.0	5.7
30.10.98	0.0	16.3
6.11.98	0.1	26.9
12.11.98	1.9	52.8
18.1198	5.8	76.2
26.11.98	8.9	86.6
2.12.98	9.7	89.7
10.12.98	14.0	91.5
18.12.98	10.8	92.5

* Average of 10 observations

Observations on microclimatic parameters recorded outside and underneath tree canopy in different fields showed that outside canopy temperature was higher during the day, whereas under tree canopy temperature was more during night. The reverse trend was recorded for relative humidity. Under tree canopy RH was high during night and less during daytime. Shifts in above-mentioned parameters were recorded during morning and evening hours. Differences in magnitude of temperature and RH during daytime were more as compared to night hours. Radiation intensity was less under the canopy than outside it. Consequently, free water on leaves of under-story crops was available for longer duration under tree canopy, during morning hours.

Effect of Microclimate

Details of some microclimatic parameters viz., temperature, RH, lux meter reading, presence of free water on surface of inter-crop etc., in black gram field under and outside neem shade at different hours are shown in Table 9. As mentioned earlier, among three diseases of black-gram, shade increased the severity of only powdery mildew. Based on above mentioned data and information available in literature on powdery mildews, an attempt was made to workout the effect of microclimatic parameters on the disease development, under and outside tree canopy by calculating values of some derived parameters (Table 10). Parkash and Jhooty, 1987 have reported that, optimum temperature for development of powdery mildews range from 20-30°C. Progress of the disease becomes very slow above 35°C and below 15°C. Powdery mildews can develop in absence of free water on host surface. However, RH below 32% is deleterious to the disease development. Observations indicated that in black-gram field maximum temperature under tree canopy was below 35°C, whereas in open, maximum temperature was above 35°C for two hours. Temperature in open exceeded, maximum underneath canopy temperature, by four hours. During morning hours, free water on under-story crops was available for longer duration, under the tree canopy. Outside lux meter reading exceeded corresponding reading under tree canopy for seven hours. Length of period (hours) for other conditions viz., temperature below 15°C, temperature in the range of 20-30°C, RH below 32% was at par under and outside neem shade (Table 10). Thus, reduced availability of sun radiations, reduction in maximum temperature and availability of increased moisture under tree canopy as compared to open, accounted for increase in severity of powdery mildew of black-gram under shade. Above mentioned results and analysis of similar observations made in other fields indicated that changes in microclimate parameters, brought about by introduction of trees in crop fields, generally affected the disease severity of component crops if these operated in cardinal range of the parameters for the disease development.

Table 9. Details of some microclimate parameters in black gram field under and outside neem shade at different hours.

Time (Hours)	Temp. (°C)	Micro-climate parameters						
		Under tree canopy			Outside tree canopy			
		RH	Lux meter reading x 100	Presence of free water on leaves	Temp. (°C)	RH	Lux meter reading x 100	Presence of free water on leaves
October 30, 1998								
10.00	31.0	83.0	42	-	33.0	81.0	254	-
11.00	31.5	80.5	61	-	34.5	78.0	652	-
12.00	33.0	79.0	74	-	36.0	76.5	713	-
13.00	33.5	75.5	114	-	36.0	74.0	721	-
14.00	34.0	75.5	129	-	35.5	73.5	692	-
15.00	32.0	77.5	32	-	33.0	78.0	158	-
16.00	28.5	81.0	21	-	29.0	83.0	42	-
17.00	28.0	82.0	17	-	29.0	83.0	35	-
18.00	28.0	84.0	5	-	27.5	85.0	9	-
19.00	27.0	85.5		-	26.5	86.0		-
20.00	26.5	87.5		+	26.5	88.5		+
21.00	25.5	88.5		+	24.0	89.5		+
22.00	25.0	90.0		+	24.0	91.0		+
October 31, 1998								
4.00	22.0	92.0		+	21.0	93.5		+
5.00	23.5	90.5		+	22.5	91.5		+
6.00	25.5	89.0	11	+	26.5	86.5	42	+
7.00	27.0	86.0	14	+	27.0	86.5	72	+
8.00	27.5	85.5	24	+	28.5	83.0	146	+
9.00	28.5	83.0	148	+	29.5	81.5	419	-

Table 10. Summary of some derived microclimate parameters, important for development of powdery mildew of black-gram, under and outside tree canopy on basis of data given in Table 7.

Conditions	Length of period (hours)	
	Under tree canopy	Outside tree canopy
Maximum temperature above 35° C	0.0	2.0
Minimum temperature below 15° C	0.0	0.0
Temperature in range 20-30° C	13.0	13.0
Outside temperature exceeds maximum under canopy temperature	-	4.0
Relative humidity below 32%	0.0	0.0
Outside lux meter reading exceeds maximum reading under tree canopy	-	7.0
Free water available on under-story crop	9.0	8.0

Conclusions

1. Alterations in disease severity of under-story crops in agroforestry systems are possible due to sharing of diseases among its component crops (primary disease interactions) or modification of the microclimate by the tree species (secondary disease interactions). Primary interactions i.e. sharing of diseases among component crops of a given agroforestry system, were uncommon because of dissimilarities of plant and pathogen species involved.
2. Incidence of certain diseases viz., loose smut and alternaria leaf blight of wheat, alternaria leaf spot of mustard, cercospora leaf spot and yellow mosaic of black gram etc., did not increase under heavy shade of tree species as compared to open area.
3. Incidence of some diseases was more under tree canopy during certain years as compared to open area when the climatic conditions were just suitable for the disease development, otherwise intensity of the disease remains the same. For example, incidence of white rust and downy mildew of mustard is more under tree canopy during wet winters while it remains at same level during dry winters.
4. Incidence of certain diseases like powdery mildew of black gram was always more under shade as compared to open area.
5. Cultivation of an inter-crop in an agroforestry system also affected the disease status of the tree species in some cases. Build up of powdery mildew of ber was faster in a ber-based agroforestry system with wheat as compared to ber fields without any inter-crop.
6. Changes in microclimatic parameters, brought about by introduction of trees in crop fields, generally influenced the disease severity of component crops if these operated in the cardinal range of the parameters for the disease development.
7. Important diseases in various agroforestry systems could be kept under check by using recommended control measures for pure crops.

References

Bains, S.S. and Jhooty, J.S. 1979. Mixed infections by *Albugo candida* and *Pernospora parasitica* on *Brassica juncea* inflorescence and their control. *Indian Phytopathology*, 32: 268-273.

Huxley, P.A. and Greenland, D.J. 1990. Pest management in agroforestry systems. *Agroforestry Abstracts*, 2: 37-46.

Kolte, S.J. 1985. Diseases of annual edible oilseed crops, vol. 2. CRC Press, Florida, USA.

Kumar, A., Solanki, K.R. and Singh, R. 1998. Effect of wind-break on white rust and downy mildew of mustard (*Brassica compestris* L.). *FACTRR*,3: 4-7.

Kumar, A., Solanki, K.R. and Ajit. 1999. Effect of wheat as inter-crop on incidence of powdery mildew of ber (*Zizyphus mauritiana*). *FACTRR*, 4: 121-124.

Kumar, A., Solanki, K.R. and Singh, R. 2000. Effect of rain-fall on incidence of powdery mildew of ber. *Annals of Arid Zone*, 39: 89-92.

Kumar, A., Rai, P. and Handa, A.K. 2003. Effect of some agroforestry situations on incidence of wheat diseases under Bundelkhand conditions. *Indian Journal Agroforestry*, 5 (1&2): 88-93.

Kumar, A., Rai, P., Handa, A.K. and Rizvi, R.H. 2004. Effect of tree shade on black-gram diseases. *Annals of Arid Zone*, 43 (2): 213-216.

Kumar, A., Chaturvedi, O. P. and Singh, R. 2007. Incidence of green-gram diseases and alternaria leaf blight of *Albizia procera* in agro-forestry systems. *Indian Journal of Agroforestry*, 9 (1): 53-55.

Parkash, V. and Jhooty, J.S. 1987. Epidemiology of powdery mildew of *Ziziphus mauritiana* caused by *Microshaera alphitoides* f.sp. *ziziphi*. *Indian Phytopathology*, 40 (4): 491-494.

Schroth, G., Krauss, U., Gasparotto, L., Duarte Aguilar, J.A. and Vohland, K. 2000. Pest and diseases in agroforestry systems of the humid tropics. *Agroforestry Systems*, 50 (3):199-241.

Sehgal, J.L., Mandal, D.K., Mandal, C. and Vadivelu, S. 1990. Agro-ecological regions of India. National Bureau of Soil Survey & Land Use Planning, Tech. Bull., Nagpur, Pp. 25.

❑❑❑

Agroforestry: Natural Resource Sustainability, Livelihood & Climate Moderation © 2009
Editors: OP Chaturvedi, A Venkatesh et. al., pp. 757-775
Satish Serial Publishing House, Delhi

Chapter 54

MEETING CHALLENGES IN FORAGE PRODUCTION THROUGH AGROFORESTRY IN INDIA

S.A. Faruqui and Sunil Kumar
Indian Grassland and Fodder Research Institute, Jhansi-284 003 (U.P.)

Introduction

Livestock production is backbone of Indian agriculture contributing 7% to national GDP and source of employment and ultimate livelihood for 70% population in rural areas. They not only provide products for human sustenance but also serve as a major energy source for draught power in agricultural operations. The country accounts world's 15 percent of the total world livestock population dependent upon 2 per cent of geographical area. Among other factors, the low productivity of livestock is mainly attributed to poor fodder and feed resources. It is predicted that meat and milk consumption will grow at 2.8 and 3.3 per cent per annum, respectively, in developing countries like India. Urbanization has brought a marked shift in the lifestyle of people in feeding habits towards milk products, meat and eggs with resultant increase in demand of livestock products. Peri-urban livestock farming and emerging fodder markets are indicators of fast changing economic scenario in livestock sector. Livestock population is around 500 million and is expected to grow at the rate of 1.23% in the coming years. Production efficiency of the Indian livestock in most of the regions of the country is very low as compared to some of the states like Gujarat, Punjab and Haryana. Balanced nutrition through feed and fodder to livestock is available in selected milk shed areas of the country where intensive fodder production systems are practiced. Rest of the farming community maintains uneconomic large herds on grazing supplemented with stall feeding. As per current estimate, the availability of forage (green and dry fodder) is nearly 60 percent (Anonymous, 2001). To meet this challenge, concentrated efforts are to be made for bridging the large gap in demand and supply of the fodder in this country.

Forage Production Scenario

The requirement for fodder is mainly governed by the density of livestock in an area and secondly by management levels and climatic conditions. The growth of livestock population between the periods of 1951 to 1992 is about 61 per cent, which works out to be 1.5 per cent annual growth rate. According to an estimate, current fodder requirement is 1006 and 560 million tones for green fodder and dry matter, respectively (Anonymous, 2001). The present availability of green fodder from cultivated areas and pastures is 462.05 million tonnes, which includes supplementation from sugarcane tops and seasonal weeds. Similarly, in case of dry fodder, total availability is 393.88 million tonnes, which constitutes straw of cereal crops like paddy, wheat, barley, maize, sorghum, pearl millet and few legumes like pigeon pea and chickpea and dry grass from grazing lands and forests (Table 1). Pathak and Sunil Kumar (2004) indicated that projection of demand and supplies of forages reflects the deficit of green fodder and dry matter will be more than 65 and 25 per cent by 2025. The data are undoubtedly indicative of high pressure of livestock on available forages especially green fodder.

Forage Resources

The major fodder and feed for livestock of our country are the variety of herbage from cultivated forages, i.e. cultivated forage crops, crop residues from crop lands, dry fodder from self owned or community land and grazing from wasteland, degraded land and forest grazing lands. In India, cattle of intensive cropped areas obtain only about 25 percent of their feed from grazing in nearby forests and other uncultivated lands, the balance comes from crop residues unsuitable for human consumption. The total area under cultivated fodder is only 8.4 mha on individual crop basis which is static since last two decades (Hazra, 1998). The scope for further increase in area under forages seems to be very less due to demographic pressure of human population for food crops. In the diverse climate of India, variety of forage species, range grasses and legumes and trees as well shrubs finds their place in the natural vegetation. Various agro-climatic regions possess certain opportunity and threats to different flora of fodder value. The suitability of forage species is visible in their production potential at specific zone (Faruqui *et al.*, 2006). Apart from cultivated forages, range grasses supplements the fodder demand in Himalayan region (Zone 1,2) and western region (Zone 13,14). In coastal and Island region (Zone 11, 12 & 15), forages are grown under plantation crops in multi-tier system. Fodder trees and shrubs also contribute substantially to the fodder requirement in lean period in almost all 15 agro-climatic zones of the country.

Table 1: Availability of feed and fodder in India

Feed Resources	Availability (Million Tonnes)
Greens	
Forest Area	89.37
Fallow lands	23.21
Permanent pastures, grazing area	28.70
Cultivable wastelands, misc. tree tops	17.51
Cultivated fodders	303.26
Total	462.05
Crop Residues	
Coarse Straw-from coarse grains, sugarcane tops	154.83
Fine straw (Paddy, wheat straws)	194.11
Leguminous straws (pulses, others)	44.94
Total	393.88
Concentrates	
Oil cakes	15.76
Brans	13.26
Grains for feeding livestock	5.74
Chwins	0.53
Total	35.29

Source: XI Plan Proposals: A Report of the Sub-Committee on feed, fodder and pasture development to the Planning Commission.

Grazing Resources and Grazing Pressure

India's huge livestock population still sustains to a larger extent on the grazing resources. Grazing based animal husbandry plays significant role in rural economy of an agrarian country like ours. The grazing activity is mainly dependent on the availability of the grazing resources from pastures and other grazing lands, *viz.* forests, miscellaneous tree crops and groves, cultivable waste lands and fallow lands. According to Wasteland Atlas of India (2000), the total area available for grazing accounts for about 40 per cent of the country' geographical area. The area available for grazing is highly variable (Table 2). Among the states, Rajasthan (13.04 mha) followed by Madhya Pradesh (9.05 mha), Andhra Pradesh (7.47 mha) and Maharashtra (6.83 m ha) occupied major share of total grazing area of the country. States like Jammu & Kashmir (6.82 mha) and Himachal Pradesh (4.05 mha) in Western Himalayan region have more area under grazing. However, the percentage area under grazing to that of total geographical area lies in states like Nagaland, Manipur, Meghalaya and Sikkim in Eastern Himalayan zone.

Table 2: State-wise grazing resources and grazing pressure based on livestock units (ACU)*

State	Geographical Area (m ha)	Total grazing area (m ha)	Livestock (m)	Livestock density	Grazing pressure (ACU/ha)	ACU/ha (Net sown area)
Andhra Pradesh	27.51	7.47	32.91	1.20	2.63	2.58
Arunachal Pradesh	8.37	2.19	0.85	0.10	0.14	5.78
Assam	7.84	2.53	16.07	2.05	3.98	4.49
Bihar	17.39	3.42	47.93	2.76	7.71	3.42
Goa	0.37	0.07	0.25	0.67	1.91	1.23
Gujarat	19.60	4.89	18.60	0.95	2.39	1.68
Haryana	4.42	0.52	9.14	2.07	12.11	2.56
Himachal Pradesh	5.57	4.05	5.12	0.92	0.69	6.27
Jammu&Kashmir	10.14	6.82	8.71	0.86	0.56	6.71
Karnataka	19.18	2.92	29.57	1.54	5.73	1.99
Kerala	3.89	0.21	5.84	1.50	17.20	1.92
Madhya Pradesh	44.34	9.05	46.74	1.05	3.72	2.14
Maharashtra	30.77	6.83	36.39	1.18	3.19	1.54
Manipur	2.23	1.36	1.29	0.58	0.55	6.42
Meghalaya	2.24	1.35	1.19	0.53	0.45	3.60
Mizoram	2.11	0.44	0.21	0.10	0.14	1.14
Nagaland	1.66	1.00	1.08	0.65	0.34	4.80
Orissa	15.57	3.14	22.75	1.46	4.53	2.74
Punjab	5.04	0.27	9.45	1.88	28.64	2.60
Rajasthan	34.22	13.04	48.41	1.41	1.57	1.72
Sikkim	0.71	0.46	0.39	0.55	0.41	2.38
Tamil Nadu	13.01	3.28	25.01	1.92	3.66	2.63
Tripura	1.05	0.19	1.59	1.52	4.77	3.80
Uttar Pradesh	29.44	4.26	64.80	2.20	10.18	3.38
West Bengal	8.88	0.69	35.09	3.95	25.77	3.95
Union Territories	1.10	0.09	0.77	0.70	6.05	
Total	316.64	80.51	470.15	1.48	3.42	2.52

* ACU includes cattle, buffaloes, sheep and goats, Source: Anonymous (1997 & 2000 b)

Most of the states falling in Indo-Gangetic Plains are poor in respect of grazing resources which otherwise are potentially rich in fodder production from arable lands. Population density of livestock per unit geographical area of India is 1.48, which is indicative of high pressure on land. Among the states, West Bengal harbours highest livestock population density (3.95) followed by Bihar (2.76), Uttar Pradesh (2.20), Haryana (2.07) and Assam (2.05), much above the national average. States with moderate population density are Tamil Nadu, Punjab, Karnatka, Kerala and Tripura. All the North East hill states except Assam bear less pressure of livestock on their respective total geographical area.

Forages through Agroforestry

Agroforestry is the set of land-use practices involving the deliberate combination of trees, agricultural crops and/or animals on the same land management unit in some form of spatial arrangement or temporal sequence. Cultivating trees in combination with crops and livestock is an ancient practice. However, several factors have contributed to a rising interest in agroforestry since the 1970s: the deteriorating economic situation in many parts of the developing world; increased tropical deforestation; degradation and scarcity of land because of population pressures; and growing interest in farming systems, intercropping and the environment. Main agroforestry practices include improved fallows, taungya (growing annual agricultural crops during the establishment of a forestry plantation), home gardens, alley cropping, growing multipurpose trees and shrubs on farmland, boundary planting, farm woodlots, orchards or tree gardens, plantation/crop combinations, shelterbelts, windbreaks, conservation hedges, fodder banks, live fences, trees on pasture and apiculture with trees (Nair, 1993).

Agroforestry has received much attention in India from researchers, policy makers and others for its perceived ability to contribute significantly to economic growth, poverty alleviation and environmental quality, that is why today agroforestry is an important part of the 'evergreen revolution' movement in the country (Puri and Nair, 2004). On the other hand, the success stories of wasteland reclamation and poplar-based agroforestry show that the technologies are widely adopted when their scientific principles are understood and socio-economic benefits are convincing. Presence of animal is considered essential to agroforestry system since it provides a channel for the crop biomass through animals to human food and nutrition on one side and increase in soil fertility through manure on the other. Animal is thus an essential component of the agroforestry system.

Based upon the four-five decades of research on Agroforestry and dryland agriculture, various options suitable for fodder production from culturable waste/marginal lands were suggested as : These are farming/Wood lot, Range/Pasture management, Silvipastoral systems, Timber and Fiber systems. In a comparison of the energy fixation by various land use systems on rainfed marginal lands, it was found that the silvipastoral farming produced much higher biomass energy, new coppice farming model for such lands obtained optimized land productivity by more than 2.5 times (Table 3). If biologically it is possible, the same could be harnessed for several other purposes of human use. Pathak and Daggar (1998) advocated agroforestry as a central theme for rehabilitation/bioremediation of degraded and polluted habitats. In other study on the Sub-Himalayan rocky gravelly degraded lands. The peak B/C value of 3.43 under *Acacia catechu* plantation with bhabhar grass (*Eulaliopsis binata*). In case of ravines, they observed that silvipasture or silviculture of fast growing species is the best option with B/C ratio of 1.9 (Table 4).

Table 3 : Comparative estimate of energy fixation (in harvestable biomass) different systems of land use.

Land use system	Aerial biomass ($t\ ha^{-1}\ yr^{-1}$)		Energy ($k\ cal\ m^{-2}\ yr^{-1}$)	
	Mean	Range	Mean	Range
Grazing lands	5.4	2.8-8.0	605.0	250-960
Managed pastures	4.5	3.5-5.5	1560.0	1120-2000
Rainfed agriculture(with inputs & management)				
Sorghum (105 days)	5.9	-	2597.0	-
Pearlmillet (67 days)	7.3	-	3197.0	-
Maize (Diara land)	1.7	-	782.0	-
Silvipasture (with intensive managementpossibilities)	11.0	7.0-15.0	4775.5	3040-6511

Table 4 : Economics of utilization of degraded lands (Dhruvanaravana & Ram Babu 1984)

Land use activity (Fuel + Fodder plantation)	B/C Ratio
Hilly region	
Dalbergia sissoo + *Chrysopogon fulvus*	2.75
Acacia catechu + *Eulaliopsis binata*	3.43
C.fulvus (Sole)	3.02
Eucalyptus sp. (Sole)	1.68
Ravines	
Acacia nilotica + *Prosopis juliflora* + *Dicanthium annulatum* + *Cenchrus ciliaris* at Kota (Rajasthan)	1.55
Acacia nilotica + *D. annulatum* at Vasad (Gujarat)	1.38
Acacia nilotica + *P. juliflora* at Agra (Uttar Pradesh)	1.80
Semi-arid black soils	
Eucalyptus tereticornis (Sole)	1.90

Components of Agroforestry for Fodder Supplementation

Silvipastoral System

Silvipasture as an agroforestry practice involves grasses and legumes with multi-purpose trees used initially under cut and carry system and latter on grazed *in-situ* (Table 5). This system is important for increasing fodder production from the marginal, sub-marginal and other wastelands, which comprise of about 50 per cent of the total land area in the country. Such modified and adaptable systems provide quality forage/grazing resource for a longer period of the year normally more than the situation with local/natural grasses of the marginal and sub-marginal lands. Silvipastoral land use has been identified as a holistic approach to rehabilitate the degraded lands and rangelands for meeting the needs of fodder and fire wood by rural masses and conservation of natural resources (Paroda, 1990). Productivity of some multipurpose trees in silvipastoral systems in highly degraded lands in semi-arid regions was worked out and found that *Albizia amara* based silvipastoral system produced 7.95 t ha^{-1} yr^{-1} fodder followed by *Acacia tortilis* with 7.60 t ha^{-1} yr^{-1}. The production of *Hardwickia binata* based silvipastoral system was quite low (2.32 t ha^{-1} yr^{-1}) on account of slow growth of the tree (Roy *et al.*, 1998).

Leucaena, Acacia, Albizia, Hardwickia and *Dalbergia; Cenchrus, Panicum, Sehima, Dichanthium, Borthiocloa* and *Heteropogon; Stylo* and *Siratro* are the most successful trees, grasses and legume species respectively, for silvipasture on different types of degraded lands in the Central India. As against less than 1 t ha^{-1} yr^{-1} biomass production, the improved technology has yielded 10 t ha^{-1} yr^{-1} in a rotation of 10 years. The year-round grazing by the mixed herd of goat, sheep and cattle has given >75 kg ha^{-1} yr^{-1} secondary production without loss of vegetation vigour. Grasses like Panicum maximum have shown their high adaptability to shade and maintained carbohydrate and crude protein level, indicating potential for optimizing quality forage production to support the animal production (IDRC Report, 2002).

The productivity and adaptation of *Cenchrus ciliaris* and *Panicum maximum* under shade were studied under the canopies of fifteen-year old trees of *Acacia tortilis* and *Leucaena leucocephala* and in open field in degraded lands in semi-arid tropical environment at IGFRI, Jhansi (Bhatt *et al.*, 2005). On the average of seven years, *C. ciliaris* and *P. maximum* produced 4.20 t ha^{-1} yr^{-1} and 6.84 t ha^{-1} yr^{-1} dry matter under tree canopies of *A. tortilis* and *L. leucocephala* respectively. The total above ground biomass productivity of the systems increased from 1.5 to 2.25 times when compared to only pasture land use system. *Panicum maximum* has shown its high adaptability to shade and maintained carbohydrate and crude protein level, indicating potential for optimizing quality forage production to support the animal production.

Silvipastoral Model

Silvipastoral model has shown their potential for increasing fodder production from marginal lands for sustaining livestock. In one of the model, the degraded lands (shallow red gravelly soil) in Bundelkhand region under semi-arid tropic producing < 1 t ha^{-1} yr^{-1} have been improved to produce more than 10 t ha^{-1} yr^{-1} at a 10 years rotation through silvipastoral system (Pathak, 2002; Pathak *et al.*, 1995). The comparative studies of different silvopasture system in natural grasslands revealed that *Acacia tortilis, Albizia amara, A. lebbeck, H. binata* and *L. leucocephala* along with grasses at 10th years of establishment produced 19.83, 13.88, 22.01, 17.81 and 14.11 t ha^{-1} yr^{-1} biomass respectively. The *L. leucocephala* + natural grass based silvipasture systems produced 5.34 t ha^{-1} yr^{-1} (2.95 t from pasture grasses and 2.39 t from tree leaves) forage biomass, 1.44 t ha^{-1} branches as fuel wood and 7.33 t ha^{-1} as timber. A technology package developed at NRC for Agroforestry, Jhansi included silvipasture consisting of *Albizia amara* + *Leaucaena leucocephala* as tree component, *Dichrostachys cinerea* as bush planted in alternate row spaced at 5 m x 5m and *L. leucocephala* in between *A. amara* or *D.cinerea.* The grasses (*Chrysopogon, Stylosanthes hamata* + *S. scabra)* were planted at spacing of 100 cm x 50cm in the interspaces of tree components. This silvipastoral system was found to sustain 12 small ruminants (sheep and goats) per hectare as against natural grassland with carrying capacity of 6 small ruminants. There was 33-39 percent gain in body weight of ruminants under silvipastoral system than natural grasslands (Rai *et al.*, 1998). In a study in ravines, the assessment of production of different tree species showed that *Acacia nilotica* is the most promising species under ravine area for producing the fuel wood. In ravine areas, the natural grasslands are overgrazed and degraded with poor productivity. The main grasses found in gullies are *Dichanthium, Cenchrus* and *Sehima*. By protection, sowing and planting of these grasses along with *Stylo*, the green fodder yield can be achieved in 2–5 years. This practice also reduces run-off and soil loss considerably to the tune of 6–10 t ha^{-1} yr^{-1} (Prajapati, 1979).

Table 5: Suitable species for silvipasture systems in different regions of India

Trees/Shrubs	Grasses/Legumes
(1) Hot Arid Desert (Western Rajasthan, parts of Gujarat and Haryana)	
Acacia nilotica, A.tortilis, Albizia amara, Azadirachta indica, Dichrostachys cinerea, Eucalyptus tereticornics, E.camaldulensis, Prosopis cinerarea, P. juliflora	Grasses – *Cenchrus ciliaris, C.setigerus, Lasiurus sindicus* Legumes – *Atylosia scarabaeoides, Stylosanthes hamata, Clitoria ternatea*
(2) Semi-arid, Rocky & Gravely (Parts of UP, MP, AP and Karanataka)	
Acacia catechu, Ailanthus excelsa, Albizia lebbek, Cassia siamea, D. sissoo, Dendrocalamus strictus, D. cinerea, H. binata, P. juliflora	Grasses – *Cenchrus ciliaris, C.setigerus, Panicum antidotale, Chrysopogan fulvus, Dicanthium annulatum* Legumes – *Macroptelium atropurpureum, Desmodium trifolium, Stylosanthes spp.*
(3) Cold Desert (Laddakh – J&K)	
Hippophae spp., Juniperus communis J. wallichiana, Populas spp., Salix spp.	Hay species – *Pranges pabularia*
(4) Ravine lands (UP, MP, Rajasthan & Gujarat)	
Acacia nilotica, A.catechu, A.tortilis, Albizia lebbek, Dalbergia sissoo, Dendrocalamus strictus, Dichrostachys cinerea, Eycalyptus spp., Prosopis juliflora	Grasses – *C. ciliaris, C.setigerus, P. pedicellatum, Chrysopogon fulvus* Legumes –*Atylosia scarabaeoides, Stylosanthes gracilis, S. humilis*
(5) Cultivable wastelands	
Acacia spp., Albizia spp., Hardwickia binata, Leucaena leucocephala, Sesbania spp.	Grasses – *Cenchrus ciliaris, C.setigerus, Chrysopogon fulvus, Dicanthium annulatum, Pennisetum polystachyon, Sehima nervosum* Legumes – *Clitoria ternatea, Glycine javanica, Macroptelium atropupureum, Stylosanthes spp.*

Agrisilviculture

In agri-silvicultural systems, MPTs including fodder-cum-fuel trees can be grown in association with field crops. Trees are normally pruned annually yielding fodder as well as fuelwood. In arid and semi-arid regions, hardy trees like *P. cineraria, T. undulata, Eucalyptus* sp., *A. tortilis, H. binata, Azadirachta, Acacia* sp., *A. excelsa, Z. jujube*, etc. could be grown along with dry land crops. In Rajasthan, Gujarat, Haryana; Khejri and Rohida are grown along with crops of pearlmillet /sorghum, moth, til, etc. without affecting crop production. Even crops under and near the trees produce more biomass under rainfed situation (Kumar, 1998). The timber-cum-fodder trees produced fuel, fodder and timber besides the crop production. Such systems on arid and semi-arid areas have been considered highly economic in long run and also provide better employment opportunities (Gupta and Mohan, 1980). Poplar and berseem combination gave 50 t/ha green fodder in aspan of 5-6 months under irrigated situation in tarai belt of U.P. (Chaturvedi, 1983).

Hortipastoral Systems

There is considerable scope of introducing multipurpose trees preferably nitrogen fixing trees in hortipastoral systems with the aim of generating additional resources in terms of forage, fuel wood and monetary returns. Attempts have been made to identify suitable fruit trees for different agro-ecological zones of the country for their use in hortipastoral and agri-horticultural plantations. Investigations conducted at Jhansi under semi-arid conditions on marginal land with low fertility, revealed that average forage yield of natural grass in association with guava, lemon and mango was 21.2 t ha^{-1} and 5.2 t ha^{-1} for green and dry forage, respectively. In the degraded arid and semi-arid rangeland regions number of overgrazed plants of wild ber (Jherberi) (*Z. nummularia*) are found which could be successfully budded with improved var. of ber viz. Gola, Seb, Umran, Banarsi *Karaka* besides planting MPTS like anjan, subabul, khejri along with grasses and legumes like *Cenchrus, Lasierus, Chrysopogon, Stylosanthes, Siratro* etc.

Studies of the production potential of pasture alone(C. ciliaris), fodder trees alone (*H. binata* + *C. mopane*) and silvipastrol system (*C. ciliaris* + *H. binata* + *C. mopane*) were done at CAZRI, Jodhpur for nine years. It vwas observed that silvipastrol system was better for higher average forage production (1.73 t ha^{-1} yr^{-1}) and livestock maintenance 4.1 ACU ha^{-1}) followed by pure pasture and tree blocks (Harsh *et al.,* 1992). Experimental results at IGFRI and NRCAF, Jhansi; CAZRI, Jodhpur and elsewhere revealed that budded ber started fruiting even in the 2nd year (3 kg/tree) to 10 kg in the 4th year thereby improving the economy of the poor farmers (Rs. 1600 and Rs. 9880 in 2nd and 4th year, respectively) (Deb Roy, 1994, 1995). Productivity of both Panchayat as well as private sand dunes in rangeland will get a boost for the poor people besides other agroforestry/silvipastoral technology for fixation of sand dunes. *Anacardium occidentale, Annona squamosa, Emblica officinalis* (*Phyllanthus emblica*), *Psidium guajava* and *Zizyphus mauritiana* along with grasses and Stylosanthes can be grown in wastelands under such hortipastoral systems. Production of grasses and legumes has been found to be between 2.5 and 6.0 t ha^{-1} which sustain livestock industry. Other environmental benefits are a bonus. Sunil Kumar and Choubey (2008) recently reported that Aonla- *Dicanthium* based hortipasture realize higher dry matter (3.38 t ha^{-1}) compared with pure pasture(DMY- 3.07 t ha^{-1}). The B: C ratio was also higher for hortipasture (1: 3.70) than pure pasture (1: 1.85). This study signifies that in semiarid situations, understorey of Aonla can be suitably utilized by growing suitable grasses with optimum nutrient management.

Forages under Plantation Crops

Forage production under plantation crops has given encouraging results from different agro-climatic regions particularly in coastal belts. In Kerala, the land is not available for fodder cultivation, growing of forages under partial shade may be a viable option. In coastal areas, in many situations, sole cropping is not feasible. Under such conditions, the forages are to be interwoven as intercrops in the existing systems (Table 6). Guinea grass evaluated under partial shade of coconut at AICRP on Forage crops centres viz., Vellayani, Coimbatore, Bhubaneswar and Jorhat (*Machillus bombicini)* shown that it can be grown with increased net return per unit area. In south zone, the reduction in forage yield of guinea grass was 16-17 % over open. In North East zone, at Jorhat 16 % reduction was found under *Machillus bombicini.* However at Bhubaneswar condition, the forage yield increased by 6 % under coconut than open (Faruqui and Sunil Kumar, 2006). It indicates that guinea grass and other shade tolerant grasses like congosignal grass, signal grass provides unique opportunity for quality fodder production under different resources (open area or with trees / orchards having moderate canopy) using recommended input management options in diverse agro-climatic situation of the country for sustainable livestock production.

Table 6 : Production potential of forage crops under plantation crops

Crop / system	Green fodder yield (t ha^{-1})
A. Open area	
Guinea grass	80-120
Congosignal grass	50-75
Napier bajra hybrid	200-250
Cowpea	30-35
Stylosanthes	25-30
B. Intercrop in coconut	
Guinea	80-100
Congosignal grass	50-60
Napier bajra hybrid	150-200
C. Intercrop in tapioca	
Cowpea	20-25
Stylosanthes	15-20
D. Rice fallow cropping	
Cowpea	20-30
Sesbania rostrata	20-25

Agroforestry Systems for Problem Areas

Waterlogged Areas

Under continuous water stagnation *Tamarix articulate, Eucalyptus rostrata, Salix balyonia, Morus alba, Delbergia sissoo, Casurina equesetifolia, Bischofia javanica* can be planted under agroforestry system in sub tropical India. *Lagerstroemica fleriginae, Syzium cumini* and *Bischofia javanca* are suitable species for tropical climate. Besides this the *Bracharia* sp., *Vitivaria zizanodis, Medicago denticulate, Vicia* sp., *Lathyrus* sp. and *Pisum* sp. can be grown on raised sunken emerged bunds as well as skip and staggered planting of vegetative species. There are large river bedaffected areas in Uttar Pradesh and Bihar called Diara lands. These areas should be provided with permanent vegetation cover of economic fuel-cum-fodder trees along with suitable grasses (Table 7).

Table 7: Suitable species for fodder in diara lands and river – bed affected areas

Components	Suitable species
Grasses	*Pennisetum pedicelatum, Chrysopogon fulvus, Dicanthium annulatum, Eulaliopsis binata (Bhabhar grass)*
Legumes	*Stylosanthes gracilis, Calapogonium mucunoides*
Trees	*Delbergia sisoo, Acacia catechu, A. nilotica, Salix spp. Zizyphus spp., Psidium guajava*

Salt Affected Lands

There are about 8 million ha of land affected by saline and alkali soils in the country. Utilization of these problematic lands can be considered by growing salt tolerant species.

Grasses : *Cynodon dactylon, C. plectostachyon, Paspalum notatum, P. dilatatum, Choloris gayana, Lasiurus sindicus, Brachiaria mutica, Sporobolus marginatus, Urochloa mossambicensis*

Legumes : *Glycine javanica, Macroptilium spp., Phaseolus junatea, Stylosanthes humilis, Lotononis bainesil*

Trees : *Acacia tortilis, A. nilotica, P. juliflora, Dendrocalamus strictus, Salvdora spp., Zizyphus spp., Sesbania spp. Albizia amara.*

Acidic Soils

The acid soil can also be suitable utilized by growing tolerant species.

Grasses : *Pennisetum polystachyon, P. pedicellatum, Paspalum notatum, Pennisetum clandestianum.*

Legumes : *Centrosema pubesens, Desmodium intortum, Stylosanthes guianensis*

Trees : *Ficus numeralis, Albizia chiensis, Morus serrata, Alnus nepalensis, Schima wallichi, Bucklandia pupulnea.*

Ravine Lands

In India, there are about 4.0 million ha of ravine lands, a major portion of which is confined to Utter Pradesh, Madhya Pradesh, Rajasthan and Gujarat. These lands come under capability class VI and VII and are far below their economic utilization. It is suggested that small to medium gullies can be put to silvipasture or hortipasture.

Grasses : *Cenchrus ciliaris, C. setigerus, Pennisetum pedicellatum Chrysopogon fulvus*

Legumes : *Stylosanthes humilis, S. gracilis, Stizolobium deeringeanum, Macroptilium atropurpureum, Atylosia scarabaeoides, Alysicarpus monilifer*

Trees : *Acacia tortilis, A. nilotica, A. catechu, Albizia lebbek, A. amara, Dalbergia sissoo, Dendrocalamus strictus, Zizyphus spp., Emblica officianlis, Eugenia jambolana.*

Lands Affected by Shifting Cultivation

In the hilly terrain of Assam, Orissa, Madhya Pradesh and Rajasthan, shifting cultivation has been practiced for centuries. This results in mass soil erosion and loss of land productivity. This has been recognized as a complex problem and specialized research efforts are urgently needed to tackle this challenging problem on priority. Introduction of agro-silvipastoral management will not only reduce the soil erosion hazards but will also be able to improve the socio-economic status of the tribal people.

North East region

Grasses : *Cenchrus ciliaris, C. setigerus, Pennisetum pedicellatum, Chrysopogon fulvus*

Legumes : *Clitoria ternatea, Lablab purpureus*

Trees : *Leucaena leucocephala, Hardwickia binata, Acacia nilotica, A.tortilis, Dalbergia sissoo, Albizia spp., Anogeissus pendule.*

Assam and Orissa

Grasses	:	*Pennisetum pedicellatum, Dichanthium annulatum, Cynodon plectostachyon.*
Legumes	:	*Stylosanthes hamata, Macroptilium atropurpureum.*
Trees	:	*Albizia lebbek, A. procera, Leucaena leucocephala, Ficus spp.*

In a field experiment conducted at Kumarganj, Faizabad, Uttar Pradesh, India on a sodic wasteland (silt clay loam soil) with sub-humid tropical climate during 1992-96, to investigate comparative forage yield potential of four stress tolerant perennial grasses (*Pennisetum purpureum, Brachiaria mutica, Cenchrus ciliaris* and Panicum maximum) raised in association with *Dalbergia sissoo* (tree species). *P. purpureum* recorded the highest fodder yield (434.3 q ha^{-1}), followed by *B. mutica* (361.2 q ha^{-1}) and *P. maximum* (195.5 q ha^{-1}). The data on the average tree height of *D. sissoo* in combination with different grasses indicated that the maximum tree height (3.74 m averaged over years) and circumference at breast height (20.55 cm tree^{-1}) was attained in association with *P. purpureum* followed by *C. ciliaris.* A gradual change in soil physico-chemical properties was observed (Rana *et al.*, 2000).

Role of Fodder Trees in Forage Supplementation

Tree leaves are called top feed due to their high nutritive value and quality. More than 200 species of leguminous trees are described to have forage value of which most of the species are tropical and subtropical in nature. It has been now established that tree legumes have enormous potential for supplementing the low quality diets. Fodder trees provide are very important component of any mixed farming situation like India (Table 8), as they provide-

i. High quality, high digestibility biomass resource.
ii. Supplement to provide nutrients deficient in the diet, an enhancement of microbial growth and digestion of cellulose biomass in the rumen of cattle, sheep and goats.
iii. Source of protein that escapes rumen degradation to be digested in the intestines and enhance the protein status of the animal.
iv. Source of vitamins and minerals to complement deficiencies in the basal feed resource.

The dry matter content of the various trees leaves ranges from 20 to 40 percent (Rekib *et al.*, 1991). There is the large variation in the chemical composition of shrub and trees leaves, the range being CP 10-28, EE 1.8-7.4,

CF 10-33, NFE 36-66, Ca 1.07-6.90, P 0.05-0.31, DCP 4-17, TDN 38-63 percent respectively.Several workers have been studied the leaf fodder production potential of different shrubs and trees and found satisfactory result in almost all the experiments. The most promising exotic and indigenous fodder trees and shrubs are *Acacia tortilis, Dichrostachys cinerea,Brasilettia mollis, Pittosporum phillyraesides, Schirueus mole, Atriplex sps., Cordia mopane, Leucaena leucocephala, Zizyphus spinacristi and Albizia amara, C. rothi, A. lebbek, Acacia nilotica, Hardwickia binata, Azadirachta indica, A. excelsa, P. cineraria , Sesbania sesban, Dalbergia sissoo, A.procera,* respectively (Raghavan, 1992). Most of the potential tree leaves fodder are reasonably good in palatability. However, there are certain tree fodders, which are not liked by the animals. Tree fodder like *L. leucocephala, Sesbania sp., G. optiva, A. lebbek , Ficus sp., M. alba, C. calothyrsus* etc., are much liked by animals.

Table 8: Production potential of selected fodder trees under semi arid situation

Species	Age (years)	Trees (per ha)	Leaf fodder production (kg tree^{-1})
Acacia tortilis	7	440	3.2
Albizia amara	7	440	13.5
Albizia lebbek	6	440	5.3
Albizia procera	6	440	11.3
Dichrostachys cinerea	6	440	0.8
Leucaena leucocephala	3	5000	0.8
Hawaiian giant	1.5	5000	1.5
Sesbania grandiflora	3.5	5000	0.9
Sesbania sesban	3.5	5000	0.3
Azadirachta indica	5	300	0.09
Anogeissus pendula	5	00	0.03
Dendrocalamus strictus	5	300	0.3
Pongamia pinnata	5	300	0.32
Zizyphus mauritiana	5	300	0.12
Acacia nilotica sub sps.,cupressiformis	5	300	0.39
Acacia nilotica sub sps., indica	5	300	0.07
Dalbergia sissoo	5	300	0.70
Emblica officinalis	5	300	0.85
Hardwickia binata	5	300	0.15
Melia azedarch	5	300	0.60
Terminalia arjuna	5	300	0.18

Challenges in Adoption of Agroforestry

Much has been learned about how to promote agroforestry and increase benefits to farmers and others through research, extension and policy reform. For example, the effectiveness of alley farming practices to improve soil fertility and crop yields helped refocus strategies on growing trees and crops in rotation rather than together (Pathak and Ram Newaj, 2003). Some trees, such as *Leucaena leucocephala*, have become invasive in some areas, and this has helped researchers to recognize the importance of screening species. Some of the issues are outlines which are seen as constraint or a factor for slow adoption of agroforestry measures.

Adaptation to Local Conditions: Successful efforts to introduce agroforestry often combine modern science and traditional knowledge. Experience has also shown that individual preferences, adaptations and entrepreneurial skills make a big difference and that communities need help to document and spread innovations of farmers.

Availability of Information and Training: Farmers need more information and training for agroforestry relative to other agricultural activities, which limits the spread of some practices. When starting operations, they often lack skills to establish tree and shrub nurseries, pre-treat the seeds and carry out tree pruning activities. However, extension strategies, including field schools, exchange visits and farmer training, are effective ways of disseminating needed information.

Government and Project Support: Lack of financial credit is not a major constraint to adopting agroforestry practices because of the small size of farms and scale of operations, the incremental approach that farmers use to plant trees and the desire of most farmers to avoid risks. In many instances, offering free inputs or paying farmers to plant trees encourages dependency and acts as a disincentive to planting when a project ends. Once farmers start planting on a small scale and see the benefits, they are usually able and willing to continue. On the other hand, government and project interventions are needed to promote tree planting, provide information and technical assistance and fill other gaps such as supplying tree seeds where they are not available

Linking Farmers to Markets: Assessing demand before planting trees is a critical first step in adopting agroforestry, as looking for a market only in times of surplus is problematic. It is also more advantageous to assist farmers to sell their produce locally before they attempt to enter a more competitive export market, and to help them strengthen their links with the private sector as part of market development. In addition, training in entrepreneurship and business skills has proven highly beneficial to farmers, and farmer organizations can have an important role in assembling produce, bargaining collectively and reducing transaction costs.

Decentralized, Community-based Germplasm Strategies: The most successful approaches to supplying and distributing planting material arethose involving community-based seed stands and nurseries managed by individual farmers or groups. Seed and nursery enterprises can also help to increase incomes.

Future Thrust

- Systematic and holistic R & D efforts for augmentation of forage from degraded lands.
- Integration of livestock of suitable types and species in different land use patterns for getting potential benefits.
- Appropriate understanding of livestock grazing and management of top feed species.
- Studies related to nutritive value, palatability, extent of anti-nutritional factors of ground and top feed species.
- Documentation of research on system based animal production.
- Due consideration for cost effective, socially acceptable and ecologically safe production systems for different agro-climatic regions of India.

References

Anonymous, 2001. Annual Report, Kharif & Rabi, AICRP on Forage Crops, IGFRI, Jhansi

Anonymous, 1997. Basic Animal Husbandry Statistics. Ministry of Agriculture, Department of Animal Husbandry and Dairing, New Delhi, Pp. 43-52.

Anonymous, 2000b. Wastelands Atlas of India, Ministry of Rural Development, Department of Land Resources Environment of India, New Delhi, Pp. 10.

Chaturvedi, A.N. 1983. Poplar farming in Utter pradesh. Forest Bulletin No. 47 Government of Uttar Pradesh , Lucknow.

Deb Roy, R., 1994. Agroforestry – Sustainable agriculture and environment. In: *Natural Resource Management for Sustainable Agriculture and Environment* (Ed., (DebB.L.). Angkor Pub. Ltd., New Delhi.

Deb Roy, R., 1995. Agroforestry – Principles and practices. In: *Sustainable Development of Dryland Agriculture in India* (Ed. Singh, R.P.). Sci. Pub., Jodhpur.

Faruqui, S.A. and Sunil Kumar 2006. Guinea: an excellent tropical fodder grass, AICRP(FC), Tech. Pub. No. 4/2006 , AICRPFC, IGFRI, Jhnasi.

Faruqui, S.A., Singh, D.N. and Sunil Kumar 2006. Technological innovation for augmenting forage resources for dry regions. In : Pathak, P.S. and Kundu, S.S. (eds.) *Livestock feeding strategies for dry regions*. International Book Distribution Co. Lucknow, Pp. 549-573.

Gupta?T. and Mohan?D?, 1980. Economics of afforestation in hot arid zones of India. In: *Proc. Second Forestry Conference*, Dehradun.

Harsh, L. N., Tiwari, J.C., Burman, U. and Sharma, S.K., 1992. Agroforestry for arid region. *Indian farming*, 42: 32-37.

Hazra, C.R. 1998. Advances in fodder production systems. In : Strategy for maximization of forage production Ed (s) A.K. Mukherjee, S. Maiti, & M.K. Nanda, Directorate of Research, BCKV, Kalyani, West Bengal, Pp. 40-56.

IDRC, 2002. *Final Report of Silvipasture Operational Research Project for the Bundelkhand Region*, IGFRI, Jhansi.

Kumar, Rabi., 1998. Evaluation of alternate tree and bushes for farm forestry in sand dune areas. *Forest Report*. Deptt. of Agroforestry, CCS Harayana Agricultural University, Hissar.

Nair, P.K.R. 1990. The Prospects for Agroforestry in the tropics. World Bank Technical Paper No. 131. Washington , DC: The World Bank (IBRD).

Pathak, P.S. and Ram Newaj, 2003. Agroforestry: Potentials and Opportunities, Agrobios (India), Jodhpur, Pp. 1-337.

Paroda, R.S., 1990. Silvipastoral approach for wasteland development: Present options and future strategy. In: *Silvipastoral Systems in India* (Eds., Pathak, P.S., Singh, P.) Pp. 1–3. Range Management Society of India, Indian Grassland and Fodder Research Institute, Jhansi.

Pathak, P.S., Gupta S.K. and Singh, P., 1995. *IGFRI Approaches to Rehabilitation of Degraded Lands*. Bulletin, IDRC-IGFRI, Jhansi, Pp. 1–23.

Pathak, P.S. 2002. Potential technological and management inverventions for improving the productivity of grasslands. In : Brithal, P. and Parthasarathy. Rao (eds). *Technology options for sustainable livestock production in India,* Pp. 164-182.

Pathak, P.S. and Dagar, J.C. 1998. Agroforestry: A strategy for rehabilitation of degraded lands. In : (Gopal, B. and Pathak, P.S. eds.) Ecology Today : An Anthology of Contemporary Ecological Research International Scientific Pub., New Delhi, Pp. 333-371.

Pathak, P.S. and Sunil Kumar 2004. Forage and grazing resources in different agro-climatic regions in India In : S.S. Kundu, A.K. Mishra and P.S. Pathak (eds.). *Buffalo production under different climatic regions.* International Book Distribution Co. Lucknow, Pp. 1-41.

Patil, B.D. and Pathak, P.S., 1981. Transformation of tropical wet lands habitat for forage and fuel production in India. *International J. Ecology and Environmental Sciences*, 1: 151–56.

Prajapati, M.C., 1979. Silvipastoral system of production in ravines. In: *Proc. Summer Inst. Silvipasture Production for Marginal and Sub-marginal Land*. June, 4 - July, 3, IGFRI, Jhansi.

Puri, S. and Nair, P.K.R., 2004. Agroforestry research for development in India: 25 years of experiences of a national programme. *Agroforestry Systems*, 61–62: 437–52.

Rai, P., Solanki, K.R. Deb Roy, R., and Singh, R. 1998. Performance of lambs and kids on Silvipastoral system and effects of grazing on constituent vegetation . *Indian Journal of Animal Science*, 68: 973-975.

Rana, B.S., Parihar, A.K.S. and Singh, B.P., 2000. Herbage yield of four grasses in sodic soil under silvipastoral production system. *Range Management and Agroforestry*, 21 (2): 213–217.

Rekib, A., Upadhyay, V.S. and Pathak, P.S., 1991. Trees for fodder. *Indian farming* 11: 56-61.

Roy, M.M., NigamG. and Kumar, V, 1998. Productivity of semi-multipurpose trees in silvopastoral system on highly degraded lands an semi-arid region. *Indian Journal of Forestry*, 2 (1): 4–8.

Singh, P. 1990. Forage Production for Sustainable Mixed Farming In: *Agronomic Research Ttowards Sustainable Agriculture*; Indian Society of Agronomy, IARI, New Delhi, Pp. 96-118.

Sunil Kumar and Choubey, B.K., 2008. Performance of aonla (*Emblica officinalis*)- based hortipastoral system in semi-arid region under rainfed situation, *Indian Journal of Agricultural Sciences*, 78 (9): 748-51.

❑❑❑